The World Economy

Geography, Business, Development

Sixth Edition

Frederick P. Stutz
Barney Warf

Pearson

ISBN 9789332536470
First Impression, 2014
Seventh Impression, 2024
Eighth Impression, 2025

Published by Pearson India Education Services Pvt. Ltd, CIN: U72200TN2005PTC057128.

Head Office: 1st Floor, Berger Tower, Plot No. C-001A/2, Sector 16B, Noida - 201 301, Uttar Pradesh, India.
Registered Office: Featherlite, 'The Address' 5th Floor, Survey No 203/10B, 200 Ft MMRD Road, Zamin Pallavaram, Chennai – 600 044.
Website: in.pearson.com; Email: companysecretary.india@pearson.com

Printed in India at: Manipal Technologies Limited, Manipal

Table of Contents

GLOSSARY

Absolute advantage. The ability of one country to produce a product at a lower cost than another country.

Acid rain. Acid rain, snow, or fog derives from the combustion of coal, releasing sulfur and nitrogen oxides that react with water in the earth's atmosphere.

Affluenza. A play on the term *influenza*, it refers to the psychology of consumerism in which more is never enough, although surveys show that once basic needs are satisfied, further purchases do little to raise overall happiness.

Agglomeration economies. The benefits gained by firms by clustering near other firms, including reduction of transport costs of inputs and outputs, access to specialized labor and ancillary services, access to specialized information, and ability to access a particular type of infrastructure.

Agribusiness. Food production by commercial farms, input industries, and marketing and processing firms that contribute to the total food sector.

Agricultural subsidies. Government payments to producers of agricultural products who simultaneously sing the praises of the free market.

Animate sources of energy. Energy derived from living human or animal muscle power, typically the dominant form in preindustrial societies.

Aquaculture. Commercial harvesting of fish or aquatic species for food.

Baby boom. The generation created by the dramatic rise in the U.S. birth rate following World War II, between the years 1946 and 1964.

Back offices. Segments of services work that involve low-wage, unskilled functions such as data entry.

Backward integration. The process of purchasing productive capacity "upstream" in the production process, that is, in the creation of inputs, through establishing a unit in-house or purchasing an existing company.

Balance-oriented lifestyle. A mind-set which insists that because resources are finite, they must be recycled and input rates slowed down to prevent ecological overload.

Basic sector. The part of an urban or regional economy engaged in the export of goods or services to clients located elsewhere.

Behavioral geographers. A school of thought that analyzes space from the perspective of individual perception, cognition, and psychology.

Biomass. The sum of mass of living beings, mostly plants, in a given place at a particular moment in time.

Birth rate (crude). The number of live births per 1000 population per year.

Brain drain. The process whereby less developed countries lose talented people to industrially advanced nations through emigration.

Bubonic plague. The massive epidemic that swept through Europe from 1347 to 1351, killing roughly 25 million people and forever changing the continent.

Burghers. The emerging merchant class of early capitalist Europe, which grew wealthy and powerful from the expanding markets and trade routes.

Business services. Service functions that cater primarily to corporations rather than households, including financial and producer services.

Call centers. Offices established by corporations that handle routine activities over the telephone, including reservations, customer assistance, and trouble-shooting.

Capital. A factor of production, including tools, buildings, and machines used by labor to fashion goods from raw material.

Capital flight. The investment of their monies by local individuals and whole countries in overseas ventures and in foreign banks for safekeeping.

Capital-intensive production. Forms of industry in which a high proportion of capital is used relative to the amount of labor employed per unit output.

Capital markets. Financial markets in which money in various forms is bought and sold, including banks, securities markets, public and private debt refinancing, and so forth.

Capitalism. The social system in which markets and production for profit are the primary (but not only) means of organizing resources. Ownership of private property is a key institutional characteristic. The importance of markets varies over time and space, giving rise to many different forms of capitalism, and the state always plays a role.

Carrying capacity. The maximum population an ecosystem can support.

Cartel. An organization of buyers and sellers, capable of manipulating price and/or supply.

Cattle ranching. Commercial raising of cattle, often on ranges and sometimes on public lands, for the production of beef.

Central business district. The downtown of an urban area, typically given over to firms that need centralized locations (often to take advantage of agglomeration economies there) and that pay high rents to do so.

Chlorofluorocarbons (CFCs). Chemicals implicated in the destruction of the atmospheric ozone layer.

Chronic malnutrition. Malnutrition that is essentially permanent in a given area, usually attributable to lack of protein.

Colonialism. The global system of empires, largely European, that emerged in the sixteenth century and dominated most of the world until the mid-twentieth century, in which foreign powers conquered other societies and restructured their economies, societies, and landscapes to fit foreign interests.

Commercial agriculture. Agricultural goods produced for sale in the city or on the international market.

Commodification. The process by which goods and services that were not commodities become produced and consumed through market transactions.

Commodities. Goods and services that are produced for sale on a market (i.e., to make a profit). Most goods under capitalism are commodities, but not all are (e.g., public-sector goods or those that have not been commodified, such as air).

Commodity chains. A perspective on economic activity that views commodities via a series of stages from producer to consumer.

Common market. A form of regional economic integration among member countries that disallows internal trade barriers, provides for common external trade barriers, and permits free factor mobility.

Comparative advantage. The theory that stresses relative advantage, rather than absolute advantage, as the true basis for trade. Comparative advantage is gained when countries focus on exporting the goods they can produce at the lowest relative cost.

Competitive advantage. The theory of international competitiveness that stresses the production of high-value, high-profit goods using skilled labor, agglomeration economies, and constructive state policy.

Conservation. Careful management and use of resources to assure continuing availability in the longer run.

Conspicuous consumption. Consumption well above one's needs, designed to show off social status.

Consumer services. Services oriented toward households and individuals rather than firms, such as retail trade and personal services (e.g., haircutters).

Consumerism. The systematic creation of consumer wants via advertising, and the culture and values that result in privileging shopping over other activities.

Consumption. The purchase and use of commodities (goods and services) to satisfy human needs and desires.

Contingent labor. Workers who are employed part-time or temporarily rather than full-time.

Core and periphery. An economic and spatial relationship between regions and countries where those on the outside export raw materials to industrialized regions at the center. Core regions are self-sustaining, whereas peripheral areas are dependent on the core.

Cost–space convergence. The reduction of travel costs between places as a result of transport improvements.

Current account. Measure of U.S. government trade that includes both merchandise trade and intangibles (services, interest payments, etc.).

Customs union. A group of countries that enters into an agreement to reduce or eliminate trade barriers among themselves in order to promote trade with one another.

Dairy farming. Commercial raising of cows for the production of milk, butter, and cheese.

Death rate (crude). Annual number of deaths per thousand population.

Deforestation. The clearing and destruction of forests (especially tropical rain forests) to make way for expanding settlement frontiers and the exploitation of new economic opportunities.

Deindustrialization. The loss of manufacturing employment through plant closures, in part due to movements of firms overseas.

Delayed gratification. The postponement of consumption in order to generate savings.

Demographic transition. The historical shift of birth and death rates from high to low levels in a population.

Dependency theory. The perspective that the economies of less developed countries were purposely underdeveloped via colonialism and transnational corporations to facilitate the development and expansion of the world's wealthier economies.

Deregulation. The reduction of government controls over economic activity within a country.

Desertification. The expansion of desert lands and noncultivatable regions by overuse, overpopulation, and drought.

Development. A historical process that encompasses the entire economic and social life of a country, resulting in change for the better. Development is related to, but not synonymous with, economic growth.

Diminishing marginal returns. The decline in unit output per each extra unit of input, that is, lower increases in productivity.

Direct effects. Employment and income generated at the site of a basic sector establishment, not including subcontracts or expenditures of worker incomes.

Diseconomies of scale. The disadvantages arising from attempts to produce at a scale larger than the optimum point on the long-run average cost curve, including overcrowded facilities, production pressure that leads to equipment breakdowns, and rising costs of inputs.

Distance decay. The decline in the level of interaction between two places with an increase in distance.

Distance learning. Virtual instruction, class attendance, or participation via telecommunications allowing the instruction source and the student to be separated from one another.

Diversification. A strategy by which a firm enters a different product market from the one in which it has traditionally been engaged.

Division of labor. The specialization of production within or among firms, regions, and countries by occupation or by region (in a spatial division of labor); because specialization raises productivity and thus profitability, more specialized divisions of labor tend to be more competitive.

Double cropping. Use of an agricultural area for more than one harvest in a given year.

Doubling time. The time in years required for the population of a region or country to double.

Ecological footprint. A measure of how each individual's consumption level and patterns contribute to the annihilation of the earth's ecosystems.

E-commerce. Electronic commerce, that is, electronic transactions between businesses and consumers or among businesses.

Economic base analysis. A model of local economies in which the basic, or export, sector is analytically privileged as the motor of local growth.

Economic geography. The branch of the discipline of geography concerned with the spatial organization of economic activity, including production, consumption, and transportation of goods and services, raw materials, people, and information. Economic geography includes a broad array of topics ranging from corporate location to international trade and development, and several different conceptual approaches, such as quantitative modeling and feminist, Marxist, ecological, and poststructural approaches.

Economic union. A form of regional economic integration having all the features of a common market as well as a common central bank, unified monetary and tax systems, and a common foreign economic policy.

Economies of scale. The cost savings derived from producing goods in large volumes, that is, by spreading fixed costs over a higher quantity of output.

Edge city. Communities in the exurbs, or suburbs distant from downtown.

Electronic data interchange (EDI). The electronic movement of standard business documents between and within firms.

Electronic funds transfer systems (EFTS). The network of telecommunications that allows financial institutions to move large volumes of funds almost instantaneously around the world.

Euro The currency used in most, but not all, member countries of the European Union.

Exchange rate. The value of one currency in terms of another.

Exchange value. The market price of a commodity, in contrast to the use value.

Export-led industrialization. The development strategy that relies upon encouraging foreign investment, developing a comparative or competitive advantage internationally, and generating large trade surpluses by exporting as much as possible.

Export-processing zones. Areas designated within countries by their governments in order to attract foreign firms and promote export-oriented activity and thus enhance foreign revenues, typically with tax breaks, subsidies, infrastructural developments, and labor training programs.

Export-restraint agreement. A nontariff barrier whereby governments coerce other governments to accept voluntary trade export restraint agreements.

Export subsidies. Payments made by governments that lower the final cost of goods and services to importers.

Externalization. The purchase of a service by a firm rather than producing it in-house, usually through subcontracting.

Exurbs. Residential areas on the outermost fringes of urban areas.

Feedlots. Large commercial areas in which cattle are stored and fattened before slaughter.

Feudalism. The social system that preceded capitalism in Europe, as well as some other places such as Japan. The class relations, culture, and geography of feudalism differed markedly from those of capitalism. The system was organized primarily through the power of the state, not the market.

Filtering model. The view of housing change that holds that lower-income households will move "up" through the housing stock as older homes are made available by the movement of the more well-to-do into newer houses.

Financial capital. Liquid or floating capital in the form of savings, loans, stocks, and other monetized commodities that can move rapidly through space, in contrast to fixed capital.

FIRE. Finance, insurance, and real estate.

First World. The economically advanced countries of Europe, Japan, the United States, and Canada.

Fixed capital. Capital investments that are set in one place and difficult to change, including buildings and equipment. Fixed capital can remain constant over small expansions in output, as opposed to variable capital, which rises as a function of output.

Flexible production. Also known as post-Fordism, this term refers to the form of capitalism that took shape in the 1970s and is characterized by vertical disintegration, computer technologies (e.g., just-in-time inventory systems), and lack of reliance on economies of scale.

Flow resources. Resources that are continually replenished, such as sunlight or water.

Food stamp program. A federal system for supplementing the incomes of low-income families by providing them with vouchers that may be used for the purchase of food.

Footloose industries. Firms that possess considerable locational mobility, i.e., relatively little inertia; such companies are generally labor-intensive with few barriers to entry and exit.

Fordism. The system of industrial production attributed to Henry Ford, who pioneered the moving assembly line and a specialized division of labor in factories that mass produced goods cheaply and profitably.

Foreign aid. Assistance given by one country to another, which takes many forms ranging from military aid, grants and loans, to disaster relief. Contrary to widely held opinion, foreign aid comprises a minuscule part of the government budgets of all countries, including the United States.

Foreign direct investment (FDI). Investing in companies in a foreign country, with the purpose of managerial and production control.

Forward integration. The process of purchasing productive capacity "downstream" in the production process, that is, in the creation of outputs, through establishing a unit in-house or purchasing an existing company.

Fossil fuels. Fuels, including oil, coal, and natural gas, that are formed from plant and animal remains.

Four-field rotation system. Rotating three crops among four fields over a period of years, while allowing a fourth rotated field to remain fallow, thus resting the soil for that year.

Four Tigers. South Korea, Taiwan, Hong Kong, and Singapore. See *newly industrializing countries (NICs)*.

Free-trade area. A form of regional economic integration in which member countries agree to eliminate trade barriers among themselves but continue to pursue their independent trade policies with respect to nonmember countries.

Friction of distance. Friction exerted on movement and flow by time and cost factors associated with movement across space.

General Agreement on Tariffs and Trade (GATT). An international agency, headquartered in Geneva, Switzerland, supportive of efforts to reduce barriers to international trade; replaced by the World Trade Organization in 1995.

Gentrification. The growth in incomes and property values in low-income, inner-city neighborhoods associated with either commercial investments or the influx of relatively wealthy professional households.

Geothermal energy. Energy produced from deep inside the earth as water interfaces with heated rocks from the earth's core, producing steam.

Global city. A city that is a preeminent international location for business decision making and corporate services.

Global warming. The rise of average atmospheric temperatures over the past two centuries, often attributed to human-induced quantities of carbon dioxide from automobiles and manufacturing.

Globalization. The set of economic, political, and cultural processes that generate increases in the volume, scope, and velocity of international transactions and linkages.

Green Revolution. A popular term for the greatly increased yield per hectare that followed the introduction of new, scientifically bred and selected varieties of such food crops as wheat, maize, and rice.

Greenhouse effect. The warming of the atmosphere due to increased amounts of carbon dioxide, nitrous oxides, methane, and chlorofluorocarbons.

Growth-oriented lifestyle. A mind-set that insists on maximum production and consumption. It assumes an environment of unlimited waste and pollution reservoirs and indestructible ecosystems.

Guilds. The medieval system of handicraft production according to which craft workers were organized around different types of goods (e.g., paper, leather, iron) to restrict competition. Master artisans or craftsmen ran workshops in which apprentices worked and learned the trade.

Hanseatic League. A group of trading cities located mostly around the Baltic Sea that engaged in extensive trade relations during the late medieval period.

Hegemonic power. In international political economy, the most powerful country in the world, which "sets the rules" that others follow. Examples include Britain in the nineteenth century and the United States in the twentieth century.

Homo economicus. The model of human behavior widely used in neoclassical economics, that is, an all-knowing, self-interested individual who behaves rationally.

Horizontal integration. A business strategy to increase a firm's scale by buying, building, or merging with another firm at the same stage of production of a product, leading toward oligopoly.

Hub-and-spoke networks. Hubs are major cities that collect passengers from small cities, in the local vicinity, via spoke lines. Hubs redistribute passengers between sets of original major cities.

Human capital. The sum of skills, education, and experience that makes labor productive.

IMF conditionality. The set of restrictions the International Monetary Fund imposes on countries to qualify for IMF loans or debt restructuring.

Import-substitution industrialization. A trade strategy, now largely discredited, that puts high tariffs on imports as a way to stimulate domestic production of goods. The opposite of export promotion.

Inanimate sources of energy. Energy derived from sources other than living muscle power or remains of living tissue (e.g., firewood), including solar energy, nuclear power, wind, geothermal energy, and fossil fuels.

Income elasticity. The percentage change in demand that accompanies a change in income.

Indirect effects. Employment and income effects of a commercial establishment that are generated by its backward linkages and subcontracts to suppliers of goods and services.

Induced effects. Employment and income effects of an establishment that arise from expenditures of the wages and salaries that it pays its workers.

Industrial inertia. The resistance of some types of firms to changing their spatial location, often due to heavy fixed capital investments.

Industrial restructuring. A term used to refer to the alternating phases of growth and decline in industrial activity. It emphasizes changes in employment between regions and links these with changes in the world economy.

Industrialization. The movement from an agricultural economy to a manufacturing-based, export-oriented economy.

Infant industry. A young industry that, it is argued, requires tariff protection until it matures to the point where it is efficient enough to compete successfully with imports.

Infant mortality rate. Number of deaths during the first year of life per 1000 live births.

Informal economy. The part of the economy that is essentially untaxed and unregulated, including, but not limited to, many illegal activities (e.g., the black market) but also including casual labor, street vendors, and a variety of similar occupations; comprises a large share of the economy in the developing world.

Information technology. Communications technologies based on microelectronics, including microprocessors, computers, robots, satellites, and fiber-optic cable.

Infrastructure. The transportation and communication systems and other public goods (e.g., dams, sewers) necessary for an economy to function.

Intangible output. The output of services, which cannot be directly observed or measured.

Integration. The process of expanding either vertically (upstream or downstream) in the production process or horizontally (within the same product market).

Intellectual property rights. Establishing and policing patent, copyright, and trademark rights on an international basis.

Intensive subsistence agriculture. A high-intensity type of primitive agriculture practiced in densely populated areas of the developing world.

International currency markets. The internationalization of currency, banking, and capital markets.

International division of labor. The global system of geographically differentiated production set into motion under colonialism that concentrated high-wage, high-profit activities in the First World and relegated colonies and developing countries to low-wage, low-profit activities largely in the primary economic sector.

International economic order. The placement of countries within the world economy based on capital, trade, and production.

International economic systems. The system of the world economy based on flow across international boundaries.

International Monetary Fund (IMF). An international financial agency that attempts to promote international monetary cooperation, facilitate international trade, make loans to help countries adjust to temporary international payment problems, and lessen the severity of international payments disequilibrium, often by imposing Structural Adjustment Policies.

Internet. The global network of computer networks that allows data, video, and other information to be shared electronically among users.

Intracorporate trade. Trade among subsidiaries of the same corporation, typically using shadow pricing rather than market prices.

ISDN. Integrated Services Digital Network, the technical format that allows data to be exchanged on the Internet.

Journey-to-work. Travel by individuals to work; in American households, yielding the largest proportion of travel.

Just-in-time inventory systems. Quick response and delivery of parts and inventory from component plants to final assembly operations.

Knowledge worker. The class of laborers who work in information-intensive professional occupations such as management and producer services.

Kondratiev cycles. Named after the Russian economist who discovered them, these refer to long-term oscillations (roughly 50–75 years) in the capitalist economy linked to major waves of technological change, as measured by fluctuations in prices, output, profits, productivity, and employment.

Labor. An input factor of production that consists of living human beings and their capacity to generate value.

Labor force. Those in society who work, including both the employed and unemployed.

Labor-intensive production. Forms of industry in which a high proportion of labor is used relative to the amount of capital employed per unit of output.

Labor migration theory. An explanation of the process of changing residences from one geographic locale to another due to economic factors.

Land. A factor of production that includes not only a geographic portion of the earth's surface but also the raw materials from this region.

Land degradation. Decline in the usable quality of a landscape via deforestation, soil erosion, and diminished soil fertility.

Less developed countries (LDCs). The Third and Fourth worlds, encompassing Latin America, Africa, and most of Asia, characterized by relatively high rates of population growth and low per capita income.

Limits to Growth. The opinion by the Club of Rome that the optimum population size for the world shows that growth must be limited; a gloomy forecast by Paul Ehrlich suggesting worldwide famine and war as the inevitable results of continued increases in world population.

Line-haul costs. Costs involved in moving commodities along a route.

Location theory. A compilation of ideas and methods dealing with questions of accessibility.

Locational factors. Major elements that shape the decision of the firm to locate in some places and not others, including cost and productivity of labor, land, and other inputs. .

Maglev. A magnetically levitated train that operates with a linear induction engine and cruises on a cushion of air at high speeds on a detached right-of-way; heralded as the state of the future in ground transportation systems.

Malthusianism. The ideology originating with Thomas Malthus that holds overpopulation to be the major, or even only, cause of most world problems.

Maquiladoras. Assembly plants in Mexico, usually foreign-owned, for the production of textiles, electronics, automobiles, and other goods, mostly for export.

Mariculture. Commercial growth of maritime species for sale on the market.

Market. An institution composed of buyers and sellers of commodities. Just as there is a large array of different types of producers and consumers, there are many kinds of markets for different goods and services, ranging dramatically in size and sophistication.

Maximum sustainable yield. Maximum production consistent with maintaining future productivity of a renewable resource.

Mediterranean agriculture. A type of agriculture that produces specialty crops because of mild climates, including citrus, grapes, nuts, avocados, tomatoes, and flowers.

Mercantilism. The economic ideology that holds the state's primary responsibility is to maximize a country's wealth by discouraging imports and promoting exports; it was widely popular from the sixteenth to the early nineteenth centuries.

Microelectronics. Semiconductors, integrated circuits, and electronic components and parts.

Migration. A change in residence intended to be permanent, frequently across international boundaries.

Minerals. Natural inorganic substances that have a definite chemical composition and characteristic crystal structure, hardness, and density.

Mixed crop and livestock farming. The raising of beef cattle and hogs as a primary revenue source, with the crops fed to the livestock.

Mode of production. A Marxist term that refers to the basic forces and social relations of production. Slavery comprised one mode, feudalism another, capitalism yet a third. Each is typified by an ensemble of class relations, culture, technologies, and geographical landscape.

Modernization theory. The approach to development that maintains countries should embrace global capitalism, reduce trade barriers, invite foreign investment, and diffuse markets to stimulate growth.

Multiplier. The effect on total employment (or output, wages, and profits) generated by changes in an industry, including interindustry linkages and expenditures resulting from changes in personal income (wages and salaries).

NAFTA. The North America Free Trade Agreement signed by Mexico, Canada, and the United States that began in 1994 and gradually removed all tariffs and nontariff barriers among them.

Nation-state. A nation of people who enjoy an independent territorial state.

Natural growth rate (NGR). Population growth measured as the excess of live births over deaths per 1000 individuals per year; it does not reflect emigration or immigration.

Negative population growth. A falling level of population where out-migration and death exceed in-migration and births.

Neocolonialism. The state of being economically independent on paper (*de jure*), but not in practice (*de facto*), that is, the domination of a country's economy by foreign corporations.

Neo-Malthusian. The perspective that holds that despite the fact that Malthus' original analysis and predictions were flawed, the essence of his views remains correct in the long term.

Net migration rate (NMR). The difference between in-migration (or immigration, if international) and out-migration (or emigration) rates.

Newly industrializing countries (NICs). Rapidly growing economies in the less developed world, mostly in Asia, that have experienced sustained growth and rising levels of prosperity. See also *Four Tigers*.

Nonbasic sector. The part of an urban or regional economy that caters to local demand (i.e., it is not export-oriented), including retail sales, real estate, and consumer services.

Nondirect production workers. Workers in a manufacturing firm who are not directly involved in the production process (e.g., management, administration, research, marketing, and sales).

Nonpoint sources. Sources of pollution that do not lie in one point but are broadly spread out; typically this refers to agricultural sources.

Nonprofit services. Services provided for reasons other than making a profit, including those provided by churches, charities, and civic organizations.

Nonrenewable resources. Resources that are fixed in amount—that cannot regenerate—such as fossil fuels and metals.

Nontariff barriers. Restrictions other than tariffs that limit entry into an industry by competitive firms or countries, including quotas and a variety of legal limitations such as licensing requirements.

North American Free Trade Agreement (NAFTA) See *NAFTA*.

North American Manufacturing Belt. The core manufacturing region of the United States extending from Boston westward through upstate New York, southern Ontario, Pennsylvania, Ohio, southern Michigan, and southeastern Wisconsin.

Nuclear energy. The energy released either through nuclear fission or fusion; for humans, this includes nuclear weapons and power plants.

Nuclear fission. Energy released when hydrogen atoms are fused; while this powers the sun, and has been generated artificially by people, it is not a commercially feasible energy source.

Nuclear fusion. Energy released when large, usually uranium, atoms are split; the basis for nuclear power plants.

Offshore assembly. An arrangement whereby firms based in advanced industrial countries provide design specifications to producers in underdeveloped countries, purchase the finished products, and then sell them at home or abroad.

Offshore banking centers. Banking centers in less regulated parts of the world, often in small island states that offer generous tax benefits to attract foreign financial firms.

Oligopoly. The control of a market by a small number of firms or producers that can affect the market price.

Organization of the Petroleum Exporting Countries (OPEC). The international cartel of oil-producing countries.

Outsourcing. The subcontracting and shifting of work to other locations and firms outside the principal corporation.

Overpopulation. A level of population in excess of the "optimum" level relative to the food supply or rate of consumption of energy and resources.

Ozone layer. Layer of O_3 (trioxygen) in the atmosphere that protects life from excessive ultraviolet radiation.

Parity price. Equality between the prices at which farmers could sell their products and the prices they need to spend on goods and services to run the farm.

Passive solar energy. Means of harvesting solar energy through panels.

Pastoral nomadism. Animal herds used for subsistence, moved from one forage area to another, in a cyclical pattern of migration.

Peasant agriculture. Subsistence agriculture, using little mechanical equipment and producing labor-intensive crops.

Plantation. Sites of intensive cultivation of commercial crops grown largely for export; developed during the colonial era, plantations are still used to grow both food and nonfood crops.

Point sources. Sources of pollution that are concentrated in one location, such as a factory.

Political economy. The approach to studying society that views social relations as a unified whole organized along lines of class, gender, ethnicity, and other lines of power.

Population density. The average number of people per unit area, usually per square mile or square kilometer.

Population pyramid. A special type of bar chart indicating the distribution of a population by age and sex.

Post-Fordism. Also called flexible production, the type of production that emerged in the late twentieth century characterized by automated production, just-in-time inventory systems, and the capacity to produce goods profitably in small quantities, offsetting the advantages formerly generated by economies of scale.

Postindustrial economy. A theory that modern society is dominated by services and maintains that they form a historically new type of economic and social system and landscape in which a class of knowledge professionals will end traditional scarcity through rising productivity.

Poststructuralism. An approach to philosophy and social science that questions the ordered appearance of the world and the idea of objectivity, asserting instead there are many language-based interpretations that rival one another.

Price floor. A guaranteed price above the market price set by the government as the basis for agricultural subsidies.

Primary sector. Economic activities associated with the extraction of raw materials, including farming, fishing, mining, and forestry.

Privatization. The process by which government-owned assets are transferred to private ownership and management.

Producer services. Services that sell their output (primarily expertise and specialized information) to other firms rather than to households, including financial services and business services (e.g., legal services, advertising, accounting, public relations, etc.).

Product life cycle. The typical sequence through which a product passes, from its introduction into the market to when it is replaced by a new product.

Product market. The market where households buy and firms sell the products and services they have produced.

Production linkages. Purchases and sales of tangible inputs and outputs by firms, as opposed to other types of interfirm linkages such as specialized services and information.

Productivity. The ratio of outputs to inputs; productivity growth is the rise in productivity over time.

Profit. The difference between gross revenues and production costs.

Projected reserves. Estimated future supply of a given finite resource.

Protectionism. An effort to protect domestic producers by means of controls such as tariffs and quotas on imports.

Push-and-pull factors. The conditions in the source area that tend to drive people away and the perceived attractiveness of the destination that simultaneously stimulates migration flows.

Quota. A restriction on imports imposed by one country against another as a form of protectionism; the most common form of nontariff barrier.

Raw materials. A substance in the physical environment considered to have value or usefulness in the production process.

Recycling. The reuse of discarded materials after they have passed through some form of treatment (e.g., melting down glass bottles to produce new bottles).

Regional economic integration. The international grouping of sovereign nations to form a single economic region.

Renewable natural resources. Resources, such as water, timber, or fish, that can be replenished relatively easily.

Renewable resources. Resources capable of yielding output indefinitely if used wisely, such as water and biomass.

Reserve. A known and identified deposit of resource materials that can be tapped profitably with existing technology under prevailing economic and legal conditions.

Residential location decision. The process whereby households decide where to live, typically involving trade-offs between rent and commuting costs and subject to budget constraints.

Resource. A naturally occurring substance that can be extracted under prevailing conditions and be of potential profit.

Reverse commuting. Daily commuting from city center to suburbs, rather than the opposite direction, which is more common.

Scale. As opposed to spatial scale, scale in production refers to varying levels of output over which firms can spread their fixed costs and achieve economies of scale.

Second World. A term used during the Cold War but now obsolete; it included the former Soviet Union and its client states in Eastern Europe, Mongolia, and Cuba.

Secondary sector. The processing of materials to render them more directly useful to people; manufacturing.

Serfs. The basis of the rural agricultural labor force under feudalism. Unlike slaves, serfs were not owned by a master but were bound to the land by law and custom. Unlike wage workers, they did not receive payment for their work; rather, they paid their lord rent and kept any surplus that might remain.

Service linkages. Purchases and sales of intangible inputs and outputs, such as information and expertise by firms, as opposed to production linkages in tangible goods.

Services. Economic activity associated with the buying and selling of intangibles, including expertise and information.

Shifting cultivation. Also called swidden or slash-and-burn, it is the temporary use of rain forest land for agriculture by cutting and burning the overgrowth.

Slash-and-burn agriculture. Also called swidden or shifting cultivation, it is the temporary use of rain forest land for agriculture by cutting and burning the overgrowth.

Social relations of production. A Marxist term that refers to ownership and control of the means of production, that is, the pattern of class in a given mode of production.

Solar energy. Radiation from the sun, which is transformed into heat primarily at the earth's surface and secondarily in the atmosphere.

Spatial integration. Linkages between a city or firm and the members of its economic environment.

Spatial interaction. The movement, contact, and linkage between points in space, for example, the movement of people, goods, traffic, information, and capital between one place and another.

Spatial mismatch. The mismatch between the supply and demand for skills in a particular region, usually meaning the need for skilled workers and the supply of unskilled ones.

Special Economic Zones (SEZs). The export processing zones set up by the Chinese government along its Pacific Coast in the late twentieth century.

Spring wheat. Wheat that is planted in the spring and harvested in the fall, in contrast to winter wheat.

Squatter settlements. Residential areas that are home to the urban poor in underdeveloped countries. The various terms used to identify squatter settlements include the following: *callampas, tugurios, favelas, mocambos, ranchos,* and *barriadas* in Latin America; *bidonvilles* and *gourbivilles* in North Africa; *bustees, jhoupris, jhuggis, kampongs,* and *barung barong* in South and Southeast Asia.

Stage model of agriculture. Developed by Boserup, it refers to the intensification of agricultural activity and rising productivity that accompany growing population levels.

Stock resources. Resources with fixed total supplies, such as mineral ores or petroleum, or ones that are replenished only very slowly, such as soil or timber.

Strategic minerals. Those minerals deemed critical to the economic and military well-being of the nation.

Structural adjustment. The policies advocated by the International Monetary Fund as requirements for debt restructuring, typically including currency devaluation, reductions in government subsidies, and privatization.

Subsistence agriculture. Peasant agriculture, using little mechanical equipment, producing meager and labor-intensive crops.

Suburbanization. The movement of people and economic activity from inner-city regions to the outer rings of a city.

Surplus value. A Marxist term for the value of output a worker generates but for which he or she is not compensated. Based on the labor theory of value, this view holds that workers ultimately create all wealth but are not compensated for their efforts by the market price of their labor. All employment relations are thus held to be inherently exploitative.

Sustainable agriculture. Agricultural practices that do not do long-term damage to the environment and that minimize ecological disruption.

Sustainable development. Economic development that aims at long-term environmental sustainability rather than short-term profit maximization.

Swidden agriculture. See *shifting cultivation*.

Target pricing. Direct payment by the government to a farmer of the difference between the market selling price and the price that the government has set artificially.

Tariff. A schedule of duties placed on products. A tariff may be levied on an *ad valorum* basis (i.e., as a percentage of value) or on a specific basis (i.e., as an amount per unit). Tariffs are used to serve many functions—to make imports expensive relative to domestic substitutes; to retaliate against restrictive trade policies of other countries; to protect infant industries; and to protect strategic industries, such as agriculture, in times of war.

Tax-haven country. A typically small nation, often an island, that gives extraordinary tax breaks to lure foreign capital and corporations.

Tax havens. Local governments that give tax-related inducements to influence banking or industrial location.

Telework centers. Offices organized by firms where telecommuters may work, facilitating telework but not from the home.

Teleworking. Work performed via electronic lines of communication, including the Internet, either at home or via telework centers.

Terminal costs. Costs incurred in loading, packing, and unloading shipments and preparing shipping documents.

Terms of trade. The relative prices of exports to imports for a country.

Tertiary sector. Economic activity that produces intangibles, that is, services.

Third World. The set of countries, mostly former European colonies and mostly economically underdeveloped, located in Latin America, Asia, and Africa.

Time–space compression or convergence. The reduction in travel time between places that results from transportation and communication improvements.

Total fertility rate. The number of live births per 1000 people in a country per year.

Township and Range System. Initiated by Thomas Jefferson, it is the primary system of land demarcation west of the Mississippi River; using a series of latitude and longitude lines, it not only was critical in the creation of property parcels in the nineteenth century but became the basis of many county and state borders too.

Trade deficit. The excess of a country's imports over exports for any specific year.

Tragedy of the commons. The frequent overuse rather than conservation of public resources by the cumulative isolated actions of individuals, thus ruining resources held in common.

Transmaterialization. Change in the form of a product into a consumable form, for example, sugar cane into refined sugar.

Transnational corporations (TNCs). Companies that operate factories or service centers in countries other than the country of origin; also known as multinational corporations.

Transport costs. The alternative output given up when inputs are committed to the movement of people, goods, information, and ideas over geographical space.

Truck farming. Farms on the periphery of urban areas, mostly vegetables, with low transport costs to market.

Underconsumption. Consumption levels inadequate to meet basic human needs.

Underdevelopment. The state characterized by poverty, low rates of investment, high unemployment and rates of population growth, and low per capita incomes.

Underemployment. A shortage of job opportunities that forces people to accept less than full-time employment or being employed well beneath their training and ability.

Undernutrition. A state of poor health in which an individual does not obtain enough calories.

Unemployment. The state of actively seeking but unable to find employment.

Unequal exchange. The argument that an artificial division of labor has made earning a good income from free trade difficult for most less developed countries.

Uneven development. The persistent tendency of capitalism to generate social and spatial inequality, manifested geographically in rich and growing regions on the one hand, with abundant life opportunities, and poor or stagnant regions on the other, with widespread unemployment and poverty.

Uneven spatial development. The existence of geographic disparities in terms of living standards, quality of life, and opportunities.

Urban hierarchy. The system that ties cities together via various tiers stratified by population size and economic significance.

Urban sustainability. Urban growth that can occur over long periods by minimizing environmental impacts.

Urbanization economies. The benefits firms accrue from locating in large cities, thus gaining access to other firms, information, labor, and infrastructure.

Use value. The usefulness of a commodity to the person who possesses it, i.e., its subjective ability to satisfy wants.

Utility maximization. In neoclassical models of consumption, the tendency of consumers to achieve the most satisfaction possible by allocating their incomes among goods and services that generate different use values.

Value added by manufacturing. The difference between the revenue of a firm obtained from a given volume of output and the cost of the input (the materials, components, services) used in producing that output.

Vertical integration. Expansion of a firm's in-house productive capacity "upstream" (i.e., of inputs) or "downstream" (i.e., of outputs).

von Thünen model. A famous nineteenth-century land use model that revealed how land markets reflect the interaction of agricultural prices, production costs, and transportation as users seek to maximize their income or rent.

Wheat belts. Areas, such as those in North America, in which wheat is the predominant agricultural product.

Wholesale and retail services. Services that act as intermediaries between producers and consumers of goods, including warehouses and stores that sell to the public.

Wind farm. Capturing wind energy with wind turbines and converting it to electricity.

Winter wheat. Wheat that is planted in the fall, lies dormant through the winter, and is harvested in the spring.

World Bank. A group of international financial agencies that includes the International Bank for Reconstruction and Development, the International Finance Corporation, and the International Development Association.

World economy. A multistate economic system created in the late fifteenth and early sixteenth centuries by European capitalism and, later, its overseas progeny.

World Trade Organization (WTO). The world trade union that came into existence following the Uruguay Round of the GATT Treaty. The WTO enforces trade rules and assesses penalties against violators.

World-systems theory. A theory that holds that countries and regions practice economic activities and succeed based on their ability to produce needed goods and services for the world economy.

Zero population growth (ZPG). A stable level of population in a country, not rising or falling from year to year, as a result of the combination of births, deaths, and migration.

Economic Geography: An Introduction

OBJECTIVES

- To acquaint you with the discipline of geography and the subfield of economic geography
- To discuss five major analytical themes useful in comprehending social and spatial issues
- To summarize the major paradigms for approaching economic geography
- To introduce capitalism as a system that forms the major focus of this volume
- To note the various dimensions of globalization
- To situate economic geography within the context of world development problems

Capitalist development, often expressed most intensely in the built environment of the city, reflects the constellations of forces that produce landscapes in different places and times. In Manhattan, flows of capital, labor, energy, raw materials, and information interact with the local physical environment to generate a unique combination that is both global and local simultaneously.

Jean-Pierre Lescourret Photolibrary

Economic Geography: An Introduction

GEOGRAPHIC PERSPECTIVES

Everything that happens on the earth's surface is geographic. All social processes, events, problems, and issues, from the most local—your body—to the most global, are inherently geographic; that is, they take place in space, and where they are located influences their origins, nature, and trajectories over time. Everything that is social is also spatial, that is, it happens someplace. Where you are sitting now, how you got there, where you live and work, the patterns of buildings and land uses in your school or city, transport routes, and the ways people move through them all are different facets of geography; so are the distributions of the world's cultures, the patterns of wealth and poverty, the flows of people, goods, disease, and information.

Geography is the study of space, of how the earth's surface is used, of how societies produce places, and how human activities are stretched among different locations. In many respects, geography is the study of space in much the same way that history is the study of people in time. This conception is very different from simplistic popular stereotypes that portray geographers as a boring bunch concerned only with drawing boundaries and obsessed with memorizing the names of obscure capital cities. Essentially, the *discipline of geography examines why things are located where they are*. Simply knowing where things are located is relatively simple; anyone with a good atlas can find out, say, where bananas are grown or the distribution of petroleum. Geographers are much more interested in explaining the processes that give rise to spatial distributions, not simply mapping those patterns. Much more interesting than simply finding patterns on the earth's surface is the *explanation* linking the spatial outcomes to the social and environmental processes that give rise to them. Thus, geographers examine not only where people and places are located but how people understand those places, give them meaning, change them, and are in turn changed by them. Because this issue involves both social and environmental topics, geography is the study of the distribution of both human and natural phenomena and lies at the intersection of the social and physical sciences.

All social processes and problems are simultaneously spatial processes and problems, for everything social occurs somewhere. More important, *where* something occurs shapes *how* it occurs. Place is not some background against which we study social issues, but it is part of the nature and understanding of those issues. Geographers ask questions related to location: Why are there skyscrapers downtown? Why are there famines in Africa? How does the sugar industry affect the Everglades? Why is Scandinavia the world's leader in cell phone usage? How is the North American Free Trade Agreement (NAFTA) reshaping the U.S., Canadian, and Mexican economies? Why is China rapidly becoming a global economic superpower? How has the microelectronics revolution changed productivity and competitiveness and the global locational dynamics of this sector? What can be done about inner-city poverty?

To view the world geographically is to see space as socially produced, as made rather than simply given, that is, as a product of social relations, a set of patterns and distributions that change over time. This means that geographic landscapes are social creations, in the same way that your shirt, your computer, your school, and your family are also social creations. Geographers maintain that the production of space involves different spatial scales, ranging from the smallest and most intimate—the body—to progressively larger areas, including neighborhoods, regions, nations, and the least intimate of all, the global economy.

Because places and spaces are populated—inhabited by people, shaped by them, and given meaning by them—geographers argue that all social processes are embodied. The body is the most personal of spaces, the "geography closest in." Individuals create a geography in their daily life as they move through time and space in their ordinary routines. Societies are formed by the movements of people through space and time in everyday life. In local communities, neighborhoods, and cities—the next larger scale—these movements form regular patterns that reflect a society's organization, its division of labor, cultural preferences and traditions, and political opportunities and constraints. Geographies thus reflect the class, gender, ethnicity, age, and other categories into which people sort themselves. Spatial patterns reflect the historical legacy of earlier social relations; political and economic organization of resources; the technologies of production, transportation, and communications; the cultures that inform behavior and guide it; and legal and regulatory systems. The global economy itself—an intertwined complex of markets and countries—involves planet-wide patterns of production, transportation, and consumption, with vast implications for the standard of living and life chances of people in different areas.

Geographers study how societies and their landscapes are intertwined. To appreciate this idea, we must recognize that social processes and spatial structures shape each other in many ways. Societies involve complex networks that tie together economic relations of wealth and poverty, political relations of power, cultural relations of meanings, and environmental processes as well. Geographers examine how societies and places produce one another, including not only the ways in which people organize themselves spatially but also how they view their worlds, how they represent space, and how they give meaning to it. Divorcing one dimension, say the economic, from another, such as the political, is ultimately fruitless, but to make the world intelligible we must approach it in manageable chunks. This text centers upon only one aspect of this set of phenomena, economic landscapes.

Economic geography is a subdiscipline concerned with the spatial organization and distribution of economic activity (production, transportation, communication, and consumption); the use of the world's resources; and the geographic origins, structure, and dynamics of the world economy. Economic geographers address a wide range of topics at different spatial scales using different theories and methodologies. Some focus on local issues such as the impacts of waste incineration facilities, while others study global patterns of hunger and poverty. Conceptual approaches found in economic geography include models of supply and demand, political economic analyses focused on class and power, feminist theorizations centered on gender, and views that deliberately blur the boundaries between the "economic" and other spheres of society such as culture, consumption, and politics. Methodologically, economic geographers use a range of tools that includes geographic information systems, mathematical models, and qualitative assessments based on interviews and field work.

FIVE ANALYTICAL THEMES FOR APPROACHING ECONOMIC GEOGRAPHY

One means of starting a comprehensive analysis of economic geography is through five analytical themes. These broad generalizations are designed to encourage you to think about economic landscapes and include: (1) the historical specificity of geographies; (2) the interconnectedness of regions, particularly with the rise of the global capitalist economy; (3) the interpenetration of human and biophysical systems; (4) the importance of culture and everyday life in the creation of social and spatial relations; and (5) the centrality of comprehending social structures and their spatial manifestations.

1. The study of space is inseparable from the study of time. If one accepts the discipline of geography as the study of human beings in space and history as its temporal counterpart, then this theme implies that geography and history are inseparable, indeed indistinguishable. It is the accumulated decisions of actors in the past—firms, individuals, organizations, governments, and others—that created the present, and it is impossible to explain the contemporary world meaningfully without continual reference to their actions and the meanings they ascribed to them. Historical awareness undercuts the common assumption that the present is the "typical" or "normal" way in which human beings organize themselves, for it is history as much as geography that teaches us the full range and diversity of human behavior, cultures, and social systems. All geographies are constructed historically, and all histories unfold spatially. Such an emphasis leads directly to the question of *how* histories and geographies are produced, particularly through the everyday lives of ordinary individuals. Historical geography—a redundancy, for *all* geography is inescapably historical—is thus much more than simple reconstructions of past worlds; it is the analysis of the reproduction of social systems over space-time as they are transformed into the present.

But there is a broader meaning to unveil here. Taking history seriously means acknowledging that geographies are always changing, that they are forever in flux, that landscapes are humanly created and therefore plastic and mutable. History is produced through the dynamics of everyday life, the routine interactions and transient encounters through which social formations are reproduced. "Time" is thus not some abstract independent process; it is synonymous with historical change (but not necessarily progress) and the capacity of people to make, and change, their worlds. There is, for example, no need to accept the geography of poverty (at any spatial scale) as fixed and inevitable, whether in New York City or Bangladesh. Like landscapes or buildings, poverty is socially constructed, the outcome of political and economic forces. To understand how geographies are produced historically, therefore, is to focus on the dynamics that underpin their creation. Views that purport to represent a "snapshot in time," therefore, are more deceiving than illuminating; it is the *process* that underlies the creation of places that is central, the social dynamics at work, not their appearance at one instant in time.

2. Every place is part of a system of places. Unlike traditional approaches to geography, which studied regions in isolation, this theme notes that all regions are interconnected, that is, they never exist in isolation from one another. Indeed, places are invariably tied together to a greater or lesser extent by the biophysical environment (e.g., winds and currents, flows of pollution), flows of people (migration), capital (investment), and goods (trade), and the diffusion of information, innovations, and disease. Places are inevitably part of a network of places because contemporary social relations stretch across regions. It follows that what happens in one place must affect events in others; the consequences to action are never purely local. For example, the Chernobyl meltdown in Ukraine in 1986 led to clouds of radioactive emissions that crossed Scandinavia and entered North America; the North American Free Trade Agreement (NAFTA) links regions from southern Mexico to Quebec; and the jet airplane made the contemporary world vulnerable to new diseases such as AIDS and swine flu.

While this theme holds in the study of many places throughout history, it is especially relevant since the rise of capitalism on a global basis beginning in the sixteenth century. More broadly, the global system of nations and markets has tied places together to an unprecedented degree, including international networks of subcontracting, telecommunications, transnational firms, and worldwide markets, as any trip to the grocery store will attest.

3. Human action always occurs in a biophysical environment. The biophysical environment (or in common parlance, "nature," a term that suffers from its popularity and unfortunately carries connotations of the nonsocial or "natural") includes the climate, topography, soils, vegetation, and mineral and water resources of a region, and affects everything from the length of a growing season to transport costs to energy supplies. It is important to acknowledge that these factors *affect* the construction of histories and geographies. But the interpenetration of people and nature is a two-way street. Everywhere, nature has been changed by human beings, for example, via the modification of ecosystems; annihilation of species; soil erosion; air, ground, and water pollution; changed drainage patterns; agriculture; deforestation; desertification; disruptions of biogeochemical cycles; and more recently, alterations in the planetary atmosphere (e.g., global warming). Indeed, human beings can't live in an ecosystem without modifying it. More recently, political ecology has much to say about the interactions of capitalism, culture, and nature. In short, the formation of geographies is neither reducible to the biophysical environment nor independent of it.

This theme points to how geographies are produced in the context of particular biophysical environments and how those environments are always and everywhere changed through human actions. For example, think about human modifications of the New World prior to the Columbian encounter, which dispels the myth that native peoples left their world in a state of untouched virginal innocence. Political conflicts over, say, water and petroleum in the Middle East, or diamonds in Africa, illustrate the role of nature in current geopolitics. The spatial structure of the Industrial Revolution may be seen as profoundly preconditioned by the location of the large coal deposits in the north European lowlands stretching from Wales to Silesia.

4. Culture—the shape of consciousness—is fundamental to economic geography. This theme begins with the recognition that human beings are sentient beings; that is, they have consciousness about themselves and their world, as manifested in their perceptions, cognition, symbolic form, and language, all of which are fundamental to any understanding of the human subject. Social science is thus fundamentally different from analyses of the nonhuman world, in which the consciousness of what is studied is not at issue (except, perhaps, in the behavior of some animals). Moving beyond the usual elementary definitions of culture as the sum total of learned behavior or a "way of life" (religion, language, mores, traditions, roles, etc.), social theory allows for an understanding of culture as what we take for granted, that is, common sense, the matrix of ideologies that allow people to negotiate their way through their everyday worlds. Culture defines what is normal and what is not, what is important and what is not, what is acceptable and what is not, within each social context. Culture is acquired through a lifelong process of socialization: Individuals never live in a social vacuum, but are socially produced from cradle to grave.

The socialization of the individual and the reproduction of society and place are two sides of the same coin, that is, the macrostructures of social relations are interlaced with the microstructures of everyday life. People reproduce the world, largely unintentionally, in their everyday lives, and in turn, the world reproduces them through socialization. In forming their biographies every day, people reproduce and transform their social worlds primarily without meaning to do so; individuals are both produced by, and producers of, history and geography. Everyday thought and behavior hence do not simply mirror the world, they constitute it. Such a view asserts that cultures are always intertwined with political relations and are continually contested, that is, dominant representations and explanations that reflect prevailing class, gender, and ethnic powers are often challenged by marginalized discourses from the social periphery. This theme is useful in appreciating how the "economy" is not sealed off from other domains of social action; culture enters deeply into economic and political behavior. For example, the ideology of nationalism was vital to the historical emergence of the nation-state. Many industries that rely on face-to-face interaction, such as investment banking, are heavily conditioned by cultural norms of trust and behavior. Ethnicity and gender roles are critical to knowing how many economies operate.

5. Social relations are a necessary starting point to understanding societies and geographies. Social relations, of course, are only one of several ways with which to view the world; other perspectives begin and end with individual actors. However, to those who view societies as structured networks of power relations and not just the sum of individual actions, the analysis of social relations is indispensable. Social relations, studied through the conceptual lens of political economy, include the uneven distributions of power along the lines of class, gender, ethnicity, and place. A focus on power brings to the fore the role that class plays in determining "who gets what, when, where, and why," that is, the ways in which social resources are distributed, as a central institution in shaping labor and housing markets, as a defining characteristic of everyday life, and as one of the fundamental dimensions of political struggle.

Political economy's dissection of the labor process, and the value-added chains that bring goods and services into our daily lives, allows for a penetration of what Marx called the "fetishization of commodities," the fact that they hide the social relations that go into their making. Given the importance of consumption in contemporary societies, such a perspective allows even the most ordinary of objects (e.g., a can of Coke) to become a vehicle for the illustration of social and spatial relations that stretch out across the globe. Further, the emphasis on social relations allows for an understanding of capitalism as one of many possible types of society, of the specific characteristics of capitalist society, and of the rich insights to be gained from recent investigations into its structural and spatial dynamics, including the periodic restructuring of regions, uneven development, the ways new technologies are incorporated into social systems, boom-and-bust cycles, the service economy, and so forth.

Similarly, feminists have shown how social and geographic reality is pervasively *gendered*, that is, how gender relations are intimately woven into existing allocations of resources and modes of thought in ways that generally perpetuate patriarchy. To ignore gender is to assume that men's lives are "the norm," that there is no fundamental difference in the ways in which men and women experience and are constrained by social relations. A wealth of feminist scholarship on everything from employment to housing to the family has made this view an essential part of economic geography. Thus, spatial patterns of work and daily life are constructed around gender relations, including spatial differences between men and women in housing, work, and commuting patterns; how such relations typically favor men and disadvantage women; as well as how gender-based meanings saturate particular places.

MODES OF THEORIZING IN ECONOMIC GEOGRAPHY

Different generations of economic geographers have sought to explain local and global economic landscapes in different ways at different moments in time. In short, economic geography is an evolving discipline whose ideas are in constant flux. There is no "one" economic geography; there is only a large array of different economic geographies from which to choose. Three principal schools of thought that have long played key roles in this subdiscipline are examined here: location theory, political economy, and poststructuralism.

Location Theory

In the 1950s and 1960s, the introduction of computers and statistical techniques provided a framework for analyzing location decisions of firms and individuals and spatial structures (e.g., land-use patterns, industrial location, settlement distributions). This approach is called logical positivism, which emphasizes the scientific method in the analysis of economic landscapes, including the formulation of hypotheses, mathematical analysis, and predictive models.

An important part of this perspective, **location theory**, attempts to explain and predict geographic decisions that result from aggregates of individual decisions, such as those that underlie the locations of companies and households. Many location theorists modeled **spatial integration** and **spatial interaction**, the linking of points through transport networks and the corresponding flows of people, goods, and information, including commuting and migration fields and shopping patterns. Others sought to uncover the location of the elements of distribution with respect to each other, such as the hierarchy of cities. Spatial structures limit, channel, or control spatial processes; because they are the result of huge amounts of cumulative investment over years and even centuries, large alterations to the spatial structures of towns, regions, or countries are difficult to make, and thus change slowly. Spatial structure and social process are circularly causal: Structure is a determinant of process, and process is a determinant of structure. For example, the existing distribution of regional shopping centers in a city will influence the success of any new regional shopping center in the area.

Location theorists developed and applied a variety of models to understand economic and demographic phenomena such as urban spatial structure, the location of firms, influences of transportation costs, technological change, migration, and the optimal location of public and private facilities such as shopping centers, fire stations, or medical facilities. Models distill the essence of the world, revealing causal properties via simplification. A good model is simple enough to be understood by its users, representative enough to be used in a wide variety of circumstances, and complex enough to capture the essence of the phenomenon under investigation. Typically, models were developed, tested, and applied using quantitative methods.

All models are simplifications of the world based on assumptions, and location theory tended to assume a world of pure competition in which entrepreneurs are completely rational and attempt to maximize profits with perfect knowledge of the cost characteristics of all locations. This image of an entrepreneur became known as ***Homo economicus*** ("economic person"), an omniscient, rational individual who is driven by a single goal—to maximize utility (or happiness, for consumers) or profits (for producers). Essentially, location theory reduced geography to a form of geometry, a view in which spatiality is manifested as surfaces, nodes, networks, hierarchies, and diffusion processes.

Critics of spatial analysis note that this approach emphasizes form at the expense of process and tends to portray geographies as frozen and unchanging. The positivist approach is silent about historical context and politics, class, gender, ethnicity, struggle, power, and conflict, all of which are absolutely central to how the world works. By not taking history seriously, this approach fails to explore the origins of contemporary processes and patterns in the past. Location theory tends to represent people as simply points on a map, abstracting them from their social worlds, as if they did not think and feel about their surroundings. Critics questioned the relevance of overly abstract mathematical models based on questionable assumptions that failed to capture the richness of political and social life. **Behavioral geographers** challenged the simplistic view of behavior as represented by *Homo economicus* and pointed to the complex ways in which spatial information is acquired perceptually and interpreted cognitively in a world of imperfect information. Others noted that location theory tends to reflect the status quo and is incapable of providing a comprehensive explanation of how geographies are tied to social, not simply individual, behavior. Location theory tends to have an inadequate understanding of inequality and how it is produced and reproduced.

Political Economy

Political economy is a way of viewing societies and geographies as integrated totalities, that is, as unified wholes with a structure that exceeds the sum of individual behaviors. In this view, *social* relations cannot be reduced to individual actions. As the term implies, this school includes both the political and economic realms and refuses to separate them: Economies are thoroughly political entities, and politics and power are inseparable from economics in many forms. Political economy is further closely related to the field of institutional economics, which analyzes the importance of formal and informal rules of behavior for economic outcomes, for instance, norms of trust and cooperation, private property rights, courts, parliamentary systems, and constitutions. Political economy is focused on the interactions between political agents; their institutional frameworks; the structure of class, power, and inequality; and social and economic constraints to individual behavior.

Because political economy embraces an enormous set of topics, it is useful to decompose the term into its constituent parts. Broadly speaking, economics may be defined as the study of the allocation of resources, including the production, distribution, and consumption of goods and services. More bluntly, it is the analysis of "who gets what, when, where, and why." Some people, lacking historical depth, erroneously assume that *economy* is synonymous with *market*, that is, that supply and demand and profit-maximizing behavior are universal phenomena throughout all space and time. A historically sensitive perspective, however, reveals that markets are only one possible way in which economic systems are organized, and fairly recent ones at that, emerging as the world's predominant mode of production only in the sixteenth century. Hunting and gathering, slavery, feudalism, and socialism are other, albeit largely extinct, forms.

The other component of political economy is politics, which may be loosely defined as the struggle for power. Power is a fundamental characteristic of all societies and, of course, takes many forms, including violence, personal charisma, status, the capacity to withhold favors, control over information, bureaucratic rank, the self-policing of ideology, the role of the state, and so on. Any time two or more individuals are gathered together, power relations exist in one form or another. One of the strengths of political economy is how it shows the multiple ways in which politics and economics are deeply interconnected, that is, as two indivisible sides of the same coin.

Political economists, many of whom were influenced by Marxism, charged that traditional theories of spatial organization obscure more than they reveal. In their view, location theories are narrowly conceived and blind to historical process—thus they are designed primarily to serve the goals of those who wield power. This approach maintains that a focus on the political organization of society and space—the ways in which power is organized and exerted to control resources—is fundamental to understanding space. Power is a fundamental part of how any social system is organized, and the economy and politics cannot be divorced, for power and wealth are always closely linked. Power is always unequally distributed among and within societies, and for political economists, therefore, social and spatial inequalities figure front and center in their analysis. Any understanding of economic geography, of who is relatively rich and powerful and who is poor and holds less power, must therefore invoke some notion of economic class, as well as gender, ethnicity, and other types of social relations. Political economists argued that the positivist views of human behavior were seriously undersocialized; that is, they ignored the social context in which people live and which deeply shapes what and how they think.

In contrast, political economy maintains that social relations cannot be reduced to individual behavior, that societies are more than the sum of their parts. Political economists dismiss the notion of the "free market" as a myth with little basis in reality; instead, there is capitalism, which is simultaneously economic, political, cultural, and spatial. As we shall see, government intervention is a hugely important part of how economic landscapes are created, unrealistic notions of the "free market" notwithstanding. Capitalist societies are defined by a particular configuration of economic relations centered on profit and accumulation, which arose in the sixteenth and seventeenth centuries, gradually coming to take over most of the world and uniting it today in a single, global division of labor. Thus, to understand economic landscapes we must understand their historical development, the class structure of a society, its relations of gender and ethnicity, and how these are tied to culture and ideology.

For political economists, economic landscapes are the products of changing social relations of power and wealth that organize space in a broad array of historically distinctive forms. To understand the developing world, for example, political economists maintained that one must examine the long history of colonialism, the dynamics of the contemporary world that perpetuate poverty and injustice, the behavior of transnational corporations (TNCs), and government policies. As more economic geographers delved into political economy, it gradually became the primary mode of analysis, an important distinction between the contemporary disciplines of geography and economics.

Poststructuralist Economic Geography

More recently, **poststructuralists** in geography and other disciplines have initiated yet another change in how we view the economy and economic landscapes. This perspective includes a wide diversity of views, but essentially poststructuralists maintain that the dynamics of capitalism cannot be understood independently of the modes of thought used to conceive, represent, and understand them. Capitalism thus does not simply exist outside of people's minds, but also inside of them. Thus, capitalism is as much "cultural" as it is "economic" and "political," and these distinctions are arbitrary; there is no reason to privilege economic relations over cultural ones. Rather, how we know the world shapes how we behave: Social discourses (e.g., maps, the news, popular conceptions) don't just reflect reality, but enter into its making. Poststructuralists thus put great emphasis on the nature of language and representation, on symbolic signification. This view tends to emphasize the complexity and randomness of social and spatial behavior. Rather than view a society and geography as a neatly organized totality, poststructuralists argued that there are instead multiple, overlapping networks of people and activities that cannot be neatly captured by a single worldview, and that we should accept the inherent complexity and messiness of the world we try to understand.

Poststructuralists initiated a "cultural turn" in geography that holds that the economy must be embedded within culture (i.e., that economic relations are always ones among people, emphasizing the role of signs and language in the production process). This view opened up areas for study that had long been ignored, such as geographies of consumption. In this view, there is no single, objective view of the world, only multiple, partial perspectives, each of which is tied to different power interests. The dominant views that naturalize the world thus tend to be those of the powerful, although there is always room to challenge them.

Economic geography has thus been characterized by major changes in thinking, and today several schools of thought coexist, often with heated debates among them. While the subdiscipline retains its long-standing interest in location theory and quantitative modeling, it has also steadily reduced the boundaries between analyses of the economic and the political, between economy and culture, between society and nature. Such bifurcations often distort more than they clarify, and economic geography today borrows freely from many points of view. Students of economic geography can learn from all of these perspectives and combine them in creative ways.

Because the reality of the world is inevitably understood from and through a particular worldview, it is essential that we are aware of different theoretical systems, their assumptions, strengths, limitations, and conclusions. For this reason, this text uses a comparative approach in which different perspectives are explained and contrasted. Looking at the world through different ideological lenses better enables us to meet the challenge of world development problems. The way in which a society answers the central questions of economic geography depends on its historical context, class and gender relations, the role of the state, its position in the world system, and cultures and ideologies.

CAPITALISM

Capitalism is the economic, social, political, and geographic system characterized by the private ownership of the economic means of production (the resources, inputs, tools, and capital necessary to produce goods and services). Because capitalism dominates the world today, its origins, structure, and changes are a central theme of this text: In many ways, economic geography today is the study of capitalist landscapes in various ways. Capitalism arose in Western Europe in the late fifteenth and sixteenth centuries, and, in the form of colonialism, ultimately came to be spread over most of the contemporary world.

The fundamental (but not the only) institution involved in the organization of factors of production in capitalist economies is the market, by which buyers and sellers interact through supply and demand on the basis of price. The guiding imperative in capitalist economies is **profit**, the difference between revenues a firm receives and its production costs. Profit dictates how capitalists behave as a class and how the market operates, and usually pushes other concerns aside. Only profitable products will be produced, based on market demand and price. Prices reflect the utility and value of goods, based on consumers maximizing their own interests, although demand is created through advertising. How and where goods are produced is based on labor and technology efficiency and the spatial distribution of production costs. In competitive market economies, the most efficient producers are the survivors; their production processes and locations will dictate how and where goods will be produced.

Capitalism features two major groups of decision makers—private households (and individuals) and businesses or corporations. The mechanisms that operate to bring households and businesses together are the resource market and the **product market**, which refer to the supply and demand for the inputs and outputs of the production process, respectively (Figure 1). Thus, resource markets organize **capital**, **land**, and **labor** to produce goods and services; product markets consist of buyers and sellers of those outputs. These markets are tied together through flows of capital (between businesses and resource markets), labor and wages (between households and resource markets), consumption expenditures for goods and

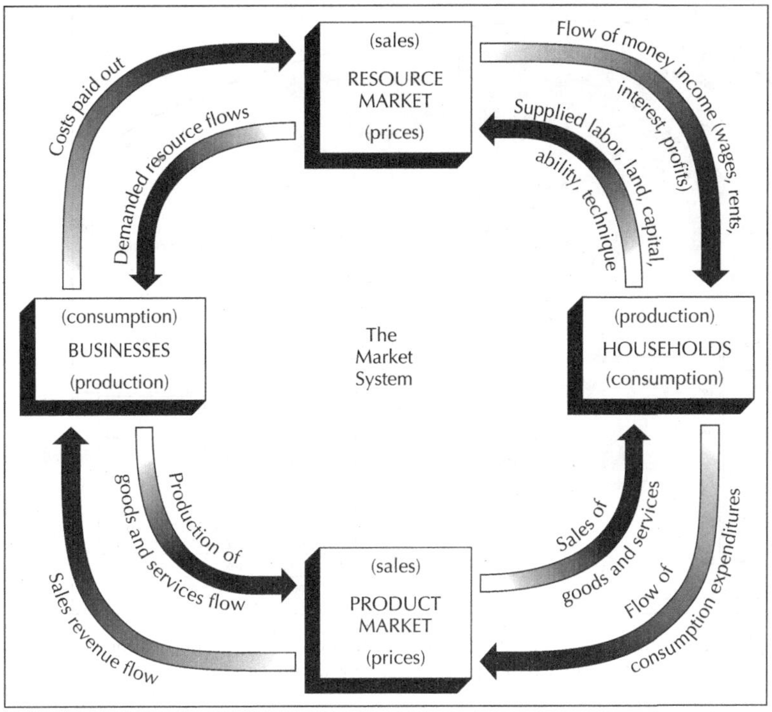

FIGURE 1 Circular flows in the capitalist economy. The circular flow in the capitalist economy involves a resource market where households supply resources to businesses and where businesses provide money income to households. It also consists of the product market where businesses manufacture and produce goods and services for households, while households provide money revenue from their wages and income to consume such goods and services. In the resource market, shown in the upper half of the diagram, households are on the supply side and businesses are on the demand side. The bottom half of the diagram shows the product market; households are on the demand side and businesses are on the supply side.

services (between product markets and households) (Figure 2), and sales revenues and profits (between product markets and businesses).

By adding the value of all the goods and services produced in a given country in one year, we can estimate its gross domestic product (GDP). (A similar measure, gross national product, GNP, includes the value of the activities of domestic companies in countries outside their borders.) Dividing each country's GDP by its population yields per capita GDP, a frequently used yardstick of quality of life (Figure 3). It is important to remember that maps and tables of abstract numbers reflect real-world conditions in which people live, work (Figure 4), find meaning and happiness, often suffer, and die. The United States, with a GDP of roughly $14 trillion, is the world's largest economy (Figure 5), followed by China and Japan. As economies grow and decline, the relative sizes of their GDPs change over time. However, although it has waxed and waned

Monkey Business Images Ltd/Photolibrary

FIGURE 2 Systems of advanced commodity production offer consumers an enormous variety of goods and services from which to choose. However, sales in America are weak, with the economic slowdown. Millions of workers are unemployed, and others have cut spending in order to reduce consumer debt, further slowing the economy. The result has been a persistent combination of weak demand and slowing supply.

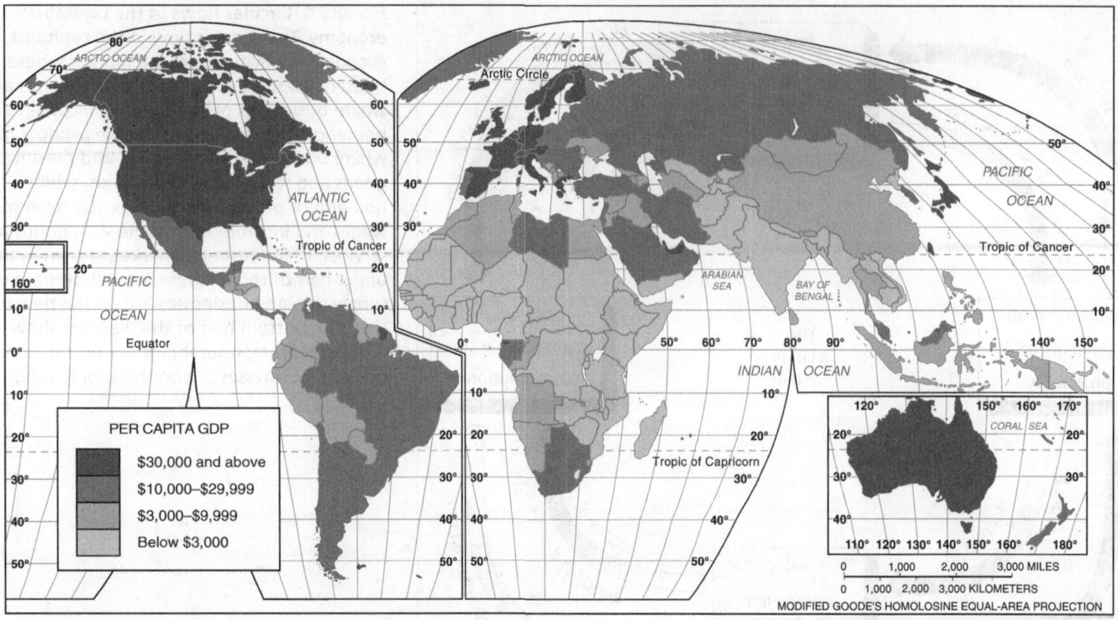

FIGURE 3 Gross domestic product (GDP) per capita—the total of the value of a country's output divided by its population—is the most commonly used measure of wealth and poverty in the world economy, and varies considerably around the world.

over time, the share of total world output produced by the United States, which has about 5% of the world's people, today stands at roughly 25% (Figure 6).

The popular understanding of capitalism holds that it consists just of markets. A commonly held view of capitalism is that it is synonymous with free markets and minimal governmental intervention, a system sometimes called *laissez-faire*. However, historically, truly free markets (with zero government rules) have never really existed; since there has been capitalism there has been a government of some form or another to shape markets. Governments have always been central to creating infrastructure, protecting property rights, providing public services such as education, and shielding producers from foreign competition, including immigrant labor. Indeed, the argument can be made that markets could not exist without some state role. This means that the various forms of capitalism are mixed systems in which both markets

UN/DPI PHOTO

FIGURE 4 Egyptian farmer tilling the soil. This field is being prepared for growing cotton, to meet a worldwide demand for cotton clothing. In the future, the poorer countries of the world will have to rely on agriculture to raise their standards of living and to supply the capital they need to create industries. Agricultural production, therefore, must be increased. Some developing countries, such as Egypt, grow a disproportionate amount of nonfood crops for the export revenue it generates.

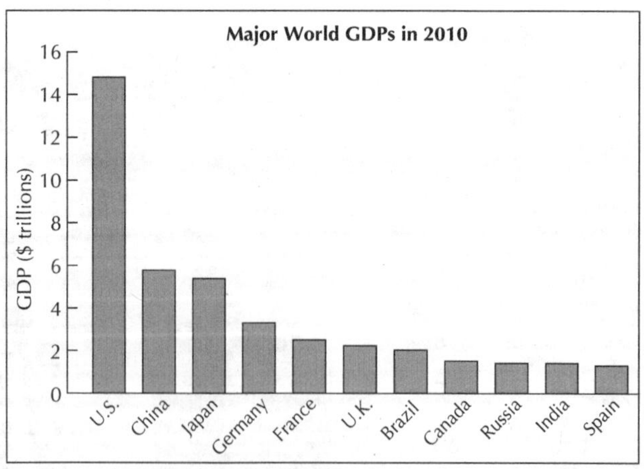

FIGURE 5 Major world GDPs in 2010. The United States, which generated roughly $14 trillion in output in 2010, is by far the world's largest economy and exerts a disproportionate influence over the rest of the world. China, the world's second largest, is rapidly growing, however. Japan and several European states form an important third tier.

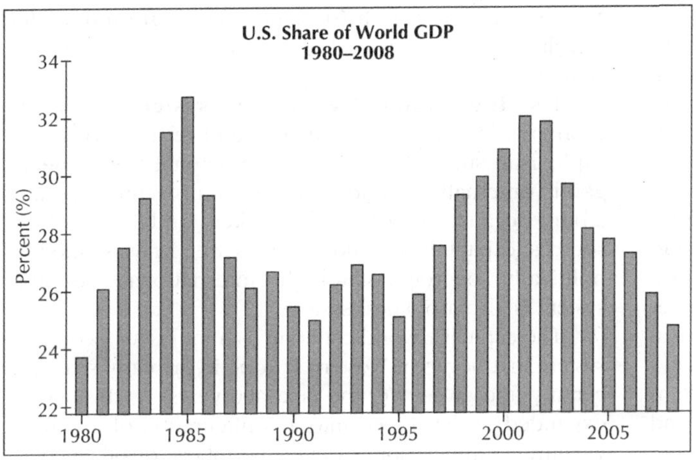

FIGURE 6 The United States, with 5% of the planet's population, has generated between one-fourth and one-third of global output over time. The large share of the planet's output (and consumption) attributable to the United States reflects its abundant resources, high rates of productivity, and political leverage within the world system. Because of globalization and good policies, almost all developing countries are starting to catch up with their rich neighbors. In 2002–2010 more than 85% of developing countries grew faster than the United States, compared with less than a third between 1960 and 2000, and none the century before that.

and governments are important decision makers, including in such vital domains as transportation, education, and health care. The balance of power between markets and governments varies widely among countries and over time, and gives rise to many national forms of capitalism, ranging from those with high levels of government intervention, such as in Scandinavia, to those with relatively little, such as in the United States.

ECONOMIC GEOGRAPHY OF THE WORLD ECONOMY

The focus of this text is the **world economy**, the networks, processes, and institutions that shape the planetary system of resource distribution, create wealth and poverty in different parts of the globe, and contribute to the rise and fall of different national powers. The global scale is only one way in which economic geography can be studied, but given the massive processes of globalization that have been at work, particularly over the last 30 years, it is highly appropriate for understanding what goes on around the world around you. The world economy links far-flung people and places so that what happens in one place shapes what happens in another through networks of interdependence. For example, it is highly probable that the clothes you are wearing now were made in a developing country such as China; that the gas you put in your car came from a foreign source such as Nigeria; that your cell phone or television was made in Southeast Asia; and that financial decisions being made in London or New York City shape your access to credit and the interest rates you pay for loans, mortgages, and credit cards. Every trip to the grocery store is a window on the global economy and an act of participation in it. Seen in this light, the global economy and everyday life are two sides of the same coin.

The world economy is constantly being transformed by a combination of technological and geopolitical forces, which in turn generate a globalization of culture, of the economy, and of environmental issues. Around the world, countries have witnessed the steady growth of large private corporations, the rising role of markets and a diminished role for the state, and lower barriers to trade. Technological changes—improvements in transportation and communications—are reducing the friction of distance and barriers to worldwide exchange. The principal instruments of the globalization of culture are worldwide television, music, and consumption patterns. The principal instruments of globalization of the economy are TNCs, which are producing new efficiencies and new geographies in production, distribution, and the use of the world's resources. The collapse of communism around the world in the 1990s, the implementation of trade alliances such as NAFTA and the European Union, the explosion of banking and finance via telecommunications systems, the rising power of corporations domestically and internationally, and the worldwide reduction of government roles via privatization and deregulation all marked a new round of globalization by removing many institutional barriers to investment and trade. This increased pace of globalization has enormous implications for countries, states, and regions.

Changes in the world economy are simultaneously cultural, technological, political, and environmental. Reductions in transportation costs, for example, have improved exchanges of people and goods. Advancements in telecommunications, including fiber-optic networks and the Internet, have rapidly increased the ease, speed, quantity, and quality of information transactions. Worldwide political-economic changes, ranging from the collapse of communism to widespread deregulation to the declining power of the United States internationally, have diminished the role of the state and increased the power of corporations, often with dire consequences for the poor and powerless. Rising populations in the developing world, and stagnant demographic growth in the developed world, have altered the global supply, demand, and cost of labor, shaping migration patterns. Globalization has accelerated international economic, political, and cultural ties, ranging from corporate investment to trade, to tourism, to terrorism, to the spread of disease. And cultural changes, including the commodification and Westernization of the world's many cultures, simultaneous secularization and growing religious fundamentalism in different places, and mounting awareness of international issues, have played a role in reshaping local and global social movements, consumption, and civil society.

Because transportation and communication costs have fallen rapidly, many local services and goods are becoming available internationally. Worldwide communication systems now allow for companies to subcontract their production and financial operations across continents, wherever prices are the cheapest and quality is the best. The global economy today is spectacularly information-intensive and relies heavily on digital technologies, corporate consultancies, cable television, Internet information services, and software systems design, programming, and application. International finance has also become both global and computerized, and capital markets are now highly mobile for all forms of marketable equities and securities, stocks, bonds, and currency transactions. The globalization of finance has been accelerated by financial deregulation—the removal of state controls over interest rates, tariffs, barriers to banking, and other financial services.

The world is full of problems—debt, unemployment, poverty, inadequate access to health care, food shortages, and environmental degradation—that have serious consequences for the lives of every person on the planet, including you. Such problems are rooted in the structure and development of the world economic system. Understanding the reasons for such problems begins by recognizing the long domination of the world system by developed countries and the existence of an **international economic order** established as a framework for an international economic system.

The **international economic system**, or world economy, includes the institutions and relations of global capitalism, such as global flows of capital (investment), goods (international trade), information, technology, and labor. Because international markets and flows of resources, capital, labor, and products are always shaped by politically sovereign states, the international economic system is also a political system.

The capitalist world economy is a multistate economic system that began in Western Europe in the early sixteenth century and grew over the next 400 years. As this system expanded, it developed into a configuration of a core of wealthy countries dominating a periphery of other countries. One common division of countries is into First, Second, and Third Worlds, a categorization that was a product of the politics of the Cold War. The **First World** includes the economically developed countries of Europe, the United States, and Canada, Australia, New Zealand, and Japan (Figure 7). The defining feature of these countries, which comprise about one-quarter of humanity, is their relatively high standard of living, characterized by a large middle class. The **Second World** was represented by the Soviet Union and Eastern Europe, a designation that lost its meaning in the post–Cold War era (the 1990s and since), when the Second World disappeared, to be divided between the First and Third. Important to understanding this division are differential rates of economic growth, which vary over time and space. Generally, the world's

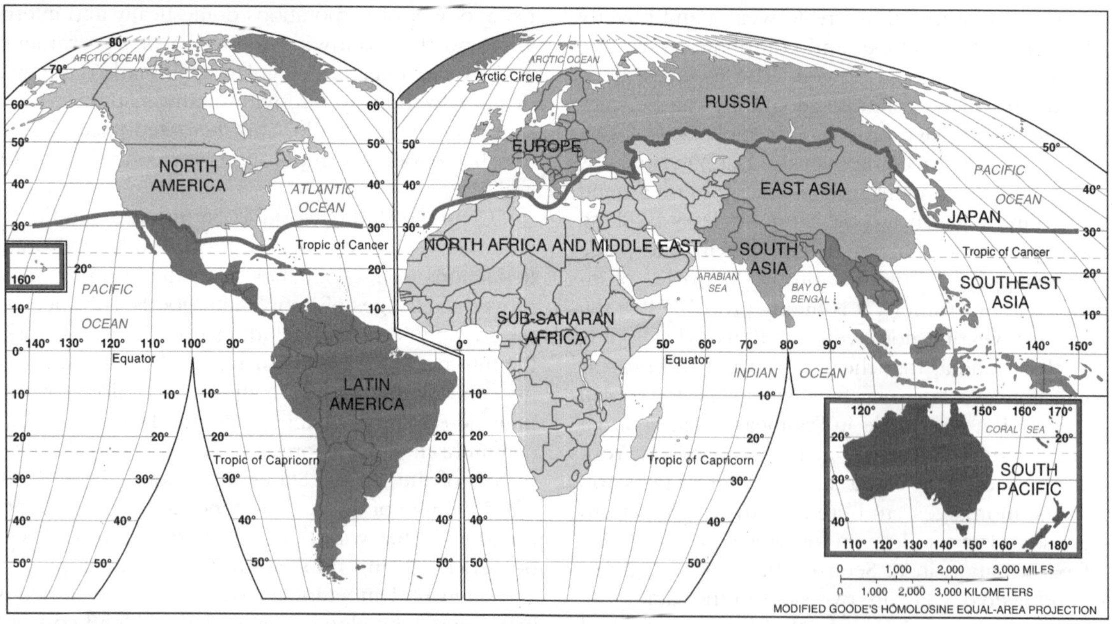

FIGURE 7 Major world regions can be divided into the First and Third Worlds, also called the global North and South. The former Second World, which consisted of the Soviet Union and its client states, expired with the collapse of communism in the 1990s and has been divided between the two. The First World, or global North, includes the developed states in Europe, Russia, Japan, and North America, as well as Australia and New Zealand. The Third World, or global South, includes everyone else, the relatively less developed countries in Latin America, Africa, the Muslim world, South Asia, East Asia, and Southeast Asia.

economies collectively grow between 3% and 5% annually (Figure 8). When a country's economy grows more rapidly than does its population, the average standard of living is likely to rise, although this depends heavily on how wealth is distributed by class, ethnicity, gender, and region; growth in which new wealth is accumulated at the top does little for the bulk of people. Conversely, when population growth exceeds that of the economy, the average standard of living is likely to decline, although there are other drivers of falling social mobility such as economic crises.

The poorest and generally weakest countries are in the underdeveloped **Third World**, sometimes also called *developing* or, a bit more accurately, *less developed* countries. The Third World consists of Latin America, Africa, the Middle East, and Asia, a broad set of diverse societies with a great range of cultures, historical backgrounds, and standards of living. A few countries have climbed out of the Third World, such as Singapore and South Korea, to enjoy standards of living that rival those in the First World. Whether Russia should be considered a First or Third World country is open to debate; its GNP per capita, after all, is lower than that of Mexico. Some observers even identify a Fourth World as a subset of the Third World, the poorest countries on earth (located mostly in sub-Saharan Africa).

At any given time, the world economy is dominated by one or more core states. In the nineteenth century, the era of the *Pax Britannica*, or period of peace dominated by the British Empire, Britain was the world's only economic and political superpower. The British navy ruled every ocean in the world, and the sun never set on the British Empire. By 1900, however, the United States overtook Britain as the world's largest national economy, and after World War II the United States displaced Britain as the world's leading superpower, a status it still enjoys today, even if its dominance is gradually eroding in the face of mounting competition. The United States is still unquestionably the largest economy in the world, although its standard of living is not the highest (falling below several countries in northern Europe). It used its wealth and power to mold the international economy to reflect its interests and those of its allies, setting up, for example, the World Bank, the International Monetary Fund (IMF), and the World Trade Organization (WTO). As the **hegemonic power** in the world, the United States created institutions that were required to establish international economic order in tune with its ideals of free trade and investment (although critics allege that "free trade" is a smokescreen for powerful countries to economically invade less powerful ones).

By the 1970s, the relative power of the United States began to decline in the face of intense competition from rival core states such as Japan and Germany. By the late 1970s, the world order created by the United States after World War II began to come to an end. One major factor in generating this change was the Organization of the Petroleum Exporting Countries (OPEC)–induced petroleum crises of 1973 and 1979, which dramatically increased the price of a critical input into industrialized economies and plunged them into recession. The "petro-shocks" dealt a significant blow to the world economy, driving up heating and transportation costs, exacerbating unemployment, accelerating deindustrialization, and curtailing many people's standards of living, essentially ending the post–WWII

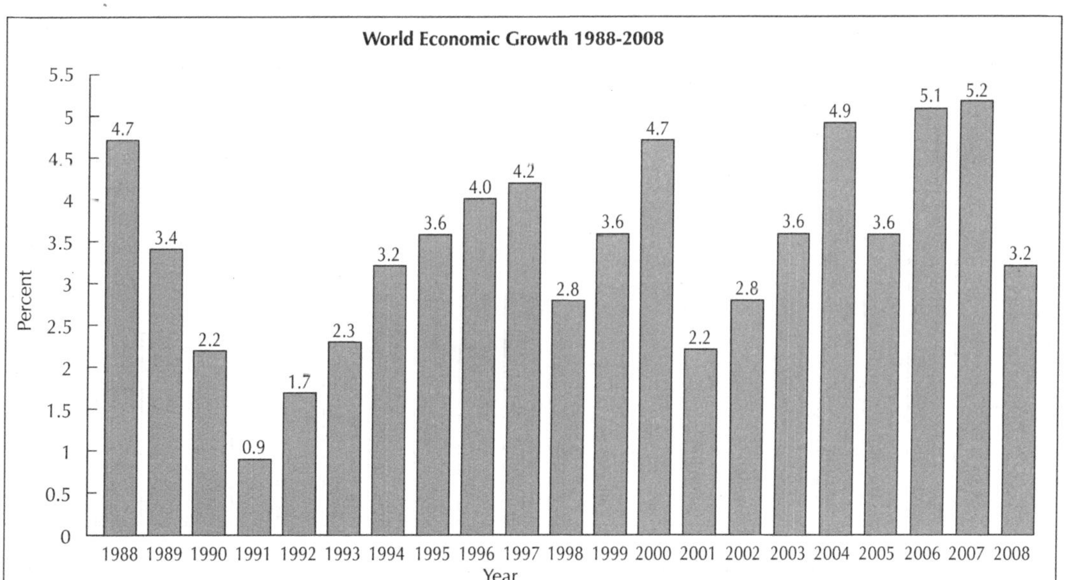

FIGURE 8 Average annual world economic growth rates vary as business cycles create boom and bust periods, but average between 3% and 5% annually. Fluctuations in average growth reflects international recessions, the price of oil, productivity growth (including the impacts of new technologies), catastrophic events such as the Indian Ocean tsunami, changes in government policies, and political turmoil or the lack thereof.

economic boom. A major reason for the breakdown of the postwar world was a decline in the rate of profit of many firms in the industrial West. Faced with intense global competition, firms had to restructure themselves and make organizational and technological changes as well as relocate parts of their operations to the developing world. Some firms went out of business, but others responded to the challenge to automate and to "go international," something they could do in part due to the rising speed of travel and new, digital information technologies.

Out of the old order came the birth of a new one in the 1980s and 1990s. This new world system, in which the Soviet bloc disappeared, left the United States as the world's only superpower. However, U.S. economic hegemony has been increasingly challenged by the rise of the newly industrializing countries (NICs), particularly those in East Asia and especially China. This global order is characterized by highly developed international markets dominated by TNCs, many of which have larger gross output than some countries (Figure 9). States and national governments also play in the global system, managing trade through protectionism, limiting movements of labor, or by reducing trade barriers (e.g., through the World Trade Organization). Like most economies around the world, the American economy has become progressively more globalized, partly as a result of the influx of foreign investment from a variety of nations, mostly in Europe, and from Canada, China, and Japan. Simultaneously, the microelectronics revolution unleashed an enormous wave of change that dramatically affected all domains of production and consumption, particularly in telecommunications and finance, accelerating the globalization of services as well as manufacturing.

GLOBALIZATION

Globalization refers to a complex set of worldwide processes that make the world economy and the various societies that comprise it more integrated and more interdependent. Globalization is essentially an expansion in the scope, scale, and velocity of international transactions. It is a useful way to explain the movements of capital, people, goods, and ideas within and among various regions of the world and their cultural, political, and environmental systems. Among other things, globalization is a process that shrinks the world by reducing transport and communication times and costs among different places. This process has exposed different people in the world to an increasingly homogeneous global culture (largely American in origin), a global market in which more goods and services are freely available everywhere than ever before, and global environmental changes on a scale never before seen.

Globalization should not be simply seen as inherently beneficial or inherently negative in character. Rather, it is a mixture of both sets of qualities that varies widely by place. In some regions, social, political, and economic problems have resulted from a tension between the processes promoting global culture, economy, and environment on the one hand, and the practice and preservation of local economic isolation, cultural tradition, and the localization of environmental problems on the other hand.

We now take a brief look at some of the most important dimensions of globalization that are occurring at an ever-increasing rate in the world today: globalization of culture and consumption, telecommunications, and economic activity, including TNCs, foreign investment, work, services, and information technology.

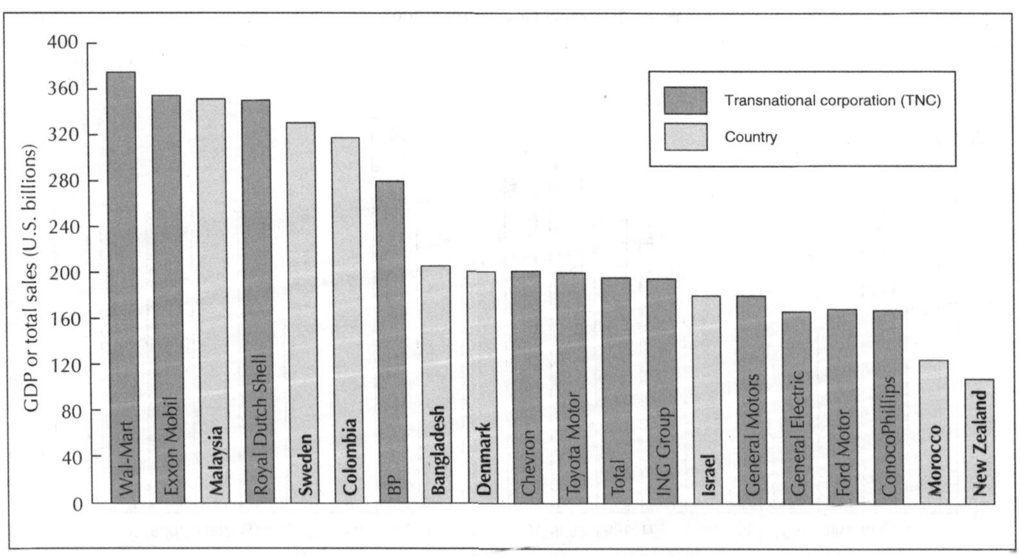

FIGURE 9 Many transnational corporations (TNCs) are larger than some national economies. The relative size of a TNC is important to small countries whose economies are often drastically affected by decisions a global corporation makes. Wal-Mart seems to be a microcosm of the U.S. economy in that shoppers are focusing on savings in this age of austerity. With high gas prices and unemployment rates in America, Wal-Mart's sales are booming.

Globalization of Culture and Consumption

Culture is the total learned way of life of a society. Culture can be defined as that body of beliefs, customs, traditions, social forms, and material traits constituting a distinct social tradition of a group of people. Cultural practices include religion, attitudes toward family size, as well as language, which is the transmission of ideas through languages, symbols, and signs. Historically, different cultures were distinct from one another, but contemporary capitalism has increasingly homogenized cultures around the planet, largely by exporting American culture. Go to any shopping mall in Brazil, South Africa, or Indonesia, and you are likely to hear American music and see American movies, American clothes, and American fast food. For example, large numbers of the world's young people enjoy wearing blue jeans and Nike shoes, consuming Coca-Cola and Pepsi, smoking Marlboro cigarettes, eating McDonald's hamburgers, listening to Lady Gaga, or watching American action movies. (That these examples all involve purchases speaks volumes about the commodification of culture under capitalism; little American culture diffuses elsewhere that does not involve buying.) Thus, for many people globalization is essentially synonymous with Americanization, a fact that often generates resentment against what they view as cultural imperialism. Students of globalization observe an increasingly homogenized global landscape of office towers, stores, restaurants, and service stations. The recognizable logos and visual appearance of retail chains do not vary from one region to another, and customers recognize these logos and building designs in whatever landscape or part of the world they may find themselves. However, the penetration of global culture in different regions across the earth is taking place at different rates; some societies have enthusiastically adopted Western culture, others have walled themselves off from it, and in most countries one finds a mix of many levels of adoption.

Telecommunications

The growth of a global digital telecommunications network greatly enhances the globalization of culture. Because of cable television and international news services, we know a great deal about political and economic events happening anywhere in the world within a few hours. Far-away places are less remote and more accessible now than they were just 10 years ago. Through television, cell phones, and the Internet, we can reach people who live far away, interact with them, and receive pictures and messages from around the world at the click of a mouse.

Citizens in developed countries take such telecommunication innovations as cell phones and cable television for granted. But the entire world is being wired into global networks of millions of personal machines interconnected by fiber-optic and satellite links, which allows essentially instantaneous communications with anyone on the Net. That interchange can include mail, documents, books, pictures and photographs, voice and music, video and television images, and programs and film. The largest of these networks, the *Internet*, includes over 1.6 billion people, or one-fourth of the planet.

The spread of telecommunications is not ubiquitous however, and it generates its own geographies. The world contains significant handfuls of people who have never seen television, used a phone, or ridden in a motor vehicle. Access to the communications of the information age and modern transportation is restricted by an uneven division of wealth worldwide. Even within countries, access may be restricted because of uneven distribution of wealth or because of discrimination against a tribe, an ethnic group, or women.

Globalization of the Economy

Companies, societies, and individuals that were once unaffected by events and economic activity elsewhere now share a single economic world with other companies, societies, and workers. The fate of an aerospace worker in Los Angeles is tied to political changes in Eastern Europe. The job of an auto worker in Detroit is related to the value of the Mexican peso and the auto industry's investment in production plants along the Mexican border. The globalization of the economy has meant that national and state borders and differences between financial markets have become much less important because of a number of trends: (1) international finance; (2) the increasing importance of TNCs; (3) foreign direct investment from the core regions of the world—North America, Western Europe, and East Asia; (4) global specialization in the location of production; (5) globalization of the tertiary sector of the economy; (6) the globalization of office functions; and (7) global tourism.

Globalization involves international financial flows. In the deregulated, hypermobile, electronic world of international banking today, telecommunications has allowed a single global capital market. Computers allow traders to monitor and trade instantaneously in national currencies, stocks, bonds, and futures listed anywhere in the world. Banks, financial houses, and corporations can operate worldwide partly because of the decision centers that control the global economy. Consequently, banks and corporations can react immediately to changes in the value of commodities or gold on the world market and the rate of exchange between the dollar and the euro, the Japanese yen, the Chinese yuan, and other currencies.

Transnational Corporations

The globalization of the economy has been spearheaded by **transnational corporations (TNCs)**, sometimes referred to as multinational enterprises (MNEs). A TNC may conduct research, operate industries, and sell products in many countries, not just where its headquarters are based. Most TNCs maintain their headquarters, offices, and factories in one of the three regions of the core countries—North America (United States and Canada), Western Europe (especially Germany, France, Italy, the United Kingdom, and the Netherlands), and Japan. In 2000, TNCs employed 100 million people in the core regions, and 20 million elsewhere.

In 1970, the world's 15 richest nations were host to the headquarters of 7500 TNCs. However, by 2000 these same countries hosted 25,000 TNCs. According to the World Investment Report by the United Nations Conference on Trade and Development, there are some 53,000 TNCs in the world today, controlling about 40% of all private-sector assets and accounting for a third of the goods produced for the world's market economies. They employ 100 million people directly, which is 4% of the employment in developed regions and 12% in developing regions. In some countries, TNCs are responsible for extremely high proportions of total domestic production—the mining, manufacturing, and petroleum sectors of the Canadian economy, for example. TNCs also play a disproportionately dominant role in other developed countries, such as Belgium, France, the Netherlands, Italy, Britain, and Japan. Sales of goods and services by very large TNCs exceed $100 billion annually, and the sales of the largest TNCs are larger than most countries' total economies, making the decisions of a global corporation important to a small country's economy. Today, TNCs control more than half of all international trade simply via intracorporate transfers of components, services, investments, profits, and managerial talent among their scattered plants and offices in various countries. Most of this intrafirm trade is not finished products and services, but components, subassemblies, parts, and semifinished products.

Thus TNCs, not countries, are the primary agents of international trade, largely between and within their organizations. In effect, TNCs change countries' reserves of resources by moving human and physical capital and technology from one part of the world to the other, creating a new asset base, and allowing production and manufacturing to occur in outsourced locations where they may not have happened otherwise. A TNC will operate in a country where a set of characteristics taken together is more attractive: location, resource endowments, size and nature of market, and political environment. Further, the TNC is able to use transfer pricing, the practice of setting prices for goods and services provided by subsidiaries so as to transfer taxable profits to countries that have the lowest corporate tax rates.

TNCs are able to compete on a world scale due to their transnational communications ability that allows them to share information, via the Internet and satellite and fiber-optic communication systems, with their subsidiaries and branches throughout the world. This is a tremendous advantage in that all components of the TNC can stay aware of markets, products, labor, and business opportunities. TNCs also have the advantages of large stores of capital, technological and managerial skill, and overall economies of scale.

Foreign direct investment (FDI) refers to investment by foreigners in factories that are operated by the foreign owners of a TNC. TNCs in the United States are most likely to invest in Europe, Canada, and Latin America. Western European TNCs are most likely to invest in eastern European, Russian, and African markets, as well as in North America. Japanese transnationals are most likely to invest in Asia and in North America.

Since the 1980s, governments in the three core regions where TNCs are based—North America, Europe, and East Asia—have made changes to accommodate international corporate capital, altering tax codes and regulations that formerly hindered transnational operations. Other countries where TNCs wish to invest, especially developing countries, have also modified their laws, taxes, and regulations to encourage transnational operations within their borders. These accommodations, often labeled "neoliberalism," have changed the relationships between countries and corporations, favoring the latter over the former.

Globalization of Investment

The direction of the world economy is centered in the core regions—North America, Western Europe, and Japan—as well as the Pacific Rim. From the three major world cities, or command centers, in New York, London, and Tokyo, orders are sent instantaneously to factory shops and research centers around the world because manufacturing production and assembly lines and lower-cost offices have been located outside the high-cost core countries. For example, most U.S. sportswear companies, which are centered in New York City and Los Angeles, have moved their production to Asian countries. Latin America, Africa, and Asia contain three-fourths of the world population and almost all of its population growth. These countries find themselves on the periphery of the world economy, suffering a sustained lack of foreign investment; this pattern is the result of centuries of colonialism and a world system in which the rules often work against them.

Three trends are apparent in foreign direct investment (FDI) in developing countries. First, the proportion of FDI that core countries are allocating to periphery countries is declining. Core countries increasingly invest in one another. Second, FDI is becoming more geographically selective. Countries that attract the greatest FDI from the core countries are those that have chosen the export-led strategy of economic growth. Countries welcome foreign investment in order to build factories that will manufacture goods for international markets and employ local labor. Export-led policies rely on global capital markets to facilitate international investment and global marketing networks to distribute the products. The countries that have grown the fastest in recent decades have generally followed the export-led approach as opposed to the alternate approach, import substitution.

Locational Specialization

In the global economy, every location plays a distinctive role based on its particular combinations of assets and weaknesses that have evolved over time, and TNCs assess the economic and locational assets of each place. The original factors of production in global development—

population and resources—are declining in importance and being replaced by specialization. Today, brain power has largely replaced muscle power as the primary source of wealth in the world and transmaterialization (substitutability among inputs) has changed the nature of resources. Input factors and components move intrafirm, final goods are fabricated close to the point of consumption, and national boundaries count much less than they did in the global economy of the past.

In the new global economy, TNCs maintain a competitive edge by correctly identifying optimum geographic factors and locations for each of its activities, including engineering systems, **raw material** extraction, production, storage, office functions, marketing, and management. Suitable places for each activity may be clustered in one country or may be disbursed in countries around the world. The resulting globalization of the economy has increased economic differences among have and have-not places in the world. Factories are closed in some locations and reopened in other countries. Some countries become centers of technical research, whereas low-skilled manual tasks are concentrated in others. Changes in the geography of production have created a spatial division of labor in which regions specialize in particular functions. TNCs decide where to locate in response to the characteristics of the local labor force, its skill level, the prevailing wage, and attitudes toward unions, tariffs, and transportation rates. A TNC may close factories in regions with high wage rates and strong labor unions.

Globalization of Services

The globalization of services and consumption also plays an important economic role. For example, U.S. business service exports generate one-third of the nation's foreign revenues, dwarfing auto exports. Business services are essential inputs to TNCs as they expand into the world arena. This international sector includes legal counsel, business consulting, accounting, marketing, sales, advertising, billing, and computer services. Many professionals—architects, software designers, business consultants—market their skills throughout the world. The sector also includes tourism, education of foreign students, and entertainment—TV, music recordings, and movies. By 2000, 60% of the gross revenues of the five largest U.S. motion picture studios came from outside the United States. As with manufacturing, the globalization of services operates in a world of a declining role for the nation-state but a continuing emphasis on cultural differences at both the national and regional levels.

The influence of the United States is reflected in the global transmission of television shows as intercontinental information networks allow international subscribers access to huge amounts of American culture, its most powerful export. But this will have both positive and negative effects on the culture and the disposition of the world's peoples. It is likely to broaden the common links among the younger generations of the developed world, especially those who are savvy about the Internet and the World Wide Web. At the same time, it threatens to alienate the more conservative elements in those cultures, many of whom have turned to religious fundamentalism.

Globalization of Tourism

Tourism is one the world's fastest-growing industries, employing 230 million people in the year 2009 and contributing about 12% of the world's gross domestic product. The tourism industry is already one of the leading export sectors and is expected to grow at an annual rate of 3% worldwide. The highest rate of growth will take place in some developing countries, especially tropical regions and areas with picturesque scenery and those that provide both natural and cultural attractions for their visitors, along with pleasant climates, good beaches, and attractive social and political milieus. Political stability is critical to this industry.

Information Technology and Globalization

Improvements in communication mean that globalization of the world economy is moving forward rapidly, to a point where many people in any location can receive and send information to others elsewhere at almost any time. Increasingly, the world economy depends on moving information instead of people. This export and import not of products but information will allow innovations to sweep the world at a rapid rate and will exacerbate and increase the disparity between the have and have-not nations. In the future, the designation of "fast" and "slow" societies will refer to the effect that information technology has on the tempo of human affairs. The information-based economy means that the relative success of individuals or groups is based on access to information, more than on money or products and more than on natural resources, labor pools, and other traditional metrics of power and wealth. A global information network will allow a **knowledge worker** in the global economy to mine the databases and other knowledge bases of the world. The world will be interdependent and the interchange of information among researchers via information systems will be facilitated as never before.

Real-time information systems are those that make information available as it happens, or at least as soon as software programs process it and make it available, so that everyone can seek critical information by accessing a computer. This is one of the essential differences between the world economy of the future and the world economy of the past. With real-time information systems, more people will make more decisions in a customized world economy as people who interface with customers become part of a self-managing business unit. Individuals, companies, and TNCs need feedback about their decisions as soon as it becomes available so that they can adapt faster and continuously to customers' needs and thus compete more effectively. Real-time information systems allow business decisions to be made with the minimum of bureaucracy.

The communications and information technology (IT) revolution has come about through the networking of individual computers, which are linked to global networks of personal computers and information databases. These communication networks, which include the Internet, allow instantaneous communication with anyone else on the network. Communication can include photographs, voice and music, videos, television images and programs, films, documents, books, pictures, mail, and spreadsheets. High-speed Internet connections allow users to shop online at tens of thousands of stores, make reservations at hotels in almost any country in the world, buy airplane tickets, monitor the weather and stock market, pay bills, and read, comment on, and even contribute to newspapers, magazines, and encyclopedias.

GLOBALIZATION VERSUS LOCAL DIVERSITY

Globalization has affected different regions in different ways, and therefore must be understood geographically. Generally, it has damaged but not completely destroyed unique local diversity. Many current political, social, and economic problems arise from the tension between forces promoting globalization of the culture and economy versus those striving to preserve local cultural traditions and economic self-sufficiency. The desire to retain traditional economies and cultural preferences in the face of increasing globalization has led to political conflict, social chaos, and market fragmentation in more traditional regions of the world.

Globalization and local diversity will coexist and shape each other, a development some geographers call "glocalization." The hypothesis of uneven fragmentation—the world economy produces different results in different places—accommodates continuing antagonism between globalizing and localizing tendencies that will, even if unevenly, coexist with each other. For this to take place, individuals must appreciate that they can advance both local and global values without damaging either and that multiple loyalties to different local, national, and transnational affiliations need not be mutually exclusive. People can be loyal to family, community, country, and the world's people simultaneously. In a globalized world, more and more people become aware of the extent to which their well-being is dependent on events and trends elsewhere.

PROBLEMS IN WORLD DEVELOPMENT

The structure, behavior, and impacts in time and space of the world economy are highly uneven. Temporally, world economic growth rates have waxed and waned, dropping during recessions and rising during years of prosperity, and rates of economic growth are very uneven among different world regions. As growth plays out differentially over different regions it generates new geographies of wealth and poverty. High growth rates, such as those that have occurred for decades in East Asia, pull people out of poverty and create a middle class. Low economic growth rates, such as those in Latin America and especially Africa, mean that people's standards of living increase slowly, or not at all, depending on labor markets, rates of inflation, unemployment, and population growth.

Despite the economic progress in many parts of the world, there are still vast areas of the planet in which billions of people remain mired in deep poverty. Much of the world has not benefited from globalization. Economic development, and the lack of it, are thus important questions for economic geographers. Development is a concept full of hope, even though the jolts and dislocations can be horrendous when long-standing traditions and relationships are broken down. The purpose of **development** is to improve the quality of people's lives—to provide secure jobs, housing, adequate nutrition and health services, clean water and air, affordable transportation, and education. Whether development takes place depends on the extent to which social and economic changes and a restructuring of geographic space help or hinder in meeting the basic needs of the majority of people.

Problems associated with the development process occur at every level, ranging from a Somalian villager's access to food and a health clinic to international trade relations between rich and poor countries. Our attempts to understand development problems at the local, regional, and international levels must consider the principles of resource use as well as the principles surrounding the exchange and movement of goods, people, and ideas. Two critical issues require immediate attention. One is the challenge to economic expansion posed by the environmental constraints of energy supplies, resources, and pollution. The other element is the enormous and explosive issue of disparities in the distribution of wealth between rich and poor countries, urban and rural areas, wealthy and poor people, dominant and subordinate ethnic groups, and men and women.

Environmental Constraints

The world environment—the complex and interconnected links among the natural systems of air, water, and living things—is caught in a tightening vise. On the one hand, the environment is being stressed by the massive overconsumption and wasteful consumer culture of the developed world. On the other hand, the environment is being squeezed by the poor people in developing countries who must often destroy their resource base in order to stay alive. The constraints of diminishing energy supplies, resource limitations, and environmental degradation are three obstacles that threaten the possibility of future economic growth.

There is a significant energy problem in much of the developing world. Oil is an unaffordable luxury for much of the world's population, who cook and heat with fuelwood, charcoal, animal wastes, and crop residues. In countries such as India, Haiti, Indonesia, Malaysia, Tanzania, and Brazil, fuelwood collection is a major cause of deforestation—one of the most severe environmental problems in the underdeveloped world.

The fragility of the environment poses a formidable obstacle to economic growth. Are there limits to growth? Is the world overpopulated? Some of our present activities, in the absence of controls, may lead to a world that will be uninhabitable for future generations. Topsoil, an irreplaceable resource, is being lost because of overcultivation, improper irrigation, grassland plowing, and deforestation. Water tables are falling, including in the United States, where, for example, the Ogallala water basin under the Great Plains is in increasing danger of being rapidly depleted. Forests are being torn down by lumber and paper companies and by farmers in need of agricultural land and wood to keep warm or cook their food. Water is being poisoned by domestic sewage, toxic chemicals, and industrial wastes. The waste products of industrial regions are threatening to change the world's climate. Accumulated pollutants in the atmosphere—carbon dioxide, methane, nitrous oxide, sulfur dioxide, and chlorofluorocarbons—are said to be enhancing a natural **greenhouse effect** that may cause world temperatures to rise. El Niño events, or periodic warming of the Pacific Ocean by just 0.25 degrees Fahrenheit, caused violent weather disruptions worldwide, with billions of dollars worth of damage from floods, mudslides, and loss of life. Chlorofluorocarbons, which were used as aerosol propellants and coolants and in a variety of manufacturing processes, are blamed for damaging the earth's ozone layer, which protects living things from excessive ultraviolet radiation from the sun. Yet another hazard to the environment is the fallout from nuclear bomb tests that took place in the 1950s and 1960s and from nuclear power reactor accidents such as those at Three Mile Island, Pennsylvania, and Chernobyl, Ukraine.

Disparities in Wealth and Well-Being

The world economy generates great variations in economic structures, standards of living, and quality of life around the globe. There are enormous differences between the world's richest and poorest nations in wealth and standard of living as measured by economic statistics such as GNP per capita and paralleled by social and demographic measures such as life expectancy, infant mortality rates, literacy, and caloric consumption. In short, maps of economic measures are simultaneously maps of other dimensions of people's lives, including how long they live, the chances that their babies will grow into adults, their ability to read and write, and the quality of the food they eat. These numbers point to the multifaceted nature of poverty and development, which is not just economic but also social and political.

Poverty afflicts relatively few people in economically developed countries, although there are nonetheless disturbingly large numbers of poor in wealthy societies such as the United States, including hunger and malnutrition among families in Appalachia or on Native American reservations, bankrupt farmers on the Minnesota prairie, unemployed factory workers in Detroit, and single mothers on welfare in New York. Deeply entrenched, institutionalized poverty confines billions of people to lives of inadequate food, shelter, health care, transportation, education, and access to other resources. Mass poverty is the single most important world development problem of our time. You cannot doubt this assertion when you see maimed people on the streets of Bombay, begging children in Mexico City, desperate farm laborers in Brazil, emaciated babies in Mali, or women and children carrying firewood on their backs in the countryside north of Nairobi. Mass poverty is ethically intolerable and a critical issue that we must try to overcome.

Who are the world's poor? They are the 15 million children in Africa, Asia, and Latin America who die of hunger every year. They are the 1.5 billion people, or 24% of the world's population, who do not have access to safe drinking water. They are the 1.4 billion without sanitary waste disposal facilities. They are the 3 billion people—50% of the world's population—who live in countries in which the per capita income was less than \$400 in 2005. Half of the world (largely in Africa) earns \$2 per day or less. These numbers are characteristic of impoverished countries in which much economic activity takes place outside of the market. These people are caught in a vicious cycle of poverty (Figure 10), often with few ways out, and lead lives of quiet desperation and hopelessness. Their life expectancies tend to be short, infant mortality levels high, and access to energy, medical care, transportation, and education often minimal. The economic geography of the world is at its core concerned with these social and spatial discrepancies among and within countries.

The poor of the world overwhelmingly live in developing countries, most of them former European colonies, which failed for one reason or another to keep up with the economic levels of the West over the past 500 years. During the worldwide economic boom that occurred in the three decades following World War II (1945–1975), the GNP of the developed countries more than doubled. Although per capita real income in developing countries also rose, incomes in developed countries rose much more quickly. Developed countries enjoyed 66% of the world's increase, whereas half of the world's population in underdeveloped countries (excluding China) made do with one-eighth of the world's income. By 1982, the national income of the United States (then 235 million people) was about equal to the total income of the Third World (more than 3 billion people). In short, over the past half century the rich have become richer and the poor have gained only slightly.

The developing world is far from a homogeneous entity; that is, there are enormous differences among and within developing countries in terms of their historical background, cultures, economies and standards of living, and when and how they were incorporated into the world system. So great are the variations among countries, and often within, that it is simplistic to speak of a single developing world without immediately acknowledging its differences. To lump, say, South Korea, which has a standard of living similar to southern Europe, together with Mozambique, one of the world's poorest states, is to fail to understand the profound differences that separate them.

With the debt crisis of the 1980s, the United States finally discovered it had a real stake in the prosperity of the developing world. The inability of some countries to make payments on their debt placed the financial structures of

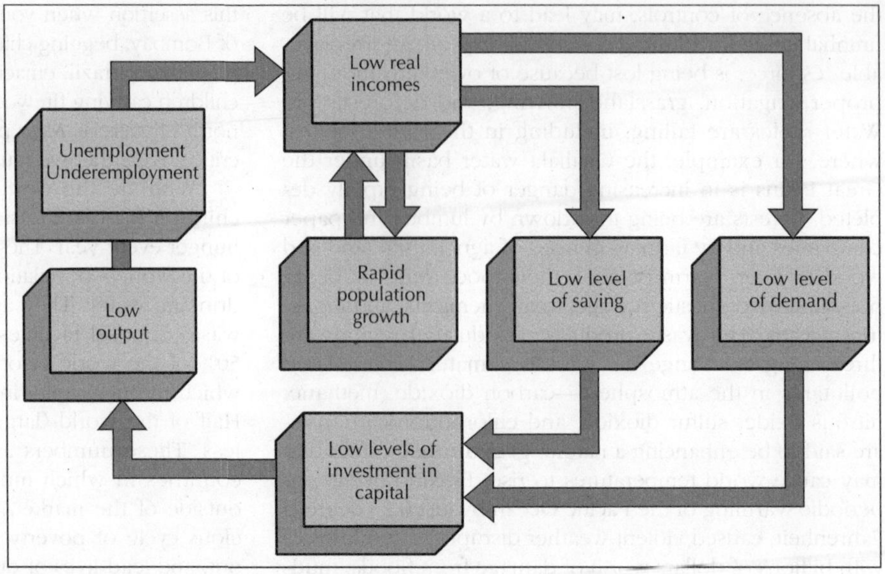

FIGURE 10 The cycle of poverty in Third World countries. Most Third World nations have low per capita income, which leads to a low level of saving and a low level of demand for consumer goods. This makes it very difficult for these nations to invest and save. Low levels of investment in physical and human capital result in low productivity for the country as a whole, which leads to underemployment and low per capita income. In addition, many of these countries are faced with rapid population growth, which contributes to low per capita incomes by increasing demand without increasing supply or output. Yet, the number of people going hungry in the world, as of 2011, has been dropping since 2007, partly due to a recession-fueled drop in world food prices.

the United States and some European nations in jeopardy. Many U.S. banks, including some of the largest, would technically have been insolvent if their loans to developing nations had been declared in default. This led to enormous pressure to resolve the immediate problems of the debt crisis, many of which were directly related to the poor performance of the economies of the debtor nations.

Unfortunately, for many debtors, the solution often proved to be more painful than the problem itself. Under strict rules imposed by the IMF and other international agencies, which believed in market fundamentalism (the narrow notion that only free markets can alleviate social problems), stringent limits were placed on the economic policies of debtors, with the result that a majority of citizens in these nations often found themselves worse off. The goals of **IMF conditionality**, as it came to be called, were to restore growth, reduce central government involvement in the economy, and expand the exports of goods and services while reducing imports so that the debtors would have sufficient earnings of foreign revenue to make payment on the interest and principal of their debt.

There is little evidence that these policies helped to restore economic growth, and they even lowered many people's quality of life, as the former chief economist of the World Bank, George Stiglitz (2002), noted, by forcing drastic cutbacks in necessary government services and forcing currency devaluations that drove up the cost of imports. However, such changes did result in export surpluses that made debt servicing easier. As a consequence, the 1990s saw a remarkable reversal in the flow of financial resources—instead of the flow from rich nations to poor nations to assist in development efforts, there was a flow from poor to rich. But debt repayments have become a serious obstacle to further economic development in poor countries where capital and financial resources are scarce and every dollar lost has repercussions throughout the economy.

Summary and Plan

This text explores the economic geography of capitalism, especially on a global scale. Although it is important to understand the local and national levels of economic activity, the rapid growth of the world economy has increasingly focused attention on processes, problems, and policies at the international scale. In this chapter, we present the geographer's perspective. We provide a definition of the field and introduce the main concepts geographers use to interpret and explain world development problems at a variety of scales, ranging from small areas and regions to big chunks of the world.

Key Terms

behavioral geographers
capital
capitalism
development
economic geography
First World
foreign direct investment (FDI)
globalization
greenhouse effect
hegemonic power
Homo economicus
IMF conditionality
international economic order
international economic systems
knowledge worker
labor
land
location theory
political economy
poststructuralism
product market
profit
raw materials
Second World
spatial integration
spatial interaction
Third World
transnational corporation (TNC)
world economy

Study Questions

1. What defines the geographic perspective?
2. Define *economic geography.*
3. Why are geography and history inseparably linked?
4. Why can't places be studied in isolation from each other?
5. What are some ways in which nature shapes, and is shaped by, the economy?
6. How is the economy related to culture?
7. Why are social relations an important place to begin understanding economic landscapes?
8. Define the term *globalization* and list reasons why it has occurred.
9. What are four ways in which globalization is manifested?
10. What is location theory?
11. What is the political economy approach to geography?
12. How have poststructuralists contributed to the analysis of economic issues?

Suggested Readings

Coe, N., P. Kelly, and H. Yeung. 2007. *Economic Geography: A Contemporary Introduction.* Oxford: Blackwell.

Dicken, P. 2010. *Global Shift: Mapping the Changing Contours of the World Economy.* 6th ed. New York: Guilford Press.

Florida, R. 2004. *The Rise of the Creative Class.* New York: Basic Books.

Friedman, T. 2005. *The World Is Flat: A Brief History of the Twenty-first Century.* New York: Farrar, Straus and Giroux.

Harvey, D. 2006. *A Brief History of Neoliberalism.* Oxford: Oxford University Press.

Knox, P., J. Agnew, and L. McCarthy. 2008. *The Geography of the World Economy,* 5th ed. London: Edward Arnold.

Scott, A. J. 2006. *Geography and Economy.* Oxford: Oxford University Press.

Shepard, E., and T. Barnes, eds. 2002. *A Companion to Economic Geography.* New York: Wiley.

Stiglitz, J. 2002. *Globalization and Its Discontents.* New York: WW Norton.

Web Resources

Association of American Geographers
http://www.aag.org

Summary of economic geography emphasizing career possibilities.

NationMaster
http://www.statemaster.com/encyclopedia/Economic-geography

Short, efficient summary of economic geography, its history and diversity.

U.S. Bureau of the Census
http://www.census.gov

Vast collection of data and reports about the U.S. economy and society, with some international data as well.

Wikipedia
http://en.wikipedia.org/wiki/Economic_geography

Good but brief overview of economic geography.

The Historical Development of Capitalism

OBJECTIVES

- To explore the historical context of capitalism, including its feudal origins
- To provide an overview of the characteristics of capitalist economies
- To document the importance of the Industrial Revolution and its impacts
- To shed light on the relations between colonialism and global capitalism

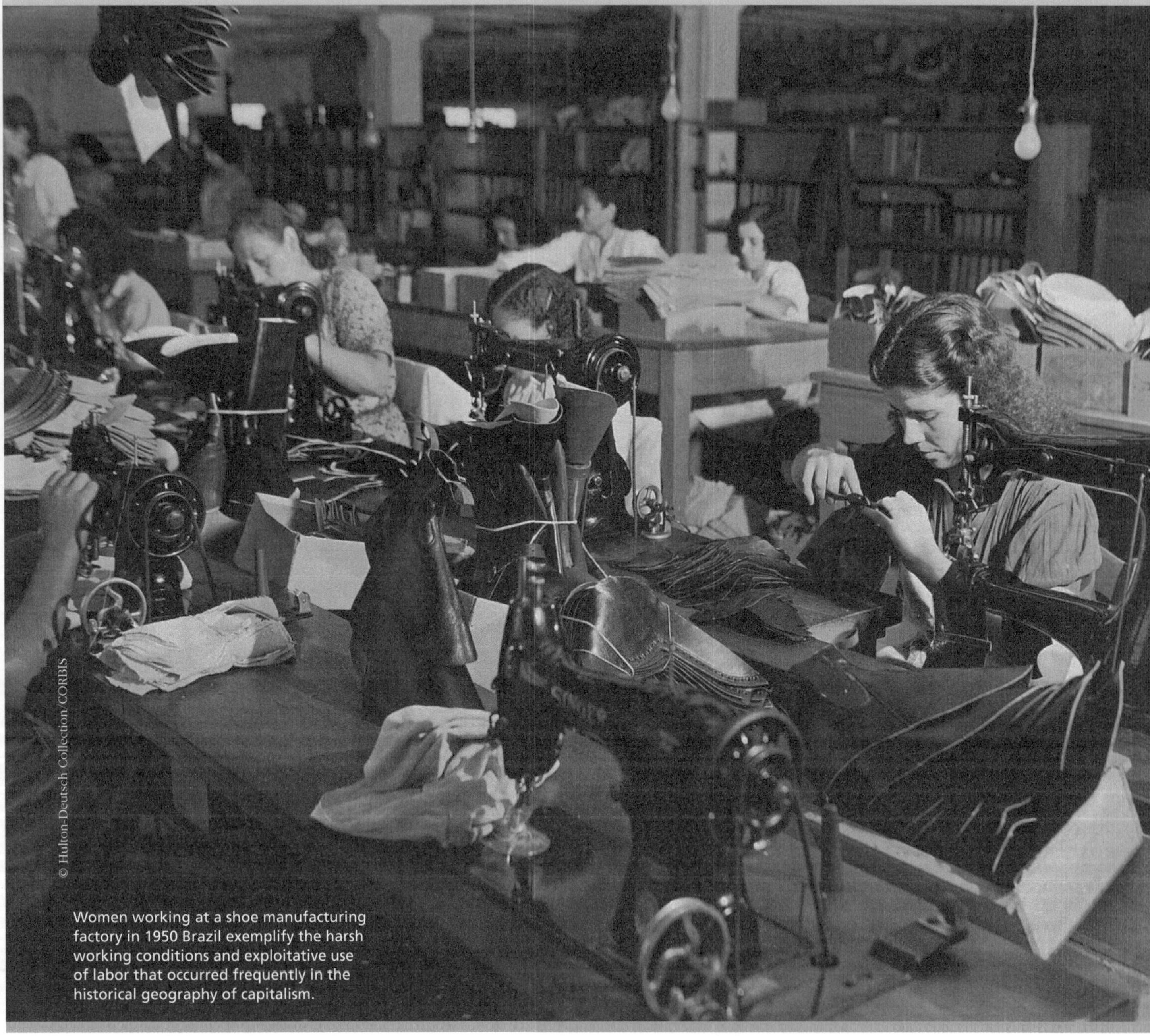

Women working at a shoe manufacturing factory in 1950 Brazil exemplify the harsh working conditions and exploitative use of labor that occurred frequently in the historical geography of capitalism.

The Historical Development of Capitalism

Geographies are not created overnight. The spatial distribution of people and economic activities reflects the imprint of processes that take years, even centuries, to unfold. For this reason, a historical understanding of economic landscapes is absolutely essential for understanding the contemporary geographies of the world. Because the present is produced out of the past, and shaped by it in countless ways, any serious understanding of economic geography must include an appreciation of how the contemporary world came to be. A historical appreciation reminds us that the construction of the modern world took a long time to occur and that the landscapes of the present are constantly changing.

This chapter provides a historical appreciation of capitalism in several ways. First, it delves into the context in which capitalist economies and societies were born and developed, particularly feudalism. Second, it explores the characteristics of capitalism, the features that make it unique. **Capitalism**—the dominant form of production and consumption around the world—is not the only way in which human beings have organized themselves, but came into being in the sixteenth and seventeenth centuries, mostly in Western Europe. Third, this chapter describes the Industrial Revolution, which began in the eighteenth century and marked an exponential increase in the scale and speed of capitalist activities. Finally, it addresses the relations between capitalism and colonialism, the process by which capitalism "went global," spilling out of Europe and effectively conquering the rest of the globe.

FEUDALISM AND THE BIRTH OF CAPITALISM

Human beings have developed many ways of organizing resources and production systems and providing for themselves over time. For the vast bulk (95% or more) of human existence, we were hunters and gatherers, food collectors depending on nature for food and other necessities. The agricultural revolution that began roughly 10,000 years ago saw a major transition in the ways in which people worked and lived, including the establishment of settlements and the first class-based societies. Many early agricultural societies were based on slavery; in Europe, this process culminated in the Roman Empire, which ended in the fifth century A.D.

Prior to capitalism, the prevailing form of economic and social relations was **feudalism**, which lasted for more than a millennium, approximately from the fifth to the fifteenth centuries. Sometimes called the Middle Ages or Dark Ages, the feudal system was deeply entrenched and relatively stable for a long period. Feudalism was not unique to Europe, as other places had a similar social organization, including Japan and to some extent India. Politically, this type of system manifested itself in Europe as a changing series of empires, including the Frankish kingdoms, the Normans, the Holy Roman Empire, tsarist Russia (which lasted until World War I), and the Austro-Hungarian Empire, which ended only in World War I. Indeed, one of the major differences between Europe and the United States is the impact that feudalism had in Europe: In North America, capitalism emerged on a landscape that had not been shaped by more than a millennium of feudalism, as was Europe, including its land use and property systems, cities, and class and gender relations.

Characteristics of Feudalism

Feudalism was marked by a distinct set of interlocking characteristics that made it qualitatively different from capitalism, reminding us of the uniqueness of the economy and social system in which we live and work today.

Compared to the dynamic, ever-changing world in which we exist, feudalism comprised a remarkably stable and conservative world that changed relatively little. To an observer of feudal France in the eighth century and Poland in the eighteenth century, there would appear to be relatively few differences. Most people's lifestyles—working the land in a cycle of endless drudgery—would be the same from one generation to another. Almost everyone lived like their fathers and mothers before them and their grandparents before them. Tradition gave the dominant shape to human experience and to everyone a sense of where they fit into the world. In this sense, feudalism actively discouraged experimentation and change. Under capitalism, in contrast, novelty is the norm, for it helps to sell goods and services in the market.

However, it is erroneous to think of the feudal era as completely static. Indeed, this view arose during the Renaissance, when historians sought to contrast the changes of their day with the alleged stasis of the past. In fact, during the later feudal period there was significant change: Universities were established, new types of farming and new technologies introduced, wetlands drained, forests cleared, plagues and diseases spread, and political conflicts caused enduring changes. The introduction of the longbow and guns in the fourteenth century, for example, made knights essentially obsolete.

In Europe, the church (that is, the Catholic Church until the Protestant Reformation of the sixteenth century) was by far the predominant political/ideological institution. Most people were extremely religious, and their belief in god informed every aspect of their behavior and everyday life. The population fatalistically accepted its lot in life, and the idea of progress, of change for the better, was largely unthinkable. In most towns, the cathedral was the largest and most impressive building, its size and design a testimony to the wealth and power of the church (Figure 1). Local priests, who were often the only ones who could read and write (and even many of them were illiterate), were important actors in the community's spiritual and intellectual life, serving as judges and teachers and officiating at weddings and funerals. Education and schooling emphasized the Bible. In Rome, the Pope exerted great power over kings and nobles throughout the continent, often appointing leaders and threatening to excommunicate those who did not obey. Popes were masters of politics, wealthier than anyone else, and often corrupt. The church owned farmland and hunting estates, raised taxes, and even had its own armies. and the Pope could excommunicate recalcitrant kings.

FIGURE 1 Notre-Dame de Reims, France. The size and beauty of medieval cathedrals, which often took more than a century to construct, testify to the power and wealth of the church during that era.

Under the feudal system, an aristocratic nobility made up the ruling class, whose power lay in the ownership of land, which was the basis of wealth and political power. There were many tiers within this ruling class, including a variety of lords, dukes, earls, barons, and others. Ownership of land was the basis of wealth and political power. Aristocrats typically owned vast estates of farmland, and under the manorial system that characterized feudalism, the extraction of surplus value occurred through the payment of rent by tenant farmers who paid tribute to their local lords, who owned the land. Often the farmers paid one-half or more of their output as rent, in exchange for protection. Thus, rent payments were the primary form of wealth transfer from the poor to the wealthy, which occurred through the state rather than through a market system, for unlike the capitalist system, there was no effective division between public and private property. The aristocracy controlled the reins of government, including the military and penal system, and their private interests were synonymous with those of the state. Knights and the military existed to enforce the rule of aristocratic law and to protect local communities from brigands, robbers, and invaders.

In an overwhelmingly rural society, in which the productivity of agriculture was comparatively low, the vast majority of people were peasants and farmers. Farming under feudalism was based on animate sources of energy, that is, living human and animal muscle power. Peasants and draft animals worked the fields, collected firewood, drew water from wells, and performed the innumerable other tasks necessary to keep their society working. Child labor on the farms was the norm, birthrates were high, and most people lived in large, extended families in small hamlets and villages, many of which were self-sufficient, producing their own food, clothing, and other necessities. Peasants were almost entirely illiterate, ignorant even of events a few miles away, unaware of what century they lived in. Although

Christian, most also believed in witches, spirits, goblins, and other supernatural beings.

Markets existed under feudalism but typically were small and poorly developed, and only the wealthy had the income required to buy luxury goods. Typically, markets consisted of seasonal fairs where itinerant merchants (often Jews) sold metal goods, silks, or jewelry. Thus, feudalism was not a type of society in which markets were the central institution that governed the allocation of resources; rather, this function belonged to the state.

A substantial share of the rural population, but not everyone, consisted of **serfs** (Figure 2). Serfdom was a uniquely feudal institution that differed from economic systems based both on slavery and on capitalism. A serf was not a slave, that is, he or she was not owned by a master. Rather, serfs were bound to the land by feudal law and custom. The standard of living for virtually everyone except the aristocrats and some merchants was very low. Serfs and other farmers (including pools of "freemen") lived a monotonous life in which each day was identical to the day before, doing the same chores, eating the same food, and seeing the same people. Most people lived very simply. Diets were typically inadequate, and malnutrition was common. Famines broke out every few years. Life expectancy in feudal Europe was typically under 50 years. Many women died in childbirth, and infant mortality rates were high. Water supplies were often infected by bacteria, and diseases such as cholera, plague, and tuberculosis took an enormous toll in human lives and suffering.

The British Library/Topham-HIP/The Image Works

FIGURE 2 Serfs were the mainstay of the feudal labor force, producing the agricultural surplus that supported the aristocracy and the state. Serfs lived monotonous lives, paying a large share of their output as "rent" to their local lord. They exemplify a noncapitalist form of labor organization (i.e., without labor markets).

Agricultural work was organized around the rhythms of the seasons, with different tasks for the spring, summer, fall, and winter. Winters, for example, might be spent indoors, weaving or fixing farm implements. Spring was a time of planting. Summers involved tending to the crops and livestock. And the fall was the time of the harvest, for many the central event of the year. The extended families that included several generations often cohabited in one dwelling; grandparents, parents, nieces, nephews, cousins, and infants lived in crowded rooms, often without paved floors. A family sleeping together in one bed, sometimes including dogs and pigs for warmth, was not unusual. Because the vast majority of people were illiterate (even many kings, queens, aristocrats, and priests were unable to read and write), peasants and farmers passed lessons, including superstitions, on to their children through poems and stories.

Although feudal society was predominantly rural and agricultural, there were a few cities and towns. Urban areas under feudalism were very different from those of today. Because agricultural productivity rates were low, and the ability of farmers to support urbanites correspondingly limited, cities were small. Most farming hamlets did not exceed 200 or 300 people, and cities over 10,000 people were rare. Of course, there were a few metropolitan areas, such as Constantinople or London, but these were few and far between. Feudal cities were densely populated, with the inhabitants crowded together, often in very unsanitary conditions. There was no running water or sewer system, and the streets were often covered with mud and animal waste. The centers of feudal cities often consisted of a walled fortress, often with a small palace located within where the local lord lived. As the town grew, new walls would be constructed, leading to concentric rings (Figure 3). Because land was not a commodity to be bought and sold, but allocated on the basis of power, there was little differentiation among land uses. Commercial and residential land uses were mixed together, and there was no effective distinction between home and work.

Within the cities, feudal **guilds**, or associations of craft workers and artisans, produced a variety of goods. Guilds consisted of skilled workers with years of experience and were organized by the type of good they produced. There were, for example, blacksmiths' guilds, weavers' guilds, goldsmiths' guilds, and guilds for bakers, leather workers, paper makers, glass workers, and shoemakers. Young men who were chosen to work in the guild spent years as apprentices learning the trade before becoming craftsmen in their own right.

The End of Feudalism

The feudal period in Europe saw relatively few changes from the sixth to the fourteenth centuries compared to the much more dynamic system of capitalism that followed. Innovation and change were discouraged, and feudal society

FIGURE 3 Carcassonne, France, offers an excellent example of feudal urbanization, including the concentric walls that often surrounded such communities.

was remarkably stable. However, the late medieval period, starting around the eleventh century, witnessed a gradual agricultural revolution based on the introduction of the heavy plow, waterwheels, the horseshoe, stirrup, the three-field system of farming, and several other innovations, which were introduced from other, more advanced societies such as the Arab, Indian, and Chinese. Other imports included cotton, the compass, sugar, rice, silk, paper, printing, the needle, the concept of zero, and the windmill. The Arab conquest of Spain made the Iberian peninsula a primary point of entry for new ideas and technologies.

Feudal Europe was the western terminus of a much larger world system that stretched across the Mediterranean, the Middle East, the Indian Ocean, and into Eastern Asia (Figure 4), connecting most of the Old World. The Arab world's strategic location astride Asia, Africa, and Europe placed it at the center of the enormous trade networks that linked places as far flung as China, Mozambique, and Belgium. Straddling the center of the feudal world system, the Caliphates centered in Damascus and Baghdad guaranteed safe passage between two critical worlds—the Mediterranean and the Indian Ocean—that had been separated since the collapse of Rome. A vital part of the Sung economy, as well as of the dynasties that came before and after it, was the Silk Road (a name coined by geographer Ferdinand von Richthofen in 1877), the umbilical cord that connected Europe, the Middle East, Central Asia, South Asia, Tibet, and China with ceaselessly flowing caravans of goods, innovations, ideas, merchants, missionaries, and armies. Few individuals traveled the entire distance of the Silk Road; rather, it was served by networks of intermediaries. The name "Silk Road" is somewhat misleading: It suggests a continuous journey, whereas goods were in fact transported by a series of routes, by a series of agents, passing through many hands before they reached their ultimate destination.

For 2000 years, the Silk Road caravans formed the primary artery of Eurasian commerce, linking ports, trading cities, oases, and innumerable different cultures. From China came jade, paper, the compass, gunpowder, printing, porcelain, lacquer ware, silk, pearls, peaches, apricots, citrus fruits, cherries, and almonds; moving in the other direction, at various times, China acquired horses, hides, furs, dyes, amber, pistachios, saffron, castor beans, sesame, peas, onions, coriander, cucumbers, grapes, sugar beets, kohlrabi, ivory, tortoise shells, rhinoceros horns, and, oddly, the chair. The Chinese established customs posts on the Silk Road to minimize smuggling and to tax goods as they passed and controlled Chinese merchants with the first known passports. So, too, did religions flow along this highway, including Manicheism, Zoroastrianism, Buddhism, and Islam. The Silk Road was therefore as important for the flows of cultures and ideas it expedited as for the flows of goods.

The introduction of innovations constituted a commercial revolution of sorts for they improved European agricultural productivity. The supply of agricultural land expanded as peasants and farmers in the late Middle Ages cut down forests and drained swamps to make room for new farmland. The heavy plow opened up the thick soils of northern Europe and expanded the amount of arable land for farming. This set of circumstances led to a gradual increase in the urban population. By the fifteenth century, much of Western Europe was carpeted by a growing network of cities, called "newtowns" in Britain and "villanovas" in Spain. The increase in productivity was commensurate with the undertaking of the construction of cathedrals.

It must be emphasized, however, that compared to much of the rest of the world, feudal Europe was relatively primitive. Standards of living and rates of innovation in Europe from the fifth to the fifteenth centuries were much lower than in the wealthier, more powerful, and more

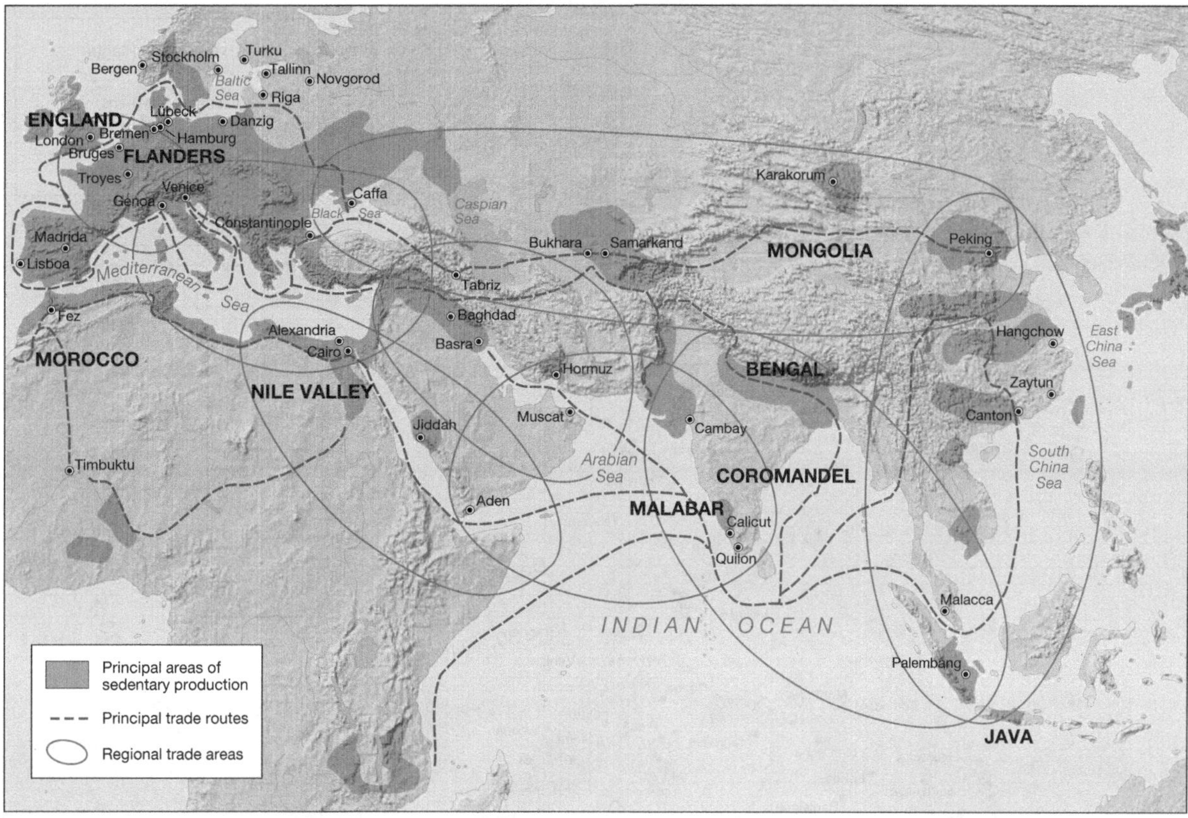

FIGURE 4 The precapitalist world system of the fourteenth century. A diverse array of societies stretching from feudal Europe to the Middle East formed one network embedded within a much broader trade system. Across the Indian Ocean, southern India, Indonesia, and China under the Sung and Ming dynasties constituted another set of networks. Overland trade along the Silk Road routes formed yet a third circuit. Unlike the capitalist world system that arose later, in the sixteenth century, this one lacked a distinct core.

sophisticated societies of the Arab world, India under the Mughals, and China under the Sung and Ming dynasties. Indeed, many of the luxury goods and innovations that Europeans imported came from these wealthier societies.

Among the other things introduced to Europe was a bacterium, *yersinia pestis,* which causes a deadly disease, bubonic plague, common in rodent populations in the central Asian steppes, or grasslands. Among humans, **bubonic plague** is highly contagious and was called the Black Death for the dark, swollen lymph glands it produced. In 1347, a ship carrying plague landed in Genoa, Italy, which was part of the expanding trade network between Europe and the Middle East and Asia. Within four years, one-quarter of Europe's population—more than 25 million people—was dead. The plague raced through the crowded, unsanitary cities, annihilating the majority of inhabitants in places. From southern Europe, it spread north, to Germany, Britain, Scandinavia, and Russia (Figure 5). Since the notion of germs was nonexistent, feudal Europe had no means of understanding or controlling the plague: frequently used "remedies," such as burning Jews or witches, did not halt its spread.

Several historians have speculated that the plague played a major role in upsetting the foundations of feudal Europe, destabilizing it and opening the door to a new type of society. Within a few years, much of the continent experienced severe labor shortages; Europe went from being a land-poor, people-rich to a people-poor, land-rich group of societies. The disintegration of legal systems allowed serfs to run away without fear of being caught and returned. Others have argued that feudalism was suffering from numerous problems anyway and would have collapsed without the plague. For example, in some cities, such as Florence, Italy, capitalist social institutions like banks were already emerging before the fourteenth century. In any case, combined with other changes, including a mini-Ice Age that reduced growing seasons from the sixteenth to the nineteenth centuries, the Hundred Years War between France and England (1337–1453), and the impacts of the Crusades, feudalism in Europe slowly began to crumble, and from its ashes a new economic, political, and social system emerged: capitalism.

THE EMERGENCE AND NATURE OF CAPITALISM

Over several hundred years, from the fifteenth to the nineteenth centuries, feudalism in Europe was gradually replaced by a new kind of society based on capitalism.

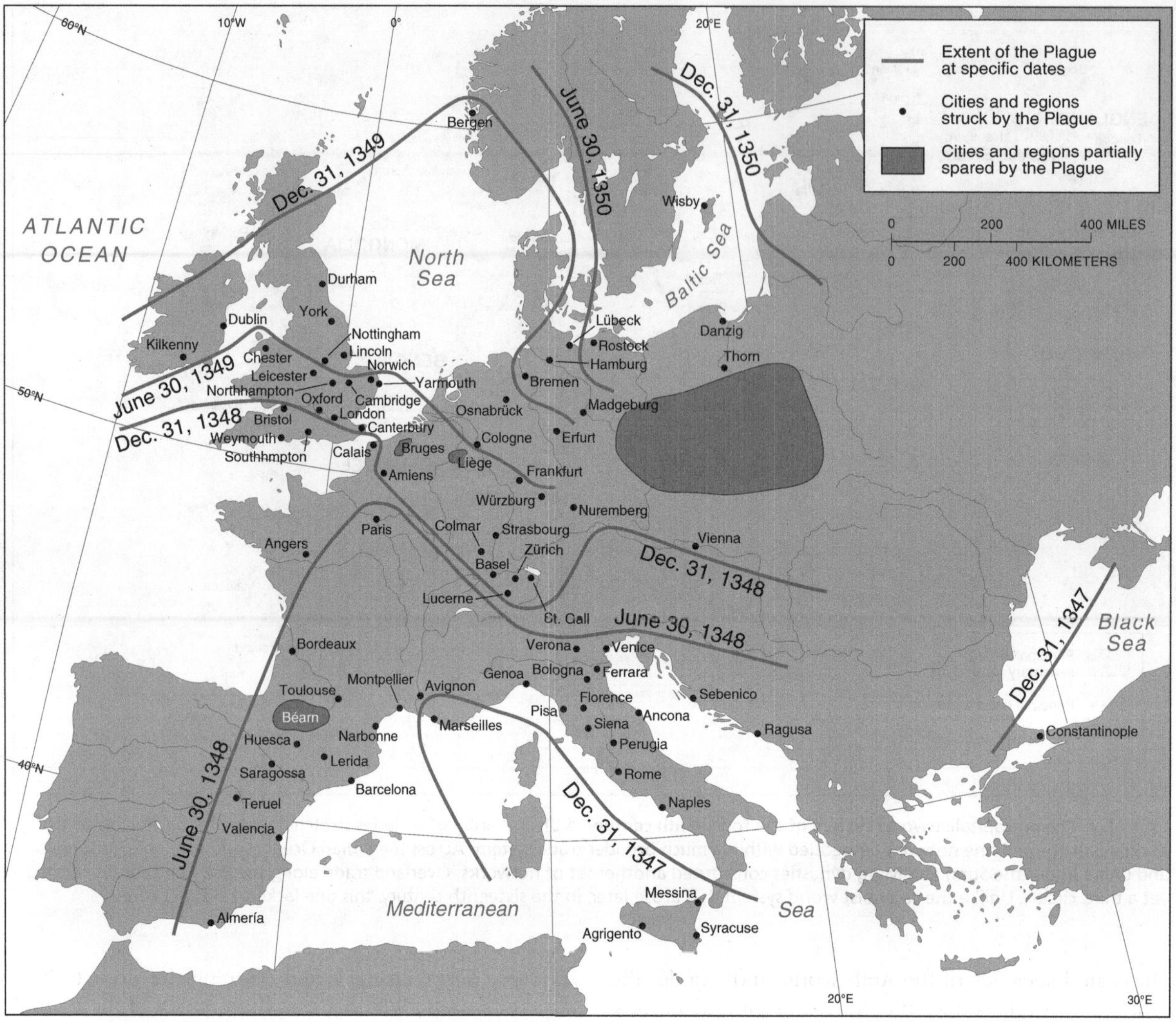

FIGURE 5 Diffusion of the bubonic plague through fourteenth-century Europe. The plague's devastation disturbed feudal Europe's equilibrium, creating labor shortages and destabilizing the social structure. Some historians argue that the plague facilitated the emergence of capitalism.

Thus, feudalism served as the womb that incubated what would ultimately become the most powerful type of economic, cultural, and political system in the world. Of course, to those who lived through it, this transition would have appeared very slow and perhaps almost invisible.

If capitalism can be said to have a birthplace, the most likely location would be in northern Italy. The city-states of this peninsula, among them Florence, Venice, Pisa, and Genoa, played a key role in creating the new kind of society. They had large groups of wealthy merchants, including the famous Medici family of Florence, who had active commercial ties to the Middle East (Figure 6) and vast holdings in silver mines, silk production, and banking. Northern Italian city-states had flourishing trade networks across the Mediterranean, including with the Arabs in Egypt and the Middle East. In northern Europe, a network of cities formed the **Hanseatic League**, which stretched from Russia and Scandinavia across northern Germany and to the North Sea (Figure 7). The Hanseatic League united a disparate series of proto-capitalist urban centers (it eventually collapsed as trade moved across the Atlantic Ocean). In these centers, the rising groups of **burghers**, or merchants, accumulated wealth and power that would make them the dominant figures in a new social order.

Like feudalism, capitalism possesses a distinct set of characteristics that define it and give it its unique form. The major features of capitalism are described next.

Markets

Unlike feudalism and slavery, in which resources are allocated principally through the political power of the state, under capitalism the most important (but not only) institution of allocation is the **market**. Markets consist of buyers and sellers of **commodities**, which are goods and services bought and sold for a price. Not everything is a commodity

FIGURE 6 Italian city-states in 1494. Capitalism began in northern Italy, and for much of its history (prior to unification in mid-nineteenth century) its political geography was characterized by city-states, not nation-states. Italian states such as Florence, Genoa, and Venice had extensive trade relations with the rest of the Mediterranean world, and their wealthy merchant families rose to power.

under capitalism (e.g., air); only scarce goods (i.e., those that command a price and can generate a profit for producers) can be classified as such. The expansion of market societies saw the steady commodification of different goods and services, including food, housing, clothing, transportation, education, entertainment, medical care, and other domains. Thus, things that used to be produced on a subsistence basis or bartered came increasingly to be made for sale on a market, their value determined by the amount of money they commanded.

Based on the type of commodity being produced and consumed, as well as the amount and nature of competition, markets come in a huge variety. They range from free-wheeling and highly competitive, with many small producers who must accept the price the market hands them, to large markets dominated by a few major producers, or oligopolists, who can shape the market price. In market-based societies, private property and the right to own it are key requirements for production; after all, a good cannot be sold unless the seller owns it. Markets thus do not exist or function well without appropriate legal guarantees to protect property rights, such as the right to own real property or intellectual property. The incentive of producers to sell goods and services is **profit**, the difference between gross revenues and production costs. Thus, markets involve production for exchange, rather than subsistence or use. In this sense, capitalism involved the triumph of the private sphere over the public one.

Because markets involve competition among different producers, there is a strong incentive to produce goods and services cheaply and efficiently and to please consumers. Generally, the more competitive a market is, the more dynamism and innovation it exhibits. Thus, competitive markets contain a powerful incentive to innovate, which is largely responsible for making capitalist societies so conducive to change. In this sense, capitalist societies differ considerably from noncapitalist ones in that capitalism rewards innovation, change, and risk taking.

Markets did not begin with capitalism. For example, there were markets in slave-based societies like the Roman Empire. There were occasional markets in feudal Europe in the form of annual fairs and festivals, which brought traveling merchants and local populations together. Typically, precapitalist markets were poorly developed, however, and involved relatively few luxury goods and centered on the needs and wants of the elites. However,

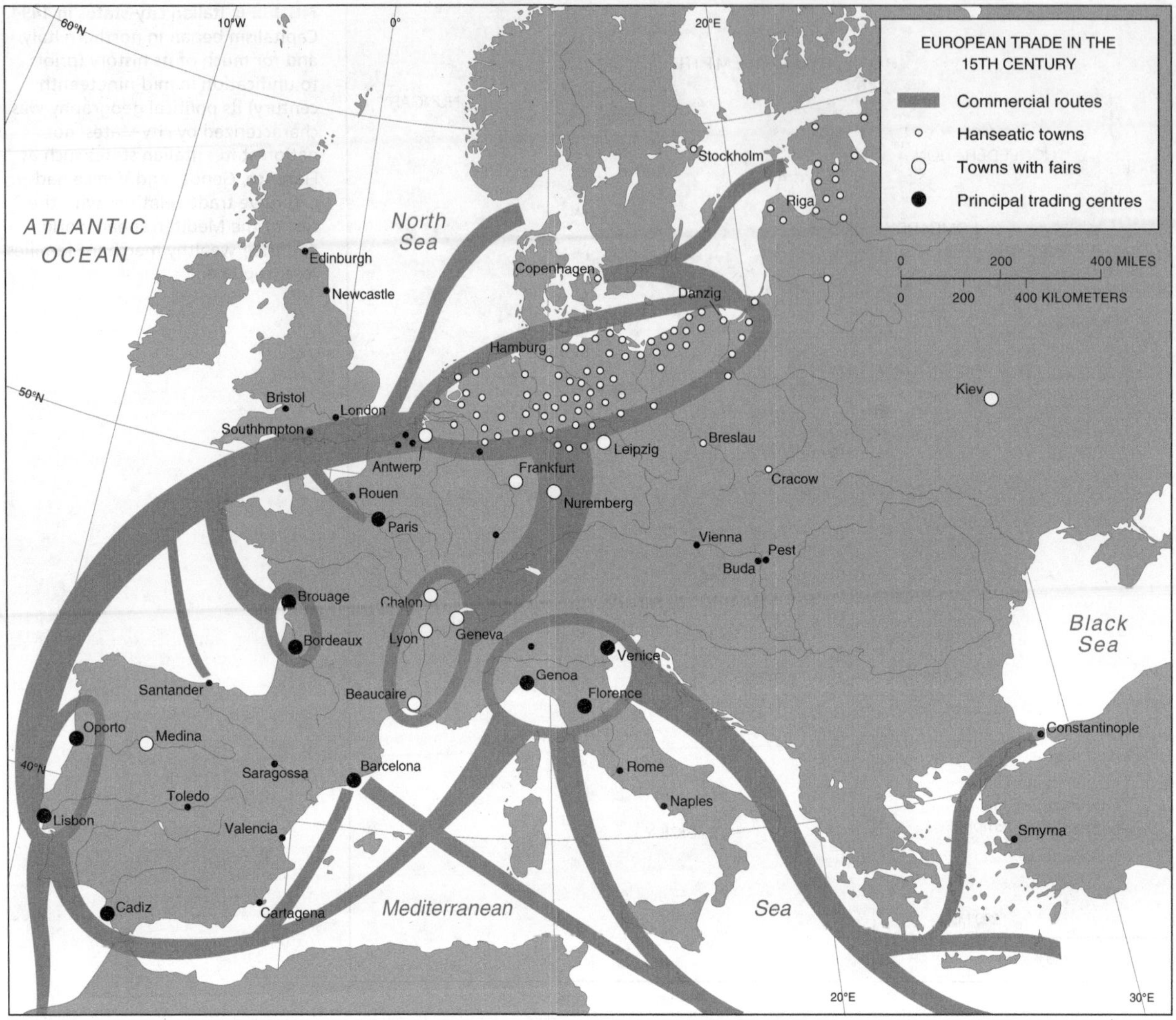

FIGURE 7 The Hanseatic League of fifteenth-century Europe was a loose connection of cities and principalities whose trading connections extended from the Baltic Sea to the North Sea. In those places, the emerging burgher or merchant class would have been at the forefront of the new class relations that accompanied the emergence of capitalism.

markets are unique in their *importance* to capitalism: Only under capitalism are markets the major way in which resources are allocated. It is worth emphasizing that markets are not the *only* way in which resources are organized, for even in ostensibly "free-market" societies (which do not exist in fact), the state, or government, plays a key role, including protecting property rights, building the infrastructure, providing public services, and protecting firms from foreign competition.

Class Relations

Although capitalist societies are sometimes depicted as being devoid of class (e.g., as consisting only of individual buyers and sellers), market-based systems in fact do have social classes of different kinds. In contrast to feudalism, the rise of the capitalist class system reflected a broad-based shift from a hierarchy based on tradition to one based on money, from status born of rank to status earned, and the ascendancy of the merchant class, or what is sometimes called the bourgeoisie (the middle class of feudalism, the ruling class of capitalism).

As the merchants and burghers of Europe gained wealth, power, and prestige in the fifteenth through seventeenth centuries, they came to enjoy increasing political and cultural power (Figure 8). Essentially, the merchant class became a capitalist class, that is, the owners of capital. The feudal aristocracy, which correctly perceived the newcomers to be a lethal threat to their centuries of rule, resisted the rise of the new class and its associated power, wealth, and prestige. Many kings, earls, and dukes held merchants in low regard, as money-grubbers who were not motivated by ancient (and increasingly quaint) notions of god and honor. Conflicts between the merchant class

Bildarchiv Preussischer Kulturbesitz/Art Resource, NY

FIGURE 8 Rembrandt's painting *Sampling Officials of the Drapers' Guild* (1662) exemplified the wealth and prestige of the new bourgeoisie in early capitalist Europe.

and its interests on the one hand and the feudal elite on the other grew steadily in number and intensity, revolving around issues like levels of taxation, the freedom to open markets at particular hours (e.g., on holidays), and merchants' loans to kings, largely for purposes of waging war. The demise of the feudal aristocracy came gradually in some places, such as England (e.g., during the civil wars of the seventeenth century), and suddenly in others, such as France, where the aristocracy collapsed abruptly in the Revolution of 1789.

In addition to changing the role and nature of the ruling classes, capitalism changed those of the workers as well. In particular, labor itself became a commodity, that is, bought and sold for a price (wages) in labor markets. Workers under capitalism, unlike those of feudal Europe, must sell their labor power to survive. Thus, the process of commodification extended to include the capacity to labor. Over several centuries, the peasants and serfs of Europe were gradually forged into a working class. This was not a simple process; it involved driving rural workers off the land and inculcating new norms, such as working the hours dictated to them by their new bosses, the capitalists. In the context of industrial Europe, the working class became known as the proletariat.

Finance

The growth of capitalism brought about a deep and fundamental change in the role of money, which increasingly became the measure of all worth. In noncapitalist societies, *barter* plays a major role in organizing economic relations: Goods may be traded for one another, or labor traded for goods. This approach has the advantage of providing transparency in the exchange process. In the context of barter-based economies, money is relegated to a relatively small role. Obviously, money existed before capitalism—the Romans, for example, had vast quantities of coins—but under market-based societies money assumed a new level of importance. As the cash system replaced barter, money became standardized and ubiquitous as a measure of value. The time, space, and labor were measured in monetary terms, so that even human life came to have a financial value. Wealth and power were increasingly defined along monetary lines, that is, economically rather than politically. Many traditional social roles that were defined by kinship, religion, friendship, and trust became depersonalized and formalized as they were mediated by money. Money thus acquired important social and political as well as economic roles. For this reason, we must see money as a social product, that is, it cannot exist or have meaning outside of society.

The organization and control of money became an industry in its own right in the emerging capitalist system. Large, complex societies cannot function without well-established financial systems, which not only reflect systems of production but also shape them. The turnover rate of money—the pace with which it changes hands—is important to the process of capital accumulation. Banking arose primarily among the goldsmiths of feudal Europe, who stored gold for their customers and then loaned it out to borrowers, for a price (interest). By the seventeenth century, commercial credit became widespread, and with it, came different types of banks and insurance firms. Starting with small savers who pooled their funds to purchase ships to trade with Asia, joint stock companies, the nucleus of modern stock markets, spread the risks of large investments over many small investors. Accounting became an important profession. By the nineteenth century, financial systems were increasingly regulated by the state through central banks, which sought to control money supplies and thus interest, inflation, and exchange rates. Modern banking has become a huge and complex industry linking savers and borrowers of different types and with different needs.

Territorial and Geographic Changes

If capitalism fundamentally changed the rules of societies, it also reshaped how they were organized geographically. Because capitalism is overwhelmingly the most significant economic and political system worldwide, economic geography is primarily the analysis of how capitalism produces landscapes. Not surprisingly, the geographies of capitalism are unique to the logic of profit-maximizing societies.

An idea shared by many economic geographers is that capitalism creates **uneven spatial development**, that is, varying levels of economic growth, wealth, and poverty in different locations. Uneven development is reflected in the simultaneous existence and interaction of rich and poor places, those with high and low unemployment and regions and countries with happy, prosperous residents and those with large numbers of impoverished, hungry ones. Those who believe in neoclassical economics (the mainstream economics taught in the United States) often hold that uneven development is a temporary phenomenon that markets eliminate in due course. Others, working from the viewpoint of political economy, believe there are several mechanisms through which uneven development occurs, but the most significant is that of capital investment and disinvestment. As capital seeks out the highest rate of profit, it flows into some regions and out of others, in the process simultaneously creating prosperous places (e.g., New York City) and abandoning others to economic decline (e.g., Detroit). In this light, wealthy regions and poverty-stricken ones are intimately connected—two sides of the same coin. The creation of uneven development, which we will explore at greater depth throughout this text, occurs at different spatial scales.

At the global level, capitalism, through the mechanism of colonialism, created a worldwide system of commodity production and consumption, with Europe at the center and its colonies on the periphery, a process we will examine in more detail shortly. Essentially, Europe became wealthy, in part due to its extraction of cheap raw materials from its colonies in Latin America, Asia, and Africa. Prior to colonialism, Europe was poor and backward compared to competing regions; by the sixteenth century, when colonialism was undertaken in earnest, Europe had begun to surpass many other regions of the planet. Thus, early globalization had highly uneven impacts—positive and negative—on different parts of the world.

Within Europe, capitalism unleashed a division between the relatively prosperous northwestern region and a comparatively poorer southern and eastern region. Historically, southern Europe, including Greece, Italy, and Spain, had been the wealthiest and most powerful region. With the ascendancy of capitalism, particularly in the form of the Industrial Revolution in the late eighteenth and nineteenth centuries, however, northwestern Europe became much more advanced economically, particularly following the defeat of Napoleon in 1815 (Figure 9). To this day, the societies of Western Europe remain wealthier than those elsewhere on that continent.

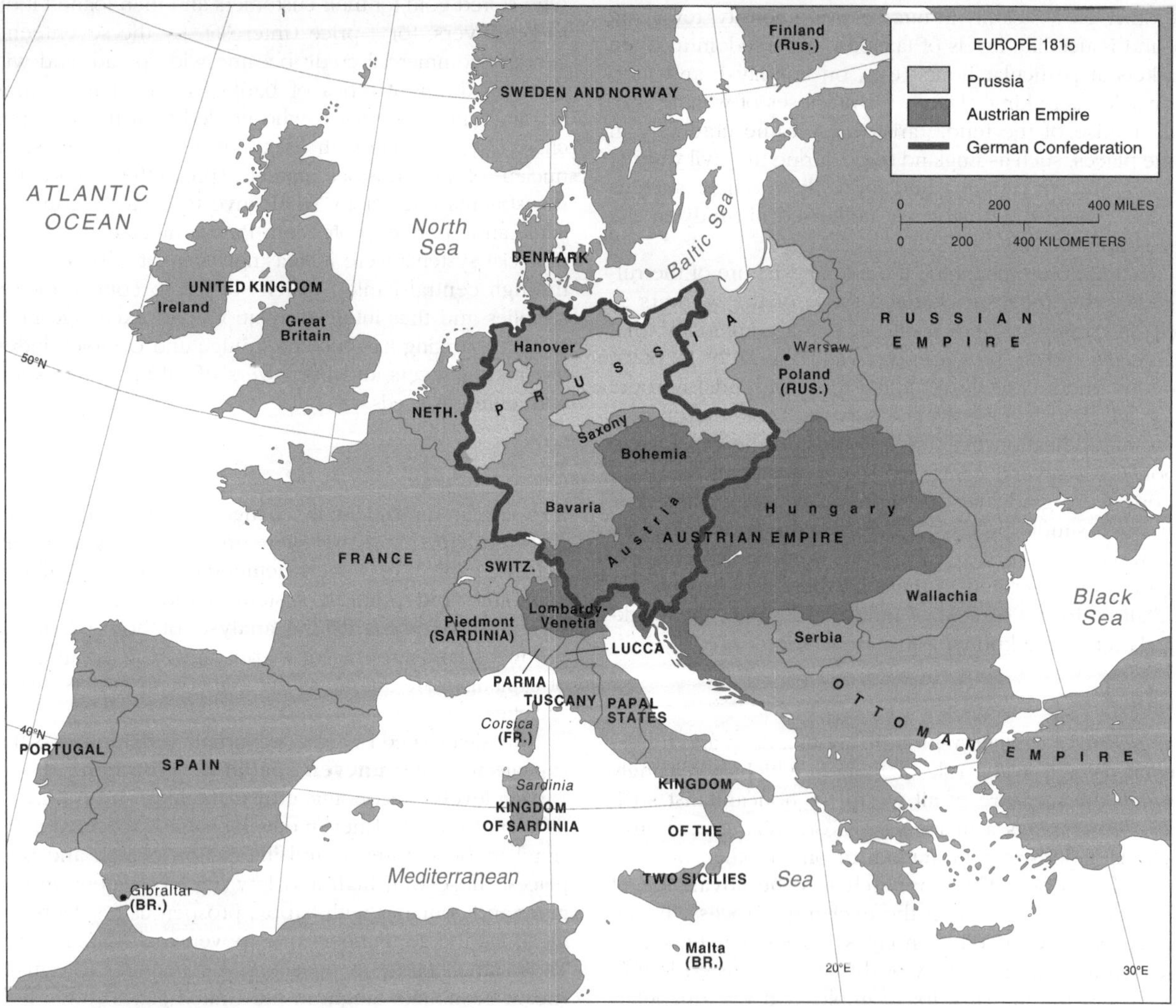

FIGURE 9 Europe in 1815, at the end of the Napoleonic wars. The turbulence in Europe after the French Revolution of 1789 saw the nation-state emerge as the primary political entity on the continent.

Within the individual countries of Europe, as well as within other emerging nation-states dominated by capitalism, there arose a division in wealth between cities and the countryside. Feudal Europe, with its tiny cities, had a relatively small urban-rural schism and there were relatively few differences in standards of living between urban and rural areas. Under capitalism, as rural areas were reshaped by waves of enclosures of farmland and the commodification of agriculture in the form of cash crops, large numbers of people migrated to the urban areas, which became wealthier. Port cities in particular thrived, given the maritime basis of the world economy. Today, almost everywhere, cities have higher incomes, more jobs, and more opportunities than do rural areas.

Finally, within capitalist cities, urban land, like labor, became a commodity, organized through land markets. In the process, profit became the mechanism for establishing different uses of land, including separating home and work, areas of production and social reproduction. Gradually, as cities grew larger, particularly under the Industrial Revolution in the nineteenth century, the distances between home and work were stretched to the point where workers engaged in mass commuting. Thus, an intraurban division of labor complemented the division between cities and the countryside.

The key for the economic geographer is to see all of these scales as different versions of the same process (i.e., uneven development manifested at the global, continental, national, and local scales).

Long-Distance Trade

The ability to buy and sell goods over long distances is a fundamental part of capitalist societies. Trade reflects the geographic organization of exchange, linking producers and consumers who generally never see one another. As capitalism took hold and became entrenched across the European continent, trade networks proliferated in diversity and extent. Of course, there was trade prior to capitalism. In feudal Europe, trade with the Muslim world, and via the Silk Road, with East and South Asia, allowed the influx of many goods that Europeans did not produce for themselves. Indeed, Europe was the western terminus of a much larger fourteenth-century trading system that stretched across the Middle East, India, and into Southeast Asia and China (see Figure 4). Yet prior to capitalism, trade was largely confined to luxury goods, such as spices, silks, porcelain, and precious metals. Only aristocrats who had the means to purchase such luxuries were the consumers.

If long-distance trade was peripheral to feudalism—those societies could have survived without such goods—it is central to capitalism. In market-based societies, trade occurs in all sorts of goods, from luxuries to those for everyday use. The expansion of trade networks was the major incentive for the formation of networks of land and sea routes that tied different parts of Europe together and tied the continent to the rest of the world. Within Europe, a vast expansion in roads, canals, and, somewhat later, railroads was undertaken in the sixteenth, seventeenth, and eighteenth centuries that stitched places together into an increasingly interdependent spatial division of labor. The increased speed with which people, capital, goods, and information circulated is commonly called time-space compression. New ships allowed Europeans to sail long distances relatively quickly, and in the process they created new maps and charts for navigation, learning the behavior of winds and currents all over the planet.

If trade reflects differences among places in the nature of production, it also helps to shape those places. In economics and economic geography, this idea is reflected in the concept of comparative advantage, the specialization of production that occurs when places begin to trade extensively with one another. The growth of long-distance trade within Europe, and between Europe and its colonies, increased competitive pressures on local producers and helped to fuel declines in production costs and associated increases in standards of living. By the seventeenth century, for example, an upper-middle-class family in Britain could purchase salted cod from Newfoundland, furs from Russia, timber from Scandinavia, wines from France, blown glass from what is now the Czech Republic, and olive oil and citrus from Spain or Greece. In short, capitalist trade relations made consumers better off as countries became increasingly interdependent on one another.

New Ideologies

Capitalism, it should be emphasized, is not simply an economic, political, cultural, or geographic set of relations—it is all of these simultaneously. Just as feudalism consisted of multiple dimensions, including the ideological dimension of religion, so too does capitalism exist both in the economy and in the domains of culture and ideology. Market-based systems reflected the ways in which people were organized and interrelated, but they also were reflected in how people perceived the world. The emergence of capitalism brought with it a vast panoply of ideological changes that revolutionized the ideas, science, and culture of the modern world.

After its invention in 1450, the printing press made the production of books easier, faster, and cheaper (Figure 10). Of course, Europeans were acquainted with printed textiles, money, and playing cards long before they encountered printed books. And the paper used to produce handwritten copies of books and manuscripts had been imported into Europe by Arab merchants in Spain during the tenth century, who in turn acquired it from China in the eighth century. From Spain, paper spread to Sicily and Italy in the eleventh century, and to France in the twelfth century. Printing had a huge impact on the societies of Europe. It allowed large quantities of materials to be produced cheaply and distributed quickly, and the effects of this revolution, in conjunction with the numerous other massive changes criss-crossing the face of Europe, were monumental.

FIGURE 10 Guttenburg's invention of the printing press in 1450, using the Chinese innovation of movable type, revolutionized the way in which ideas could circulate through late medieval and early capitalist Europe. It is no coincidence that the cheap books made possible by the printing press played a big role in the European Enlightenment of the seventeenth and eighteenth centuries.

The new communications environment made possible by the printing press accelerated the decline of the feudal order by destabilizing traditional society. Printing brought literacy to adults—especially males, for female literacy lagged far behind—who now had more ready access to texts. The first major step in the mechanization of communication, printing accelerated the diffusion of information by packaging it conveniently, democratizing books in much the same manner that cheap clocks and maps democratized time and space, respectively, widening access for the literate to people, places, and events far removed from them historically or geographically. Similarly, printing undermined the centrality of the clergy in the production of knowledge, and unlike handwritten monastic copies, printed books gave their audiences identical copies to read, experience, and discuss; it made censorship more difficult as well. By helping to break the monopoly on learning held by monasteries and universities, printing encouraged the growth of a lay intelligentsia. The technology therefore did much to enlarge the domain of the "political." Ideas of many sorts began to circulate around the continent, and larger numbers of people learned to read and write. Printing and rising rates of literacy facilitated the Italian Renaissance, the Protestant Reformation, European expansionism, and the rise of modern capitalism and science.

In the sixteenth and seventeenth centuries, starting in Italy, Europe witnessed the explosion of artistic and scientific knowledge known as the Renaissance (a term not coined until the nineteenth century). Leading Italian scholars such as Leonardo da Vinci exemplified the rise of secular knowledge, as did Erasmus and the Humanists in northern Europe. Although most of the intellectuals of the Renaissance were religious, the movement marked a dramatic shift from the god-centered view prevalent under feudalism to one that increasingly emphasized the role of human beings in the making of the world.

The sixteenth century also witnessed the Protestant Reformation, the second great schism in Christianity (the first being the division between the Catholic and Orthodox branches around the year 1000) (Figure 11). Beginning with Martin Luther in Germany, Protestantism offered a

FIGURE 11 Europe in the seventeenth century was gripped by severe religious wars following the Protestant Reformation. The rise of Protestantism in northern and northwestern Europe—the second great schism in Christianity—stressed individualism, one of several lines of thought that comprised the new ideological universe of emerging capitalism.

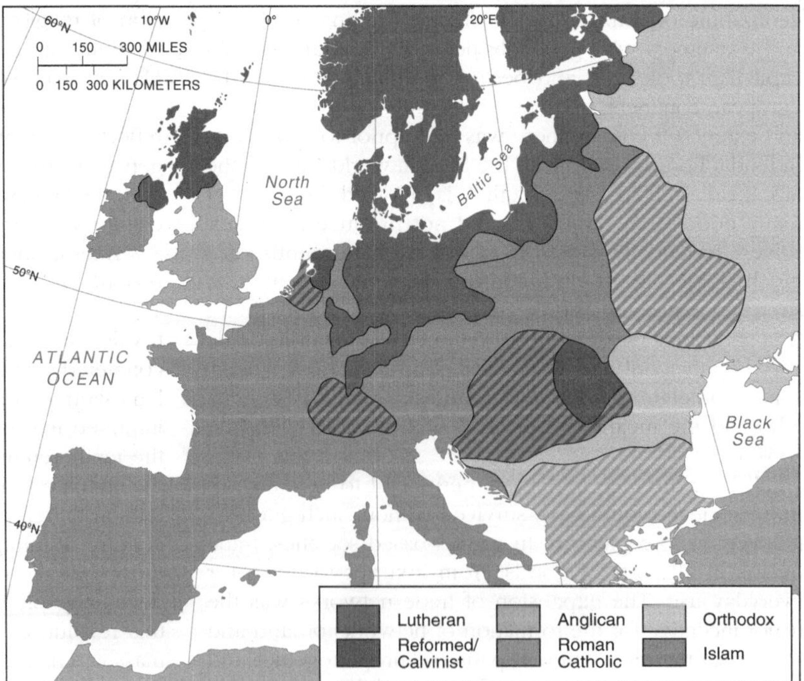

different view of life and god than that of Catholicism. In particular, Protestantism emphasized the role of the individual and the individual's direct relation to god, bypassing priests as intermediaries between individuals and god. Protestantism was one facet of a much broader growth of individualism in different social spheres and spread with the growth of literacy in the aftermath of the printing press. The famous sociologist Max Weber, who studied the relationship between Protestantism and the development of industrial capitalism in northwest Europe, argued that the "Protestant ethic," which stressed delayed gratification, savings, and material success as a sign of god's grace and one's potential entry into heaven, was instrumental in the development of capitalism. He held that Protestantism elevated work to the status of a moral obligation, making profit a reward rather than a sin, paving the way for capital accumulation. This perspective has been challenged by those who maintain that Protestantism followed in the wake of market relations rather than causing them, or that at least Protestantism and capitalism co-evolved. Others note that the origins of capitalism in Catholic Italy undermine the claim that it began in predominantly Protestant northern Europe.

Science in the sixteenth and seventeenth centuries played a critical role in restructuring how people viewed the world and their place in it. The Copernican revolution, augmented by Galileo, led to a heliocentric view of the universe rather than the older, Aristotelian geocentric one, which placed the earth in the center (Figure 12), a shift that accentuated the gradually emerging secularism of the time. Modern science emerged as scientists like Francis Bacon, Isaac Newton, Boyle, Pascal, Linnaeus, and Lavoisier made enormous contributions to physics, astronomy, and chemistry and their applications to gravity, optics, and other fields.

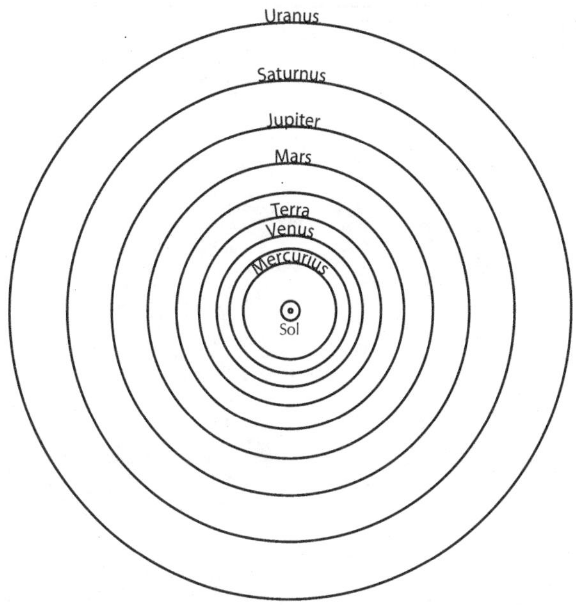

FIGURE 12 The heliocentric Copernican universe, as distinct from the older, Aristotelian geocentric one, formed an important component in the increasingly secularized view of nature that accompanied the scientific and political revolutions of early modernity. Note that the shift from a geocentric to heliocentric perspective on the universe caused an enormous intellectual and political uproar.

By the seventeenth, eighteenth, and nineteenth centuries, Western societies were undergoing the *Enlightenment*, an important explosion of scientific knowledge and secular political thought. In Britain, France, Germany, and elsewhere, advances were made in geology, chemistry, physics, and biology, including the discovery of atoms, electromagnetism, and bacteria, leading to the germ theory of disease. The publication of *On the Origin of Species* by Charles Darwin in 1859 revolutionized our understanding of evolution and ecosystems. These discoveries demystified nature by subjecting it to scientific law, a process that was accelerated by the proliferation of institutions of higher learning and academic societies. A gradual, widespread secularization of culture was underway. In political thought, theorists such as John Locke, David Hume, and Adam Smith developed a worldview that stressed secularism, individualism, rationality, progress, and democracy. Indeed, democracy, articulated by Thomas Jefferson among others, may be seen as a largely American contribution to the Enlightenment.

The Nation-State

Capitalism is not simply an economic system for organizing the production and consumption of resources through markets; it is also a political system that involves the state, that is, governments and governance. Thus, the emergence of capitalism witnessed a series of political changes that accompanied the rise of market-based societies. One of the most important of these was the rise to prominence of the **nation-state**.

A nation is a group of people who share a common culture, language, history, and territory, often manifested in a common ethnic identity. Feudal empires had many nations within their borders. The Holy Roman Empire (Figure 13) and the Austro-Hungarian Empire, for example, contained dozens of different ethnicities. Individual and collective identity was largely defined in religious terms (e.g., "Christiandom"), ancestral lineage, or as allegiance to a particular king. By the eighteenth century, however, as these empires began to disintegrate, many peoples had increasingly come to identify with a nation. *Nationalism* is a term generally used to describe either the attachment of people to a particular nation or a political movement by groups to achieve statehood (or national self-determination). As the state became the primary locus of sovereignty in the eighteenth and nineteenth centuries, it became increasingly important for ruling elites to construct a narrative that provided the often culturally heterogeneous populations that inhabited the territory of the state with a single identity. The nation and the state thus came to merge via the impetus of nationalism, self-determination, and sovereignty. But even the most classic, text examples of the nation-state lacked homogeneous ethnicity, such as France, which had different ethnic, linguistic, and

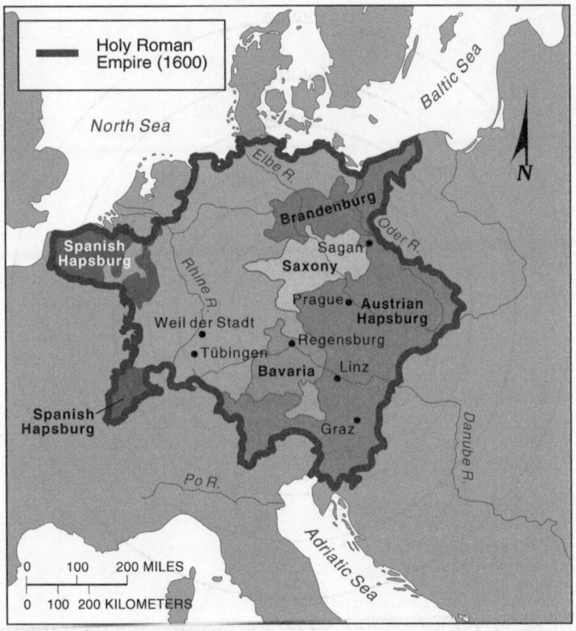

FIGURE 13 The Holy Roman Empire in 1600 was characteristic of the ethnically diverse states of medieval Europe. Although it was centered in Germany, at various times it included parts of what are today France, Italy, Austria, Poland, and other places. This social formation exemplifies how ethnicity was not a primary unit of political identity until the emergence of the nation-state.

religious minorities, including, for example, French Basques, Bretons, and Corsicans.

As national and ethnic identities displaced older feudal ones, the political geography of Europe was steadily redefined along ethnic lines. Key events like the Peace of Westphalia in 1648 effectively put an end to the Holy Roman Empire and legitimated the nation-state as the primary unit of international law and relations (Figure 14). The concluding document to the Thirty Years' War (1618–1648), the Peace of Westphalia (recognized as the first "modern" international treaty) guaranteed the independence of the Netherlands and Portugal and allowed the individual states comprising the Holy Roman Empire to choose their own religion, a power that had been usurped by the Holy Roman Emperor. The treaty ensured a state's inalienable right to universal authority, particularly the use of force within its boundaries, and the recognition of such authority and boundaries by the international community. Indeed, the modern world system of nation-states, with clear, sacrosanct borders, is essentially Westphalian in nature (although now being challenged by globalization).

The French Revolution of 1789 was another major turning point in the breakdown of the old, feudal social order and the rise of the new, modern one, creating new forms of political identity—citizen, for example. The Napoleonic wars that ended in 1815 likewise laid much of the geographic basis for the nation-states of the continent.

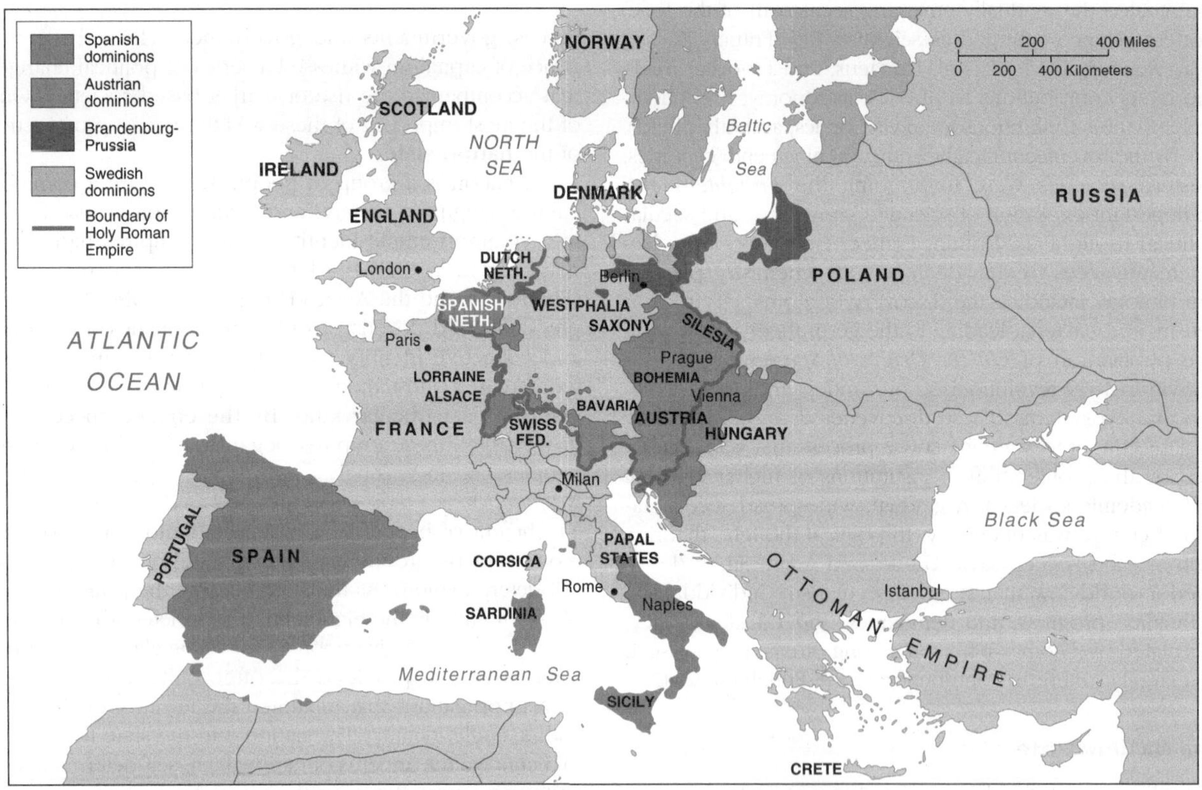

FIGURE 14 Signed in 1648, the Treaty of Westphalia not only set the boundaries among the nation-states of early modern Europe, it also legitimized the nation-state, and the principle of noninterference, in international politics. The emerging geography of capitalism was as much political as economic in nature.

In Western Europe, the centralized monarchies of feudalism were gradually replaced by constitutional republics; in Eastern Europe, this did not occur until after World War I. By the twentieth century, the nation-state had become securely entrenched as the dominant form of political organization throughout the world, including in many former European colonies.

The emergence of market societies facilitated the growth of nation-states in several ways. Rising wealth, mass literacy, growing cadres of the intelligentsia, and political parties that demanded a role in the newly democratic societies were all part of this process. Other institutions were also important, such as national banks and currency, a military draft, the media, and national rail systems, which tied together the diverse parts of the emerging nation-states and reinforced their sense of being a community of like-minded people. Thus, capitalism is not simply a system that produced markets; it made *both* states and markets. But looked at another way, the new nation-states also facilitated the establishment of markets, including, for example, the construction of public infrastructure, the provision of public services (e.g., schools, transportation), the establishment of national monetary supplies, and the protection of domestic producers from foreign competition. These relations have led some social scientists to discard the conventional view that markets were born free of the shackles of the state in favor of the idea that markets *required* the state to survive.

It is important to remember that capitalism long preceded the nation-state and that there is no necessary correspondence between the two, which is an observation with important implications in the current age of globalization. Capitalism began in the city-states of northern Italy in the fifteenth and sixteenth centuries, not in the nation-state, which is largely a product of eighteenth- and nineteenth-century industrialization. The ascendancy of the nation-state marked an expansion of the scale upon which capitalist social relations were to be managed, including trade and production networks. Capitalism thus brought about a national scope of operations at the same time as it did the international. Indeed, the political geography of capitalism has become the interstate system, at whatever scale, which is a complex of intersections between economic and political relations.

THE INDUSTRIAL REVOLUTION

Capitalism has been a dynamic force since its inception. The growth and development of market-based societies was, by historical standards, very rapid. During the Industrial Revolution in the eighteenth and nineteenth centuries, the pace of change accelerated greatly. It would ultimately become a normal feature of the modern world. It is worth emphasizing that the Industrial Revolution occurred long *after* capitalism began; indeed, for most of its history, capitalism involved preindustrial forms of production, including labor-intensive artisanal and household manufacture in the period known as **mercantilism**. Beginning in the mid-1800s, however, the speed and output of capitalist production in Europe, North America, and Japan exploded, transforming the worlds of work, everyday life, and the global economy. Somewhat later, in the twentieth century, industrialization spread to Eastern Europe and the Soviet Union; since the 1970s, it has spread to the newly industrializing countries in selected parts of the developing world.

Industrialization is a complex process that involves multiple transformations in inputs, outputs, and technologies. The three dimensions that are particularly important here are discussed next.

Inanimate Energy

Preindustrial societies relied upon **animate sources of energy** (i.e., human and animal muscle power) to get things done. Industrialization can be defined loosely as the harnessing of **inanimate sources of energy**, a major milestone in human economic evolution. Historically, several types of inanimate energy have been tapped. The first involved running water, or water in a particular stage of the hydrolic cycle when it is moving overland from higher to lower elevations. This source of energy had been used in the late Middle Ages to grind corn and flour. By the early eighteenth century, some producers of textiles began to use water-powered mills to run their machines, and it was a major source of energy in the earliest stages of the Industrial Revolution. Many textile plants in southern New England, for example, used this strategy. But it constrained firms to locating near streams and rivers. Many streams are annual, meaning they may dry up in the summer if there has been insufficient rainfall in their catchment area. Also, locating on a stream might put the producer inconveniently far away from the market.

A more efficient source of inanimate energy involved coal and the steam engine, the designs for which were laid out by Thomas Newcomen in 1712 (Figure 15). The first operating steam engine was built by the Scottish engineer James Watt in 1769; it marked a turning point in the process of industrialization. The steam engine, originally designed to pump seawater out of coal mine shafts that penetrated under the ocean, used relatively cheap and abundant fuels and could do the work of dozens of men far more efficiently. This invention required heating water into steam in order to drive the engine's pistons, and wood was the first major source of fuel. Producers began to cut down forests in Britain in large numbers, deforesting much of the country, wood supplies began to dwindle, and the rising cost eroded profits.

As wood became scarce, producers switched to coal, which could be mined in large quantities. Thus, as Britain industrialized, several areas became major coal-producing centers, including Wales and Newcastle. As the Industrial Revolution spread across the face of Europe, the large coal deposits of the northern European lowlands—northern France, Belgium, the Ruhr region in Germany, and Silesia in southern Poland and the Czech Republic—became increasingly important.

Dave King © Dorling Kindersley, Courtesy of The Science Museum, London

FIGURE 15 The steam engine, designed by Thomas Newcomen in 1712 but first built by James Watt in 1769, was the key invention of the Industrial Revolution. It was the first device to harness inanimate energy on a mass basis and revolutionized both production and transportation for two centuries.

In the United States as well, coal deposits in Appalachia played a key role in the industrialization of the nation.

In the nineteenth century, coal was joined by other fossil fuels, particularly petroleum, and to a lesser extent natural gas. The abundance of cheap energy was the lifeblood of industrialization, and production processes became increasingly energy-intensive as a result. This substitution of inanimate for animate energy both freed tens of millions of people from drudgery and allowed for large numbers to live relatively comfortable lives in the expanding middle class that industrialization brought about.

Technological Innovation

As we have seen, capitalism is a very dynamic economic system. Firms, under the lure of profits and threat of ruin, engage in frequent innovation as a way to reduce costs and increase revenues. New technologies certainly emerged prior to industrialization, but the Industrial Revolution witnessed an explosive jump in the number, diversity, and applications of new technologies (Table 1). A *technology* is a means of converting inputs to outputs. These can range from extremely simple to sophisticated. An otter using a rock to open an oyster is employing a technology, as you are when you use a pen or a computer. The increasingly sophisticated division of labor under industrialization rapidly led to opportunities for new inventions. These were employed in agriculture, in manufacturing, in transportation and communications, and in services. Figure 16 illustrates a variety of technologies that accompanied the long-term increases in productivity in transportation and production.

A major reorganization in the nature of work occurred in the Industrial Revolution with the development of the *factory* system. Prior to this era, industrial work was organized on a small-scale basis, including home-based work. The early textile industry, for example, used the "putting out" system of independent workers and contractors. By the late eighteenth century, however, firms in different industries were grouping large numbers of workers together under one roof. Some of the largest factories held thousands of workers. This process effectively created the

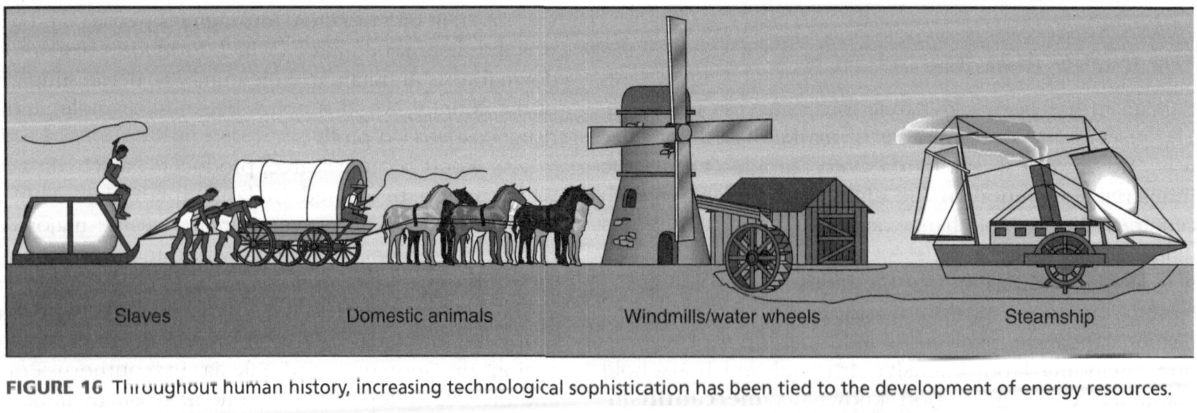

FIGURE 16 Throughout human history, increasing technological sophistication has been tied to the development of energy resources.

TABLE 1 Some Major Innovations of the Industrial Revolution

1708	Mechanical seed sower
1712	Steam engine
1758	Threshing machine
1765	Spinning jenny
1787	Power loom
1793	Cotton gin
1807	Steamboat
1828	Railroad
1831	Electric generator
1834	Reaper
1839	Photography, vulcanized rubber
1844	Telegraph
1846	Pneumatic tire
1849	Reinforced concrete
1850	Refined gasoline
1851	Refrigeration; sewing machine
1857	Pasteurization
1859	Gasoline engine
1866	Open hearth furnace
1867	Dynamite
1873	Typewriter
1876	Telephone
1877	Phonograph
1878	Microphone
1879	Electric light bulb
1884	Rayon
1886	Hydroelectric power plant
1888	Camera; radio waves
1892	Diesel engine
1895	X-rays
1896	Wireless telegraphy
1899	Aspirin
1900	Zeppelin
1903	Airplane
1906	Vacuum tube
1925	Television

industrial working class. Never before in human history had so many workers been concentrated on a permanent basis, a feature that changed how they lived and viewed each other—and themselves. Inside factories, workers used vast amounts of capital (i.e., many types of machines), that is, work became much more capital-intensive. The introduction of interchangeable parts, a concept invented by American gun maker Eli Whitney, made machines more reliable and easier to fix. In the early twentieth century, Henry Ford introduced the moving conveyor belt, which further accelerated the tempo of work and the ability of workers to produce.

Productivity Increases

As a consequence of the massive technological changes of the Industrial Revolution, productivity levels surged. **Productivity** refers to the level of output generated by a given volume of inputs; productivity increases refer to higher levels of efficiency, that is, greater levels of output per unit of input (e.g., labor hour or unit of land), or, conversely, fewer inputs per unit of output.

As Figure 17 indicates, productivity levels rose exponentially in the nineteenth century. This process had several important repercussions. As the cost of producing goods declined, standards of living rose. Most workers labored long hours under horrific conditions and endured standards of living still quite low compared to those we enjoy today. But nonetheless, over several decades, industrialization saw many kinds of goods become increasingly affordable. Because wage rates have been linked historically to the productivity of labor, the working class gradually became better off. Clothing, for example, which was scarce before the Industrial Revolution, became relatively cheap and ceased to be as accurate an indicator of class status as it had been previously.

The industrialization of agriculture was most important in this regard. As machine after machine was introduced into farming, including mechanical reapers, harvesters, threshers, and tractors, food became progressively cheaper and diets improved; more people ate more and better food than ever before. With the notorious exception of the Irish Potato Famine of the 1840s, hunger and malnutrition

gradually declined throughout Europe, although they did appear in the aftermath of wars. As diets improved, so did resistance to disease, and life expectancies rose as well.

The Geography of the Industrial Revolution

Like all major social processes, the Industrial Revolution unfolded very unevenly over time and space. Whereas capitalism had its origins in Italy, industrialization was very much a product of northwestern Europe. Some scholars locate the first textile factories in Belgium, in cities such as Liege and Flanders. However, without question it was Britain that became the world's first industrialized nation. By the end of the eighteenth century, Britain stood virtually alone as the world's only industrial economy, a fact that gave it an enormous advantage over its rivals. For example, Britain's industrial base allowed it to triumph over France in their eighteenth-century rivalry for global hegemony and to flood its competitors' markets with cheap textiles. Cities in the Midlands of Britain, such as Leeds and Manchester, centers of the British textile and metal-working industries (Figure 18), were known as the workhouses of the world for their high concentrations of workers, capital, and output. Given the lack of government regulation over labor practices, the use of brutally exploitative child labor was common (Figure 19). Others, such as London, Glasgow, and Liverpool, became centers of ship building, which, in a maritime-based world economy, was a major industry in its own right. In many cities, networks of producers of guns, watches, shoes, metals, and light industry formed dense industrial districts of small firms with intricate linkages of inputs and outputs.

Why Britain? There are no simple answers to this question. Britain had already enjoyed a network of long-distance trade relations with its colonies in North America and elsewhere. Agriculture in Britain was well advanced in

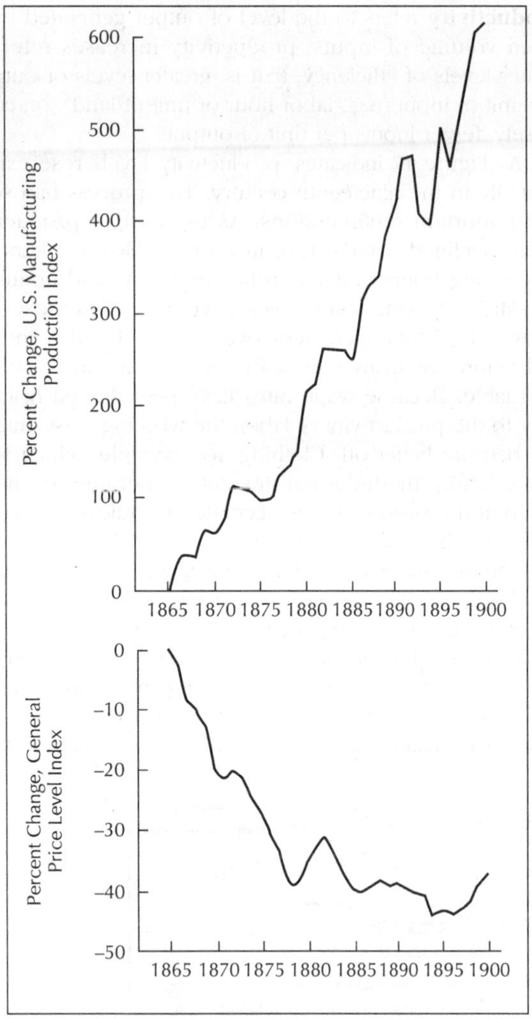

FIGURE 17 Manufacturing productivity in the United States rose exponentially in the latter part of the nineteenth century. After the Civil War, the Industrial Revolution began in earnest. As productivity increased, the prices of goods dropped accordingly, and standards of living rose. Geographically, this period saw the emergence of the Manufacturing Belt along the southern shores of the Great Lakes.

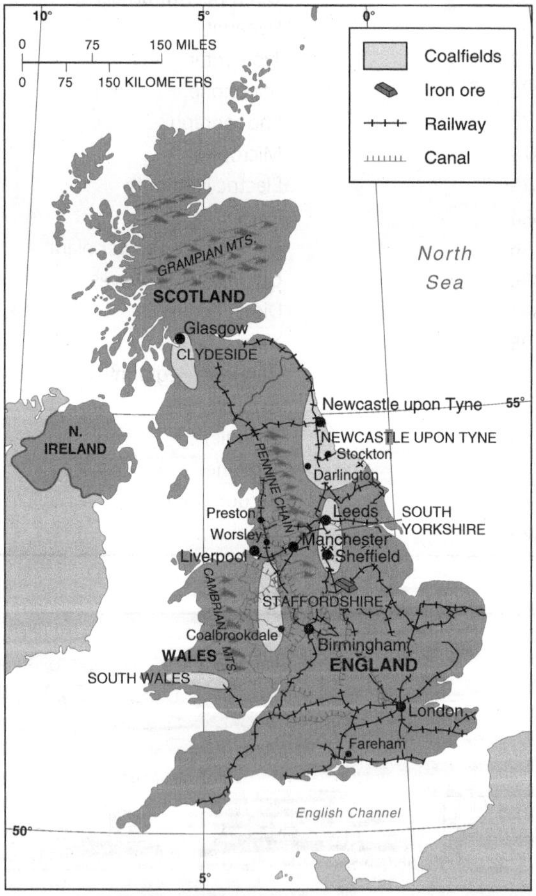

FIGURE 18 **Britain's industrial areas, the sources of the Industrial Revolution.** Coal from Wales and Newcastle fueled the development of the factory system, particularly in the Midlands cities such as Manchester, Sheffield, Birmingham, Leeds, and Liverpool.

Hulton Archive/Getty Images

FIGURE 19 The use of children as workers was common during the Industrial Revolution, and persisted until the implementation of child labor laws in the l9th and early 20th centuries. Similar conditions apply to many children in the developing world today.

the process of commodification compared to the European continent. And Britain enjoyed large deposits of coal and was the locale where the steam engine was invented.

A half-century after it began in Britain, the Industrial Revolution diffused over the European continent during the nineteenth century (Figure 20). In France, industrial complexes formed on the lower Seine River and in Paris. In Italy, the Po River valley became a major producer of textiles and shoes. In Scandinavia, cities such as Stockholm became major ship-building centers. In Germany, which was relatively late to industrialize (following its unification in 1871), the Ruhr region became a global center of steel, automobile, and petrochemical firms. By the early nineteenth century, the revolution also began to spread worldwide (Figure 21), leapfrogging across the Atlantic and igniting the industrialization of southern New England, and in the 1870s, spreading to Japan, which became the first non-Western country to join the club of industrialized nations as the old feudal order there collapsed following the Meiji Restoration of 1868. Russia, which flirted with industrialization under Peter the Great, did not become fully industrialized until the 1920s, when the Soviet Union under Stalin leaped forward to become the world's second-largest economy in the span of a decade. In the twentieth century, the process of industrialization spread to many developing countries, particularly in East Asia, where it has had profound consequences for millions of people.

In a sense, then, the industrialization of the developing world, which is still partial and incomplete, is a continuation of a long-standing historical process. Although the process changed over the years, its broad outlines have remained the same. The industrial complexes formed by the diffusion of the Industrial Revolution remain highly important to the global economy today.

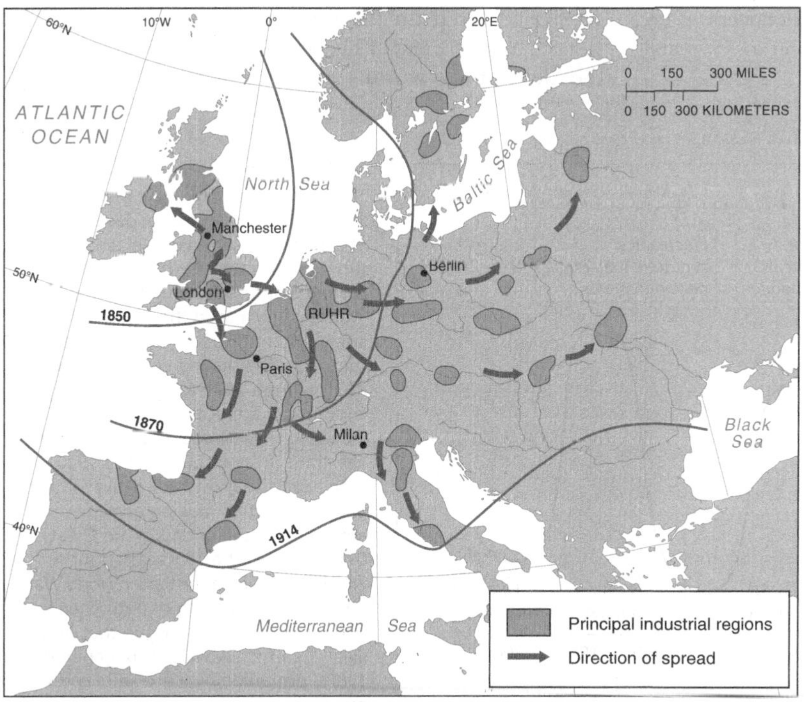

FIGURE 20 Spread of the Industrial Revolution across the European continent. Well after Britain had industrialized, the new form of manufacturing led to the formation of industrial complexes in France, then later in Germany and Italy.

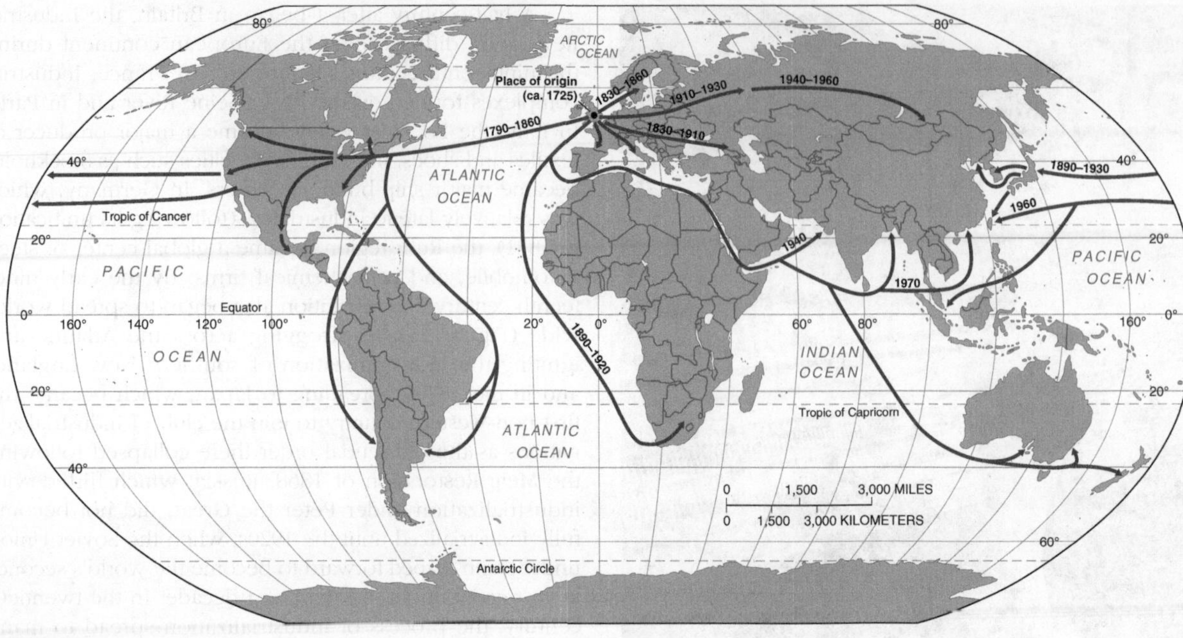

FIGURE 21 The global diffusion of the Industrial Revolution. By the early nineteenth century, the process had become entrenched in New England and later took root across the rest of North America. Japan emerged from a long period of isolation in 1868 and became the first non-Western industrial power shortly thereafter. Eastern Europe lagged well behind Western Europe, and Russia industrialized only in the 1920s under the Stalinist government in the Soviet Union. The newly industrialized countries (NICs) of East Asia started the process in the 1960s, and it continues today in selected parts of the developing world.

Cycles of Industrialization

Just as the process of industrialization occurs in different places at different times, so too did the nature and form of industrialization vary in successive historical periods. Capitalism is prone to long-term cyclical shifts in its industries, products, labor markets, and geographies, a concept often referred to as Kondratiev waves of roughly 50 to 75 years' duration (Figure 22). For our purposes, this means that industrialization saw the rise of different industries at different times. Industrialization was thus not one process but a series of them that varied over time and space.

The first wave of the Industrial Revolution (1770s–1820s) centered on the textile industry (Figure 23). In Britain, as in the rest of Europe, North America, Japan, and the developing world today, textiles have *always* led

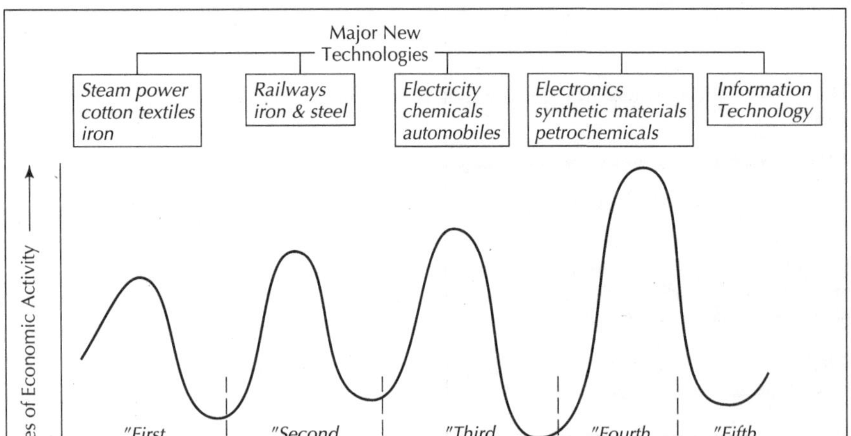

FIGURE 22 Long waves of economic activity, named after their discoverer, Kondratiev. The historical development of capitalism is often organized into four such waves, each of which was centered around a different technology. The textile industry dominated the first, from the late eighteenth to the early nineteenth centuries. The second wave, lasting roughly from 1820 to 1880, saw the widespread application of steam power, including the railroad and the steamship. The third, from about 1880 to 1930, witnessed the growth of heavy industry, particularly steel making and the automobile, and ended in the Great Depression of the 1930s. The fourth, which began in earnest after World War II, saw the rapid growth of petrochemicals and aerospace. Many argue that we are living in a fifth wave that began in the 1980s, propelled by electronics and business services.

Lewis W. Hine/Library of Congress/Time Life Pictures/Getty Images

FIGURE 23 A factory during the Industrial Revolution. This photograph illustrates the exploitative labor relations that accompanied and underpinned the growth of modern capitalism, including the frequent use of child labor.

industrialization. Easy to enter, with few requirements of capital or labor skills, this sector initiated the industrial landscapes of most of the world. Because this wave was first centered in Britain, it catapulted that nation to prominence as the leading economic power in the world, initiating the period of the *Pax Britannica*.

The second wave, from the 1820s to the 1880s, was a period of heavy industry. In the nineteenth century, the most important sectors were those like ship building and iron manufacturing. Large-scale and capital-intensive, these types of firms differed markedly from the light industry of textiles. They required massive capital investments, were difficult to enter, and moved toward forming an oligopoly rather than a competitive market. This was the period in which the U.S. Manufacturing Belt began to form, although most of its growth occurred after the Civil War of the 1860s.

In the third wave of industrialization, from the 1880s to the 1930s, numerous heavy industries appeared, including steel, rubber, glass, and automobiles. This was a period of massive technological change, capital intensification, and automation of work, as well as economic changes. As local markets gave way to national markets, most sectors experienced a steady oligopolization, or concentration of output and ownership in the hands of a few large firms. Many companies became multiestablishment corporations. Not surprisingly, this wave saw the rise to power of the "robber barons": Carnegie (steel), Rockefeller (oil), Duke (tobacco), Dupont (chemicals), J. P. Morgan (banking), and John Deere (agricultural machinery).

The primary growth sectors in the fourth wave of industrialization, which started during or immediately after the Depression of the 1930s and lasted until the oil shocks of the 1970s, were petrochemicals (including plastics) and automobiles. With a relatively stable global economy, the world system was dominated by the United States, which produced a huge share of the planet's industrial output.

The electronics industry led the fifth wave of industrialization, often thought of as beginning after the oil shocks of the 1970s. Powered by the microelectronics revolution, and by the explosive growth of producer services, this era experienced rapid productivity growth in household electronics and information-processing technologies.

It is important to note that during each era, the major propulsive industry was commonly featured as the "high-tech" sector of its day. Thus, just as electronics is often celebrated at this historical moment for its innovativeness and ability to sustain national competitiveness, so too were the textile industry in the eighteenth century and steel industry in the nineteenth century associated with high levels of productivity and wages. What was a leading industry at one moment would become a lagging one in the next historical epoch, as high-wage, high-value-added sectors replaced low-wage ones in the world's core and as low-wage, low-value-added sectors dispersed to the world's periphery.

Consequences of the Industrial Revolution

The Industrial Revolution permanently changed the social and spatial fabric of the world, particularly in the societies that now form the economically developed world. No part of their social systems, economy, technology, culture, or everyday life was left untouched. Within a century of its inception, industrialization changed a series of rural, poverty-stricken societies into relatively prosperous, urbanized, and cosmopolitan ones. Some of the major changes included those described next.

CREATION OF AN INDUSTRIAL WORKING CLASS As we noted earlier, a significant part of industrialization was the reorganization of work along the lines of the factory system. For the first time in human history, large numbers of workers labored together using machines. These conditions were quite different from those facing agricultural workers, who were dispersed over large spaces and relied on animate sources of energy. Industrialization gave rise to organized labor markets in which workers were paid by the hour, day, or week.

In short, as firms created a new form of labor, they created a new form of laborer, a proletariat, or industrial working class, which became socialized according to the new conditions of work. This process was not easy, given how brutally exploitative working conditions were during this time. Workers typically labored for 10, 12, or even 14 hours per day, six days per week, for relatively low wages. (The 8-hour day and Saturdays free from work were products of workers' movements in the 1930s.) Often the work was unsanitary and dangerous, even lethal, as workers were subjected to accidents, poor lighting, and poor air quality. Child labor was also common, subjecting those as young as 4 or 5 to horrendous and exploitative conditions like those now found in the developing world.

As a result, time—like space, and so much else—became a commodity, something bought and sold. The transition from agricultural time to industrial time was important. Prior to the Industrial Revolution, people experienced time seasonally and rarely felt the need to be conscious of it. Time was simply lived, without worry about the precise beginnings and endings of events. With industrialization, however, time was measured and divided into discrete units, as signaled by the factory whistle, bell, and stopwatch. This change marked the commodification of time via the labor market.

Industrialization produced a working class; it also produced labor unions. The first resistance to the industrial system arose in the early nineteenth century among the British Luddites, named after their leader, "General" Ned Ludd. Luddites blamed their miserable working conditions on the machines they used and often destroyed them as a way of halting their exploitation. In France, workers wearing large wooden shoes, or "sabots," jammed them into the machinery in an act of "sabotage." By the late nineteenth century, labor had formed a number of powerful unions; in the United States these included the Knights of Labor, the American Federation of Labor, and, in the twentieth century, the Industrial Workers of the World and the Congress of Industrial Organizations. Industrialization thus often led to considerable class conflict.

URBANIZATION Geographically, the Industrial Revolution was closely associated with the growth of cities. Almost everywhere, industrialization and *urbanization* have been virtually simultaneous. Manufacturing firms concentrated in cities, especially large ones, such as the British Midlands (e.g., Leeds, Manchester) or throughout the U.S. Manufacturing Belt (e.g., Pittsburgh, Cleveland, Detroit, Chicago, and Milwaukee). The reasons firms concentrated in cities are important. Often, this tendency is attributed to the presence of workers in urban areas. But which came first, firms or workers? Cities were clearly centers of capital as much as they were centers of labor, which poses something of a "chicken or egg" problem. Yet cities were very small when the Industrial Revolution began, but through agricultural mechanization (which reduced rural job opportunities) and rural-to-urban migration, the urban labor supply was created.

There are powerful reasons for firms to concentrate, or agglomerate, in cities. Most firms benefit by having close proximity to other firms, including suppliers of parts and ancillary services. Concentration allows firms to share a specialized infrastructure, information, and labor force. The cities of the nineteenth century were composed of dense webs of industrial firms, with intricate input and output relationships tying them together. Industrialization changed societies from predominantly rural to predominantly urban in character. In Europe, North America, and Japan, for the first time in history, the majority of people lived in cities. The growth of cities in industrial societies is frequently depicted using an urbanization curve (Figure 24), which illustrates the percentage of people living in urban areas over time. In the United States, for example, the first national census of 1790 showed that 95% of Americans lived in rural areas. This rural proportion decreased throughout the nineteenth century, and by 1920, 50% of the nation's population lived in cities. Today, it is roughly 85%. The pattern is similar in every other country that has industrialized.

POPULATION EFFECTS Industrialization changed more than simply the geographic distribution of people (i.e., concentration in cities); it also shaped growth rates and demographic composition. The Industrial Revolution unleashed drastic changes in both the rate of growth and the health of the population.

On the eve of the Industrial Revolution, the famous theorist Thomas Malthus predicted that rapid population growth

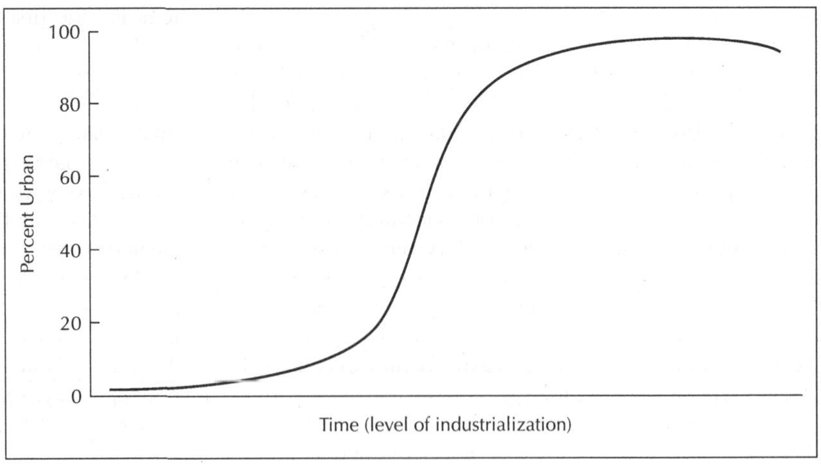

FIGURE 24 An urbanization curve expresses the proportion of a country's population that lives in cities at different stages in industrialization. Preindustrial societies are agricultural and rural; because manufacturing is concentrated in urban areas, industrialization causes cities to grow more quickly than the countryside.

would create widespread famine. Yet Malthus was soon proved to be wrong, at least in the short run. The industrialization of agriculture generated productivity increases greater than the rate of population growth, and the creation of a stable and better food supply improved most people's diets. As a result, life expectancy rose. Industrialization also lowered death rates, particularly as malnutrition declined and infant mortality rates dropped. Eventually, public health measures and cleaner water helped to control the spread of most infectious diseases. As death rates dropped, the populations of industrializing countries increased dramatically (Figure 25). Accompanying this change was a shift from the extended to the nuclear family. Eventually industrialization also led to a decline in the birth rate, the number of children per family, and growth rates.

GROWTH OF GLOBAL MARKETS AND INTERNATIONAL TRADE The Industrial Revolution had an impact on the existing global economy as well. Under capitalism, a loose network of international trade had formed well before the eighteenth century. Indeed, as we saw earlier, there were extensive linkages from Europe across Eurasia and the Indian Ocean as early as the fourteenth century. The harnessing of inanimate energy for transportation dramatically accelerated the speed of both land and water transportation, significantly compressing time and space. Sailing ships and horse-drawn transport traveled at roughly 10 miles per hour (mph), for example, but steamships could reach speeds of 40 mph and railroads more than 65 mph. Moreover, the new, industrialized forms of transportation were not only faster but also cheaper, resulting in cost-space convergence as well.

These changes dramatically lowered the barriers to trade, and the international volume of imports and exports began to soar. Europe, starting with Britain, could import unprocessed raw materials, including cotton, sugar, and metal ores, and export high-value-added finished goods, a process that generated large numbers of jobs in Europe and contributed to a steady rise in the standard of living. It is worth noting that classical political economists like Adam Smith and David Ricardo began to attack the philosophy of mercantilism, which preached state protection against imports, at precisely this historical moment.

Finally, the industrial world economy saw explosive growth in international finance. British banks, largely concentrated in London, for example, began to expand their activities internationally, lending to clients and investing in markets overseas. Much of the capital that financed the American railroad network came from Britain. The globalization of production was thus accompanied by the steady globalization of money and credit.

The timing of industrialization was significant to different nations as well. There existed an important difference between early and late industrializers in this regard. Early industrializers (e.g., Britain, the United States) faced little international competition. Thus, their light industries (i.e., textiles) associated with the first wave of industrialization could develop relatively slowly, with minimal government intervention. Firms in sectors such as textiles—with few barriers to entry, low infrastructural demands, and quite competitive internationally—were important in shaping the national political climates that were characterized by laissez-faire politics and minimal government intervention.

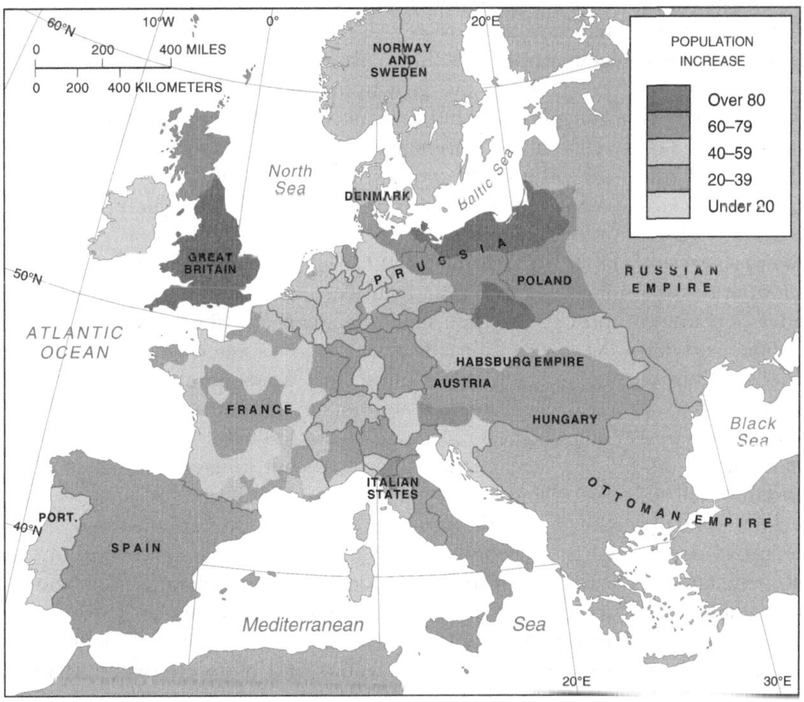

FIGURE 25 Population growth in Europe from 1800 to 1850. The capacity of industrialized societies to support large, dense nucleations of people led to higher rates of population growth and larger numbers of people in northern Europe than in the south, which industrialized later and less completely.

Case Study

Railroads and Geography

Railroads, one of the central technologies of the Industrial Revolution, along with steamships, gave birth to radically new urban and economic geographies. Railroads were central to the rise of modern capitalism and the opening up of continental interiors to rapid, cheap, and dependable transportation. In addition to their ability to shuttle people, railroads could move heavy loads over long overland stretches, reducing land transport costs by as much as 95%.

British railroads, the world's first, began in the early nineteenth century. In 1814, the first steam locomotive was introduced, and the British system expanded rapidly. In 1830, the first interurban railroad connected Liverpool and Manchester. Early railroads achieved speeds between 20 and 30 mph, or three times that achieved by stagecoaches; later ones achieved speeds up to 70 or 80 mph. The enormous costs of constructing and maintaining railroad networks required, however, high volumes of traffic to amortize expenses over numerous clients.

In Russia, railroads helped to forestall a decline into Third World status. With the opening of the Moscow–St. Petersburg line in 1851, Russia's railroads zoomed from 700 total miles in 1860 to 36,000 in 1900. The first leg of the Trans-Siberian railroad opened in 1903, stretching 6000 miles from Moscow to Vladivostok; by reducing journey times between Europe and Asian Russia from months to days, it brought the vast resources of Siberia into the tsarist spatial division of labor. Similarly, the Japanese rail system first linked Tokyo and Yokohama in 1872, part of the rapid unification of Japanese space following the Meiji Restoration of 1868. In Italy, the railroad became an instrument of national unification, a process achieved in the face of the growing economic and political ascendancy of the north at the expense of the peninsula's southern districts. The French railroad system, centered, naturally, upon Paris, soon made even the capital accessible to people throughout the nation; equally important, it made French peripheral territories accessible to Parisian capitalists.

Within cities, too, railroads were potent in restructuring urban land markets, initiating a series of changes in the urban rent structure that amplified some land values and eroded others. Railroads simultaneously intruded upon older cities, destroying the traditional fabric of urban spaces and giving rise to new ones. As the commodification of urban space intensified, specialized districts of production and social reproduction materialized apace within divisions of labor that became spatially extended and more integrated. Outside urban areas, railroads allowed the rhythms of the city to penetrate the countryside. By making distant lands accessible, railroads dramatically extended the rent structures of urban areas, commodifying land in hitherto inaccessible regions, encouraging farmers, settlers, and planters to assault distant, relatively untouched ecosystems.

The American rail experience paralleled that of Britain in generating an increasingly ubiquitous transport surface across national space. Beginning with the first line in Quincy, Massachusetts, in 1826, Boston became the country's first rail hub, with three radial lines. It took only 43 years to move from the first rail line in 1826 to the famous transcontinental connection of the Union Pacific and Central Pacific lines at Promontory Point, Utah, on May 10, 1869, that stitched together the multitude of farms, towns, and cities stretched across North America. Thus, distances from New York that took 6 weeks to traverse in 1830 could be crossed in 1 week by 1857. American railroads had profound impacts on the country's social, urban, and economic geography, opening the coal mining districts of Appalachia and the meat and grain belt of the Midwest alike. The time-space effects of the railroad system were geographically highly uneven; the Northeast and Midwest were integrated far more than were the Northeast and the South. Because American labor was expensive and land was cheap—the exact opposite of Europe—railroads in the United States tended to curve much more around hills and valleys.

American railroads were instrumental in opening up the vast wilderness of the continental interior, and train travelers across the prairies compared railroads to ships on land. The completion of national rail networks also facilitated the specialization of individual cities, which often benefited from comparative advantages based on local resources, strategic locations, and contingent pools of skilled labor. Chicago in particular witnessed explosive growth in the 1840s due to its role as a transportation hub. The accelerated speed and improved reliability of shipments brought the Midwest steadily into the orbit of Eastern capital. Shortly after the Civil War, the Rocky Mountains and West Coast were incorporated into the American space-economy, which soon thereafter became the largest in the world, surpassing Britain in the 1890s. Far from resulting from some mythical process of "free market" expansion, the growth of the American railroads was actively abetted by the federal government, including the land grants introduced in 1850. The U.S. Interstate Commerce Act of 1887 standardized rail gauges.

One of the most significant long-term impacts of the relational geographies that the railroad fostered was the rise of dramatically expanded markets on the scale of the nation-state. Prior to the Industrial Revolution, markets were highly localized and self-sufficient and long-distance trade was comparatively rare. In 1817, for example, it took 52 days to ship goods from Cincinnati to New York using wagons and rivers. Between 1830 and 1857, most of the country was within a few days' travel of New York City. Interregional commerce tended to be export-oriented and controlled by merchants in large, East Coast cities. As long as production and consumption were confined to regional markets, goods retained the identity of their origin. Railroads, however, allowed large-scale commercial providers to lower costs, increase productivity, and raise profits while using the new technologies to conquer space in ever more effective ways, expanding their operations by diminishing geographic barriers. Within the new, larger markets, prices tended to equalize quickly. This process generated "economies of speed" in which maximization of throughput set the stage for the emerging industrial regime of production. All of this was central to the transition from small, isolated, localized markets to larger, national ones, a new geographic formation that was highly conducive to the rise of larger, capital-intensive, vertically integrated, multiestablishment firms and oligopolies as companies internalized their commodity chains, that is, produced many inputs for themselves rather than purchased them from suppliers.

One of the most explicit repercussions of railroads was the standardization of time. England, first to build railroads, was also the first to adopt a standard railroad time: In 1847, the British Railway Clearing House suggested that all rail stations adopt Greenwich Mean Time (GMT), the special preserve of the Royal Observatory located in Greenwich and founded in 1675. By 1855, GMT became the legal standard throughout the country.

In contrast, relatively late industrializers faced a significantly different international climate, one dominated by early industrializers. Countries such as Germany and Japan, which did not begin industrializing until the late nineteenth century—long after Britain and the United States—faced significant competition in industries like textiles. Consequently, these nations tended to experience relatively short periods in which their economies were dominated by light industry and moved rapidly into heavier sectors. Germany, for example, developed a comparative advantage in steel, armaments, and automobiles. In countries in which heavy industry dominates and places significantly higher demands on the state for labor training, infrastructure, and trade protection, national political cultures that look favorably upon state intervention are more likely to develop. This is true of newly industrializing countries (NICs) today. Thus, the internal political culture within countries was strongly affected by the timing of their entry into the international division of labor.

COLONIALISM: CAPITALISM ON A WORLD SCALE

Intimately associated with the development of capitalism in Europe was Europe's conquest of the rest of the globe. This process, euphemistically called the "Age of Exploration," can be viewed as the expansion of capitalism on a global scale. Just as the geographies of capitalism are typified by uneven spatial development, as noted earlier, colonialism involved uneven development on a global scale, with Europe at the center and its colonies on the world periphery. Many theories of world development, such as world-systems theory, incorporate this theme.

Colonialism was simultaneously an economic, political, and cultural project. It was also an act of conquest, by which a small group of European powers came to dominate a very large group of non-European societies. Culturally, under colonialism the distinction between the "West" and the "Rest" emerged. In conquering the "Orient" and encountering cultures very different from their own, Europeans discovered themselves as Westerners, often in contrast to other people whom they represented in highly erroneous terms.

Everywhere, colonized people fought back against colonial rule. Examples include the Inca rebellions against the Spanish, Zulu attacks on the Dutch Boers, the great Indian Sepoy uprising of 1857, and the Boxer Rebellion of 1899–1901 in China. Yet Western powers, armed with guns, ships, and cannon, effectively dominated the entire planet. While a few countries remained nominally independent, such as Thailand, the only one to escape colonialism substantively was Japan, which, under the Tokugawa Shogunate, closed itself off from the world until 1867.

Colonialism had profound implications for both the colonizers and the colonized, which is why a sophisticated understanding of economic geography must include some understanding of this process. Globally, colonialism produced the division between the world's developed and less developed countries. Colonialism changed European states too, strengthening the formation of capitalist social relations and markets as well as the nation-states in Western Europe. Prior to colonialism, Europe was a relatively poor and powerless part of the world, compared to the Muslim world, India, or China; afterward, Europe became the most powerful collection of societies on the planet.

The Unevenness of Colonialism

It is important to note that colonialism did not occur in the same way at different historical moments and in different geographic places. Temporally, there were two major waves of colonialism (Figure 26), one associated with the preindustrial mercantile era and the other with the Industrial Revolution. From the sixteenth century, when colonialism began, until the early nineteenth century,

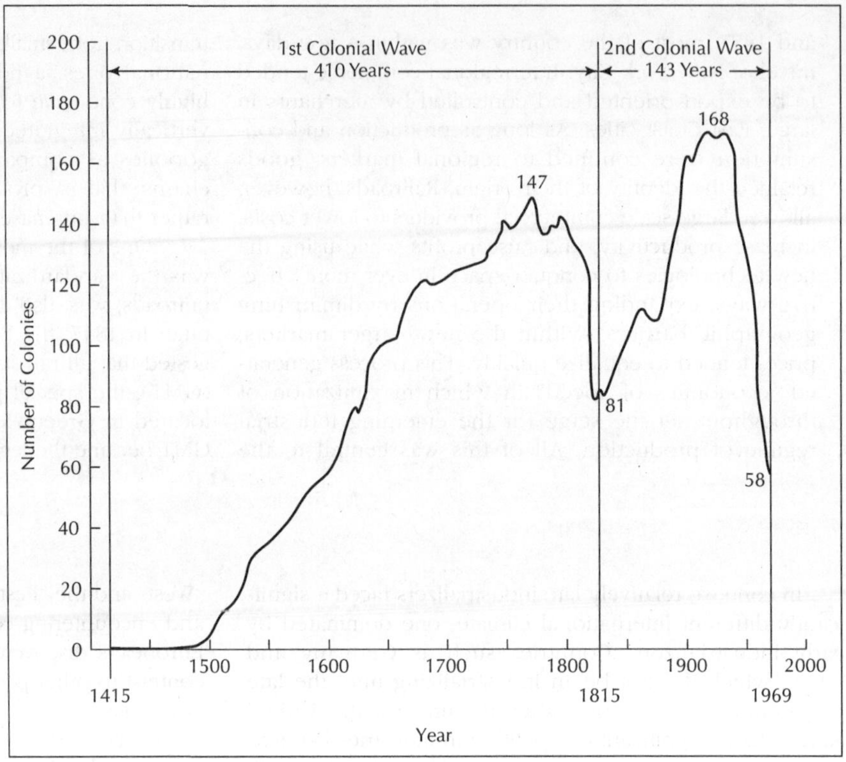

FIGURE 26 Waves of European colonialism. The first major wave, lasting from the sixteenth century until 1815, was dominated by the Spanish and Portuguese conquest of the New World. The Napoleonic wars, however, weakened Spain, and its colonies in the Americas broke away, leading to a decline in the total number of colonized countries. The second wave, from 1815 until the 1960s, was the European conquest of Africa, Asia, and the Middle East, particularly by the British and French. After World War II, the colonial empires broke up and the number of nominally independent countries in the world increased.

Western economic thought was characterized by mercantilism, in which state protection of private interests was justified as necessary for the national well-being. During this period, the largest colonial powers were Spain and Portugal, and their primary colonies were in the New World and parts of Africa.

Following the Napoleonic wars, which ended in the Treaty of Vienna in 1815, European powers were relatively weak. This provided an opportunity for nationalists in Latin America, led by Simón Bolívar, to break away and become independent countries. Thus, the number of colonies declined sharply in the early nineteenth century. However, during the subsequent phase of industrialization, characterized economically by the ideology of free trade, the number of colonies grew again. This time, Britain emerged as the world's premier power, and along with France, colonized large parts of Africa and Asia. Finally, as Figure 26 illustrates, the number of colonies declined rapidly after World War II amid the era of decolonization.

Colonialism was also uneven spatially. Different colonial empires had widely varying geographies, as illustrated by the distribution of empires at their peak in 1914, the eve of World War I (Figure 27). The British Empire, which encompassed one-quarter of the world's land surface, stretched across every continent of the globe, including large parts of western Africa, the Indian subcontinent, and parts of Southeast Asia. The French ruled in parts of western Africa and in Indochina as well as Madagascar. The Portuguese had Brazil, chunks of Africa such as Angola and Mozambique, Goa in India, parts of Indonesia such as Timor, and Chinese Macau. The Belgians possessed Congo, the private holdings of Leopold II. Portugal retained control over vast swaths of Africa. Italy was in Libya and Ethiopia. Even Germany, late to unify and to industrialize, controlled parts of Africa (Togo, Namibia, Tanganyika) and New Guinea.

How Did the West Do It?

What allowed Europe to conquer the rest of the planet? The answers to this question are not simple. Obviously, they do not lie in any innate superiority of Europeans. Indeed, for much of history, Europe was relatively weaker and poorer than the countries it conquered. By the sixteenth century, however, Europe did come to possess several technological and military advantages over its rivals.

Jared Diamond (1999), in his Pulitzer–prize-winning book *Guns, Germs, and Steel,* maintains that Western societies enjoyed a long series of advantages by virtue of geographical accident. Agriculture in the West, centered on wheat, was productive and could sustain large, dense populations. Old World societies were often stretched across vast East-West axes, or regions with common growing seasons. There was a long Western history of metalworking, which not only increased economic productivity but also led to such weapons as guns and cannons. By the Renaissance, Europeans had become highly skilled at building ships and navigating the oceans. And by the late

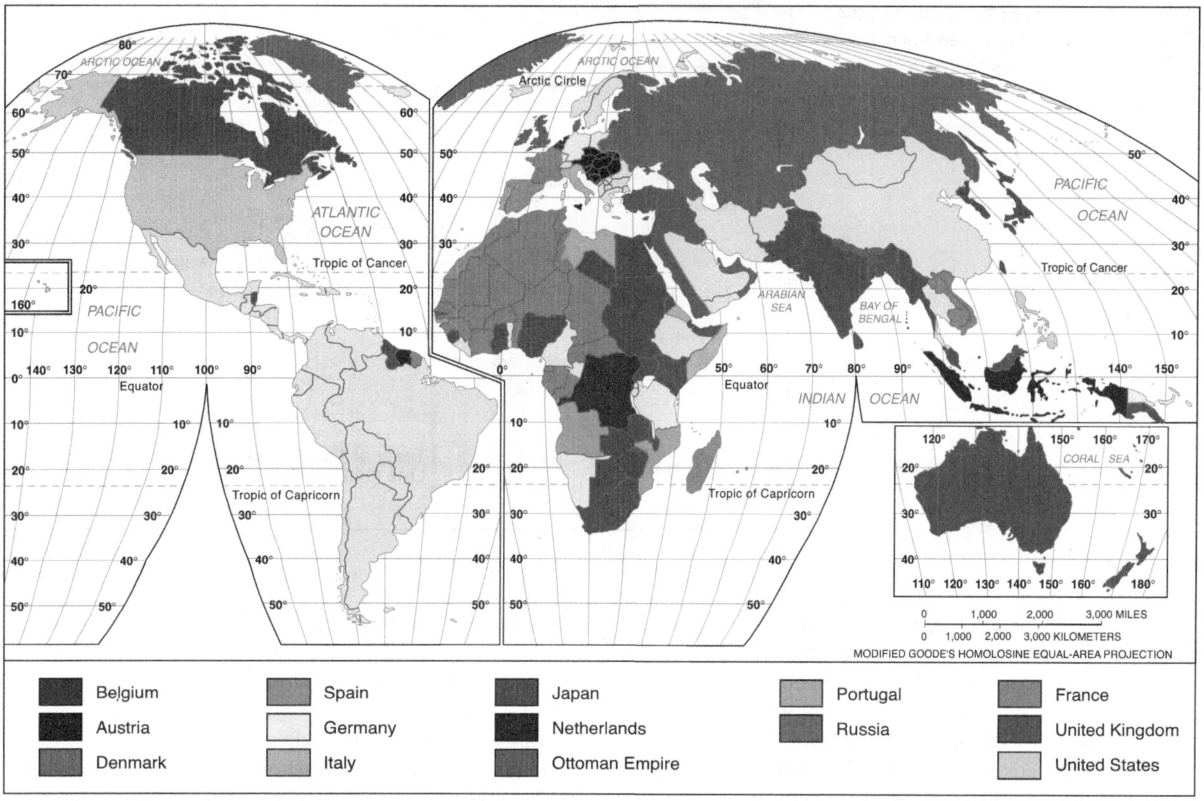

FIGURE 27 The geography of colonial empires in 1914, on the eve of World War I and the peak period of European influence. The British Empire encompassed vast areas in Africa, South Asia, and Australia. France ruled over most of western Africa and Indochina. Portugal retained its hold over Angola and Mozambique. Belgium controlled Congo. Indonesia belonged to the Dutch. Even late-developing Italy and Germany controlled parts of Africa. Japan was becoming a new colonial power in Asia, and the United States had become both a colonial power, in the Philippines and Puerto Rico, as well as an emerging neocolonial one.

eighteenth century, the West had discovered inanimate energy, which offered numerous economic and military advantages. In addition, the Europeans unleashed diseases (if unintentionally), particularly smallpox and measles, on the New World, which provided them with an unintended advantage but also led to labor shortages.

Others maintain that the West's advantages were not simply technological, but political. The Western "rational" legal and economic system stressed secular laws and the importance of property rights. Yet others point out that unlike the Arabs, Mughal India, or China, Europe was never united politically. Indeed, every time one European power attempted to conquer the others, it was defeated, as exemplified by France in the early nineteenth century and Germany twice in the twentieth. The lack of centralized political authority created a climate in which dissent and critical scholarship was tolerated. For example, the French Huguenots, Protestants in a predominantly Catholic country, could flee persecution by moving to Switzerland, where they started the Jura district watch industry. Similarly, when Columbus failed to obtain financing for his voyages from the Italians, he could switch to Spain, whose king ultimately consented.

A Historiography of Conquest

To appreciate colonialism, it is necessary briefly to delve into its specifics in different times and places. This short review is intended to demonstrate that colonialism meant quite different things under different contexts (i.e., it was historically and geographically specific).

LATIN AMERICA Home to wealthy and sophisticated civilizations such as the Inca, Mayan, and Aztec cultures, Latin America was one of the first regions to be taken over by Europeans. Two years after Columbus arrived, Spain and Portugal struggled over who owned the New World, a contest settled by the Pope in the Treaty of Tordesillas in 1494 (Figure 28). The conquistadors who spread out over Mexico and Peru annihilated, respectively, the Aztecs and Incan states. In large part, this was accomplished through the introduction of smallpox, which killed 50 to 80 million people within a century of Columbus's arrival, the greatest act of genocide in human history.

Under the philosophy of mercantilism, which stressed bullion, or precious metals, as the key to national wealth, the Spanish took home large quantities of silver from the

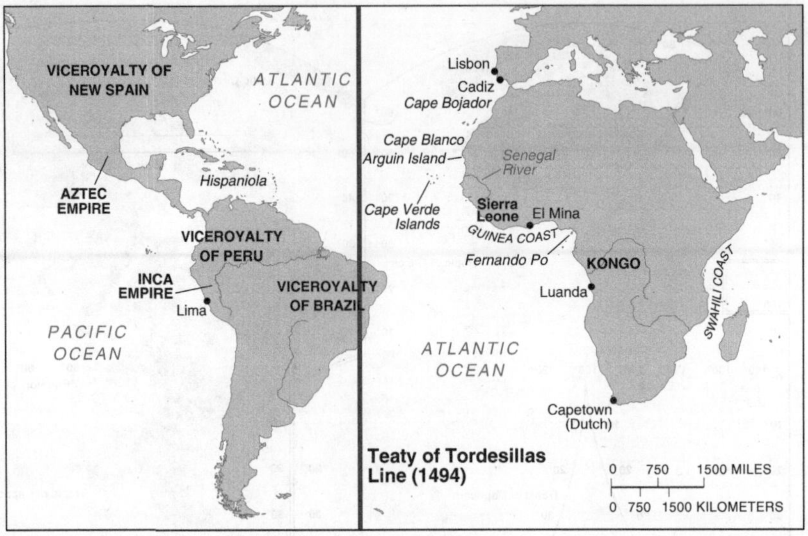

FIGURE 28 The Treaty of Tordesillas in 1494 divided the New World between Spain and Portugal, which is why Brazilians speak Portuguese rather than Spanish.

New World. The silver mines in central Mexico were among the largest in the world, and in the Potosi mines in Bolivia, 2 million Aymara Indians perished digging silver. Argentina takes its name from the Latin word for silver and is home to the Rio Plata (or "silver river"). Most of this metal was taken back to Spain in galleons and provided an enormous base of capital that financed economic activities throughout Europe.

Spain also introduced the land grant system into the New World, giving large tracts of lands to potential rivals for the Spanish throne to remove any threat they might have posed. This process was an extension of the latifundia system practiced in Iberia, which itself had roots going back to the Roman Empire. As a small landed aristocracy consolidated its hold, the distribution of farmland became highly uneven, with a few wealthy landowners and large numbers of landless *campesinos*. This pattern continues to the present in Latin America, indicating how colonialism still profoundly shapes the geographies of the contemporary world.

The Spanish empire in the New World was largely ended by the independence movements of the 1820s that followed the Napoleonic wars (although it took the Spanish-American War of 1898 to finish it off). The Spanish empire broke up into a series of independent countries stretching from Mexico to Argentina. The Portuguese empire in Brazil did not fragment in the same way, leaving that nation as the giant of Latin America.

NORTH AMERICA The colonialization of North America proceeded along somewhat different lines. Here, the Spanish were active in Florida and in the Southwest (Figure 29). A century before the Pilgrims arrived at Plymouth Rock, the Spanish had control of what is now Texas and California. The French took over Quebec and the St. Lawrence River valley, only to lose it to Britain in the eighteenth century, and the Mississippi River valley, with the key port of New Orleans, only to sell it to the United States in the Louisiana Purchase of 1803. The Russians crossed the Bering Straits and seized Alaska but sold it to the United States in 1867. The Dutch established colonies in New Amsterdam, including Haarlem and Brueklyn, but the British captured the area in 1664 renaming it New York. Britain emerged as the dominant power in North America, controlling New England and the Piedmont states along the eastern seaboard. Canada was colonized largely through the famous Hudson Bay Company, then the world's largest, which controlled much of the fur and fish trade.

British colonialism in North America began with a series of port cities on the east coast, typically at the mouths of rivers (e.g., Boston, New York City, Philadelphia). After these colonies achieved independence in the Revolutionary War, settlers moved west, across the Appalachians in the early nineteenth century and across the Great Plains and Rocky Mountains somewhat later. Railroads opened up the region to even further settlement. As in Latin America, this process involved the wholesale eradication of Native American peoples, theft of their land, and commodification of territory.

AFRICA As it was everywhere else, colonialism in Africa was unique. The process included slavery, the kidnapping of roughly 20 million people and exporting them to the New World (Figure 30), typically under brutally inhumane conditions, where they were used to compensate for the labor shortages brought on by the European annihilation of Native Americans. The capture of slaves robbed these societies of young adults in their prime working years and sometimes occurred with the assistance of local chiefs who sought to profit from the trade. Slaves were generally taken from western Africa, and the largest numbers worked the sugar plantations of Brazil and the Caribbean. Others worked the cotton and tobacco planta-

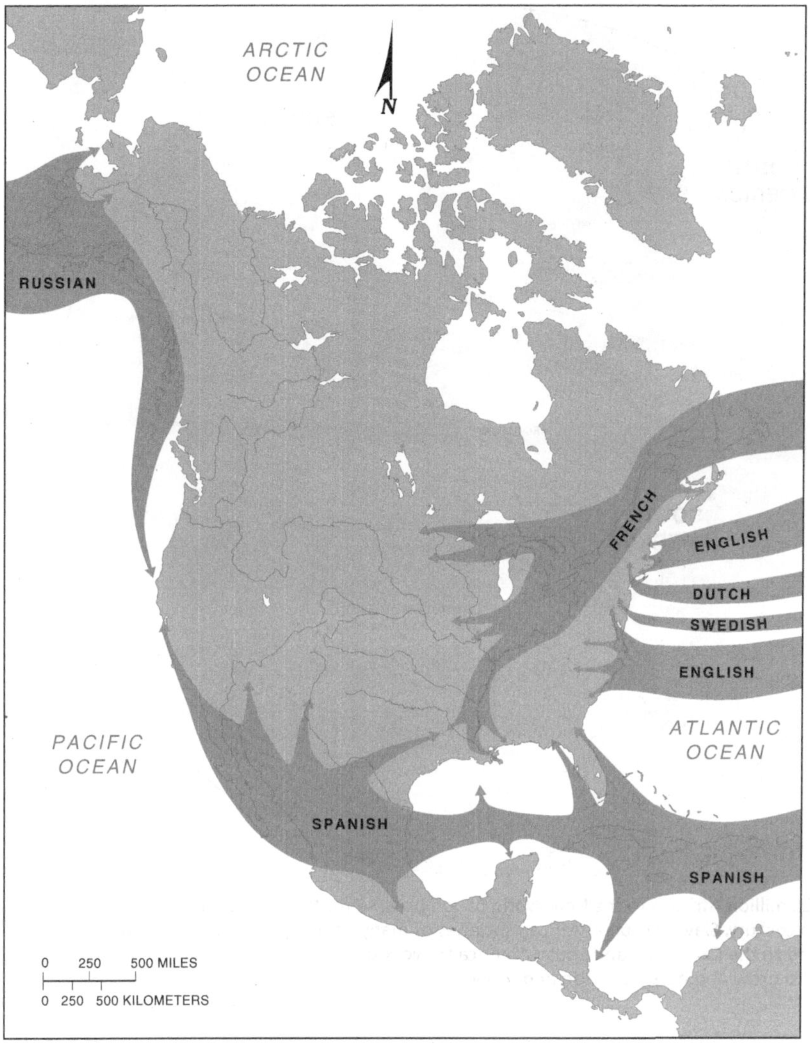

FIGURE 29 The colonial conquest of North America involved a variety of European powers. The Spanish were in Florida and the southwestern United States long before any other Western power. The French occupied the St. Lawrence River valley and the Mississippi, before selling the territory between the Mississippi and the Rockies to the United States in 1803 (the Louisiana Purchase). Russia occupied Alaska, sending explorers as far south as California. Britain had colonies in New England and the Piedmont and ruled Canada through the Hudson Bay Company.

tions of the American South. In eastern Africa and across the Sahara desert, there long existed a smaller system of slavery operated by Arab traders.

Africa is exceptionally rich in minerals, and colonialists were quick to take advantage of that fact. In the nineteenth century, when European powers penetrated from the coastal areas into the interiors, copper, gold, and diamond mines soon opened, often using slave labor. These activities remain the basis of many African economies today.

Perhaps the most important colonial impact on Africa was the political geography that Europe imposed upon that continent. Following the famous Berlin Conference of 1884, when European powers drew maps demarcating their respective areas of influence, the colonies of Africa bore no resemblance whatsoever to the distribution of indigenous peoples. Roughly 1000 tribes were collapsed into about 50 states (Figure 31). Some groups were separated by colonial boundaries; many others, with widely different cultures and economic bases, were lumped together. Not surprisingly, since independence in the 1950s and 1960s, many African states, including Angola, Congo, Rwanda, Liberia, and Sudan, have been wracked by civil wars and tribal conflicts in which tens of millions have perished.

THE ARAB WORLD The Arab world, one of the most powerful and sophisticated centers of world culture from the seventh through the fifteenth centuries, had been colonized long before the Europeans arrived. The expansion of the Ottoman Empire over several centuries led one group of Muslims—Turks, who are not Arabs—to dominate another group of Muslims, the Arab peoples of the Middle East and North Africa (Figure 32). Gradually, starting in the nineteenth century, European powers encroached on this vast domain. In 1798, the French seized Egypt from the Ottomans, only to lose it to Britain four years later. The British used Egypt as a source of cotton, establishing large plantations along the Nile River and building the Suez Canal in 1869. In the nineteenth century, the French seized Morocco, Algeria, and Tunisia.

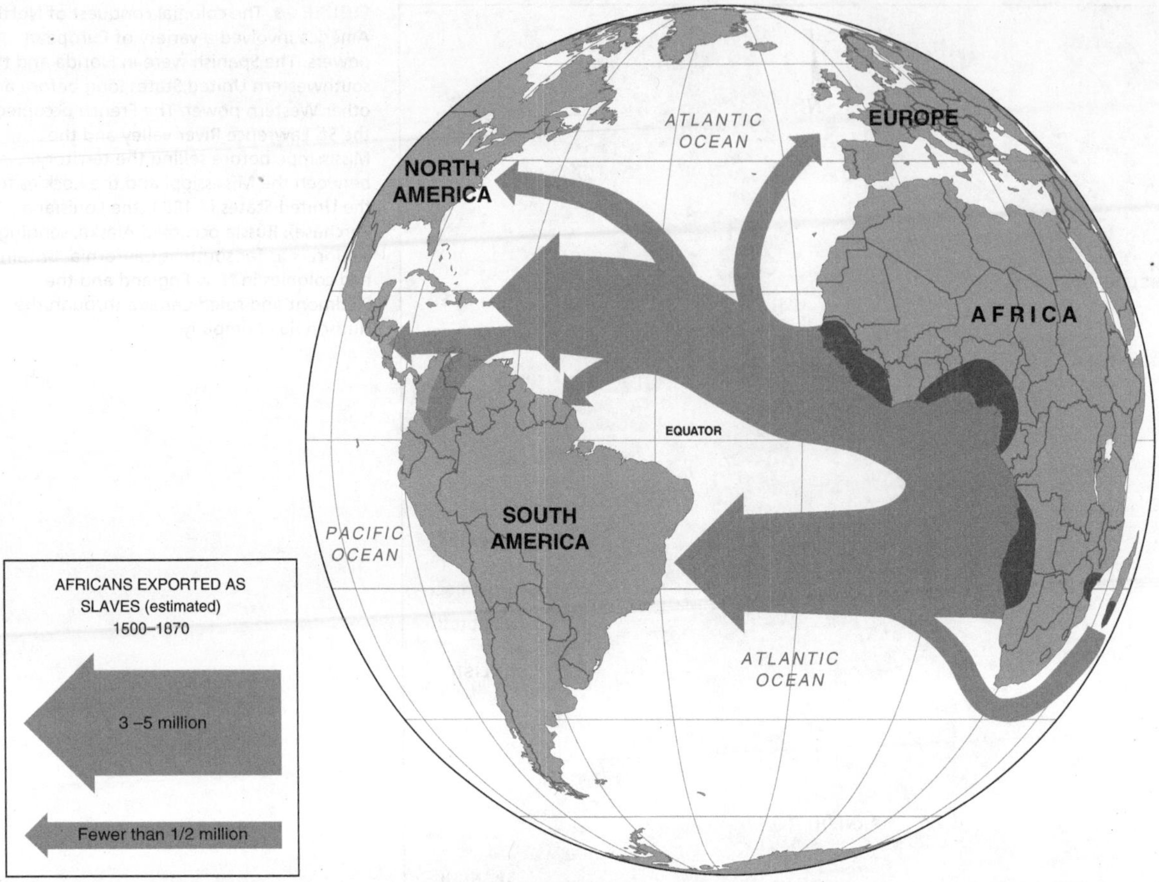

FIGURE 30 The slave trade brought roughly 20 million Africans to the New World to compensate for the labor shortages induced by the great smallpox epidemics of the sixteenth century. Slavery robbed African societies of many of their young people in the prime working years. Most slaves were shipped to the Caribbean and South America to work on sugar and fruit plantations; in the southern United States, slaves were used to grow and harvest cotton and tobacco.

After World War I, the Ottoman Empire collapsed, and the French and British seized its Arab colonies. The French took over Syria and Lebanon. The British assumed control over Palestine, much of which became Israel in 1948, as well as Iraq and the sheikdoms of the Persian Gulf. Many Arabs, who initially welcomed the Europeans as liberators, quickly learned that their new, European boss was very similar to the old, Ottoman one in its suppression of their liberties.

SOUTH ASIA The Indian subcontinent came to be the jewel in the crown of the British Empire. Starting in the seventeenth century, the British East India Company established footholds in this domain, founding the city of Calcutta in 1603. India, which is predominantly Hindu, had long been controlled by the Muslim Mughals, whose power was gradually usurped by the foreigners. A vast land that stretched from Muslim Afghanistan in the west through Hindu India to Buddhist Burma, the Indian colony was the largest colonial possession on earth.

Britain made an enormous economic impact on this land. In the nineteenth century, British textile imports flooded India, destroying the Mughal textile industry in a classic example of colonial deindustrialization. This event later became symbolically important in the independence movement after World War II, when Gandhi called upon Indians to boycott British textiles. Indian laborers were exported throughout the British Empire, including the Caribbean, eastern Africa, and parts of the Pacific such as Fiji. Tea plantations were established in Bengal and Ceylon (now Sri Lanka). To facilitate the extraction of Indian wealth, Britain built railroads.

In 1857, a mass uprising against the British, known as the Sepoy Rebellion, took place. Encouraged by the local raj, or Mughal ruler of Calcutta, the revolt claimed the lives of tens of thousands of Indians when it was eventually crushed. Although it failed, the revolt is often considered India's first blow for independence because it forced the British crown to assume direct control over this land rather than administer it through the East India Company, which was abolished as a consequence.

EAST ASIA East Asia, comprising China, Korea, and Japan, also had a unique colonial trajectory. Japan, as

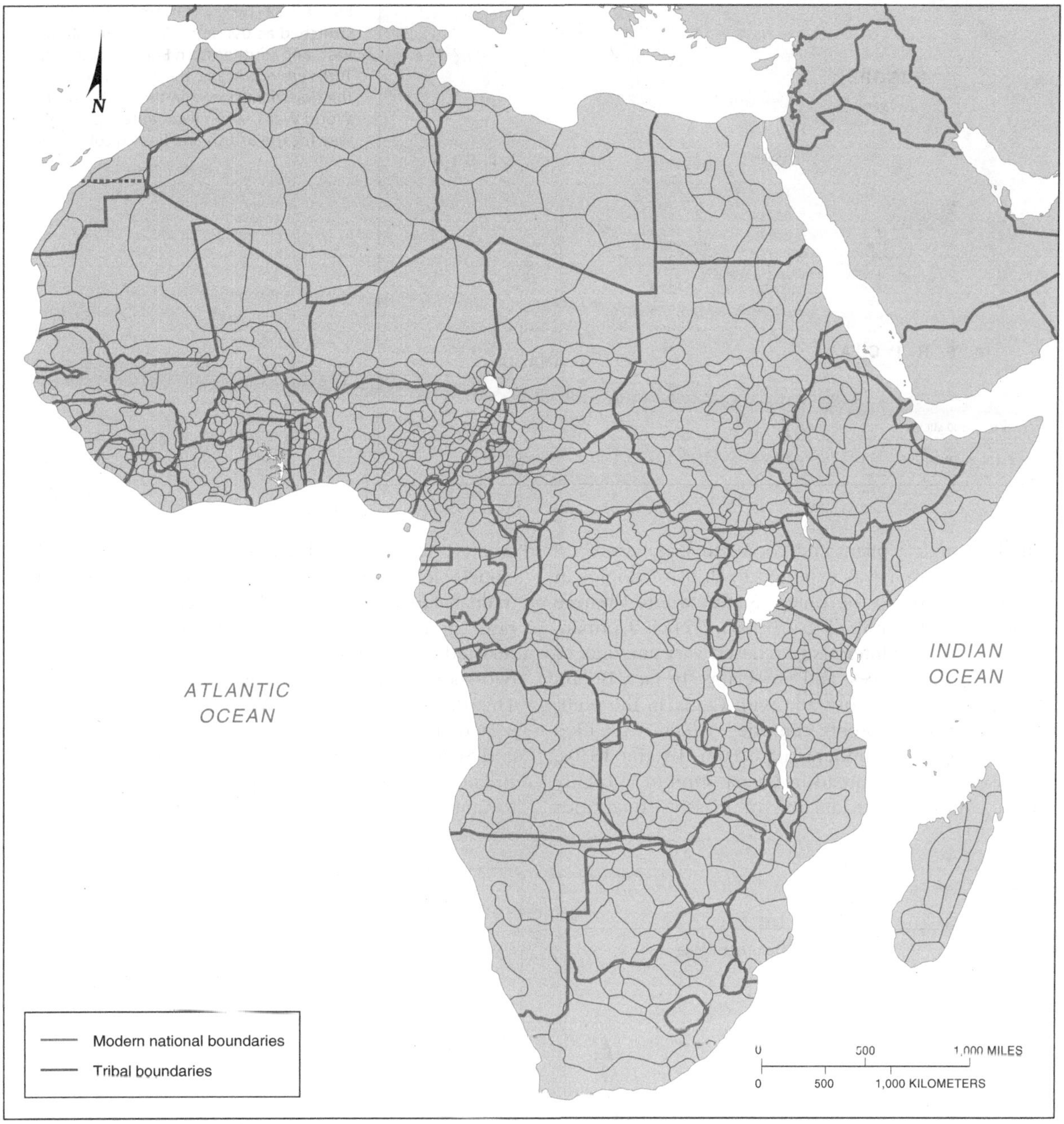

FIGURE 31 The Berlin Conference of 1884 redrew the map of Africa, collapsing roughly 1000 tribes into 50 different states. Because the boundaries of what would become Africa's states bore no resemblance to the geography of the people who lived there, many tribes were split in two, while others, without any cultural or economic similarities, were lumped together. The political geography of Africa has played out disastrously in the form of numerous tribal conflicts and civil wars that have claimed the lives of tens of millions of people.

noted earlier, was never colonized. When it emerged from a long period of isolation that began in the early seventeenth century and lasted until 1853, Japan rapidly westernized and industrialized and became the only non-Western power to challenge the West on its own terms, gradually expanding its power through northeast Asia. Korea, opened up under the threat of force in the 1870s, was taken over by Japan in 1895 and annexed in 1905, and Japan held it until the end of World War II. A similar situation occurred in Taiwan.

China, however, was a different story. Under the rule of the Manchus, foreigners from the north (Manchuria) who ruled the Chi'n dynasty from 1644 until the revolution of 1911, the Chinese government was weak and corrupt. Except for a few cities along the coast, China was never formally colonized; rather, European control

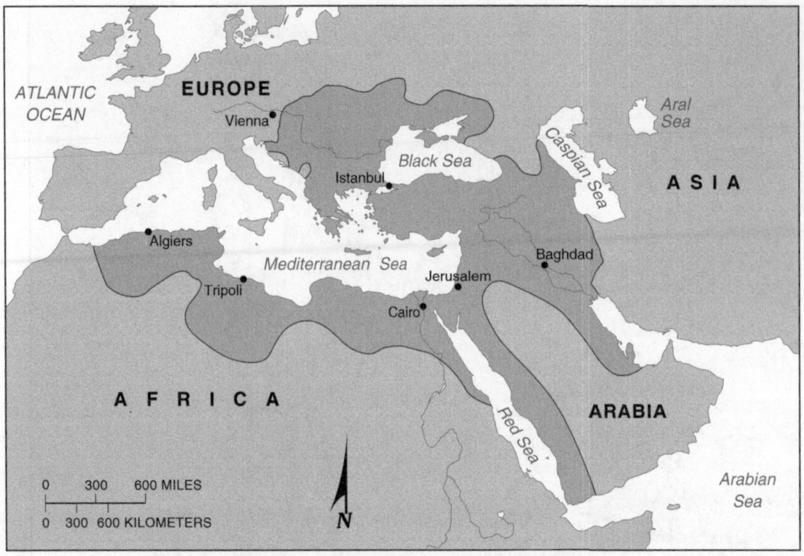

FIGURE 32 Turkey's Ottoman Empire stretched across North Africa, the Middle East, and southeastern Europe. Although Turks are not Arabs, the Turkish empire was the dominant power in the Arab world until World War I, when it collapsed, paving the way for the British and French colonialists.

operated through a pliant and cooperative government. Chinese coolie labor was exported to British colonies in Southeast Asia and to the United States, where Chinese labor built railroads in the West. British, French, German, and American trade interests purchased vast amounts of Chinese tea, silks, spices, and porcelains. In fact, Britain ran a negative balance of trade with China in the eighteenth and early nineteenth centuries, which it rectified by introducing large amounts of opium. By the 1830s, opium use was widespread in China. Opium is highly addictive, and the introduction of massive quantities created severe social disruptions in Chinese society. Nonetheless, profits were more important, and the British trade balance was restored (Figures 33a and 33b). In a rare moment of defiance, the Manchu government resisted opium imports, and Britain and China fought two short, nasty conflicts, the Opium wars of the 1840s. The British won easily. As compensation, the British seized "treaty ports," coastal cities such as Hong Kong and Shanghai, where Western, not Chinese, law applied. Britain held Hong Kong until 1997, when it was returned to China.

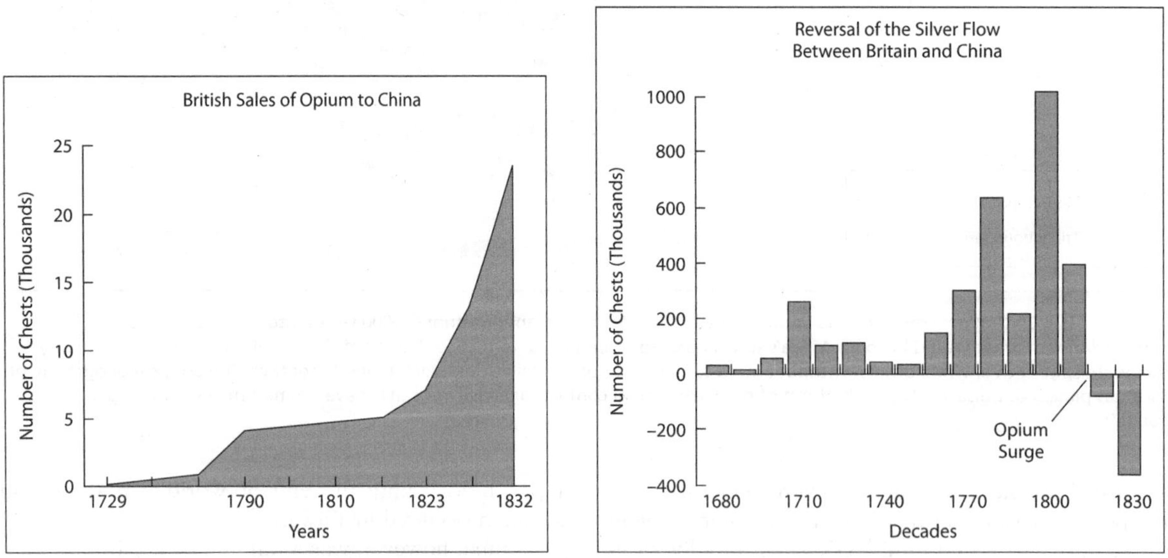

(a) (b)

FIGURE 33 British sales of opium to China in the nineteenth century represented a response to Britain's trade deficit with that nation. Economically, opium had exactly the intended effect, reversing the balance of payments, as indicated by silver flows. Socially, it was catastrophic: Opium is highly addictive and as much as 12% of China's adult population became hooked on it as a result of cheap British imports. The Manchu regime's opposition prompted the Opium wars of the 1840s, which Britain won handily, opening China to the establishment of foreign treaty ports.

Chinese resentment against the Manchus culminated in the Taiping Rebellion, a huge uprising in the southern part of the country led by Chinese Christians. Lasting from 1851 to 1864, this rebellion led to the deaths of more than 20 million people but was ultimately crushed by the Chinese government with Western backing. The shorter Boxer Rebellion of 1899–1901 was more explicitly anti-Western. These revolts set the stage for the successful nationalist revolution of 1911, which ended Manchu rule and initiated the Republic of China.

SOUTHEAST ASIA The peninsula of Indochina and the islands of Southeast Asia, long home to a rich and diverse series of peoples and civilizations, were conquered by a variety of different European powers (Figure 34). The Philippines was Spain's only Asian colony and served as the western terminus of the trans-Pacific galleon trade. Administered as part of Mexico, the Philippines was heavily shaped by Spanish rule, which affected land-use patterns (including sugar plantations), the region's language, and made it the only predominantly Catholic country in Asia. Spanish rule ended in 1898 when the United States took over; the Philippines finally became independent after World War II.

The French controlled much of Indochina, including Vietnam, Laos, and Cambodia. French rule shaped the design and architecture of cities such as Saigon, and large numbers of Vietnamese converted to Catholicism. French domination did not end until 1954, with the defeat at Dien Bien Phu an event that laid the foundations for American military involvement in Vietnam later.

Britain was also a major power in Southeast Asia; it controlled Burma and, only informally, the economy of Thailand. These countries were made into rice exporters for other parts of the British Empire. The British controlled the colony of Malaya (later Malaysia), including the strategically critical Malacca Straits. They founded the city of Singapore as a naval station and commercial center to exercise control over this region. Malaya, like other colonies in the area, became a major producer of rubber products as well as timber and tin.

Indonesia, now the fourth most populous country in the world, was dominated by the Dutch for several hundred years. Dutch rule, starting with the founding of the Batavia colony on Java in the eighteenth century, gradually expanded to include the other islands. The primary institution involved was the Dutch East Indies Company, a chartered crown monopoly similar to the British East India

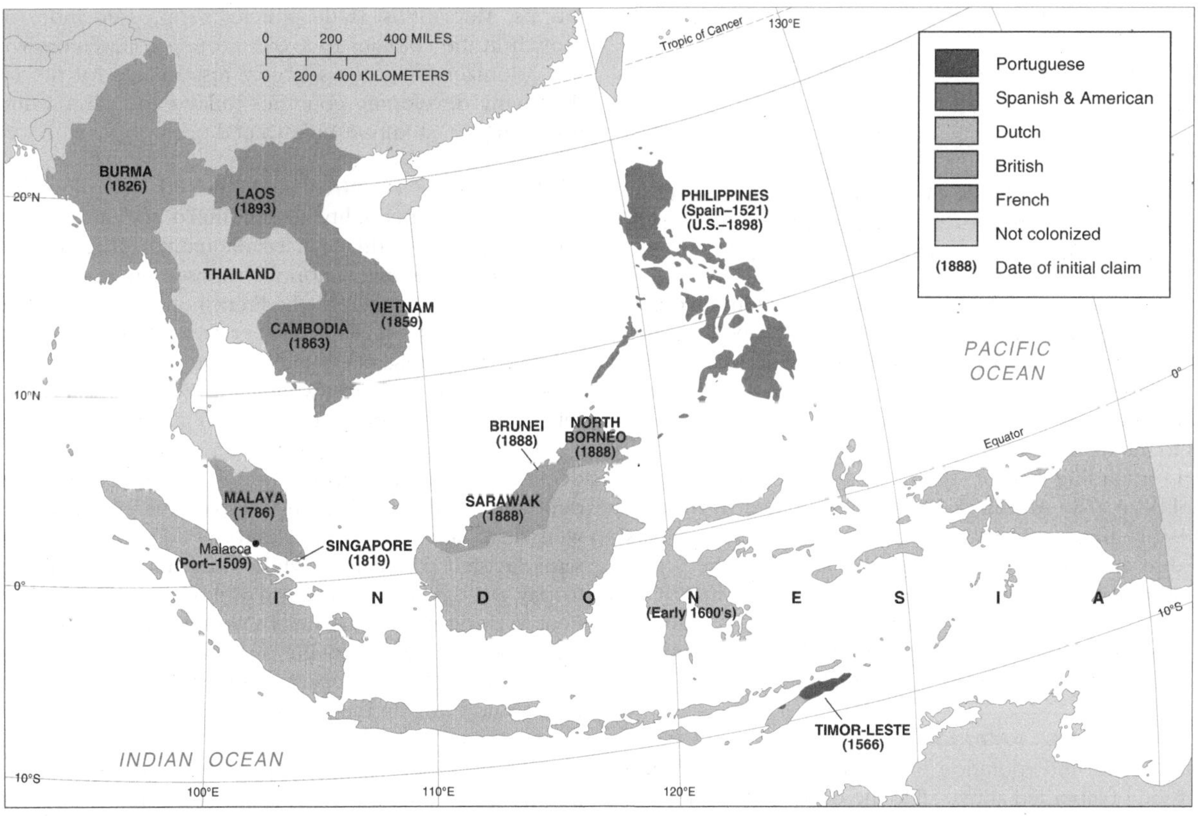

FIGURE 34 Colonialism in Southeast Asia involved a variety of powers. The French occupied Vietnam, Laos, and Cambodia. Britain ruled Burma, Malaya, and Singapore and indirectly controlled Thailand. The Spanish held the Philippines until 1898, when the United States took it away. The Dutch ruled Indonesia until 1947, when it finally achieved independence.

and British West India companies. Indonesia became a significant source of spices, tropical hardwoods, rubber, cotton, and palm oils. In all of Southeast Asia, Chinese immigrants came to play major roles in the economy as bankers and shopkeepers.

OCEANIA Australia, New Zealand, and the Pacific Ocean islands, which are commonly grouped as the region of Oceania, formed yet another domain of colonialism. Australia was inhabited for thousands of years by indigenous aborigines, most of whom were eradicated by the British as they exerted control. The native Tasmanian population was completely exterminated. The continent served originally as a penal colony for criminals. In the nineteenth century, it became a significant exporter of wheat and beef.

New Zealand, also a British colony, had a much larger native population, the Maori, who continue to have a significant presence there. The introduction of refrigerated shipping in the late nineteenth century turned this island nation into a major producer of lamb and dairy products.

Finally, the countless islands of the Pacific Ocean, home to varying groups of Polynesians, Micronesians, and Melanesians, were conquered by the British and French. Captain Cook sailed through in the seventeenth century, paving the way for followers. In the nineteenth century, fishing and whaling interests used these islands to refuel. Following World War II, when the United States drove Japan out of the Pacific, America became the leading political force in the area.

The Effects of Colonialism

By now, it is abundantly evident that colonialism introduced enormous changes in colonized places. It is worth recapitulating these in a more systematic form.

ANNIHILATION OF INDIGENOUS PEOPLES Often, colonialism imposed traumatic consequences on the people who were conquered. At times, this meant open genocide, such as in Australia, in which more than 90% of the aboriginal population was exterminated. In the New World, disease led to the deaths of tens of millions, or 95% of the population, including all of the native inhabitants of the Caribbean. The African slave trade devastated tribal societies on that continent. In other cases, brute force was used to enforce Western dominance, as during the Sepoy Rebellion in India and the Taiping Rebellion in China. These examples are the severest expressions of colonial control, but they serve as a reminder that the European conquest of the world was often brutally violent.

RESTRUCTURING AROUND THE PRIMARY ECONOMIC SECTOR The incorporation of colonies into a worldwide division of labor led above all to the development of a primary economic sector in each of them. Primary economic activities are those concerned with the extraction of raw materials from the earth, including logging, fishing, mining, and agriculture. In Latin America, Africa, South Asia, and Southeast Asia, cash crops such as sugar, cotton, tea, coffee, fruits, rubber, and tobacco were grown under the plantation system for sale abroad (Figure 35). Silver, tin, gold, copper, diamonds, iron, and other ores were mined using slave labor or peasants working under slavelike conditions. Mercantilist trade policies suppressed industrial growth in the colonies to avoid creating competitors with the colonizers, a process largely responsible for the fact that many developing countries today export low-valued goods and must import high-valued ones.

FIGURE 35 A sugar plantation in the Caribbean Antilles. Plantations used cheap labor, often slaves, working under deplorable conditions. They represented the first wave in the worldwide commercialization of many crops.

FORMATION OF A DUAL SOCIETY With it colonialism brought great inequality to colonized societies. Often, colonial powers utilized a small, native elite, typically drawn from an ethnic minority, to assist them in governing the colonies. For example, the French utilized the Alawites in Syria, a sect of Islam neither Sunni nor Shiite. The Germans and Belgians favored lighter-skinned Tutsi over darker-skinned Hutu in Rwanda and Burundi. The British relied on the Muslim Mughal rulers to govern Hindu India.

For the bulk of the population, colonialism entailed declining economic opportunities, a theme central to dependency theory. Traditional patterns of agriculture were disrupted, often with disastrous effects. Land-use patterns favored colonialists, while indigenous peasants had to pay for the cost of their own exploitation with taxes. People living in dry climates in western Africa, for example, coped well with drought until the British forced them into a system of cash cropping. Huge famines struck India, Egypt, and China in the nineteenth century.

POLARIZED GEOGRAPHIES As colonial societies became polarized, so too did the spaces they comprised. Ports, which were central to European maritime trade and control, became important centers of commerce, often to the detriment of traditional capitals further inland. For example, in

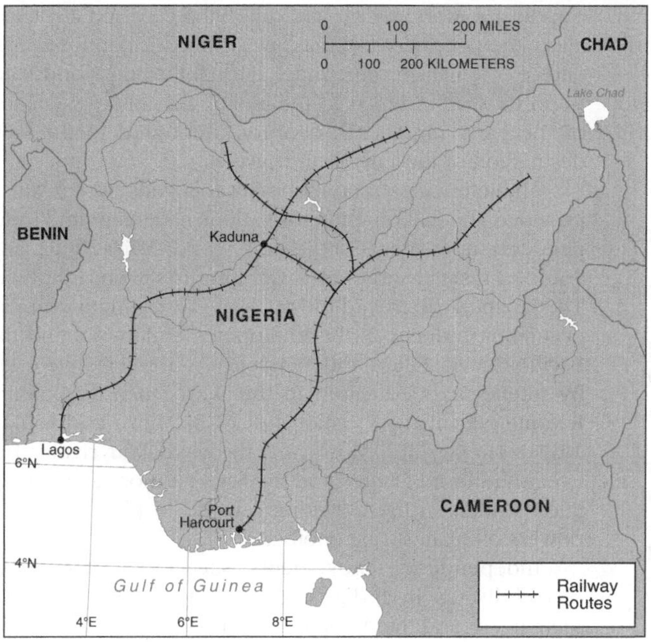

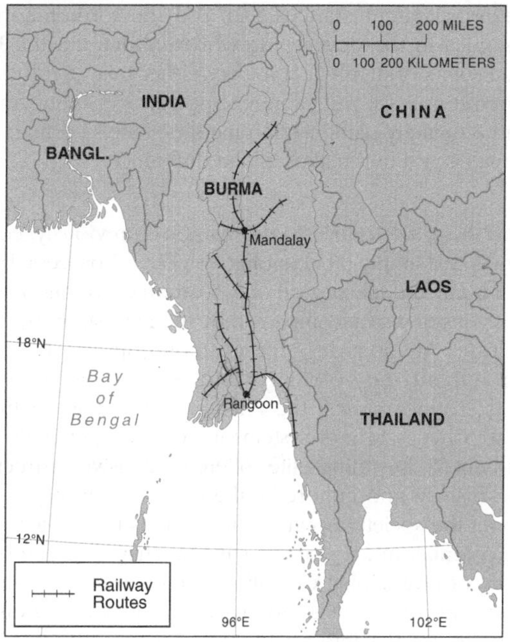

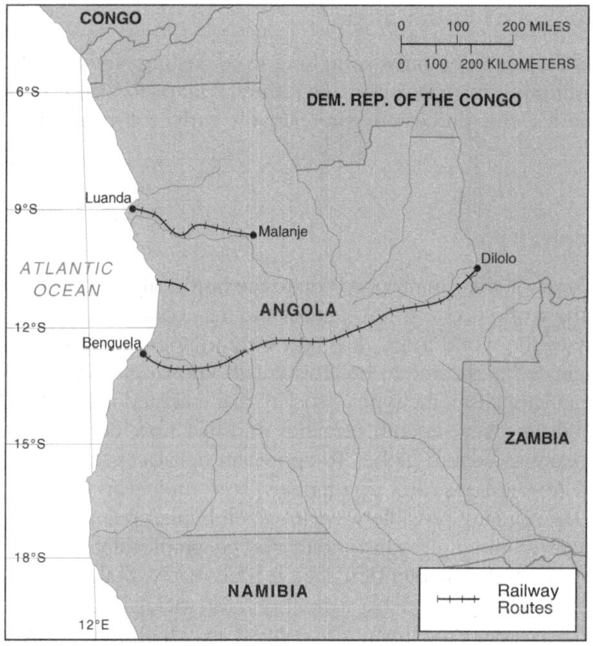

FIGURE 36 Railroads in Nigeria, Burma, and Angola reveal the colonial propensity to establish a transportation infrastructure that linked coastal port cities with the resources of the interior, which were typically mines or plantations.

Peru, Lima displaced the mountainous Incan city of Cuzco; in western Africa, the famous inland trade center of Timbuktu declined as new Atlantic maritime routes flourished; in India, coastal cities such as Calcutta, Mumbai (formerly Bombay), and Madras displaced the inland Mughal capital of Delhi; in Burma (now Myanmar), Mandalay in the interior fell far behind coastal Rangoon; in Vietnam, the imperial capital Hue declined in the face of Saigon; and in Indonesia, the traditional center of Jogjakarta in central Java was marginalized by the Dutch port city of Batavia, later Jakarta. As cities grew and offered more opportunities, millions of people left the poorer rural areas in waves of rural-to-urban migration.

From the coasts, railroads extended colonial control into the interior (Figure 36), often reaching into mineral-rich regions or plantations. A long-term consequence of this design is that the road and rail networks of developing countries often bear little resemblance to the distribution of the population that lives there and their needs; rather, they are constructed to facilitate the export of raw materials to the colonizing country, and today, to the global economy.

TRANSPLANTATION OF THE NATION-STATE The nation-state, as we observed earlier, was fundamentally a European creation. Nonetheless, it was widely dispersed around the world as colonies were made into states. In Africa, where this process was the most notorious, it led to the formation of unstable states with highly artificial borders. Similarly, Burma and Afghanistan were creations of the British. Even India, which is more culturally diverse than all of Europe, was stitched into one country; surprisingly, it divided into only three countries after independence (including Pakistan and Bangladesh). Unlike Europe, where states were centered to some degree on ethnic similarity, the states of much of the developing world were often too diverse to be understood in the same terms. In such societies, where local religious, ethnic, and tribal loyalties supersede nationalism, political conflicts can impair economic growth and development.

Of course, the United States is also very diverse ethnically and culturally, but it emerged under very different historical circumstances than did the former European colonies in the developing world. Above all, the United States was primarily a colonizer rather than a colony, able to exert military and economic power over other countries and ultimately rising to become the globe's premier superpower—and that makes all the difference.

CULTURAL WESTERNIZATION As noted previously, colonialism is not simply an economic or political process, but also a cultural and ideological one. Western economic and political control was accompanied by the imposition of Western culture. Missionaries, for example, spread Christianity throughout the colonial world, sometimes successfully (e.g., Latin America, the Philippines) and sometimes not (e.g., China). School systems in colonies, generally set up to benefit the ruling elite, offered extensive instruction in the history and culture of the colonial country but little about the society in which the students lived. More broadly, colonialism may be seen as one chapter in the broader story of how global capitalism homogenizes lifestyles, values, and role models around the world, turning disparate sorts of peoples into ready consumers.

The End of Colonialism

The European empires were long-lived, lasting almost half a millennium. Yet ultimately, they collapsed. In Latin America, this process began relatively early, following the Napoleonic wars, which weakened the core sufficiently to allow the periphery to break away. In Africa and Asia, the end of colonialism came much later, following World Wars I and II, which, similar to the global geopolitical situation of the early nineteenth century, amounted to the self-destruction of the European powers.

The international environment following World War II provided an ideal opening for various nationalist and independence movements in the Arab world, Africa, India, and Southeast Asia. Sometimes Communists were involved. The Japanese destroyed the myth of European invincibility occupying Indochina, the Philippines, Indonesia, and the Pacific Ocean. Often, independence movements were led by intellectuals educated in the West, such as Ghana's Kwame Nkrumah, Vietnam's Ho Chi Minh, and India's Mohandas Gandhi. Moreover, the Cold War rivalry between the United States and the Soviet Union allowed political leaders in the developing world to play the superpowers off against one another.

Independence movements succeeded, sometimes peacefully, as in India, and often violently, as in the Vietnamese and the Algerian defeat of the French. As a result of this process of decolonization, the number of independent states multiplied rapidly in the 1950s, 1960s, and 1970s. Today, very few official colonies remain. Whether colonialism is truly dead, however, is another matter.

Summary

This chapter has introduced you to the historical foundations of the world economy. To understand the economic geography of the world today, it is necessary to appreciate how it originated and came to be. This entails knowing something about capitalism, the type of society that emerged in the sixteenth and seventeenth centuries and that now dominates virtually the entire globe. To appreciate what capitalism is, and how it constructs geographies, is to understand that it is one of many possible ways in which human beings have organized themselves historically and geographically.

First, the chapter described feudal society, both to outline in some detail what a noncapitalist society looks like and to sketch the historical context in which capitalism emerged. Then we proceeded to describe the fundamental features of capitalism, which, uniquely, is dominated by markets as the primary way in which resources are organized. Markets are not unique to capitalism, but their importance is. However, this does not mean that markets are the only way in which economic activity is shaped, for the state plays a role even in the "freest" of market societies. Capitalism also entails a specific set of class relations, including the commodification of labor and a working class. Capitalism also creates landscapes of uneven spatial development.

Next, we delved into the Industrial Revolution, the period in the eighteenth and nineteenth centuries that saw a fantastic transformation in capitalist social relations. The ability to harness inanimate energy led to a wave of technological innovations and increases in productivity. Industrialization created the factory system, and with it, the working class. In commodifying time and space, the Industrial Revolution produced radically new geographies. Industrialization, which began in Britain, catapulted that country into becoming the most powerful in the world. It also unleashed a new international economy characterized by significantly higher levels of trade and investment, which were achieved in part by the application of the steam engine to land and ocean transportation. The rising speed of transportation and communication created great waves of time-space compression. Yet industrialization was a complex process that varied over time, with the rise of different industries, products, and technologies.

The latter part of this chapter explored the multiple dimensions of colonialism, the expansion of capitalism on a global scale. We traced some of the advantages that Europeans enjoyed that allowed them to dominate many societies much larger than themselves. The chapter examined major world regions conquered by different European powers, noting that colonialism produced different effects in different parts of the globe. Finally, we summed up the impacts of colonialism, which produced the division between the world's economically developed and underdeveloped nations.

Key Terms

animate sources of energy
bubonic plague
burghers
capitalism
colonialism
commodities
feudalism
guilds
Hanseatic League
inanimate sources of energy
industrialization
market
mercantilism
nation-state
productivity
profit
serfs
uneven spatial development

Study Questions

1. Why is historical context so important to the analysis of contemporary economic geography?
2. What is feudalism, and how did it differ from capitalism?
3. What was the fourteenth-century plague, and what impacts did it have?
4. How is capitalism a unique type of economy and social order?
5. Are there classes under capitalism? Which ones?
6. Is capitalism the only type of economic system to have markets? Why or why not? How does the role of markets in capitalism differ from their role in noncapitalist societies?
7. What territorial changes accompanied the emergence of capitalism?
8. How did capitalism unleash new ideologies and ways of looking at the world?
9. How is the nation-state related to the growth of capitalism?
10. When and where did capitalism begin?
11. What is industrialization?
12. When and where did the Industrial Revolution begin?
13. What are some major economic, social, and geographic impacts of the Industrial Revolution?
14. How did Europe manage to colonize the rest of the world?
15. How did colonialism differ among Latin America, Africa, and Asia?
16. What are five ways in which colonialism affected the societies and geographies of thé colonies?
17. When did colonialism come to an end, and why?

Suggested Readings

Abu-Lughod, J. 1989. *Before European Hegemony: The World System A.D. 1250–1350*. New York: Oxford University Press.

Chirot, D. 1985. "The Rise of the West." *American Sociological Review* 50:181–195.

Diamond, J. 1999. *Guns, Germs, and Steel*. New York: W. W. Norton.

Hindess, B., and P. Hirst. 1975. *Pre-Capitalist Modes of Production*. London: Routledge and Kegan Paul.

Jones, E. 1981. *The European Miracle*. Cambridge, UK: Cambridge University Press.

Landes, D. 1969. *The Unbound Prometheus: Technological Change and Industrial Development in Western Europe from 1750 to the Present*. Cambridge, UK: Cambridge University Press.

Mann, M. 1986. *The Sources of Social Power*. Cambridge, UK: Cambridge University Press.

Mielants, E. 2007. *The Origins of Capitalism and the "Rise of the West."* Philadelphia: Temple University Press.

Perelman, M. 2000. *The Invention of Capitalism: Classical Political Economy and the Secret History of Primitive Accumulation*. Durham, NC: Duke University Press.

Thompson, E. P. 1966. *The Making of the English Working Class*. New York: Vintage Press.

Wallerstein, I. 1979. *The Capitalist World-Economy*. Cambridge, UK: Cambridge University Press.

Web Resources

The Agrarian Origins of Capitalism
http://www.monthlyreview.org/798wood.htm

Argues that capitalism had rural, not urban, origins.

History of Capitalism
http://www.historyworld.net/wrldhis/PlainTextHistories.asp?historyid=aa49

Timeline of major processes that led to the rise of capitalism.

Internet Modern History Sourcebook
http://www.fordham.edu/halsall/mod/modsbook14.html

Good overview of the Industrial Revolution.

Industrial Revolution
http://www.42explore2.com/industrial.htm

Lots of good links to detailed accounts of the Industrial Revolution.

Population

OBJECTIVES

- To describe and account for the world distribution of human populations
- To examine the economic causes and consequences of population change
- To describe the Malthusian argument, its extensions, and weaknesses
- To describe the major demographic and economic characteristics of a population
- To outline the demographic transition
- To discuss the growth and impacts of the baby boom
- To describe and explain economic migrations, past and present

High birth rates, large families, and massive rural-to-urban migration both reflect and shape urban and economic systems in the developing world, including the structure of cities and the quality of daily life.

© Alison Wright/Corbis

Population

Human beings are the most important element in the world economy. People are not only *the* key productive factor, but their welfare is also the primary objective of economic growth and analysis. People are the producers as well as the consumers of goods and services. As the world's population continues to grow, we face the critical question of whether there is an imbalance between producers and consumers. Does population growth prevent the sustainability of development? Does it lead to poverty, unemployment, and political instability?

To help answer these questions, this chapter examines the determinants and consequences of population change for developed and developing countries. It analyzes population distributions, characteristics, and trends. It also reviews competing theories on the causes and consequences of population growth.

GLOBAL POPULATION DISTRIBUTION

There is a widespread belief that there are "too many" people in the world; the presence of large numbers of human beings is a relatively recent phenomenon (Figure 1). For the vast bulk of human existence—upwards of 95%—people existed as hunters and gatherers, collecting food from various wild plants and stalking animals. This mode of production yields relatively few calories per unit area and supports only low population densities. Further, it is demographically very stable over time: Because births equaled deaths, there was virtually no increase in the number of people in the world. Following the agricultural revolution of roughly 8000 B.C., the capacity of many societies to support more people increased somewhat. However, the really big gains in population did not occur until the onset of the Industrial Revolution of the eighteenth and nineteenth centuries, when the transformation of agriculture freed millions from lives of rural toil and allowed for large, dense urban settlements. Thus, exponential population growth is largely a product of modernity and the modern world.

In 2010, there were approximately 6.9 billion people in the world. The diverse populations that inhabit the world are very unevenly distributed geographically. Most people are concentrated in but few parts of the world (Figure 2), particularly along coastal areas and the floodplains of major river systems. Four major areas of dense settlement are (1) East Asia, (2) South Asia, (3) Europe, and (4) the eastern United States and Canada. In addition, there are minor clusters in Southeast Asia, Africa, Latin America, and along the U.S. Pacific coast. Figure 3 is a cartogram, a map that deliberately distorts areas in proportion to a variable, in this case, population size, revealing the large masses of humanity in Asia. Asia's population is the largest, as it has been for several centuries. In 2008, Asia contained 4.1 billion people, or 60% of the world's population. Six of the top 10 countries in population size—China, India, Indonesia, Japan, Bangladesh, and Pakistan—are in East, South, or Southeast Asia, which, with 4.0 billion people, is home to 59% of the planet's inhabitants. Europe (including Russia) had 729 million residents, about 11%; other continents included Africa (1 billion, or 14%), Latin America (577 million, or 8.6%), North America (337 million, or 5%), and Oceania (34 million, less than 1%). The populations of the developing world—Africa, Asia (excluding Japan), and Latin America—accounted for three out of every four humans.

National population figures show even more variability. Ten out of the world's nearly 200 countries account for two-thirds of the world's people (Table 1). Five countries—China, India, the United States, Indonesia, and Brazil—contain half of the world's population. With 1.3 billion people, China is the world's most populous country. India, with 1.17 billion, is second, but is growing more quickly, and will surpass China in population in roughly 30 years. The United States, with 310 million in 2010, is third, and is by far the most populous of the developed nations. Indonesia, the world's largest Muslim nation, is fourth, with 243 million. Other nations with large populations include

FIGURE 1 **World population growth throughout history.** For most of human existence, population levels were low and growth rates were zero. Only with the Industrial Revolution that created the modern age did growth rates begin to rise, leading to an exponential increase in the numbers of people.

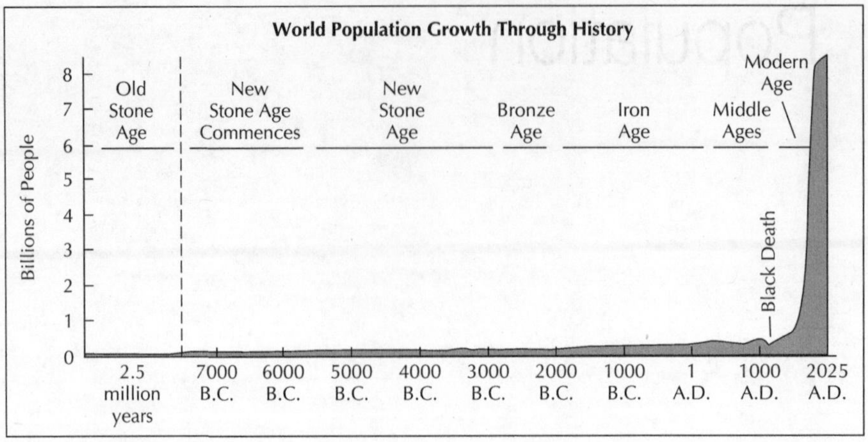

Brazil (201 million), Pakistan (184 million), Bangladesh (156 million), Nigeria (152 million), Russia (139 million and declining), Japan (126 million and declining), and Mexico (112 million). By way of comparison, other countries with significant populations include the Philippines (100 million), Vietnam (90 million), Ethiopia (88 million), Germany (82 million), Egypt (80 million), Turkey (79 million), Iran (78 million), the Democratic Republic of the Congo (71 million), Thailand (67 million), France (65 million), the United Kingdom (62 million), and Italy (58 million). Only 3 of the 10 most populous nations are considered to be economically developed (the United States, Russia, and Japan).

Population Density

Because countries vary so greatly in size, national population totals tell us nothing about crowding. Consequently, population is often related to land area. This ratio is called **population density**—the average number of people per unit area, usually per square mile or square kilometer.

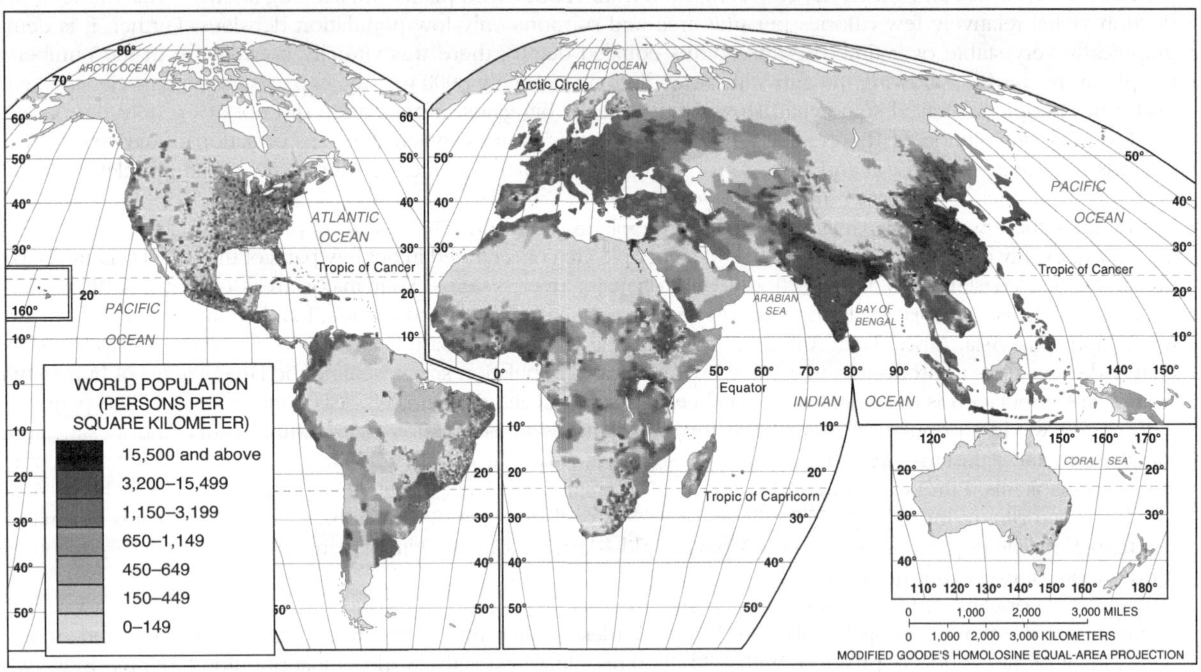

FIGURE 2 **Population map of the world, with one dot representing 10,000 people.** Population density in East Asia, notably China and Japan, as well as in South Asia, including Bangladesh, India, and Pakistan, is extremely high. Population density in Northern Asia, Africa, and South and North America is quite low, comparatively speaking. Three major and two minor areas of world population concentration occur. These are (1) East Asia; (2) South Asia; (3) Europe; (4) Northeastern United States and Southeastern Canada; (5) Southeast Asia, especially the country of Indonesia and the island of Java.

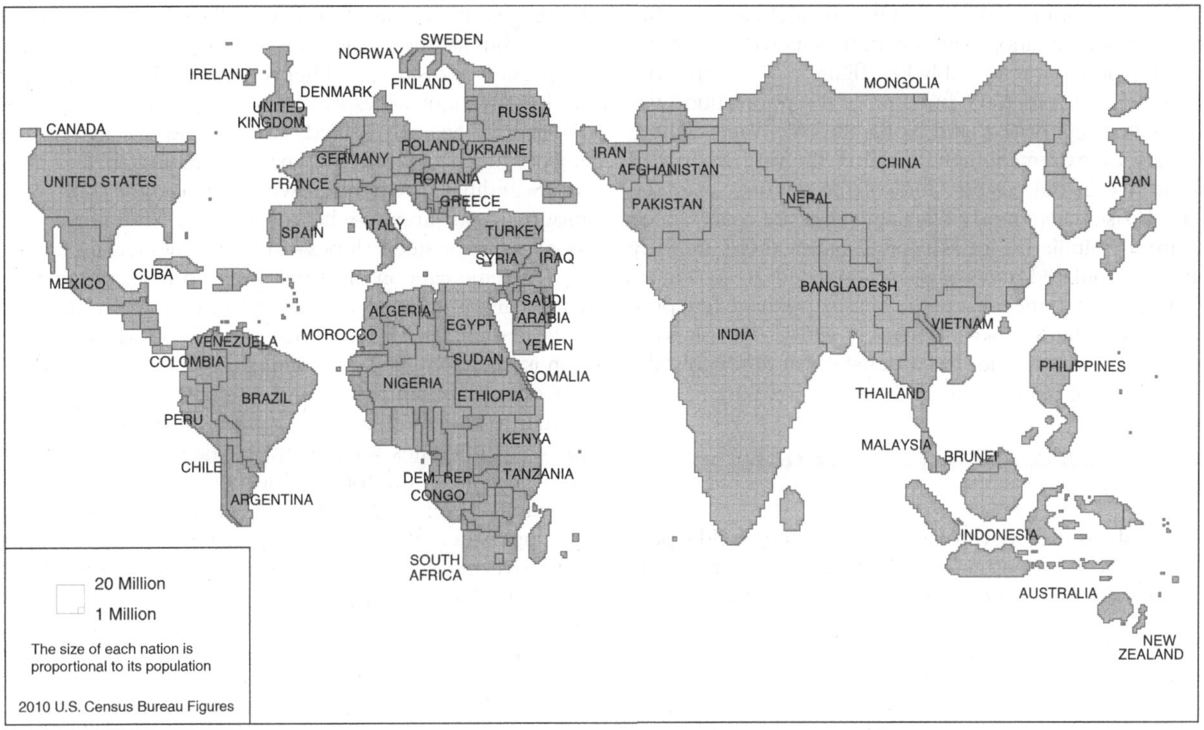

FIGURE 3 Cartogram of world population. This map shows the area of each country in proportion to its population. Geographic space has been transformed into population space. Asia dominates the map, especially China, with 1.3 billion people, and India, with 1.1 billion people. Europe is much larger than on a "normal" map of territories. Both South America and Africa show up much smaller than their territorial size because their populations are relatively small.

TABLE 1 The World's 10 Most Populous Countries, 2010

Country	Population 2010 (millions)	% Annual Growth Rate	Estimated Population 2050 (millions)
China	1320	0.6	1437
India	1173	1.4	1755
United States	310	0.6	438
Indonesia	243	1.4	343
Brazil	201	1.0	260
Pakistan	184	2.0	295
Bangladesh	156	2.1	215
Nigeria	152	2.4	282
Russia	139	−0.4	119
Japan	126	−0.1	101

Source: World Population Data Sheet.

Several countries with the largest populations have relatively low population densities. For example, the United States is the fourth most populous country, but in 2005 it had a population density of only 84 people per square mile. Although they have significant and dense metropolitan areas, the United States and Canada form one of the more sparsely populated areas of the world.

Excluding countries with a very small area (e.g., Singapore), Bangladesh is the world's most crowded nation, where more than 148 million people are crowded into an area the size of Iowa. Three of the 10 most densely populated countries—the Netherlands, Japan, and Belgium—are economically developed, whereas another three—South Korea, Taiwan, and Israel—are newly industrializing countries (NICs). The remainder are clearly less developed nations, reminding us that there is no clear relationship between population density and economic development.

Contrary to popular opinion, not all crowded countries are poor. In fact, in the Sahel states of Africa, population densities are very low. But what explains the fact that many people in the Netherlands or Singapore live well on so little land? Part of the explanation lies in their historical development and position within the colonial and contemporary world systems. Part lies in their industrious people and their ability to adapt to change. Part lies in the policies of their governments, which encourage economic growth. And part of the explanation lies in their history of trade or their relative locations. Singapore is on one of the great ocean crossroads of the world. But being on a crossroads has worked no similar miracle for Panama. In 2008, Singapore had a per capita income ($29,700) that was four times that of Panama ($7300).

National population densities are abstractions that conceal much variation within countries as well as among them. Egypt had a reasonably low figure of 71 people per square kilometer in 2005, but 96% of the population lives on irrigated, cultivated land along the Nile Valley where densities are extremely high. In China, the vast majority of people live in the eastern third of the country, near the Pacific Coast, where most of the large cities are concentrated (Figure 4). Similarly, in the United States there are very densely populated and sparsely populated areas (Figure 5). Large areas to the west of the Mississippi have few people, whereas the Northeast is densely settled. The island of Manhattan, for example, has a density that is roughly the same as that of Hong Kong.

FACTORS INFLUENCING POPULATION DISTRIBUTION

What explains the uneven distribution of the world's people? One factor among many is the physical environment. Most of the world's people tend to be concentrated along the edges of continents, in river valleys, at low elevations, and in humid midlatitude and subtropical climates. Lands deficient in moisture, and hence inhospitable to agriculture (at least without widespread irrigation), such as the Sahara Desert, are sparsely settled. Few people live in very cold regions, such as northern Canada, arctic Russia, and northern Scandinavia, where growing seasons are short. Many mountainous areas—whether because of climate, thin stony soils, or steep slopes—are also low-density habitats.

Extreme caution must be exercised in ascribing population distribution to the natural environment alone. To hold that climate or resources control population distribution is environmental determinism, a view long discredited because it is simplistic and often factually incorrect. Certainly climatic extremes, such as insufficient rainfall, present difficulties for human habitation and cultivation. However, given the forces of technology, the deficiencies of nature increasingly can be overcome. Air-conditioning, heating, water storage, and irrigation are examples of the extensive measures that technology offers to residents of otherwise harsh environments.

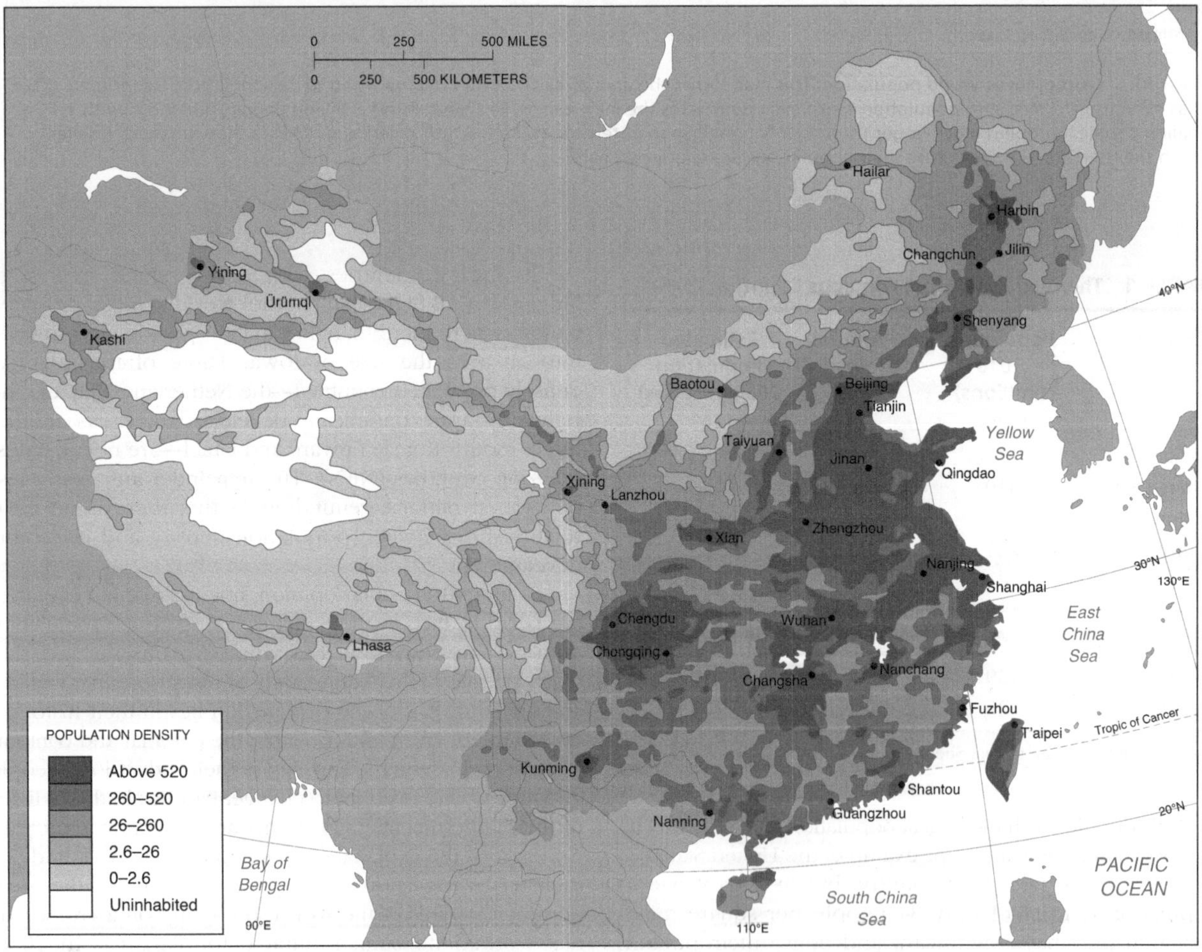

FIGURE 4 The distribution of China's population. With 1.3 billion people, most of China's population is clustered in the eastern half of the country, where moister climates allow for an agricultural base. The western half, in contrast, is mountainous or very dry.

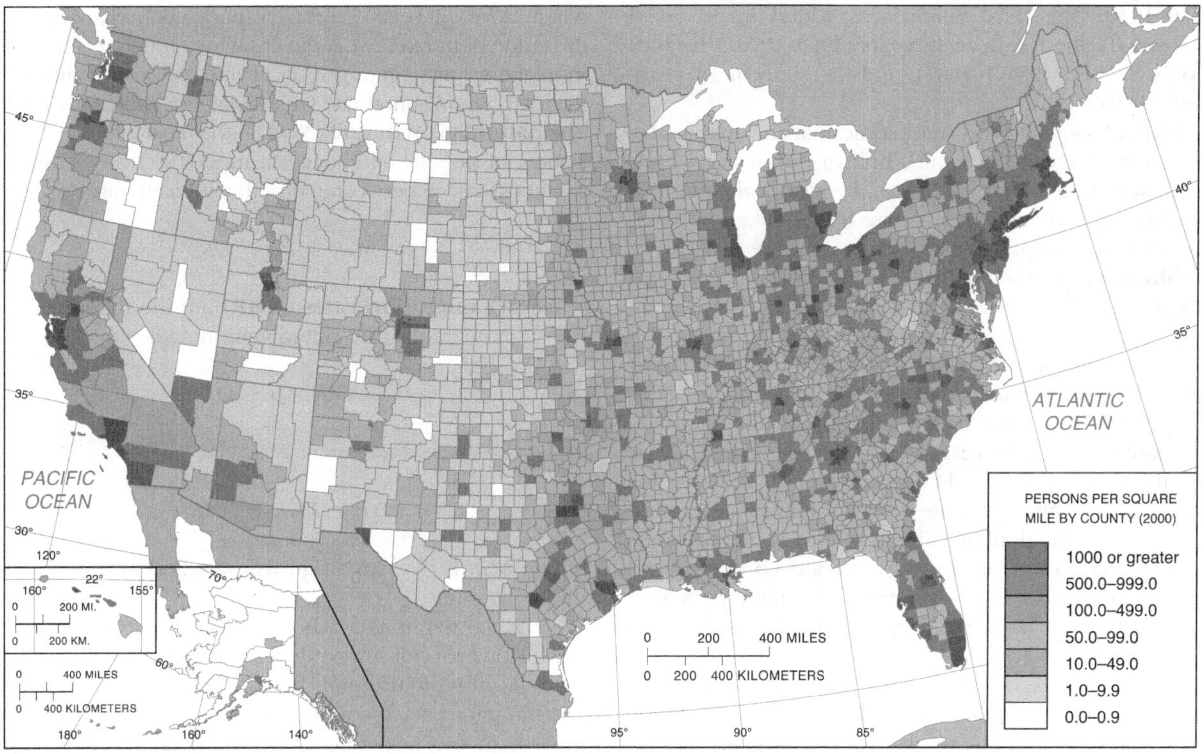

FIGURE 5 Distribution of the U.S. population, 2008. The densest regions of the United States include the metropolitan areas in the Northeast, Midwest, and California. Much of the Midwest and intermontane West, in contrast, consists of low-density agricultural and ranching areas.

If physical environments alone cannot explain the world's population distribution, what other factors are involved? Human distributions are molded by the organization and development of economic and political systems. It is the geography of economic activity—the labor markets, job opportunities, and infrastructures of urban areas—that generate large, dense populations in developed and, increasingly, underdeveloped countries. Population sizes and distributions are influenced by the demographic components of fertility, mortality, and migration. Social disasters such as war or famine may alter population distribution on any scale. Policy decisions, such as tax policies or zoning and planning ordinances, are eventually reflected on the population map.

None of these factors, however, can be considered without reference to historical circumstance. Present population distribution is explicable only in terms of the past. Geographies are never created instantaneously, and the location of the world's people is the accumulation of forces operating at the global, national, and local scales for centuries or longer. For example, the high population densities of Europe or the northeastern United States reflect the accumulated impacts of the Industrial Revolution and its associated waves of urbanization. China's large population was centuries in the making, reflecting long periods of fertile agriculture, irrigation, and a social system stretched over vast areas. The distribution of people in the developing world is largely a reflection of the centuries of colonialism, which focused growth on coastal areas, the locations of the port cities that were the centers of maritime trade in the colonial world economy.

POPULATION GROWTH OVER TIME AND SPACE

The geography of the world's population is never static but is in constant change. The world's population is increasing, albeit at a decreasing rate. Each year an additional 76 million people inhabit the earth, which means the planet adds 208,000 people daily, about 2.4 per second. Many countries in Europe, Russia, and Japan, are losing population as their deaths exceed births. The major locus of world population growth is in the developing countries, in which more than three-fourths of humankind dwells. With 6.9 billion people already and another billion expected by the year 2015, how will the developing world manage? How will the vast population increase affect efforts to improve living standards? Will the developing world become a permanent underclass in the world economy? Or will the reaction to an imbalance between population and resources be waves of immigration and other spillovers to the developed countries?

Population Change

The current rapid growth rate of the world population is a recent phenomenon. It took from the emergence of humankind until 1850 for the world population to reach 1 billion. The second billion was added in 80 years

(1850–1930), the third billion was added in 30 years (1931–1960), the fourth in 16 years (1960–1976), the fifth in only 11 years (1977–1987), and the sixth in 12 years (1987–1999).

Like all living things (and some that are inanimate, such as your savings account), human populations have the capacity to grow rapidly. Thus, the historical pattern of population growth shown in Figure 8 looks like an explosion. We can express this idea using the notion of **doubling times**—the number of years that it takes a population to double in size, given a particular rate of growth. In general, the doubling time for a population can be determined by using the *rule of 70*, which means that you divide 70 by the average annual rate of growth. For example, at 1.2%, the average rate of world population increase, the doubling time is 58 years, which means that in the year 2068 the world may have 14 billion people *if* current rates of growth continue unchanged. At an annual increase of 0.6% per year, the doubling time for the U.S. population is 116 years (meaning, if the current rate continues unchanged, the United States will have 600 million people in 2126). As growth rates increase, doubling times decrease accordingly.

The vast bulk of the world's population growth is occurring in the developing world. Of the continents, Africa has the fastest rate of growth. In 2008, the population of Africa was growing by 2.4% per year. Malawi, with a population growth rate of 3.2% per year, the highest in the world, will see its population double in just 22 years. Rapidly declining death rates and continued high birth rates are the cause of this explosion. Death rates have been falling to fewer than 10 deaths per 1000 people each year in Asia and Latin America, and to about 13 per 1000 in Africa. Crude birth rates are changing less spectacularly. They are highest in Africa (37 births per 1000 people annually), Latin America (21 per 1000), and Asia (19 per 1000). These latter figures compare with crude birth rates of 11 per 1000 in Europe and 14 per 1000 in North America.

The world population was 6.9 billion in 2010, and the United Nations projects it will reach 9 billion in 2050. Almost all of this increase will occur in the developing countries. The largest absolute increase is projected for Asia, reflecting its huge population base. Future population growth will further accentuate the uneven distribution of the world's population. In 2010, 80% of the world's population lived in the developing world, but by the year 2050, the proportion will increase to 90%.

The rate of natural increase for a country or a region is measured as the difference between the birth rate and the death rate. Births and deaths represent two of the three basic population change processes; the third is migration. Every population combines these three processes to generate its pattern of growth. We can express the relationship among them using the equation

$$\text{Population change} = \text{Births} - \text{Deaths} + \text{In-migration} - \text{Out-migration, or,}$$

$$\Delta P = BR - DR + I - O,$$

where ΔP represents the rate of population change, BR is the crude birth rate, DR is the crude death rate, I is the total in-migration rate (immigration, if internationally), and O is the total out-migration rate (or emigration, if internationally). The **natural growth rate (NGR)**—the most important component of population change in most societies—is defined as the difference between the birth and death rates:

$$NGR = BR - DR,$$

while **net migration rates (NMR)** are the difference between in-migration and out-migration rates:

$$NMR = I - O.$$

Thus,

$$\Delta P = NGR + NMR,$$

where NGR is the natural growth rate and NMR is the net migration rate.

For the world as a whole, net migration is obviously zero. However, for any scale smaller than the globe, both natural growth and net migration must be included. Natural increase accounts for the greatest population growth in most societies, especially in the short run. However, in the long run, migration contributes significantly because the children of immigrants add to the population base.

Fertility and Mortality

The immediate cause for the surge in the growth of the world population is the difference between the crude birth rate and the crude death rate. The crude **birth rate** is the number of babies born per 1000 people per year, and the crude **death rate** is the number of deaths per 1000 per year. For example, the U.S. birth rate in 2010 was 14 and the death rate was 8; the growth rate was therefore 14 minus 8, or 6 per 1000, which is a natural growth rate of 0.6%.

Birth rates fluctuate over time, in response to changing economic and political circumstances and cultural values about having children. In the United States, crude birth rates rose sharply after World War II, producing the baby boom. The children of the baby boom, Generation X, were born largely in the 1980s and 1990s.

MALTHUSIAN THEORY

One of the first social scientists to tackle the matter of population growth and its consequences was the British Reverend Thomas Robert Malthus (Figure 6). Malthus's ideas, contrived in the early days of the Industrial Revolution in the late eighteenth century, had an enormous impact on the subsequent understanding of this topic. He offered his most concise explanation in his 1798 book, *Essay on the Principle of Population Growth*. Malthus was concerned with the growing poverty evident in British

Photos.com/Jupiter Images

FIGURE 6 Thomas Robert Malthus was an influential theorist who started the idea of overpopulation. His pessimistic views of food and demographic growth influenced many early political economists, earning the discipline the nickname "the dismal science."

cities at the time, and his explanation was largely centered on the high rates of population growth that he observed, which are common to early industrializing societies. Thus, it is with Malthus that the theory of overpopulation originates. His pessimistic worldview earned economics the label of the "dismal science" and stood in sharp contrast to the utopian socialism emanating from France in the aftermath of the French Revolution of 1789.

The essence of Malthus's line of thought is that human populations, like those of most animal species, grow exponentially (or in the parlance of his times, geometrically). A geometric series of numbers increases at an increasing rate of time. For example, in the sequence 1, 2, 4, 8, 16, 32, and so on, the number doubles at each time period, and the increase rises from 1 to 2 to 4 to 8 and so forth. Exponential population growth, in the absence of significant constraints, is widely observed in bacteria and rodents, to take but a few examples from zoology. Note that there is an important assumption regarding fertility embedded in Malthus's analysis here: He portrayed fertility as a biological inevitability, not a social construction. This argument was in keeping with the large size of British families at the time and the excess of fertility over mortality. In short, in Malthus's view, humans, like animals, always reproduced at the biological maximum; they were, and are, portrayed as prisoners of their genetic urges to reproduce. It is worth noting that Reverend Malthus's argument carried with it a strong moral dimension: It was not just anyone who reproduced rapidly, he observed, but most particularly the poor.

Second, Malthus maintained that food supplies, or resources more generally, grew at a much slower rate than did population. Specifically, he held that the food supply grew linearly (or arithmetically, in his terminology). An arithmetic sequence of numbers, in contrast to an exponential one, grows at a constant rate over time. For example, in the sequence 1, 2, 3, 4, 5, and so on, the difference from one number to the next is always the same. Malthus's view that agricultural outputs increased linearly over time reflected the preindustrial farming systems that characterized his world. In such circumstances, without economies of scale, an increase in outputs is accomplished only with a proportional increase in inputs such as labor, reflective of what economists call a linear production function. However, this view of agricultural output is actually rather optimistic by Malthus's reckoning. He argued that in the face of limited inputs of land and capital, agricultural output was likely to suffer from **diminishing marginal returns**. For example, as farmers moved into areas that were only marginally hospitable to crops, perhaps because they are too dry, too wet, too cold, or too steep, they would need increases in inputs that are proportionately much larger than the increases in output. Diminishing returns, he held, would actually lead to increases in agricultural output that were smaller than a linear production function would generate (Figure 7).

When one plots the exponential growth of population against the linear growth of food supplies (Figure 8), it is clear that sooner or later, the former must exceed the latter. Thus, in the Malthusian reading, populations always and inevitably outstrip their resource bases, and people are condemned to suffering and misery as a result. Malthus blamed much of the world's problems on rapid population growth, and subsequent generations of theorists influenced by his thoughts have invoked overpopulation to explain everything from famine to crime rates to deviant social behavior. Malthus himself entered into a famous debate with his friend David Ricardo over whether the British government should subsidize food for the poor, Malthus maintaining that

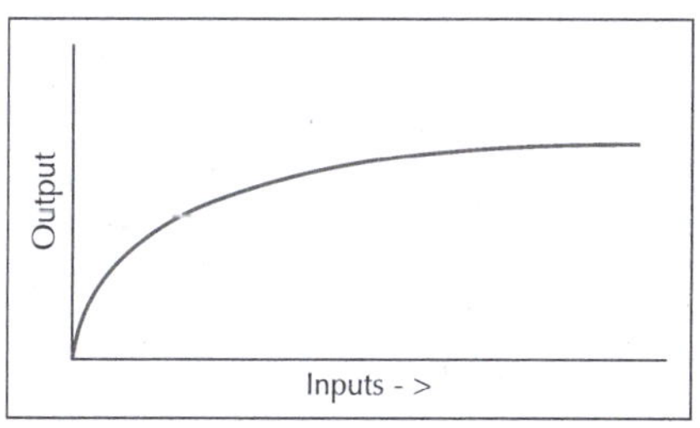

FIGURE 7 Diminishing marginal returns, proposed by Malthus, set in when increases in inputs (e.g., land, labor) fail to generate equal increases in outputs. This process leads to declining productivity growth. To some extent, diminishing returns can be offset by technological change.

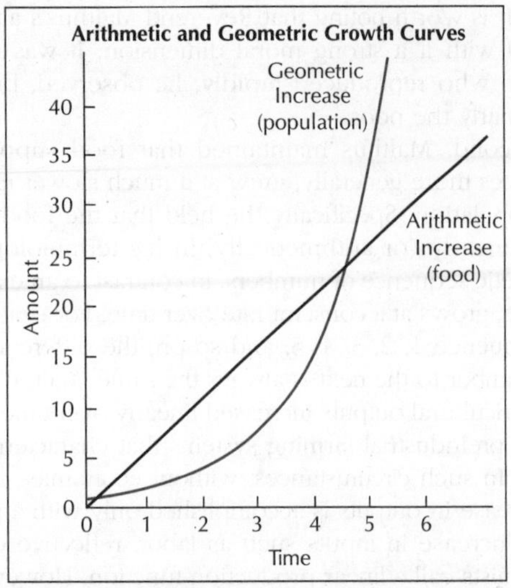

FIGURE 8 Malthus's argument contrasted the geometric (exponential) increases in population with arithmetic (linear) increases in the food supply. Eventually, the demand for food, or resources more generally, must exceed the supply, leading to famine. However, critics noted that Malthus's conceptions were flawed by their simplistic treatment of production, as well as the determinants of fertility.

such subsidies only encouraged the poor to have more children and thus exacerbated poverty in the long run. Indeed, he was contemptuous of the poor, blaming them for their poverty, and even advocating mass death to keep their numbers in check:

> Instead of recommending cleanliness to the poor, we should encourage contrary habits. In our towns we should make the streets narrower, crowd more people into the houses, and court the return of the plague. . . . The necessary mortality must come.

Malthusianism thus attributes to rapid population growth a variety of social ills, including poverty, hunger, crime, and disease.

Malthus refined his argument to include checks to population growth. Given that natural population growth is the difference between fertility and mortality, "preventative checks" are factors that reduce the **total fertility rate**. Contraceptives are an obvious example, although Malthus objected to their use on religious grounds, advocating instead moral restraint or abstinence. Other preventative checks include delayed marriage and prolonged lactation, which inhibits pregnancy. Should preventative checks fail, as he predicted they would, population growth would ultimately be curbed by "positive checks" that increased the mortality rate, particularly the familiar horsemen of the Apocalypse—death, disease, famine, and war.

Malthus's ideas became widely popular in the late nineteenth century, particularly as they were incorporated into the prevailing social Darwinism of the time, which represented social change in biological terms, often naturalizing competition as a result. However, to many observers it became increasingly apparent that his predictions of widespread famine were wrong. The nineteenth century saw the food supply improve, prices decline, and famine and malnutrition virtually disappear from Europe (except for the Irish potato famine of the 1840s). By the early twentieth century, Malthusianism was in ill repute.

Critics noted that Malthus made three major errors. First, he did not foresee, and probably could not have foreseen, the impacts of the Industrial Revolution on agriculture; the mechanization of food production simply rendered the assumption of a linear increase untenable (Figure 9). Indeed, the world's supply of food has consistently outpaced population growth, meaning that productivity growth in agriculture has been higher than the rate of increase in the number of people. This observation implies that there is plenty of food to feed

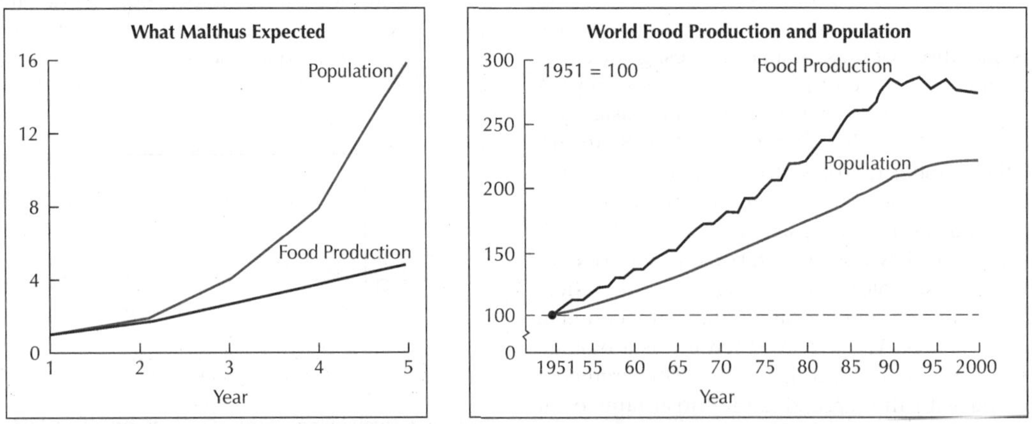

FIGURE 9 Malthus's predictions of catastrophe were belied by the productivity gains and declining fertility unleashed by the Industrial Revolution. Since World War II, the world's food supply has increased more quickly than its population, indicating that the causes of hunger are not simply reducible to population growth, but involve complex political questions about colonial legacies, uneven development, corrupt governments, and war and conflict.

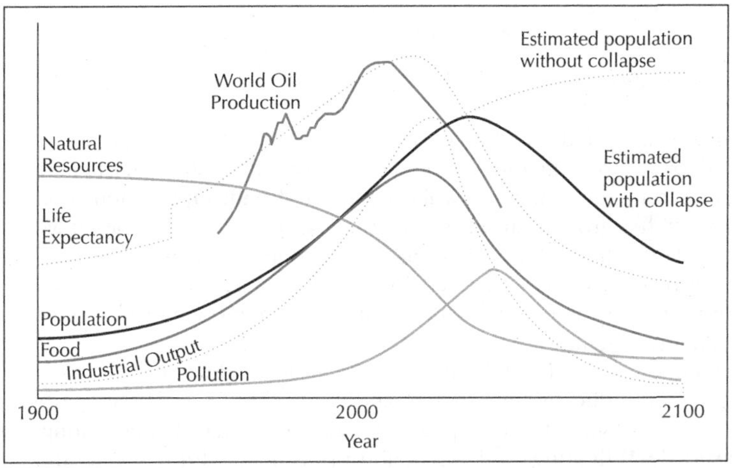

FIGURE 10 Neo-Malthusians, such as the Club of Rome, revived Malthus's arguments in the 1960s by using computer models of the world economy, population growth, and resource usage. They included ecosystems in their analysis and predicted that in the long run, his predictions still had merit. Unlike Malthus, however, neo-Malthusians advocate the use of birth control.

everyone in the world and that hunger is not simply caused by overpopulation, but by a variety of other factors, including politics.

Second, Malthus did not foresee the impacts of the opening up of midlatitude grasslands in much of the world, particularly in North America, Argentina, and Australia, which increased the world's wheat supplies during the formation of a global market in agricultural goods. Third, and perhaps most important, Malthus's analysis of fertility was deeply flawed. During the Industrial Revolution, total fertility rates declined and family sizes decreased. Thus, contrary to his expectation, humans are not mere prisoners of their genes and the birth rate is a socially constructed phenomenon, not a biological destiny.

In the 1960s, when the world experienced average population growth rates in excess of 2.6% annually, Malthusianism underwent a revival in the form of **neo-Malthusianism**. Neo-Malthusians acknowledged the errors that Malthus made but maintained that while he may have been wrong in the short run, much of his argument was correct in the long run. In keeping with the growing environmental movement of the times, neo-Malthusians also added an ecological twist to Malthus's original argument. The most famous expression of neo-Malthusian thought was the Club of Rome, an international organization of policy makers, business executives, scholars, and others concerned with the fate of the planet. The Club of Rome funded a famous study of the planet's future, published as **The Limits to Growth** (1972), which modeled the earth's population growth, economic expansion and resource consumption, and energy and environmental impacts (Figure 10). It concluded that the rapid population and economic growth rates of the post–World War II boom could not be sustained indefinitely and that ultimately there would be profound worldwide economic, environmental, and demographic crises. Much of this argument was framed in terms of the exhaustion of nonrenewable resources such as oil and ecological catastrophe. Indeed, the last half century has indeed witnessed massive degradation of ecosystems around the world, and some resources such as petroleum have already passed their peak production year. Unlike Malthus, neo-Malthusians advocated sharply curtailing population growth through the use of birth control and had an important impact on international programs promoting contraceptives and family planning, such as the Peace Corps and Agency for International Development.

While neo-Malthusianism retains a credibility that the original Malthusian doctrine does not, it, too, suffered from a simplistic understanding of how resources are produced (e.g., when the price of oil rises, corporations find more oil). Blaming overpopulation for all the world's problems is simplistic, skips over the historical forces such as colonialism that generate poverty, and ignores major issues such as government policy. For example, is the crowding of a train in a developing country (Figure 11) the result of overpopulation or the lack of investment in transportation services? In addition, family-planning programs in the

The World Bank

FIGURE 11 The trains are crowded in Bangladesh. Here, in one of the world's most fertile lands, the combination of high population density and the concentration of land into fewer and fewer hands contribute to continuing poverty and hunger.

Case Study

Population and Land Degradation

Human population growth is a leading force in global land degradation and environmental change. Global population increased from 5 billion in 1987 to 6.9 billion in 2010, an average annual growth rate of 1.4%, with Africa recording particularly high growth rates. This population increase, together with the gap between rich and poor and discrepancies in income-earning opportunities, has increased the demand for food and energy, putting pressure on available environmental resources such as fresh water, fisheries, agricultural land, and forests.

Population growth has resulted in increased demand for agricultural productivity, higher incomes, and changes in consumption patterns. Inequitable land distribution, a legacy of colonialism and political conflicts, has exacerbated the problem. Land degradation is especially acute in developing countries where a significant portion of the population is dependent on subsistence farming.

Poverty also is a major cause of land degradation, with population growth and poverty reinforcing each other to bring about a spiral of decline in soil fertility. Neo-Malthusians warn that population growth will outstrip the natural supplies of food, water, and shelter. The neo-Malthusian theory further regards population growth and the environment to be in conflict, with the quest for food security coming at huge inevitable environmental costs, ultimately leading to an escalation in land degradation, a decline in agricultural productivity, and greater food insecurity. The argument effectively states that the world would literally run out of food unless drastic steps are taken to protect the environment from people. Simultaneously, common property tenure systems also exacerbate the utilization of common property resources, leading to overutilization and degradation from elements such as overgrazing. People (especially the rural poor) have been blamed for misusing the resources at their disposal for short-term gains. Such land misuse is accentuated by the coping strategies employed in the face of food insecurity.

In contrast to the Malthusian view, food security and the environment are considered to be complementary and interdependent, with a healthy natural resource base providing food security. It is argued that population increase and income growth drive technological inventions such as agricultural intensification to solve food production constraints, thus achieving environmental and economical sustainability. It is further argued that the neo-Malthusian perception disregards the potential for economic development within sustainable environments when confronted with rapid population growth, and assumes that agro-ecological zones have limited carrying capacities. Population growth, food security and a healthy natural resource base are viewed as interdependent and mutually reinforcing for positive gain to both.

Land degradation as a global development issue directly impacts the natural resource base that supports subsistence livelihoods, agriculture, and manufacturing, limiting development opportunities for current and future generations. For example, deforestation and biodiversity loss affect income from tourism and contribute to food insecurity, challenging the eradication of extreme poverty and hunger. Land degradation occurs through various processes, including loss of vegetation, soil erosion leading to loss of fertility, declining soil biodiversity, and soil compaction, which leads to reduced infiltration and increased runoff salinity. Other effects of land degradation processes include water pollution, siltation of watercourses and reservoirs, and loss of animal and plant diversity, leading to loss of ecological functions. Land degradation is often viewed as a trigger for disasters such as landslides.

Land-use change has both negative and positive effects on human well-being and on the provision of ecosystem services. Positive changes include more food and forestry products that have resulted in increased income and secure livelihoods. Negative changes include biodiversity loss and disturbances of biophysical cycles (e.g., water and nutrients) that impinge on human welfare in many regions. Soil erosion, particularly in the rural regions where the majority of the population resides, may reduce agricultural yields, resulting in increased food insecurity, famine, and poverty, as well as forced migration, especially for impoverished people and countries. Demand for more food production contributes to overexploitation of good agricultural soil and expansion into wooded and environmentally marginal areas that are susceptible to degradation. Processes such as clearing of woodlands, logging, firewood collection, and charcoal production lead to deforestation. The highest rates of deforestation occur in areas where hunger is prevalent. Land degradation consequences can induce declines in forest products and wild foods and worsen levels of poverty and malnutrition, especially since these resources are harvested often as coping strategies in the face of droughts and floods.

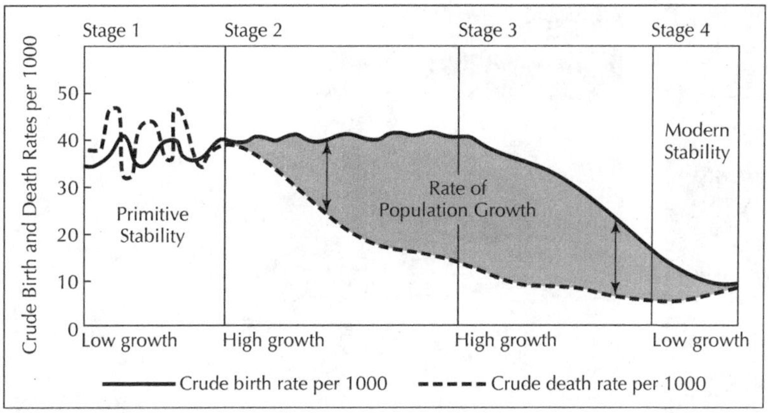

FIGURE 12 The demographic transition. Four stages of demographic change are experienced by countries as they develop from preindustrial to modern. Stage 1: Birth rates and crude death rates are both high; the resulting natural increase is quite low. Stage 2: A declining death rate, but a continually high birth rate, result in an increasing population growth rate. Stage 3: Death rates remain low, while birth rates start to decline rapidly, resulting in a rapid but declining growth rate. Stage 4: Both death rates and birth rates are low, resulting in a low natural increase. The important issue here is *why* these rates decline (i.e., the social forces that change birth and death rates).

developing world have often failed to live up to expectations, often for the simple reason that simply advocating contraception to curb population growth ignores the fundamental economic reasons *why* people in impoverished countries have large families and many children. In short, whatever its merits, neo-Malthusianism must be viewed in light of other models of population growth that originate from different premises and often arrive at different conclusions.

DEMOGRAPHIC TRANSITION THEORY

Developed by several American demographers in the 1950s, the demographic transition theory stands as an important alternative to Malthusian notions of population growth. Essentially, this is a model of a society's fertility, mortality, and natural population growth rates over time. Because this approach is explicitly based on the historical experience of Western Europe and North America as they went through the Industrial Revolution, time in this conception is a proxy for industrialization and all of its economic, social, and geographic characteristics. In short, the **demographic transition** examines how birth, death, and growth rates change, and, more important, why they change, as a society moves from a rural, impoverished, and traditional context into a progressively wealthier, urbanized, and modern one. This approach can be demonstrated with a graph of birth, death, and natural growth rates over time that divides societies into four major stages (Figure 12). Each stage is discussed here in detail.

Stage 1: Preindustrial Society

In the first stage, a traditional, rural, preindustrial society and economy, total fertility rates are high; families are large, and most women are pregnant much of the time (Figure 13). Thus, impoverished countries such as Rwanda have exceptionally large families (8.5 children per mother, in contrast to wealthy ones in Europe, North America, and some of the NICs in Asia, where women have on average fewer than 2 children apiece).

Traditionally, total fertility rates in preindustrial societies have been very high for a variety of reasons. In agrarian economies, children are a vital source of farm labor, helping to plant and sow crops, tend farm animals, perform chores, carry water and messages, and watch over younger siblings (Figure 14). Econometric studies reveal that even children as young as 4 can generate more income than they consume. Even in North America, summer school vacations were created so that school kids could help their families on the farm. Families in this context are typically extended, with several generations

Fertility Rates Around the World

Lowest		Highest	
Macau	0.91	Niger	7.68
Hong Kong	1.04	Uganda	6.73
Singapore	1.10	Mali	6.54
Taiwan	1.15	Somalia	6.44
Japan	1.20	Burundi	6.25
South Korea	1.22	Burkina Faso	6.21
Montserrat	1.23	Dem. Rep. of the Congo	6.11
Lithuania	1.24	Ethiopia	6.07
Belarus, Czech Republic	1.25	Angola	6.05

Averages	
Industrialized Countries	1.9
Developing Countries	3.6
World	3.3

FIGURE 13 Fertility rates around the world vary widely. The average number of children each woman may expect to have ranges from as little as 1.3 in Italy to as high as 8.5 in impoverished nations such as Rwanda. These patterns reflect the dynamics of the demographic transition and the social forces that lead people in different societies to have large or small families.

FIGURE 14 The use of child labor is widespread throughout the developing world and is a major driver of high fertility rates. Children are used for a variety of tasks, including planting, weeding, and harvesting crops; tending to livestock; carrying water and firewood; and caring for one another. In urban areas, children often work in factories, on the streets selling items, or as sex workers. Often, such children are highly vulnerable and are brutally exploited, and may be paid virtually nothing.

living together. In addition, children are important resources for taking care of their elderly parents; in the absence of institutionalized social programs such as Social Security, the aged depend on their offspring for assistance. Finally, in such societies with high infant mortality rates, having many children is a form of insurance that some proportion will survive until adulthood. In short, there are very clear reasons why poor societies have high crude birth rates. In contrast to Malthusianism, which explains this observation by an appeal to human genetics, the demographic transition portrays high total fertility rates as a rational strategic response to poverty.

Thus, a map of crude birth rates around the world (Figure 15) reveals that the poorest societies have the highest rates in the world, particularly in Africa and most of the Middle East. In contrast, for reasons we shall soon see, crude birth rates in North America, Europe and Russia, Japan, Australia, and New Zealand are relatively low. The world's lowest birth rates are found in Spain and Italy. In societies with high birth rates, the age distribution of populations tends to be young. Thus, the proportion of the population aged less than 18 (the median age in many developing countries) is a reflection of high total fertility rates (Figure 16).

However, in preindustrial societies, mortality rates are also typically quite high, which means that average life expectancy is relatively low. The primary causes of death in poor, rural contexts are the result of inadequate diets, particularly protein, which weaken the immune system, as well as unsanitary drinking water and bacterial diseases. The most common diseases in this context are diarrheal ones which lead to dehydration, including cholera, as well as others such as dengue fever, schistosomiasis, bilharzia, malaria, tuberculosis, plague, and measles, although historically smallpox was also important. Table 2 lists the most dangerous infectious diseases in the world in 2009, including respiratory infections brought on by pneumonia and influenza (which kill 3.9 million annually); AIDS, which takes 2.8 million lives annually (especially in

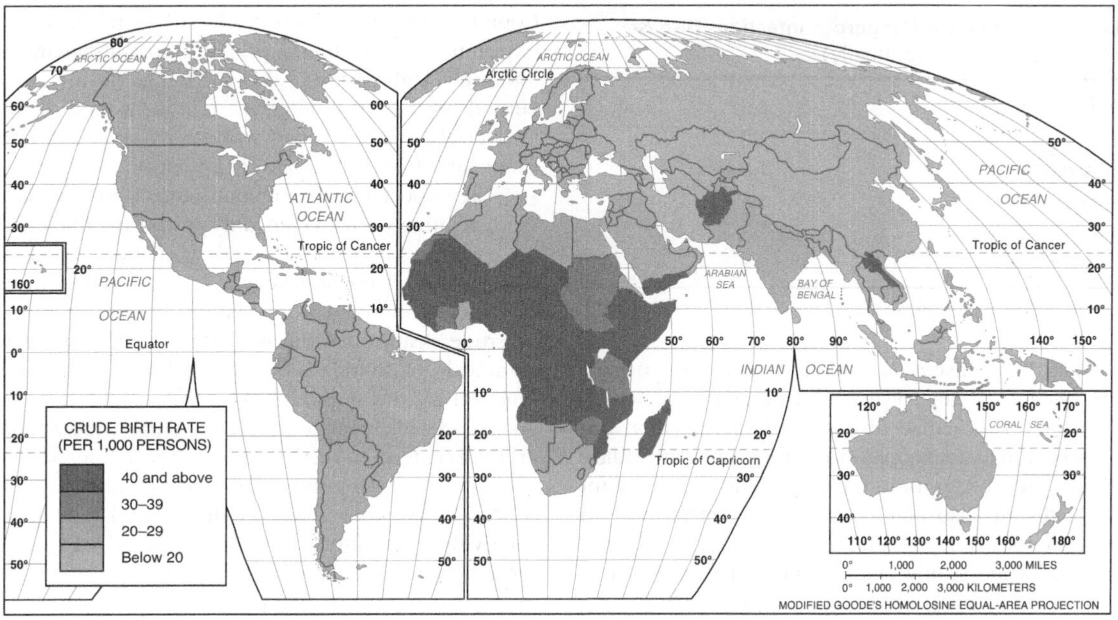

FIGURE 15 The geography of crude birth rates around the world closely reflects the level of economic development. Generally, the poorest societies have the largest families. Africa and much of the Muslim world tend to have the highest fertility levels. In these circumstances, the need for child labor, particularly on farms, is the predominant motive. In economically advanced societies, this need is mitigated, and the opportunity costs of raising children rises, so families are smaller. Thus, birth rates in Europe, Japan, the United States and Canada, and Australia are low.

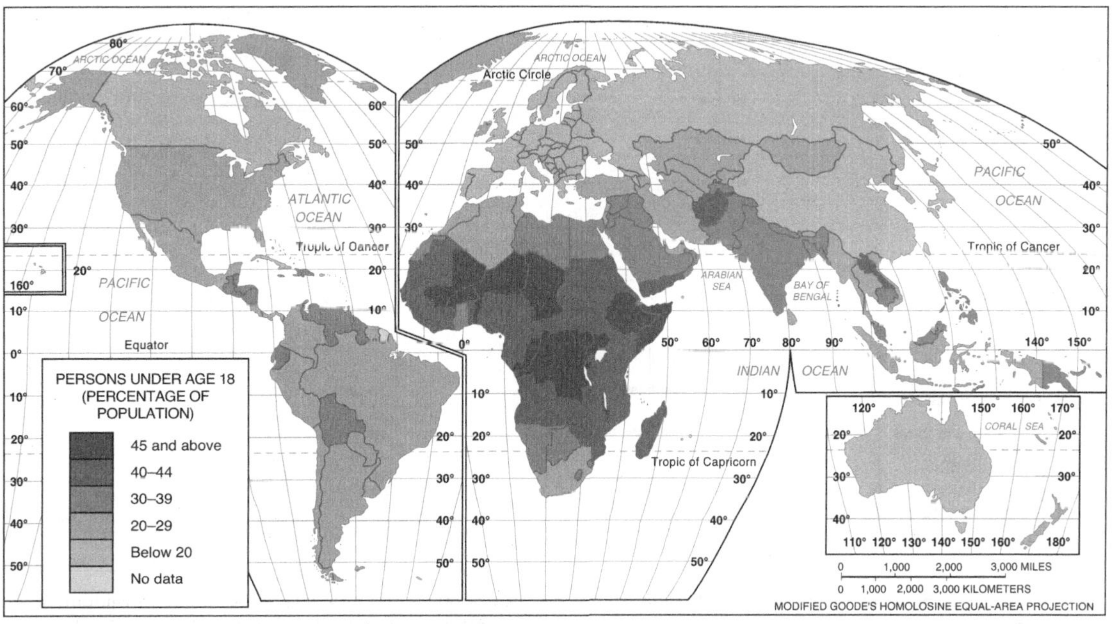

FIGURE 16 The proportion of each country's population below the age of 18 is closely correlated with total fertility rates. In poor, rural societies, where birth rates are high and families large, a large share of the people (often more than half) are younger than 18, particularly in Africa and parts of the Arab world. In contrast, in more developed states with lower birth rates and smaller families, the median age rises; in the First World, the fastest-growing age groups are the middle-aged and elderly.

TABLE 2 World's Most Dangerous Infectious Diseases, 2009 (millions of victims annually)

Respiratory infections	3.9
AIDS	2.8
Diarrheal diseases	1.8
Tuberculosis	1.6
Malaria	1.3
Measles	0.6

Source: World Health Organization.

Africa); diarrheal diseases, which deplete the body's nervous system of electrolytes; and tuberculosis, which kills 1.8 million. Because disease and malnutrition are ever-present threats to people in poor societies, including particularly the very young, the most vulnerable, infant mortality rates are also high, and a significant proportion of babies do not live to see their first birthday.

The world geography of death rates (Figure 17) thus closely reflects the wealth or poverty of societies (including their historical development and role in the global economy), which in turn is manifested in a variety of issues that shape national mortality rates: the amount, quality, and consistency of adequate food; access to health care; the public health infrastructure; care for expectant mothers, babies, and young children; smoking rates; and several other factors. Countries with the highest death rates—and thus lowest life expectancies—are found primarily in sub-Saharan Africa, although Afghanistan, Pakistan, Bangladesh, Russia, and central Asian states also have relatively high death rates. Conversely, the developed world, as well as Latin America, China, and India, has relatively low death rates.

Life expectancy throughout most of human history has been relatively low, often only in the twenties, although once people survived infancy their chances of living to old age improved considerably. The geography of life expectancy around the world (Figure 18) closely reflects that of crude death rates but is also shaped by differences in standards of living. Living for a long time is a luxury enjoyed by the populations of economically developed societies, while people in most of Africa, the Middle East, and Russia tend to die before they reach age 70.

Because both fertility and mortality rates are high, the *difference* between them—natural population growth—is relatively low, often fluctuating around zero. Thus, although families are large and parents have many children, growth rates are curtailed by malnutrition, disease, and infant mortality. For this reason, for thousands of years, human growth rates worldwide have been very slow, occasionally even negative, and new arrivals to a community were welcomed (Figure 1). Indeed, prior to Malthus, rapid population growth was celebrated as a way to increase the local **labor force**, diversify the division of

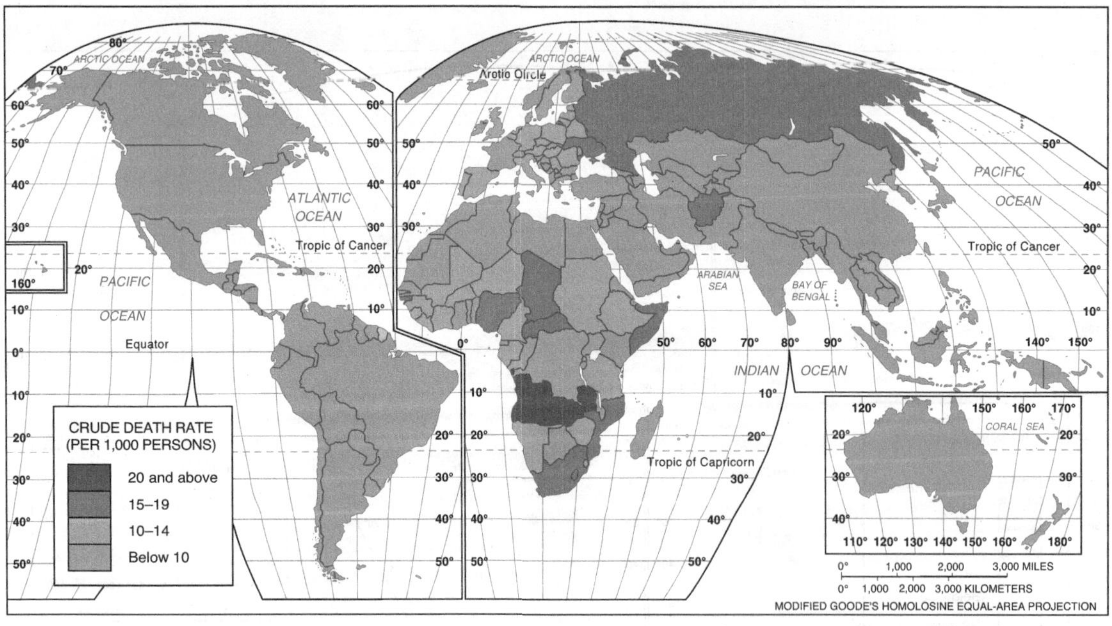

FIGURE 17 Like birth rates, the geography of crude death rates also closely mirrors the level of economic development. The poorest countries, especially in sub-Saharan Africa, have high death rates, largely due to inadequate diets, infectious diseases such as malaria, and contaminated water supplies. Russia, which is relatively more developed, has a remarkably high death rate due to the economic collapse it suffered in the 1990s and the spread of diseases such as tuberculosis. In much of the rest of the world, in contrast, including even in poor countries like China and India as well as the First World, death rates have gone down and life expectancy has increased due to adequate food supplies and public health infrastructures.

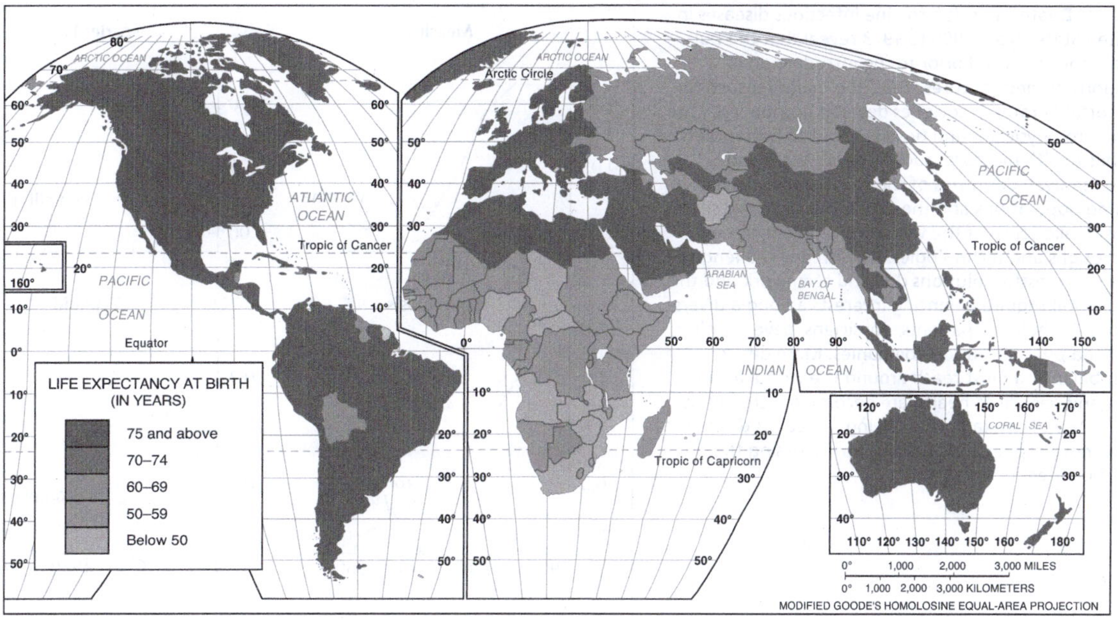

FIGURE 18 A world map of life expectancy at birth reveals that people in wealthier countries live longer than those in poorer ones. Whereas the inhabitants of impoverished societies such as Mali or Chad cannot expect to live beyond age 50, the average age at death for those in Europe, Japan, Australia, the United States, and Canada is over 75. Life expectancies are thus closely correlated with people's access to adequate nutrition, public health, and health care.

labor, and raise standards of living. While relatively few societies in the world live in the circumstances described here—that is, few people today live isolated from the world economy and its demographic aftermath—Stage 1 may be held to describe certain tribes in parts of central Africa, Brazil, or Papua New Guinea.

Stage 2: Early Industrial Society

The second stage of the demographic transition pertains to societies in the earliest phases of industrialization, when manufacturing jobs are growing in urban areas. Such conditions pertain, for example, to the Britain of Malthus's day, the United States in the early nineteenth century, or selected countries in the developing world today, such as Mexico or the Philippines (although these countries have experienced fertility declines). In this context, economic changes in the labor market as well as in consumption, particularly diet and public health care, have important demographic consequences.

Early industrial societies retain some facets of the preindustrial world, particularly high total fertility rates. Because most people still live in rural areas, children remain an important source of farm labor. The major difference is the decline in mortality rates, which leads to longer life expectancies. Why do mortality rates decline as societies industrialize? One often-claimed reason is access to better medical care, particularly access to hospitals and vaccinations from diseases, an assertion frequently advocated by those in the health care occupations.

However, the historical evidence does not sustain this view; early hospitals were often filthy, and patients may have been more likely to die in them than if they stayed home! Moreover, the introduction of vaccinations often came *after,* not before, declines in deaths due to many diseases occurred. For example, in the United States, the vaccines for measles, scarlet fever, typhoid, diphtheria, tuberculosis, and pneumonia, which were invented in the 1920s, 1930s, and 1940s, all occurred after the majority of the declines in the death rate from each disease in the first two decades of the twentieth century (Figure 19). Indeed, rather than private health care providers, it was public health measures, particularly clean drinking water and sewers, that played a significant role in lowering diseases. Better housing was also important. Finally, the industrialization of agriculture and cheaper food, which led to better diets, was vital in improving immune systems and raising life expectancies, including lowering **infant mortality rates** (percent of babies who die before their first birthday).

Death rates for different demographic groups do not drop evenly as an economy develops over time. Infant mortality rates tend to drop earliest and most quickly. It was not uncommon in premodern societies to find an infant mortality rate in excess of 200 infant deaths per 1000 live births—20% or more of all babies died before reaching their first birthday. Nowhere in the world today are rates that severe, but the highest rates are found in sub-Saharan Africa (Figure 20), where poverty, lack of adequate diets, disease, inadequate drinking water, and insufficient health care services conspire to kill 10% of all infants before age 1.

FIGURE 19 Death rates due to nine infectious diseases in the United States from 1900 to 1973 reveal that the bulk of the declines occurred prior to the introduction of inoculations or medical cures. Thus, the major reasons for lower mortality rates as societies develop economically are not linked to physicians and hospitals, but to better diets and public health measures such as clean water and garbage removal. Countries of the world must address expanding populations and the explosions of new diseases and must make sense of the vast array of scientific new breakthroughs and technologies. As with any challenge of this scope, successful solutions demand teamwork and this teamwork will require balancing interests among a diverse group of stakeholders: patients, physicians, payers, policy makers, and pharmaceutical companies. More difficult still, this balancing must reach around the world, as the business and science of health are becoming increasingly a global concern. For example, evolving infectious diseases can now move rapidly between countries, raising the threat of pandemic.

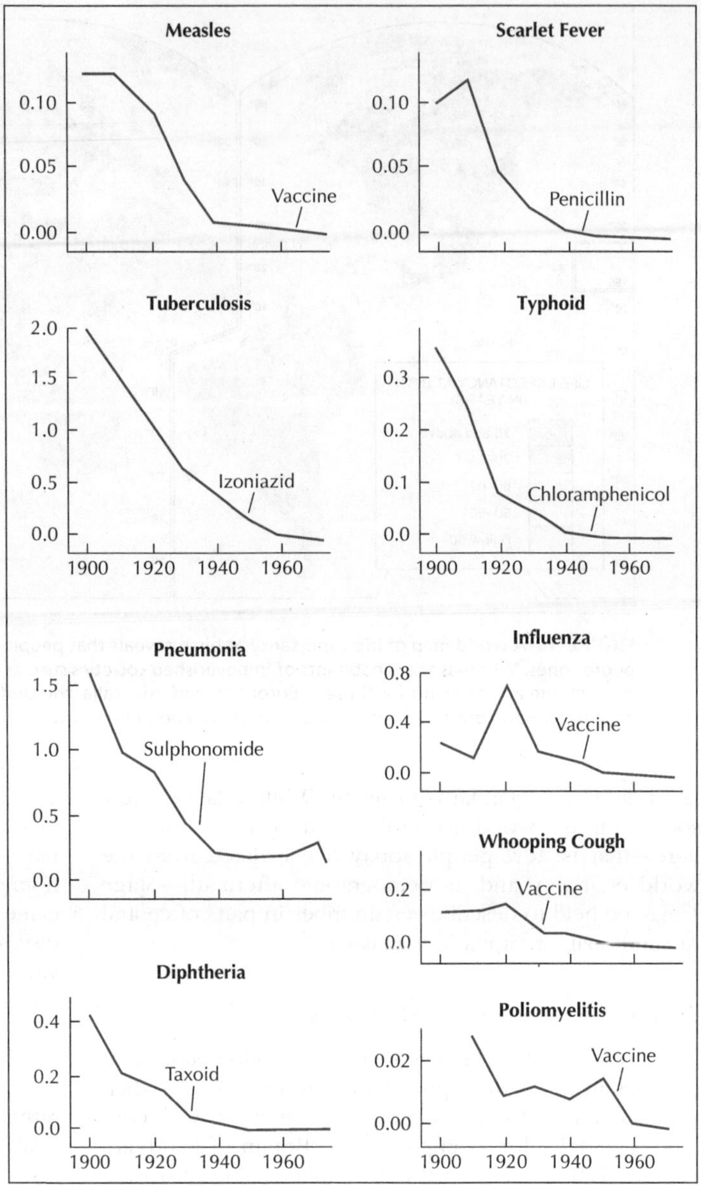

Because the drop in the death rate disproportionately affects the very young, it acts much like an increase in the birth rate—more babies survive to grow to adulthood. Life expectancies likewise increase. One of the reasons the very young are more affected is that, as the death rate drops, it does so initially because communicable diseases are brought under control, and the very young are particularly susceptible to such diseases. The control of communicable disease has the serendipitous economic side effect of reducing the overall illness level in society, thus promoting increased labor productivity. Workers miss fewer days of work, are healthier when they do work, and are able to work productively for more years than when death rates are high. Eventually, as death rates drop, the timing of death shifts increasingly to the older ages, to the years beyond retirement when the economic impact on the labor force is minimal. Although death rates have declined throughout the world, mortality is usually, but not always, lowest in the developed world (especially in northwest Europe and in Japan) and highest in the underdeveloped world (especially in sub-Saharan Africa). Variations in fertility tend to be more pronounced, with much higher levels in the developing world.

In early industrial societies, because the death rate has dropped but the birth rate has not, the natural growth rate grows explosively. This situation characterized the world Malthus observed at the end of the eighteenth century and is evident in a wide number of countries in the developing

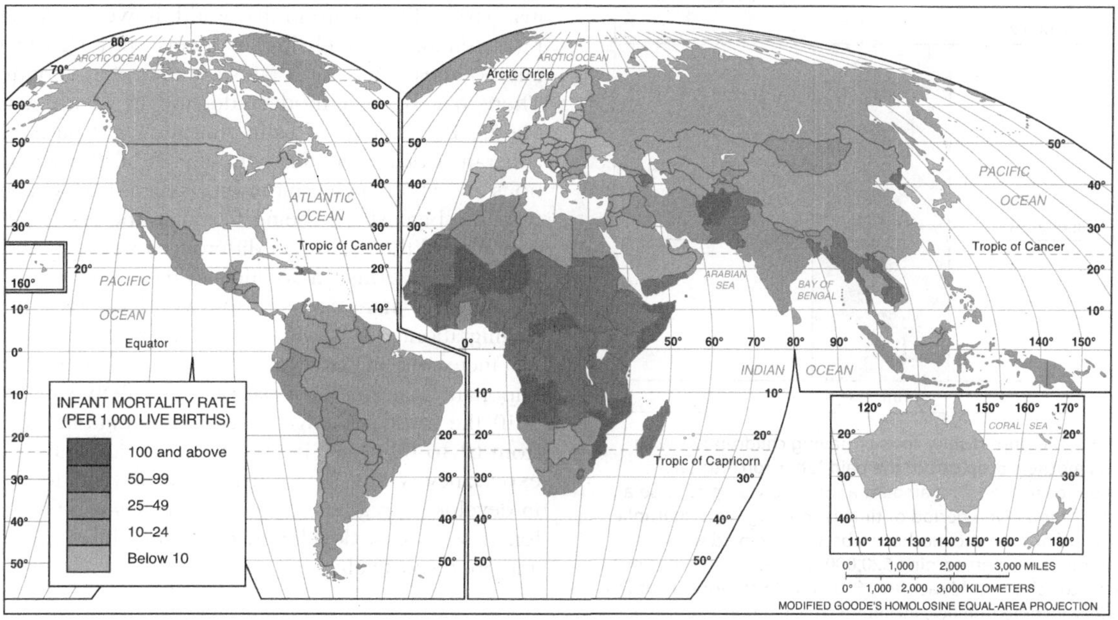

FIGURE 20 The geography of infant mortality rates is perhaps the most sensitive and optimal measure of economic development. The capacity of societies to protect their most vulnerable members—babies—reflects their parents' access to nutrition and clean water and exposure to infectious diseases. Poor countries suffer from high rates of infant deaths, often exceeding 100 deaths per 1000 births (e.g., much of Africa, Afghanistan, Iraq); people in the economically developed world, in contrast, can expect that the vast majority of babies will live past their first birthday.

world today. In short, poor countries have rapid increases in population not because they have more children than before, but because fewer people die earlier.

Stage 3: Late Industrial Society

Societies in the throes of rapid industrialization, in which a substantial share, if not the majority, lives in cities, exhibit a markedly different pattern of birth, death, and growth rates than those earlier in the transition. Death rates remain relatively low, for the reasons discussed earlier. What is different is that total fertility rates also exhibit a steady decline. It is important to note that declines in crude birth rates almost always occur *after* declines in the death rate; societies are much more amenable to death control than birth control.

Why do crude birth rates fall and families get smaller as societies become wealthier? The answer to this important question lies in the changing incentives that people face as their worlds shift from primarily rural to primarily urban, with a corresponding increase in the size and complexity of labor markets. For many people, the decision to have, or not to have, children is the most important question they will ever face, with profound consequences not only for their personal well-being but also for society at large. Essentially, urbanization and industrialization lead to smaller families because the benefit/cost ratio of children changes over time. This assertion is not meant to reduce children to simple economic commodities. However, in societies in which a large number of women enter the paid labor force—become commodified labor outside of the home, rather than unpaid workers inside of it—the constraints to child rearing become formidable. For women, who typically have primary responsibility for child care, working outside of the home and taking care of young children in urbanized societies pose extraordinarily difficult obstacles. Many women understandably drop out of the labor market, if only temporarily, to take care of their kids. As a result, they do not earn an income while staying at home, relying on their husband's income for support. Economically, this process generates an opportunity cost to having children: The more children a couple has, or the longer a mother refrains from working outside of the home, the greater the opportunity cost she faces or they face as a family (Figure 21). As women's incomes rise, either over time or comparatively within a society, the opportunity cost of children rises accordingly, leading to lower total fertility rates.

In distinct contrast to neo-Malthusian family planning, which tends to ignore the social circumstances pertaining to fertility, in this model there is a clear link between labor markets and fertility behavior. Getting women to work outside of the home in commodified labor markets is the surest form of birth control. As total fertility rates decline, so too does the natural growth rate. In short, relatively prosperous societies tend to have smaller families, and there is frequently a corresponding shift from extended to nuclear families in the process.

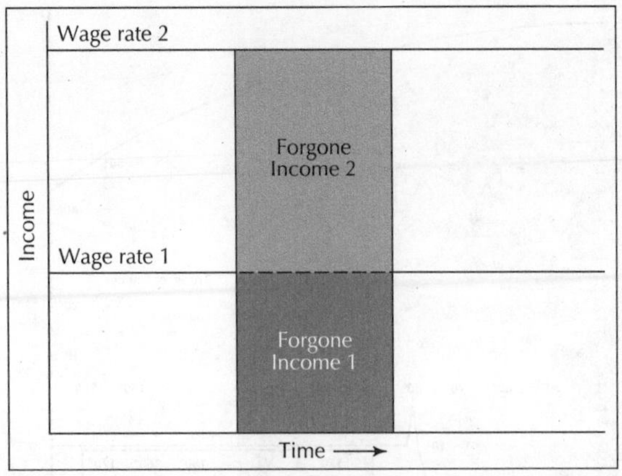

FIGURE 21 The opportunity costs of having children rise with income as mothers drop out of the paid labor force to take care of young children. The amount a family gives up to raise a young child for a given period of time is the wage rate multiplied by the period of time the mother is absent from paid work. (For example, a woman earning $20,000 per year who drops out for 5 years to raise a child sacrifices $100,000 in forgone income.) Among low-income families, including those in poor countries or the economically disadvantaged in the First World, the amount forgone is relatively low (hatched area) because incomes are low. As societies develop and family incomes rise, however, the amount forgone rises (shaded area). Thus, the economic cost of raising a child rises with the level of economic development; for this reason, poor families tend to be large families, both among and within countries, and wealthier families tend to be smaller ones.

Historically, fertility levels fell first in Western Europe, followed quickly by North America, and more recently by Japan, and then the remainder of Europe. In all of those areas of the world, reproductive levels are below the level of generational replacement; the United States is the only major country still above that level, and just barely. Elsewhere in the world, however, crude birth rates remain at much higher levels, although in China and Southeast Asia birth rates are dropping very quickly. There has been a modest decline in South Asia, the Middle East, much of Latin America, and parts of sub-Saharan Africa. Overall, then, almost all the world's nations are experiencing more births than deaths each year, with the biggest gap being found in the less developed nations and the narrowest difference existing in the more developed nations. In addition to these patterns of natural increase, many areas of the world are also impacted by migration.

Increasing life expectancy in a region or country is an important indicator of social progress. Between 1980 and 2010, the world's average life expectancy at birth increased from 61 to 68 years. In developed regions, average life expectancy is 74 years for males and 81 years for females; in developing regions, 65 years for males and 68 years for females; in least developed countries, the values are 53 and 56, respectively.

Stage 4: Postindustrial Society

The fourth and final stage of the demographic transition, postindustrial society, depicts wealthy, highly urbanized worlds with their own configurations of birth, death, and growth rates. In this context, indicative of Europe, Japan, and North America today, death rates are very low and life expectancies correspondingly high. Do death rates ever reach zero? No, for that would mean life expectancies become infinite! Even the declines in the death rates will not exhibit much improvement, and they face diminishing marginal returns: The easy causes of death have been largely eliminated, and overcoming the remaining ones will be much harder.

As societies industrialize and become progressively wealthier, the causes of death change from infectious

FIGURE 22 Deaths in economically advanced societies such as the United States are less due to hunger and infectious diseases and much more likely to result from behavioral or lifestyle choices. The primary drivers of death in the United States are smoking, which kills roughly 435,000 Americans annually; poor diets coupled with physical inactivity, which kill 400,000; and alcohol abuse, which takes 85,000 lives each year. Illegal drugs, in contrast, kill many fewer people (less than 30,000).

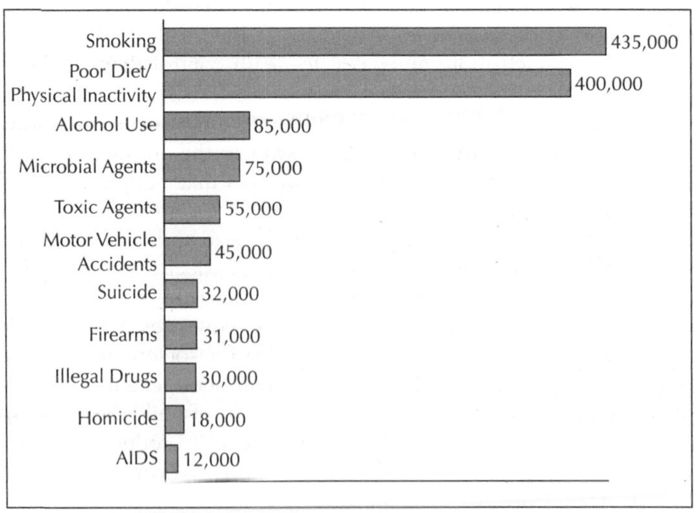

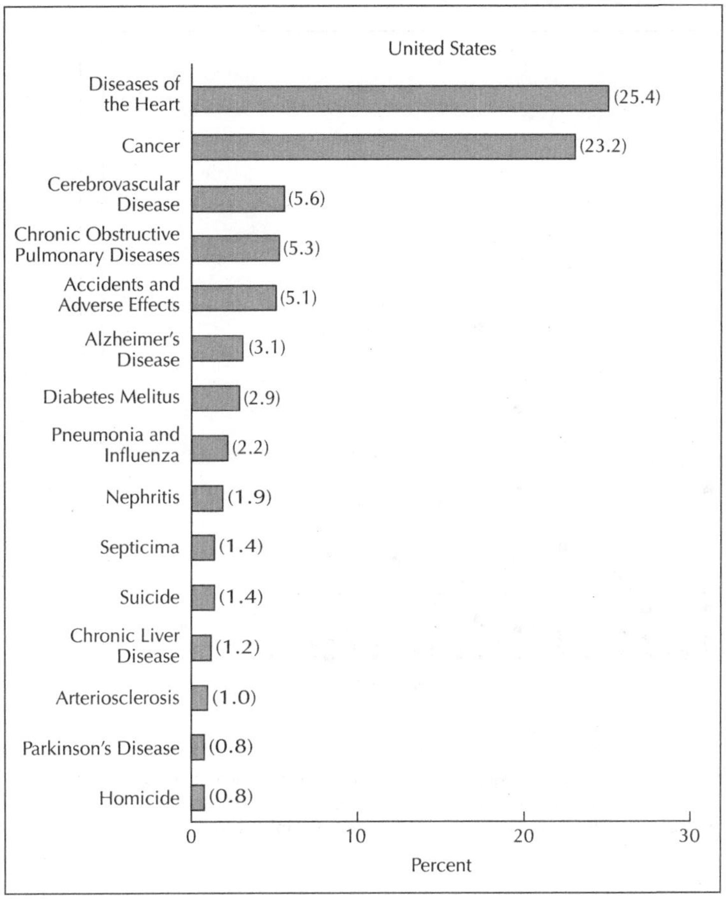

FIGURE 23 The drivers of death are manifested in different biological outcomes. Heart disease, which is closely related to smoking, alcohol abuse, and obesity, is the largest single killer in the United States, responsible for more than 28% of all deaths, followed by cancers of all forms combined (22.7%). Cerebrovascular disease (strokes) is third. In today's world, governments have big underlying structural gaps in budgets and large deficits that will not be easily filled by economic recovery anytime soon. Rising health care costs, especially in European welfare states, and also in the United States, with new health care legislation, plus the burden of pension and Medicare spending, will put relentless pressure on government debt. Eventually, the world's rich economies should return to full employment, and when they do, public borrowing to finance the health-conscious postindustrial society will hurt growth and may lower standards of living.

diseases, the bane of the preindustrial world, to lifestyle-related ones, particularly those associated with smoking and obesity, as well as, to a lesser extent, car accidents, suicides, and homicides (Figure 22). The leading causes of death in the United States today, for example, are heart disease, cancers of all forms, and strokes (Figure 23).

Birth rates too, continue to fall in such contexts, as families grow smaller and many couples elect to go childless or have only one. Around the world, national income and population growth rates are inversely related (Figure 24). When crude birth rates drop to the level of crude death rates, a society reaches zero population growth (ZPG). When birth rates drop below death rates, as they have in virtually all of Europe, the society experiences negative population growth. Japan, the oldest major society in the world (older demographically than Florida), will see its population decline by 30% or more in the next 50 years. Such situations are characterized by large numbers of the elderly, a high median age, and a relatively small number of children, all of which have dramatic implications for public services. Often in such contexts, governments take steps to increase the birth rate with ample rewards for childbirth (e.g., subsidized child care and long paid parental leaves) and publicity campaigns. In societies with extremely low or negative population growth, the major cause of demographic growth is often net immigration.

Globally, uneven economic development—the legacy of colonialism, different national policies, position in the global system, and national cultural systems—generates uneven patterns of natural population growth; the geography of natural growth rates is the difference between the geographies of birth and death rates (Figure 25). The most rapid rates of increase are found throughout the poorer parts of the developing world (i.e., in Africa and in the Arab and Muslim worlds). The economically developed nations, including North America, Japan, Europe, Australia, and New Zealand, in contrast, have low rates of population growth, often hovering around zero or even lower.

These patterns have significant implications for the nature and future of the world's people. Although the world's average natural growth rate has been slowly declining, it still adds roughly 100 million people per year (Figure 26), roughly the population of Mexico. Projecting into the future, declining fertility levels are believed to lead to slower rates of demographic growth throughout the twenty-first century (Figure 27). However, because there are so many people of child-bearing age in the developing world, the total population of the planet is projected to rise to roughly 10 billion

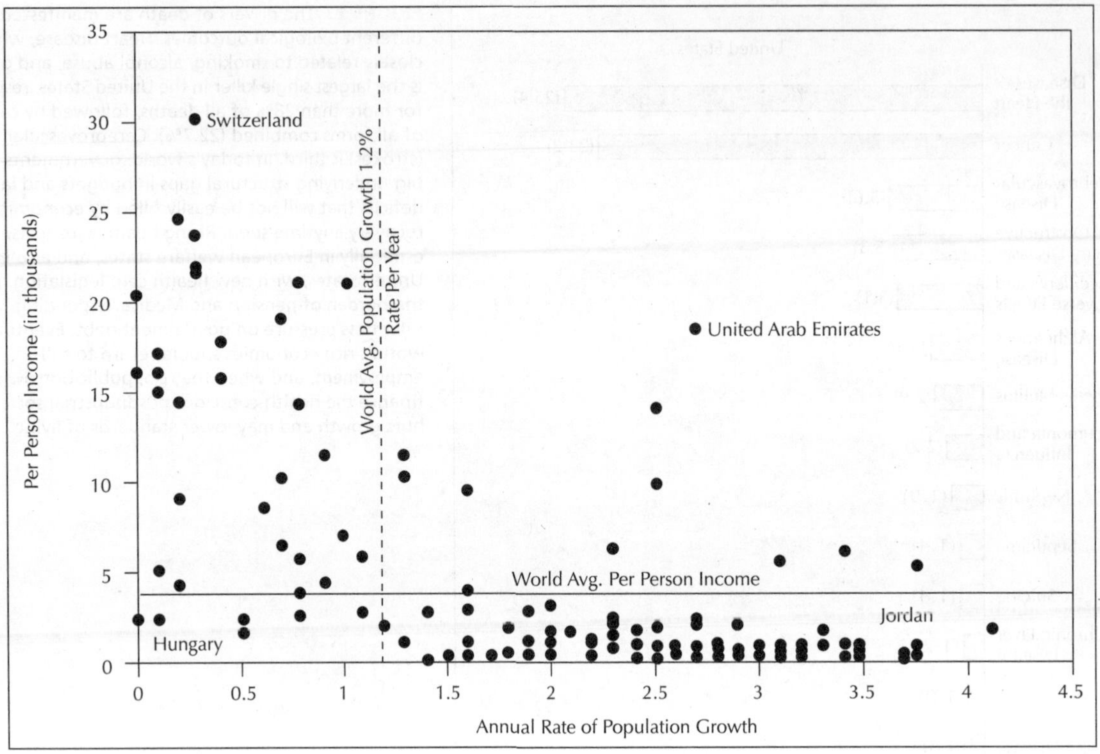

FIGURE 24 The rate of population growth in different societies is closely associated with their average income. Countries with high rates of demographic growth tend to have incomes below the world average; conversely, those with high per capita incomes tend to have low rates of growth. This pattern reflects the dynamics of the demographic transition discussed earlier.

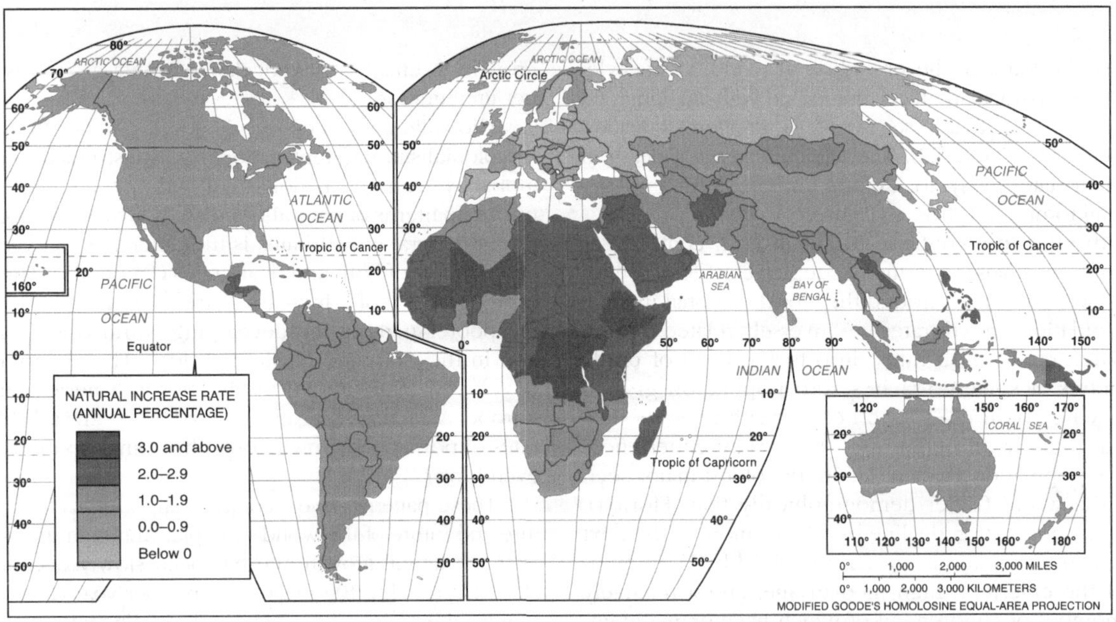

FIGURE 25 Rates of natural population change (birth rates minus death rates). The fastest-growing areas of the world include Africa, Central America, and Southwest Asia. Here growth rates exceed 2.5%, with a number of countries in Central America and Africa actually exceeding 3% per year. Three percent growth per year does not seem like a high growth rate; however, it indicates total population doubling time for a country of only 23 years. With a 2% growth rate, a country would double in 35 years. For a 1% growth rate, a country may double in 70 years. Natural increase in Europe as a whole is only 0.2%, and several countries have negative rates of natural growth.

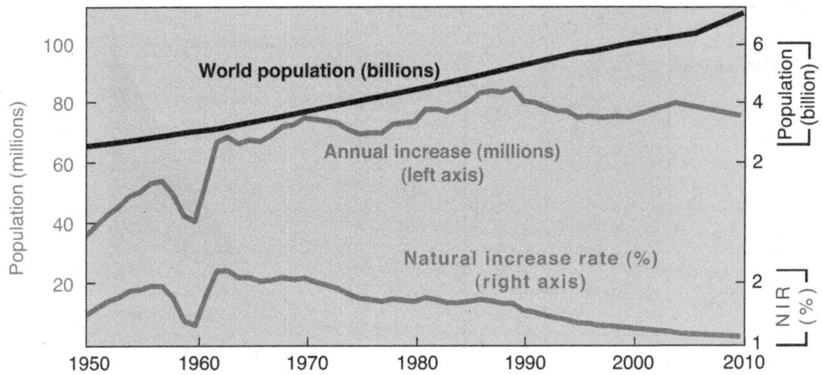

FIGURE 26 World population growth rates have slowly declined as fertility levels gradually dropped. Despite lower rates of increase, the world still adds roughly 100 million people each year.

by the year 2100 (Figure 28). The vast bulk of these additions will be in the Third World. However, these projections are based on different assumptions about the future of fertility (Figure 29). Should total fertility rates decline more rapidly than expected, the increase in the world's people may not be as dramatic as some believe.

However, the population explosion in the developing world will have enormous impacts. Not all of the world's problems are reducible to population growth, but many are not independent of it either. Rapid population growth will accelerate, among other things, agricultural overcultivation and soil depletion, overfishing, poaching, deforestation, depletion of water resources, loss of biodiversity, and rural-to-urban migration (Figure 30). Further, generating an infinite pool of poor people keeps wages low, not only in the developing world but also in the developed, as globalization pits the labor forces of countries against one another. Thus, it is important to keep the dynamics of population growth in mind in relation to environmental degradation, economic development, and international issues.

Contrasting the Demographic Transition and Malthusianism

By now you may have realized that the assumptions, analyses, and conclusions offered by Malthusian theory and the demographic transition are markedly different. Whereas Malthusianism tends to take fertility for granted—arguing that people are prisoners of biological imperatives to reproduce uncontrollably—the demographic transition reveals fertility is socially constructed (i.e., families have children or don't have them, as the case may be, depending on the costs and benefits that children offer). Moreover, whereas Malthusian scenarios inevitably depict the population as growing uncontrollably, to the point of resource exhaustion, the demographic transition predicts steadily declining levels of world population growth as crude birth rates converge upon death rates. The evidence supports this view. After accelerating for two centuries, the overall rate of world population growth is slowing down. In the 1960s, it reached a peak of 2.6%, declining to 1.7% a year in the 1990s, and dropping further to 1.2% in 2010. However, the absolute size of population will continue to increase because the size of the base population to which the growth rate applies is so large.

Criticisms of Demographic Transition Theory

Although the demographic transition has wide appeal because it links fertility and mortality to changing socioeconomic circumstances, it has also been criticized on several grounds. Some critics point out that it is a model derived

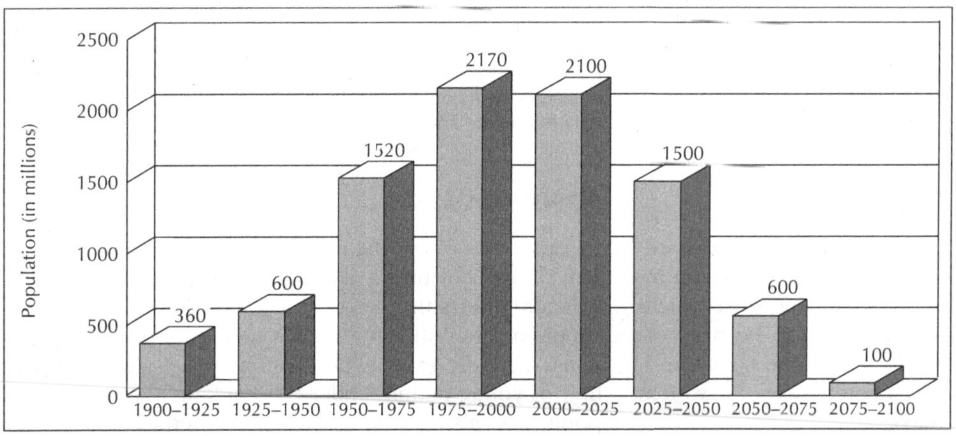

FIGURE 27 Net additions to the world's population between 1900 and 2100. The world's population increased exponentially in the twentieth century. The period from 1975 to 2025 shows the largest half-century of increase, with more than 4 billion people being added to the world's population. From there, the world's population should slow down in rate of increase as fertility levels decline everywhere.

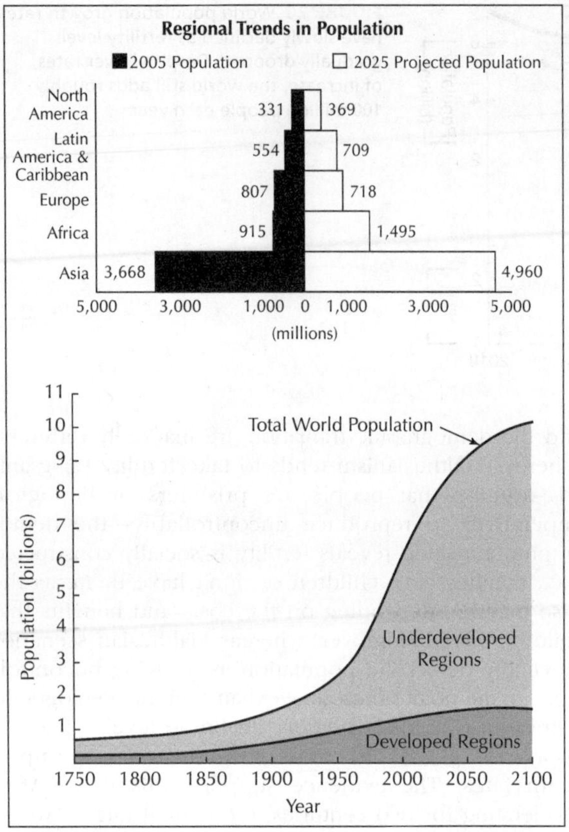

FIGURE 28 The vast bulk of additions to the world's population has occurred, and will continue to occur, in the world's economically underdeveloped regions. Natural growth rates in the First World are low, often negative, whereas they tend to be relatively high in poorer countries.

from the experience of the West and then applied to many non-Western societies, as if they are bound to repeat the exact sequence of fertility and mortality stages that occurred in Europe, Japan, and North America. There is no inevitable logic that assures the developing world must meekly follow in the footsteps of the West. Some have pointed out that the developing world is in many ways qualitatively different from the West, in no small part because of the long history of colonialism. Further, demographic changes in the developing world have been much more rapid than in the West. For example, whereas it took decades, or even centuries, for mortality rates in Europe to decline to their modern levels, in some developing countries the mortality rate has plunged in only one or two generations. Because mortality rates do not vary geographically as much as fertility rates, most of the spatial differences in natural growth around the world are due to differences in fertility. These caveats are useful in cautioning us to examine the historical context in which all theories and explanations emerge and to be wary of blindly importing models developed in one social and historical context into radically different ones for which they were not originally intended.

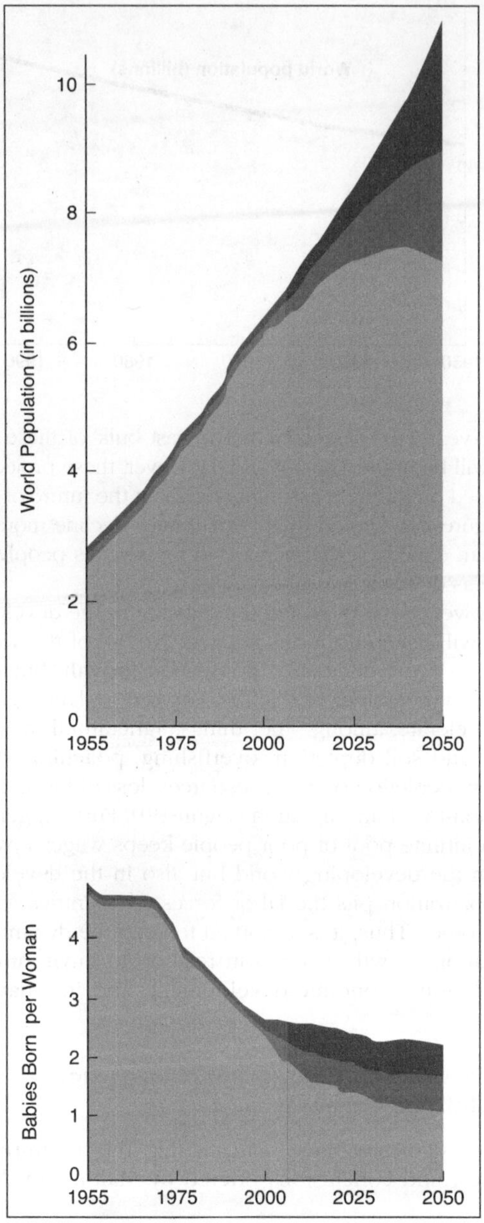

FIGURE 29 The future of the world's population size depends on how fertility levels change. If current fertility levels are maintained, the number of people in the world could exceed 10 billion by 2050. However, if fertility levels drop, as they have been doing, the rate of increase will slow down; at a minimum, the world can expect to hold more than 7.5 billion people in 50 years.

POPULATION STRUCTURE

Except for total size, the most important demographic feature of a population is its age-sex structure. The age-sex structure affects the needs of a population as well as the supply of labor; therefore, it has significant policy implications. A rapidly growing population implies a high proportion of young people under the working age. A youthful population also puts a burden on the education system.

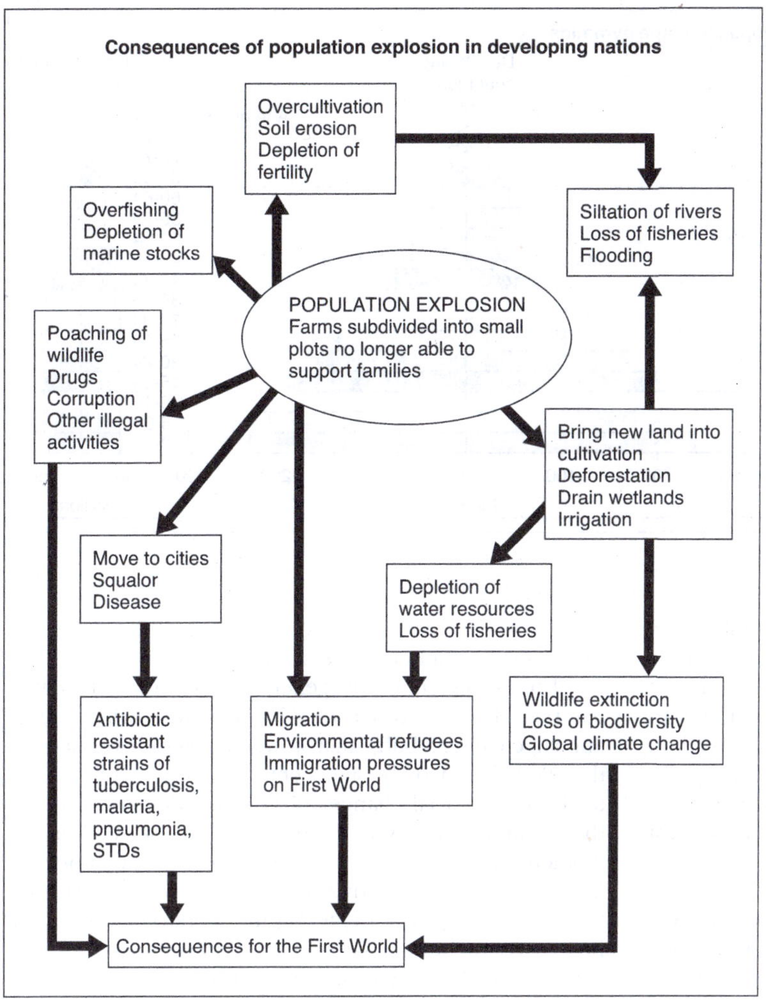

FIGURE 30 The population boom in developing countries will have numerous repercussions. Not all of the world's problems can be blamed on population growth, but rapid rates of increase are likely to exacerbate environmental degradation, such as soil erosion and wildlife loss, and generate large numbers of people seeking better opportunities in cities and in the economically developed world.

When this cohort enters the working ages, a rapid increase in jobs is needed to accommodate it. By contrast, countries with a large proportion of older people must develop retirement systems and medical facilities to serve them. Therefore, as a population ages, its needs change from schools to jobs to medical care.

The age-sex structure of a country is typically summarized or described through the use of **population pyramids**. They are divided into 5-year age groups, the base representing the youngest group, the apex the oldest. Population pyramids show the distribution of males and females of different age groups as percentages of the total population. The shape of a pyramid reflects long-term trends in fertility and mortality and short-term effects of baby booms, migrations, wars, and epidemics. It also reflects the potential for future population growth or decline.

Two basic, representative types of pyramid may be distinguished (Figure 31). One is the squat, triangular profile. It has a broad base, concave sides, and a narrow tip. It is characteristic of developing countries having high crude birth rates, with a young average age, and relatively few elderly. Natural growth rates in such societies tend to be high.

In contrast, the pyramid for economically developed countries, including the United States, describes a slowly growing population. Its shape is the result of a history of declining fertility and mortality rates, augmented by substantial immigration. With lower fertility, fewer people have entered the base of the pyramid; with lower mortality, a greater percentage of the births have survived until old age. In short, the structure of the population pyramid closely reflects the stage of the demographic transition in which a country is positioned. Like all developed countries, the U.S. population has been aging, meaning that the proportion of older persons has been growing. The pyramid's flattened chest reflects the baby dearth of the Great Depression in the 1930s, when total births dropped from 3 million to 2.5 million annually. The bulge at the waist of the pyramid is a consequence of the **baby boom** that followed World War II. By 1976, the total fertility rate had fallen to 1.7, a level below replacement. Members of the

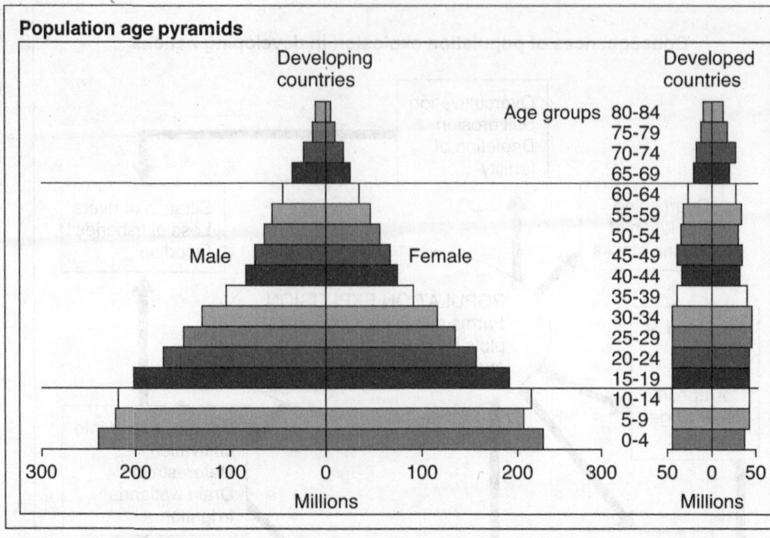

FIGURE 31 Population pyramids reflect the age and sex structure of a society. Poor countries, with high birth rates and a large share of the population at early ages, have pyramids with a broad base and narrow top. Economically wealthier societies, with low rates of natural growth, have older populations; their pyramids have proportionately smaller bases and a larger share of people in the middle aged and elderly groups.

baby-boom generation, however, were having children in the 1980s and 1990s, driving the total fertility rate up to 2.0, almost at replacement level and the highest in the developed world. Thus, the U.S. population continues to grow from natural increase as well as from immigration. Because different parts of the United States have very different socioeconomic conditions, cultures, and migration patterns, various places in the country have very different population pyramids (Figure 32).

A few developed countries have very low rates of population growth—in some cases **zero population growth (ZPG)** or **negative population growth**. They have low crude birth rates, low death rates, and, in some cases, net out-migration. France is an example. Because of very low fertility, the country is experiencing negative population growth, and although there is a steady stream of foreigners (especially Arabs and Africans) into the country, France tries to limit immigration. Population decline is an economic concern to many European countries, as well as Japan, the world's oldest nation demographically. Who will fill the future labor force? Is the solution the immigration of guest workers from developing countries? In these respects, demographic changes have profound influence on immigration policies and the size and nature of the labor force.

THE BABY BOOM, AN AGING POPULATION, AND ITS IMPACTS

The so-called baby boom—everyone born between 1946 and 1964—is the largest generation in world history, 90 million strong, and comprises roughly 40% of Americans today. Baby boomers are the children of the "greatest generation," those who lived through the Great Depression and fought World War II. After the war, with the American economy booming and standards of living rising rapidly, birth rates increased dramatically, giving rise to a flood of children in the schools.

The movement of this generation through its life cycle has been likened to that of a python swallowing a large animal, creating a bulge that slowly tapers off over time. In the 1960s, the baby boomers entered college, contributing to the rapid social changes of that decade. In the 1970s, they entered the workforce, including more working women than ever before, which increased the supply of labor and helped to drive up unemployment rates for a while. This large swell entering the workforce required a huge investment in capital stock and infrastructure—office space, desks, training programs, computer terminals, parking garages, not to mention cafeterias and clothing stores. However, because they swelled the labor supply, income growth for boomers was relatively low, and the number of children they had per family was also much smaller than that of their parents, although the large size of this generation meant that the absolute number of children they produced (Generation X) was considerable. Economic opportunities for boomers varied considerably in terms of their birth year: Those born in the early years of the boom encountered labor shortages and high wages, while those born after the boomer peak year in 1957 faced markedly less positive circumstances.

Expenditures by baby boomers in the midst of their prime earning years helped to fuel the consumer and housing booms of the 1990s and 2000s. The result of the influx of baby boomers was new products, new services, and new technologies in niche markets, improving service and reducing service delivery times. Now that the baby boom is preparing to retire, labor force growth is low. As they leave the workforce, the numbers of retirees will rise dramatically, and demands on Social Security and Medicaid will grow exponentially, with serious challenges for public financing. Indeed, an aging

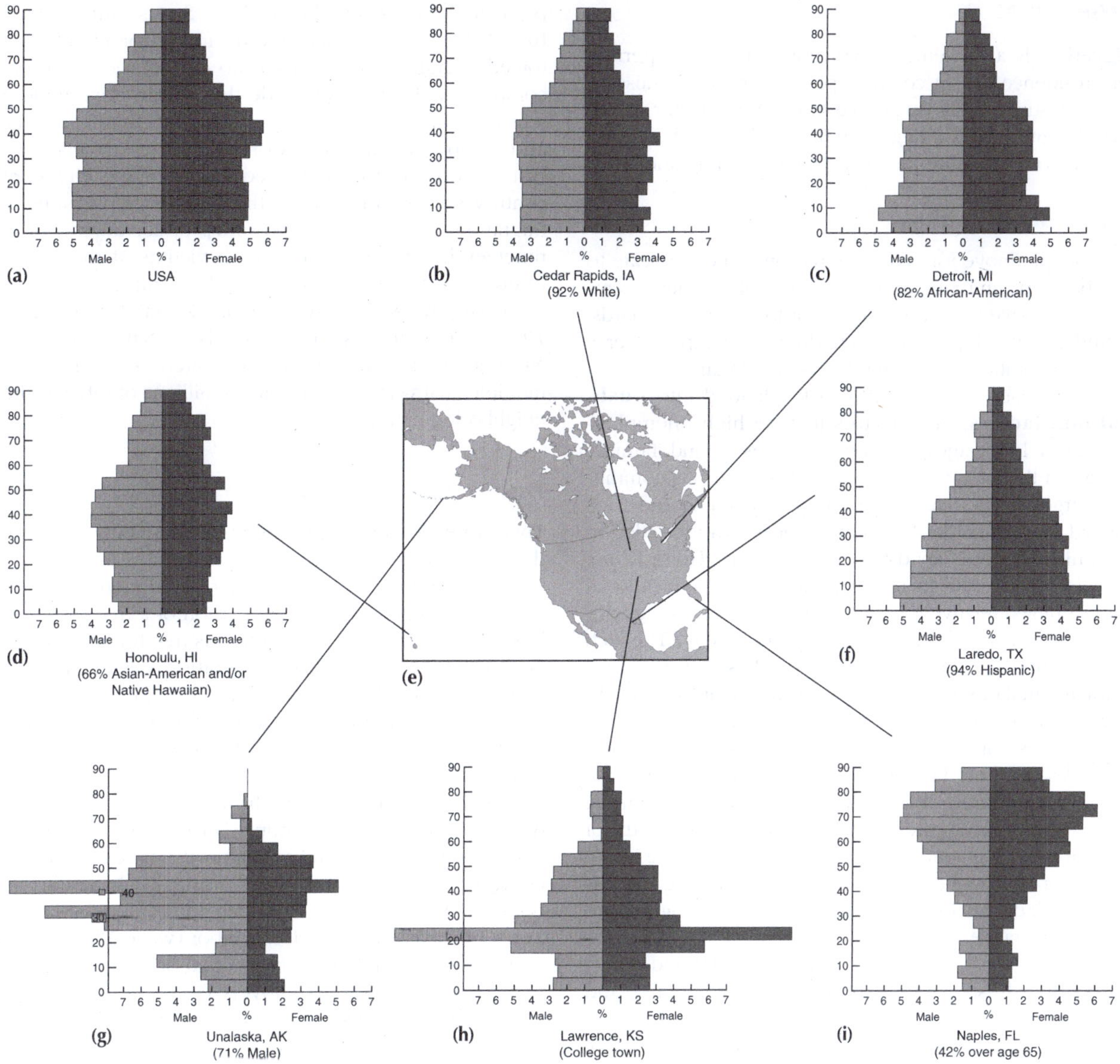

FIGURE 32 Population pyramids for selected communities in the United States reflect a diverse suite of demographic conditions. In college towns, a disproportionately large number of young adults is evident. In retirement communities, such as Naples, Florida, there are far more elderly than young people. Note the imbalance between males and females in Unalaska, a reflection of the labor market there.

population is one of the major driving forces behind the growth of health-related services, which have grown steadily and in tandem with the rising numbers of the elderly. As the baby boom retires—a process already underway—expenditures for pensions and Social Security will also rise rapidly, setting the stage for intergenerational conflicts over the allocation of resources. When the boomers begin to die en masse, starting around the year 2020, death-related industries (e.g., funeral homes) will experience a surge in demand.

The aging of the baby boom is symptomatic of the demographics in most economically developed countries, all of which tend to have low birth rates and high average life expectancy. In Europe, Russia, Japan, the United States, Canada, Australia, and New Zealand, the most rapidly growing age groups are the middle-aged and the elderly. As these societies age, the ratio of nonworking to working people, the so-called dependency ratio, increases steadily, meaning there is a declining ratio of working adults to nonworking people, with associated problems concerning who will pay for the services required by children and the elderly. Aging societies change in many other ways as well: The elderly tend to move more slowly and require more assistance. These observations reflect how economic and demographic changes are closely intertwined.

MIGRATION

Migration is a movement involving a change of permanent residence. It is a complex phenomenon that raises a lot of questions. Why do people move? What factors influence the intensity of a migratory flow? What are the effects of migration? What are the main patterns of migration?

Causes of Migration

Most people move for economic reasons. They relocate to take better-paying jobs or to search for jobs in new areas. They also move to escape poverty or low living standards, to find a better life for their children, to escape adverse political conditions, or to fulfill personal dreams.

The causes of migration can be divided into **push-and-pull factors**. Push factors include high unemployment rates, low wages, poverty, shortages of land, famine, or war. In the late 1970s and early 1980s, various communist purges in Vietnam, Cambodia/Kampuchea, and Laos pushed approximately 1 million refugees, who resettled in the United States, Canada, Australia, China, Hong Kong, and elsewhere. Pull factors include job and educational opportunities, relatively high wages, the hope for agricultural land, or the "bright lights" of a large city. The rich oil-exporting countries in the Middle East act as a pull factor for millions of immigrants seeking employment. In Kuwait and the United Arab Emirates, nearly 80% of the workforce is composed of foreigners, mostly drawn from South Asia and the Philippines. Spatial differences in economic opportunities, therefore, go far toward explaining why young people often leave rural areas, the influx of Mexicans into the southwestern United States, or the immigration of non-Westerners into Europe, including Indians and Bangladeshis into Britain, Turks into Germany, and Arabs into France.

Migrations can be voluntary or involuntary and reflect the historically specific matrix of cultural, economic, and political circumstances in which migrants live. Most movements are voluntary, such as the westward migration of pioneer farmers in the United States and Canada. Involuntary movements may be forced or impelled. In forced migration, people have no choice; their transfer is compulsory. Examples include the African slave trade, in which 20 million people were stolen from their homelands and shipped to the New World, and the deportation of British convicts to the United States in the eighteenth century and to Australia in the nineteenth. In impelled migration, people choose to move under duress. In the nineteenth century, many Jewish victims of the Russian pogroms elected to move to the United States and the United Kingdom. Civil wars in Central America in the 1980s led hundreds of thousands of Salvadorans and Nicaraguans to emigrate to the United States. In Africa, multiple civil wars have displaced millions of refugees into neighboring nations.

The Economics of Migration

This subsection examines migration of a voluntary nature due to economic incentives, which comprises the largest single category of human migrations throughout history. Unfettered migration is an expression of a free market for labor, although markets are never as free for labor as they are for capital. Unrestricted migration is generally only feasible within the borders of the same nation-state. Among states, restricting the flow of immigrants is one of many ways states as well as markets shape economic landscapes.

Consider two regions as shown in Figure 33. Region A on the right, is a highly industrialized country (e.g., the United States or Germany). Region B is a less developed country. A labor market exists within each nation. The quantity of labor supplied and demand is measured on the horizontal axis and the price of labor (wages) is measured on the vertical axis. As the price of labor increases, the demand decreases (i.e., employers face a demand curve), and as the price decreases, the demand increases. The supply curve S_1 increases upward to the right, suggesting

FIGURE 33 Migration and wage differentials. The quantity of labor is measured on the horizontal axis, and the price of labor is measured on the vertical axis. As the price increases on the vertical axis, the demand for labor decreases in both the developed country and the less developed country. The supply curves of both developed and undeveloped countries slope upward, suggesting that a greater supply of labor is available at higher prices. Because the equilibrium price of labor is higher in the developed country than in the undeveloped country, labor migrations occur from less developed to developed, or from B to A, to take advantage of higher wages. The greater the wage differential, both the greater the flow and the longer the distance of flow. The wage rate differential between San Diego and Tijuana, for example, is 8 to 1 for comparable occupations.

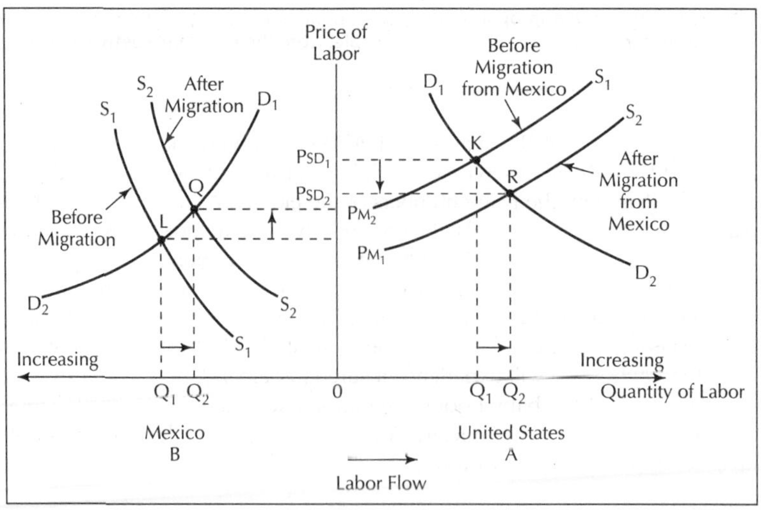

that a greater supply of labor is available at higher prices. As the price for labor increases, individuals who would not care to work at lower wages now come into the market. They substitute work at higher prices for staying home and taking care of children, going to school, or being in retirement.

In order to facilitate the analysis for the less developed country, such information is shown on the left side of Figure 33. Instead of measuring the quantity of labor from zero to the right for the less developed country, we now measure it from zero to the left. Price remains on the vertical axis. The supply curve in the less developed country slopes upward to the left and the demand curve slopes downward to the left. In this manner, we get a back-to-back set of supply and demand curves for a less developed country and a developed country.

Observe the equilibrium position of the supply and demand for labor in the developed country before migration occurs, which occurs at point K. Also note the equilibrium position for the supply and demand of labor in the less developed country. This equilibrium position occurs at L. In classical **labor migration theory**, there is an assumption that information about job availability and wage differentials is widely available and held by workers (i.e., there is little uncertainty). Labor in the less developed country finds out about jobs available in the developed country at higher wage rates. Because the equilibrium price in Region A is higher than that in Region B, labor migrates from Region B to Region A to take advantage of higher wages (assuming there are no barriers to migration). The greater the differential in wages, the greater the flow of labor will be. In Region A, extra labor is now coming into the region, which is used to working for lower wages. Because the extra labor is supplied over and above the indigenous supply and the labor is used to working at lower wages, the supply curve moves downward toward the right to the new equilibrium R. In addition, because the labor pool has left Region B, the supply of labor is reduced. The supply curve moves upward and to the right in Region B, thus raising the equilibrium price to Q. The new equilibrium price in Region A, at R, is at a lower level than it was prior to labor migration. Thus, migration will continue as long as there is a difference between the wages of Region A and Region B, which exceed a cost associated with migration. In the case of flows from Mexico to the United States, most categories of employment are paid two to five times the rate in Mexico. Consequently, the flows both on a daily basis and on a longer-term basis continue to occur at high levels, including many illegal or undocumented workers (see Figures 32 and 33).

In classical migration theory, transportation costs and other costs associated with moving an individual or family are included, such as selling a property and purchasing one in the new region. The costs of labor in the different regions will not be exactly equalized. But classical trade and migration theory tell us that the long-run price of labor in the two regions should come into close harmony with one another. Relocation and similar migration costs should be split up over the period of work remaining in the life of the mover.

However, when we observe real-world labor movements and price differentials, we find that wage rates do not seem to be converging among regions. Major discrepancies occur in the wages paid in and among various regions of the United States and countries of Europe, as well as in South America and India. If neoclassical theory held, there would be less difference in national averages of per capita incomes. One reason that labor differential rates exist is because of the imperfections in the availability of knowledge about opportunities. Many workers in the less developed countries do not know that jobs they may be qualified for in more highly developed countries even exist. As an individual contemplates changing locations and even countries, he or she is beset by a series of social factors, including lack of friends and knowledge and the feeling of uneasiness in the new setting. Consequently, the largest numbers of international labor migrants are young males who do not have families to relocate.

Cultural differences are also important to understanding migration. The cultural shock of living in a new environment, especially when one does not have the resources to live adequately or does not speak the native language, presents social problems. Institutional barriers also exist, such as the status of immigration or the length of time allowed in the host country. African Americans, Puerto Ricans, Chinese immigrants, Mexican Americans, and women have encountered such resistance in the past in the United States in their search for improved working conditions and wages, often in the face of xenophobic racism and ostracism.

Consequently, we cannot expect that economic forces alone will lead to a total eradication of wage inequities throughout a country or throughout the world. Barriers to migration, including legal obstacles and immigration restrictions, imperfect information, lack of skills, inability to afford transportation, and the powerful bonds that hold people in place work to prevent a free flow of labor among and often within countries. At best, only a small portion of the population in a low-wage region has the ability to gain access to higher pay in developed nations. Therefore, there will continue to be a discrepancy in per capita earnings between less developed countries and developed countries and between depressed regions and economically healthy regions within countries. The demand curves for labor also shift down and to the right as the more highly qualified, productive labor leaves, and up and to the right as new workers arrive. Migration is thus intertwined with local and national labor markets in complex ways that shape average incomes and unemployment rates.

The availability of work and wage rates account for major labor flows throughout the world, from countries lacking in jobs and with low wages to countries with jobs available at relatively higher wage rates. Major labor flows occur (1) from Mexico and the Caribbean to the United States and Canada; (2) from South American countries to

Argentina, Venezuela, and Peru; (3) from North Africa and southern European nations to northwestern Europe; (4) from Africa and Asia to Saudi Arabia; and (5) from Indonesia to Malaysia, Singapore, and Australia. Migrants vary by age. Young adults are most likely to be migrants because of their desire for an improved life and greater ability to travel and overcome hardships.

Barriers to Migration

All countries regulate the flow of immigration. The United States limits legal immigration to approximately 600,000 people annually, although a total of roughly 1.1 million enter the United States legally or illegally every year. Altogether, about 33.1 million immigrants live in the United States, comprising 11% of the population. Of this group, an estimated total of 5 million people live in the country illegally, often under constant threat of being caught and deported. Billions of dollars are spent annually to police the U.S. borders, much of which is used to try to keep Mexicans and other Latin Americans out. The status of illegal immigrants is a significant political issue often arousing passionate feelings: Some argue that by paying income and sales taxes, immigrants generate more wealth than they consume, while others argue that they compete with unskilled American residents, many of whom are ethnic minorities, and drive down wages in the bottom rungs of the labor market.

Characteristics of Migrants

Some countries have higher rates of migration than do others—both into and within the country. In general, the countries that have long histories of migration, such as the United States, Canada, and Australia, have higher migration rates in the modern world than do other countries, such as China, where migration is far less common. Recently, China has witnessed an enormous surge of people leaving rural areas for the more prosperous coastal cities; with more than 100 million such migrants, this stream is one of the largest migrations in history.

When people do move, they are far more likely to be young adults than any other group. Young adults have the longest working life ahead of them and thus stand to gain more than the elderly from the accumulated benefits of moving to a relatively higher-wage region. Many migrants send part of their wages to their home country or village. These remittances often form an important source of income for the recipients back home. In the developed world, migration rates tend to drop significantly by the time people have established families and purchased homes, typically in their thirties.

Consequences of Migration

Migration has demographic, social, and economic effects, especially due to the fact that migrants tend to be young adults and are often the more ambitious and well-educated members of a society. Obviously, the movement of people from one region to another causes the population of the country of origin to decrease and of the destination country to increase. Because of migratory selection, the effects are more complicated. If the migrants are young adults, their departure increases the average age, raises the death rate, and lowers the birth rate in the country of origin. For the destination region, the opposite is true (i.e., their arrival tends to lower the average age and the death rate but increase the fertility rate). If migrants are retirees, their effect is to increase the average age, raise the death rate, and lower the birth rate. Arizona and Florida, for example, have attracted a large number of retirees, resulting in higher-than-average death rates.

Social conflict is a fairly frequent social consequence of migration. It often follows the mass movement of people from poor countries to rich. There were tensions in Boston and New York after the Irish arrived in the 1840s; fleeing the potato famine, they were the first Catholics to arrive in the United States in large numbers. Similar tensions have come with recent migrants, such as Cubans settling in Miami. Social unrest and instability also follow the movement of refugees from poor countries to other poor countries. Many immigrants are subject to racist xenophobia and become scapegoats for all the problems in their new country, especially during economic downturns. In much of Europe, for example, nationalists blame Turks, Arabs, Pakistanis, and other immigrants for unemployment. Generally, poor migrants have more difficulty adjusting to a new environment than the relatively well educated and socially aware.

The economic effects of migration are varied. With few exceptions, migrants contribute enormously to the economic well-being of places to which they come. For example, guest workers from Turkey and Yugoslavia were indispensable to the economy of Germany in the 1960s and 1970s. Without them, assembly lines would have closed down, and patients in hospitals and nursing homes would have been unwashed and unfed. Without Mexican migrants, fruits and vegetables in Texas and California would go unharvested and service in restaurants and hotels would be much more expensive. Migrants to the United States also pay income and sales taxes, but illegal ones do not reap the benefits of programs such as Social Security.

In the short run, the massive influx of people to a region can cause problems. The U.S. Sunbelt states have benefited from new business and industry but are hard pressed to provide the physical infrastructure and services required by economic growth. In Mexico, migrants to Mexico City accelerate the competition for scarce food, clothing, and shelter. Despite massive relief aid, growing numbers of refugees in the developing world impoverish the economies of host countries.

Emigration can relieve problems of poverty by reducing the supply of labor. External migration reduced poverty somewhat in Jamaica and Puerto Rico in the 1950s and 1960s. However, emigration can also be costly. Some of the most skilled and educated members of the population of Third World countries migrate to developed countries. Each year, the income transferred through the "brain drain" to the United States amounts to significant sums, although billions of dollars are also sent back home in the form of remittances to family members who stayed behind. Indeed, remittances are often a major source of income for impoverished villages in the developing world.

Patterns of Migration

In examining patterns of migration, it is helpful to consider migration internationally or within a country separately. It is also convenient to subdivide external migration into intercontinental and intracontinental, and internal migration into interregional, rural-urban, and intrametropolitan. International migrations are greatly exceeded by internal population movements, especially to and from cities.

The great transoceanic exodus of Europeans and the Atlantic slave trade to the New World are spectacular examples of intercontinental migration. In the five centuries before the economic depression of the 1930s, these population movements contributed greatly to a redistribution of the world's population. One estimate is that between 9 and 10 million slaves, mostly from Africa, were transported by Europeans into the sparsely inhabited Americas. The importance of the "triangular trade" among Europe, Africa, and the Americas can hardly be exaggerated, especially for British economic development. Africans were purchased with British manufactured goods. They were sent to work on plantations where they undertook the production of sugar, cotton, indigo, molasses, and other tropical products. These commodities were then sent to Britain for processing, which created new British industries. Plantation owners and slaves became a new market for British manufacturers whose profits helped finance Britain's Industrial Revolution.

The Atlantic slave trade, however, was dwarfed by the voluntary intercontinental migration of Europeans. Mass emigration began slowly in the 1820s and peaked on the eve of World War I, when the annual flow reached about 900,000 (Figure 34). At first, migrants came from densely

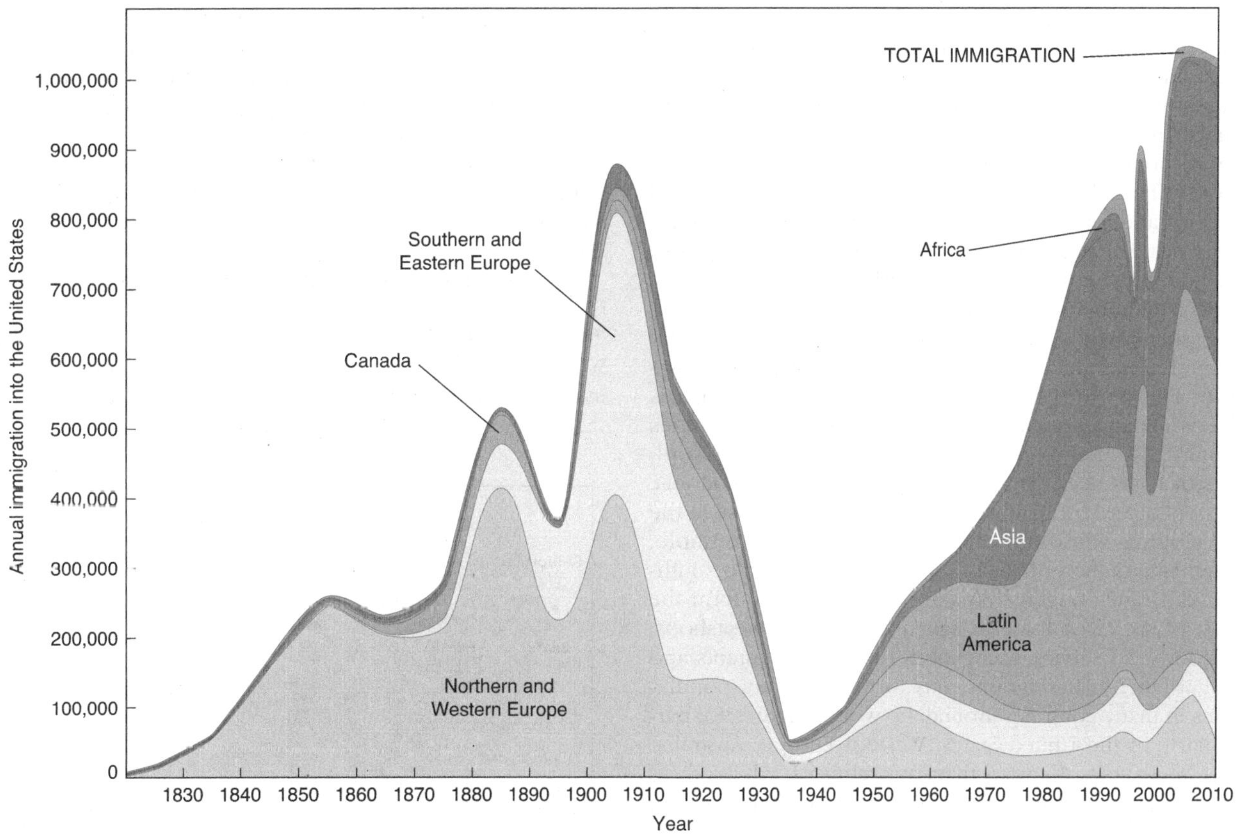

FIGURE 34 Immigration to the United States by region of the world. European countries provided more than 90% of all immigrants to the United States during the 1800s, and up until 1960, Europeans continued to provide more than 80% of the total migration. But since the 1960s, Latin America and Asia have supplanted Europe as the most important source of immigrants to the United States. During the Great Depression and World War II years, immigration was at an all-time low. Add the taxes that an immigrant and his or her descendants are likely to generate over their lifetimes; then subtract the cost of the government services they are likely to consume. The result is that each new immigrant yields a net gain to the government of $80,000. States and cities lose $25,000, while the impact on the federal treasury is $105,000.

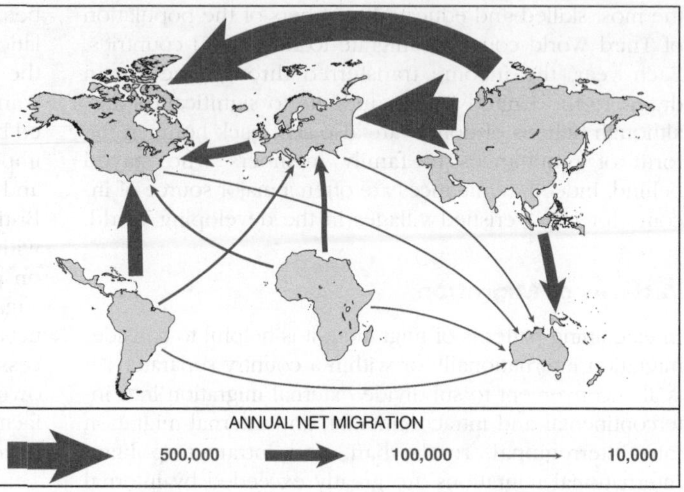

FIGURE 35 Intercontinental migration flows are dominated by people leaving the developing world in search of opportunities in the developed world, including immigrants from Asia and Latin America to the United States and Canada and Africans and Asians to Europe. In the long run, there is a drawback to the rich country low fertility rates: an impending slowdown in the labor supply, which will threaten productivity and economic growth. With millions of workers unemployed in the rich countries, an impending labor shortage might not seem much of a problem. But these demographic shifts set the boundaries for a country's future, including their ability to maintain their standard of living and their ability to service their public debt. Unless more immigrants are allowed in, or a larger proportion of the working-age population joins the labor force, or people retire later, or their productivity rapidly accelerates (all of which are unlikely events), the aging population will translate into permanent slower potential growth. America is debating this thorny issue now.

populated northwestern Europe. Later, they came from poor and oppressed parts of southern and eastern Europe. Between 1840 and 1930, at least 50 million Europeans emigrated. Their main destination was North America, but the wave of migration spilled over into Australia and New Zealand, Latin America (especially Argentina), and southern Africa. These new lands were important for Europe's economic development. They relieved population pressure and provided new sources of foodstuffs and raw materials, markets for manufactured goods, and openings for capital investment. Another large-scale intercontinental migration was the Chinese diaspora of the nineteenth and early twentieth centuries, especially into Southeast Asia.

Since World War II, the pattern of intercontinental migration has changed. Instead of heavy migratory flows from Europe to the New World, the tide of migrants is overwhelmingly from developing to developed countries (Figure 35). Migration into industrial Europe and to North America has been spurred partly by widening economic inequality and by rapid rates of population increase in the developing world. In the United States, for example, roughly 5 million Mexican immigrants live in the southwestern states (Figures 36 and 37), where they form the bulk of the labor force in agricultural labor, sweatshops, and low end service occupations (e.g., in restaurants and as gardeners). Immigrants thus form significant populations in many countries around the world (Figure 38), particularly in the United States, Western Europe, Australia, and South Asia. Some of them are refugees, while others are unskilled workers seeking jobs outside of their native lands.

The era of heavy intercontinental migration is over. Mass external migrations still occur, but at the intracontinental scale. In Europe, forced and impelled movements of people in the aftermath of World War II have been succeeded by a system of migrant labor. In the United States, thousands of people from Latin America, particularly Mexicans, many of whom are illegal aliens, arrive each year to find work (Figure 39). Similarly, the most prosperous industrial countries of Europe attract workers from the agrarian periphery (Figure 40). France and Germany are the main receiving countries of labor migrants to Europe. France attracts workers from North Africa. During the post-WWII boom years (1945–1975), West Germany drew many "guest workers" from Italy, Yugoslavia, Greece, and Turkey. Migrant workers from southern Europe usually have low skills and perform jobs unacceptable to indigenous workers.

The system of extraterritorial migrant labor also exists in the developing world. In Africa, laborers move great distances to work in mines and on plantations. In West Africa, the direction of labor migration is from the interior to coastal cities and agricultural export areas. In

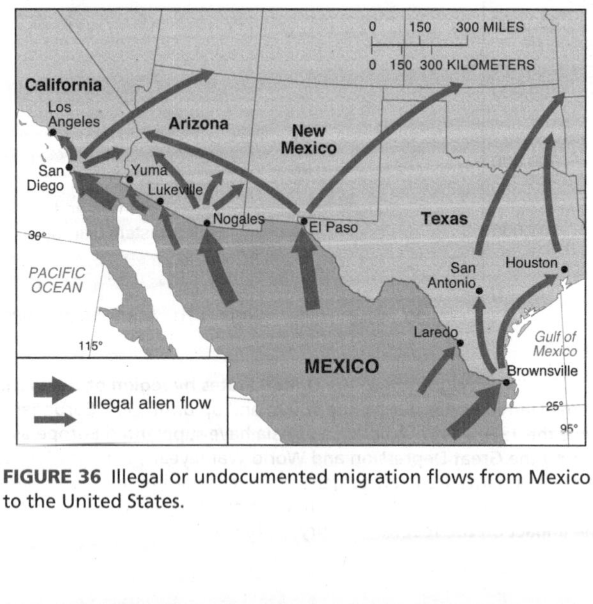

FIGURE 36 Illegal or undocumented migration flows from Mexico to the United States.

Courtesy of the Library of Congress

FIGURE 37 Flows of millions of illegal immigrants from Mexico to the United States take place at a variety of locations along the border. Often migrants undertaking such journeys are exposed to very dangerous conditions, and many die in transit. Such desperation is driven by the enormous differences in standards of living and opportunities on either side of the border, that is, uneven spatial development.

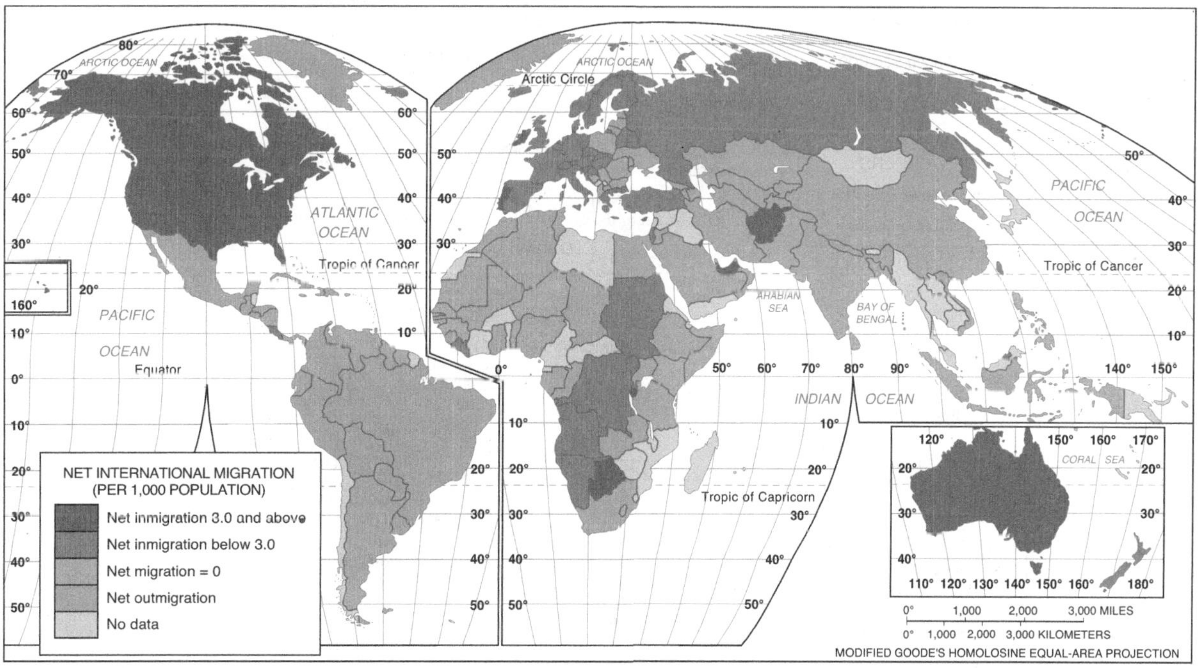

FIGURE 38 Net international migration rates around the world. The United States has the largest group of immigrants, although other populations are found in France, Britain, Germany, and Australia, as well as in India, Pakistan, Iran, and Saudi Arabia. In the latter group of countries, immigrants include many unskilled workers from Asia.

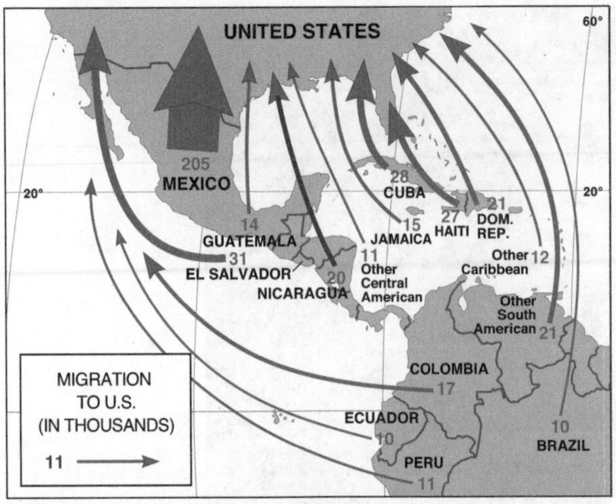

FIGURE 39 Migration flows to the United States from Latin America. The largest source of emigrants is Mexico, with smaller streams from various parts of the Caribbean, Central America, and northern South America.

East Africa, agricultural estates attract extraterritorial labor, typically refugees (Figure 41). In southern Africa, migrants focus on the mining-urban-industrial zone that extends from southern Zaire in the north, through Zambia's Copper Belt and Zimbabwe's Great Dyke, to South Africa's Witwatersrand in the south.

Today, refugee-generating and receiving countries are concentrated in Africa (6 million people), Southeast Asia (4 million), and Latin America (2 million). The causes of refugee movement typically include wars (e.g., Vietnam, World War II, Afghanistan), racial and ethnic persecution (e.g., South Africa, Bosnia-Herzogovina), economic insufficiency increased by political turmoil (e.g., Sudan), and natural and human-caused disasters (e.g., Central American hurricanes).

Colonizing migration and population drift are two types of interregional migration. Examples of colonizing migration include the nineteenth-century spontaneous trek westward in the United States and the planned eastward movement in Russia beginning in 1925. General drifts of population occur in almost every country, and they accentuate the unevenness of population distribution. Between the two World Wars, there was a drift of African Americans from the rural South to the cities of the nation's industrial heartland in the Northeast and Midwest. Since the 1950s, there has been net out-migration from the center of the United States to both coasts and a shift of population from the Rustbelt states to the Sunbelt (Figure 42). Today, the majority of Americans live in the South and West, as opposed to the North and Midwest, although the vast expanses of land in the Sunbelt states generate lower population densities than in the Northeast.

The most important type of internal migration is rural-urban migration, which is usually for economic motives.

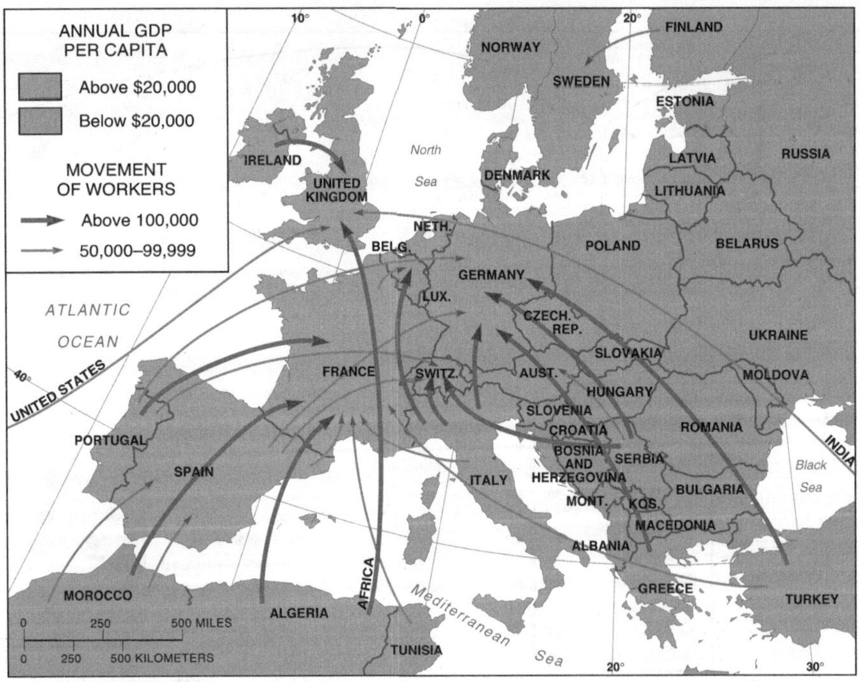

FIGURE 40 Migrant flows to and within Europe are generally from poorer countries with a labor surplus to wealthier ones with better employment opportunities. For example, Turks have long served as "guest workers" in Germany, and France hosts a growing population of Arabs from Algeria and Morocco. There are also net migration streams out of Spain, Portugal, Greece, and Italy to France, Switzerland, Germany, and other states.

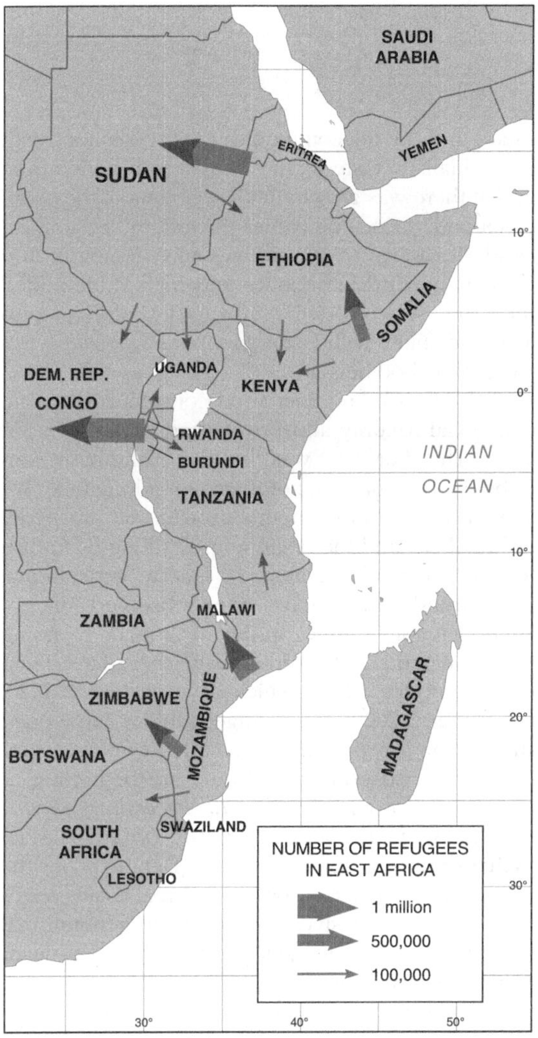

FIGURE 41 Involuntary migrations in East Africa in the late twentieth century. Eritreans, Ethiopians, Somalis, and Sudanese have been forced to move because of raging civil wars. To escape civil war, Hutus have been forced to migrate from Rwanda, and residents of Mozambique have fled to Malawi and other neighboring states.

The relocation of farm workers to industrial urban centers was prevalent in developed countries during the nineteenth century. Since World War II, migration to large urban centers has been a striking phenomenon in nearly all developing countries. Burgeoning capital cities, in particular, have functioned as magnets attracting migrants in search of "the good life" and employment.

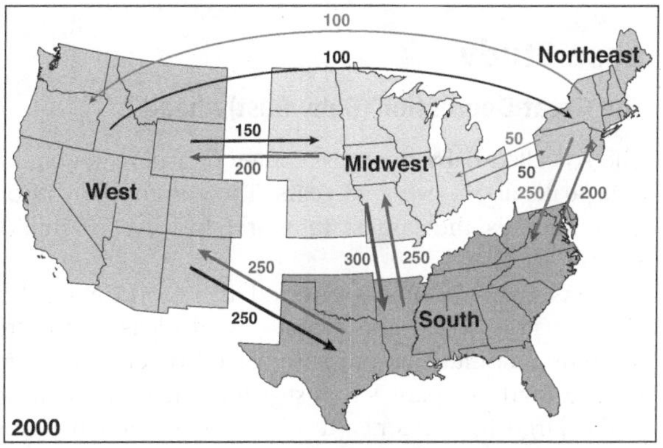

FIGURE 42 Migration flows among the four major census regions of the United States. While all regions exchange people with each other, the largest flows are to and from the South. These flows are one factor in the growth of the Sunbelt (South and West), which is now more populous than the Snowbelt or Rustbelt (Northeast and Midwest).

In highly urbanized countries, intermetropolitan migration is increasingly important. Although many migrants to cities come from rural areas and small towns, they form a decreasing proportion. Job mobility is a major determinant of intercity migration. So, too, is ease of transportation, especially air transportation. For intermetropolitan migrants from New York, the two most popular destinations are Miami and Los Angeles.

Internally Displaced Persons (IDPs)

An internally displaced person (IDP) is a person or group of people who, as a result of conflict usually characterized by violence, is either strongly compelled or forced to leave their home. Other compelling factors that could cause displacement are natural or human-made disasters or abuses of human rights. IDPs are not considered refugees per se because they do not leave their country's borders when seeking sanctuary. The majority of IDPs are on the continent of Africa. Sudan presently has 5 to 6 million IDPs, and they are located in the eastern provinces of the Darfur region where government forces have been attacking local villages. Both Iraq and Afghanistan have more than 2.5 million IDPs due to sectarian violence and war. The Democratic Republic of the Congo in central Africa has several million IDPs, mainly in the eastern provinces, due to civil war and unrest, while Colombia continues to house millions of IDPs due to civil war between government forces and the FARC (Revolutionary Armed Forces of Colombia).

Case Study

The Great Depression (Baby Bust) Ahead

The so-called baby boom is a surge of a new generation that peaks about every 40 years. The current baby-boom generation is the largest in world history, 90 million strong.

As baby boomers entered the workforce, they brought new social and technological ideas and entrepreneurial spirit. This pressure forced older, mature industries and companies to restructure themselves and to make large investments at all levels to train the new workers. Initially, this new boomer generation was not very productive and had low earnings and savings rates. In the 1960s and 1970s, the baby boomers entered the workforce, including more working women than ever before. This large swell entering the workforce required a huge investment in capital stock and infrastructure: office space, desks, training programs, computer terminals, parking garages, not to mention cafeterias and clothing stores.

The baby boomers redefined the workplace, causing social and technological change, although their conformist, civic-minded bosses were often not accustomed to such change. A few of the upstart new entrepreneurs included Bill Gates (Microsoft), Michael Dell (Dell Computers), Steven Jobs (Apple), Larry Ellison (Oracle), and Eric Anderson (Netscape). The result of the influx of baby boomers was new products, new services, and new technologies in niche markets, improving service and reducing service delivery times.

The baby-boom spending wave began in the 1980s, peaking around 2006. According to the U.S. Bureau of Labor Statistics, family spending follows a predictable life cycle that results in maximum spending between ages 46 and 50. The great number of baby boomers moving as consumers in and out of their peak spending years causes booms and busts in the economy. The peak in 2006 created the biggest business boom in the history of the world.

The baby-boom spending swell changed the demand for products and services and changed how we work and live. The innovation of the baby-boom generation required a new *customized/flexible economy* requiring creativity in products and services and an unprecedented surge in productivity. This business trend was propelled by the advancement of baby boomers into their power years when they had the business decision-making capacity to change organizations and accommodate new technologies to service their different skills and lifestyles. The combination of the microcomputer revolution and the telecommunications revolution, with the individualistic taste of the baby-boom generation, meant that there was growth in all segments of the economy, especially those companies that offered customization and flexibility to an individual's demands and needs. This meant high-quality and high-value-added products and services delivered rapidly, and custom-produced flexibility with fast response and delivery.

The baby boomers are now retiring, and there is a new dearth of spending, increased consumer debt, and strained Social Security and private and public pension programs. The year 2009 was the end of the baby economic boom, one that will not return for a long time. We are now at the end of economic growth and prosperity in most markets, and the beginning of a long downturn in the U.S. and world economies. Due to generational spending trends, there have been observed 40-year growth cycles, for example, in the stock market, ending in 1929, 1968, and 2009. Additionally, there have been 30-year commodity cycles, which have peaked in 1920, 1950, 1980, and 2009. Three huge bubbles have been expanding for the past three decades—stocks, commodities, and real estate—and they are beginning to deflate. In his book, *The Great Recession Ahead: How to Prosper in the Debt Crisis of 2010–2012* (**www.hsdent.com**), forecaster Harry Dent explains the "Perfect Storm" as peak oil prices and real estate prices collide with the massive downturn in generational spending and the stock market—or the baby bust—leading to more severe downturns for the global economy and individual investors alike. He argues that the next broad-based global bull market will not occur until 2020–2025, because of the dearth of consumer spending and this baby bust. It will take until this time for the children of the baby boomers, the generation Xers, to enter their peak spending years and drive the market upward once again.

Migration to Exurbs: Counterurbanization

Another major spatial change is occurring due to the changing demographics in the workplace. A geographical shift of the population is occurring from the urban area as the workforce begins moving to the exurbs, or outer suburbs, and back to smaller towns and communities. This trend, which began in the 1970s, is called *counterurbanization*. The move to the exurbs is being propelled by (1) retirees, (2) those seeking relief from the high-cost suburbs, and (3) those desiring a greater nature or native lifestyle. These shifts will be made possible by the *electronic cottage*—the increased power of computers and telecommuting in ways that have not been used in the past. This power will be employed to decentralize today's firms through the communication revolution—the moving of information rather than people. As baby boomers moved into their peak spending years, bolstered by two-income family earnings, they spent more than ever before on high-quality durable goods and convenience items, and less on the standardized, mass-produced items and services of the past. Just as in flexible manufacturing, baby-boomer consumer demand centered on the high-quality, high-choice, niche markets. Labor patterns of the future will focus on exurban personal freedom and an ability to control one's time schedule and work location, health, and environment, rather than only securing high income and retirement benefits, which in themselves are disappearing fast in the new post-boom economy. Such new workers in the companies of the year 2015 will not so much be "doing their own thing" as "controlling their own time and space" in exurbia.

The X generation (or gen X, children of the baby boomers) will provide business growth opportunities for housing, office buildings, restaurants, all manner of personal and business services, home conveniences, entertainment, and more. Small town exurbs will become boom towns, and as many as 80 million North Americans could shift outside metropolitan areas by the year 2015.

Such boom towns will provide a slashed cost of living, especially for food and real estate needs, but ample opportunity to start up a *carbon copy business*, which is a successful business idea transplanted from a major city to the exurbs that share common consumer demographics and lifestyles with the city. Starbucks, Red Box Videos, Arco, Charles Schwab Investment, 24-Hour Fitness, and Papa John's Pizza are examples. The *carbon copy business* will be an important exurb model for small firms, while improving the quality of life for individuals. Other exurbanites will connect to cities and markets remotely through the Internet, while leveraging information technologies to exploit the growth of small towns.

Retirement exurbs on the East Coast include coastal Maine; Cape Cod; the Green and White Mountains of Vermont and New Hampshire; coastal areas of New Jersey, Delaware, and Maryland; southern coastal zones of South Carolina and Georgia; and the Florida peninsula (see map available at **www.FrederickStutz.com**). Inland from the East Coast are retirement exurbs in southern Quebec, central Pennsylvania, the Blue Ridge in central Virginia, the Piedmont of western North Carolina, South Carolina, and the Smoky Mountains of eastern Tennessee. Favorite Midwestern exurbs include northern Michigan and Wisconsin, Kentucky Lake, and the Ozarks of southern Missouri. Western exurbs are booming due to many rural amenities and available land for development at a low cost. South Texas, Padre Island, eastern New Mexico, the western Colorado Rockies, western Arizona, Las Vegas, and the northern Rockies of western Montana and northwestern Wyoming are growing into such regions. The West Coast is lined with such exurban centers as British Columbia, Victoria Island, the Puget Sound and coastal Washington and Oregon, northern California and the Sierra Nevada Mountains, the central California coast, and San Diego County, including coastal Baja California, in Mexico.

Summary

Because people are the single most important element in the world economy, it is essential to learn about population distribution, qualities, and dynamics. In this chapter, we began by examining the uneven distribution of people over the earth's surface. The vast majority of the 6.7 billion people alive today live in the developing world, particularly in Asia, where more than half reside. We emphasized that the distribution of people is a reflection of centuries, or even millennia, of uneven economic development, particularly following the Industrial Revolution and the formation of global colonial empires in the eighteenth and nineteenth centuries, which dramatically concentrated populations in cities and along the coasts. We noted population density is not an adequate variable to account for economic well-being: Some of the world's poorest countries are sparsely inhabited, and some of the wealthiest, such as the Netherlands, are very dense.

We also examined the processes of population change. The two major components of population change are migration and natural increase. The principal force affecting world population distribution used to be migration; now it is natural increase, the difference between fertility and mortality rates. This chapter explored the nature of population structures, particularly the age-sex distribution portrayed by population pyramids. This device is useful in contrasting the distribution of people by age and sex among different countries or the same country over time, particularly to reveal how changing economic circumstances, by changing fertility and mortality rates, create larger or smaller pools of young and elderly. This line of thought is useful in forecasting the future population status of regions or countries, including, for example, the changing size and composition of labor forces.

Malthusian conceptions of population change held that the growth in the number of people must inevitably outstrip the resource base of the planet, or parts thereof. While this position was useful in noting that unconstrained population growth cannot go on unchecked indefinitely, it also was undermined by its simplistic views of why people have children and derailed by the growth in productivity of the Industrial Revolution. Indeed, by focusing on population growth as the source of the world's complex problems, Malthusianism tends to blame the victim (i.e., the poor) and ignore other, more important political and economic forces.

The demographic transition offers a superior model of population growth in that it links fertility and mortality rates, and thus natural growth, to the economic dynamics of industrialization, urbanization, and expanding capitalism. This perspective offers a compelling explanation as to why crude birth rates are high in poor countries and why in the premodern context natural growth rates were low. Further, it explains the declines in death and birth rates associated with economic development, including the essential question of why couples have fewer children as their incomes rise. Thus, unlike Malthusianism, it embeds fertility in its historical and economic context, and its conclusions about the future of the world's population growth are markedly different. Although the population growth rate is falling, the world's population is projected to increase for decades to come, due to the large momentum coming from the vast and youthful population of the developing world.

In addition to fertility and mortality, migration is a major force in shaping the geography of population. Excluding involuntary migration such as slavery or impelled migration (typically from wars), we noted that spatial discrepancies in economic opportunities are the major forces driving migration. The causes include both push and pull factors, but typically center on unemployment rates and average incomes. The great transcontinental migration streams of the nineteenth century have given way to flows from the developing to the developed world. Young people, particularly males, constitute the largest group of migrants. Migration has important impacts on local labor markets, affecting the supply of labor and thus wage rates. Thus, economic and demographic forces are fused unevenly over the geographic landscape.

Key Terms

baby boom
birth rate
death rate
demographic transition
diminishing marginal returns
doubling time
infant mortality rate
labor force
labor migration theory
Limits to Growth
Malthusianism
migration
natural growth rate (NGR)
negative population growth
neo-Malthusianism
net migration rate (NMR)
population density
population pyramid
push-and-pull factors
total fertility rate
zero population growth (ZPG)

Study Questions

1. Summarize the spatial distribution of the world's population.
2. Define *crude fertility* and *death rates.*
3. What are the mathematics of the four major components of population growth?
4. Why did Malthus have such a gloomy view of the future?
5. How do neo-Malthusians resemble and differ from Malthusians?
6. What are the four stages of the demographic transition? Give examples of each.
7. What explains the world map of total fertility rates?
8. Did doctors alone reduce death rates?
9. What are the major causes of death in the United States?
10. Why are the poorest countries growing the most rapidly?
11. What is a population pyramid and how does it vary between developed and developing countries?
12. Summarize the major causes of international migration.
13. What are some consequences of international migration?
14. Where will the bulk of the world's population growth occur in the twenty-first century? Why?
15. When were the two largest periods of immigration in the United States?

Suggested Readings

Kirk, D. 1996. "Demographic Transition Theory." *Population Studies* 50:361–388.

Newbold, K. 2006. *Six Billion Plus: Population Issues in the Twenty-First Century*. Boulder, CO: Rowman and Littlefield.

Peters, G., and R. Larkin. 2005. *Population Geography: Problems, Concepts and Prospects*. New York: Kendall/Hunt.

World Resources Institute. 2004. *World Resources*. New York: Oxford University Press.

Web Resources

U.S. Population Estimates
http://www.census.gov/population/www/projections/popproj.html

The U.S. Census Bureau, in association with the Federal-State Cooperative Program for Population Estimates (FSCPE), has recently released updated population estimates at the national, state, and county levels.

The Census Bureau
http://www.census.gov

The Census Bureau Web site was designed to enable "intuitive" use and is intended to be visually appealing, concise, and quick-loading. It was designed so users can effectively locate and utilize the resources the site has to offer, such as the "Population Clock" and its small search engine.

Population Reference Bureau
http://www.prb.org/Publications/Datasheets/2010/2010wpds.aspx

Up-to-date population estimates of the world and countries, searchable database, webcasts, and more.

World Health Organization
http://www.who.int/en/

Various data and reports about the health of the world's population, including diseases.

Resources and Environment

OBJECTIVES

- To describe the nature, distribution, and limits of the world's resources
- To examine the nature of world food problems and the difficulties of solving them
- To describe the distribution of strategic minerals and the time spans for their depletion
- To consider the causes and consequences of the energy crisis and to examine alternative energy options
- To examine the major causes of environmental degradation

Open-pit mining operations, such as Kennecott's Bingham Canyon Mine in Utah, are huge, efficient, and capital-intensive enterprises, but also do enormous damage to the local environment.

Resources and Environment

Economic growth and prosperity depend partly on the availability of natural resources and the quality of the environment. There is growing concern that the consumption of inputs and goods in developed countries, and increasingly in developing countries, is depleting the world's stock of resources and irreparably degrading the natural environment. What can be done to effectively manage resources and protect the environment?

Optimists believe that economic growth in a market economy can continue indefinitely; they see relatively few limits in raw materials and great gains in technological productivity. In contrast, pessimists assert that there are inherent **limits to growth** imposed by the finiteness of the earth—by the fact that air, water, minerals, space, and usable energy sources can be exhausted or ecosystems overloaded. They believe these limits are near and, as evidence, point to existing food, mineral, and energy shortages and to areas now beset by deforestation and erosion.

How can we create a habitable and sustainable world for generations to follow? One solution is to transform our present **growth-oriented lifestyle**, which is based on a goal of ever-increasing production and consumption, to a **balance-oriented lifestyle** designed for minimal environmental impact. A balance-oriented lifestyle would include an equitable and modest use of resources, a production system compatible with the environment, and appropriate technology. The aim of a balance-oriented world economy is maximum human well-being with a minimum of material consumption. Growth occurs, but only growth that truly benefits all people, not just the elite few. However, what societies, rich or poor, are willing to dismantle their existing systems of production to accept a lifestyle that seeks satisfaction more in quality and equality than in quantity and inequality?

This chapter deals with the complex components of the population–resources issue. Have population and economic growth rates been outstripping food, minerals, and energy? What is likely to happen to the rate of demand for resources in the future? Could a stable population of 10 billion be sustained indefinitely at a reasonable standard of living utilizing currently known technology? These are the critical questions with which this chapter is concerned.

RESOURCES AND POPULATION

Popular opinion in the industrial West generally appreciates the need to reduce population growth but overlooks the need to limit economic growth that exploits resources. Most people in the economically developed world suffer from a view that resources are limitless and do not appreciate that our rapid consumption of them ultimately threatens our affluent way of life. The First World is, in short, liquidating the resources on which our way of life was built. The growth of some developing countries is aggravating the situation. Their growing populations put increasing pressure on resources and the environment, and many aspire to affluence through Western-style urban industrialization that depends on the intensive use of resources. Poor countries generally do not have the means for running the high-energy production and transportation systems manifest in the industrial West. The production of a middle-class basket of luxury goods (e.g., cars) requires six times as much in resources as a basket of essential or basic goods (e.g., food). The expansion of gross domestic product (GDP) through the production of middle-class baskets means that only a minority of people in poor countries would enjoy the fruits of economic growth. Resource constraints prevent the large-scale production of consumer goods for the growing populations of the developing countries.

However, numerous measures of material well-being (e.g., per capita incomes, calories consumed, life expectancy) show that people in most, but not all, countries are better off today than their parents were. But there are problems with this optimistic assessment. These improvements are based on averages; they say nothing about the distribution of material well-being. Another difficulty is that the world may be achieving improvements in material well-being at the expense of future generations. This would be the case if economic growth were using up the world's resource base or environmental carrying capacity faster than new discoveries and technology could expand them.

Carrying Capacity and Overpopulation

The population–resources problem is much debated, particularly during periods of economic shortages and rising prices. Neo-Malthusian pessimists believe that the world will eventually enter a stationary state at **carrying capacity**, which is the maximum population that can be supported by available resources. They point to recurring food crises and famines in Africa as a result of **overpopulation**. However, carrying capacity, an idea borrowed from ecology, is simplistic in that it ignores the historical, political, and technological context in which the production and consumption of goods occurs. Human beings are not mindless products of an unchanging nature and are capable of modifying their environment and altering the constraints and opportunities it presents. On the other hand, optimists believe in the saving grace of modern technology. Technological advances in the past 200 years have raised the world's carrying capacity, and future technical innovations as well as the substitution of new raw materials for old hold the promise of raising carrying capacity still further.

The answer to the population–resources problem also depends on the standard of living deemed acceptable. To give people a minimal quality of life instead of one resembling the American middle class would require vast quantities of additional resources. The establishment of an economy that provides for the basics of life—sufficient food, housing, education, transportation, and health care—depends on our capacity to develop alternatives to the high-energy, material-intensive production technologies characteristic of the industrial West. Already, there are outlines of a theory of resource use suited to the needs of a basic goods economy. Some of the main ideas are: (1) the adoption of organic agriculture; (2) the use of renewable sources of energy; (3) the use of appropriate technology, labor-intensive methods of production, and local raw materials; and (4) the decentralization of production in order to minimize the transport of materials and their associated carbon footprints. These productive forces would minimize the disruption of ecosystems and engage the unemployed in useful, productive work. Typically, economies that produce essential goods for human consumption face neither excessive unemployment nor overpopulation. Moreover, secure supplies of basic goods provide a strong motivation for reducing population size, as families no longer require many children to ensure economic prosperity.

TYPES OF RESOURCES AND THEIR LIMITS

All economic development comes about through the use of human resources (e.g., labor power, skills, and intelligence). In order to produce the goods and services people demand in today's global economy, we need to obtain natural resources. What are natural resources and what are their limits?

Resources and Reserves

Natural resources have meaning only in terms of historically specific technical and cultural appraisals of nature and are defined in relation to a particular level of development. **Resources**, designated by the larger box in Figure 1, include all substances of the biological and physical environment that may someday be used under specified technological and socioeconomic conditions. Because these conditions are always subject to change, we can expect our determination of what is useful to also change. For example, petroleum was not considered a resource until the mid-nineteenth century, when the Industrial Revolution led to rising demand for fuels. Uranium, once a waste product of radium mining of the 1930s, is now a valuable ore. Taconite ores became worthwhile in Minnesota only after production from high-grade, nonmagnetic iron ores declined in the 1960s.

At the other end are reserves, designated by the box in the lower left corner of Figure 1. **Reserves** are quantities of resources that are known and available for economic exploitation with current technologies at current prices. When current reserves begin to be depleted, the search for additional reserves is intensified. Estimates of reserves are also affected by changes in prices and technology. **Projected reserves** represent estimates of the quantities likely to be added to reserves because of discoveries and changes in prices and technologies projected to occur within a specified period, for example, 50 years.

Renewable and Nonrenewable Resources

There is a major distinction between nonrenewable and renewable resources. **Nonrenewable resources** consist of finite masses of material, such as fossil fuels and metals, which cannot be used without depletion. They are,

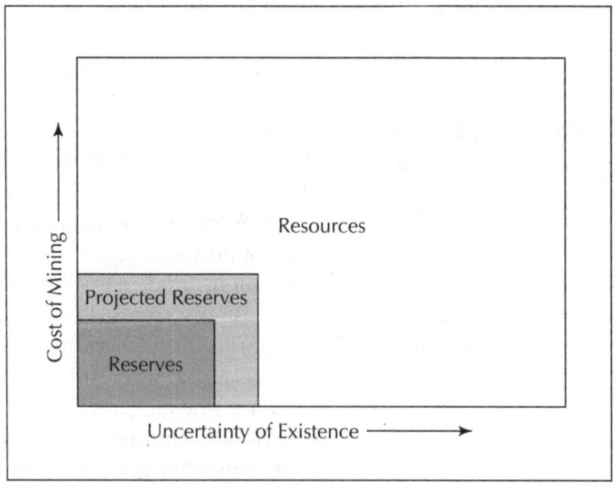

FIGURE 1 Classification of resources. Resources include all materials of the environment that may someday be used under future technological and socioeconomic conditions. Reserves are resources that are known and available with current technologies and at current prices. Projected reserves are reserves based on expected future price trends and technologies available.

for all practical purposes, fixed in amount, or in some cases, such as soils, they form slowly over time. Consequently, their rate of use is important. Large populations with high per capita consumption of goods deplete these resources fastest. Many nonrenewable resources are completely altered or destroyed by use; petroleum is an example. Other resources, such as iron, are available for recycling. **Recycling** expands the limits to the sustainable use of a nonrenewable resource. At present, these limits are low in relation to current mineral extraction.

Renewable resources are those resources capable of yielding output indefinitely without impairing their productivity. They include **flow resources** such as water and sunlight and **stock resources** such as soil, vegetation, fish, and animals. Renewal is not automatic, however; resources can be depleted and permanently reduced by misuse. Productive fishing grounds can be destroyed by overfishing. Fertile topsoil, destroyed by erosion, can be difficult to restore and impossible to replace. The future of agricultural land is guaranteed only when production does not exceed its maximum sustainable yield. The term **maximum sustainable yield** means maximum production consistent with maintaining future productivity of a renewable resource.

In our global environment, the misuse of a resource in one place affects the well-being of people in other places. The misuse of resources is often described in terms of the **tragedy of the commons**, a term coined by biologist Garrett Hardin in 1968. This metaphor refers to the way public resources are ruined by the isolated actions of individuals, which occurs when the costs of actions are not captured in market prices. Originally it referred to the tendency of shepherds to use common grazing land; as each one sought as much of the commons as possible, it became overgrazed (Figure 2). Similarly, people who fish are likely to try to catch as many fish as they can, reasoning that if they don't, others will. Thus, the tragedy of the commons exemplifies a market failure, a problem generated by individual actors who behave "rationally" but collectively create an irrational and self-destructive outcome. Similarly, dumping waste and pollutants in public waters and land or into the air is the cheapest way to dispose of worthless products. Firms are generally unwilling to dispose of these materials by more expensive means unless mandated by law.

Sometimes resources are unavailable, not because they are depleted but because of politics. Resources are under the control of sovereign nation-states. Many wars in the twentieth century have been resource wars. For example, Japan invaded Korea and Taiwan in the 1890s largely in order to obtain arable land and coal. The Iraqi invasion of Kuwait in 1990 and the U.S. invasion of Iraq in 2003 were largely motivated by concerns over the region's oil supplies. In the Middle East, fierce national rivalries make water a potential source of conflict: While some parts are blessed with adequate water supplies, most of the region is insufficiently supplied. Some observers predict that political tension over the use of international rivers, lakes, and aquifers in the Middle East may escalate to war in the next few years.

Food Resources

Thanks to scientific advances in farming, world food production has been increasing faster than population (Figure 3). While there is sufficient food to feed everyone in the world, there are huge geographical variations in people's access to a sufficient number and quality of calories (Figure 4). The populations of the industrialized world are generally well fed; indeed, in the United States, the major dietary problem is an overabundance of calories and an epidemic of obesity. In the developing world, in contrast, hundreds of millions of people worldwide still go hungry daily. With demand for food expected to grow at 4% per year over the next 20 to 30 years, the task of meeting that need will be more difficult than ever before.

The World Bank

FIGURE 2 The Kenyan rangelands on which these herders' cattle graze are in jeopardy. With growing grazing pressures, more than 60% of the world's rangelands and at least 80% of African, Asian, and Middle Eastern rangelands are now moderately to severely desertified. About 65 million hectares of once productive land in African have become desert during the past 50 years.

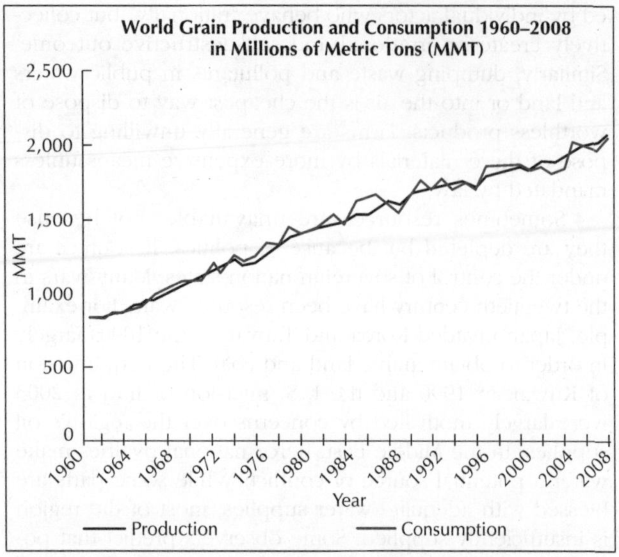

FIGURE 3 World grain production, 1960–2007. While total output has risen steadily, the amount of grain per person has remained relatively constant.

A record explosion in the world's population coupled with the problem of poverty threatens the natural resources on which agriculture depends, such as topsoil. To make matters worse, environmental degradation perpetuates poverty, as degraded ecosystems diminish agricultural returns to poor people.

The gulf between the well fed and the hungry is vast. Average daily calorie consumption is 3300 in developed countries and 2650 in developing countries. But these are average figures. There are people in the developed world who go hungry and some people in Africa with plenty to eat. Averages mask the extremes of **undernutrition**—a lack of calories—and overconsumption. Even with a high calorie satisfaction, people may suffer from **chronic malnutrition**—a lack of enough protein, vitamins, and essential nutrients. The most important measure in assessing nutritional standards is the daily per capita availability of calories, protein, fat, calcium, and other nutrients. In the world today, the sharpest nutritional differences are not from country to country or from one region to another within countries. They are between rich and poor people. The poor of the earth are the hungry, and those with the least political power often suffer in terms of an insufficient food supply.

Hunger among the poor of the world is often attributed to deforestation, soil erosion, water-table depletion, the frequency and severity of droughts, and the impact of storms such as hurricanes. Although the environment does have a bearing on the food problem, it has limited significance compared to the role of social conditions such as war and a world economy whose rules are tilted against the impoverished. Subsidized agricultural exports from the United States, for example, have bankrupted millions of farmers in the developing world, reducing those countries' ability to feed themselves.

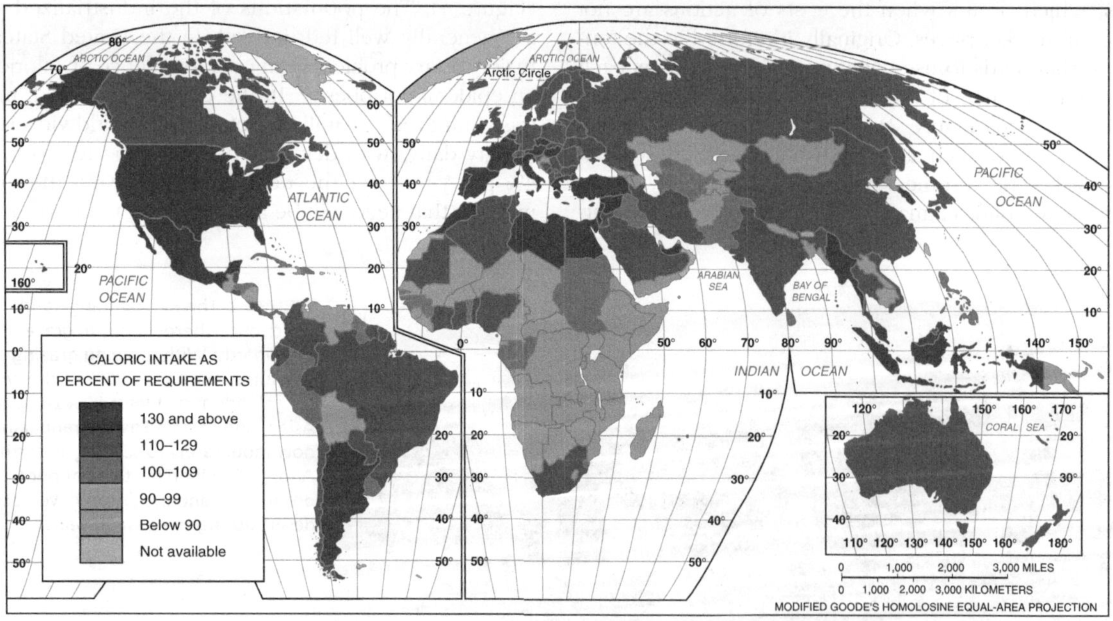

FIGURE 4 Caloric intake as a percentage of adult daily requirements. Highly developed regions of the world receive, on average, 130% of the daily caloric requirements (2400 calories per day) set by the United Nations Food and Agricultural Organization (FAO). Some countries in South America, South Asia, and many countries in Africa receive less than 90% of the daily caloric requirements needed to sustain body and life. Averages must be adjusted according to age, gender, and body size of the person and by regions of the world. Although it appears from this map that the great majority of the world is in relatively good shape with regard to calories per capita, remember that averages don't reveal destitute groups in each country that receive less than their fair share. Again, the situation is most severe in the Sahel, or center belt of Africa.

Population Growth

Population growth is one of many causes of the food problem, and Malthusian views often influence the public's opinion of this issue. However, presently, at the global level, there is no food shortage. In fact, world food production grew steadily from 1961 to 2008. Even over the next several decades, production increases, assuming continuing high investments in agricultural research, are likely to be sufficient to meet effective demand and rising world population. However, some are more pessimistic about future world food production. They argue that food production will be constrained by the limits to the biological productivities of fisheries and rangelands, the fragility of tropical and subtropical environments, massive overfishing of the world's oceans, the increasing scarcity of fresh water, the declining effectiveness of additional fertilizer applications, and social disintegration in many developing countries.

The success of global agriculture has not been shared equally. In Africa, per capita food production has not been able to keep up with population growth. By contrast, Asia, and to a lesser extent Latin America, have experienced tremendous successes in per capita terms. The reasons for this are complex and have to do with the relative equality in patterns of land ownership, government policies toward farmers (e.g., price ceilings on agricultural crops), the respective ability of countries to build infrastructures and extend credit to small farmers, and the role of different states in the world economy.

The food and hunger problem is most severe in sub-Saharan Africa, a region that has long suffered from centuries of colonial misrule, corrupt and uncaring governments, artificial political boundaries that fuel tribal conflicts and secessionist wars, rapid population growth, and lack of foreign investment. Fifteen countries are experiencing exceptional food emergencies. Of the 28 countries with food-security problems, 23 are in sub-Saharan Africa (Figure 5). Indeed, famine, the most extreme expression of poverty, is now mainly restricted to Africa. The fact that famine has been declining for decades in Latin America and Asia suggests that famine can be eliminated. But how? Certainly, bringing an end to Africa's multiple civil wars would go a long way toward eradicating famine. Africa has witnessed countless brutal conflicts that have killed tens of millions of people, most recently in the Congo. Such conflicts divert resources from civilian use, interrupt the production of crops, terrorize populations, destroy the infrastructure, destabilize markets, and complicate the stability of the governments, creating famine and prohibiting the flow of development aid. But peace is not in itself a sufficient condition for removing acute hunger. Appropriate policies and investments are needed to stimulate rural economic growth that underpins food security and to provide safety-net protection for the absolute poor. Rural infrastructure development, credit to farmers, and land redistribution are also necessary steps in this regard (Figure 6). Price controls on food crops create disincentives to produce, and heavily subsidized food imports from the developed world, especially the United States, bankrupt farmers. Often elites in the developing world care more about their foreign bank accounts than the well-being of their own populations.

The pace of urbanization in the developing countries has also contributed to the food problem. In recent decades, hundreds of millions of people who previously lived in rural areas and produced some food have relocated

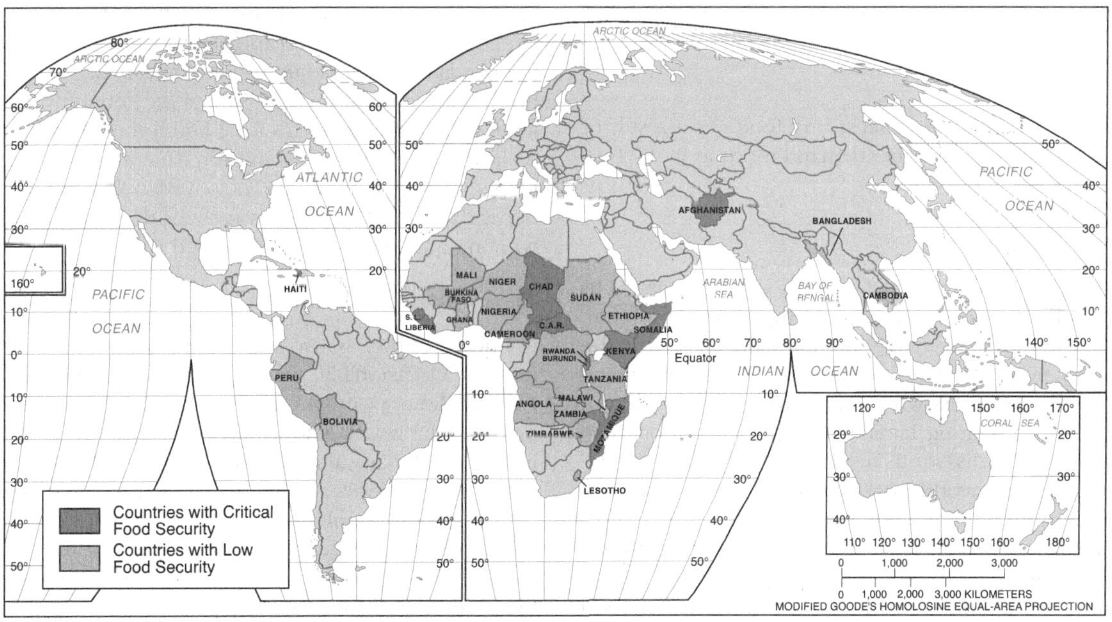

FIGURE 5 Developing countries with low or critical food security indexes. Africa remains the continent most seriously challenged by food shortages. Fifteen countries in the region are facing critical food emergencies. Of the 28 countries with household food security problems, 23 are in sub-Saharan Africa.

FIGURE 6 Third World farmers, such as these in Indonesia, depend on high rice yields. Rice is the staple food for more than one-half the world's population. While rice and other grains supply energy and some protein, people must supplement grains with fruits, nuts, vegetables, dairy products, fish, and meat in order to remain healthy.

to urban areas, where they must buy food. As a result of urbanization, there is a higher demand for food in the face of lower supply.

Poverty

The inequitable allocation of food is directly related to poverty, the single greatest cause of the hunger problem. Hungry people are inevitably poor people who lack the purchasing power to feed themselves. Under capitalism, food goes to customers who can afford it, not to where it is needed most. During famines, the prices of foods rise dramatically, with disastrous results for the poor. From the perspective of the world market, where food is produced is immaterial as long as costs are minimized and a profitable sale can be made. Thus, in the midst of hunger, food may be exported for profit. Since the populations of the developed world can afford to pay much more for food than their counterparts in less developed countries, it is not surprising that the market fails to include the poor. Solving the world food problem is ultimately a matter of alleviating poverty in developing nations. This is no easy task, and while parts of the developing world have made great economic strides over the past 40 years (e.g., East Asia), much of Africa and parts of India and Latin America remain mired in poverty and hunger.

Maldistribution

The problem of world food distribution has three components. First, there is the problem of transporting food from one place to another. Although transport systems in developing countries lack the speed and efficiency of those in developed countries, they are not serious impediments under normal circumstances. The problem arises either when massive quantities of food aid must be moved quickly or when the distribution of food is disrupted by political corruption and military conflict.

Second, serious disruptions in food supply in developing countries are traceable to problems of marketing and storage. Food is sometimes hoarded by merchants until prices rise and then sold for a larger profit. Also, much food in the tropics is lost due to poor storage facilities. Pests such as rats consume considerable quantities, and investments in concrete storage containers can help to minimize this loss.

A third aspect of the distribution problem is in the inequitable allocation of food. Only North America, Australia, and Western Europe have large grain surpluses. But food grain is not always given when it is most needed. Food aid shipments and grain prices are inversely related. Thus, U.S. food aid was low around 1973, a time of major famine in the Sahel region of Africa, because cereal prices were at a peak.

Closely associated with poverty as a cause of hunger in developing countries is the structure of agriculture, including land ownership. Land is frequently concentrated in the hands of a small elite. In Bangladesh, less than 10% of households own more than 50% of the country's cultivable land; 60% of Bangladesh's rural families own less than 2%. A similar situation applies in Latin America. Many rural residents own no land at all. They are landless laborers who depend on extremely low wages for their livelihoods. But without land, there is often no food.

Civil Unrest and War

Political conflict is an important cause of hunger and poverty. Occasionally, governments withhold food to punish rebellious populations. Devastating examples of depriving food to secessionist areas include the government in Nigeria starving the Biafrans in the 1970s and the government in Ethiopia starving the Eritreans into submission, with 6 million people dying in the process. In Sudan, the Arab government's genocide against the African population in Darfur has led to the starvation of millions. In Zimbabwe, the government of Robert Mugabe has systematically denied food to his political opponents in order to quash domestic opposition. Civil wars, which are frequent in developing countries whose political geographies were shaped by colonialism and which have unstable governments, devastate agricultural production. Without a stable political environment, the social mechanisms necessary to produce and distribute food to the hungry cannot operate.

Environmental Decline

As population pressure increases on a given land area, the need for food pushes agricultural use to the limits, and marginal lands, which are subject now to **desertification** (Figure 7) and **deforestation**, are brought into production. Removal of trees allows a desert to advance, because the windbreak is now absent. The cutting of trees also lowers the capacity of the land to absorb moisture, which diminishes agricultural productivity and increases the chances of drought. Desertification and deforestation are symptoms as well as causes of the food problem in developing countries (Figure 8). Natural resources are mined by the poor to meet the food needs of today; the lower productivity resulting from such practices is a concern to be put off until tomorrow.

Government Policy and Debt

In many developing countries, government policies have emphasized investment in their militaries and cities at the expense of increasing agricultural production. In addition, some governments in Africa have provided food at artificially low prices in order to make food affordable in cities. While this practice keeps labor affordable for multinational corporations and placates the middle class, it robs farmers of the incentive to farm. Farmers cannot make a living from artificially low commodity prices.

The average debt of many developing countries runs into the billions. In 2008, aggregate debt of African countries stood at $260 billion. Simply put, African countries have no surplus capital to invest in their infrastructure or food production systems. Instead, they have to enforce austerity, reducing levels of government services in support of economic growth, particularly agricultural growth. Debt repayments subsume a large share of foreign revenues, decreasing funds available for investment.

In recent decades, agriculture in developing countries has expanded. This expansion is in the export sector, not in the domestic food-producing sector, and it is often the result of deliberate policy. Governments and private elites have opted for modernization through the promotion of export-oriented agriculture. The result is the growth of an agricultural economy based on profitable export products and the neglect of those aspects of farming that have to do with small farmers producing food for local populations.

Imports from the developed world, particularly the United States, also exacerbate food problems. For example, after the passage of the North American Free Trade Agreement (NAFTA) in 1994, massive U.S. exports of government-subsidized corn caused the price of corn in Mexico to fall by 70%, bankrupting 2 million Mexican farmers. All over the world, farmers protest subsidized U.S.

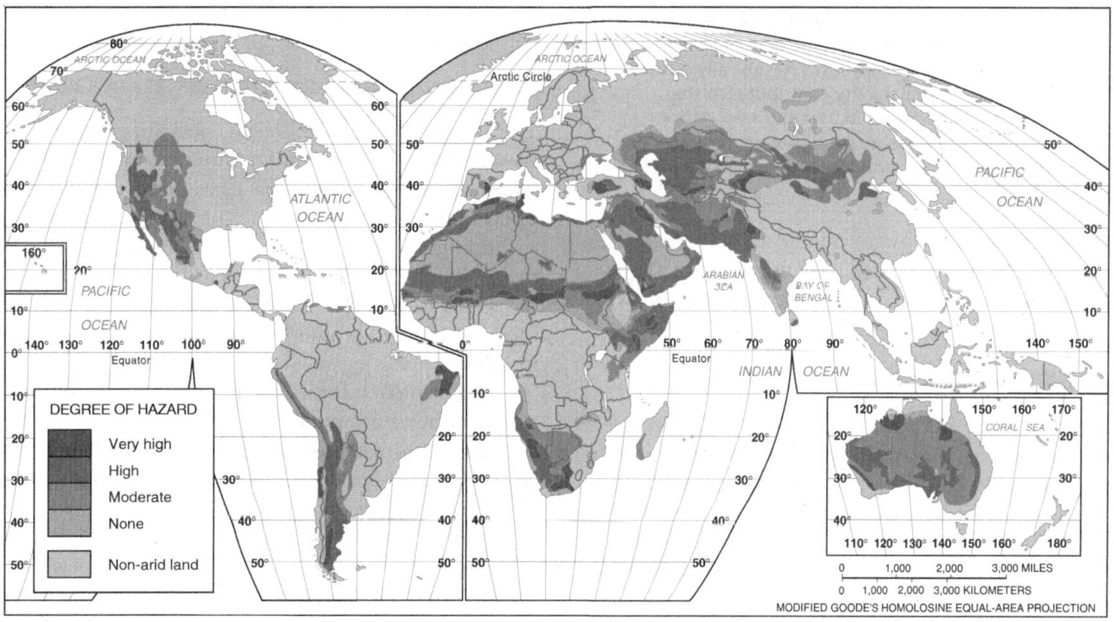

FIGURE 7 World desertification. The main problem is overuse by farmers and herders. Approximately 10% of the earth's surface has lost its topsoil due to overuse of lands by humans, creating desertification. An additional 25% of the earth's surface is now threatened. Topsoil is being lost at a rate of approximately 30 billion metric tons per year. Approximately 20 million acres of agricultural land are lost every year to desertification by agricultural overuse. When plants are uprooted by overplowing or by animals, the plants that stabilized shifting soil are removed. When the rains come, water erosion can wash away the remaining topsoil.

UN/DPI PHOTO/Ruth Massey

FIGURE 8 Water for Chad. Water is an important ingredient to sustain human life. Fifty percent of the world's people do not have adequate, clean water. Villagers in Chad are delighted as the water pours out of a new water system they have worked together to construct. The system is part of an antidesertification project funded by the United Nations Development Program and the U.S. government. Acute water shortage in many parts of the world requires solutions that will be costly, technically difficult, and politically sensitive. Water scarcity contributes to the impoverishment of many countries in east and west Africa, threatening their ability to increase food production fast enough to keep pace with modern population growth.

grain exports (produced by a country that celebrates the "free market") for undermining local food-producing systems. The issue has also become a major obstacle in world trade negotiations.

INCREASING FOOD PRODUCTION

Yield increases will be the major source of future food production growth. These can be achieved through the expansion of arable land and increased crop intensity. The result of these methods of increasing food supply would be to put additional pressures on land and water resources and contribute significantly to human-made sources of greenhouse gases.

Expanding Cultivated Areas

The world's potentially farmable land is estimated to be about twice the present cultivated area. Vast reserves are theoretically available in Africa, South America, and Australia, and smaller reserves in North America, Russia, and Central Asia. However, many experts believe that the potential for expanding cropland is disappearing in most regions because of environmental degradation and the high cost of developing infrastructure in remote areas. About half of the world's potentially arable land lies within the tropics, especially in sub-Saharan Africa and Latin America. Much of this land is under forest in protected areas, and most of it suffers from soil and terrain constraints as well as excessive dryness. In Asia, two-thirds of the potentially arable land is already under cultivation; the main exceptions are Indonesia and Myanmar. South Asia's agricultural land is almost totally developed.

The expansion of tropical agriculture into forest and desert environments contributes to deforestation and desertification. Since World War II, half of the world's rain forests in Africa, Asia, and Latin America have disappeared. Conversion of this land to agriculture has entailed high costs, including the loss of livelihoods for the people displaced, the loss of biodiversity, increased carbon dioxide emissions, and decreased carbon storage capacity. Desertification—the growth of deserts due to humanly caused factors, typically on the periphery of natural deserts—threatens about one-third of the world's land surface and the livelihood of nearly a billion people. Many of the world's major rangelands are at risk. The main factor responsible for desertification is overgrazing, but deforestation (particularly the cutting of fuel wood), overcultivation of marginal soils, and salinization caused by poorly managed irrigation systems are also important influences. Deforestation and desertification are destroying the land resources on which the development of the developing countries depends.

Raising the Productivity of Existing Cropland

The quickest way to increase food supply is to raise the productivity of land under cultivation. Remarkable increases in agricultural yields have been achieved in developed countries through the widespread adoption of new technologies. Corn yields in the United States are a good example. Yields expanded rapidly with the introduction of hybrid varieties, herbicides, and fertilizers. Much of the increase in yields came through successive improvements in hybrids.

The approach for increasing yields in developed countries has been adopted in developing countries. One approach is known as the **Green Revolution**, started by Nobel-laureate American agronomist Norman Borlaug in the 1960s, in which new high-yielding varieties of wheat, rice, and corn are developed through plant genetics, including crops that grow more quickly, perhaps yielding several harvests per year, are more pest and drought resistant, and have higher protein content. The Green Revolution has had enormous impacts in Asia and Mexico, increasing the food supply, but it is not a panacea. It depends on machinery, for which the poor lack sufficient capital to buy. It depends on new seeds, which poor farmers cannot afford. It depends on chemical fertilizers, pesticides, and herbicides, which have contaminated underground water supplies as well as streams and lakes. It depends on large-scale, one-crop farming, which is ecologically unstable because of its susceptibility to pestilence. It depends on controlled water supplies, which have increased the incidence of malaria, cholera, schistosomiasis, and other diseases. It is confined largely to a group of 18 heavily populated countries, extending across the tropics and subtropics from South Korea to Mexico (Figure 9). It is also benefiting countries that include half of the world's population. This approach involves the widespread application of artificial fertilizers, an increasingly common practice throughout the developing world (Figure 10).

Politically, the Green Revolution promises more than it can deliver. Its sociopolitical application has been largely unsatisfactory. Even in areas where the Green Revolution has been technologically successful, it has not always benefited large numbers of hungry people without the means to buy the newly produced food. It has benefited mainly Western-educated farmers, who were already wealthy enough to adopt a complex integrated package of technical inputs and management practices. Farmers make bigger profits from the Green Revolution when they purchase additional land and mechanize their operations. Some effects of labor-displacing machinery and the purchase of additional land by rich farmers include agricultural unemployment, increased landlessness, rural-to-urban migration, and increased malnutrition for the unemployed who are unable to purchase the food produced by the Green Revolution.

The Green Revolution generated substantial increases in agricultural output worldwide. However, world hunger remains a serious problem, indicating that the problem is not so much one of food production, but of food demand in the economic sense (i.e., purchasing power). Unfortunately, the Green Revolution does nothing to increase the ability of the poor to buy food. Hunger is a complex and intractable problem in large part because it is so closely tied to questions of poverty and economic development, not simply increasing agricultural productivity.

The Green Revolution has helped to create a world of more and larger commercial farms alongside fewer and smaller peasant plots. However, given a different structure of land holdings and the use of appropriately intermediate technology, the Green Revolution could help developing countries on the road toward agricultural self-sufficiency and the elimination of hunger. *Intermediate technology* is a term that means low-cost, small-scale technologies intermediate between primitive stick-farming methods and complex agroindustrial technical packages.

Creating New Food Sources

Expanding cultivated areas and raising the productivity of existing cropland are two methods of increasing food supply. A third method is the identification of new food sources. There are three main ways to create new food sources: (1) cultivating the oceans, or mariculture; (2) developing high-protein cereal crops; and (3) increasing the acceptability and palatability of inefficiently used present foods.

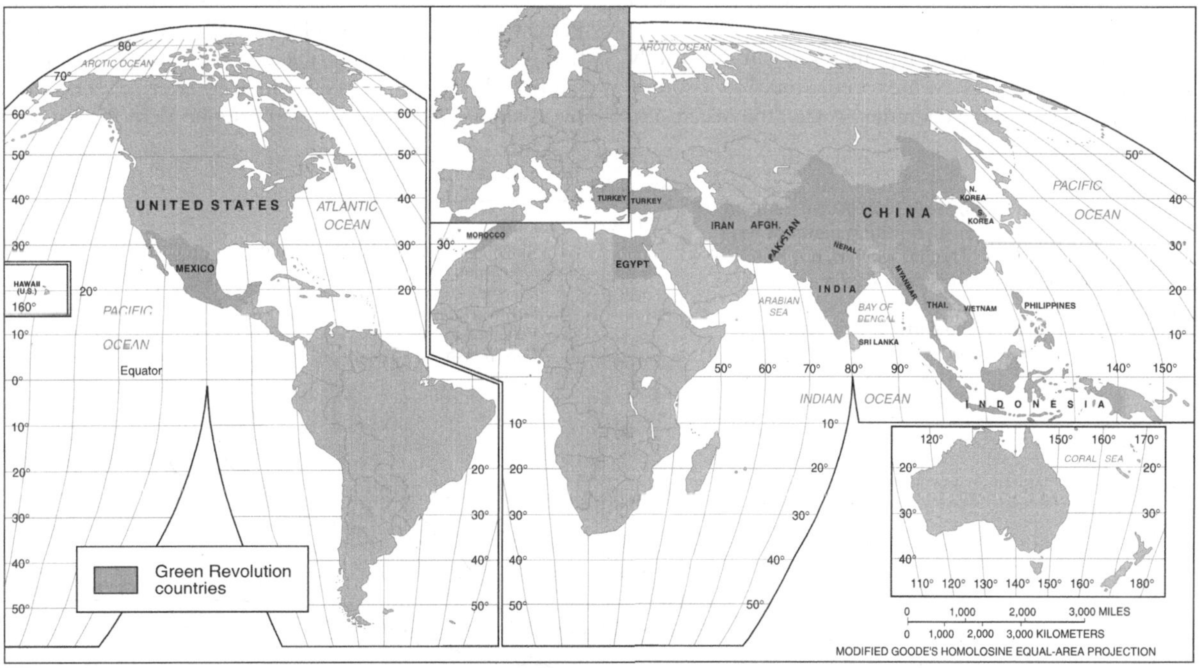

FIGURE 9 The chief countries of the Green Revolution. The Green Revolution was the result of plant scientists genetically developing high-yielding varieties of staple food crops such as rice in East Asia, wheat in the Middle East, and corn in Middle America. By crossing "super strains" that produced high yields with more genetically diverse plants, both high yield and pest resistance were introduced.

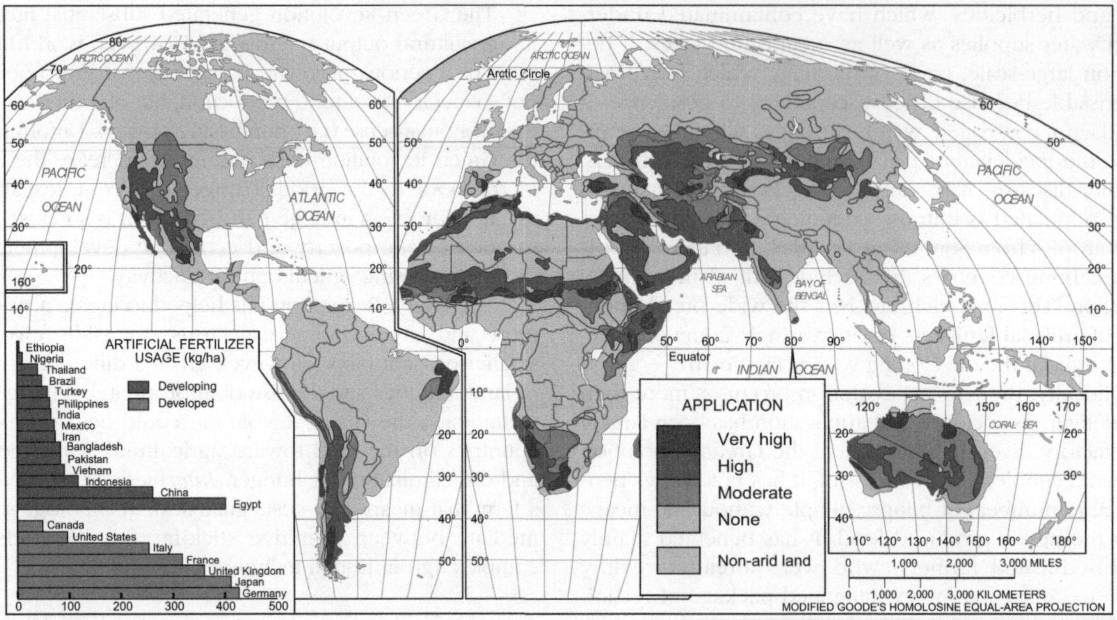

FIGURE 10 Artificial fertilizer usage. The application of artificial fertilizers, as opposed to natural ones obtained from people and animals, may enhance agricultural productivity but also makes economies more dependent on petroleum.

Cultivating the Oceans

Fishing and the cultivation of fish and shellfish **(aquaculture)** from the oceans is not a new idea. At first glance, the world seems well supplied with fisheries because oceans cover three-fourths of the earth. However, fish provide a very small proportion—about 1%—of the world's food supply.

World fish consumption has increased more rapidly than the population, and even exceeded beef as a source of animal protein in some countries. Today, the oceans are in dire threat of being heavily overfished, with catastrophic implications for marine ecosystems as well as the future world food supply. Over the past 30 years, the total tonnage of fish caught by commercial fishing fleets has leveled off and declined as a result of overfishing (Figure 11). Overfishing has been particularly acute in the North Atlantic and Pacific oceans. Countries such as Iceland and Peru, whose economics rely heavily on fishing, are sensitive to the overfishing problem. Peru's catch of its principal fish, the anchovy, has declined by over 75% because of overfishing. The Peruvian experience demonstrates that

FIGURE 11 Global fish catch, 1950–2008. Rising demand and increasingly efficient industrial fishing methods have not only yielded dramatically higher catches, but have increasingly depleted the world's oceans of many species. Pelagic fish comprise the world's largest fishery and are found in the vast oceans of the earth at all levels. See Richard Ellis, *The Empty Ocean.*

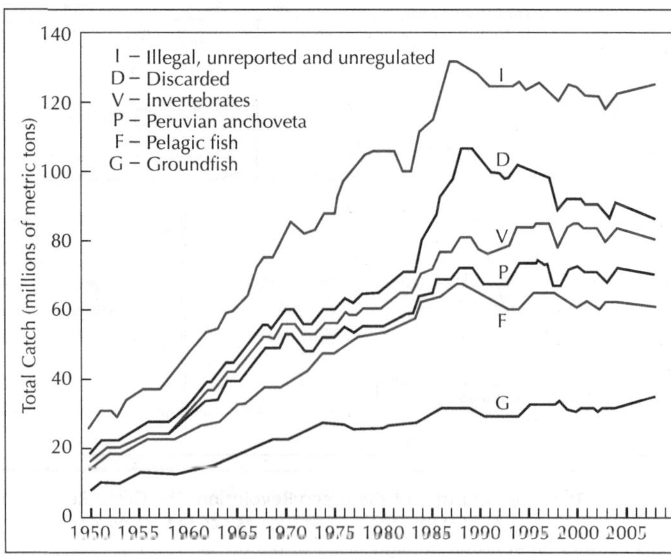

the ocean is not a limitless fish resource, as did the quest for whales a century earlier. Indeed, if current fish consumption levels continue, the ocean's ecosystems are likely to experience severe stress, or collapse, with dire consequences not only for the world environment but for the human food supply as well.

An alternative to commercial fishing fleets, which employ sophisticated techniques but catch only what nature has provided, is fish farming. **Mariculture** is now expanding rapidly and accounts for 11% of the world's fish caught yearly. The cultivation of food fish such as catfish, trout, and salmon is big business in the United States, Norway, Japan, and other fishing countries.

High-Protein Cereals

Another source of future food production rests in higher-protein cereal crops. Agricultural scientists seek to develop high-yield, high-protein cereal crops in the hope that development of hybrid seeds will be able to help the protein deficiency of people in developing countries who do not have available meats from which to gain their protein needs, as do people in developed countries.

Fortification of present rice, wheat, barley, and other cereals with minerals, vitamins, and protein-carrying amino acids is an approach that also deserves attention. This approach is based on the fortified food production in developed countries and stands a greater chance of cultural acceptance because individual food habits do not necessarily need to be altered. But developing countries rely on unprocessed, unfortified foods for 95% of their food intake. Large-scale fortification and processing would require major technological innovation and scale economies to produce enough food to have an impact on impoverished societies.

More Efficient Use of Foods

In many developing countries, foods that satisfy consumer preferences as well as religious taboos and cultural values are becoming limited. The selection of foods based on social customs should be supplemented with information concerning more efficient use of foods presently available. An effort should be made to increase the palatability of existing foods that are plentiful.

Fish meal is a good example. Presently, one-third of the world's fish intake is turned into fodder for animals and fertilizer. Fish meal is rich in omega-3 fatty acids and amino acids necessary for biological development. However, in many places, the fish meal is not used because of its taste and texture.

Another underused food resource is the soybean, a legume rich in both protein and amino acids. Most of the world's soybeans wind up being processed into animal feed or fertilizer and into nondigestible industrial materials. World demand for tofu and other recognizable soybean derivatives is not large. By contrast, hamburgers, hot dogs, soft drinks, and cooking oils made partially from soybeans are more acceptable.

A Solution to the World Food Supply Situation

As we have emphasized, there is a widely shared belief that people are hungry because of insufficient food production. But the fact is that food production is increasing faster than population, and still there are more hungry people than ever before. Why should this be so? It could be that the production focus is correct, but soaring numbers of people simply overrun these production gains. Or it could be that the diagnosis is incorrect—scarcity is not the cause of hunger, and production increase, no matter how great, can never solve the problem.

The simple facts of world grain production make it clear that the overpopulation/scarcity diagnosis is incorrect. Present world grain production can more than adequately feed every person on earth. Ironically, the focus on increased production has compounded the problem of hunger by transforming agricultural progress into a narrow technical pursuit instead of the sweeping social task of releasing vast, untapped human resources. We need to look to the policies of governments in developing countries to understand why people are hungry even when there is enough food to feed everyone. These policies influence the access to knowledge and the availability of credit to small farmers, the profitability of growing enough to sell a surplus, and the efficiency of marketing and distributing food on a broad scale.

The fact is that small, carefully farmed plots are more productive per unit area than large estates because the families that tend to them have more at stake and invest as much labor as necessary to feed themselves when they can. Yet, despite considerable evidence from around the world, government production programs in many developing countries ignore small farmers. They rationalize that working with bigger production units is a faster road to increased productivity. In the closing years of the twentieth century, many agricultural researchers, having gained respect for traditional farming systems, agree with this conclusion.

NONRENEWABLE MINERAL RESOURCES

Although we can increase world food output, we cannot increase the global supply of minerals. A mineral deposit, once used, is gone forever. A **mineral** refers to a naturally occurring inorganic substance in the earth's crust. Thus, silicon is a mineral, whereas petroleum is not, since the latter is organic in nature. Although minerals abound in nature, many of them are insufficiently concentrated to be economically recoverable. Moreover, the richest deposits are unevenly distributed and are being depleted.

Except for iron, nonmetallic elements are consumed at much greater rates than metallic ones. Industrial societies do not worry about the supply of most nonmetallic minerals, which are plentiful and often widespread, including nitrogen, phosphorus, potash, or sulfur for chemical fertilizer, or sand, gravel, or clay for building purposes. Those commodities the industrial and industrializing countries do worry about are the metals.

Location and Projected Reserves of Key Minerals

Only five countries—Australia, Canada, South Africa, the United States, and Russia—are significant producers of at least six **strategic minerals** vital to defense and modern technology (Figure 12). Of the major mineral-producing countries, only a few—notably the United States and Russia—are also major processors and consumers. The other major processing and consuming centers—Japan and western European countries—are deficient in strategic minerals.

How large is the world supply of strategic minerals? Most key minerals will be exhausted within 100 years and some will be depleted within a few years at current rates of consumption, assuming no new reserves. The United States is running short of domestic sources of strategic minerals. Its dependence on imports has grown steadily since 1950; prior to that year, the country was dependent on imports for only four designated strategic minerals. When measured in terms of percentage imported, U.S. dependency increased from 50% in 1960 to over 82% in 2008. Minerals projected to be future needs by the United States are unevenly distributed around the world. Many of them, such as manganese, nickel, bauxite, copper, and tin (see Figure 12), are concentrated in Russia and Canada and in developing countries. Whether these critical substances will be available for U.S. consumption may depend less on economic scarcity and more on international tensions and foreign policy objectives.

Solutions to the Mineral Supply Problem

Affluent countries are unlikely to be easily defeated by mineral supply problems. Human beings, the ultimate resource, have developed solutions to the problem in the past. Will they in the future? Although past experience is never a reliable guide to the future, there is no need to be unduly pessimistic about the exhaustion of minerals as long as we develop alternatives.

If abundant supplies of cheap electricity ever became available, it might become possible to extract and process minerals from unorthodox sources such as the ocean. The oceans, which cover 71% of the earth, contain large quantities of dissolved minerals. Salt, magnesium, sulfur, calcium, and potassium are the most abundant of these minerals and amount to over 99% of the dissolved minerals. More valuable marine minerals also include copper, zinc, tin, and silver. Some minerals such as bromine and magnesium are being obtained electrolytically from the oceans at the present time. Finally, improved efficiency of production has reduced the demand for various minerals per unit of output (Figure 13).

Much more feasible than mining the oceans is devoting increased attention to improving mining technology, especially to reducing waste in the extraction and processing of minerals. Equally feasible is to utilize technologies that allow minerals to be used more efficiently in manufacturing. Also, if social attitudes were to change, encouraging lower per capita levels of resource use, more durable products could be manufactured, saving not only large amounts of energy but large quantities of minerals too.

Reusing minerals is still another option for our mineral problems. Every year in the United States and other

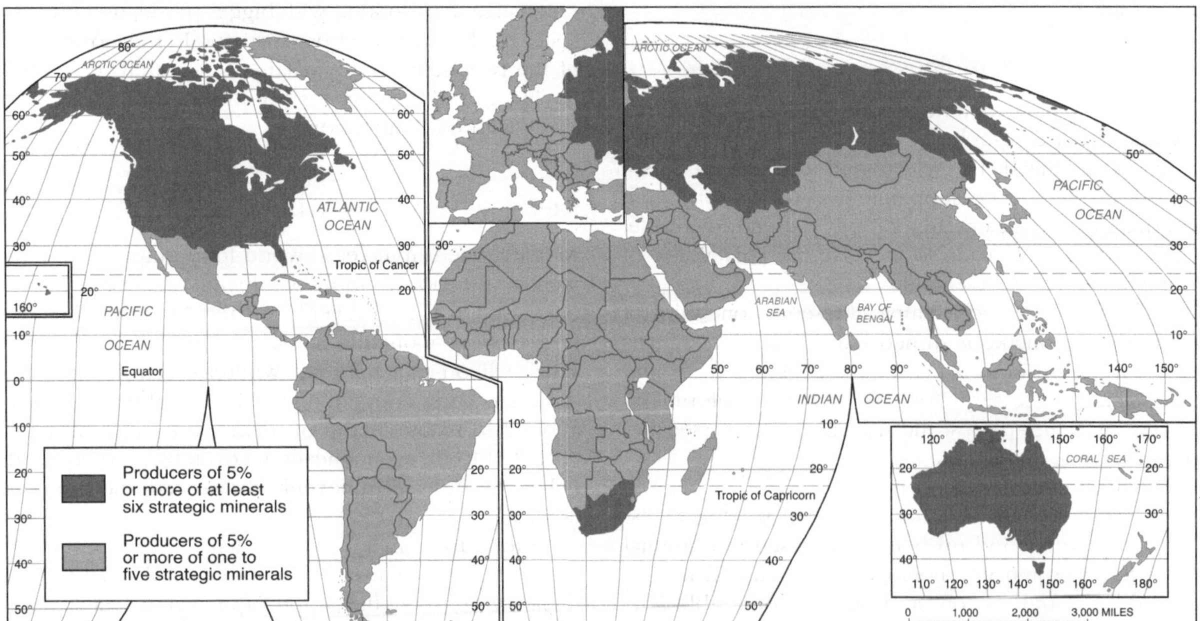

FIGURE 12 Major producers of strategic minerals.

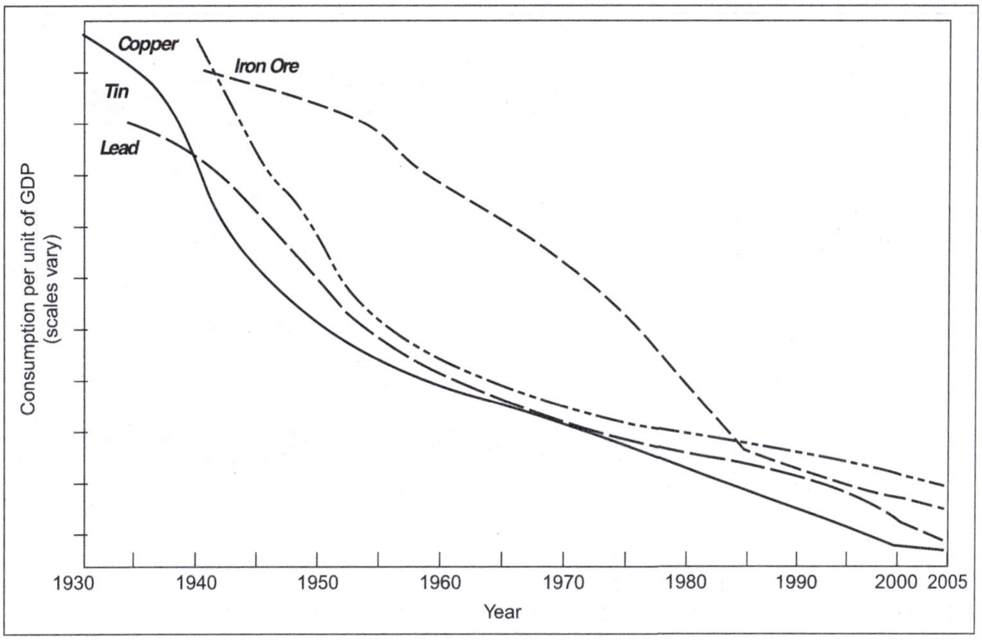

FIGURE 13 The consumption of lead, tin, copper, and iron ore per unit of GDP for the United States, 1930–2005. Transmaterialization is the process whereby natural materials from the environment are systematically replaced by higher-quality or technologically more advanced materials linked to new industries—glass fibers, composites, ceramics, epoxies, and smart metals.

affluent countries, huge quantities of household and industrial waste are disposed of at sanitary landfills and open dumps. These materials are sometimes called "urban ores" because they can be recovered and used again. For years, developed countries have been recycling scarce and valuable metals such as iron, lead, copper, silver, gold, and platinum, but large amounts of scrap metals are still being wasted. Although we could recover a much greater proportion of scrap, this is unlikely when prices are low or when virgin materials are cheaper than recycled ones.

Environmental Impacts of Mineral Extraction

Mineral extraction has a varied impact on the environment, depending on mining procedures, local hydrological conditions, and the size of the operation. Environmental impact also depends on the stage of development of the mineral—exploration activities usually have less of an impact than mining and processing mineral resources.

Minimizing the environmental impacts of mineral extraction is in everyone's best interest, but the task is difficult because demand for minerals continues to grow and ever-poorer grades of ore are mined. For example, in 1900 the average grade of copper ore mined was 4% copper; by 2000, ores containing as little as 0.4% copper were mined. Each year more and more rock has to be excavated, crushed, and processed to extract copper. The immense copper mining pits in Montana, Utah, and Arizona are no longer in use because foreign sources, mostly in the developing countries, are less expensive. As long as the demand for minerals increases, ever-lower quality minerals will have to be used and, even with good engineering, environmental degradation will extend far beyond excavation and surface plant areas.

ENERGY

The development of energy sources is crucial for economic development. Today, commercial energy is the lifeblood of modern economies. Indeed, it is the single biggest item in international trade. Oil alone accounts for about one-quarter of the volume (but not value) of world trade. The U.S. economy consumes vast amounts of energy, overwhelmingly consisting of **fossil fuels** (Figure 14). With roughly 5% of the world's people, the United States consumes 25% of its fossil fuels. These form the inputs that, along with labor and capital, run the economic machine that feeds, houses, and moves the population. As Figure 15 indicates, the primary uses of petroleum are transportation and industrial purposes, whereas the major uses of coal are for electrical power generation.

Until the energy shocks of the 1970s, commercial energy demands were widely thought to be unproblematic, that is, always there to generate rising affluence. Suddenly, higher prices brought energy demands in the industrial countries to a virtual standstill, generating inflation, unemployment, and accelerating deindustrialization (Figure 16). Thousands of factories were shut down, and more than 3 million workers were laid off. They learned firsthand that when energy fails, everything fails in an urban-industrial economy. During the 1980s and 1990s, oil prices decreased from $30 per barrel in 1981 to $14 per barrel in 1999. **OPEC (Organization of the Petroleum Exporting Countries)**, once considered an invincible cartel, saw its share of world oil output drop steadily as non-OPEC countries expanded production. Oil-consuming countries, including developing countries strapped by heavy energy debts, were relieved to see prices falling. Oil-exporting developing countries, such as Mexico, Venezuela, and Nigeria, which came to depend on oil revenues for an important source of income, were hurt the worst. By 2008,

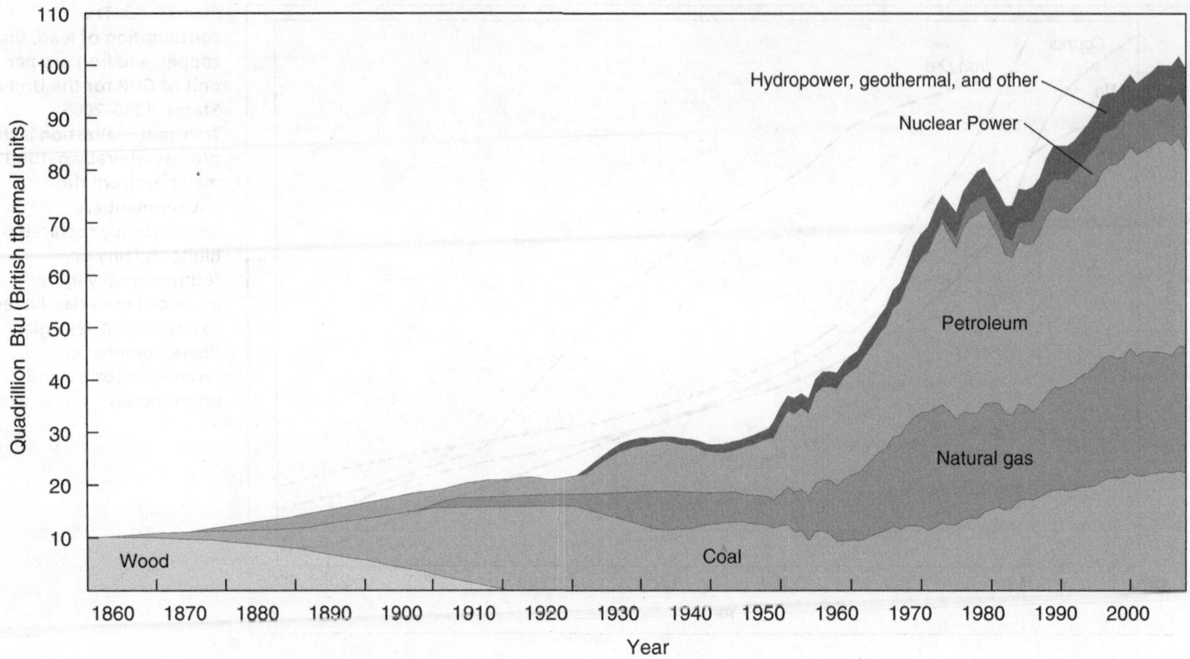

FIGURE 14 U.S. energy consumption, 1850–2008. The U.S. economy contains 5% of the world's people but uses one-third of its energy. The three principal sources of fossil fuels are coal, natural gas, and petroleum. After World War II, petroleum and natural gas surpassed coal as the chief source of energy in the United States. Hydro and nuclear have also increased recently.

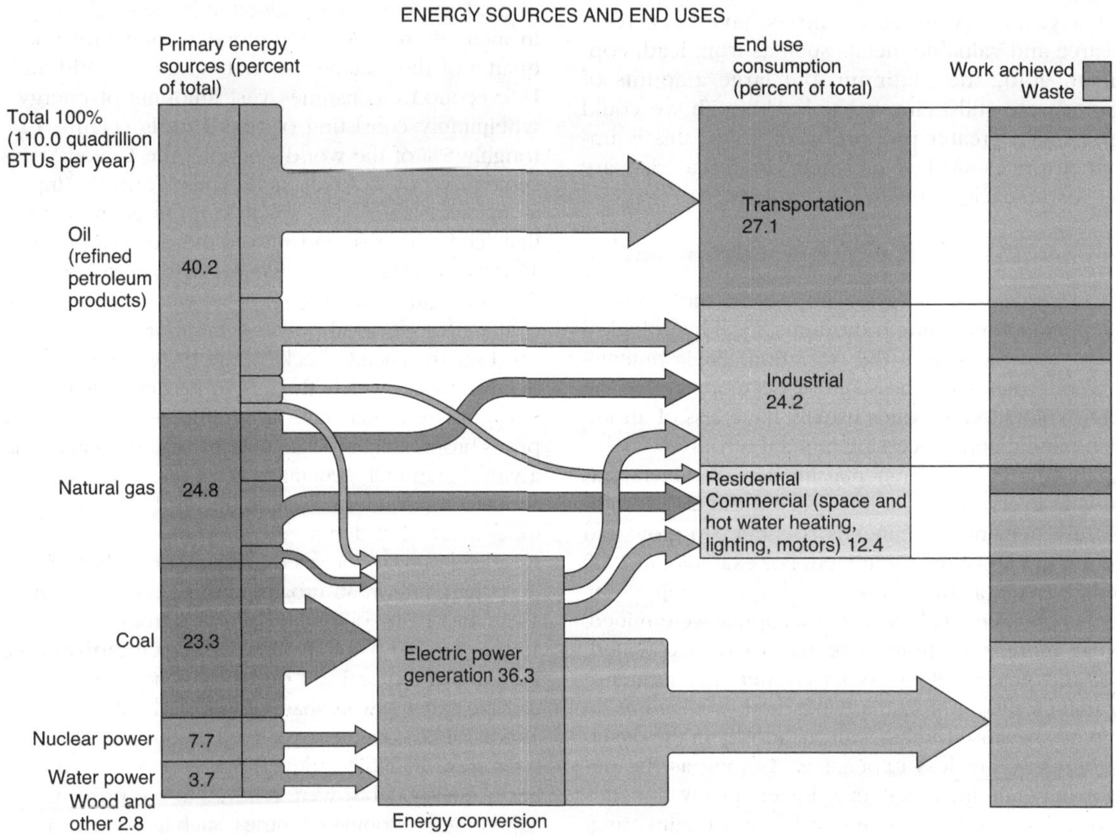

FIGURE 15 U.S. energy sources and end uses. Different energy inputs are applied to different uses. While coal is still widely used for electrical generation, petroleum is the most common energy source for transportation and industrial production.

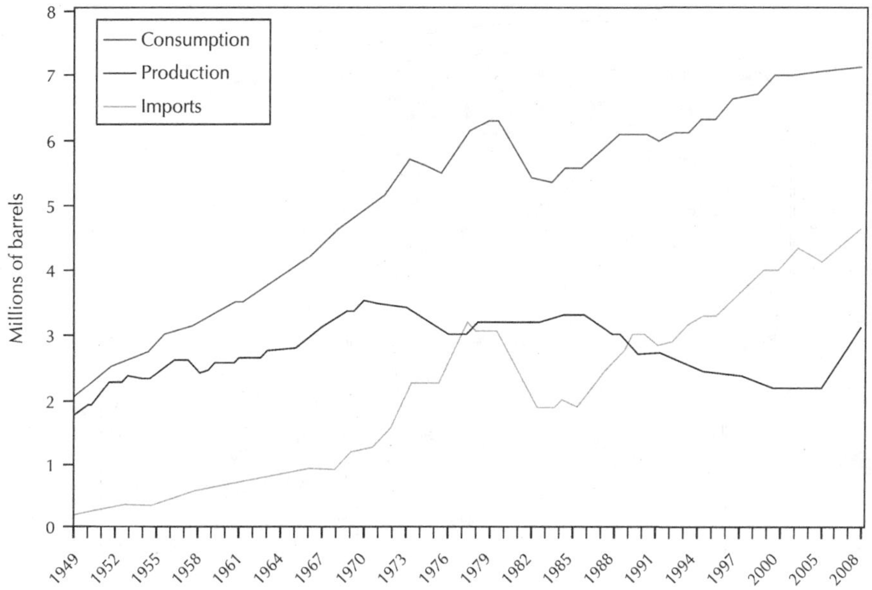

FIGURE 16 Oil production, consumption, and imports in the United States, 1950–2008. In the late 1970s and early 1980s, high prices created increased production and lower consumption; also, the Alaskan oil fields came into production. From 1980 onward, oil prices declined sharply due to the decline in OPEC's oligopolistic power. Imports comprise over one-half of all oil consumed in the United States. By 2010, total petroleum consumption had reached 19.5 million barrels per day (mbpd) in America, with 57% of its crude oil being imported. The top oil-importing country was Canada, importing 2.5 mbpd to the United States. Yet, the combined OPEC countries imported over 6 mbpd into the United States. Texas was the leading oil-producing state in the United States, with Prudhoe Bay, Alaska, the top oil field. Saudi Arabia is the leading oil-producing country in the world, with 11 mbpd, while the United States is the leading oil consumption country.

however, world oil prices had risen to $140 per barrel, only to fall again in 2009 and 2010 as the world's financial crisis and recession reduced global demand. Such oscillations point to the cyclical nature of the petroleum industry, which has huge impacts on other sectors of the world's economy.

Energy Production and Consumption

Most commercial energy produced is from nonrenewable resources. Most renewable energy sources, particularly wood and charcoal, are used directly by producers, mainly poor people in the developing countries. Although there is increasing interest in renewable energy development, commercial energy is the core of energy use at the present time. Only a handful of countries produce much more commercial energy than they consume. If we take petroleum consumption and production as an example, the main energy surplus countries include: Saudi Arabia, Iraq, Mexico, Iran, Venezuela, Indonesia, Algeria, Kuwait, Libya, Qatar, Nigeria, and the United Arab Emirates. Saudi Arabia is by far the largest exporter of petroleum and has the largest proven reserves. Nearly one-half of all African countries are energy paupers. Several of the world's leading industrial powers—most notably Japan, many western European countries, and the United States—consume much more energy than they produce, making them heavily reliant on imported oil, largely from the Middle East. This fact profoundly shapes the foreign policies of countries such as the United States.

The United States leads the world in total energy use, but leaders in per capita terms also include Canada, Norway, Sweden, Japan, Australia, and New Zealand (Figure 17). With 5% of the world's population, the United States consumes roughly one-quarter of the world's energy, largely for transportation, which consumes 40% of American energy inputs. The automobile, for all the convenience it offers, is a highly energy-inefficient way to move people. In contrast, developing countries consume about 30% of the world's energy but contain about 80% of the population. Thus there exists a striking relationship between per capita energy consumption and level of development. Most developing countries consume meager portions of energy, well below levels required with even moderate levels of economic development. Commercial energy consumption in developed countries has been at consistently high levels, whereas in developing countries it has been at low but increasing levels.

Oil Dependency

Much of the world was seriously affected by the 1973 and 1979 Arab oil embargoes. Imported oil to the United States as a proportion of total demand increased from 11% in the late 1960s to 50% in the 1970s to about 58% today. As a result, numerous presidential administrations of the United States have repeatedly called for a national policy of oil self-sufficiency to reduce U.S. dependency on foreign supplies of petroleum—without much success. Under heavy political pressure from corporations and campaign donors, air and water pollution regulations have been relaxed, and tax credits for home energy conservation expenditures were ended. The U.S. Congress toyed at times with imposing stricter fuel standards on new cars, but relaxed these under pressure from automobile producers; indeed, American fuel standards are considerably below those in Europe, Japan, and China. These conflicting policies worked against federal efforts to encourage American households and companies to conserve fossil fuels. The United States imports about 58% of the oil it consumes, but only a small proportion comes from the Middle East. Japan, Italy, and France are comparatively much more dependent on Persian Gulf oil.

Nonetheless, U.S. industry did become more energy-efficient. Manufacturing reduced its share of total U.S.

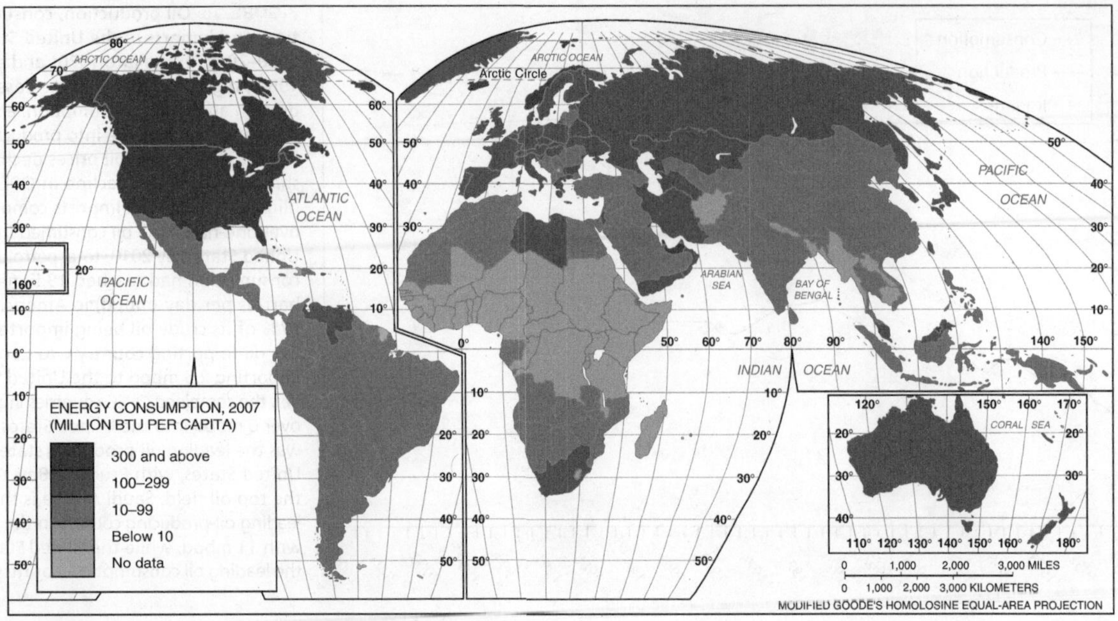

FIGURE 17 World per capita energy consumption. The United States, Canada, and the Scandinavian countries consume more energy per capita than any other countries. When the electricity usage of the United States, Canada, Europe, and Russia is combined, 75% of electricity usage in the world is accounted for, but only 20% of the people.

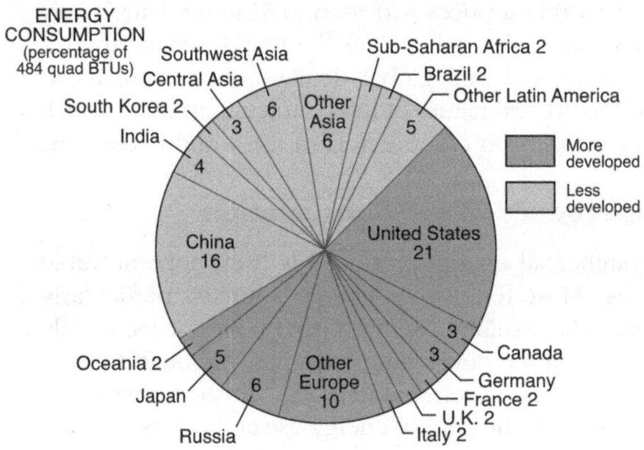

energy consumption from 40% to 36%, and the burgeoning service economy consumed relatively little energy. In terms of conservation efforts, however, the United States lags behind Japan and Europe, where energy is more expensive. Gasoline taxes in Europe, for example, help to fund more energy-efficient public transportation.

Production of Fossil Fuels

As an accident of geology, the world's fossil fuels are highly unevenly distributed around the globe. Two-thirds of the world's oil resources are located in the Middle East (Figures 18 and 19). Other large reserves are found in northern Africa, Latin America—primarily Mexico and Venezuela—and in Russia and Nigeria (Figure 20). Offshore drilling, such as in the North Sea (Figure 21), forms another supply. Natural gas, often a substitute for oil, is also unevenly distributed, with nearly 40% in Russia and central Asia and 34% in the Middle East.

The unevenness of the world's supply and demand for petroleum creates a distinct pattern of trade flows of petroleum (Figure 22), the most heavily traded commodity (by volume) in the world. Primarily, these flows represent exports from the vast reserves of the Middle East to Europe, East Asia, and North America, although the United States also imports considerable quantities from South America and Nigeria. The differences between crude oil production and consumption for each major world region are sketched in more detail in Figure 23.

Adequacy of Fossil Fuels

In the next few decades, energy consumption is expected to rise significantly, especially because of the growing industrialization of developing countries. Most of the future energy production to meet increasing demand will come from fossil energy resources—oil, natural gas, and coal. How long can fossil fuel reserves last, given our increasing energy requirements? Estimates of energy reserves have increased substantially in the past 20 years, and therefore there is little short term concern over supplies; consequently, energy prices are relatively low. If energy consumption

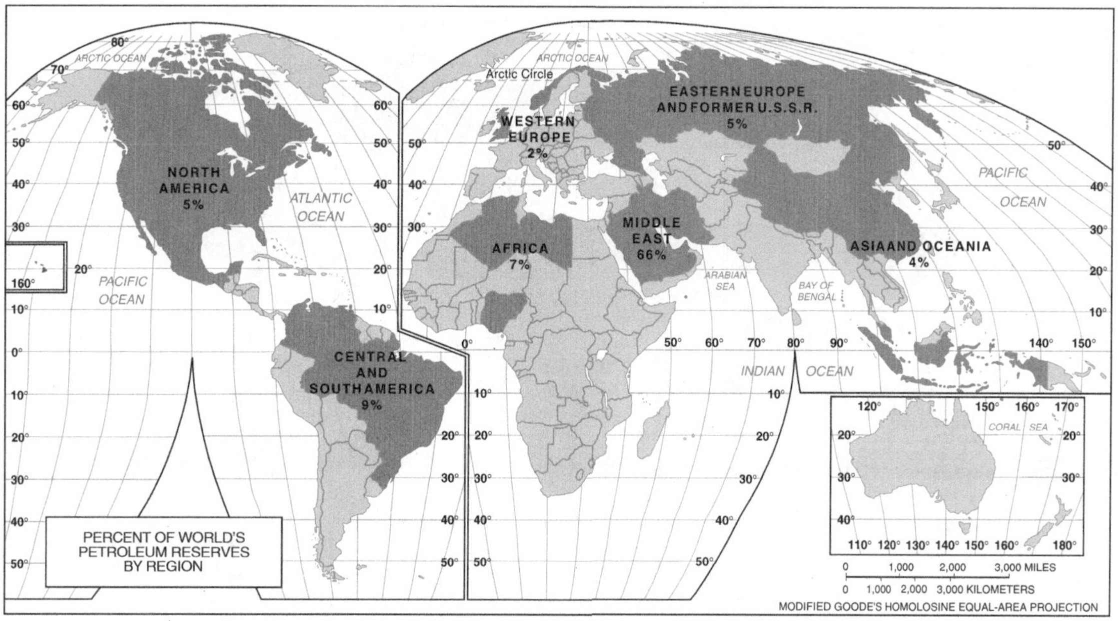

FIGURE 18 World's proven petroleum reserves, revealing the primacy of the Middle East.

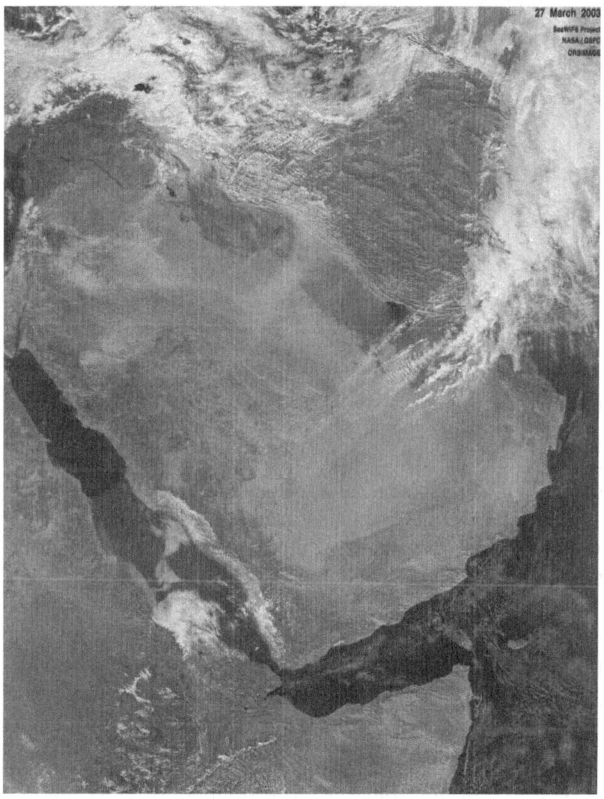

Image courtesy the SeaWiFS Project, NASA/Goddard Space Flight Center, and ORBIMAGE

FIGURE 19 The Persian Gulf (or, to the Arabs, the Arab Gulf), lies in the heart of the world's largest petroleum deposits, an immense ocean of oil that fuels the global economy.

were to remain more or less at current levels, which is unlikely, proved reserves would supply world petroleum needs for 40 years, natural gas needs for 60 years, and coal needs for at least 300 years. Although the size of the world's total fossil fuel resources is unknown, they are finite, and production will eventually peak and then decline.

Oil: Black Gold

Most of the world's petroleum reserves are heavily concentrated in a few countries, mostly in politically unstable regions. Although reserves increased by 170% between 1978 and 2003, most of this increase is attributed to new discoveries in the Middle East. Regionally, however, reserves have been declining in important consuming countries. For example, reserves in Russia declined by 9% between 1991 and 2004. They also declined by 9% in the United States during the same period. Despite new discoveries, Europe's reserves are likely to be depleted by 2050. Moreover, exports of oil from Africa and Latin America will probably cease by 2050. The Middle East will then be the only major exporter of oil, but political turmoil (e.g., wars, revolutions) there could cause interruptions in oil supplies, creating problems for the import-dependent regions of North America, Western Europe, Japan, and the East Asian Newly Industrializing Countries (NICs).

Natural Gas

The political volatility of the world's oil supply has increased the attractiveness of natural gas, the fossil fuel experiencing the fastest growth in consumption. Natural gas production is increasing rapidly, and so too are estimates of proven gas reserves. Estimates of global gas reserves have increased

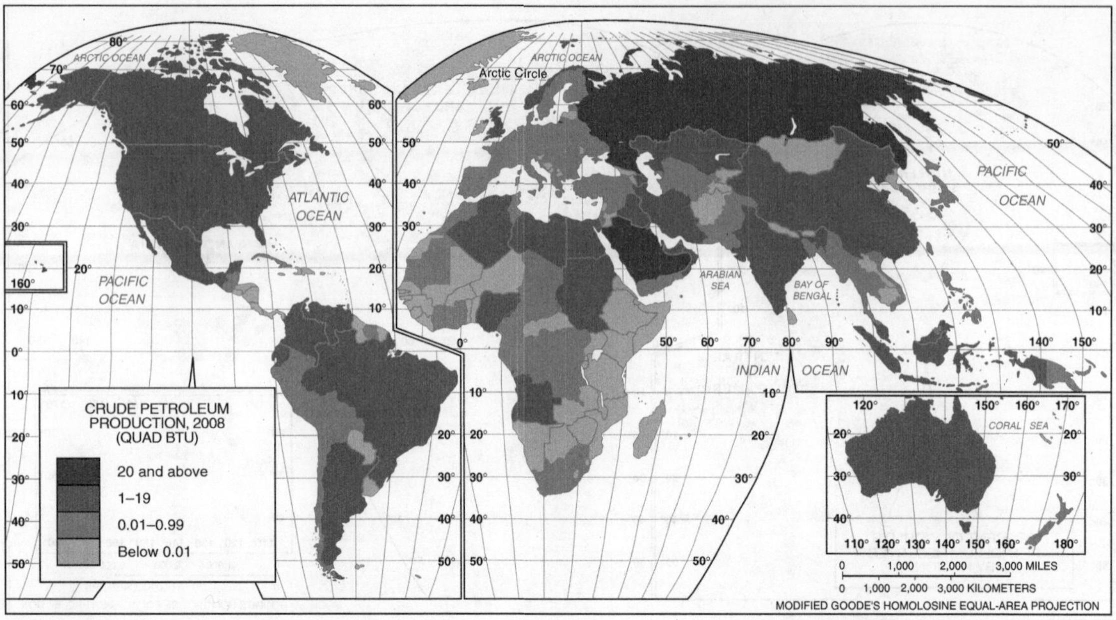

FIGURE 20 World crude petroleum production. Saudi Arabia is the largest producer and has the largest proven reserves; the United States is a major producer but with limited reserves.

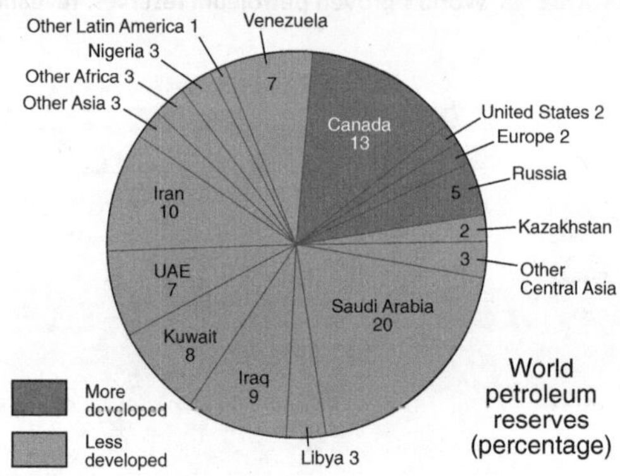

during the past decade, primarily due to major finds in Russia, particularly in Siberia, and large discoveries in China, South Africa, and Australia. Reserves have also been increasing in Europe, Latin America, and North America. Gas production will eventually peak, probably in the first two or three decades of the twenty-first century. As a result, gas supplies will probably last a bit longer than oil supplies.

The distribution of natural gas differs from that of oil. It is more abundant than oil in the former Soviet Union, Western Europe, and North America, and less abundant than oil in the Middle East, Latin America, and Africa. A comparison of Figures 24 and 25 shows that natural gas also differs in its pattern of production and consumption. Because of the high cost of transporting natural gas by sea, the pattern of production is similar to that of consumption.

Coal

Coal is the most abundant fossil fuel, and most of it is consumed in the country in which it is produced (Figure 26). Use of this resource, however, has been hampered by inefficient management by the international coal industry, the inconvenience of storing and shipping, and the environmental consequences of large-scale coal burning.

China is the world's largest consumer of coal, which is a major reason why it is the world's leading producer of greenhouse gases, such as carbon dioxide, associated with global climate change. The principal fossil fuel in North America is coal (Figure 27). With the exception of Russia, the United States has the largest proven coal reserves. Coal constitutes 67% of the country's fossil fuel resources, but only a small fraction of its energy consumption. It could provide some relief to the dependence on oil and natural gas. However, the use of coal presents problems that the use of oil and natural gas do not, making it less desirable as an important fossil fuel. These problems are as follows:

1. Coal burning releases more pollution than other fossil fuels, especially sulfur. Low-grade bituminous coal has large amounts of sulfur, which, when released into the air from the burning of coal, combines with moisture to form **acid rain** (Figure 28).
2. Coal is not as easily mined as oil or natural gas. Underground mining is costly and dangerous, and

open-pit mining leaves scars difficult to rehabilitate to environmental premining standards.

3. Coal is bulky and expensive to transport, and coal slurry pipelines are less efficient than oil or natural gas pipelines.
4. Coal is not a good fuel for mobile energy units such as trains and automobiles. Although coal can be adapted through gasification techniques to the automobile, it is an expensive conversion and it is not well adapted to motor vehicles, overall.

ENERGY OPTIONS

The age of cheap fossil fuels will eventually come to an end. As societies prepare for that eventuality, they must conserve energy and find alternatives to fossil fuels, especially alternatives that do not destroy the environment. How viable are the options?

Conservation

One way to reduce the gap between domestic production and consumption in the short run is for consumers to restrict consumption. Energy **conservation** stretches finite fuel resources and reduces environmental stress. Conservation can substitute for expensive, less environmentally desirable supply options and help to buy time for the development of other, more acceptable sources of energy.

Many people believe that energy conservation means a slow-growth economy; however, energy growth and

FIGURE 21 A Shell/Esso production platform in Britain's North Sea gas field. British oil exploration was stimulated by a dramatic increase in the price of oil in the 1970s and early 1980s. Britain's North Sea oil and gas investment may keep the country self-sufficient for the foreseeable future. However, the huge 2009 British Petroleum oil platform blowout and oil spill in the Gulf of Mexico has raised safety issues for offshore oil drilling.

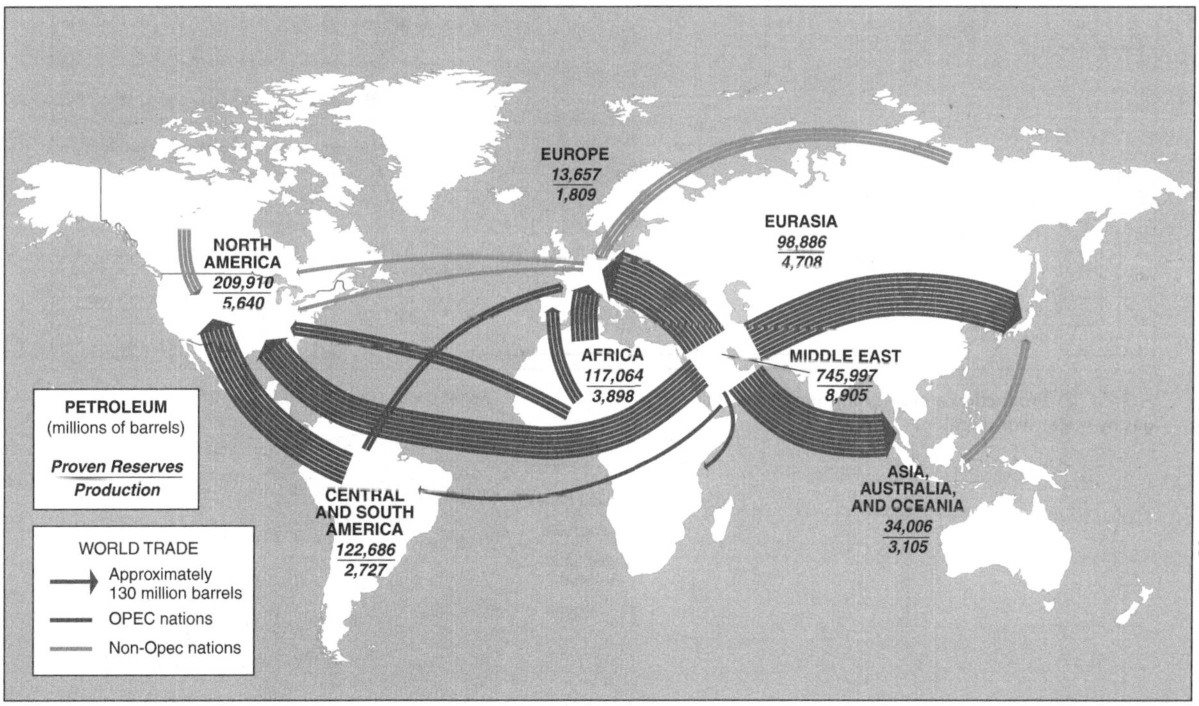

FIGURE 22 World trade patterns in petroleum. The major flows are from the Middle East to Europe, East Asia, and the United States, which also imports from Latin America and Nigeria.

FIGURE 23 The production and consumption of crude oil by major world regions, 2008. The developed market economies of Europe, North America, and Asia, especially Japan, consume a far greater proportion of energy resources than they produce. Conversely, the Middle East, especially the Persian Gulf region, produces the most crude oil of any world region but consumes only one-sixth of its production. Latin America, Africa, and the former Soviet Union consume less crude oil than they produce.

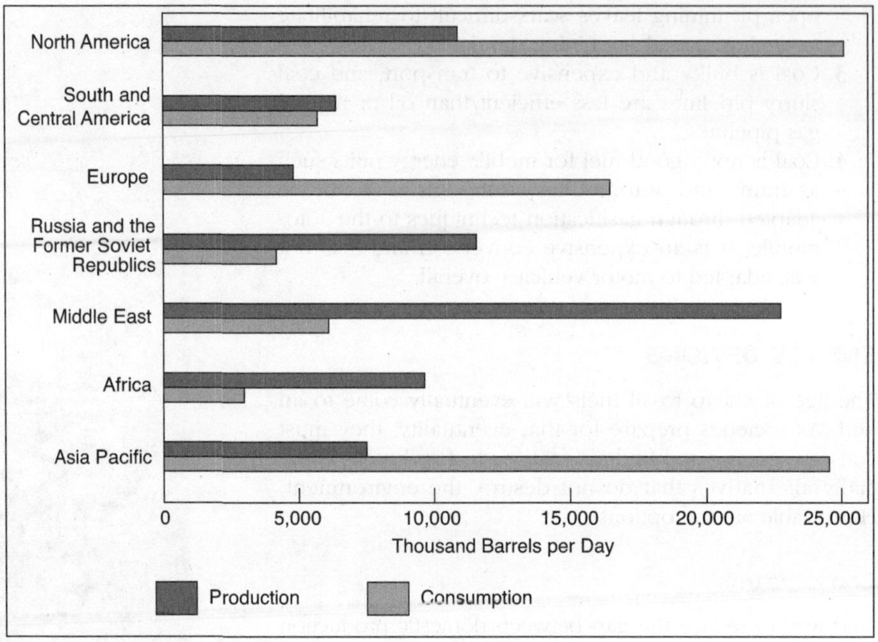

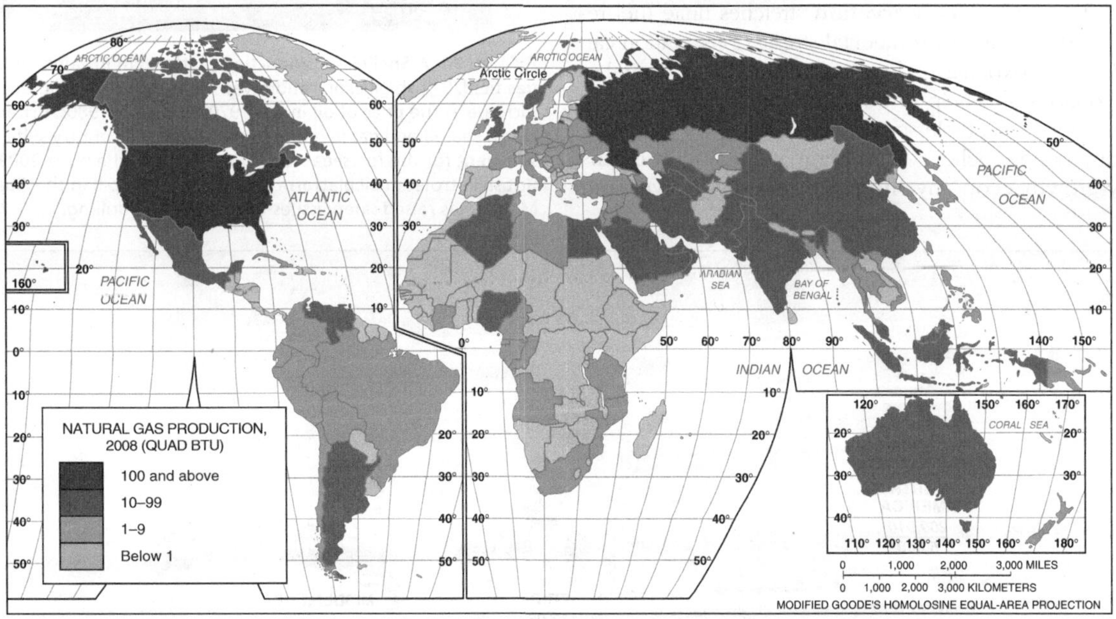

FIGURE 24 Production and consumption of natural gas by major world regions.

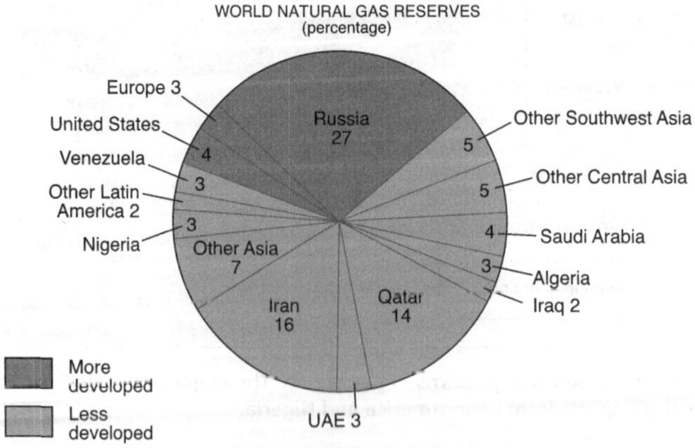

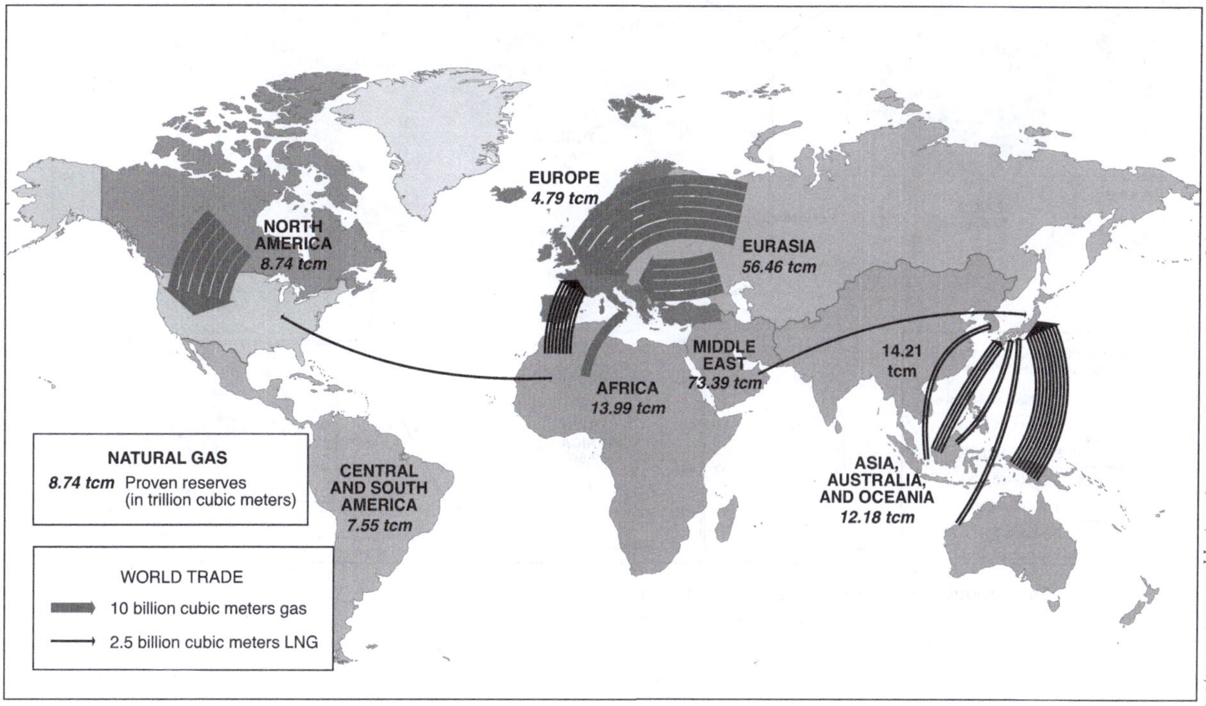

FIGURE 25 World trade patterns in natural gas. Russia is the world's largest exporter, primarily to Europe. Japan receives most of its gas from Southeast Asia, whereas the United States imports most from Canada.

economic growth are not inextricably linked. It is possible to enjoy economic progress while simultaneously consuming less energy per unit of output. In the United States, from the early 1870s to the late 1940s, gross national product (GNP) per capita increased sixfold, whereas energy use per capita only slightly more than doubled. Energy efficiency, the ratio of useful energy output to total energy input, increased steadily throughout the twentieth century, partly as a result of industries installing better equipment.

Nuclear Energy

The form of **nuclear energy** currently in use commercial ly—**nuclear fission**—involves splitting large uranium atoms to release the energy within them. But nuclear fission causes many frightening problems, which became alarmingly clear after the nuclear accidents at the Three Mile Island plant in Pennsylvania in 1979 and at Chernobyl in Ukraine in 1986. Concerns over nuclear energy range from environmental concerns caused by radiation to problems of radioactive waste disposal. Early radioactive wastes were dumped in the ocean in drums that soon began leaking. Likewise, many sites throughout the United States have contaminated groundwater supplies and leak radioactive wastes. One hotly discussed strategy currently underway is to store much of the nation's nuclear waste at Yucca Mountain, Nevada, miles away from major towns and cities. Another problem associated with the use of nuclear energy is the danger of terrorists stealing small amounts of nuclear fuel to construct weapons, which, if detonated, would wreak world havoc. Yet another is its high costs: Each plant costs billions of dollars to build and needs elaborate engineering and backup systems, as well as precautionary safety measures.

Nuclear power is less widely used in North America than in some western European countries and Japan (Figure 29). In France, one-half of energy comes from nuclear power; in Japan, 25%; Belgium, France, Hungary, and Sweden produce more than one-half of their energy from nuclear power plants, while Finland, Germany, Spain, Switzerland, South Korea, and Taiwan are also major producers and users. In the United States and Canada, countries that are less dependent on nuclear energy, the eastern portions rely on nuclear power plants more than do the western portions. For example, New England draws most of its electricity from nuclear power plants. Interestingly, some countries have decided to throw in the towel on nuclear power generation because of high risks and high costs (Figure 30). For example, Sweden began phasing out its nuclear plants in 1995, a process completed in 2010.

Nuclear fusion, the combining of smaller atoms to release their energy (the process that fuels the sun), has the potential to be a solution to the environmental concerns of nuclear fission because it does not release radioactive waste. The raw material for nuclear fusion is the common element hydrogen. Fusion is the process that powers the sun and can be made to occur artificially, but is not yet commercially viable. If this technology is ever commercially successful, nuclear fusion would provide limitless amounts of very cheap energy and pose no radiation dangers.

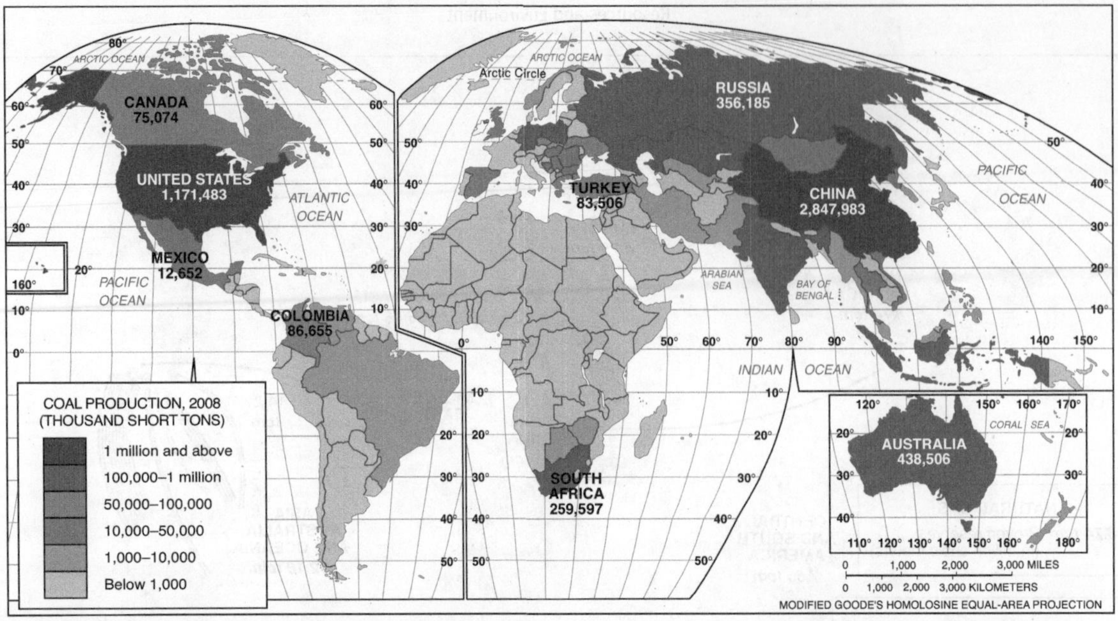

FIGURE 26 Coal production. The United States, China, and Russia lead the world in major coal deposits. Australia, Canada, and Europe also have large quantities.

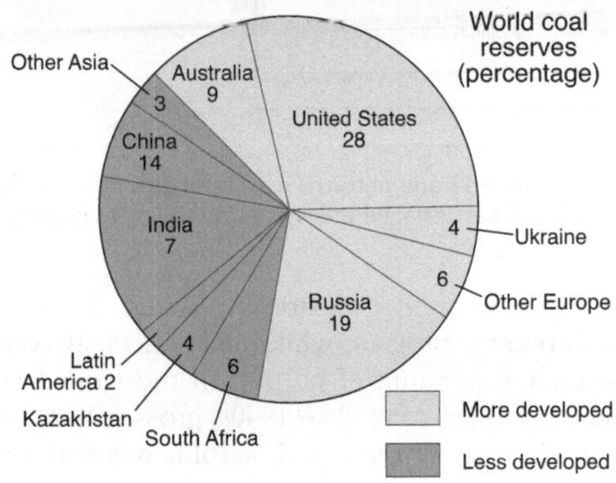

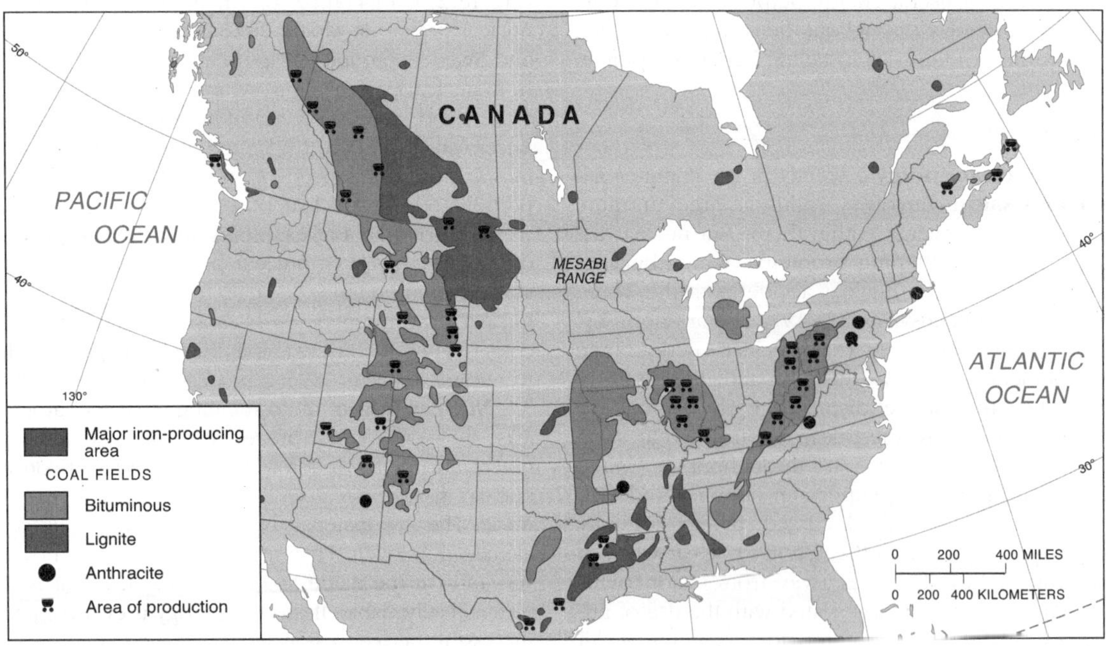

FIGURE 27 Major iron-producing areas and coal fields of North America.

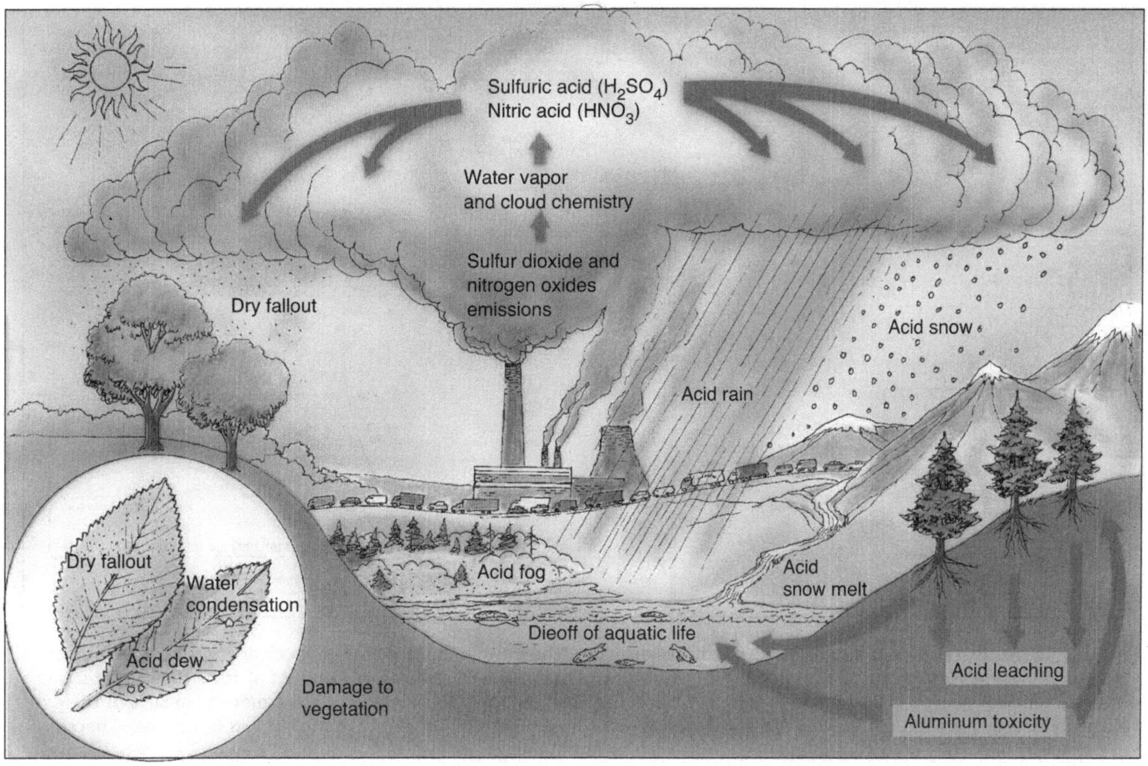

FIGURE 28 Acid rain creation. When sulfur is released into the atmosphere from the burning of coal and oil, it combines with moisture to create acid rain.

Geothermal Power

The development of geothermal power holds promise for the future in several countries that have hot springs, geysers, and other underground supplies of hot water that can easily be tapped. The occurrence of this renewable resource is highly localized, however. New Zealand obtains about 10% of its electricity from this source, and smaller quantities are utilized by other countries such as Italy, Japan, Iceland, and the United States. If the interior of the earth's molten magma is sufficiently close to the surface (i.e., 10,000 feet), underground water may be sufficiently warm to produce steam that can be tapped by drilling geothermal wells. **Geothermal energy** is most producible in giant cracks or rifts in the earth's tectonic plate structure that occur in earthquake or active volcanic areas around the Pacific Rim. Wyoming and California are noted examples.

Hydropower

Another source of electric power, and one that is virtually inexhaustible, is hydropower—energy from rivers. Developed countries have exploited about 50% of their usable opportunities, Russia and Eastern Europe about 20%, and developing countries about 7%. In developed countries, further exploitation of hydropower is limited mainly by environmental and social concerns. In developing countries, a lack of investment funds and sufficiently well-developed markets for the power are the main obstacles.

One of the main problems of constructing dams for hydropower is the disruption to the natural order of a watercourse. Behind the dam, water floods a large area, creating a reservoir; below the dam, the river may be reduced to a trickle. Both behind and below the dam, the countryside is transformed, plant and animal habitats are destroyed, farms are ruined, and people are displaced. Moreover, the creation of reservoirs increases the rate of evaporation and the salinity of the remaining water. In tropical areas, reservoirs harbor parasitic diseases, such as schistosomiasis. For example, since the construction of the Aswan High Dam in the 1970s, schistosomiasis has become endemic in lower Egypt, infecting up to one-half of the population. An additional problem is the silting of reservoirs, reducing their potential to produce electricity. The silt, trapped in reservoirs, cannot proceed downstream and enrich agricultural land. The decrease in agricultural productivity from irrigated fields downstream from the Aswan High Dam has been substantial. China is completing the Three Gorges Dam on the Yangtze River, which will be the largest hydroelectric dam on the planet and may well experience similar problems.

The long-term hydrological, ecological, and human costs of dams easily transform into political problems on international rivers. An example is Turkey's Southeast Anatolia Project, which envisages the construction of 22 dams and

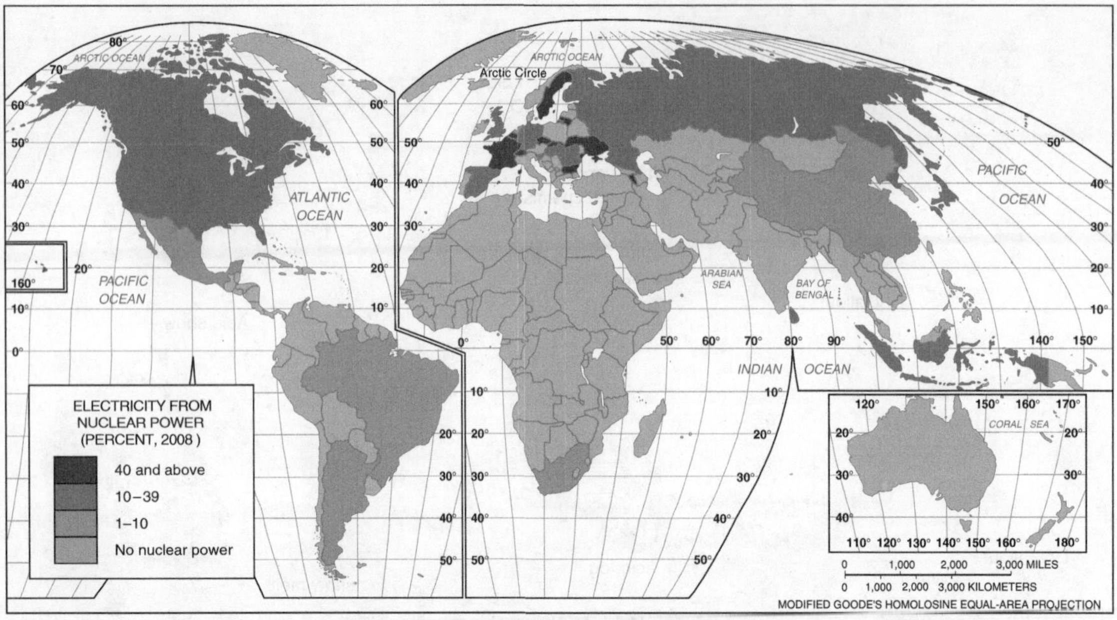

FIGURE 29 Nuclear power as a percentage of total energy use. The most important areas of nuclear energy production in the world include Western Europe and Japan. These are areas that have a relatively small amount of fossil fuels to satisfy local demand for energy. In Europe, for example, France, Germany, Sweden, and Finland produce more than 50% of their electrical energy from nuclear sources. Nuclear power is much less prevalent in the developing nations of the world because of extremely high start-up costs and the need for expensive uranium fuels.

19 hydroelectric power plants. Because the project is being developed on two transboundary rivers—the Tigris and Euphrates—problems and disputes have arisen with two downstream users—Syria and Iraq—whose interests are affected by the project.

Solar Energy

Like river power, **solar energy** is inexhaustible. During the petrocrises of the 1970s, solar energy caught the public imagination, including a few solar-powered homes and buildings. Large-scale utilization of solar energy still poses

U.S. Department of Agriculture

FIGURE 30 The large, hyperbolic cooling tower and reactor containment dome of the Trojan nuclear power plant in Rainier, Oregon. Safety issues surrounding the use of nuclear energy are fraught with turmoil. Most industrialized countries expanded their nuclear energy production during the past 20 years, with France and Japan in the lead. Expansion of nuclear capacity had slowed by 1998 because of cost concerns and the chilling effects of the accidents at Three Mile Island in Pennsylvania and at Chernobyl in Ukraine. New energy sources, such as geothermal, solar, biomass, and wind energy, have increased and now provide up to 5% of total primary energy requirements in Australia, Austria, Canada, Denmark, Sweden, and Switzerland.

Case Study

Resources: Wind Energy

The wind is a natural energy resource that has been used by humans on a modest scale for hundreds of years. Recently, however, concerns over climate change and diminishing reserves of fossil fuels have increased interest in wind as a major contributor to future energy needs. This renewed interest has not been without controversy, and major obstacles remain to realizing the full potential of wind as a modern energy resource, one of them being the significant modifications to our energy infrastructure that would be required.

Although in the past wind energy was used to propel sailing ships, pump water, and grind grain, it is currently used almost exclusively to generate electric power. Turbines used to generate electricity come in many shapes and sizes. The most common modern design uses three vertically oriented blades to drive a shaft oriented in the horizontal direction, usually termed a three-blade, horizontal-axis turbine. Other designs exist (e.g., different types of vertical-axis turbines) but are comparatively rare. Modern wind turbines are typically rated by how much electric power they can generate, from several hundred watts to several megawatts. One MW of electricity-generating capacity can serve 200 to 300 average homes. The country with the largest installation and rated capacity of wind turbines, or "wind farms," today is Germany, with the United States and Spain a distant second and third.

The theoretical potential of wind energy is one reason that it is seen as an attractive energy resource. The amount of energy contained in earth's winds is enormous in terms of human needs. Some estimates indicate that, in theory, there is enough wind energy potential in the United States to serve *all* of the country's current energy consumption, not just electricity usage. Areas of high wind energy potential are widespread throughout the world, although areas of high potential are frequently concentrated in certain regions, such as the Great Plains in the United States. Complementing its widespread availability and potential, wind energy is attractive because it is a renewable resource and because it does not produce any of the harmful emissions associated with fossil fuel electricity generation, including the carbon dioxide associated with climate change. A more recently realized benefit is that the cost of wind-generated electricity is generally lower than that of other renewable alternatives. In fact, wind even compares well to some types of fossil fuel electricity generation, particularly natural gas.

A chief drawback to using wind to generate electricity is that it is an uncontrollable, intermittent resource. Electricity cannot be effectively stored for long, and supply must balance demand at all times in order to keep the electric grid functioning. In some cases, wind power may not be available when it is needed, which requires expensive back-up generation infrastructure. Compounding this problem, electricity demand is typically highest on hot summer afternoons, a condition that often does not coincide with peaks in wind electricity generation. A further disadvantage of wind power is that wind farms must be sited in areas with good wind resources in order to be effective. Unfortunately, more often than not such areas are long distances away from population centers where the electricity is needed, making the development of electricity transmission infrastructure to transport wind power a lengthy and expensive undertaking. One final drawback is that utility-scale wind turbines are large, highly visible, and intrusive. Some people consider them ugly and argue that they destroy treasured landscapes. Others believe that they pose a threat to wildlife, particularly birds and bats. Both of these issues are often hotly debated in areas where wind farms are proposed.

technical difficulties, however, particularly that of low concentration of the energy. So far, technology has been able to convert only slightly more than 30% of solar energy into electricity; however, depending on the success of ongoing research programs, it could provide substantial power needs in the future. Solar energy's positive aspects are that it does not have the same risks as nuclear energy, nor is it difficult, like coal, to transport, and it is free of pollution. It is almost ubiquitous but varies by latitude and by season. In the United States, solar energy and incoming solar radiation are highest in the Southwest and lowest in the Northeast.

Passive solar energy is trapped rather than generated. It is captured by large glass plates on a building. The greenhouse effect is produced by short-wave radiation from the sun. Once the rays penetrate the glass, they are converted to long-wave radiation and are trapped within the glass panel, thus heating the interior of the structure or a water storage tank. The other way of harnessing solar energy is through photovoltaic cells made from silicon. A bank of cells can be wired together and mounted on the roof with mechanical devices that maximize the direct rays of the sun by moving at an angle proportional to the light received. Another type of active solar energy system is a wood or aluminum box filled with copper pipes and covered with a glass plate, which collects solar radiation and converts it into hot water for homes and swimming pools.

Cost is the main problem with solar energy. High costs are associated with the capturing of energy in cloudy areas and high latitudes. But unlike fossil fuels, solar energy is difficult to store for long periods without large banks of cells or batteries. Currently, solar energy production is often more expensive than other sources of fuel. To promote the development of innovative energy supplies when the Arab oil embargo hit in the 1970s, the U.S. government offered tax incentives, including tax deductions for solar units mounted on housetops. Although this tax deduction offset the high costs of installing solar energy systems, maintenance and reliability soon became a problem. Families that moved lost their investment, because most systems installed are rarely recoverable in the sale price of homes.

Wind Power

The power of the wind provides an energy hope for a few areas of the world where there are constant surface winds of 15 mph or more. The greatest majority of **wind farms** in the United States are in California. However, wind machines are an expensive investment, and the initial cost plus the unsightliness of the wind machine has ended most wind farm projects. Wind farm potential in California has never matched expectations, and investment in wind farming declined; however, it is currently experiencing a revival.

Biomass

Still another form of renewable energy is **biomass**—wood and organic wastes. Today, biomass accounts for about 14% of global energy use. In Nepal, Ethiopia, and Tanzania, more than 90% of total energy comes from biomass. The use of wood for cooking—the largest use of biomass fuel—presents enormous environmental and social problems because it is being consumed faster than it is being replenished. Fuelwood scarcities—the poor world's energy crisis—affect 1.5 billion people and could affect 3 billion in the future unless corrective actions are taken.

With good management practices, biomass is a resource that can be produced renewably. It can be converted to alcohol and efficient, clean-burning fuel for cooking and transportation. Its production and conversion are labor-intensive, an attractive feature for developing countries that face underemployment and unemployment problems. But the low efficiency of photosynthesis requires huge land areas for energy crops if significant quantities of biomass fuels are to be produced.

ENVIRONMENTAL DEGRADATION

On some days in Los Angeles, pollution levels reached what is called locally a *level-three alert,* although conditions there have improved recently. People are advised to stay indoors, cars are ordered off the highways, and strenuous exercise is discouraged. In Times Beach, Missouri, which is some 50 miles south of St. Louis, dioxin levels from a contaminated plant became so high that the Environmental Protection Agency required the town to be closed and the residents to be relocated. Around Rocky Flats, nuclear wastes of plutonium have degraded the soil so that radioactivity levels are five times higher than normal. In New England, acid rain has become so bad that it has killed vegetation and fish in rivers and lakes.

Environmental problems, caused mainly by economic activity, may be divided into three overlapping categories: (1) pollution, (2) wildlife and habitat preservation, and (3) environmental equity.

Pollution

Pollution is a discharge of waste gases and chemicals into the air, land, and water. Such discharge can reach levels sufficient to create health hazards to plants, animals, and humans as well as to reduce and degrade the environment. The natural environment has the capacity to regenerate and cleanse itself on a normal basis; however, when great amounts of gases and solids are released into it from industrial economic activity, recycling and purification needs are sometimes overwhelmed. From that point on, the quality of the environment is reduced as pollutants create health hazards for humans, defoliate forests, inundate land surfaces, reduce fisheries, and burden wildlife habitats.

Air Pollution

Air pollutants, the main sources of which are illustrated in Figure 31, are normally carried high into the atmosphere, but occasionally, and in some places more than others, a temperature inversion prevents this from occurring. Inversions are caps of warm air that prevent the escape of pollutants to higher levels. Under these conditions, the inhabitants of a place are under an even greater risk. These conditions promote the formation of smog that blocks out sunshine; causes respiratory problems; stings the eyes; and creates a haze over large, congested cities everywhere.

Air pollution gives rise to different concerns at different scales. Air pollution at the local scale is a major concern in cities because of the release of carbon monoxide, hydrocarbons, and particulates. Air pollution at the regional scale is exemplified by the problem of acid precipitation in eastern North America and Eastern Europe (Figure 32).

At the global scale, air pollution may damage the atmosphere's **ozone layer** and contribute to the threat of **global warming**. The earth's protective ozone layer is thought to be threatened by pollutants called **chlorofluorocarbons (CFCs)**. When CFCs such as freon leak from appliances like air conditioners and refrigerators, they are carried into the stratosphere, where they contribute to ozone depletion. As a result of the 1987 Montreal Protocol, developed countries stopped using CFCs by the year 2000 and developing countries must stop by 2010. Scientists hope that this international agreement will effectively reduce ozone depletion, and preliminary evidence points to some success, offering an example of effective global cooperation to address an important environmental crisis.

Concern about global warming centers around the burning of fossil fuels in ever greater quantities, which increases the amount of carbon dioxide in the air, which in turn makes the atmosphere more opaque, reducing thermal emissions to outer space. Heat-trapping gases, such as carbon dioxide, warm the atmosphere, enhancing a natural **greenhouse effect** (Figure 33). The vast majority of

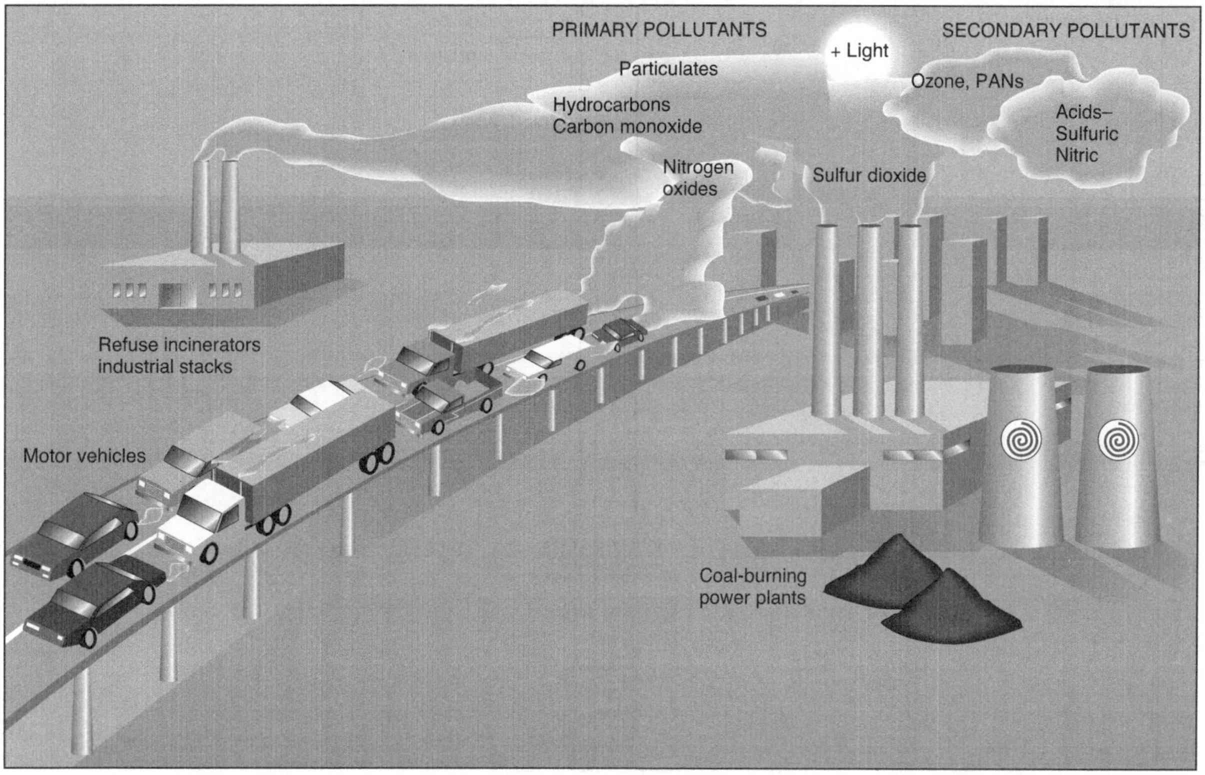

FIGURE 31 The primary sources of major air pollutants. Industrial processes are the major sources of particulates; transportation and fuel combustion cause the lion's share of the other pollutants.

these are produced by industrialized economies. Since the 1890s, the average temperature of the earth's surface has increased by 2 degrees Fahrenheit. This increase in temperature may or may not be humanly induced, however. There are divergent views on the issue. Nonetheless, even if the observed global warming is consistent with natural variability of the climate system, most scientists agree that it is socially irresponsible to delay actions to slow down the rate of anthropogenic greenhouse gas buildup. For example, continued warming would increase sea levels, disrupt ecosystems, and change land-use patterns. While agriculture in some temperate areas may benefit from global warming, tropical and subtropical areas may suffer.

Water Pollution

Although there is more than enough fresh water to meet the world's needs now and in the foreseeable future, the problem is that when we use water, we invariably contaminate it. Major wastewater sources that arise from human activities include municipal, mining, and industrial discharges, as well as urban, agricultural, and silvicultural runoff. The use of water to carry away waste material is an issue that has come into prominence because water is being used more heavily than ever before. As populations and standards of living rise, problems of water utilization and management increase. These problems are most acute in developing countries, where some 1 billion people already find it difficult or impossible to obtain acceptable drinking water. But water is also an issue in developed countries. In the United States, for example, the major water management problem through most of the twentieth century focused on acquiring additional water supplies to meet the needs of expanding populations and associated economic activities. Recently, water management has focused on the physical limits of water resources, especially in the West and Southwest, and on water quality. Passage of the Clean Water Act in 1972 resulted in improvements in water quality of streams that receive discharges from specific locations or **point sources** such as municipal waste-treatment plants and industrial facilities. Recent efforts to improve water quality have also emphasized the reduction of pollution from diffuse or **nonpoint sources** such as agricultural and urban runoff and contaminated groundwater discharges. These sources of pollution are often difficult to identify and costly to treat, and often meet with resistance from entrenched agribusiness interests.

Wildlife and Habitat Preservation

The habitats of wildlife and plants, called **renewable natural resources**, are in danger throughout the world. These natural environments are critical reserves for many species of plants and animals. Wildlife, forestlands, and wetlands, including lakes, rivers and streams, and coastal marshes, are subject to acid rain, toxic waste, pesticide

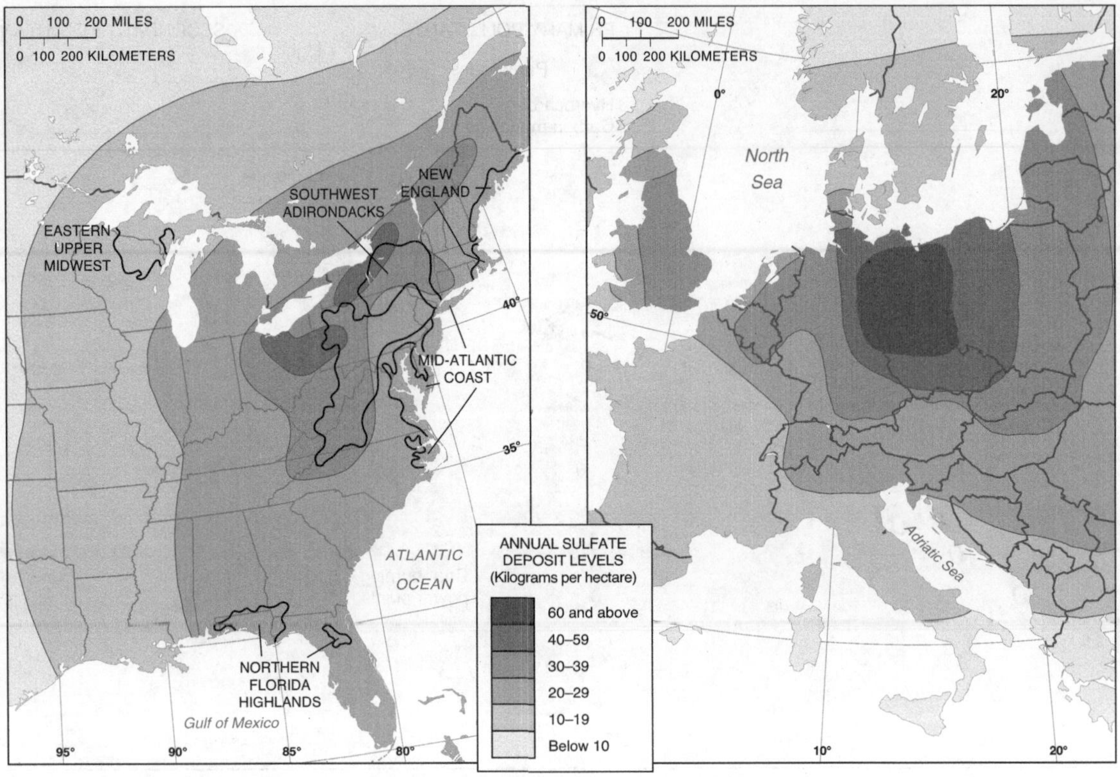

FIGURE 32 The areas worst afflicted by acid rain in North America occur downwind from the principal polluting regions of the industrialized Midwest. Ohio, western Pennsylvania, and northern New York State are the areas most heavily impacted by acid rain. Acid precipitation and snow deposits are also well documented in Europe, which is in a belt of prevailing wind coming from the west, as is the United States. Sulfur, released into the atmosphere from the burning of coal, combines with water vapor to produce sulfuric acid. Such acid creates substantial air pollution and etches away at limestone buildings, monuments, and markers on the earth. Acid precipitation can also kill plant and animal life, especially aquatic life. Literally thousands of lakes in Sweden and Norway no longer support the fish they once did due to serious acid precipitation.

discharge, and urban pollution. They are also endangered by the encroachment of development and transportation facilities worldwide. The demand for tropical hardwoods, such as teak and mahogany, has already stimulated waves of destruction in tropical rain forests. In the United States alone, expanding economic activity has consumed forests and wetlands, depleted topsoil, and polluted local ecosystems at a rapid rate. Many species of plants and animals have been reduced, including the grizzly bear, American bison, prairie dog, gray wolf, brown pelican, Florida panther, American alligator and crocodile, and a variety of waterfowl and tropical birds. All in all, the exponential increase of human beings has been closely associated with a corresponding dramatic decline in the number of species in the world (Figure 34), and future economic growth may threaten yet another catastrophic round of loss of the planet's biodiversity.

The problem of wildlife and habitat preservation is exacerbated by the need for economic gain and corporate profit. A variety of questions beset wildlife managers and environmental farmers. Should farmers be permitted to drain swamps in Louisiana to farm the land, thus destroying the habitat for alligators? Should forest fires started by lightning be allowed to burn themselves out, as has been the practice on western U.S. forests and rangelands? The trade-offs between residential lands versus wetlands, wildlife migration versus forest management, highway safety versus habitat preservation, and conservation versus real economic development and growth of the U.S. economy are problematic issues. It is difficult to select the best alternative, and policies may reflect the power of entrenched economic interests as much as the public welfare.

Regional Dimensions of Environmental Problems

Because economic structures and change are not uniform around the world, but exhibit great variation among the world's continents (as well as within them), and because population growth, cultural patterns, standards of living, and state policies regulating problems such as pollution vary widely, the environmental problems unleashed by capitalism and demographic change worldwide differ considerably. See the color insert for a spatial depiction of these issues.

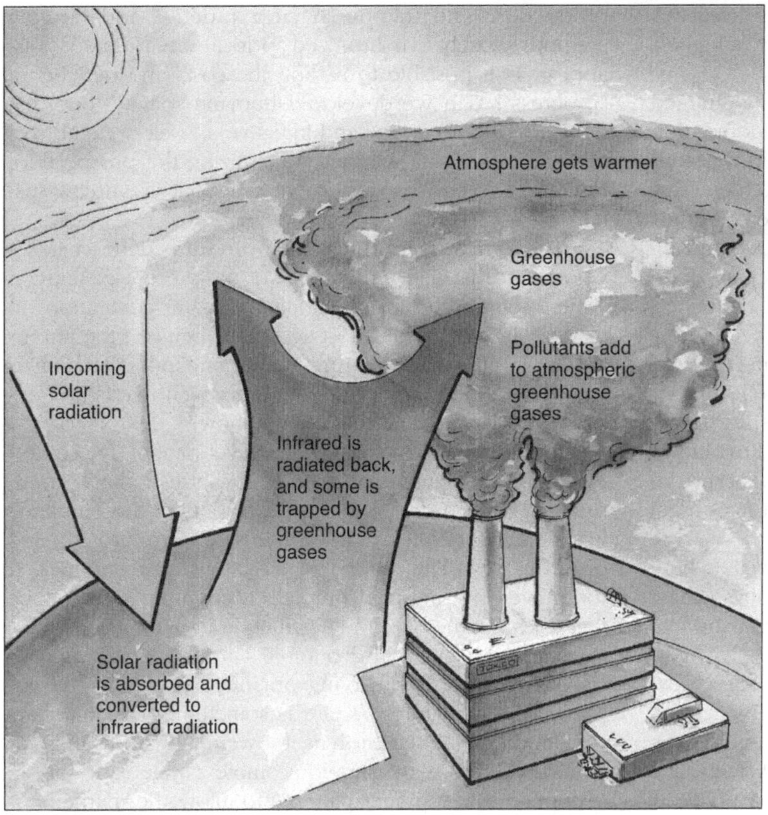

FIGURE 33 The greenhouse effect. Incoming solar radiation is absorbed by the earth and converted to infrared radiation. Greenhouse gases act as an insulator and trap the infrared heat, raising the temperature of the lower atmosphere. The overall effect is global warming. Presently, the scientific consensus on global warming is that it leads to climate change and most of this change is induced by human causes. The result has been an increase in global average temperatures over the past several decades. While some still debate whether climate change is happening, and if it has been induced by human causes, the scientific debate has largely shifted to ways to reduce human impact and to mitigate impacts that have already occurred. Of most concern with these anthropogenic (human) factors is the increase in carbon dioxide levels due to emissions from the combustion of fossil fuels. Aerosols and cement manufacture are additional problems, as well as ozone depletion, deforestation, and animal agriculture.

In North America, the major environmental problems include acid rain downwind from industrial source areas. In addition, water pollution and withdrawal of groundwater are serious issues. Because the U.S. economy is so huge and energy-intensive, the pollutants generated there are major sources of global problems, such as the depletion of the ozone layer and global warming. Across the face of North America, other issues such as wetlands destruction, saltwater intrusion, and urban air quality are serious predicaments

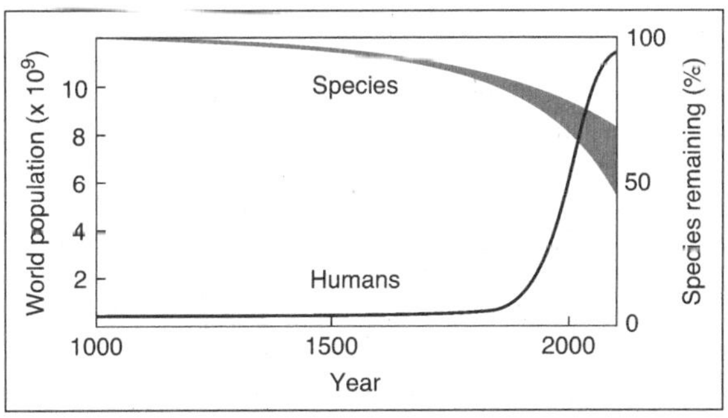

FIGURE 34 An inverse relationship exists between human population size and the survival of species worldwide. Uncertainty about the extent of decreasing biodiversity is reflected in the width of the species curve. Overall, the growth of the human species has been a disaster for most others.

In Latin America, the primary environmental issue is deforestation, particularly in the Amazon River basin, where farmers and logging companies are stripping away the forest cover of some of the world's richest ecosystems. Further, degradation of ground cover leads to mudslides that can be lethal to thousands. Overgrazing and soil erosion are other important consequences, and in many Latin American cities, such as Mexico City, air quality is very poor.

In Europe, these issues range from pollution of the Mediterranean Sea, whose coasts are populated by dense clusters of cities, as well as acid rain in Germany and Poland. Rising sea levels pose a threat to the Netherlands, much of which lies below sea level and uses dikes to hold the ocean back. Air and water pollution are constant problems requiring government intervention.

Environmental problems in Russia and its neighbors are tangled up in decades of mismanagement by the Soviet regime and an economy that collapsed in the 1990s. The Chernobyl disaster in 1985 left parts of Ukraine polluted with nuclear radiation. The extraction of water from the Aral Sea has left it nearly dead. And air pollution and acid rain take a toll on the region's enormous forests.

The predominantly Muslim world of North Africa and the Middle East has its own environmental problems. Soil erosion and overgrazing have contributed to desertification, particularly in the Sahara. Egypt's Aswan High Dam has been a mixed blessing, reducing floods on the Nile but also reducing the siltation that keeps farmlands fertile.

Heavy use of river water contributes to soil salinization. All this takes place in a region with one of the world's highest rates of population growth.

Sub-Saharan Africa's environmental problems, framed in the context of extreme poverty and rapid population growth, include widespread deforestation in western Africa. A continent in which islands of people were surrounded by oceans of wildlife has become one of islands of wildlife surrounded by oceans of people. Overgrazing and soil erosion threaten agricultural land as well as biodiversity, and the region is rocked by wars, drought, famines, and diseases such as AIDS and malaria.

In South Asia, the Indian subcontinent, home to more than 1.5 billion people, has seen widespread soil salinization and water and air pollution, which conspire to reduce the supply of agricultural land. Green Revolution farming accentuates these issues. Deforestation in the Himalaya Mountains has increased flooding downstream in Bangladesh.

East Asia, home to 1.3 billion in China as well as hundreds of millions more in Korea and Japan, exhibits similar problems as South Asia. The encroachment of farmland into forests has reduced the habitats of many species. Trying to stop periodic floods on the Yangtze River, the Chinese government recently finished the Three Gorges Dam, which has modified the basin and ecosystems of Asia's largest river. Japan supports 124 million people in a country with little arable land, and its dense cities exhibit severe air pollution levels, although the rapid growth of Chinese cities has rendered the air quality in those cities considerably worse yet.

Southeast Asia, one of three major regions in the world that sustain rain forests, has enjoyed rapid economic growth over the past two decades. Its environmental problems have grown proportionately. Deforestation in Indonesia, Malaysia, and Thailand has gone unchecked, threatening rich tropical ecosystems. Logging, farming, and international paper companies all contribute to this trend. Water pollution and soil erosion are increasingly widespread. Huge forest fires periodically carpet the region with smoke, adding to the deteriorating air quality of the region's cities, some of which are huge.

ENVIRONMENTAL EQUITY AND SUSTAINABLE DEVELOPMENT

Economic development policies and projects all too often carry as many costs as benefits. Build a dam to bring hydroelectric power or irrigation water, and fertile river bottomlands are drowned, farmers are displaced, waterborne diseases may fester in the still waters of the reservoir, and the course of the river downstream to its delta or estuary is altered forever. Dig a well to improve water supplies in dry rangelands, and overgrazing and desertification spread outward all around the points of permanent water. Mine ore for wealth and jobs, and leave despoiled lands and air and water pollution. Build new industries, shopping malls, and housing tracts, and lose productive farmland or public open space. Introduce a new miracle crop to increase food production, and traditional crop varieties and farming methods closely synchronized to local environments disappear. Is it possible to reckon the costs and benefits of "progress"? Can we develop a humane society, one that encourages both equity and initiative, a society capable of satisfying its needs without jeopardizing the prospects of future generations? How do we go about creating a sustainable society?

The term **sustainable development** is defined as development that meets the needs of the present generation without compromising the ability of future generations to meet their own needs. Most accept its focus on the importance of long-range planning, but as a policy tool it is vague, providing no specifics about which needs and desires must be met and fulfilled and how.

In the debate on sustainable development, two different emphases have emerged. In the industrialized countries of North America, Europe, and Japan, the emphasis has been on long-term rather than short-term growth, and on efficiency. The emphasis has been on economics: If today we rely on an incomplete accounting system, one that does not measure the destruction of natural capital associated with gains in economic output, we deplete our productive assets, satisfying our needs today at the expense of our children. There is something fundamentally wrong in treating the earth as if it were a business in liquidation. Therefore, we should promote a systematic shift in economic development patterns to allow the market system to internalize environmental costs. The environmental costs of automobiles, for example, should include those associated with acid rain, primarily in the form of more expensive gasoline.

This Western emphasis on the economic aspects of sustainable development has been criticized in Africa, Asia, and Latin America. Critics from the less-developed world accuse environmentalists from the industrialized world of dodging the issues of development without growth and the redistribution of wealth. While some observers in the developing world may believe in the power of markets to distribute goods and services efficiently, they argue that social and political constraints are too severe to provide answers to all our problems. Many criticize the West's excessive consumption of resources. Many advocates in the developing world put basic human needs ahead of environmental concerns. Let us work toward a sustainable future, but let us do so by ensuring food, shelter, clean water, health care, security of person and property, education, and participation in governance for all. An extension of this sentiment is a desire to protect basic values as well—to respect nature rather than dominate it, and to use the wisdom of indigenous groups to reexamine current, mostly Western, structures of government and the relationships that people have with the environment.

Is the consumerist West ready to listen to those with different values and priorities? Surely there are many paths to a sustainable future, each determined by individualized priorities of what is desired and therefore worthy of sustaining. Surely too, in following those paths, we must recognize that future growth will be constrained by resources that are finite or whose availability is difficult to determine. Finally, we must realize that no region can achieve sustainability in isolation. A desirable and sustainable future will be the result of many social and policy changes, some small and at the local level, others international and far-reaching. If we accept that the futures of the world's rich and poor are inextricably linked, perhaps we will enforce policies that lead to a more just and equitable distribution of the world's resources. However, a world that rewards only the rich at the expense of the poor—as has been the case globally for the past several decades—is doomed to social inequality and environmental destruction.

From a Growth-Oriented to a Balance-Oriented Lifestyle

Given the dynamics of the market system, it is unlikely that energy availability will place a limit on economic growth on the earth; however, ultimately, drastic changes in the rate and nature of the use of energy resources are certain. The ultimate limits to the use of energy will be determined by the ability of the world's ecosystems to dissipate the heat and waste produced as more and more energy flows through the system.

In countless ways, energy consumption improves the quality of our lives, but it also pollutes. As the rate of energy consumption increases, so too does water and air contamination. Sources of water pollution are numerous: industrial wastes, sewage, and detergents; fertilizers, herbicides, and pesticides from agriculture; and coastal oil spills from tankers. Air pollution reduces visibility; damages buildings, clothes, and crops; and endangers human health. It is especially serious in urban-industrial areas, but it occurs wherever waste gases and solid particulates are released into the atmosphere.

Pollution is the price paid by an economic system emphasizing ever-increasing growth as a primary goal. Despite attempts to do something about pollution problems, the growth-oriented lifestyle characteristic of Western urban-industrial society continues to widen the gap between people and nature. "Growthmania" is ultimately a road to self-destruction. Many argue that we must transform our present growth-oriented, economic system into a balance-oriented one that explicitly recognizes that natural resources are exhaustible, that they must be recycled, and that input rates must be reduced to levels that do not permanently damage the world's environment. A balance-oriented economy does not mean an end to growth, but a new social system in which only desirable low-energy growth is encouraged. It requires a de-emphasis on the materialistic values we have come to hold in such high esteem. If current resource and environmental constraints lead us to place a higher premium on saving and conserving than on spending and discarding, then they may be viewed as blessings in disguise.

Summary

We conclude by restating the resources-population problem. It is possible to solve resource problems by (1) changing societal goals, (2) changing consumption patterns, (3) changing technology, and (4) altering population numbers. In the Western world, much of the emphasis is on technological advancement and population control.

Following a review of renewable and nonrenewable resources, we explored the question of food resources. The food "crisis" is essentially a consequence of social relations, including war and disruptions of agricultural systems. Food production is increasing faster than population growth, yet more people are hungry than ever before. In the course of transforming agriculture into a profit base for the wealthy in the developed and in the less developed worlds, the Third World poor are being forced into increasingly inhospitable living conditions. Famine, like poverty, is a social construction, not a natural event, and its origins, like its solutions, must be found in the uneven distribution of resources among and within countries.

Unlike food, which is replenishable, nonrenewable minerals and fossil fuels, once used, are gone forever. We discuss some of the alternatives to fossil fuels and point to energy conservation as a potent alternative with potential that remains to be fully exploited. Finally, the comparison between growth-oriented and balance-oriented lifestyles underscores the importance of social concerns as they relate to economic growth. Growth and inequality are inextricably linked parts of social change and environmental protection.

Key Terms

acid rain
aquaculture
balance-oriented lifestyle
biomass
carrying capacity
chlorofluorcarbons (CFCs)
chronic malnutrition
conservation
deforestation
desertification
flow resources
fossil fuels
geothermal energy
global warming
Green Revolution
greenhouse effect
growth-oriented lifestyle
limits to growth
mariculture
maximum sustainable yield
minerals
nonpoint sources
nonrenewable resources
nuclear energy
nuclear fission
nuclear fusion
Organization of the Petroleum Exporting Countries (OPEC)

overpopulation
ozone layer
passive solar energy
point sources
projected reserves
recycling
renewable natural resources
renewable resources
reserve
resource
solar energy
stock resources
strategic minerals
sustainable development
tragedy of the commons
transmaterialization
undernutrition
wind farm

Study Questions

1. What is meant by carrying capacity?
2. Differentiate renewable from nonrenewable resources.
3. What are the major causes of Third World hunger?
4. What are three methods of expanding world food production?
5. What was the Green Revolution? Where was it largely located?
6. Summarize major flows of oil on the world market.
7. Where are the major world coal deposits located?
8. What are some alternative energy options to fossil fuels?
9. What are some environmental consequences of high energy use? Be specific.
10. What is sustainable development?

Suggested Readings

Castree, N., and B. Braun, eds., 2001. *Social Nature: Theory, Practice, and Politics.* London, UK: Blackwell.

Ellis, R. 2003. *The Empty Ocean: Plundering the World's Marine Life.* New York: Shearwear Books.

Falola, T., and A. Genova. 2005. *The Politics of the Global Oil Industry.* New York: Praeger.

Klare, M. 2002. *Resource Wars: The New Landscape of Global Conflict.* New York: Owl Books.

Zimmerer, K. 2006. *Globalization and New Geographies of Conservation.* Chicago: University of Chicago Press.

Web Resources

Conservation Databases—WCMC
http://www.unep-wcmc.org

The World Conservation Monitoring Centre, whose purpose is the "location and management of information on the conservation and sustainable use of the world's living resources," provides five searchable databases. Users can search by country for threatened animals and plants (plants are available for Europe only), protected areas of the world, forest statistics and maps, marine statistics and maps, and national biodiversity profiles (12 countries only at present). Information is drawn from several sources, and database documentation varies from resource to resource.

State of The World's Forests—FAO
http://www.fao.org/forestry/home/en/

The United Nations Food and Agriculture Organization presents information on the current status of the world's forests, major developments over the reporting period, and recent trends and future directions in the forestry sector. It provides information on global forest cover, including estimates for 1995, change from 1990, and revised estimates of forest cover change.

Environmental Protection Agency
http://www.epa.gov

This site provides everything you ever wanted to know about environment and material resources.

Theoretical Considerations

OBJECTIVES

- To present the basic factors underlying the location decisions of firms
- To summarize the Weberian model of transport costs
- To show how production technique, scale, and location are interrelated
- To illustrate how and why firms grow and change over time and space
- To reveal the geographical organization of large corporations
- To describe the relevance of the product cycle in the changing locational requirements of firms
- To depict the role of business cycles, particularly Kondratiev long waves
- To emphasize the role of the state in shaping economic landscapes

Female workers on the assembly line at a low-voltage electrical appliance factory in Taizhou, China.

Theoretical Considerations

Theory is a way of looking at the world, an explanation, a way of making sense of the relationships among variables. Theory is what separates description from explanation. A theory allows us to establish causality, to test hypotheses, to justify arguments, and to make claims to truth. Theories are simplifications about the world that allow us to gain understanding. Theories are thus indispensable to knowing how the world works, whether in the formal intellectual world or in our everyday lives at home and at work. Understanding theory is not a choice, because theory is inescapable. We all use theories every day, whether we're aware of them or not. Economic geographers use theory as well, ranging from the simple to the very sophisticated, to understand the order and chaos of economic landscapes.

There are a variety of theoretical frameworks for interpreting economic landscapes, including traditional industrial location theory; the behavioral approach; and the political economy, or structural, perspective. Industrial location theory derives from and shares the conservative ideology of neoclassical economics, using abstract models to search for best, or optimal, locations. The behavioral approach focuses on the psychology of the decision-making process: Rather than considering how decisions *should* be made in firms, it examines how decisions *are* made. This approach recognizes the possibility of suboptimal behavior. The political economy perspective emphasizes the historical context and social relations within which industrial activity takes place.

This chapter offers an overview of some of the major conceptual approaches and topics in economic geography. It begins with a discussion of the factors of location that influence industrial locations of single-plant firms. *Industrial* in this context should not be interpreted to be the same as manufacturing; every major sector of the economy is an industry, including many services, such as finance, legal services, advertising, and accounting. The factors of location include labor, land, capital, and management. This chapter delves into the first, and still highly influential, model of industrial location, introduced by Alfred Weber a century ago. It explores the interrelations among production technique, scale of output, and geographic location. It discusses how and why firms grow over time and space and then turns to the geographical organization of firms, moving from simple single-establishment ones to large, multi-establishment, multiproduct corporations with well-developed internal divisions of labor. The chapter emphasizes the social forces and contexts that firms simultaneously produce and are produced by, including conflicts between capital and labor, and then turns to the cyclicality of capitalism, its tendency toward boom-and-bust cycles, which in turn shape uneven spatial development. The chapter concludes by reciting some of the many ways the state, or government, plays a critical role in building, changing, and reproducing the economic geographies of capitalism.

FACTORS OF LOCATION

Numerous variables influence the location of firms and industries, which are aggregations of firms. The locational decision of a firm is thus quite complex, and many companies spend considerable time and effort in choosing the optimal location. After all, investments in inappropriate locations can be disastrous. Thus, firm decision making is a rational process, if an imperfect one, and is subject to the iron laws of market competition. Although personal considerations such as climate or the owner's preferences may figure in from time to time on the margins, firms cannot choose locations arbitrarily or they will be forced out of business by their more rational competitors. The major factors that shape a firm's location include labor, land, capital, and managerial and technical skills. All these are necessary for production, and all exhibit spatial variations in both quantity and quality.

Labor

For most industries (except those in the primary economic sector concerned with extractive activities), labor is the most important determinant of location, especially at the regional, national, and global scales. (At local scales, such as within a commuting field, other factors such as rents may become more significant.) When firms make location decisions, they often begin by examining the geography of labor availability, productivity, skills, and militancy. The degree to which firms rely on labor, however, varies considerably among different sectors of the economy, and even among different firms, which may adopt different production techniques.

Labor is required for all forms of economic production, but the relative contribution of labor to the cost structure and value added varies considerably among industries. For example, the contribution of labor costs is high in the automobile industry but very low in the petroleum industry. The demand for labor depends on how labor-intensive or capital-intensive a given production process is. In highly capital-intensive industries such as petroleum, labor costs may be so small a share of total costs as to be essentially irrelevant. Thus it is a mistake, but a common one, to assume that all industries seek out low-cost labor. Over time, most industries have become increasingly capital-intensive, that is, they have substituted capital for labor, particularly when production in large quantities justifies the investments involved. The use of tractors in agriculture, machinery and robots in manufacturing, or computers in services and office work are examples. Much of the productivity growth of capitalism has historically resulted from capital-intensification.

The supply of labor in a given region also greatly affects its cost. In countries with high birth rates, the supply tends to be relatively high, and labor costs are therefore low. In economically advanced countries, late in the demographic transition, the birth rate is low and labor is relatively expensive. These trends shape the willingness of firms to become more capital-intensive: When labor is cheap, there is little incentive to mechanize.

There are great variations among industries in terms of their preferences for different types of workers. Because some firms demand particular types of workers in terms of their age or sex, the demographic structure of a region also shapes the supply of certain types of employees. The fast-food industry, for example, prefers young workers who will work for minimum wage, and the supply of this demographic segment is more limited in certain regions than in others. Finally, since labor is mobile over space (but not perfectly so), migration (or internationally, immigration) also shapes the local supply of labor. In regions that can attract labor easily, wage rates will tend to be low, all else being constant. When the supply is limited by, say, immigration restrictions, wage rates tend to go up. For example, limitations placed by the U.S. Congress on immigration shortly after WWI contributed to the rise in wages in the 1920s. At the local level, housing costs can also constrain the supply of labor if they are so high that workers cannot find affordable places to live, an increasingly common problem in the coastal areas of the United States.

In theory, because labor is relatively mobile, regional differences in the supply and demand for labor should move toward equilibrium over time. A high demand for labor in one place, and thus high wages, and an excess supply in another, and thus low wages, should be brought into equilibrium by labor migration. Examples are the nineteenth-century migration of people from rural areas to the city in countries such as Britain and the United States, and the twentieth-century movement of labor from the Third World to the wealthy countries of northern Europe: from the Caribbean and South Asia to Britain, from North Africa to France, and from Turkey to Germany.

As is the case with all production factors, the response of labor to changing market conditions is not instantaneous. Labor has a relatively high degree of inertia, especially in the short run. People are generally reluctant to leave familiar places, even if jobs are plentiful in other areas. They will sit out short periods of unemployment or accept a smaller net income than could be earned elsewhere. Public policies, such as unemployment payments and workers' compensation, have reduced the plight of the unemployed and underemployed in advanced industrial countries in this century. The lack of instantaneous adjustment in labor demand and supply has resulted in variations in the cost of labor within and among countries. In the United States, wages are often higher in the more industrial states and in highly unionized environments.

It is a common myth, but simply not true, that firms always want the cheapest labor possible. Cost is only one of many factors. Equally important is productivity. Productivity is largely a function of the skills present in the local labor force, or **human capital**, which in turn are derived from formal and informal educational systems, on-the-job training, and years of experience. Firms will pay relatively high wages for skilled, productive labor if the cost is justified by higher output and profits. To illustrate this point, consider that if labor costs were central to the location of all firms, then very low-wage countries—say, Mozambique—should attract vast quantities of capital, which they don't, and high-wage countries, such as Germany or Japan, should see a rapid exodus of jobs. Similarly, low-wage states in the United States, such as Mississippi, should attract firms, while high-wage ones such as Minnesota should repel them, although in reality the opposite is the case. The reality of the geography of labor is much more complex, of course, and involves national and local labor markets in which jobs are constantly created and destroyed; skills are produced, reproduced, and changed; new technologies come into play; and other cultural, economic, and social forces interact. What firms seek as much as cheap labor is *productive* labor.

The skill level of a given occupation greatly affects the size of its labor market. Generally, skilled labor markets tend to be geographically larger than unskilled ones. Workers may migrate hundreds or thousands of miles for well-paying positions, and the market for many skilled jobs is global in reach, including corporate management, professional services, and university professors. Unskilled positions, in contrast, typically draw from a relatively small laborshed; few people would travel cross-country, for example, to take a job as a janitor or cashier.

It is worth emphasizing that labor is also important because the labor factor is saturated with politics. Labor is the only input that is sentient, and it alone is able to resist the conditions of exploitation, to go on strike, to engage in slowdowns, or to unionize (Figure 1). Unionization rates vary widely across the United States (Figure 2), adding to differentials in the cost of labor. The South is generally much less unionized than the North, in part due to "right to work" laws that forbid workers in an establishment from being forced to join a union. Thus, in addition to the cost of labor, firms must consider working conditions, health and safety standards, pensions and health benefits, vacations and holidays, worker training, subsidized housing, and the role of labor unions, all of which shape local wage rates and productivity levels.

AP Photo/Ric Francis

FIGURE 1 Labor alone, unlike other inputs to production, is sentient and capable of changing the conditions of work, as with strikes or slowdowns.

On the world scale, developed countries have higher wage rates than newly industrializing countries. One factor responsible for differential wage rates is the level of worker organization. Higher rates of unionization are associated with higher wages. Unionization is generally more prevalent in the older, established industrial countries than in the newly industrializing countries. Thus, considerable advantages can be gained by companies that relocate to, or purchase from, newly industrializing countries, especially if these countries are characterized by low levels of capital–labor conflict. The capital–labor conflict, manifested in industrial disputes, is a powerful force propelling the drift of industrial production outward from the center to the periphery of the world system.

Land

At the local scale (i.e., within a particular metropolitan commuting area, in which labor costs are relatively constant), land availability and cost are the most important locational factors affecting firms' location decisions. The cost of land reflects the highly localized supply and demand, and different types of firms require different quantities in the production process. Generally, larger firms, particularly in manufacturing, require more land and are thus more sensitive to the costs, although in some sectors, such as producer services, firms pay very high costs (in rent or by purchasing a site). Firms often engage in intensive examination of several possible sites before settling on an optimum location.

The primary determinant of the cost of land is its accessibility. Transport costs (the measure of accessibility) determine the location rent of parcels at different distances from the city. Thus, because land downtown is the most accessible, it is by far the most expensive; in most cities, land costs decline exponentially away from the city center (Figure 3).

However, not all firms necessarily seek out low-cost land. The imperative to do so depends on the trade-off between land and transportation costs that firms make to maximize their profits. Firms that must have accessible land—generally labor-intensive firms that must maximize their accessibility to labor, to each other, to specialized pools of information, and to urban services such as legal firms—will pay very high rents in order to locate near the city center. Firms that do not require access to clients, suppliers, and services, on the other hand—such as large manufacturing firms in suburban industrial parks—make a different trade-off, choosing to locate on the urban periphery where land costs are low but transport costs are higher. The demand for land and the need to agglomerate are thus inversely related.

Within metropolitan areas, a centrifugal drift of manufacturing to suburban properties has been taking place since the 1970s. Factories historically needed a large amount of land, preferably in a one-story building. Large parcels of land are more likely to be available in the suburbs than in central-city locations, where accessibility makes land relatively

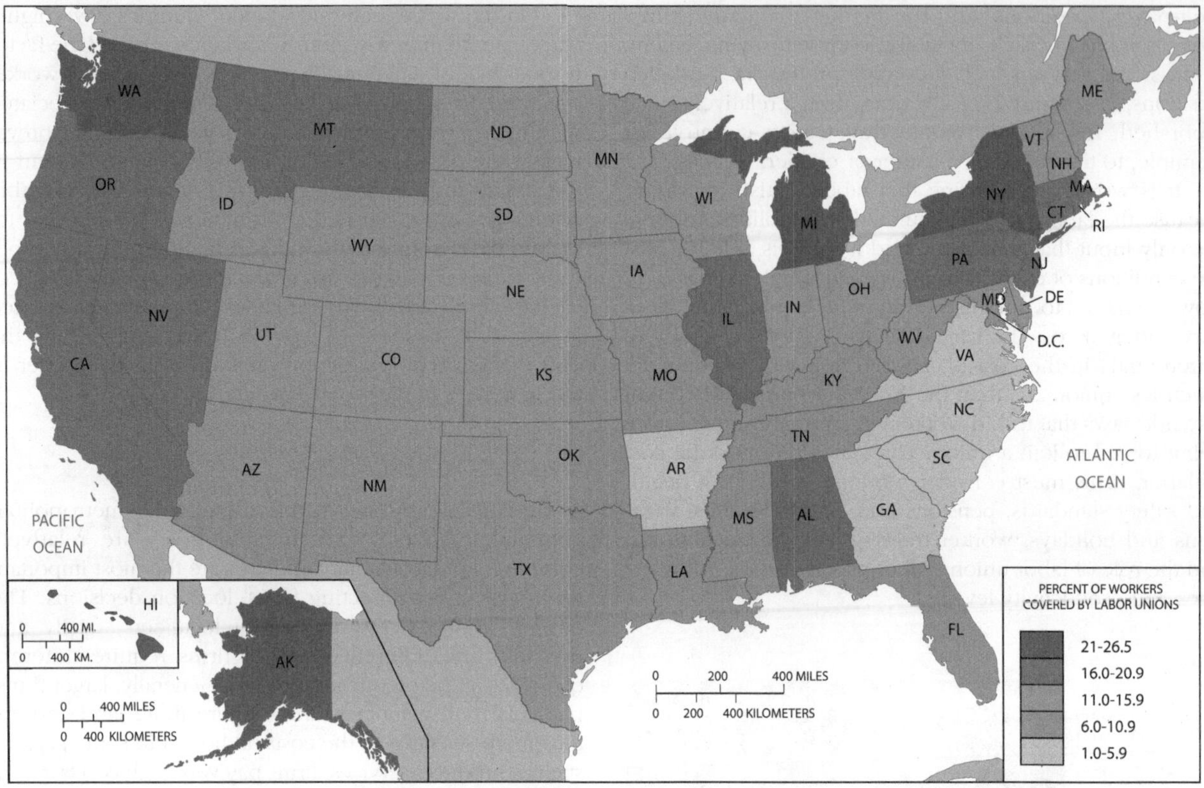

FIGURE 2 Percent of U.S. workers covered by labor unions, 2007. Unlike other inputs to the production process, labor alone can struggle to improve the conditions of its work. Unionization has declined sharply in the United States, to 15% of the labor force from a high of 45% in 1950, largely due to deindustrialization. Unions are much less well represented in services than in manufacturing and less prevalent in "right-to-work" states, particularly in the South, where they are not legally entitled to require workers at an establishment to join.

expensive. However, computerization and just-in-time inventory systems have diminished the need for land in some cases. More reasons why industrial properties have expanded into the suburbs include locations that are easily accessible to motor freight by interstate highway and beltway and access to suburban services and infrastructure, including ample sewer, water, parking, and electricity. Industries may also be attracted to the suburbs because of nearness to amenities and residential neighborhoods. Suburban locations minimize labor's journey to work, which helps to hold wages down.

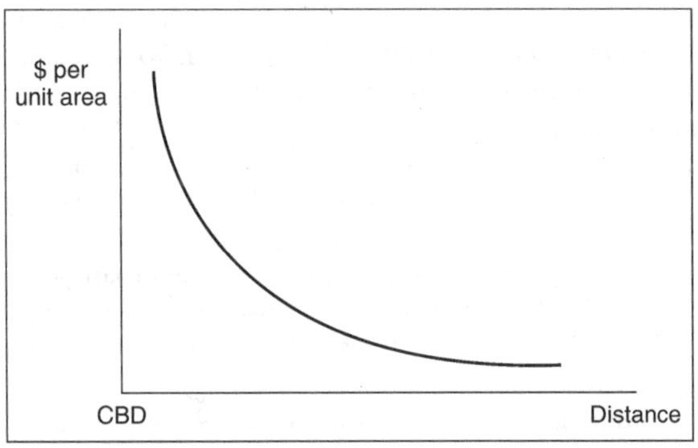

FIGURE 3 Land costs decline with distance from the central business district (CBD) in most cities because accessibility to parcels in the periphery falls and transport costs rise. Firms and households all make trade-offs between rents and transport costs depending on their specific locational needs.

Capital

Under capitalism, capital plays a dominant role in structuring the production process. Capital takes one of two major forms: **fixed capital** and **financial capital**. Fixed capital includes machinery, equipment, and plant buildings, and is often expensive. Besides the installation and construction costs, firms must budget for maintenance and repair and depreciation. The age of the capital stock of a region, or how recent the technology it deploys is, greatly affects its overall productivity levels: Places with newer fixed investments are more productive. Thus, after World War II, countries like Japan and Germany, whose capital stock had been destroyed, started over with new equipment and machinery and enjoyed high levels of productivity growth as a result. Japanese and German steel producers, for example, used electrical furnaces while American ones still relied on older Bessemer open hearth ones.

Financial capital includes intangible revenues, including corporate profits, savings, loans, stocks, bonds, and other financial instruments. The rate of capital formation reflects variables such as corporate profitability (including market prices, production costs), savings rates, interest rates, and taxation levels. Financial capital is by far the most mobile production factor. The cost of transporting liquid capital is almost zero and it can be transmitted almost instantaneously over fiber-optic lines. Fixed capital is much less mobile than liquid capital; for example, capital invested in buildings and equipment is obviously immobile and is a primary reason for industrial inertia. Any type of manufacturing that is profitable has an assured supply of liquid capital from corporate revenues or borrowing (depending on the firm's credit rating). Interest rates—the cost of capital—hardly vary within individual countries, but do vary among them.

Most types of manufacturing, however, initially require large amounts of fixed capital to establish the operation—or, periodically, to expand, retool, or replace outdated equipment or to branch out into new products. The cost of this capital must be paid from future revenues. Investment capital has a variety of sources: personal funds; family and friends; lending institutions, such as banks and savings and loan associations; and the sale of stocks and bonds. Most capital in advanced industrial countries is raised from the sale of stocks and bonds, although American firms rely on this approach more than do firms in Europe, where banks play a larger role in industrial financing. The total supply of investment capital is a function of total national wealth and the proportion of total income that is saved. Savings become the investment capital for future expansion.

Whether a particular type of industry, or a given firm, can secure an adequate amount of capital depends on several factors. One factor is the demand for capital, which varies from place to place and from time to time. Because financial capital can travel instantaneously, variations in the cost of loans within a country tend to be very small. Of course, capital can always be obtained if users are willing to pay high enough interest rates, which reflect the dynamics of capital markets and government policy. Beyond supply-and-demand considerations, investor confidence is the prime determinant of whether capital can be obtained at an acceptable rate. Investor confidence in a particular industry may exist in one area but be lacking in another. Investor confidence is expressed in several ways, including the willingness to purchase a company's stocks and bonds as well as venture capitalists' willingness to provide start-up capital for new firms.

Capital is important as well because firms can substitute capital for labor in a process of capital intensification, that is, using technology to displace workers (Figure 4). The history of capitalism is largely one of steady capital intensification in different industries, particularly in agriculture, in which only a tiny fragment of the labor force in industrialized countries now works. Capital intensification can increase productivity, but it may also leave workers without jobs. For workers, this process is beneficial only if the cost of goods drops sufficiently to increase real incomes and the rise in worker expenditures can generate job growth. Historically, this has generally been the case, although much of the job losses associated with deindustrialization are attributable to capital intensification.

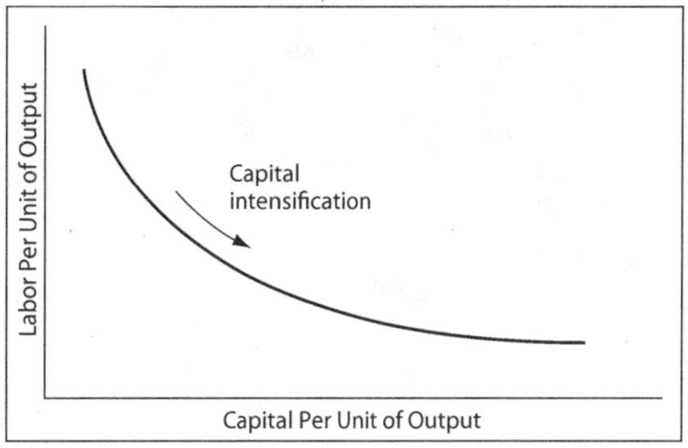

FIGURE 4 The process of capital intensification, or substitution of capital for labor, is a long-term trend under capitalism. Firms seek to reduce costs and raise productivity by using machines instead of workers per each unit of output. However, when labor costs are low, or jobs are difficult to automate, capital substitution may not occur.

One of capital's crucial advantages over labor is geographic mobility; it can make use of distance and differentiation in a way that labor cannot. Corporations take advantage of such flexibility by shifting production to low-wage regions, setting up plants in areas with low levels of worker organization, or establishing plants in areas that offer incentives. Even threatening to move can aid companies in their attempts to hold wages and benefits down.

Managerial and Technical Skills

Managerial and technical skills are also required for any type of production. Management involves the nuts and bolts of corporate decision making, including allocating resources, raising capital and negotiating financial markets, keeping abreast of the competition and government rules and policies, making investment decisions, hiring and firing, marketing and public relations, and performing numerous similar types of functions. Corporate management reflects and shapes the organizational structure of a firm, including the pattern of ownership and how decisions are made. The forms of management may range from sole proprietorships to partnerships to complex corporate hierarchies, and firms may be either publicly or privately owned (the former sell stocks and are owned by shareholders, the latter are typically family owned).

Within firms, management forms an important part of the corporate division of labor (as illustrated by the functions of corporate headquarters as compared to those of branch plants). In large firms, headquarters decides a firm's overall competitive strategy, what markets and products to focus on, labor policies, mergers and acquisitions, and types of financing. Thus, these jobs tend to be skilled, well

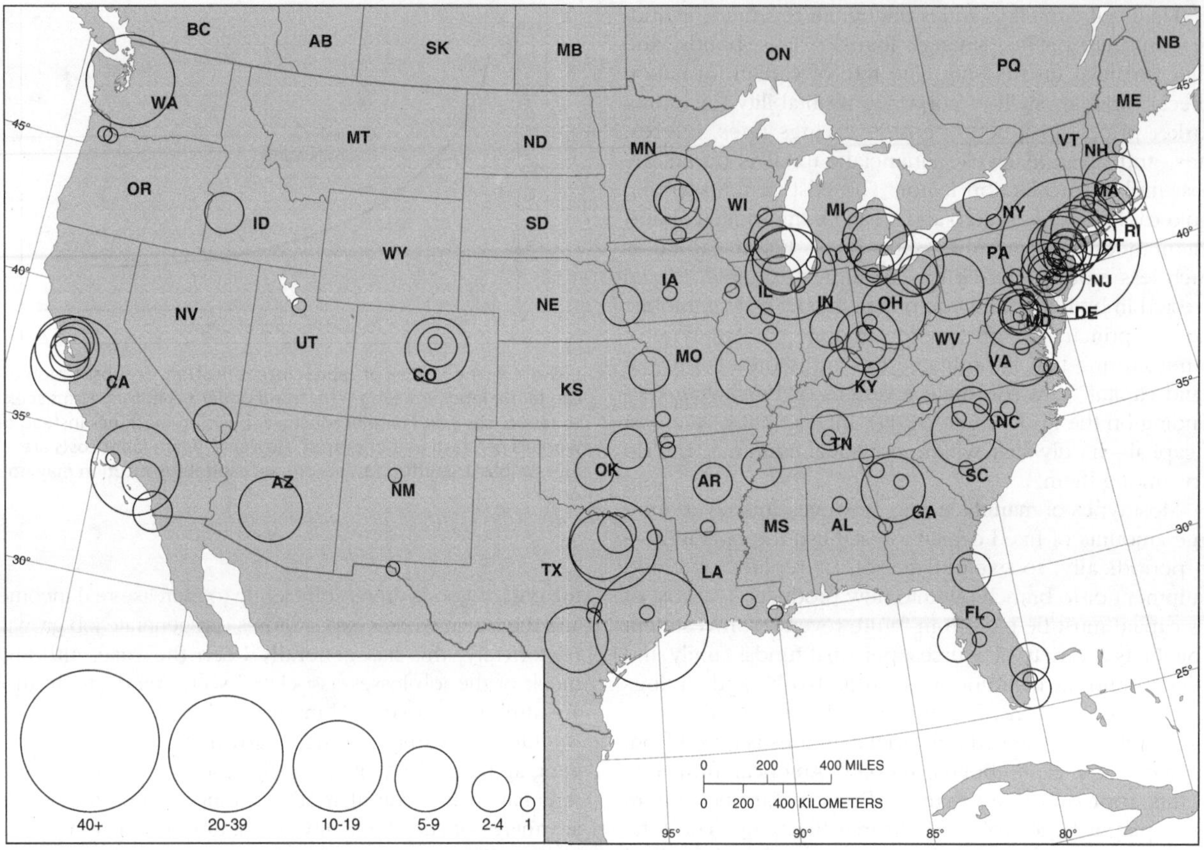

FIGURE 5 Corporate headquarters in the United States, 2008. **Headquarters are the sites for the producer services and decision-making functions required to make all companies operate effectively and profitably, including management and administration, public relations, advertising, legal services, and research and development. Because they require numerous linkages to other firms, particularly other high value-added services, headquarters generally cluster in large cities. Cities in the northeastern and midwestern United States, with skilled labor pools and dense agglomerations of ancillary services, have been the most attractive. Historically, New York City has been the location of the majority of the nation's headquarters, although there has been a gradual decentralization to other cities lower in the urban hierarchy.**

paying, and white collar. Most U.S. headquarters are in the largest urbanized areas of the country (Figure 5; Table 1). The top 10 metropolitan headquarter areas for large firms are New York City, Chicago, San Francisco, and Los Angeles, followed by Dallas, Houston, Philadelphia, Washington, DC, Boston, and Atlanta.

Technical skills are those necessary for the continued invention of new products and processes. These skills are generally categorized as research and development (R&D). In the early phases of industrialization in developed countries, product development was usually carried out in tandem with production by small firms, many of which, together with their innovations, failed to survive. Today, the R&D required for new products is typically a large and expensive process, involving long lead times between invention and production, a process that is often beyond the scope of small firms. The cutting edge of advanced industrial economies, R&D is concentrated in a few major research-university clusters and established areas of innovation, often as measured by the geography of patent applications. Three of these in the United States are Silicon Valley, the region south of San Francisco Bay in the vicinity of Stanford University (Figure 6); Boston and Route 128, home to Harvard and the Massachusetts Institute of Technology; and the Research Triangle of North Carolina, so called because of three universities located there—the University of North Carolina at Chapel Hill, Duke University in Durham, and North Carolina State University at Raleigh. Roughly equidistant from the three main cities, Research Triangle Park is home to the laboratories of IBM, Burroughs Wellcome, Northern Telecom, and other major companies conducting R&D.

For most of the twentieth century, the United States was the world's leader in technological innovation. Starting in the 1970s, however, the rate of new inventions and granted patents declined. Factors that slowed U.S. innovation included a decline in federal government support for basic research (e.g., funding for the National Science Foundation and National Institutes of Health), particularly because the bulk of federal research dollars is oriented

Table 1 Location of Major U.S. Corporate Headquarters

New York City	239
Chicago	109
San Francisco	91
Los Angeles	85
Dallas–Ft. Worth	76
Houston	70
Philadelphia	70
Washington, DC	66
Boston	66
Atlanta	53
Minneapolis	50
St. Louis	39
Cleveland	35
Detroit	34
Miami	31
Denver	27
Milwaukee	26
Nashville	25
Phoenix	23
Tampa	20
Seattle	19
San Diego	18

toward the military, because of a widespread lack of interest in science education among the public, because many organizations focus on quick returns rather than long-run growth, and because of difficulties in obtaining venture capital. Corporate outlays for R&D rose in the late twentieth century as foreign competition and high labor costs forced firms to automate in order to reduce costs and increase productivity. In 2006, U.S. expenditures for R&D represented 2.6% of the gross domestic product (GDP), compared to Japan (3.4%), South Korea (3.2%), Germany (2.5%), the United Kingdom (2.3%), France (2.1%), and China (1.4%). As a group, these countries dominate the world in the number of R&D scientists and engineers and in the amount of R&D expenditures.

THE WEBERIAN MODEL

A list of location factors is not a theory. Combining these elements in an analytical way is the task of location theory, a time-honored part of economic geography. Classical industrial location theory is founded on the work of Alfred Weber in 1929, a German economist and younger brother of the famous sociologist Max Weber.

Weber's approach emphasized the role of transportation costs in the location decisions of individual firms. He attempted to determine the patterns that would develop in a world of numerous, competitive, single-plant firms. Although the model originated in the study of manufacturing, it is applicable to other sectors, such as services, as well. Weber began by assuming that transportation costs are a linear function of distance. The model assumes that producers face neither risk nor uncertainty and that the demand for a product is infinite at a given price; producers could sell as many units as they produced at a fixed price.

Weber taught geographers to think about the distinction between material- and market-oriented industries. The first cost faced in the production process is that of assembling raw materials. Raw materials, such as coal, are found only at specific locations; their transportation costs are a function of the distance that they must be moved. For each case we must consider (1) the costs of assembling raw materials (*RM*), (2) the costs of distributing the

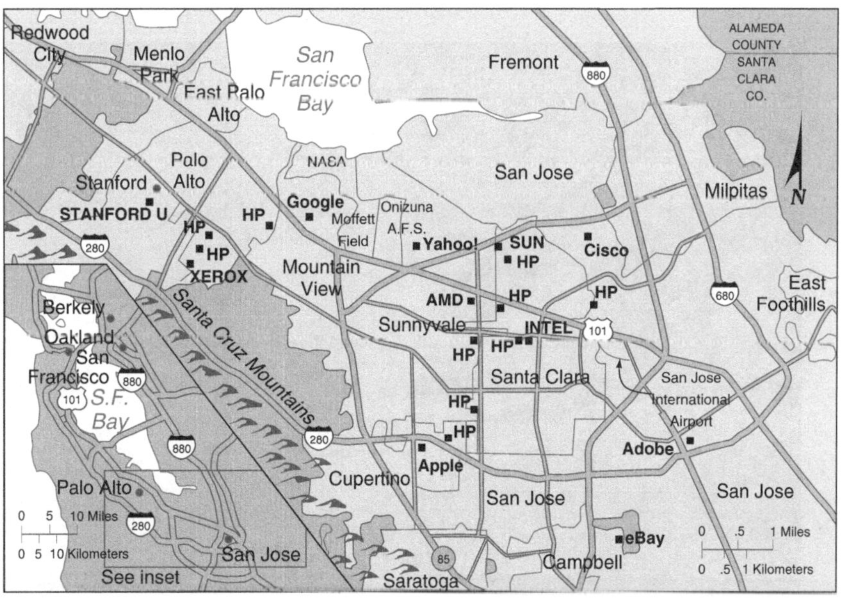

FIGURE 6 Silicon Valley, just south of San Francisco, California, is the world's largest region for producing computer software. It exemplifies the tendency of skilled, high-value-added economic activity to cluster in distinct areas, much like Italy's Emilia-Romagna or Germany's Badden-Wurtenburg. U.S. leadership in high-value economic activity is being challenged these days by China, however. China repeatedly and willfully drives its currency value down to make its exports cheaper than those produced in America. From 2000 to 2010, the U.S. trade deficit with China grew 400%. During the same time, 5.4 million American jobs in manufacturing were eliminated. Even though the United States develops most of the innovations in high value-added economic activity, once China adopts the innovations, it is tough for U.S. manufacturers to compete against China's lower wages, looser regulations, and cheaper currency. However, there is an upside to cheap Chinese imports: They lower the cost of products to world and U.S. consumers, and China uses many of its excess dollars gained by selling products to America to buy U.S. Treasury Bills, financing the $14 trillion deficit of the U.S. government, and holding down interest rates.

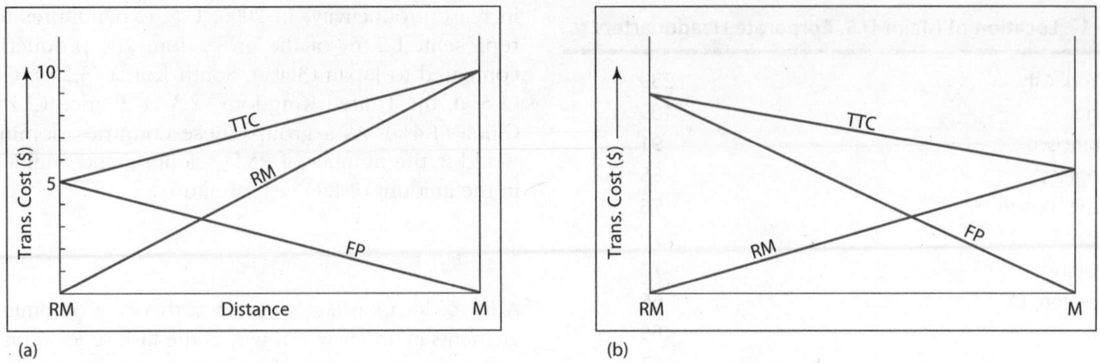

FIGURE 7 Weber's model: one localized raw material, located at *RM*. The line *RM* represents assembly costs, while the line *FP* represents finished-product distribution costs. In Figure 7a, total transport costs are lowest at *RM*, as in primary-sector activities. In Figure 7b, total costs (*TTC*) are lowest at the market (*M*); in this example an industry in which the raw material accrues weight in the production process, total transportation costs are minimized by locating the plant at the site of the market.

finished product (*FP*), and (3) the total transportation costs (*TTC*) (Figure 7). The best location for a manufacturing plant is the point at which total transportation costs are minimized.

In Figure 7a, the assembly costs for the localized raw material (*RM*) are minimized at point *RM*. Finished-product distribution costs are minimized at *M*. Thus, in cases where the cost of shipping raw materials outweighs the costs of shipping the final product, the firm will locate near the raw materials—an iron ore mining company, for example (Figure 8). In Figure 7b, raw materials, once processed, add to the weight of the finished product, so the total transportation costs (*TTC*) are minimized at the market (*M*). Bottled and canned soft-drink manufacturing exemplifies the use of one pure raw material (syrup concentrate). If the plant locates at the market, the water, which makes the largest contribution to the weight of the finished product, does not need to be moved (Figure 9).

Weber in Today's World

Weber's model was originally developed from an analysis of manufacturing firms, although it can be applied to other sectors. It is useful to explore the influence of transportation costs in a rigorous way, and many economic geographers still work in this tradition. But despite its broad appeal, several developments limit its applicability. First, not all firms need to minimize transport costs; as we saw with the rent–transport trade-off that all firms must make, some firms will accept higher transport costs and locate on the urban periphery. Second, the production process is much more complex than it was in the early twentieth century

Tim Graham/The Image Bank/Getty Images

FIGURE 8 Gravel mining operations, such as this open cast iron ore in Carajas, Brazil, reflect an industry in which transport costs for the inputs are high, which leads firms to locate near the raw material. Localized pure raw materials notwithstanding, few firms are making new locational decisions at the present time due to the recent Great Recession. New investment money is tight. All recessions are painful, but the hangovers that follow financial crises, such as the recent one from 2007 to 2009, are particularly long and grim. Growth is substantially lower than during "normal" recoveries as households and firms reduce their debt burdens. Output is sluggish, and credit is growing weakly or shrinking across much of the rich world.

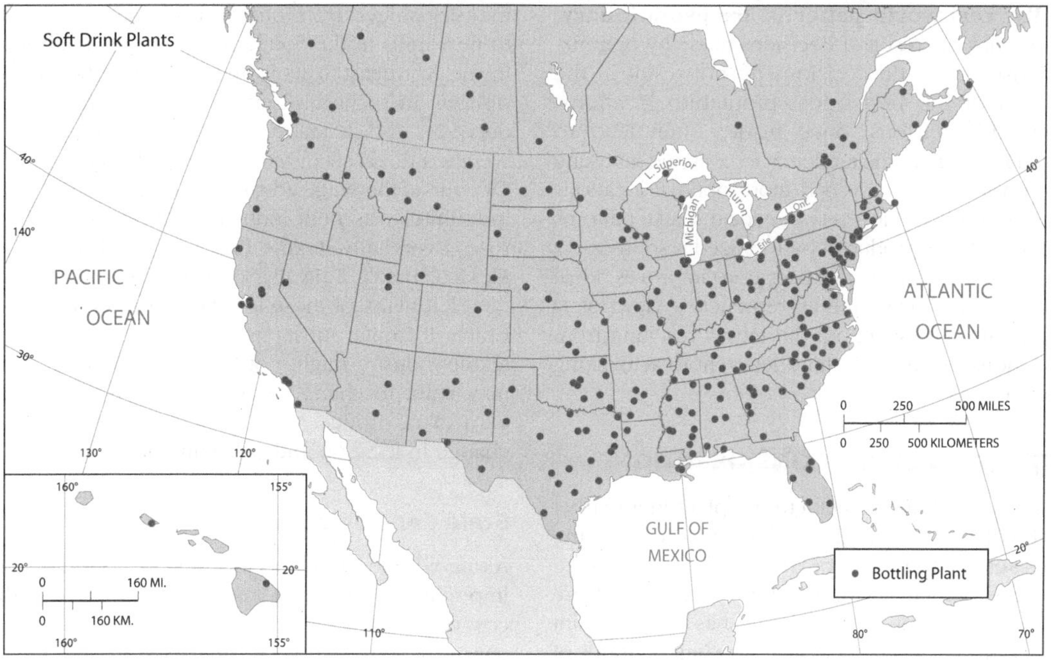

FIGURE 9 Bottled and canned soft-drink manufacturing exemplifies a form of production that is market oriented. Most concentrate on the peripheries of metropolitan areas.

when Weber developed his model. Many plants begin with semifinished items and components rather than with raw materials. Producers' goods seldom lose large amounts of weight; therefore, there is not much of a tendency to locate near raw materials.

Weber's model has also been criticized for its unrealistic view of transportation costs as a linear function of distance. Because of fixed costs, especially terminal costs, the mileage costs of long hauls are less per unit of weight than those of short hauls. Plants tend to locate at material or market points rather than at intermediate points, unless there is a forced change in the transportation mode, such as at a port. However, the disadvantages of intermediate locations have been reduced with the expansion of the modern trucking industry and its flexibility for short hauls.

Two other developments have a bearing on how choices of industrial location have changed:

1. **Transportation costs have been declining in the long run**. This decline increases the importance of other locational factors, particularly labor costs and productivity. This is most obvious in firms producing high-value and high-tech products. For these firms, transportation costs are relatively unimportant. Yet for firms that distribute consumer goods (e.g., soft drinks) to dispersed markets, transportation costs remain a significant factor (Table 2).
2. **Brainpower has been steadily displacing muscle and machine power** and transforming natural resources. Natural resources are no longer as important in the growth of economies as they were historically. Instead, a widespread transmaterialization of resources has occurred as smaller, lighter products are made from resources to which high technology and brainpower have been added.

TABLE 2 Transport Costs as a Percentage of Product Prices

Stone, clay, and glass	27
Petroleum	24
Lumber and wood	18
Food	13
Furniture	12
Paper products	11
Primary metals	9
Textiles	8
Fabricated metals	8
Transport equipment	8
Rubber and plastics	7
Tobacco	5
Machinery	5
Instruments	4
Apparel	4
Printing and publishing	4
Electronics	4
Leather products	3

Source: Compiled by authors from U.S. Department of Commerce national input-output tables.

3. **Finally, real-world patterns are evolutionary**; they are not the result of decisions made by optimizers. Most real-world decisions do not result in the selection of the best (most profitable) locations. Locational decisions, once made, often lead to **industrial inertia**, the tendency to continue investing in a nonoptimal site even if more optimal locations exist. Tensions develop between ideal spatial patterns and the patterns produced by localized resources. As technology (especially transportation) improves, ideal spatial patterns (from the entrepreneur's viewpoint) become more feasible, but the inertia resulting from past actions exerts a constant deterrent to actualizing these patterns.

TECHNIQUE AND SCALE CONSIDERATIONS

The establishment of any manufacturing plant in a market economy involves the interdependent decision-making criteria of **scale**—the size of the total output—and technique—the particular combination of inputs that are used to produce an output. Technique has an important effect on a firm's locational decision. A certain amount of land (resources), labor, and capital is needed to produce any finished product, but, within economic and technological limits, firms can substitute among inputs to minimize their costs: Capital may be substituted for labor. The greater the differences are among inputs in terms of their prices and productivity levels, the more incentive firms have to substitute among them.

The most evident trend has been substitution of capital in the form of machinery for labor. This trend is most evident in agriculture, which has progressively changed from very labor-intensive to capital-intensive over time. More and more manufacturing systems, which apply sophisticated technology, including robotics, to improve the quality and efficiency of production are replacing labor. In services, capital intensification is also evident in the computerization of the office. Whether or not substitution between production factors occurs depends on the relative costs and productivity of the two inputs and the scale and locational decisions already made by the firm. If, for example, labor costs rise at a given location, the firm may choose to substitute capital for labor at that location, or it may opt to change locations to take advantage of lower labor costs and thus maintain the same labor-to-capital ratio.

The limits to substitution among inputs vary considerably from industry to industry and are fixed for given periods of time by technological constraints and the prices of inputs. Firms must choose from an available suite of options as to how they want to produce, and the costs of labor, land, and capital will greatly affect these choices. Some industries lend themselves more to capital intensification than do others. Petroleum refining, for example, can be readily automated, whereas garment manufacture cannot. The garment industry, therefore, is much more sensitive to changes in labor costs than is petroleum. In the late nineteenth and early twentieth centuries, the U.S. textile industry shifted from old, multistoried New England mills to new mills in the Southern Piedmont as labor costs rose in the Northeast. This is an example of the influence that options in technique exert to determine the locational decision. The increased labor costs in the Northeast outweighed the costs of moving the industry to the Southeast. Of course, the wage advantage of the South did not persist indefinitely; as new industry moved south, wages there rose. Eventually, textile firms migrated farther afield—to Mexico, Brazil, Taiwan, South Korea, and Singapore, and then left many of those for even cheaper locations such as China. If capital substitution had been a viable option, the textile industry might not have moved. Many times a firm may want to change its scale to increase output and to earn extra profits. A change in scale may also require a change in location and/or technique.

Scale Considerations

Along with location and technique, economic scale is also important because producers are concerned with the unit cost of production—and adjustments in scale can produce considerable variations in unit cost. Scale is the means by which production is "tuned" to meet demand. In some economies, this tuning may be done by the state; in others, by private entrepreneurs.

Principles of Scale Economies

Along with standardization of parts, the **division of labor** is a primary feature of mass production. Workers who perform one operation in the production process over and over are much more efficient than those who are responsible for all phases. The division of labor not only increases the efficiency of production but also facilitates the use of relatively unskilled labor. A worker can learn one simple task in a short time, whereas the skills required to master the entire operation might take years to learn. This process was instrumental, for example, in the early growth of the U.S. automobile industry in the system pioneered by Henry Ford. Division of labor, however, requires a relatively large scale because a large pool of workers is necessary. A common way to measure the size of a firm is by the number of employees. Capital, once invested in machinery and buildings, becomes fixed capital and produces income only when in operation. A three-shift firm makes much more efficient use of its fixed capital than a single-shift firm does. The three-shift firm is three times larger in scale, measured by employment, yet its fixed capital investment may be no more than that of the single-shift firm.

Economies of scale, or scale economies, refer to the reductions in costs associated with the production of output in large quantities. Large firms generally pay much less for material inputs than small firms do. For example, Ford Motor Company can obtain tires for a much lower unit price than an individual dealing with the same tire company can, because Ford buys millions of tires a year.

Increasing scale, in other words, generally lowers the unit cost of inputs.

Economists portray scale economies as a curve of long-run average costs (LRACs), which graphs the unit costs as a function of scale. Several possible LRAC curves are indicated in Figure 10. Notice that unit costs decrease, reach an optimum point (O), and then began to increase. The rise in the curve is termed **diseconomies of scale** (diminishing marginal returns to scale) and occurs when a firm becomes too large to manage and operate efficiently. The optimum scale of operation is very small in Industry A, very large in Industry C, and fairly wide-ranging in industry B. Firms in Industry A should be small, they should be large in Industry C, and they can range from small to large in Industry B.

Vertical and Horizontal Integration and Diversification

Besides simply increasing plant size, two other means are commonly employed for effecting scale changes. Some firms purchase raw material sources or distribution facilities. This is called **vertical integration** (or vertical merger) in that the firm controls more steps "up and down" in the total production process. Some early large automobile firms, for example, owned their own iron and coal mines and produced their own steel; over time, however, auto producers have become more vertically disintegrated, that is, the production of different parts has been taken over by specialized firms. Large oil companies are often vertically integrated; they control exploration, drilling, refining, and retailing. In contrast, **horizontal integration** (or horizontal merger) occurs when a firm gains an increasing market share of a given niche of a particular industry. This is typical when markets become oligopolistic, that is, when they are controlled by a handful of large firms.

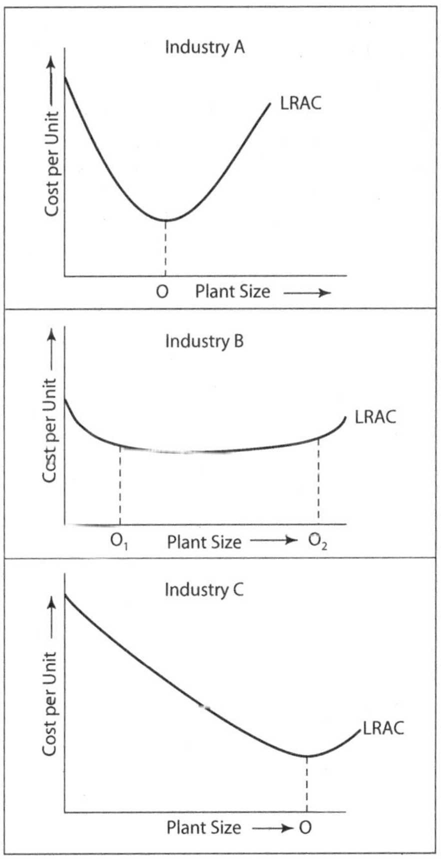

FIGURE 10 Variations in long-run average cost curves (LRACs). Organizational and technical constraints make the optimal level of output vary according to the characteristics of the industry and the firm as it tries to achieve economies of scale and avoid diseconomies of scale.

Another trend among corporations in the United States, Japan, and Western Europe has been a strategy of **diversification**. Many large corporations, through conglomerate mergers and acquisitions, control the production and marketing of diverse products. A company may produce many unrelated products, each of its operations having elements of horizontal and vertical integration. Diversification spreads risk and increases profits. Diminishing demand for the products of one division may be offset by rising sales in another.

Most industrial location theory is based on small, single-plant operations producing a single product. Large corporations are much more complex, but they deal with all the variables of location theory and must still make locational decisions. Although large enterprises may seem to be more concerned with technique and scale decisions, each locational decision has an effect on scale and technique. We should consider two points: First, large firms may be able to operate in less than optimal locations and still have a significant effect on the market through the control that they exert over government policies and the prices and sources of raw materials. Conversely, large firms may be able to make optimal locational choices through their employment of the scientists and technical personnel who help top management make more profitable decisions.

Interfirm Scale Economies: Agglomeration

So far, we have been concerned with intrafirm scale economies, that is, within one company. However, scale economies also apply to clusters of firms in the same or related industries—for example, the computer firms located in California's Silicon Valley. By clustering, unit costs can be lowered for all firms, in ways that could not occur if they were spatially separate. These economies, called **agglomeration economies**, take several forms.

Production linkages accrue to firms locating near other producers that manufacture the same basic raw materials. By clustering, distribution and assembly costs are reduced. **Service linkages** occur when enough small firms locate in one area to support specialized services. The advertising industry in New York offers an example (Figure 11). Advertising agencies must cluster within a short distance of Madison Avenue in order to avail themselves of the dense networks of interfirm linkages to be found there, including information and gossip on the latest trends, markets, clients, hires, and products. For a firm to be successful, it must be near this complex; otherwise, it might as well be on the other side of the moon.

In addition to production and service linkages, there are marketing linkages, which occur when a cluster is large enough to attract specialized services. The small firms of the garment industry in New York City have collectively attracted advertising agencies, showrooms, buyer listings, and other aspects of finished-product distribution that deal exclusively with the garment trade. Firms within the cluster have a cost advantage over isolated firms that must provide these specialized services for themselves or that must deal with New York City firms at a considerable distance and cost.

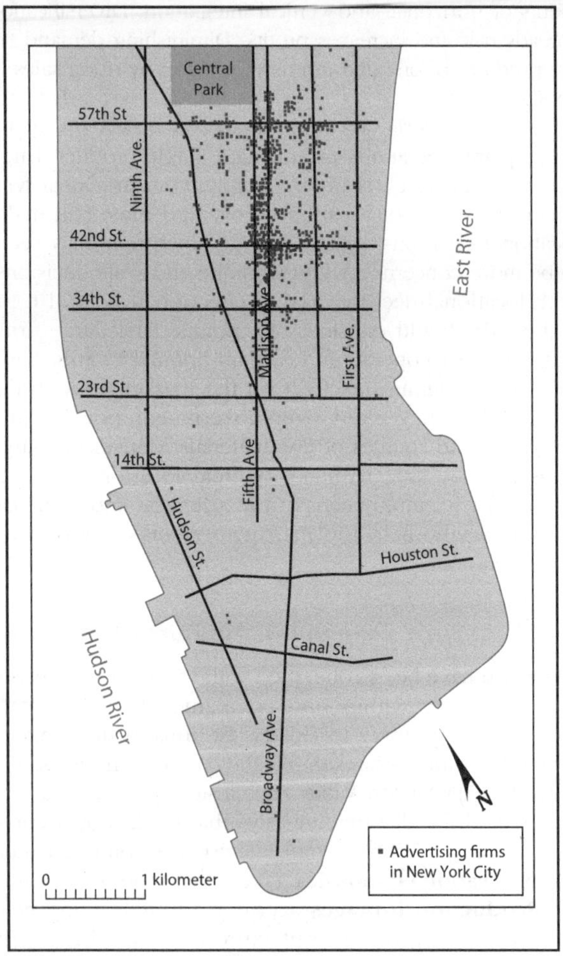

FIGURE 11 Advertising agencies in Manhattan must cluster near Madison Avenue to achieve the agglomeration economies so essential to labor-intensive, information-intensive, white-collar, high-value added functions such as producer services.

Cities provide markets, specialized labor forces and services, utilities, and transportation connections, as well as access to specialized information. Clusters of production thus arise from the numerous interconnections of people, goods, services, and information that stitch individual entrepreneurs, firms, their supplier networks, ancillary services, and related institutions together. In addition to firms, other entities in clusters may include government agencies and offices, public/private partnerships, trade associations, universities, legal services and patent attorneys, accounting firms, specialized advertising firms, and related ancillary services.

Clusters arise because the creation of new knowledge and products in a constant process of innovation typically takes place within the confines of small geographic areas. Agglomeration is essential to the creation of knowledge, typically through the interactions of skilled individuals, and to the production of synergies, the positive aspects of interaction, such as new ideas, processes, and products that would be unlikely to arise from firms or individuals acting in isolation. Organizational learning and knowledge transfer occur most effectively when firms are in close physical proximity, whether they are rivals or collaborators. Both situations are essential to the generation of an entrepreneurial and creative dynamic. The process of discovery and innovation is closely related to collaborative relationships, networking, and tight spatial linkages among firms and individuals. In essence, knowledge spillovers rely on frequent, repeated, and sustained interactions among individuals and firms in a given location, which creates synergies: interactions that generate benefits (i.e., ideas) that would not be possible if actors (firms, individuals) operated in isolation (i.e., the combined effect is greater than the sum of the parts). The synergistic benefits of agglomeration are often labeled "positive external economies of scale" (i.e., external to an individual firm) in that they lower production costs in ways that would not be possible if firms and workers were geographically dispersed. Thus, in most economic sectors, research and intellectual activity are concentrated in or near large metropolitan areas, whereas unskilled, low-wage occupations that involve little creative activity (e.g., branch plants, back offices) tend to disperse to smaller towns. **Urbanization economies**, therefore, are a combination of production, service, and marketing linkages concentrated at a particular location.

Evaluation of Industrial Location Theory

Are industrial location patterns rational? Do firms always search for optimal locations? What factors shape their decisions? Do they make mistakes?

From the perspective of behavioral geography, the main defect of normative industrial location theory is that it fails to describe what decision makers actually do. How do enterprises actually select profitable sites for branch plants? Economic geographers often find that the behavior of corporate executives does confirm the validity of the framework of locational search, learning, and choice

FIGURE 12 The classical principles of industrial location theory are evidenced in the river valley and railroad site of Bethlehem Steel Corporation's plant at Bethlehem, Pennsylvania. This huge plant, which extends for nearly 8 kilometers along the south bank of the Lehigh River, converts raw materials—Appalachian coking coal and Minnesota iron ore—into structural shapes, large open-die and closed-die forgings, forged steel rolls, cast steel and iron rolls, ingot molds, and steel, iron, and brass castings. The main market for the steel products is the American Manufacturing Belt with its abundance of metal-using industries.

evaluation (Figure 12). The first phase of the decision-making process is the recognition that a growth problem exists with respect to demand. The possible responses are in situ expansion, relocation, acquisition, or construction of a new plant. A new plant involves a three-stage search procedure, the outcome of which leads to a decision and, finally, to the allocation of resources. The process also generates feedback into learning and decision-making behaviors. Classical industrial location theory tends to be static and ignore the time horizon within which firms operate: it is important to distinguish between the organization's short- and long-term responses. The behavioral approach is much more realistic in that it recognizes that the environment in which the enterprise operates is in a constant state of flux.

Location theory may be evaluated by comparing optimal patterns against real-world ones. The final location chosen by an industry is not always determined by transportation costs, as was Weber's principal conclusion, nor by production costs at the site, including land, labor, capital, and managerial and technical skills. There are several factors that complicate locational decisions:

1. A firm may have more than one critical site or situation factor, each of which suggests a different location.
2. Even if a firm clearly identifies its critical factor, more than one critical location may emerge.
3. A firm cannot always precisely calculate costs of situation and site factors within the company or at various locations due to unknown information.
4. A firm may select its location on the basis of inertia and company history. Once a firm is in a particular community, expansion in the same place is likely to be cheaper than moving to a new location.
5. The calculations of an optimal location can be altered by a government grant, loans, or tax breaks.
6. Noneconomic factors play important roles in **footloose industries** that have gravitated to coastal areas in the Sunbelt of the United States because of recreational opportunities and other amenities.

Structuralists have also criticized industrial location theory because it focuses on individual firms as abstract entities, without embedding and contextualizing them within the rest of the economy. For structuralists, locational analysis begins at the top, with the world's capitalist system, not at the bottom, with individual firms. The actual behavior of the individual firm only takes on meaning in the broader economic context of the class relations and the historical dynamics of capitalism as an integrated system. Working up from the bottom can explain neither the individual elements nor the system as a whole. According to structuralists, then, industrial location theory is unrealistic because it focuses on only a small part of reality, that is, the firm, at the expense of the broader organization of society, power, and class. This criticism extends to the new approaches in location theory in which the simple conception of the single-plant firm has been replaced by a model of the firm as a complex organizational structure.

HOW AND WHY FIRMS GROW

Most large companies operate at the maximum scale possible on the LRAC curve. In fact, evidence of increasing returns to scale has led to a reappraisal of the theory of the firm. The tendency toward increasing scales of operation is therefore based on the motivating force of growth. Firms expand for two reasons: survival and growth. Both goals are promoted by horizontal and vertical expansion and by diversification.

The view that corporate growth is part of a natural progression is overly deterministic, however, and it flies in the face of reality. The majority of firms in an economy remain small and peripheral. Only some firms, especially those that manufacture capital goods, have the potential to develop into large corporations. Financial barriers (i.e., lack of investment capital and necessary information and management skills) prevent most firms from making successive transitions from a small, regional base to larger, national organizations and then to multinational operations. Access to finance—banking capital, venture capital, and international bond and currency markets—has become increasingly uneven, favoring some firms and not others. Because these gaps in financial access have become wider, a small firm has much less of a chance of evolving into a corporate giant today than it did a hundred years ago.

How a firm grows depends on the strategy that it follows and the methods that it selects to implement its strategy. As we discussed earlier in the chapter, growth strategies are either **integration** or diversification. In the United States, horizontal integration predominated from the 1890s to the early 1900s, vertical integration came to the fore in the 1920s, and diversification has been the principal strategy since the 1950s. This three-stage sequence provides a framework for understanding the interrelationship of the various strategies. The early growth of large enterprises involves the removal of competition by absorption, leading to oligopoly. This is followed by a period in which the oligopoly protects its sources of supply and markets by vertical integration, buying firms "upstream" and "downstream" in the production process. Once a dominant position is achieved, rapid corporate growth can proceed only with diversification.

Methods for achieving growth are internal or external to the firm. Growth can be financed internally by the retention of funds, taking out loans, or by issuing new shares of stock. Or it can be generated externally by acquiring the assets of other firms through mergers. Most large firms employ both means, but external growth is particularly important for the largest and fastest-growing enterprises in industries such as biotechnology and electronics.

Whatever strategy and method are adopted, corporate growth typically involves the addition of new factories and, thus, a change in geography. Initially, much of the employment and productive capacity of a firm concentrates in the area in which it was founded. As enterprises grow, they become more widely dispersed multiplant operations, which is sometimes accompanied by decreasing dominance of the home region. Exceptions tend to be companies confined to one broad product area and based in a region where there is historical specialization within that product area.

The choice of growth strategy affects corporate geography. Horizontal integration frequently involves setting up plants over a wider and wider area. The geographic consequences of vertical integration vary according to whether the move is backward ("down" in the production process) or forward ("up" in the production process). **Backward integration**, in which a firm takes over operations previously the responsibility of its suppliers, can lead a firm into resource-frontier areas. An example is the development of iron-ore deposits by U.S. and Japanese companies in Venezuela and Australia. Conversely, **forward integration**, in which a firm begins to control the outlets for its products, can lead a resource-based organization to set up plants in market locations. Diversification does not have such predictable consequences for the geography of large enterprises.

The method of growth also affects the geography of multiplant firms. When growth is achieved internally, enterprises can carefully plan the location of new branch plants. When growth is achieved externally, enterprises inherit facilities from acquired firms; hence, there is less control over their locations. Moreover, the attractiveness of new facilities often lies in their economic, financial, and technical characteristics. Nonetheless, geography does play a role in the decision process. Firms typically confront the uncertainty and risk of expansion by investing first in geographically adjacent or culturally similar environments.

Geographers have developed models of how firms grow. Most of these models postulate a single development path beginning with a small, single-plant operation and culminating with the multinational enterprise. As we have mentioned, this kind of evolution along a path from a local to a national and then to an international company is exceptional. Unequal access to finance makes it difficult, if not impossible, for many firms to expand beyond the subnational scale. In the late twentieth century, the size distribution of firms resembles a broad-based pyramid in which fewer and fewer firms can move from one level to another. Rather than the single developmental sequence that may have existed in the nineteenth century, today multinationals follow a distinctive path through a series of discrete development sequences.

GEOGRAPHIC ORGANIZATION OF CORPORATIONS

Multi-establishment, multiproduct corporations, which include headquarters, manufacturing plants, research laboratories, education centers, offices, warehouses, and distribution terminals, have their own distinctive geographies. To appreciate the internal geography of these systems, two issues must be considered: (1) the ways in which corporations are organized to maximize efficiency, and (2) the influence of hierarchical management structures on the location of employment.

Organizational Structure

Companies organize themselves hierarchically in a variety of ways to administer and coordinate their activities. The basic formats are (1) functional orientation, (2) product orientation, (3) geographic orientation, and (4) customer orientation. A fifth format, which is a combination of at least two of the basic formats, is called a matrix structure. Different companies may select different formats, but all formats are always subject to review and modification.

The organizational format that is based on various corporate functions—manufacturing, marketing, finance, and research and development—is illustrated in Figure 13. With this framework, all the company's functional

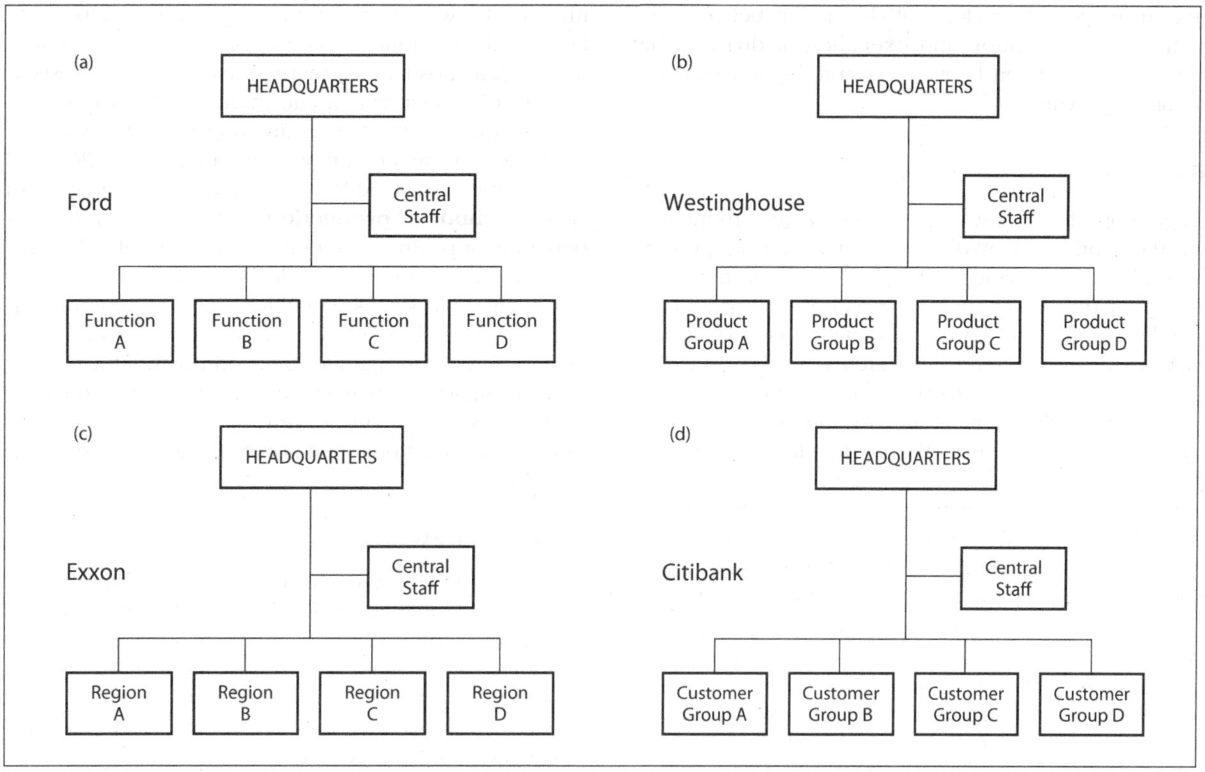

FIGURE 13 Organizational structures: (a) functional orientation, such as the Ford Motor Company; (b) product orientation, such as Westinghouse; (c) geographic orientation, such as Exxon; (d) customer orientation, such as Citibank.

operations are concentrated in one sector of the enterprise. An example of a company with this type of organizational structure is Ford Motor Company. This form of organization works well for companies with relatively narrow ranges of products.

Figure 13b illustrates the product-orientation organizational structure. For a major motor vehicle manufacturer, product groups can be cars, trucks, buses, and farm equipment. Although a central corporate staff is needed to provide companywide expertise and some degree of assistance, each group also has its own functional staff. Thus, a fairly high degree of managerial decentralization occurs. The product-orientation format works well for companies with diverse product lines. Westinghouse is an example of a company organized according to this format.

A third organizational format is based on geographic orientation—either the geographic location of customers or of the company's production facilities. The company is organized around regions rather than functions or products. Under this form of organization, most or all of the corporation's activities relating to any good or service that is bought, sold, or produced within a region are under the control of the regional group head, and each geographic region is a separate profit center. This organizational format is best suited to companies with a narrow range of products, markets, and distribution channels. It is popular among oil companies and major money-center banks.

Some companies organize according to the types of customers they want to serve rather than the locations of customers. For example, commercial banks are commonly made up of personal, corporate, mortgage, and trust departments. Alternatively, manufacturing corporations might be structured into industrial, commercial, and governmental divisions according to the prevalent type of customer for each group.

The various organizational structures all have advantages, but none is ideal for all companies. Indeed, it is safe to say that these formats have drawbacks for most or all of the companies that have adopted them. Nonetheless, a company usually chooses one basic format as the most satisfactory structure for its needs at a particular time in its evolution—or it creates a combination of two or more types.

For the small, single-plant firm, strategy and production functions are not geographically separated; hence, there is no need for an intermediate tier of coordinating activities. As firms grow to become multilocational companies, more complex functional and spatial divisions of labor develop. One of the best-known forms of spatial organization draws on the characteristics of large electronics companies. Strategic functions such as research and marketing are performed at the headquarters. Coordinating functions are dispersed to regional offices that control interdependent production facilities. This organizational structure represents a clear-cut distinction between the functions of conception and execution, a division that mirrors the distinction between nonproduction and production employment.

Administrative Hierarchies

A large proportion of the employees of large corporations, even those primarily involved in manufacturing, perform nonproduction activities. This proportion is increasing because of the substitution of capital for labor and because of the growth of various activities associated with administration, management, research, advertising, finance, legal services, and so forth. The ratio of nonproduction to production employment is less important from a geographic perspective than is the relative location of these activities.

Strategic head-office functions tend to cluster in a relatively few large metropolitan areas, especially in the case of huge firms with a financial orientation rather than a production orientation. Headquarter functions rely heavily on access to other firms, including clients, suppliers, advertisers, repair and maintenance, and specialized law firms, and the agglomeration economies of large cities. Many of these functions require face-to-face contacts and business meetings. The contribution of head-office establishments to nonproduction employment within corporations has more strategic than numerical significance.

ECONOMIC GEOGRAPHY AND SOCIAL RELATIONS

Because economic geography includes analysis of how firms locate and behave in space, it is necessary to embed firms in their social and historical contexts. From the political economy perspective, production is a *social*, not purely an individual, process.

The crucial **social relation of production** is between owners of the means of production and the workers employed to operate these means. Under capitalism, the means of production are privately owned, which divides people into owners of capital and those who must sell their labor power in order to survive. Owners of capital—the capitalist class that came to power beginning in the sixteenth century—control the labor process and extract surplus value through the exploitation of workers. Competitive relations exist among owners; cooperative and antagonistic relations emerge between owners (capital) and workers (labor).

Relations among Owners

Capitalists make independent production decisions under competitive conditions. An appreciation of capitalist development requires understanding that a raw competitive struggle for survival is fundamental to it. Competition requires producers to apply a minimum of resources to achieve the highest output. It forces companies to minimize costs, which means the extreme specialization of labor and subordination of workers to machine automation. It demands large-scale production to lower costs and to control a segment of the market. It also entails the acquisition of linked or competing companies and the investment of capital in new technology and in R&D.

Competition is the source of capitalism's immense success as a **mode of production**. But the tensions that arise between opposing elements cannot be solved without fundamental change. Consider, for example, the critical environmental issues generated by the contradiction between capitalism and the natural environment. For productive forces to continue to expand without a reduction in living standards, new values must be built into the production system. The use of renewable energy sources and the imposition of pollution controls are evidence of these new values.

Relations between Capital and Labor

The relations between capital and labor are both cooperative and confrontational. Without a cooperative workforce, production is impossible. However, cooperation is often subsumed by antagonism.

Because producers make decisions according to their desire to make profits, they try in every way possible to pay workers only part of the value produced by their labor. Value produced by workers in excess of their wages—called **surplus value**—is the basis for profit. This view emphasizes the dynamics of the workplace and the labor process rather than of supply and demand in the market. Workers try to increase their wages in order to enjoy a higher standard of living. They sometimes organize into unions and, if necessary, strike to demand higher wages. If management agrees to meet labor demands, cooperative relations may exist for a time before antagonistic relations resume.

Competition forces management to invest in technology and research to increase productivity. As production increases, the struggle between employers and employees puts higher wages into the workers' hands. Machines and low-wage labor can replace high-wage labor. Low-wage peripheral regions can sell products to high-wage center regions. Industrial migration to the periphery removes jobs in the center, which disciplines organized labor. Pressures to increase wages slacken, and mass demand decreases. A problem of underconsumption develops. Thus, in capitalism, the solution to one problem may be the breeding ground for new problems.

Competition and Survival in Space

Relations among owners and between capital and labor are sources of change in the geography of production. Competition among owners may cause a company to relocate all or parts of its operation to a place where it can secure low-wage labor. From the company's perspective, this strategy is mandatory for survival; if other companies lower their costs and it does not, it will inevitably lose in the competitive struggle. Capitalists must expand to survive, and the struggle for existence leads to the survival of the biggest. In their search for profits, giant corporations have extended their reach so that few places in the world remain untouched.

The incessant struggle of companies to compete successfully is especially evident in the entrepreneurial response to differential levels of capital–labor conflict. Old industrial regions of the core—Europe, North America—have high conflict levels. In contrast, peripheral regions have various combinations of lower conflict levels and lower wages. Organized labor in the old industrial areas induced the owners of capital to switch production and investment to countries that were not yet industrialized or to newly industrializing countries. The reason that mobile capital could avoid the demands of organized labor was the development of productive forces—an increased ability to traverse space and conquer the technical problems of production—and the emergence of a huge alternative labor force in the developing world following the colonial revolutions in Asia, Africa, and the Caribbean.

These dramatic changes in the 1970s and 1980s ended the original international division of labor that was formalized in the late nineteenth century. Under the old imperial system, the advanced powers were the industrialists and the colonies were the agriculturalists and producers of raw materials. After decolonization, light industry and even some heavy industry began to emerge in the former colonies, assisted by the advanced economies. The increasing globalization of production was accompanied by a new international division of labor. The world became a "global factory," in which the developed countries produced the sophisticated technology and the developing countries were left with the bulk of the low-skill manufacturing jobs. The emergence of the international division of labor, mainly a consequence of the activities of the footloose multinational corporations, resulted in deindustrialization in the old industrial regions of the advanced economies and a precarious export-led industrial revolution in parts of the developing world.

THE PRODUCT CYCLE

Product cycles help us to appreciate the importance of technological considerations in corporate spatial organization. The **product life cycle**, which begins with a product's development and ends when it is replaced with something better, is important geographically because products at different stages of production tend to be manufactured at different places within corporate systems. Moreover, at any given stage of the cycle, the various operations involved in the manufacture of a product such as a camera are not necessarily concentrated at a single factory. For example, production of a camera's complex components occurs at a different place from where the final product is assembled.

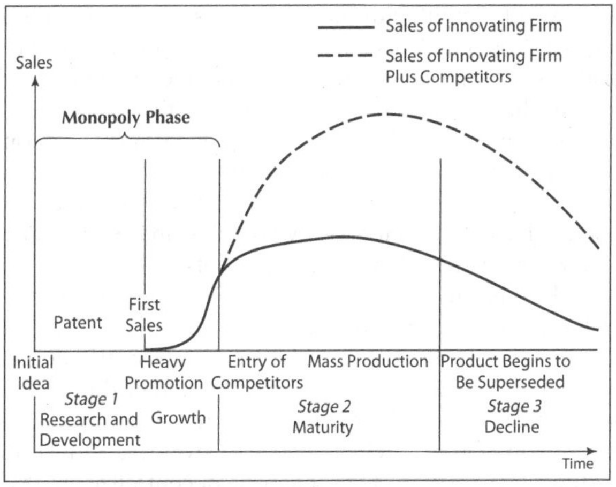

FIGURE 14 A typical product life cycle. Stage 1 is the monopolistic phase in which initial discovery and development are followed by the commercial launching of the product. Rapid sales ensue. The company may enjoy a monopoly during this period, at which time it attempts to improve the product. Stage 2 is characterized by the entry of competition. Emphasis is now on mass-produced, inexpensive items that are standardized and intended to expand the market. Competition begins to erode a large share of the innovating firm's sales. In Stage 3, a large share of the market has been lost to new products and other companies. Overall sales of the product decline as alternative products and manufacturing processes are introduced.

The famous economist Simon Kuznets developed the concept of the three-stage product cycle (Figure 14). In Stage 1, innovators discover, develop, and commercially launch a product. They also benefit from a temporary monopoly and all the special privileges—high profits—that result from it. In Stage 2, competitors buy or steal the new idea, which forces an emphasis on low-cost, standardized, mass-production technologies. Sales of the product increase for a while, but the initial high returns diminish. By Stage 3, the product begins to be superseded. Markets are lost to new products, and manufacturing capacity is reduced.

Innovation begins in an advanced industrial country. These countries have the science, the technology, and the internal market to justify R&D. As a result, they also have an international advantage, and they export their product around the world. But as the technology becomes routinized, other producers appear on the scene, first in the other advanced countries, then on the periphery. Meanwhile, back in the rich country, investment in the newest generation of sophisticated technology is the cutting edge of the economy. There is no doubt that developed countries are the innovators of the world economy and that less developed countries increasingly specialize in the task of transforming raw materials into commodities. But developed countries are also engaged in activities associated with the second and third stages of the product cycle. Indeed, Britain and Canada have expressed concern about their recipient status, a concern that has also been voiced in the United States.

Not all manufacturing operations are fragmented. Corporate branch plants are often *clones*, supplying identical products to their market areas. For example, medium-sized firms in the clothing industry often have this structure, as do many multiplant companies manufacturing final consumer products. Part-process structures tend to be associated with certain industrial sectors, such as electronics and motor vehicles, characterized by complex finished products comprising many individual components.

Labor is an important variable in the location of facilities that make components. Manufacturers seek locations where the level of worker organization, the degree of conflict, and the power of labor to affect the actions of capital are more limited than in long-established production centers such as Detroit, Coventry, and Turin. Starting in the 1970s, Fiat began to decentralize; part of the company's production was moved away from its traditional base in Turin to the south of Italy. Compared with the workers of Turin, who were relatively strong and well organized, the workers of the south were new to modern industry and had little experience of union organization. At the international level, Ford adopted a similar tactic when it invested in Spain and Portugal in the 1970s. Ford management perceived that it could operate trouble-free plants in a region of low labor costs. The labor factor is further emphasized by the practice of *dual sourcing*. To avoid total dependence on a single workforce that could disrupt an interdependent production system, companies are often willing to sacrifice economies of scale for the security afforded by duplicate facilities in different locations.

BUSINESS CYCLES AND REGIONAL LANDSCAPES

Capitalism is a society and economy notorious for its instability. The history of capitalist economies is replete with boom-and-bust periods—epochs of rapid growth, high profits, and low unemployment followed by periods of crisis, economic depression, bankruptcies, and high unemployment. Why does this instability exist? How does it affect national, regional, and local economies?

The most famous depiction of this process is that of **Kondratiev cycles**, named after the Soviet economist Nikolai Kondratiev, who first identified them in the 1920s. Examining historical data on changes in output, wages, prices, and profits, Kondratiev hypothesized that industrial countries of the world experienced successive waves of growth and decline since the beginning of the Industrial Revolution. Based on the emergence of key technologies, these long cycles have a periodicity of roughly 50 to 75 years' duration (Figure 15). The reasons that underlie the duration of these waves reflect long-term trends in the rate of capital formation and depreciation; as fixed capital investments reach the end of their useful economic life,

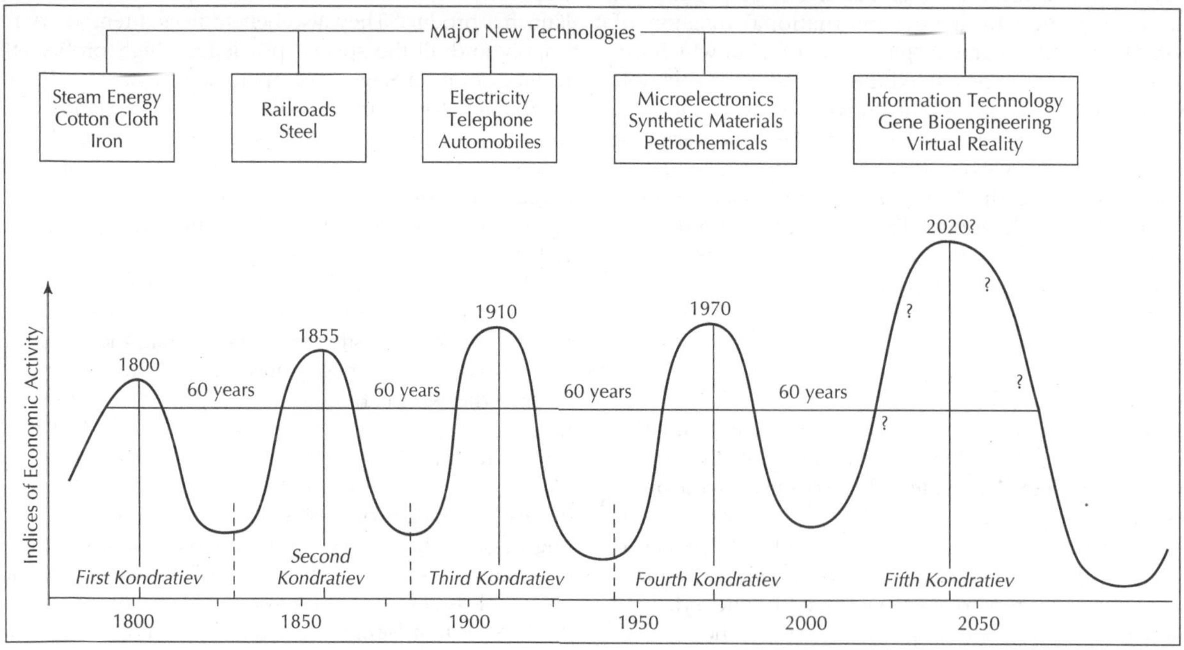

FIGURE 15 Kondratiev waves of economic activity. Kondratiev waves last approximately 50 years each and have four phases of activity, including boom, recession, depression, and recovery. Each period of economic activity has major technological breakthroughs associated with it that power economic growth and employment. On the horizon, the world appears to be entering a new boom cycle based on information technology, biotechnology, space technology, energy technology, and materials technology.

the drag on productivity they create generates incentives to search for new technologies.

The first Kondratiev wave, at the dawn of the Industrial Revolution, was centered on the textile industry and lasted from roughly 1770 to the 1820s, when the West was swept by a series of recessions, bankruptcies, and bank failures. The second wave, focused on railroads and the iron industry, originated in the 1820s, peaked in the 1850s, and ended in the great round of consolidation in the 1880s and 1890s, particularly the depression of 1893, the second worst in world history. The third wave, associated with Fordist industries such as automobiles but also including electricity and chemicals, arose at the end of the nineteenth century, peaked around World War I, and collapsed suddenly during the Great Depression of the 1930s. The fourth Kondratiev wave, which was propelled by World War II, peaked in the 1960s, corresponded to the postwar wave of growth, and included major propulsive industries such as petrochemicals and aerospace; it ended with the petroshocks of the 1970s. Many believe that we are living in the midst of a fifth wave that started in the 1980s and is centered around services.

Joseph Schumpeter, a famous German economist, explained Kondratiev's observation in terms of technical and organizational innovation. Schumpeter suggested that long waves of economic development are based on the diffusion of major technologies, such as railways and electric power. Throughout capitalist history, innovations have significantly bunched together at certain points in time, often coinciding with periods of depression that accompany world economic crises.

Simon Kuznets described Kondratiev cycles in terms of successive periods of recovery, prosperity, recession, and depression. The upswing of the first cycle was inspired by the technologies of water transportation and the use of wind and captive water power; the second by the use of coal for steam power in water and railroad transportation and in factories; the third by the development of the internal combustion engine, the application of electricity, and advances in organic chemistry; and the fourth by the rise of the chemical, plastic, and electronics industries. In the present period of world economic crisis, with higher energy costs, lower profit margins, and the growth of the old basic industries exhausted, scholars are asking whether a *fifth wave* is under way.

Information Technology: The Fifth Wave?

Some scholars argue that a fifth Kondratiev cycle began in the 1980s, and is associated with **information technology**. Information technology production is based on microelectronic technologies, including microprocessors, computers, robotics, satellites, fiber-optic cables, and information handling and production equipment, including office machinery and fax machines. Information technology, production, and use is strong in Japan, in East Asian newly industrializing countries, in the United States and Canada, and in Western Europe, notably Germany, Sweden, and France.

Information technology arises from the convergence of communications technology and computer technology. Communications technology involves the transmission of data and information; computer technology is concerned primarily with the processing, analysis, and reporting of information. A new techno-economic paradigm does seem to be emerging based on the extraordinarily low costs of storing, processing, and communicating information. From this perspective, the late twentieth century saw a prolonged period of social adaptation to the growth of this new technological system, which is now affecting virtually every part of the economy, not only in terms of present and future employment and skill requirements but also in terms of future prospects.

Leading the contemporary wave are business services such as advertising, purchasing, auditing, inventory control, and financing (**producer services**). The defining characteristic of these new services is that they create and manipulate knowledge products in almost the same way as the manufacturing industries that peaked in earlier waves transformed raw materials into physical products. Producer services have become a salient force in the new postindustrial society and the information economy. They are restructuring the geography of manufacturing, because they are the basis of increases in productivity—technological innovation, better resource allocation through expert systems, increased training and education. As a result, a new geography of world cities has emerged: a command and control economy centered on world cities such as New York, Los Angeles, London, Paris, Tokyo, Hong Kong, and Singapore.

Business Cycles and the Spatial Division of Labor

The spatial dimensions of business cycles are complex and important; they reveal that uneven development in time and space are two sides of one coin. Uneven development in space occurs through the specialization of production in different areas, including the comparative and competitive advantages that regions enjoy at different moments in time. Given the fluidity of capitalism, however, there is no reason for a region or a country to enjoy its production advantages indefinitely. Capital, labor, and information move across space, changing the conditions of profitability in different places. Put differently, uneven development in time is manifested when a region's or a nation's comparative advantage is created and lost as capital creates and destroys regions over successive business cycles. The loss of comparative advantage makes a region attractive to firms: It offers pools of unemployed, and hence cheap, labor; an infrastructure; and often other advantages as well. In short, regions abandoned by capital may be ready to be recycled for a new use.

Over different business cycles, several industries may locate in one region, each leaving its own imprint on the local landscape. Each constructs a labor force, invests in buildings, and shapes the infrastructure in ways that suit its needs (and profits). From the perspective of each region, therefore, business cycles resemble waves of investment and disinvestment. Each wave, or Kondratiev cycle, deeply shapes the local economy, landscape, and social structure and leaves a lasting imprint on a region that is not easily erased. For example, the textile industry in New England created its industrial landscape in the nineteenth century, ranging from small mills located on streams to the large factories in Lowell, Massachusetts, or Manchester, New Hampshire. These landscapes, including the people who inhabit them, persisted long after the industry abandoned New England for the South in the early twentieth century. For many years, New England was a relatively poor part of the country, with high unemployment rates. By the 1980s, however, a new wave of production—the electronics industry—had concentrated in the region. Firms producing computer hardware and software found the human resources and the Route 128 corridor of the Boston metropolitan region attractive, and the local geography of this industry was shaped to no small extent by the residues of earlier ones.

In short, each set of investment/location decisions in a region is prestructured by earlier sets of decisions. Thus, as their comparative advantage changes, regions accumulate a series of different imprints: Each wave is shaped by and transforms the vestiges of past waves. Thus, regions are unique combinations of layers of investment and disinvestment over time. Such an approach explains the unique characteristics of places through their economic histories. Because capitalism constantly reproduces spatial inequality by diverting capital from low-profit to higher-profit regions, individual places are perpetually open to the lure of new forms of investment and vulnerable to the risk of being abandoned by capital. This view allows us to integrate the specifics of regions with broader understandings of capitalist processes.

THE STATE AND ECONOMIC GEOGRAPHY

Contrary to widespread popular opinion, the economic landscapes of capitalism are not simply the products of "free markets," but also involve the role of the state (government in all its forms and functions). In noncapitalist societies, particularly feudalism, the state was the major allocator of resources; there was, effectively, no division between public and private power. Under capitalism, markets are the *primary* means through which decisions are made to hire people, use land, or determine how to utilize capital, but they are not the *only* means. The state does what no individual or firm can do, tackling problems too big for private firms and providing necessary, but unprofitable, services. Given the long involvement of the state—often obscured by notions of a mythical "free market" that has never existed in fact – capitalism could never survive without the state. Even the most unfettered of markets, for example, such as a garage sale, presupposes the existence of state-generated money, property rights, and infrastructure such as a road system.

The degree of state involvement varies historically and geographically—it waxes and wanes depending on the political forces and economic imperatives at work—but it is never zero. In the nineteenth century, particularly in the United States, state involvement was considerably less than it is today; there were few public services, for example, although the federal government did subsidize railroads and erect tariffs against imports. Starting with the Great Depression of the 1930s, when the market generated extreme suffering for millions of people, the role of the state changed; it became the so-called "welfare state," offering numerous social protections and safeguards (e.g., Social Security, minimum wage). Since the late twentieth century, under the pressure of globalization, the welfare state has retreated around much of the world, particularly in the United States, and has been replaced by a deregulated, "neoliberal" governmental concept based on the premise that the market is the optimal way of allocating public as well as private resources. Any account of how economic landscapes are produced, therefore, must include some understanding of the role of the state. Several dimensions are sketched here.

One way that states shape economies is through the creation and enforcement of a legal system. Laws, which carry the moral authority of right and wrong, also act economically to protect property rights. Without the right to buy and sell, to have assets secured against forcible appropriation, markets would simply not exist. The development of capitalism thus entailed a secularized legal system. Laws and regulations encompass a vast array of activities that both constrain behavior and protect some parties from the actions of others, including, to take but a handful, health and safety regulations (e.g., environmental protection, workplace safety rules, restaurant inspections), antidiscrimination ordinances, and antitrust laws. The state enforces laws through the police, judiciary, and the military and maintains a monopoly over the legal use of violence.

The state is also heavily involved in setting fiscal and monetary policy. Fiscal policy—which determines how governments spend their money—has enormous impacts on localities throughout a nation-state. In the United States, for example, the federal government's budget is more than $2 trillion annually. Governments collect revenues through a variety of taxes and fees, the most important of which in the United States are individual income taxes, but taxes also include the less important social insurance payroll tax and corporate income taxes (the latter generate only 12% of the U.S. government's revenues); see Figure 16. National governments also control the money supply, which in turn affects inflation and interest and exchange rates, all of which have geographically

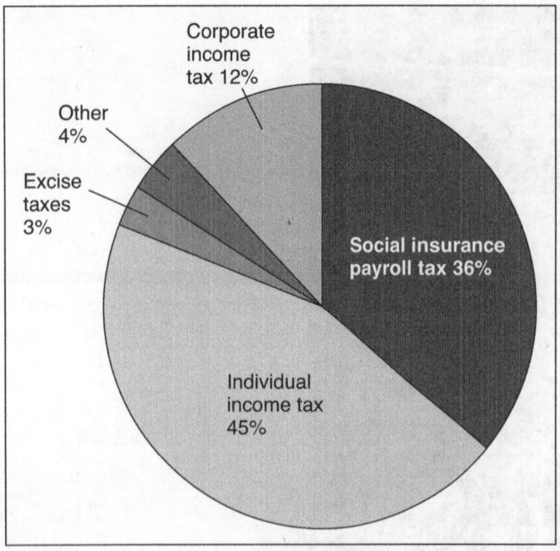

FIGURE 16 Origins of federal government tax revenues, 2008. Government receipts come in a variety of forms, but primarily from individual income taxes. Generous tax cuts for corporations have reduced their share to 12%. Although taxes are not the most important locational consideration for firms, they constitute a flow of resources among classes and regions, finance public sector activity, and represent one of many ways in which the state shapes economic landscapes.

uneven consequences, in highly complex ways. Almost all nations use a national bank to manage their money supply; in the United States, it is the Federal Reserve System.

Government expenditures, which are uneven across the landscape, have huge impacts on local areas, generating jobs, subcontracts, revenues, and profits. In some cases—Washington, DC, for example—entire metropolitan regions exist due to these expenditures. Public outlays include transfer payments. In the United States, these include Social Security, veterans' benefits, and other entitlement programs. The U.S. and other governments also subsidize producers of many goods, particularly agricultural and dairy products, spending far more on corporate welfare than on aid to poor people.

The state has a huge impact on economic landscapes through the construction of an infrastructure, including transportation and communication networks (roads, highways, bridges, airports, ports), water and electrical supply systems, hydroelectric dams, sewers, and the like. Without the infrastructure, the circulation of people, goods, and ideas that is fundamental to markets would be impossible. Thus, the U.S. Federal Interstate Highway System (Figure 17), the largest project the federal government has undertaken, plays an enormous role in facilitating the movement of

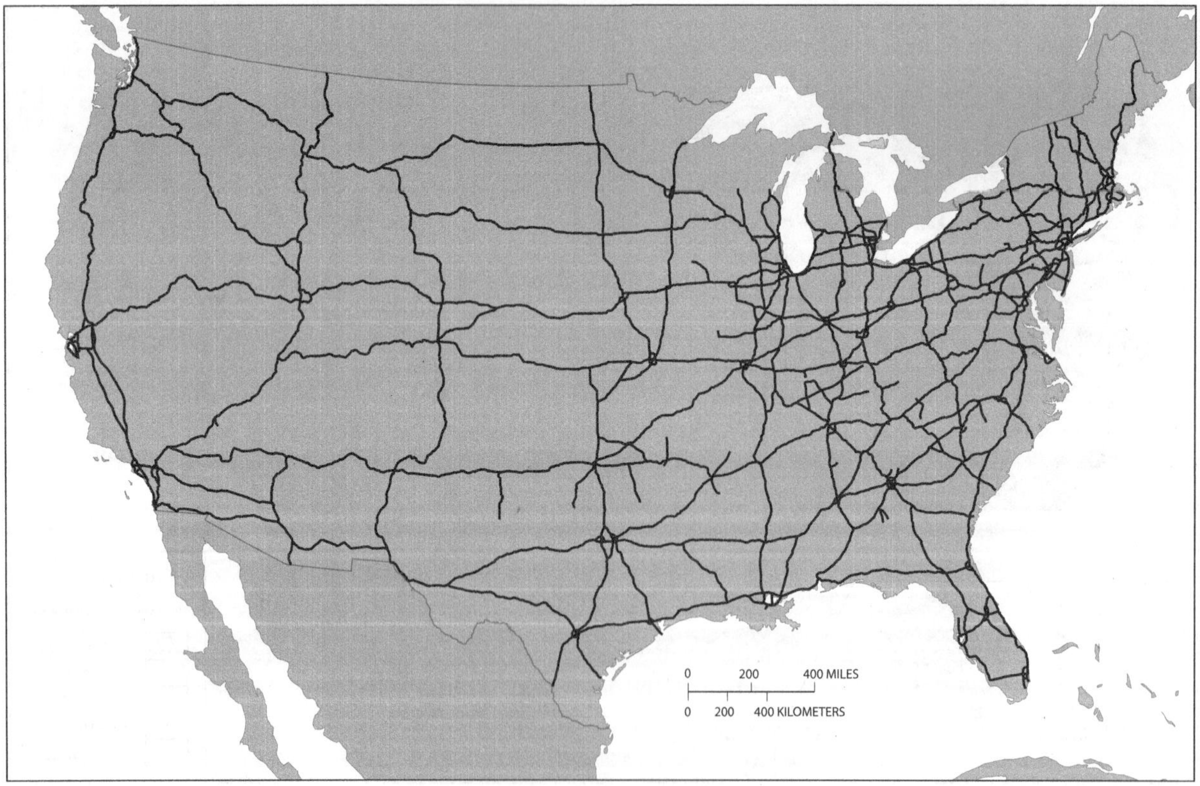

FIGURE 17 The Federal Interstate Highway System. Infrastructure is a hugely important form of state intervention in capitalist societies. The Interstate Highway System, the largest project the U.S. federal government ever undertook, has facilitated a dramatic time-space compression among American cities. Like most such projects, its costs are born publicly while the benefits are appropriated privately.

FIGURE 18 Hydroelectric dams, such as the Hoover Dam pictured here, are large, expensive projects that illustrate the state's role in the construction and maintenance of the infrastructure.

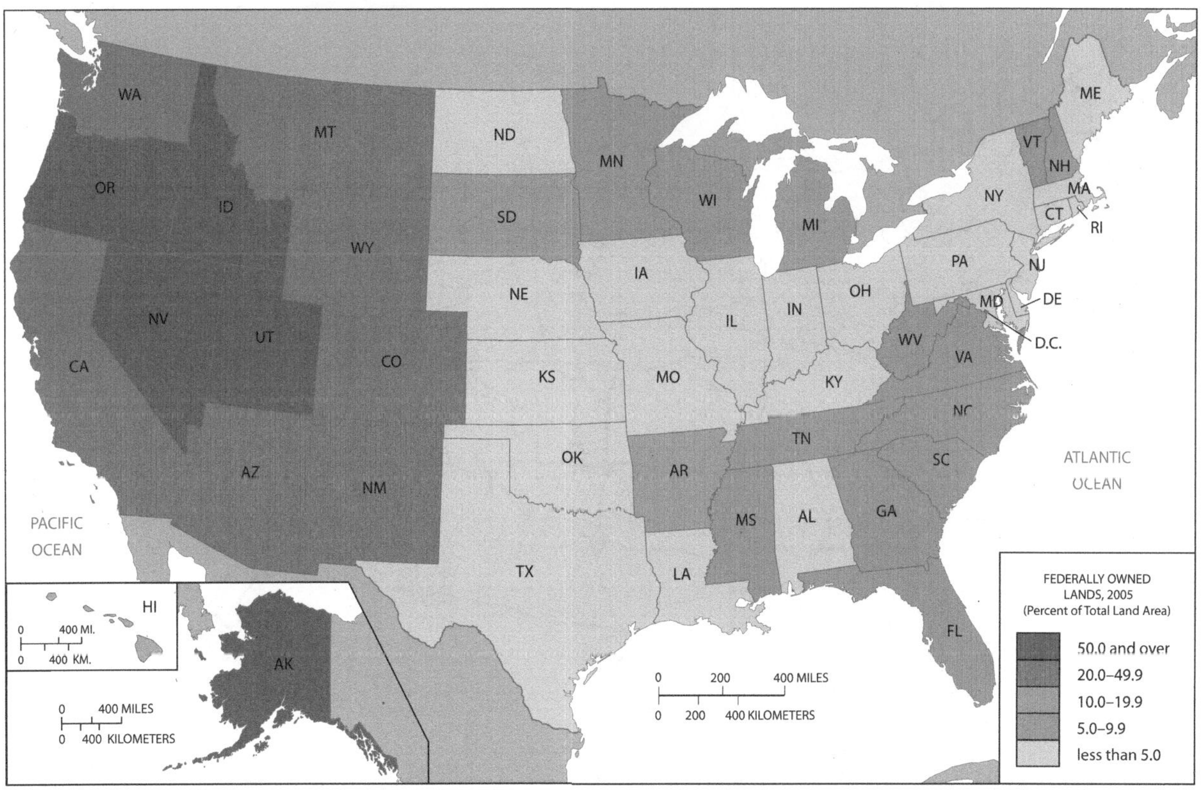

FIGURE 19 Federally owned lands. The federal government owns roughly 40% of the nation's territory, especially in the West, in many forms, including national parks, highways, forests, and military facilities.

people and goods among the nation's cities. Public services also include public education, public health services and hospitals, fire and police departments, libraries and swimming pools, trash and snow removal, public transportation and housing, parks, and so on (Figure 18). In the United States, most public services are provided at the local municipal or county level; federal government services (as opposed to transfer payments) are primarily confined to the post office and defense.

The state shapes the labor markets of capitalist societies in many ways, both directly and indirectly. In the United States, the federal government is the nation's largest single employer, with more than 2 million people in 2003 out of a national labor force of roughly 130 million. State and local government employment is even higher, amounting to approximately 20 million in 2003. Additionally, the state shapes labor markets through interventions such as the minimum wage; health and safety codes; antidiscrimination rules; and regulations concerning benefits, overtime, and vacations. The state may offer training grants and seek to generate human capital through the education system, which in turn affects private employers.

Housing markets are another area shaped by the state. Private housing markets are hugely affected by government-influenced interest and inflation rates. Public housing, which in the United States amounts to only 2% of the total housing stock, is an important resource for low-income people in many cities. Zoning codes concerning population density, minimum lot size, and architectural details affect the supply and demand for housing. In some cities, the municipal government imposes some forms of rent control, which typically create a market for housing at below-market prices. More broadly, in the United States the federal government owns 40% of the area of the whole country (Figure 19), primarily in the West, for national parks and military facilities.

Finally, the state acts as an agent in international issues. Governments shape trade in many ways (a topic we will explore in more depth later) in the form of tariffs, quotas, and nontariff barriers, as well as subsidies to exporters. Many governments attempt to manipulate exchange rates to make their exports competitive or to improve their trade balance. The state also controls the international movement of labor, putting restrictions on immigration, thus affecting the supply of workers domestically. These policies work unevenly across the national landscape, as some regions are more connected to the international economy than others.

All of these examples should serve to demolish the myth of the "free market" and demonstrate that capitalism is a system in which both markets and the state operate simultaneously, although not to an equal extent. Indeed, variations in the level of state intervention are a prime factor in explaining the different national varieties of capitalism found around the world.

Summary

In this chapter, we examined various theoretical and conceptual dimensions to the construction of economic landscapes. We began by listing the major location factors that influence the locations of firms, including labor, land, capital, and management skills. We noted that labor productivity was as important as its cost. The uneven distribution of these phenomena leads to geographically uneven patterns in their use by firms.

The chapter described in some depth the famous Weberian theory of location, which centers on minimization of transport costs. Classical location theory stresses that manufacturing patterns are caused by geographic characteristics—**locational factors**—rather than by underlying social relations. Assembly costs are incurred because the raw materials required for a particular kind of manufacturing are distributed in different places. Production costs vary because of the areal differences in the costs of labor and land; while the costs of capital investments may also vary geographically, the costs of financial capital are much more uniform due to its much greater mobility. Finished-product distribution costs are incurred when producers must sell to dispersed or widely scattered markets. Classical location theory provides a rationale for finding the points of production at which locational costs are minimized.

The chapter discussed the behavior of firms in time and space. Most geographers now question the usefulness of traditional location theory in light of the multiproduct, multiplant, multinational operations characteristic of the modern global structure of production and consumption. Accordingly, we devoted a portion of the chapter to the spatial behavior of large industrial enterprises and gave some attention to trends in industrial organization, the relationship of large firms to small ones, the reasons for corporate growth, and the internal geography of corporate systems.

The chapter illustrated how firms face a choice between their selection of a production technique, which reflects the costs of inputs, and the scale of output, which generates economies of scale. Few issues are as important as economies of scale in understanding where firms locate, the nature of the market they are in, and how they change over time. The chapter also explored the growth of firms, ranging from their strategies to the roles of vertical integration and disintegration. It also delved into the division of labor within firms, including corporate administrative hierarchies and the separation of headquarters from production functions.

We embedded firms in their social context, pointing out that they are always part of a broader nexus of capital and labor relations, which may include divisions among the owners of capital and between capital and labor. Firms are thus not isolated decision makers floating in a world without constraints but are part of the process of commodity production, transportation, and consumption. Capitalism is a tremendously dynamic society beset with constant, often wrenching, changes—in products, production processes, markets, forms of work, and locations. These changes may be examined in light of the product cycle, a metaphorical model that encapsulates the simultaneous economic, technological, organizational, and geographic changes that confront every firm in the course of its industry's evolution.

Because the economic advantages of some regions over others are always temporary, we stressed the cyclical nature of capitalism, its tendency toward boom-and-bust periods. The most famous representation of this process of long-term episodes of growth and decline is the Kondratiev wave. The periodicity of capitalism is reflected in the rise of different industries at different moments in time—textiles, steel, automobiles, electronics, producer services—each of which is the high-tech sector of its day. The Kondratiev model reminds us that the present world economic system may be in the midst of a fifth upswing, this one based on a cluster of microelectronics and information technologies. We connected business cycles to the spatial division of labor to understand how local landscapes are created through the successive imposition of different layers of investment over time.

Finally, we ended by noting that capitalism is not synonymous with the "free market" because markets are always and everywhere shaped by the state to one degree or another. The state makes and enforces laws and regulations that enforce property rights; collects and spends money; affects interest and inflation rates; generates jobs and subcontracts and regulates working conditions; regulates land use; builds the infrastructure and provides public services; shapes housing markets; and intervenes internationally in trade, exchange rates, and immigration. In all these ways, and more, the state is an actor as important in the construction of the economic geographies of capitalism as market forces.

Key Terms

agglomeration economies
backward integration
diseconomies of scale
diversification
division of labor
economies of scale
financial capital
fixed capital
footloose industries
forward integration
horizontal integration
human capital
industrial inertia
information technology
integration
Kondratiev cycles
locational factors
mode of production
producer services
product life cycle
production linkages
scale
service linkages
social relations of production
surplus value
urbanization economies
vertical integration

Study Questions

1. How is labor different from other inputs that firms use?
2. What determines the amount and type of an industry's demand for labor?
3. Do firms always pursue the cheapest labor? Why or why not?
4. What is human capital and why is it important?
5. How does land figure into the locational calculus of firms?
6. What is capital? Differentiate between fixed and liquid capital.
7. Define the Weberian model of firm location.
8. Compare and contrast scale and agglomeration economies.
9. What is the behavioral approach to industrial location?
10. Contrast horizontal and vertical integration.
11. Compare and contrast multiplant and multiproduct enterprises.
12. What is the product life cycle model and how does it affect the location of firms?
13. Define Kondratiev cycles.
14. How do business cycles shape local landscapes?
15. What are six ways the state affects economic landscapes?
16. Is there really such a thing as the "free market"? Why not?

Suggested Readings

Dicken, P. 2010. *Global Shift: Industrial Change in a Turbulent World,* 6th ed. London: Harper & Row.

Herod, A. 1998. *Organizing the Landscape: Geographical Perspectives on Labor Unionism*. Minneapolis: University of Minnesota Press.

Knox, P., J. Agnew, and L. McCarthy. 2008. *The Geography of the World Economy,* 5th ed. London: Edward Arnold.

Krugman, P. 1991. *Geography and Trade*. Cambridge, MA: MIT Press.

Porter, M. 1990. *The Competitive Advantage of Nations*. New York: Free Press.

Web Resources

U.S. Department of Commerce
http://www.doc.gov/

The government department charged with promoting American business, manufacturing, and trade. Its home page connects with the Web sites of its constituent agencies.

Bureau of Labor Statistics Web Site
http://stats.bls.gov/

Contains economic data, including unemployment rates, worker productivity, employment surveys, and statistical summaries.

Census Bureau Center for Economic Studies
http://www.ces.census.gov/

The federal government's census site contains massive amounts of data and reports about a wide variety of demographic and economic issues.

Labor History Archives
http://www.nyu.edu/library/bobst/research/tam/laborrefguide.html

Large collection of resources about the history of working men and women, including labor movements, unions, biography, and labor history.

Economic Geography Research Group
http://www.egrg.org.uk/resources.html

British Web site with information about contemporary topics in economic geography, including books, conferences, and more.

http://faculty.washington.edu/krumme/ebg/contents.html

A comprehensive portal to economic geography set up by Dr. Gunter Krumme.

Agriculture

OBJECTIVES

- To describe the world's preindustrial agricultural forms and regions
- To acquaint you with commercial agricultural practices and world regions
- To describe the agricultural policies of the United States and their shortcomings
- To summarize sustainable agriculture as an ecologically friendly alternative to contemporary forms of food production
- To analyze the von Thünen model as a means of understanding local land use patterns

A succession of combines marches over a wheat field 10 miles north of Pendleton, Oregon, and south of the little town of Helix, reflecting the enormous degree of mechanization typical of industrialized agriculture.

© Bill Stormont/CORBIS

Agriculture

Agriculture, the world's most space-consuming activity and one of humanity's leading occupations, is the science and art of cultivating crops and rearing livestock in order to produce food and fiber for sustenance or for economic gain.

The historical distribution of agriculture has been critical to the survival and success of the human species. Throughout the vast bulk of the span of our existence (95% or more of it), we were hunters and gatherers, not agriculturalists; people were food collectors, not producers. The Neolithic Revolution, roughly 10,000 years ago, witnessed the invention or discovery of agriculture, making possible a nonnomadic existence. It paved the way for a social surplus and the rise of cities, denser social forms and more refined divisions of labor, and fostered the development of new technologies such as writing, pottery, and metal working.

Until the nineteenth century, however, agriculture produced relatively little food per worker, so most of the population worked full-time or part-time on the land. The small surplus of food released few people for other pursuits. Not until the industrialization of agriculture that began in Europe during the past 200 years did large-scale employment in manufacturing and service activities become possible.

The shift of labor from the agricultural sector to other sectors constitutes one of the most remarkable changes in the world economy in modern times. In the United States and the United Kingdom, less than 2% of the economically active population now works directly in agriculture. In contrast, about 70% of the populations in a number of African and Asian countries are engaged in the agricultural sector.

Economic geographers are concerned with problems of agricultural development and change as well as with patterns of rural land use. Where was agriculture discovered? How did it diffuse? Why do farmers so often fail to prevent environmental problems? What are the characteristics of the main agricultural systems around the world? What is the effect of industrialized agriculture on farmers and the countryside? What principles can help us understand the spatial organization of rural land use? What are the consequences of government agricultural policy? In this chapter, we seek answers to these questions.

Agriculture had a long, rich history prior to the emergence of modern capitalist food production. Medieval farming methods prevailed in Western Europe until capitalism invaded the rural manor. The rise of market forces in agriculture transformed food into a commodity, something to be bought and sold for a profit. Land uses changed to crops that generated the highest rate of profit and replaced subsistence agriculture with market-oriented agriculture. Open fields were enclosed by fences, hedges, and walls. Crop rotation replaced the medieval practice of fallowing fields. Seeds and breeding stock improved. New agricultural areas opened up in the Americas. Farm machines replaced or supplemented human or animal power. The family farm came to represent the core model of commercial agriculture. This transformation resulted from a vast population increase in the new trading cities that depended on the countryside for food and raw materials. Another force that brought the market into the countryside was the alienation of the manorial holdings. Lords, who needed cash to exchange for manufactured goods and luxuries, began to rent their lands to peasants rather than having them farmed directly through feudal labor-service obligations. Thus, they became landlords in the modern sense of the term.

THE FORMATION OF A GLOBAL AGRICULTURAL SYSTEM

By A.D. 1500, on the eve of European overseas expansion, agriculture had spread widely throughout the Old World and much of the New World. In Europe, the Middle East, Africa, central Asia, China, India, and Indonesia, cereal farming and horticulture were common features of the rural economy. Nonagricultural areas of the Old World were restricted to the Arctic fringes of Europe and Asia and to parts of southern and central Africa. Agriculture had not spread beyond the Indonesian islands into Australia.

By the time of the first European voyages across the Atlantic, the cultivation of maize, beans, and squash in the New World had spread throughout Central America and the humid environment of eastern North America as far north as the Great Lakes. In South America, only parts of the Amazon Basin, the uplands of northeastern Brazil, and the dry temperate south did not have an agricultural economy. These patterns of agriculture persisted until the era of European colonial conquests. Eventually, European settlements assumed two forms: (1) farm-family colonies in the middle latitudes of North America, Australia, New Zealand, and South Africa; and (2) plantation colonies in the tropical regions of Africa, Asia, and Latin America. These two types of agricultural settlements differed considerably.

For example, farm colonization in North America depended on a large influx of European settlers whose agricultural products were initially for a local market rather than an export market. Europeans introduced the farm techniques, field patterns, and types of housing characteristic of their homelands, yet they often modified their customs to meet the challenge of organizing the new territory. For example, the checkerboard pattern of farms and fields that characterizes much of the country west of the Ohio River resulted from a federal system of land allocation (the **Township and Range System**). It involved surveying a baseline and a principal meridian, the intersection of which served as a point of origin for dividing the land into 6-by-6-mile townships, then into square-mile sections, and still further into quarter sections a half-mile long. This orderly system of land allocation prevented many boundary disputes as settlers moved into the interior of the United States.

In tropical areas, Europeans, and later Americans and the Japanese, imposed a plantation agricultural system that did not require substantial settlement by expatriates. **Plantations** are large-scale agricultural enterprises devoted to the specialized production of one tropical product raised for the market. They were first developed in the 1400s by the Portuguese on islands off the tropical West African coast. Plantations produced luxury foodstuffs, such as spices, tea, cocoa, coffee, tobacco, and sugarcane, and industrial raw materials, such as cotton, rubber, sisal, jute, and hemp (Figures 1 and 2). These crops were selected for their market value in international trade, and they were grown near the seacoast to facilitate shipment to Europe. Thus, plantations represent the first wave in the global commodification of agriculture. The creation of plantations sometimes involved expropriating land used for local food crops. Sometimes, by irrigation or by clearing forests, new lands were brought under cultivation.

Bridgestone/Firestone, Tire Sales Co.

FIGURE 1 On the world's largest rubber plantation, at Harbel, Liberia, more than 36,000 hectares, or 30% of the total land area of Liberia, are cultivated by Firestone Tire and Rubber Company. The company has also established plantations in Brazil, Ghana, Guatemala, and the Philippines. How do plantations benefit host societies?

Europeans managed plantations; they did little manual labor. The plantation system relied on forced or poorly paid indigenous labor. Very little machinery was used. Instead of substituting machinery for laborers when local labor supplies were exhausted, plantation managers went farther afield to bring in additional laborers. This practice was especially convenient because world demand for

Bridgestone/Firestone, Tire Sales Co.

FIGURE 2 A tapper on the Firestone plantation in Liberia makes an incision in a rubber tree. The latex will flow down the incision through a spout and into a cup attached to the tree. Some of the latex is carried in pails by women to collecting stations. What would this tapper do if he were not working for Firestone?

crops fluctuated. During periods of increased demand, production could be accelerated by importing additional laborers, thus making the need for installing machinery during booms unnecessary and minimizing the financial problems of idle capital during slumps.

The effect of centuries of European overseas expansion was to reorganize agricultural land use worldwide. Commercial agricultural systems have become a feature of much of the world. Hunting and gathering, the oldest means of survival, has virtually disappeared. Nomadic pastoralists, such as the Masai of Kenya and Tanzania who drive cattle in a never-ending search for pasture and water, have steadily declined in numbers, often victims of the fixed borders of the nation-state. Subsistence farming still exists, but only in areas where impoverished farmers, especially in developing countries, eke out a living from tiny plots of land. Few completely self-sufficient farms exist; most farmers, even in remote areas of Africa and Asia, trade with their neighbors at local markets.

THE INDUSTRIALIZATION OF AGRICULTURE

In the nineteenth and twentieth centuries, a third agricultural revolution took place that resolved the distinction between family and corporate models of agriculture. In other words, this revolution signified the elimination of distinct agrarian economies and communities. Industrial agriculture has become the dominant form in most developed countries and is being applied to export-oriented regions of developing countries. Key elements of industrial agriculture are extreme capital intensity, high energy use, concentration of economic power, and a quest for lower unit costs of production. Although industrial agriculture has increased output per unit of input, it has also depleted water and soil resources, polluted the environment, and destroyed a way of life for millions of farm families.

The industrialization of agriculture drastically reduced the number of farmers in North America. In the United States, the number declined from 7 million in 1935 to around 1.7 million in 2008. In Canada, 600,000 farm operators existed in 1951 but less than one-third that number were still in operation in 2008. Europe witnessed similar trends. In Britain, for example, a decline in the number of farm workers has been going on for decades. Today, the percentage of a labor force engaged in primary economic activities (agriculture, logging, fishing, mining) is a useful measure of economic development around the world (Figure 3): Economically advanced countries have a small share of their workers in agriculture, such as in

CASE STUDY

Agro-Foods

Food takes a long and complicated path to our tables that involves vast networks of inputs. For example, large-scale farming depends on industrial equipment such as tractors and combines and chemical inputs such as fertilizers, animal hormones and antibiotics, and pesticides, as well as financial networks of loans. Industrial farming is also highly dependent on energy to run the machines, pump water, produce fertilizer, and transport the finished product to the consumer. Most food that we consume in the economically developed world has been substantially modified and processed to be made durable through canning, freezing, or other methods. Such processes allow the spatial separation of production from consumption that makes long-distance trade possible. Agriculture and industry have become so intertwined that many foods have become known by the industrial process that has transformed them, such as *homogenized* milk, *pasteurized* cheese, *refined* sugar, and so on.

Industrial food producers have great flexibility in their choices of farm output and where to obtain it. For example, the manufactured food requires a sweetener, but not necessarily sugar derived from the sugarcane plant. It requires oil, yet not necessarily oil from corn. It requires a starch, but that could be derived from either potatoes or wheat or a number of other grains. The production of potato chips provides a good example of this substitution affect: Producers can fry the chips in whatever oil is cheapest at the moment of production. This illustrates why farmers are often in a disadvantaged position within the agro-food system.

Food reaches consumers via food wholesalers and retailers and the restaurant and catering industry. Powerful economic entities in food distribution can shape the agro-food system with their purchasing power, such as when a fast-food restaurant chain decides to fry its French fries in vegetable oil or lard or when large grocery chains decide to carry some food products and not others.

At the end of the agro-food system is the consumer. What we eat reflects demographic characteristics—the size and growth of the population, purchasing power, and social relations such as the structure of the family, for example. Obviously, advertising greatly influences our food choices. But more subtly, the ever-quickening pace of the economy has led to the proliferation of "fast" foods and other convenience foods meant to be consumed "on the go," in the car, or at the desk. Not only does the changing agro-food system shape our bodies, but its changing technology and consumer choices also have significant impacts on our geography.

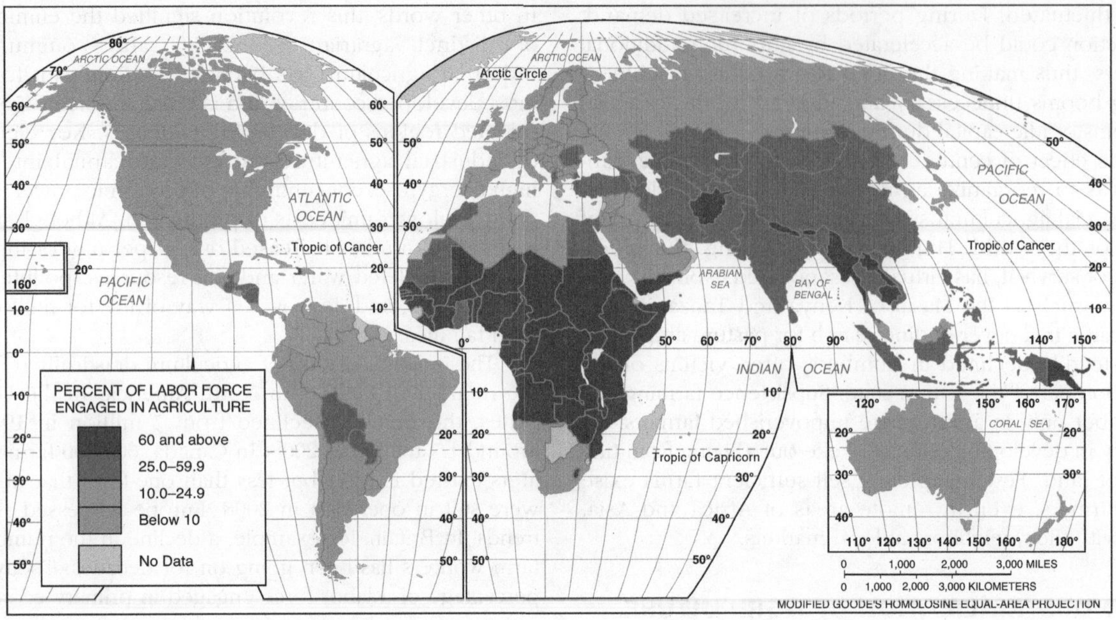

FIGURE 3 The proportion of the labor force engaged in agriculture is an important measure of economic development. In economically advanced countries, very few people are needed to generate a food supply for everyone, whereas in developing countries a much larger share of workers is engaged in farming.

North America, Europe, Japan, and Australia, whereas in much of the developing world significant proportions of workers are engaged in farming. In most of Africa and East Asia, for example, more than 60% of the employed population works as peasant farmers.

Human Impacts on the Land

The emergence of agriculture and its subsequent spread throughout the world has meant that little, if any, land still can be considered "natural" or untouched. Almost everywhere, nature has been modified extensively by human beings, making it difficult to speak of nature as a phenomenon separate from human impacts. Vegetation has been changed most noticeably. Virtually all vegetation zones show signs of extensive clearing, burning, and the browsing of domestic animals. The impoverishment of vegetation has led to the creation of successful agricultural and pastoral landscapes, but it has also led to **land degradation** or a reduction of land capability.

Hunters and gatherers hardly disturb vegetation, but farmers must displace vegetation to grow their crops and to tend their livestock. Farmers are land managers; they upset an equilibrium established by nature and substitute their own. If they apply their agrotechnology with care, the agricultural system may last indefinitely and remain productive. On the other hand, if they apply their agrotechnology carelessly, the environment may deteriorate rapidly. How farmers actually manage land depends not only on their knowledge and perception of the environment but also on their relations with groups in the wider society—in the state and the world economy.

As agriculture intensifies, environmental alteration increases. Anthropologist Ester Boserup proposed a simple but famous **stage model of agriculture**. Stage 1, forest-fallow cultivation, involves cultivation for 1 to 3 years followed by 20 to 25 years of fallow. In Stage 2, bush-fallow cultivation, the land is cultivated for 2 to 8 years, followed by 6 to 10 years of fallow. In Stage 3, short-fallow cultivation, the land is fallow for only 1 to 2 years. In Stages 4 and 5, annual cropping and multicropping, fallow periods are either very short—a few months—or nonexistent. Boserup noted that the transition from one form of agriculture to another was accompanied by an increasing population density, improved tools, increasing integration of livestock, improved transportation, a more complex social infrastructure, more permanent settlement and land tenure, and more labor specialization.

In contrast, permanent agriculture (annual cropping and multicropping) usually occurs in areas of high potential productivity and high population pressure. Under permanent cultivation, the land becomes totally transformed, yet the fertility of the land may not be impaired. For example, soils of the Paris Basin have been cultivated intensively for hundreds of years and still they remain highly productive. In many parts of East Asia, carefully terraced hillsides have maintained the productivity of valuable soil resources for thousands of years.

In general, industrialized farming practices pose the main danger to the environment. Clean tillage on large fields, monoculture (the cultivation or growth of a single crop), and the breaking down of soil structure by huge machines are a few factors that may destroy the topsoil. Repeated droughts and dust storms in the twentieth century on the Great Plains of the United States gave testimony to how nature and industrial agriculture can combine to destroy the health of a steppe landscape.

Agriculture threatens ecological balances when people begin to believe that they have freed themselves from dependence on land resources. In developed countries, there is a tendency to exploit the land as a result of pressure to maximize profits. Corporate producers want to make land use more efficient and productive; thus, farming is often viewed as just another industry. However, we must remember that land is more than a means to an end; it is finite, spatially fixed, and ecologically fragile.

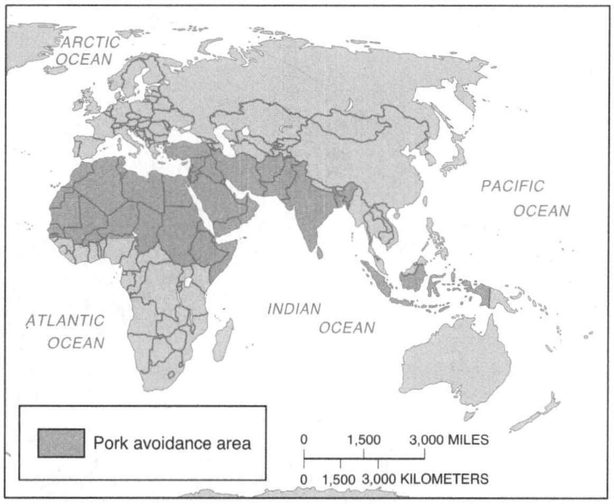

FIGURE 4 Pork-avoidance areas of the world. The pork avoidance, or pork taboo, in North Africa and the Middle East arose among the Jews during biblical times. Pigs were considered unclean because of their environmental setting. Today, the Muslim areas of the world, which subscribe to some Old Testament biblical laws, are the largest areas of pork avoidance. India is only 15% Muslim and therefore the majority of the population does not hold a taboo against pork; however, the Hindu majority practices a beef taboo in this region.

FACTORS AFFECTING RURAL LAND USE

Rural land-use patterns, which are arrangements of fields and larger land-use areas at the farm, regional, or global level, are difficult to understand. Worldwide, hundreds of farm types exist. The most interesting aspect of the world's agricultural land-use areas or regions is not their extent but the uniformity of land-use decisions farmers make within them. Given any farming region, why do farmers make similar land-use decisions? Several variables determine land use, including site characteristics and cultural preferences and perception, which are discussed next.

Climatic Limitations

Variations in rural land use depend partly on climatic characteristics, such as soil type and fertility, slope, drainage, exposure to sun and wind, and the amount of rainfall and average annual temperature. Plants require particular combinations of temperature and moisture. Absolute physical limits of the crop are "too wet," "too dry," "too cold," and "too hot." Absolute climatic limits are wide for some crops, such as maize and wheat, but narrow for others, such as pineapples, cocoa, bananas, and certain wine grapes.

Cultural Preferences and Perceptions

Food preferences, often having religious origins, are one variable affecting the type of agricultural activity at a given site. For example, many Africans avoid protein-rich chickens and their eggs. Hindus abstain from eating beef; Jews and Muslims do not eat pork (Figure 4). Most people in East and Southeast Asia abstain from drinking milk or eating milk products. In the United States, a consumer preference for meat (enhanced by the enormous lobbying power of corporate agribusiness) leads American farmers to plant a greater proportion of their land in forage crops for animal feed than do European farmers, who grow more food crops.

People interpret the environment through different cultural lenses. Their agricultural experiences in one area influence their perceptions of environmental conditions in other areas. Consider the settlement of North America. The first European settlers were Anglo-Saxons accustomed to moist conditions and a tree-covered landscape. They equated trees with fertility. If land was to be suitable for farming, it should, in its natural state, have a cover of trees. Thus, the settlers of New England and the East Coast realized their expectations of a fertile farming region. When Anglo-Saxons edged onto the prairies and high plains west of the Mississippi River, they encountered a treeless, grass-covered area. They underestimated the richness of the prairie soils, in particular, and the area became known as the "Great American Desert." In the late nineteenth century, a new wave of migrants from the steppe grasslands of Eastern Europe appraised the fertility of the grass-covered area more accurately than the Anglo-Saxons who preceded them did. The settlers from Eastern Europe, together with technological inventions such as barbed-wire fencing and the moldboard plow, helped to change the perception of the prairies from the Great American Desert to the Great American Breadbasket.

Land degradation is a function of many variables, including the type of farming system utilized and the educational levels of land managers. In the mountains of Ethiopia, where cultivation has been occurring for 2000 years with a low rate of soil loss, the cumulative erosion of good soil has resulted in a serious decline in the capability of the land. In comparison, in the hills of northern Thailand, where rates of soil loss are much higher, the local land management system has compensated for soil erosion and the capability of the land has been maintained.

However, land is sometimes devastated by land managers—not because of ignorance or stupidity but because of the social systems that keep farmers trapped in small plots of land, dependent on cash crops with low market prices and powerful merchants with a local monopoly on the collection of crops for sale in the market. Local farmers may well be aware of the causes of land degradation and attempt to combat it with fertilizers, mulching, and

Agriculture

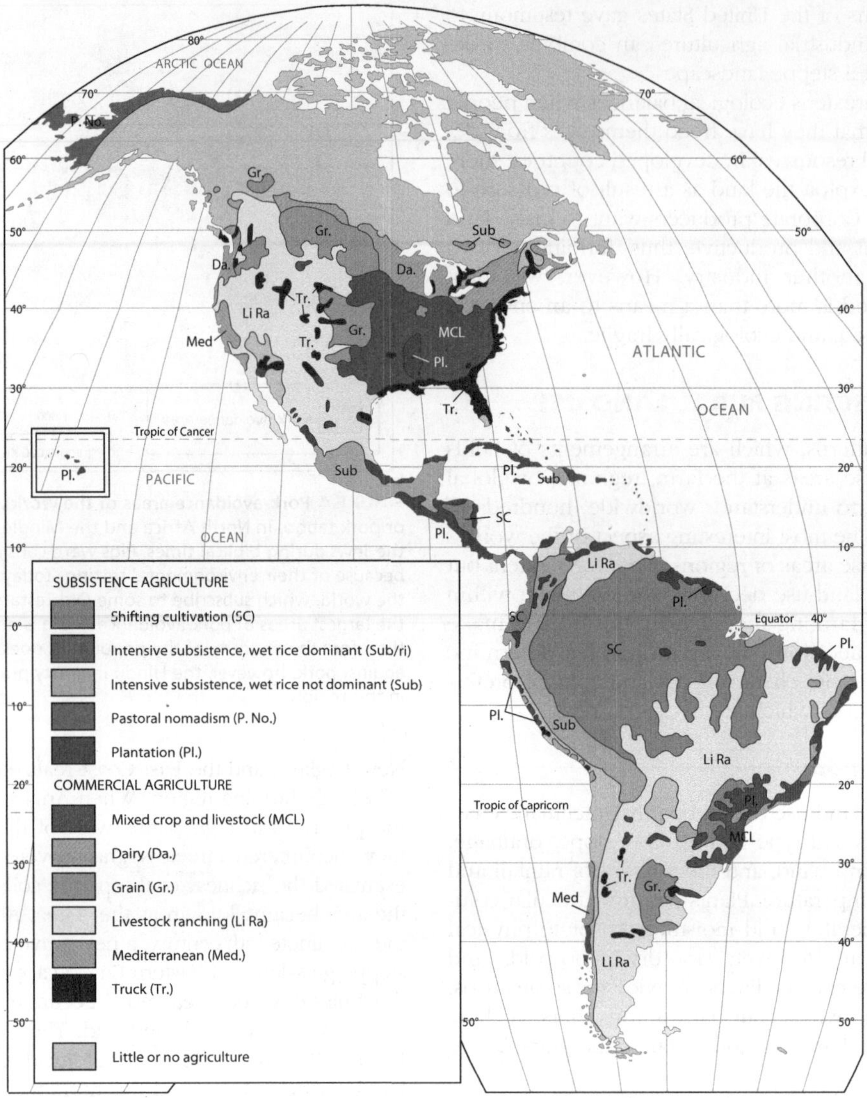

FIGURE 5 **World agricultural regions.** Africa, Southeast Asia, and the Amazon basin are the principal regions of shifting agriculture. Intensive subsistence agriculture is found in East Asia, South Asia, and Southeast Asia. Plantation agriculture is primarily in tropical and subtropical regions of Latin America, Asia, and Africa. Commercial regions include mixed crop and livestock farming, which exists primarily in the northern Unites States, southern Canada, and central Europe. Grain cultivation exists in Argentina, the Great Plains of the United States, the plains of Russia and Ukraine, and Australia. Livestock ranching includes areas too dry for plant cultivation, including western Canada and the United States; southeastern South America; central Asia; and large portions of Australia. Mediterranean agriculture specializes in horticulture and includes areas surrounding the Mediterranean Sea, regions of the southwestern United States, central Chile, and the southern tip of Africa. Finally, truck farming—commercial gardening and fruit farming—is found in the southeastern United States and in southeastern Australia.

terracing. However, without real land reform, many peasants around the developing world are trapped in desperate circumstances reflecting the dynamics of the global economy and oppressive national and local social structures.

SYSTEMS OF AGRICULTURAL PRODUCTION

Systems of agricultural production set their imprint on rural land use. Like manufacturing, agriculture is carried out according to two basic modes of production, peasant or precapitalist systems and capitalist, commodified systems. The major distinction between these is the labor commitment of the enterprise. In the peasant system, production comes from small units worked entirely, or almost entirely, by family labor. In the capitalist system, family farming is still widespread, but labor is a commodity to be hired and dismissed by the enterprise according to changes in the scale of organization, the degree of mechanization, and the level of market demand for products.

In any geographic region, one system of production dominates the others. For example, capitalist agriculture dominates parts of South America, whereas peasant agriculture dominates other parts. Capitalist agriculture finds expression in a vast cattle ranching zone extending southwest from northeastern Brazil to Patagonia; in Argentina's wheat-raising Pampa, which is similar to the U.S. Great

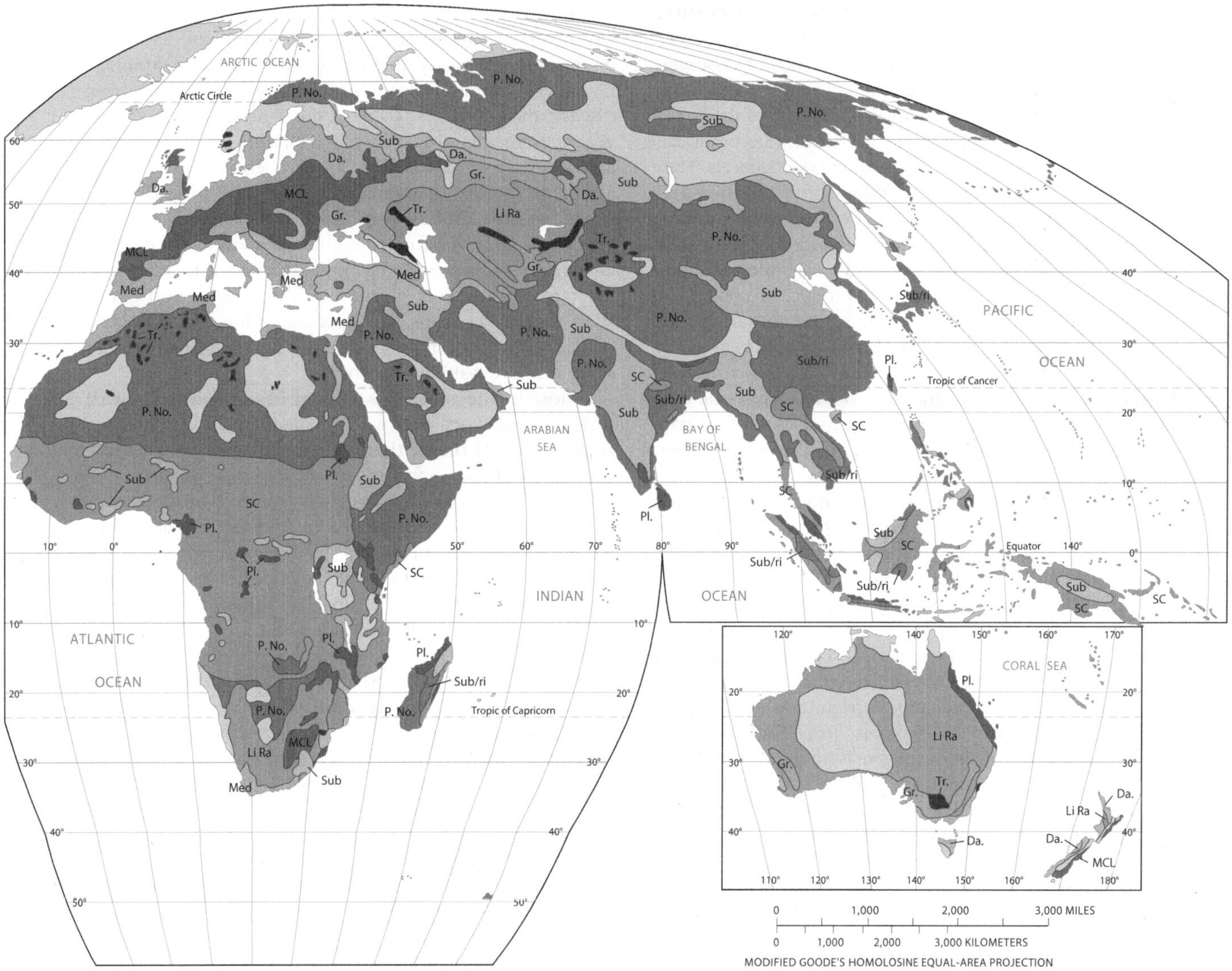

Plains; in a mixed livestock and crop zone in Uruguay, southern Brazil, and south central Chile, which is comparable to the U.S. Corn Belt; in a Mediterranean agriculture zone in central Chile; and in a number of seaboard tropical plantations in Brazil, the Guianas, Venezuela, Colombia, and Peru. Peasant agriculture dominates the rest of the continent. There is shifting cultivation in the Amazon Basin rainforest, rudimentary sedentary cultivation in the Andean plateau country from Colombia in the north to the Bolivian Altiplano in the south, and a wide strip of crop and livestock farming in eastern Brazil between the coastal plantations and livestock ranching zones.

Preindustrial Agriculture

Most of the world's farmers, including the people of Latin America, Africa, and Asia, practice subsistence agriculture on a preindustrial basis. These regions have several characteristics in common:

1. The majority of workers are engaged in agriculture instead of manufacturing or services.
2. Agricultural methods and practices are technologically primitive. Farms and plots are small in comparison with those of the developed world, labor is used intensively, and mechanization and fertilization are used only infrequently. The primary energy source is animate, that is, from living human and animal muscle.
3. Agricultural produce that is harvested on the farms is used primarily for direct consumption. The family, or the extended family, subsists on the agricultural products from the farm. Although in certain years surpluses may be produced, this is rarely the case.

There are several major categories of **subsistence agriculture**: peasant-based agriculture; shifting cultivation in the tropics; pastoral nomadism in North Africa and the Middle East; and intensive subsistence agriculture in South and East Asia, where rice is grown (Figure 5).

PEASANT MODE OF PRODUCTION Peasant agriculture occurs entirely in developing countries where market relations have not fully encompassed all domains of economic life. It is relatively **labor-intensive**, involving endless hours of backbreaking toil. In such societies, the bulk of the population lives in rural areas. Farmers are small-scale producers who invest little in mechanical equipment or chemicals. They are interested mainly in using what they produce rather than in exchanging it to buy things that they need. Food and fiber are exchanged, particularly through interaction with capitalist agriculture at the global, national, and local scales, but farm families consume much of what they produce. To obtain the outputs required to be self-supporting, peasant farmers are frequently willing to raise inputs of labor to very high levels, especially in crowded areas where land is rarely available. Highly intensive peasant agriculture occurs in the extensive rice fields of South, East, and Southeast Asia. Most of the paddies are prepared by ox-drawn plow, and the rice is planted and harvested by hand—millions of hands.

Another example of the peasant mode of production exists in the semiarid zone of East Africa. This zone includes the interior of Tanzania, northeast Uganda, and the area surrounding the moist high-potential heartland of Kenya. As in most parts of the developing world, peasant agriculture in this region has been complicated by the colonial and postcolonial experience. People earn a living by combining several activities. They eat their crops and livestock and sell or exchange agricultural surpluses at markets. They grow cash, or export, crops such as cotton. They maintain beehives in the bush and sell part of the honey and wax. They brew and sell beer. They hunt, fish, and collect wild fruits. They earn income by cutting firewood, making charcoal, delivering water, and carrying sand for use in construction. Some of them have small shops or are tailors. Most important, people sell their labor, both short term and long term, nearby and far away.

To farm and herd successfully in the semiarid zone, land managers must meet certain requirements set by the environment and by the nature of crops and animals. Livestock requires water, graze, salt, and protection from disease and predators. To meet these needs day after day, year after year, land managers must have considerable skill and knowledge. They must know a great deal not only about the ability of animals to withstand physiological stress but also about environmental management—which grass to save for late grazing and where and when to establish dry-season wells to enable the stock to withstand the rigor of the daily journey between water and graze. With respect to crops, land managers must know about plant-moisture and nutrient needs. They must also be sensitive to the variability of rainfall.

Most of the time, this system of agriculture in East Africa provides peasants with an adequate and varied food supply. In bad times, there are mechanisms for sharing hardship and loss so that those farmers who are hardest hit can usually rebuild their livelihoods after bad times end. However, the peasant mode of production has been forced to adjust to pressures from governments and the world economy during colonial and postcolonial periods, including competing with subsidized grain imports from countries such as the United States.

SHIFTING CULTIVATION Shifting cultivation, **slash-and-burn**, *or* **swidden agriculture** is practiced in three main tropical rainforest areas of the world: (1) the South American Amazon region; (2) central Africa; and (3) Southeast Asia, Indonesia, and New Guinea (see Figure 5). Rainfall is heavy in these regions, vegetation is thick, and soils are relatively poor in quality.

When shifting cultivation is practiced, the people of a permanent village clear a field adjacent to their settlement by slashing vegetation (Figure 6). After the field is cleared

UN/DPI PHOTO/P. Sudhakaran

FIGURE 6 A scene showing deforestation in the rainforest in Acro, Western Brazil. Forests have been burnt to the ground to create temporary pastures for cattle. The nation's rainforests are being cut down at a rate 50% faster today than they were 10 years ago. Rainforest loss creates or contributes to a number of intractable problems: It contributes to greenhouse warming, eliminates the cleansing of the atmosphere, creates new semideserts, increases large-scale flooding, and threatens wildlife habitats.

with axes, knives, and machetes, the remaining stumps are burned. Daily rain returns the ash and nutrients to the soil, temporarily fertilizing it. (Because of this clearing technique, shifting cultivation is sometimes called *slash-and-burn* agriculture.) The field is used for several years. At the end of this time, the soil is depleted, and the village turns elsewhere to clear another field. Eventually, the forest vegetation again takes over, and the area is refoliated. The soil is thus allowed to replenish itself.

Swidden agriculture survives in areas of the humid tropics that have low-potential environmental productivity and low population pressure. Under ideal conditions, this form of agriculture leaves much of the original vegetation intact. Farmers make small, discontinuous clearings in forests. They cut down some trees, burn the debris, and prepare the soil for a variety of crops—groundnuts, rice, taro, sweet potato. Because no fertilizer is used, soil nutrients are quickly depleted. Using hoes or knives, farmers plant the fields by hand with tubers or seeds: An indentation is made in the soil, a stem of a plant is submerged or a seed is dropped into a hole, and soil is pushed over the opening by hand. Mechanization and animals are not used for plowing or for harvesting. The most productive farming occurs in the second or third year after burning. Following this, surrounding vegetation rapidly regenerates, weeds grow, and soil productivity dwindles. The plot, sometimes called a *swidden* or *milpa*, is abandoned; then a new site is selected nearby. Usually, the village does not permanently relocate. The villagers commonly return to the abandoned field after 6 to 12 years, by which time the soil has regained enough nutrients to grow crops again. Except on steep slopes, where soil erosion can be a serious problem, shifting cultivation can be a sustainable system of agriculture. It allows previous plots to regenerate natural growth. However, shifting cultivation can lead to degradation when an increasing population demands too much of the land, reducing the fallow period.

The predominant crops grown in shifting agricultural areas include corn and manioc (cassava) in South America, rice in Southeast Asia, and sorghum and millet in Africa. In some regions, yams, sugarcane, and other vegetables are also grown. The patchwork of a swidden is quite complex and seemingly chaotic. On one swidden, a variety of crops can be grown, including those just mentioned, as well as potatoes, rice, corn, yams, mangoes, cotton, beans, bananas, pineapples, and others, each in a clump or small area within the swidden.

Only 5% of the world's population engages in shifting cultivation today. This low percentage is not surprising because tropical rainforests are not highly populated areas. However, shifting cultivation occupies approximately 25% of the world's land surface and therefore is an important type of agriculture. The amount of land devoted to this type of agriculture is decreasing because governments in these regions deem shifting cultivation to be economically unimportant. Consequently, governments in developing countries are selling and leasing land to commercial interests that destroy the tropical hardwoods and rainforests.

PASTORAL NOMADISM Shifting cultivation and **pastoral nomadism** can be classified as extensive subsistence agriculture. Areas in which pastoral nomadism is practiced include North Africa and the Middle East, the eastern plateau areas of China and Central Asia, and eastern Africa's Kenya and Tanzania (see Figure 5). Only 15 million people are pastoral nomads, but they occupy 20% of the earth's land area, areas that are climatically opposite to those of shifting cultivators. The lands of the pastoral nomads are dry; usually less than 10 inches of rain accumulate per year, and typical agriculture is normally impossible, except in oases areas.

Instead of depending on crops as most other farmers do, nomads depend on animal herds for their sustenance. Everything that they need and use is carried with them from one forage area to another. Tents are constructed of goats' hair, and milk, clothing, shoes, and implements are produced from the animals. Pastoral nomads consume mostly meat and grain. Sometimes, in exchange for the meat, other needed goods are obtained from sedentary farmers in marginal lands near the nomads' herding regions. It is common for pastoral nomads to farm areas near oases or within floodplains that they occupy for a short period of the year. Nomadic parties usually include 6 to 10 families who travel in a group, carrying bags of grain for sustenance during the drier portions of the year.

A cyclical pattern of migration is entrenched in the nomadic way of life, and it lasts for generations. Pastoral nomads are not wandering tribes; they follow a 12-month cycle in which lands most available for forage are cyclically revisited in a pattern that exhibits strong territoriality and observance of the rights of adjacent tribes. The exact migration pattern of today's pastoral nomads has developed from a precise geographic knowledge of the region's physical landforms and environmental provision.

Nomads must select animals for their herds that can withstand drought and provide the basic necessities of the herdsmen. The camel is the quintessential animal of the nomad because it is strong, can travel for weeks without water, and can move rapidly while carrying a large load. The goat is the favorite small animal because it requires little water, is tough, and can forage off the least green plants. Sheep are slow moving and require more water, but they provide other necessities: wool and mutton. Small tribes need between 25 and 60 goats and sheep and between 10 and 25 camels to sustain themselves.

Before the railroad and telegraph, pastoral nomads were the communication agents of the desert regions, carrying with them innovations and information. This is no longer the case, as nomadic societies have fallen before the territorial imperatives of the nation-state and its fixed boundaries. However, nomadic herding remains because these vast dry areas of the world cannot be used for other economic activity. Furthermore, government attempts to settle pastoral nomads have met with little success. In the future, pastoral nomads will be allowed only on lands that do not have energy resources or precious metals beneath the surface or on lands that cannot be easily irrigated from nearby rivers, lakes, or groundwater aquifers. In any case, the number of pastoral nomads is declining.

INTENSIVE SUBSISTENCE AGRICULTURE **Intensive subsistence agriculture** is practiced by large populations living in East, South, and Southeast Asia, Central America, and South America (see Figure 5). Whereas shifting cultivation and pastoral nomadism are extensive low-density, marginal operations, intensive subsistence agriculture, as the term implies, is a higher-intensity type of agriculture in the majority of the densely populated developing areas of the world. Rice is the predominant crop because of its high levels of carbohydrates and protein. Most farmers involved in intensive subsistence rice agriculture use every available piece of land, however fragmented, around their villages (Figures 7 and 8). Most often, a farm encompasses only a few acres.

Intensive subsistence agriculture is characterized by several features:

1. Most of the work is done by hand, with all family members involved. Occasionally farm animals are used, such as water buffalo or oxen. Almost no mechanization is involved because of lack of capital to purchase such equipment and because plots are tiny.
2. Plots of land are extremely small by Western standards. Almost no piece of land is wasted. Even roads through agricultural regions of intensive subsistence are made narrow so that all cultivatable areas can be used.
3. The physiological density (i.e., the number of people that each acre of land can support) is very high.
4. Principal regions that are cultivated are river valleys and irrigated fields in low-lying, moist regions in the middle latitudes.

The World Bank

FIGURE 7 In Indonesia, harvesting rice is an example of labor-intensive peasant culture.

Image copyright javarman, 2010. Used under license from Shutterstock.com

FIGURE 8 These rice fields in Yunnan, southern China, are typical of the intensive paddy rice agriculture system that feeds more than a billion people in East and Southeast Asia. When China's technocrats decide to build a farm or dam a river, build a road or move a village, the dam goes up, the road goes down, and the village disappears. The villagers may be compensated, but they are not allowed to stand in the way of progress. China's leaders make rational decisions that balance the needs of all citizens over the long term. This has led to rapid, sustained growth that has lifted hundreds of millions of people out of hunger and poverty.

Because rice is a crop that has a high yield per acre and is rich in nutrients, it is a favorite in intensive subsistence agricultural regions (Figure 9). First, the field is plowed with a sharpened wooden pole that is pulled by oxen. Next, the field is flooded with water and planted with rice seedlings by hand. Another method is to spread dry seeds over a large area by hand. When the rice is mature, having developed for three-fourths of its life underwater, it is harvested from the rice paddy. To separate the husks from the rice itself, the farmers thrash the rice by beating it on a hard surface or by trampling it underfoot. Sometimes it is even poured on heavily traveled roads. The chaff is thus removed from the seeds, and sometimes the wind blows the lightweight material far from the pile of rice itself, a process known as winnowing.

Some year-round, tropical, moist areas of the world permit **double cropping**. This means that more than one crop can be produced from the same plot within the year. Occasionally in wet regions, two rice crops are grown, but more frequently a rice crop and a different crop, which requires less water, are produced. The field crop is produced in the drier season on nonirrigated land. In the higher latitudes of East Asia, rice is mixed with other crops and may not be the dominant crop. In western India and the northern China plain, wheat and barley are the dominant crops, with oats, millet, corn, sorghum, soybeans, cotton, flax, hemp, and tobacco also produced.

Problems of Subsistence Agriculturalists

Subsistence agriculture is subjected to variations in soil quality, availability of rain from year to year, and, in general, environmental conditions that can harm crop-production levels and endanger life. In addition, subsistence agriculturalists lack tools, implements, hybrid seeds, fertilizer, and mechanization that developed nations have had for nearly 100 years. With such drawbacks, subsistence agriculturalists can barely provide for their families, and net yields have not increased substantially for many generations. These families do not have enough capital to purchase the necessary equipment to improve their standard of living.

Finally, all too often, developing countries turn to their limited sources of export revenues to generate the cash flow needed for infrastructure, public services, and the military. They must produce something that they can sell in the world market. Often, they sell mineral resources, foodstuffs, and nonmineral energy fuels. Most frequently, these countries sell cash crops on the world market to generate foreign revenue; thus, the food is not used to sustain their own population. Another category of agricultural products that can generate revenue is nonfood or not-nourishing crops, such as sugar, hemp, jute, rubber, tea, tobacco, coffee, and a growing harvest of cotton to satisfy the world's need for fabric and denim. All of these commodities, however, command very low prices on the global market. How

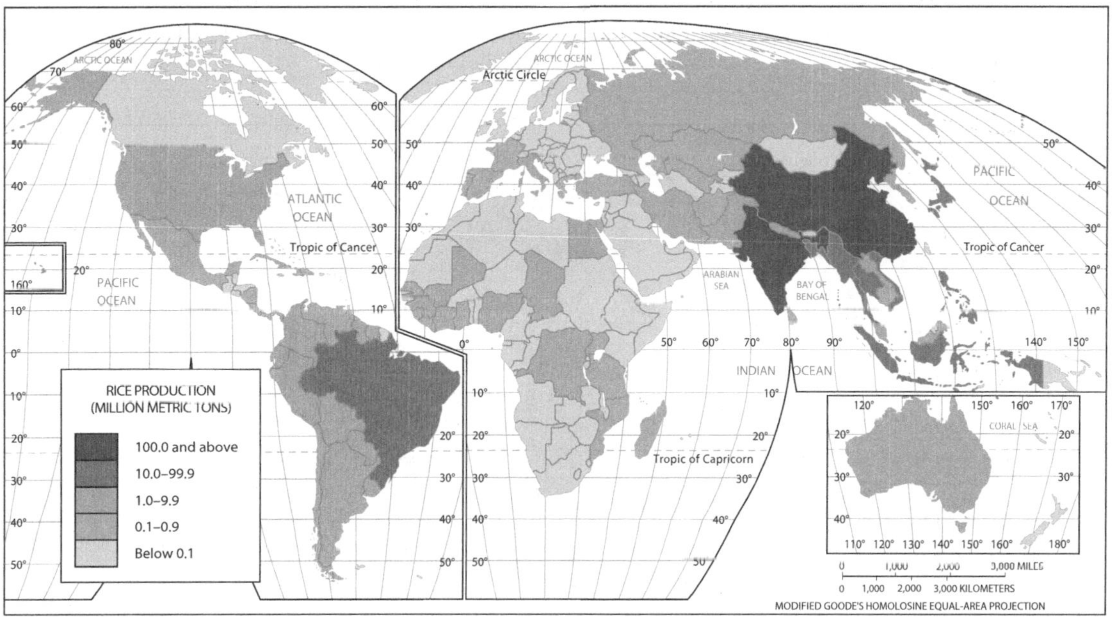

FIGURE 9 World rice production. Rice was domesticated in East Asia more than 7000 years ago. Unlike corn in the United States, which primarily goes toward animal feed, rice is almost exclusively used for human consumption. About 2 billion people worldwide are fed chiefly by rice. It is the most important crop cultivated in the most densely settled areas of the world, including China in East Asia, India in South Asia, and Southeast Asia. These areas produce more than 90% of the world's rice. Rice is an ideal crop in very humid and tropical regions because it depends on large quantities of water to develop. Rice requires a large amount of labor and tedious work, but the returns are bountiful. Rice produces more food per unit of land than any other crop; thus, it is suitable for the most densely populated areas of the world, such as China and India.

can impoverished nations feed themselves when a large proportion of their agricultural productivity and acreage is devoted to nonfood crops? This is the plight of many African, South American, Central American, and Asian countries today. As a result, sometimes alternative sources of income are inviting, even if they are illegal, such as cocoa or opium.

COMMERCIAL AGRICULTURE

Commercial agricultural areas dominated by capitalist social relations include the United States and Canada, Argentina and portions of Brazil, Chile, Europe, Russia and Central Asia, South Africa, Australia, New Zealand, and portions of China.

Agriculture in the United States epitomizes the contemporary capitalist system of food production. The American agricultural system developed in the nineteenth century as part of the unfolding of the European "frontier" across North America. Railroads and steamships dramatically lowered transport costs to the markets along the East Coast, and agricultural trading centers and ports, such as Chicago, St. Louis, and New Orleans grew rapidly. By the turn of the century, the United States had become a major supplier of wheat and other commodities to Europe. The only other major producer, Russia, effectively withdrew from world markets following the revolution there in 1917, leaving the United States as the major supplier. Consequently, the vast agricultural region stretching across the Ohio and Mississippi river basins into the Great Plains, and extending into central Canada, became the core of the North American food-producing system.

American agriculture today is a huge and very productive industry dominated by a handful of large **agribusiness** firms. Agribusiness, dominated by such giant food companies as ConAgra, Bunge, Cargill, Dole, Nabisco, General Mills, Kraft General Foods, Hunt-Wesson, Archer Daniels Midland, and United Brands, controls the whole food chain from "seedling to supermarket." Whereas the popular imagination clings to the stereotype of the small family farmer, in reality most American agriculture is organized around the needs of a small handful of large firms, which generally do not own the farmland but control the food production and processing (e.g., canning), distribution, price and cost, and marketing. The concept that describes the food companies' control of the production process from raw material to final product is **vertical integration**, which is common in **capital-intensive production**.

Agribusiness is extremely capital-intensive and energy-intensive. Farmers rely on copious quantities of chemicals, tractors, harvesters, airplanes, and other equipment—most of it very sophisticated, computer controlled, and expensive—to keep labor inputs low and productivity levels high. Only 2% of the U.S. labor force is employed in agriculture, and it not only feeds the other 98% of the populace but exports vast amounts of food as well. The very high per capita productivity has resulted in long-term rural depopulation. For example, the use of tractors worldwide (Figure 10) is a measure of the capital-intensity of agricultural production: Countries with highly commodified agricultural systems rely extensively on this technology, freeing people from the farm, whereas in developing nations with a peasant-based system of agriculture, tractors are relatively uncommon.

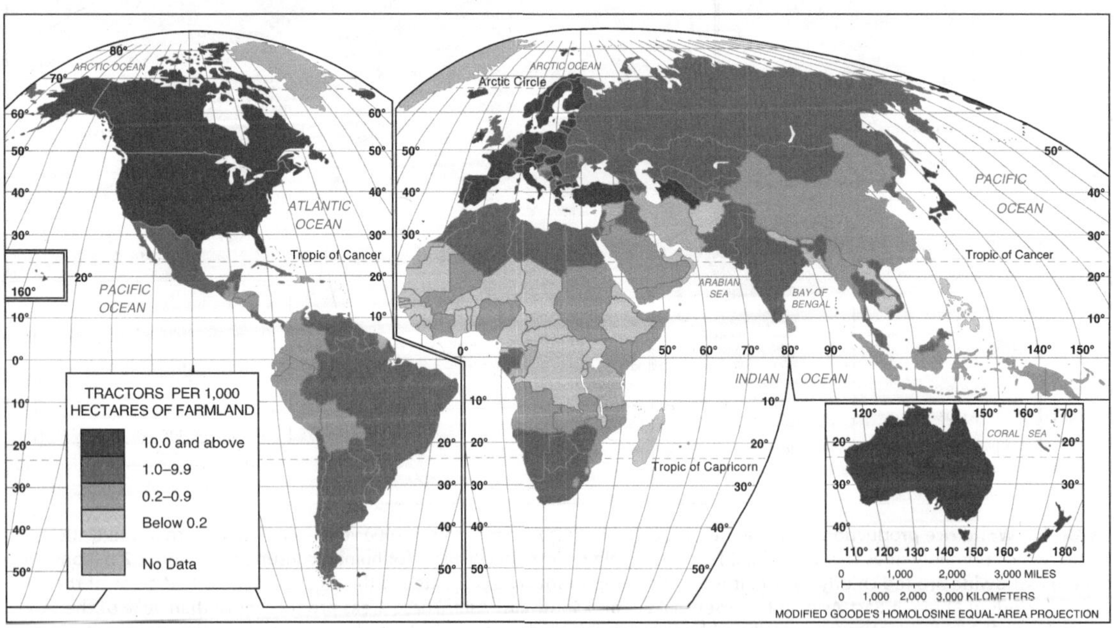

FIGURE 10 Tractors per 1000 hectares of farmland. Tractor usage is an indicator of the degree to which agriculture is mechanized. Farming systems in economically developed countries rely heavily on tractors and similar equipment—as well as fossil fuels—whereas low tractor usage is associated with labor-intensive farming in less developed countries.

The importance of corporate farming is growing in market gardening, which is sometimes called **truck farming**. Modern truck farms specialize in intensively cultivated fruits, vegetables, and vines, and they depend on migratory seasonal farm laborers to harvest their crops. In the United States, California is the epicenter of fruit and vegetable farms, although they are widespread in Florida as well.

Agribusiness has also extended livestock farming immensely; it has mechanized the raising and slaughter of cattle, often under inhumane conditions. Other examples of modern livestock production include poultry ranches and egg factories. At one time, livestock farming was associated with a combination of crop and animal raising on the same farm. In recent years, livestock farming has become highly specialized. An important aspect of this specialization has been the growth of factory-like feedlots, which raise thousands of cattle and hogs on purchased feed, generating huge quantities of animal waste. Feedlots are common in the western and southern states, in part because winters there are mild. These feedlots raise more than 60% of the beef cattle in the United States.

Thus, corporate agriculture is an industry similar to the production of other goods such as cars. Agriculture's major backward linkages—its purchases from other sectors—include petroleum and machinery; notably, labor is only a small part of the production costs, given how capital-intensive the sector is. The forward linkages of this sector—its sales to other parts of the economy—include the food processing sector and meat production; a large share of cereals, especially corn, are grown for animal feed.

Modern American farming is quick to respond to new developments, such as new production techniques. Consequently, farmers with sizable investments of money, materials, and energy can create drastic changes in land-use patterns. For example, farmers in the low-rainfall areas of the western United States have converted large areas of grazing land to forage and grass production with the use of center-pivot irrigation systems (Figure 11). Other farmers grow sugar beets and potatoes in western oases through federally subsidized water projects.

American corporate farming is also extending overseas to become a worldwide food-system model. Poultry-raising operations in Argentina, Pakistan, Thailand, and Taiwan are increasingly similar to those in Alabama or Maryland. Enormous, politically well-connected enterprises such as United Brands, Del Monte, Archer Daniels Midland, and Unilever divert food production in developing countries toward consumers in developed countries.

U.S. Commercial Agriculture: Crops and Regions

The main characteristic of **commercial agriculture** is that it is produced for sale off the farm, at the market. Following are some of the characteristics of commercial agriculture:

1. Populations fed by commercial agriculture are urban populations engaged in other types of economic activity, such as manufacturing, the services, and information processing.
2. Only a small proportion of the population is engaged in agriculture.
3. Machinery, fertilizers, and high-yielding seeds are used extensively, with high energy inputs.
4. Farms are extremely large, and the trend is toward even larger farms.
5. Agricultural produce from commercial agriculture is integrated with other agribusiness, and a vertical integration exists that stretches from the farm to the table.

Commercial Agriculture and the Number of Farmers

The percentage of laborers in developed countries working in commercial agriculture is less than 5% overall. In contrast, in some portions of the developing world where intensive subsistence agriculture is practiced, 90% of the

U.S. Department of Agriculture

FIGURE 11 The development of the center-pivot irrigation system in the 1950s enabled large-farm operators to transform huge tracts of land in sandy or dry regions of the United States into profitable cropland. Here, alfalfa is being irrigated in Montana. The alfalfa will be used to fatten cattle. The top farm products produced in the United States in order of value are corn, beef, milk, chicken, soybeans, pork, wheat, and cotton.

population is directly engaged in farming, and the average is 60% overall. Today, U.S. farmers on average produce enough food for themselves and 70 other families.

In 2008, the United States had approximately 1.7 million farms, compared with 5.7 million in 1950 (Figure 12). This reduction in the number of farm families as a percentage of the population is a result of waves of corporate consolidation and mergers and the high cost of equipment, low prices, or high interest rates that drive families off the farms. Farming is difficult, often dangerous work, and low crop prices can be ruinous. Meanwhile, the opportunity for a college education and higher-paying occupations in the cities have long lured farm children off the land. One serious problem is the encroachment of metropolitan areas onto the best farmland, which has become directly adjacent to urban areas through the expansion of housing subdivisions and shopping centers. Suburban sprawl, brought on by interstate highways that reduce the commute and penetrate into the countryside, has usurped viable topsoil and farmland around many metropolitan areas in the United States as well as around many European cities.

Machinery and Other Resources in Farming

The second aspect unique to commercial agriculture, besides the small percentage of farmers in the population, is the heavy reliance on expensive machinery—tractors, combines, trucks, diesel pumps, and heavy farm equipment—all amply fueled by petroleum and gasoline resources, to produce the large output of farm products. To this miracle seeds have been added that are hardier than their predecessors and produce more impressive tonnages. Commercial agriculture is also fertilizer-intense.

Improvements in transportation to the market have resulted in less spoilage. Products arrive at the canning and food-processing centers more rapidly than they did earlier. By 1850, many American farms were well connected to cities by rail transportation. More recently, the motor truck has supplanted rail transportation, and the advent of the refrigerator car and the refrigerator truck meant that freshness was preserved. Cattle also arrive at packing houses by motor truck as fat as when they left the farm, unlike the mid-nineteenth century, when long cattle drives were the order of the day, connecting cattle-fattening areas in Texas, Oklahoma, and Colorado with the Union Pacific rail line stretching from St. Louis to Kansas City to Denver.

Agricultural experiment stations are now located in every state and are usually affiliated with land-grant universities. These stations have made great improvements in agricultural techniques, not only in improved fertilizers and hybrid plant seeds but also in hardier animal breeds and new and better insecticides and herbicides, which have reduced pestilence. In addition, local and state government farm advisors can provide information about the latest techniques, innovations, and prices so that the farmer can make wise decisions concerning what should be produced, when it should be produced, and how much should be produced.

Types of Commercial Agriculture

We can divide commercial agriculture into six main categories: mixed crop and livestock farming, dairy farming, grain farming, cattle ranching, Mediterranean cropping, and horticulture and fruit farming (see Figure 5).

MIXED CROP AND LIVESTOCK FARMING Mixed crop and livestock farming is the principal type of commercial agriculture, and it is found in Europe, Russia, Ukraine, North America, South Africa, Argentina, Australia, and New Zealand. The primary characteristic of mixed crop and livestock farming is that the main source of revenue is livestock, especially beef cattle and hogs. In addition, income is produced from milk, eggs, veal, and poultry. Although the majority of farmlands are devoted to the production of crops such as corn, most of the crops are fed to the cattle. Cattle fattening is a way of intensifying the value of agricultural products and reducing bulk. Because of the developed world's preference for meat as a major food source, mixed crop and livestock farmers have fared well during the past 100 years.

In developed nations, the livestock farmer maintains soil fertility by using a system of crop rotation in which different crops are planted in successive years. Each type of crop adds different nutrients to the soil. The fields become more efficient and naturally replenish themselves with these nutrients. Farmers today use the **four-field rotation system**, wherein one field grows a cereal, the second field grows a root crop, the third field grows clover as forage for animals, and the fourth field is fallow, more or less resting the soil for that year.

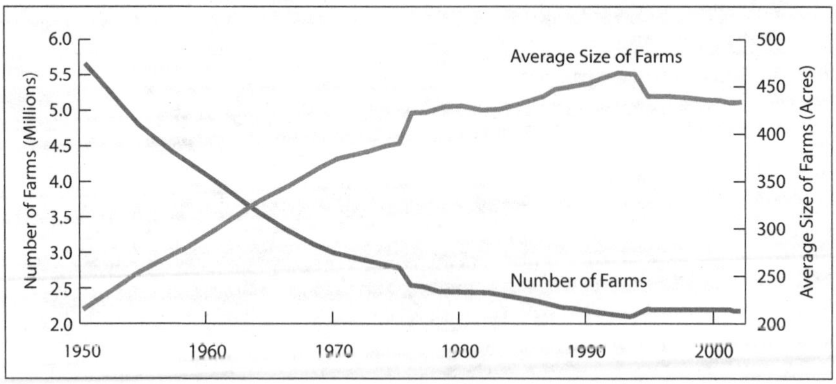

FIGURE 12 The number and average size of U.S. farms, 1950–2002. Since 1950, the number of U.S. farms has decreased steadily, while the average farm size has increased steadily. Farming today is dominated by agribusiness, not family farms.

Most cropping systems in the United States rely on corn (Figure 13) because it is the most efficient for fattening cattle. American corn is highly subsidized by the federal government, largely owing to the political clout of agribusiness, and is thus produced in much larger volumes for lower prices than would be the case if it were a "free market." Some corn is consumed by the general population in the form of corn on the cob, corn oil, or margarine, but most is fed to cattle or hogs, even though cows evolved to eat grass, not corn. Corn is widely used in corn sweeteners in a large variety of processed foods, including soft drinks and ketchup. The second most important crop in mixed crop and livestock farming regions of central North America and the eastern Great Plains is the soybean. The soybean has more than 100 uses, but it is used mainly for animal feed. In China and Japan, tofu is made from soybean milk and is used as a major food source high in protein and low in fat.

DAIRY FARMING **Dairy farming** accounts for the most farm acreage in the northeastern United States and northwestern Europe and accounts for 20% of the total output by value of commercial agriculture. Ninety percent of the world's milk supply is produced in these few areas of the world. Most milk is consumed locally because of its weight and perishability.

Some dairy farms produce butter and cheese as well as milk. In general, the farther the farm is from an urban area, the more expensive the transportation of fluid milk, and the greater proportion of production in more high-value-added commodities, such as cheese and butter. For example, the Swiss discovered ways of transforming their milk products into high-value-added chocolates, cheeses, and spreads that are distributed worldwide. These processed products are not only lighter but also less perishable. On the other hand, in the United States, the proximity of farms to Boston, New York, Philadelphia, Baltimore, and Washington, DC, on the East Coast, and to Chicago and Los Angeles in the Midwest and West, means that these farms primarily produce liquid milk. Farms throughout the remaining areas of the United States primarily produce butter and cheese. Worldwide, remote locations, such as New Zealand, for example, devote three-fourths of their dairy farms to cheese and butter production, whereas three-fourths of the farms in Britain, with a much higher population density at close proximities, produce fluid milk.

Dairy farms are relatively labor-intensive because cows must be milked twice a day. Most of this milking is done with automatic milking machines. However, the cows still must be herded into the barn and washed, the milking machines must be attached and disassembled, and the cows must be herded back out and fed. The difficulty for the dairy farmer is to keep the cows milked and fed during winter, when forage is not readily available and must be stored.

GRAIN FARMING Commercial grain farms are usually in drier territories that are not conducive to dairy farming or

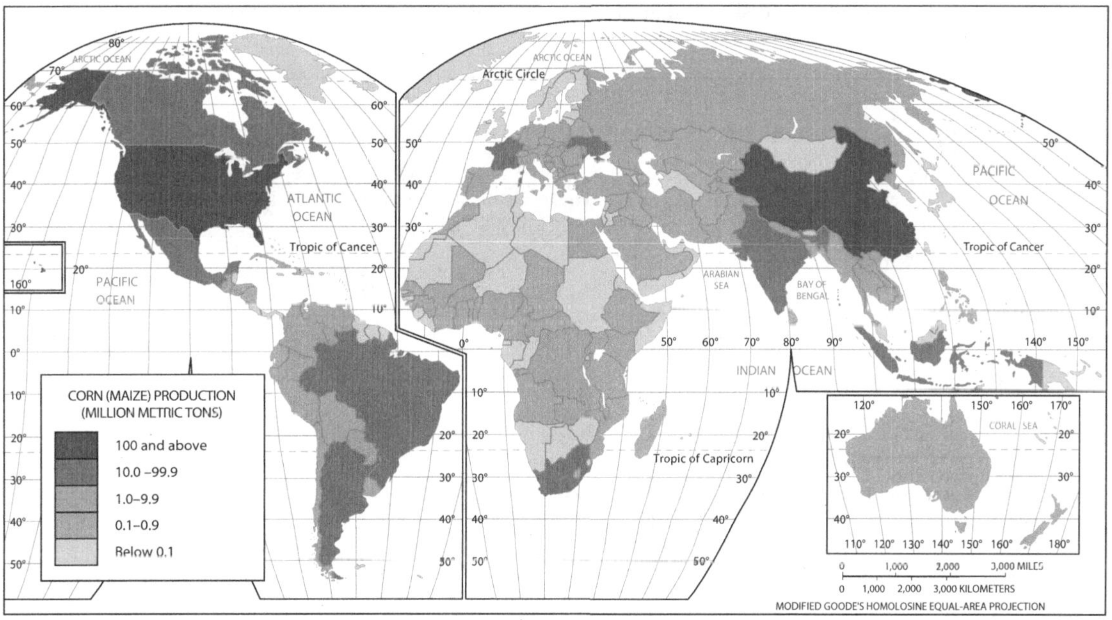

FIGURE 13 World corn (maize) production. Corn was domesticated in Central America more than 5000 years ago and exported to Europe in the fifteenth century. The United States accounts for more than 30% of the world's corn production, and 90% of this corn crop is fed to cattle and livestock for fattening and meat production. Outside the United States, corn is called maize. China is the second leading producer of corn, but the majority of the crop is consumed by humans. Because meat produces a greater market value per pound than corn by itself does, U.S. farmers convert corn into meat by feeding it to livestock on farms and feedlots. In the United States, the Corn Belt is also the livestock region of North America; it is located in the western Midwest and the eastern Great Plains. Argentina and Brazil, as well as Europe, also have sizable corn-production areas.

U.S. Department of Agriculture

FIGURE 14 Harvesting wheat by combine in the United States exemplifies capital-intensive and energy-intensive agriculture. Grain, such as the wheat shown here, is often a major crop on most farms. Commercial grain agriculture is different from mixed crop and livestock farming in that the grain is grown primarily for consumption by humans rather than by livestock. In developing countries, the grain is directly consumed by the farm family or village, whereas in commercial grain farming, output is sold to manufacturers of food products.

mixed crop and livestock farming. Most grain, unlike the products of **mixed crop and livestock farming**, is produced for sale directly to consumers. Only a few places in the world can support large grain-farming operations (Figure 14). These areas include China, the United States, Canada, Russia, Ukraine, Argentina, and Australia (Figure 15). Wheat is the primary crop used to make flour and bread. Other important grains include barley, oats, rye, and sorghum. These grains are not particularly perishable and can be shipped long distances. Wheat is the most highly valued grain per unit area and is the most important for world food production. Figure 16 shows that grain yield and production increased markedly in developing countries between 1970 and 2005, while cropland area has increased only slightly. However, production per capita is much more disappointing in developing countries, particularly in Africa. But the world's food-producing system, however constrained and imperfect, has allowed the global food supply to keep pace with world population increases (Figure 17), denying, or at least forestalling, the Malthusian prophecy.

Wheat is the leading international agricultural commodity transported among nations. The United States and Canada are the leading export nations for grains and together

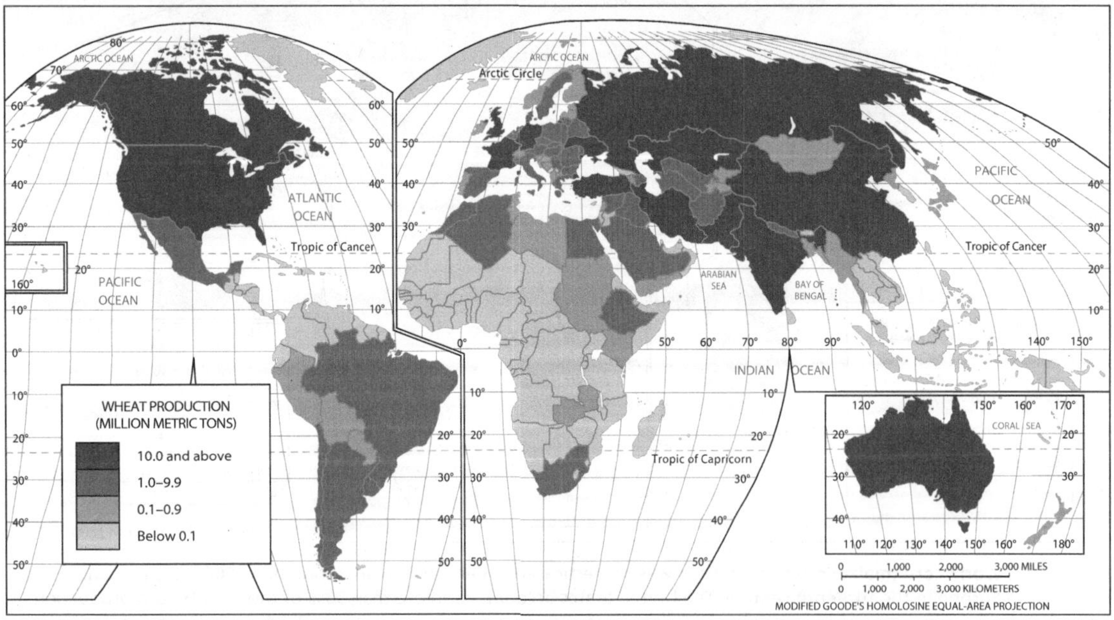

FIGURE 15 World wheat production. China is the world's leading wheat producer, followed by the United States and Russia. The United States, Canada, Australia, and Argentina are the primary wheat exporters, whereas Russia, Kazakhstan, India, and China import the most wheat. Wheat can be stored in grain elevators. Therefore, current wheat prices worldwide reflect not only growing conditions for that year but also those of supplies from commercial and subsistence operations that have been stored throughout the world.

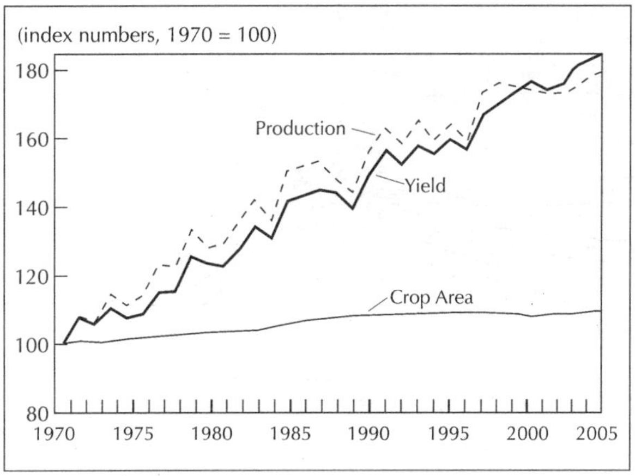

FIGURE 16 The world's output of major food crops has increased dramatically during the past 35 years, the most dramatic being cereals. Milk, meat, fish, fruit, and vegetables have also made gains in production worldwide. Every region of the world has increased production primarily as a result of an increase in yields, not an increase in cropland.

account for 50% of wheat exports worldwide. The North American wheat-producing areas have been appropriately labeled the world's breadbasket because they still provide the major source of food to many deficit areas, including several starving countries in Africa (Figure 18). The United States is the world's only agricultural superpower and plays a unique role in the global food-production system. Agricultural exports generate more than $110 billion in export revenues annually, and the United States exports 20% of all food traded internationally. As with other economic sectors, agriculture has become thoroughly globalized; for example, farmers in Nebraska and Kansas are well aware that next year's revenues will be shaped by weather and political events in markets in Europe and Asia.

In North America, the Spring **Wheat Belt** is west of the mixed crop and livestock farming area of the Midwest and is centered in Minnesota, North Dakota, South Dakota, and Saskatchewan (Figure 19). Major cities in the Great Plains, such as Minneapolis and St. Paul, were often established as centers of flour milling and distribution. Another region, just south of the Wheat Belt, is the Winter Wheat Belt,

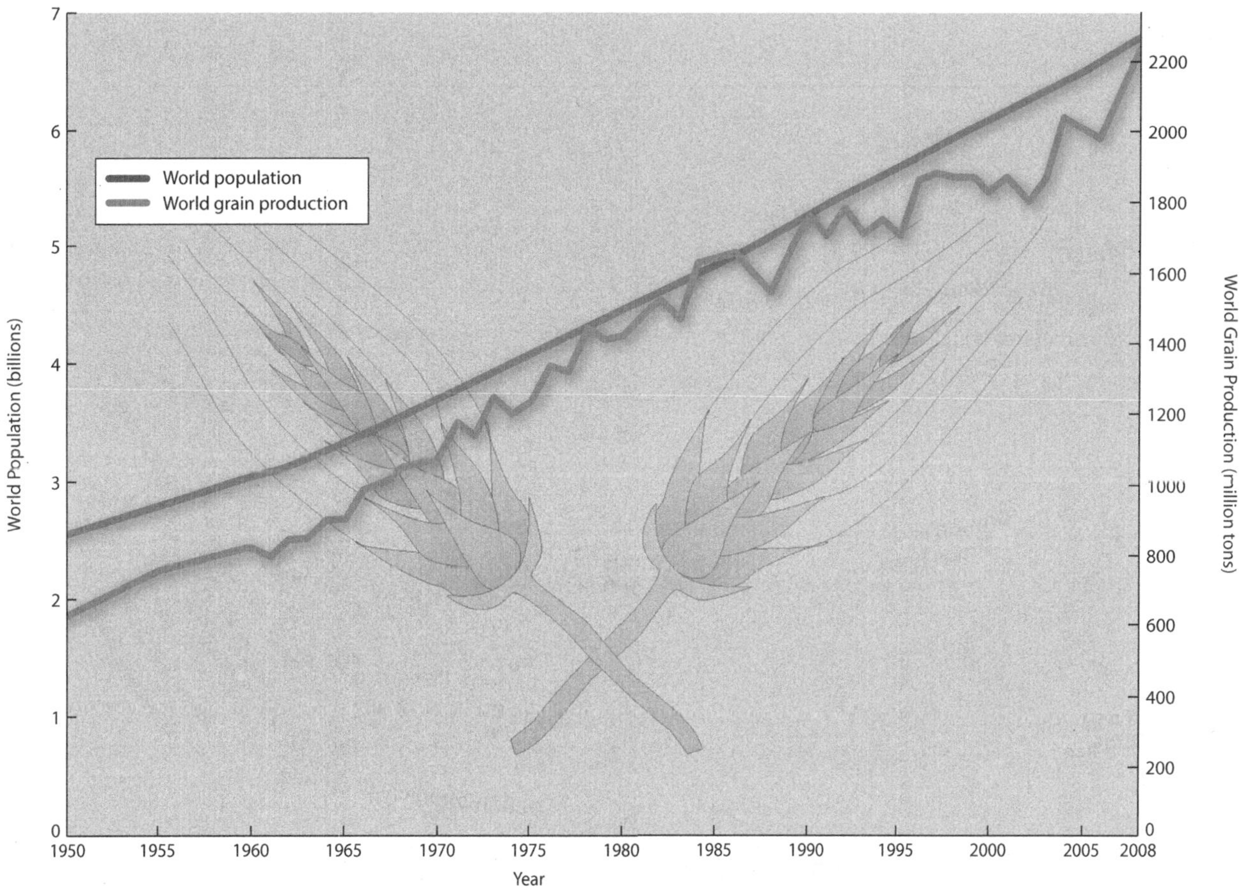

FIGURE 17 World population growth and grain production, 1950–2008. Despite persistent concerns about diminishing returns and environmental degradation, the world's food-supply system has either refuted or postponed Malthusian predictions of massive famine.

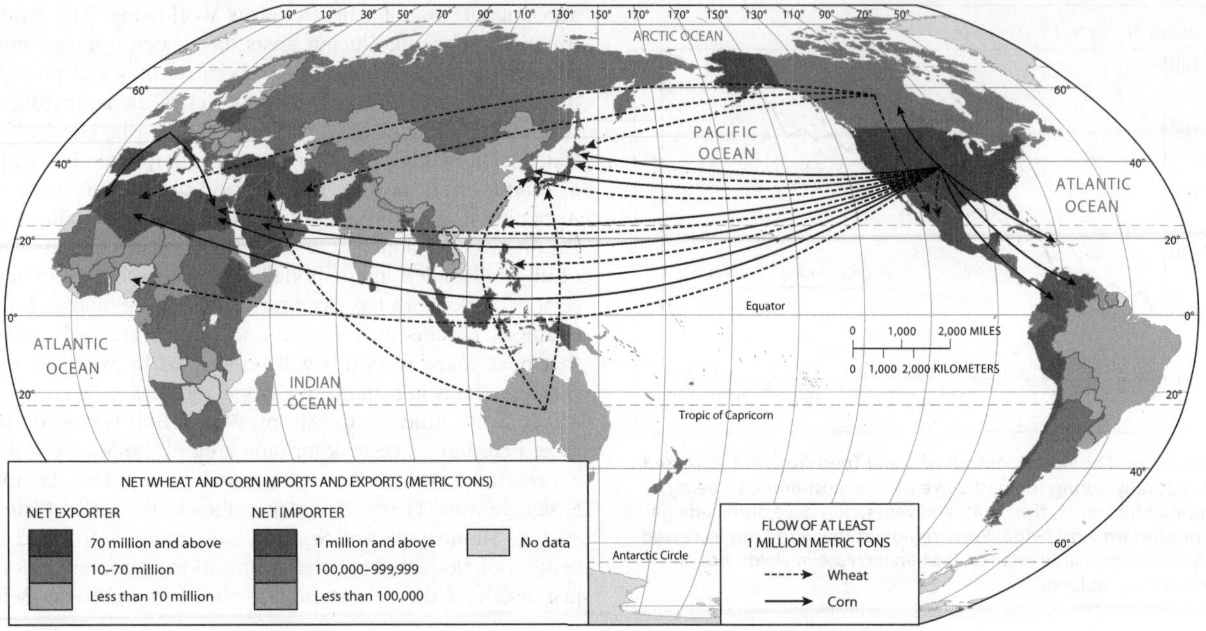

FIGURE 18 A handful of countries dominate grain exports, including the United States, Canada, and Australia.

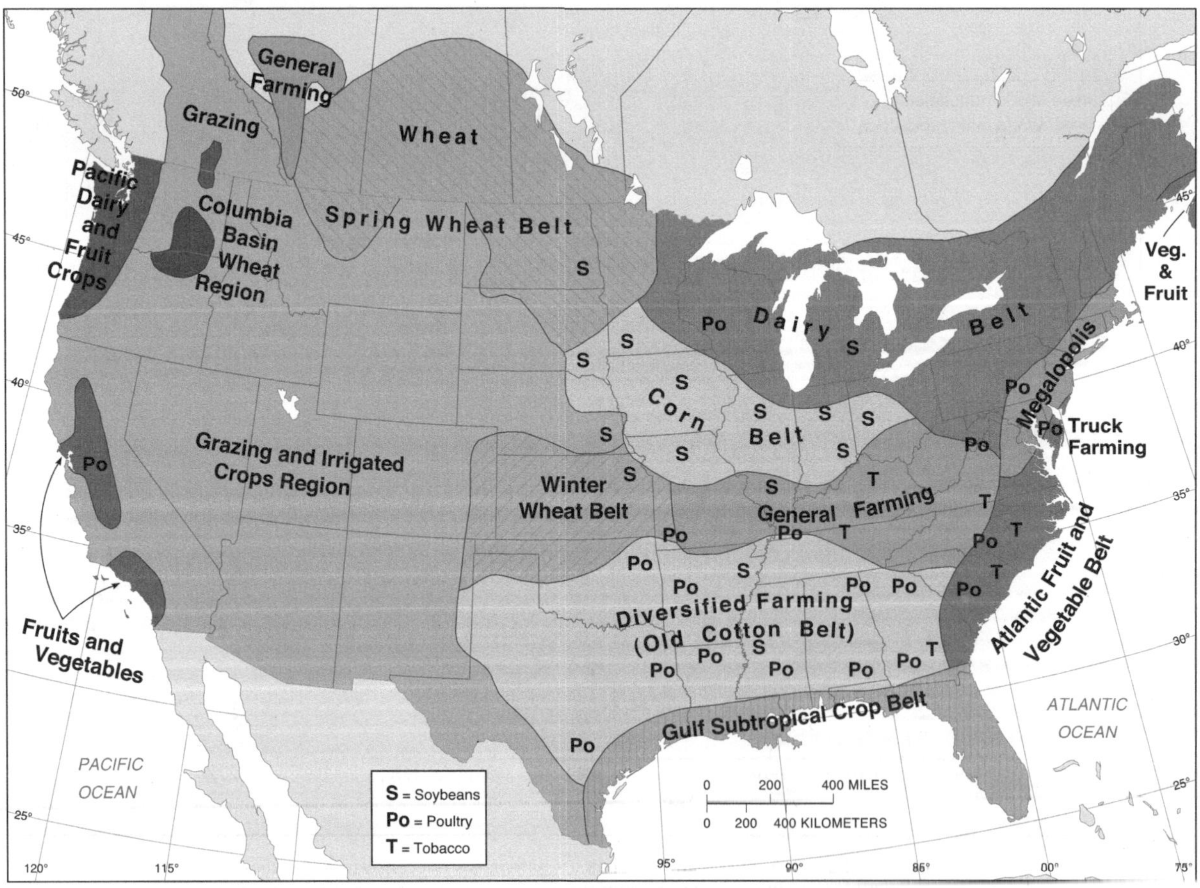

FIGURE 19 Major agricultural regions of the United States and Canada.

which is centered in Kansas, Colorado, and Oklahoma. Because winters are harsh in the Spring Wheat Belt, the seeds would freeze in the ground, so, instead, **spring wheat** is planted in the spring and harvested in the fall; the fields are fallow in the winter. **Winter wheat**, however, is planted in the fall and moisture accumulation from snow helps fertilize the seed. It sprouts in the spring and is harvested in early summer. Like corn-producing regions, wheat-producing areas are heavily mechanized and require high inputs of energy resources. Today, the most important machine in wheat-producing regions is the combine, which not only reaps but also threshes and cleans the wheat. Large storage devices called *grain elevators* are a prominent landscape feature as one traverses the Great Plains of the United States and Canada. In part, these reflect the enormous surpluses that farmers have accrued, to some degree due to government subsidies.

Joe Munroe/Photo Researchers, Inc.

FIGURE 20 Feedlots for beef cattle in California. According to the "Code of the West," cattle ranchers owned little land, only cattle, and grazed open land wherever they pleased. New cattle breeds introduced from Europe, such as the Hereford, offered superior meat but were not adapted to the old ranching system of surviving the winter by open grazing. In moist areas, crop growing supplanted ranching because it generated a higher income per acre. Some cattle are still raised on ranches, but most are sent for fattening to feedlots along major interstate highways or railroad routes. Many feedlots are owned by agribusiness and meat processing companies rather than by individuals.

CATTLE RANCHING **Cattle ranching** is practiced in developed areas of the world where crop farming is inappropriate because of aridity and lack of rainfall. Cattle ranching is an extensive agricultural pursuit because many acres are needed to raise cattle. In some instances, cattle are penned near cities and forage is trucked to cattle-fattening pens called **feedlots** (Figure 20).

Major cities grew up across the western United States partly because of the services provided by their slaughterhouses and stockyards. Denver, Dallas, Chicago, Kansas City, and St. Louis are examples. If a cattle farmer could get a steer to one of these cities, it was worth 10 times as much as it was worth on the range. Early American ranchers were not as concerned about owning territory as they were about owning heads of cattle. Consequently, the range was open and the herds grazed as they went toward market. Later, farmers bought up the land and established their perimeters with barbed-wire fences. Until about 1887, the ranchers cut the barbed-wire fences and continued to move their herds about wherever they pleased. However, after that point, the farmers seemed to win the battle, and ranchers were forced to switch from long cattle drives and wide territories of rangeland to stationary ranching. The land was divided according to the availability of water and the amount of rainfall. Farmers used the land that was productive for farm crops, such as grains and wheat, whereas the ranchers received the land that was too dry for farming. Ironically, given the frequent hostility of cattle ranchers to the state in the past, today 60% of cattle grazing occurs on land leased from the U.S. government, with ranchers paying fees well below market rates (a circumstance that again defies the conventional myth of the "free market"). Today, ranches in Texas and the West cover thousands of acres because the semiarid conditions mean that several acres alone are required to raise one head of cattle (Figure 21). Some extremely large ranches are owned by meat-packing companies that can fatten the cattle, slaughter them, and package the meat all on the same ranch.

U.S. Department of Agriculture

FIGURE 21 Corporate cattle farming in the United States.

South America, Argentina, Uruguay, and Brazil all have significant cattle-ranching industries (Figure 22). These regions, as well as Australia and New Zealand, followed a similar pattern of cattle-ranching development. First, cattle were grazed on large, open, government tracts with little regard for ranch boundaries. Later, when a conflict with farming interests occurred, cattle ranches moved to drier areas. When irrigation first began to be used in the 1930s and 1940s, farms expanded their territory and ranchers moved to even drier areas and centered much of their herds on feedlots near railroads or highways leading toward the markets. Today, ranching worldwide has become part of a vertically integrated agribusiness meat-processing industry.

MEDITERRANEAN CROPPING Mediterranean regions of the world grow specialized crops, depending on soil and moisture conditions. These regions include the lands around the Mediterranean Sea, southern California, central Chile, South Africa, and southern Australia. In these regions, summers are dry and hot and winters are mild and wet. The Mediterranean Sea countries produce olives and grapes. Two-thirds of the world's wine is produced in Mediterranean Europe, especially Spain, France, and Italy. In addition, these countries and Greece produce the world's largest supply of olive oil. In California, the crop mix is slightly different because of consumer demand and preferences. Most of the land devoted to **Mediterranean agriculture** is taken up by citrus crops, principally oranges, lemons, and grapefruit.

Unfortunately for Mediterranean farmers, these areas of the world are some of the most prized for their climates. Northern Europeans turned many Mediterranean Sea areas into tourist rivieras. No discussion is necessary to describe the land-use changes associated with the growth of Southern California. Burgeoning suburban developments around major cities, especially Los Angeles, are rapidly dwindling our Mediterranean agricultural lands. In Chile, 90% of the population lives in the Mediterranean lands in the central one-third of the country, centered on Santiago.

HORTICULTURE AND FRUIT FARMING Because of consumer preferences, purchasing power, and a severe winter season, there is a tremendous demand in U.S. East Coast cities for fruits and vegetables not grown locally. Shoppers in Philadelphia, New York City, Washington, DC, Baltimore, and Boston pay dearly for truck farm fruits and vegetables, such as apples, asparagus, cabbage, cherries, lettuce, mushrooms, peppers, and tomatoes. Consequently, a horticulture and fruit-farming industry exists as close as possible to this portion of the United States as temperature and soil conditions allow. Stretching from southern Virginia through the eastern half of North Carolina and South Carolina to coastal Georgia and Florida is the Atlantic Fruit and Vegetable Belt (see Figure 19). This is an intensively developed agricultural region with a high value per acre. The products are shipped daily to the northeastern cities for direct consumption or for fruit and vegetable packing and freezing.

As in the case of Mediterranean agriculture and subtropical cropping in southern California, the Atlantic fruit and vegetable horticulture industry relies heavily on inexpensive labor. In California, the laborers are primarily from

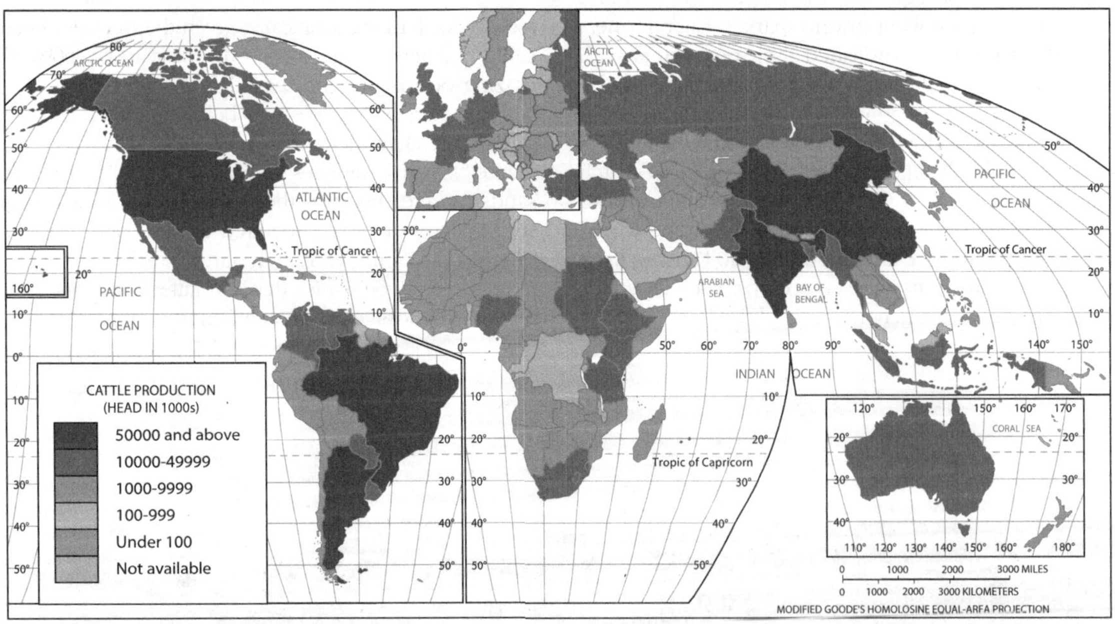

FIGURE 22 World cattle production. The developed countries produce the most beef products because a large amount of the grain crops, particularly corn, can be fed to cattle to fatten them. Poorer nations must consume all available food supplies directly or use them as revenue-producing exports. The United States, Western Europe, Russia, Brazil, Argentina, and Australia are the leading producers worldwide.

Joe Sohm/Chromosohm/Stock Connection

FIGURE 23 Farm workers in the United States, many of whom are illegal immigrants, work long hours at very low rates of pay. Agriculture is exempted from minimum wage laws, and many of these impoverished workers, who make possible fresh fruits and vegetables at low prices, are highly vulnerable to exploitative employers.

Mexico and Central America and often enter the United States illegally. On the Atlantic Coast, the laborers are primarily from the Caribbean and their immigration status is also often questionable. Farm workers often work under brutally exploitative conditions for extremely low wages, typically well below minimum wage. Inexpensive labor is the major way that specialized agriculturalists maintain profits in areas under pressure from urban growth and expansion. Often illiterate and politically powerless, the laborers suffer at the hands of unscrupulous employers. They may be cheated out of their pay, exposed to dangerous pesticides, and not be able to afford a place to live. In part, our cheap fruits and vegetables come at the expense of human misery (Figure 23).

U.S. AGRICULTURAL POLICY

Agribusiness is a long way from the mythical "free market"; active government intervention of different types has long been involved in this sector of the economy.

In the past, American farms were family owned, small, and served local markets. In those days, farm prices were relatively stable and predictable, although persistent overproduction in the late nineteenth century gradually depressed prices and bankrupted many farmers. By the early twentieth century, farms had become much larger and more highly mechanized, and technological improvements had revolutionized agriculture (Figure 24). An individual farm family could manage as many as a thousand acres with mechanized equipment. With improved transportation to the markets and between countries, the U.S. farmer now served a much wider market area.

The early twentieth century was a prosperous time for U.S. farmers, especially during World War I, when they provided a large amount of food for Allied troops. However, many farmers lost their fortunes during the 1920s and 1930s with the twin economic and environmental catastrophes of the Great Depression, and many farms ceased to operate. World War II created another upswing for agriculture as farmers once again provided food for a much larger army, the Western allies, and a hungry nation. After the war, the farmers' fortunes dwindled in the 1950s and 1960s until the U.S. government agreed to major grain trade agreements with the Soviet Union in the 1970s, which increased American exports.

Since then, world markets for U.S. grain have dwindled as many foreign countries have become better able to produce more of their own food. For example, as a result of Green Revolution technology, India, formerly a net food importer, is now a net food exporter. Relatively high interest rates made the cost of borrowing fertilizer, seeds, and equipment frequently prohibitive, and periodic increases in the value of the U.S. dollar relative to other currencies made American exports uncompetitive. At the same time, the costs of farm operation—including machinery, fertilizer, land, and transportation—increased drastically. These less profitable times for farmers, in which costs have outrun income, have continued. Compared with other sectors of the U.S. economy, the farm sector has faced higher operating costs and lower revenues, although it also enjoys subsidies.

The Farm Problem in North America

One reason that agricultural markets are currently in such desperate straits is that the demand for farm products is price-inelastic. Consumers do not demand significantly more food when farm prices are low, so the reduction in price does not lead to a substantial increase in the quantity demanded. This phenomenon is coupled with the fact that the yield from agriculture has increased manyfold during the last 100 years. Technological and mechanical improvements and hybrid seeds have increased yields so much that U.S. farm productivity is the highest in the world (Figure 25). The quantity of farm products has increased much more rapidly than has demand. These three factors have pushed down prices, as shown in Figure 26. Industrial agribusiness thus exemplifies the classic capitalist tendency toward overproduction.

In Figure 25, we see the three tendencies of American farming during the past 100 years: drastically increased supply, moderately increased demand, and

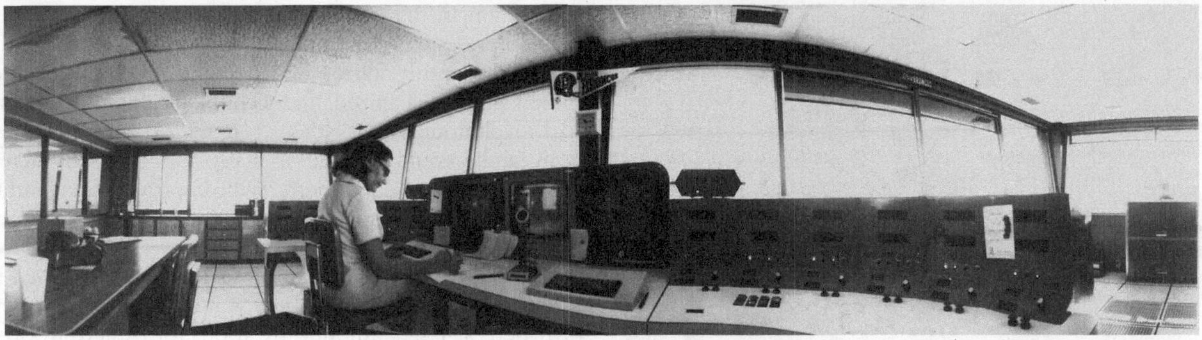

U.S. Department of Agriculture

FIGURE 24 Corporate farming in the United States. An employee watches a television monitor to see when a truck is in position to receive its computer-calculated load.

falling prices. The result has been relatively low returns to farm families and has spelled disaster for many farmers. With lower prices and increased quantities, more and more farmers cannot afford the rapidly rising costs for machinery, fertilizer, transportation, and labor. World farm prices have likewise fallen (Figure 26). The result for U.S. farmers has been a continuing reduction in return on their investment. Many farmers are seeking to apply their productivity to other, more profitable industries. However, unlike a store that can change hands and change function or a high-tech manufacturing plant that can change products, a farm is difficult to adapt to a new economic use.

As a result of persistent overproduction and low prices, there has been a large movement of farm families and farm labor away from the farm. In 1910, 35% of the U.S. population lived and worked on farms. By 2005, this figure dropped to less than 1% of the total population. However, considering present production, prices, and consumption, we still have too many farmers in America today. In a normal market situation, resources would have shifted away from agriculture into other economic activities. However, due to U.S. government price supports for agriculture, this has not been the case.

The U.S. Farm Subsidy Program

In 1933, with farming in deep crisis, Congress passed the Agricultural Adjustment Act to aid American farmers. This act was designed to help a large proportion of the population (up to 33%) who lived in rural areas. It artificially raised farm prices so that farmers could enjoy a "fair price," or **parity price**, for their products. A parity price was defined as "equality between the prices farmers could sell their products for, and the price they would spend on goods and services to run the farm." The period selected to determine parity prices was from 1910 to 1914, when farm prices were relatively high in comparison with other products.

Since 1933, however, the ratio between farm prices and all consumer goods declined until 2000, when it was approximately 30% of the original 1914 parity established in 1933. In other words, without parity, farmers could sell products and purchase only 30% of what they could in the earlier period. Admitting that markets had created widespread irrationality in agriculture, the federal government stepped in to establish a program of **agricultural subsidies**, or a **price floor**, for key agricultural commodities, a guaranteed price

FIGURE 25 Dynamics of the U.S. agricultural sector. During the past 100 years, U.S. farm production has burgeoned remarkably because of increased productivity, increased mechanization, and improved fertilizers, pesticides, and hybrid seeds. The U.S. farm output in 2000 was substantially more than it was in 1950, despite the Soil Bank program and other methods used to keep land out of production. At the same time, because food is an income-inelastic commodity, demand has increased, but not as much as supply. The result has been increased outputs and reduced prices.

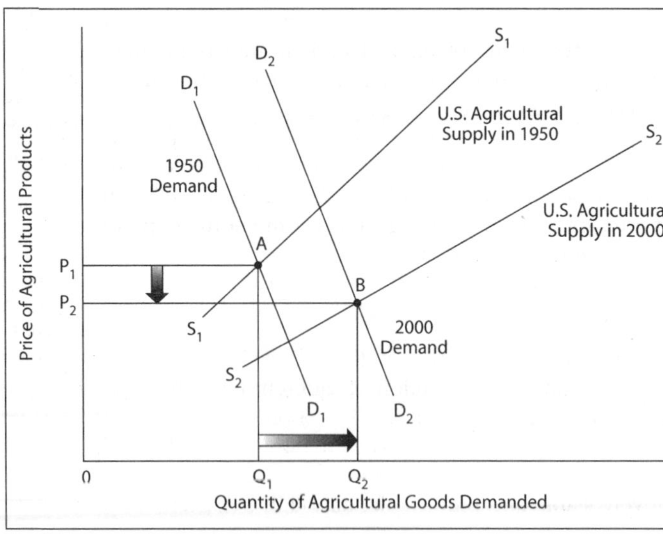

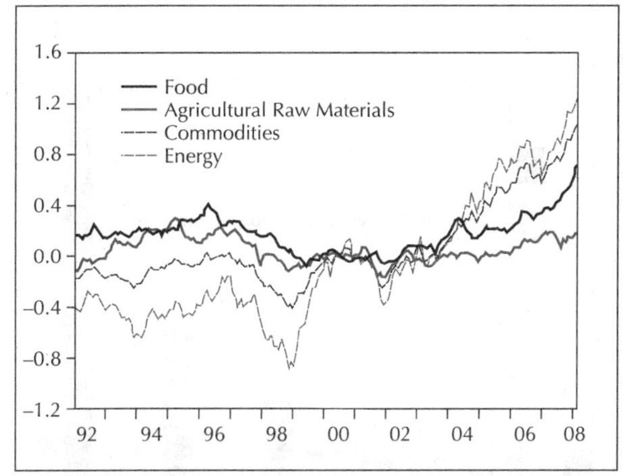

FIGURE 26 World agricultural commodity prices, 1990–2008. Whereas prices declined in the late twentieth century, there was a sharp rise in the early twenty-first, although prices for agricultural goods rose slowly.

above the market price. These supports were minimum prices that the government could assure farmers. For example, the government bought all corn and wheat from farmers and sold it at what the market would bear. It stored many of these commodities in its own storage facilities; thus public funds were used to encourage farmers to grow more than the market can consume and to store the surplus. In 1994, the U.S. government began to offer farmers **target pricing**, which is similar to the price supports of the 1950s through the 1980s. With target pricing, the government pays directly to the farmer the difference between the market selling price and the target price that the government has set; the government no longer takes control of the product.

Figure 27 shows the effects of price supports on agricultural products:

1. The market cannot arrive at an equilibrium price through the normal means of supply and demand.
2. Farmers produce a larger amount of surplus goods than consumers are willing to buy.
3. Buyers pay more and buy less than they would if market conditions prevailed.
4. Farmers' incomes are artificially raised by government subsidies, and consumers' incomes are artificially lowered.

As shown in Figure 27, with the parity price artificially high, farmers will supply the intersection of the price line and the supply curve at K for a total quantity of Q_1. However, with higher prices, the consumers will demand Q_2. The difference between Q_1 and Q_2 is the surplus that the government would purchase under the old price-support plan. Regardless of whether the subsidization is in the form of price supports or target pricing, the result is an extra cost to the taxpayer.

These price-support and target-price programs created artificially high agricultural prices for U.S. agriculture for the past 70 years. The hope, of course, was that market prices would rise to parity, and they did during World War II and during the Soviet grain trade agreements of the 1970s. However, most of the time, the price of farm products was much less than the parity price. The government also attempted to reduce production with the Soil Bank program, which paid farmers to keep acreage out of production. Initially, this approach worked, but the per-acre yields increased amazingly and completely overshadowed the lost acreage in terms of total yield.

The small American farmer as a cultural and economic institution is an endangered species. For years, the government price-support programs kept inefficient farmers in business. However, government subsidies, which amount to more than $100 billion annually, favor large corporate farms over small, family-owned ones. Because subsidies are a function of a farm's output, the subsidy program benefits the largest farmers, the corporate agribusinesses, not the small family farms as originally intended (Figure 28). The U.S. corn industry is the largest recipient of government

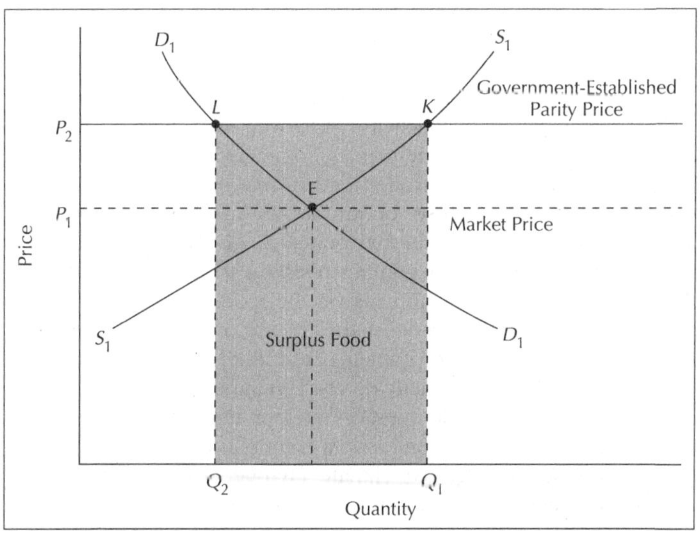

FIGURE 27 U.S. agricultural price supports. The U.S. government has supported farmers by establishing a parity price above the equilibrium price, according to supply and demand factors. For the past 60 years, the effect of the price support has been to set prices higher than they would normally be, thus producing a surplus that the government was required to purchase with tax dollars. The market cannot obtain an equilibrium at E because surplus goods are produced and too often wasted. The farmers' incomes are artificially raised, but buyers in the marketplace must pay more than the goods would warrant under normal conditions. Unfortunately, resources are artificially allocated and therefore misallocated as price shifts from P_1 to P_2 demand drops back to point L, while supply moves up to point K.

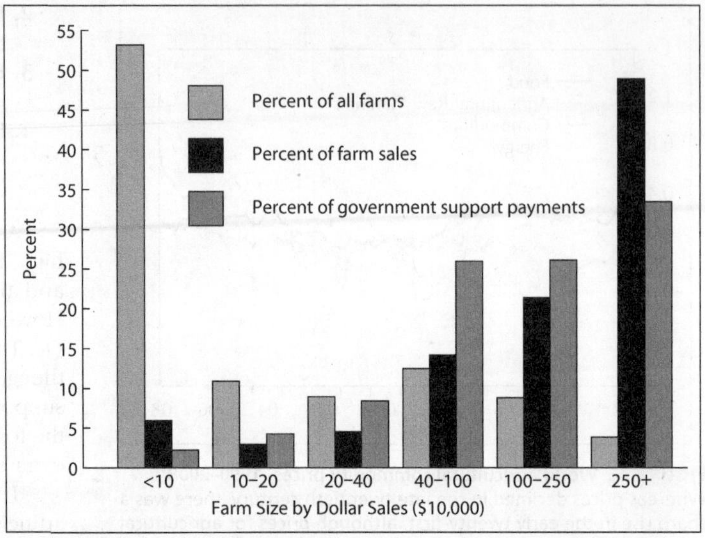

FIGURE 28 The proportion of farms, the proportion of farm sales, and the percentage of government support payments by farm size and dollar sales. Most U.S. farms are small and produce a small proportion of total farm sales. However, most government support payments go to large farms. With regard to farming and farm policy, the rich appear to be getting richer, whereas the poor or small farms, which were the original focus of price supports, are getting relatively poorer and scarcer. Oxfam International, an antipoverty group, recommends helping end hunger permanently by encouraging wealthier countries to stop subsidizing their agriculture and biofuels industries, which tend to keep world food prices high.

payments, amounting to $20 billion in 2008. In essence, the large corporate agribusiness farms have become richer and, with the lion's share of U.S. government subsidies, forced food prices even lower. This has, in effect, continued to force the small farmers off the land. Due to the political power of agribusiness, the federal government has found it politically impossible to trim agricultural subsidies, although doing so would improve the country's agricultural land, lower consumer prices, eliminate overproduction, and reduce the enmity toward the United States that subsidized agricultural exports generate.

One solution to America's farm problems is to design new uses for farm products that are not currently demanded in the United States. One example is an attempt to generate gasohol from corn and other agricultural products to fuel automobiles. A second use of the food surplus is Food-for-Peace programs, which allow agricultural surpluses to be distributed to starving nations instead of being liquidated or exterminated by dumping or destroying storehouses of food. Locally, the **food stamp program** for America's poor operates off the large agricultural surplus.

The United States is not alone in subsidizing its farmers. Half of the budget of the European Union, for example, is dedicated to the Common Agricultural Policy, which subsidizes farmers in France, Germany, and other countries. Japanese farmers are very heavily subsidized in a country with relatively little arable land, and Japanese consumers pay prices well above the world average as a result. Farming is often draped in the mantle of nationalism, and politicians everywhere find it difficult to reduce agricultural subsidies. In 2005, the world's developed countries spent more than $600 billion in agricultural subisides, flooding the world with cheap food and bankrupting farmers throughout the developing world. In any case, agriculture exemplifies the powerful role of the state in almost all market-dominated societies; there is certainly no "free market" in farming.

SUSTAINABLE AGRICULTURE

Industrialized agriculture has had profound impacts on the natural world, especially over the course of the twentieth century. Many of these impacts result from its formation of monocultures that rely heavily upon mechanical, biological, and chemical inputs (e.g., tractors, hybrid seeds, herbicides, and pesticides). Industrial agriculture requires enormous energy inputs in the form of fossil fuels for tractors and irrigation, petrochemical-derived fertilizers to increase yields, and the energy required to transport food globally. Globally, the use of pesticides has grown since World War II, costing billions of dollars annually and causing thousands of human deaths and adverse health risks due to cancer, acute poisonings, and neurological damage each year. Pesticides also harm a wide variety of wildlife, including microorganisms, fish, birds and mammals, through direct or secondary environmental exposure. Soil health, critical for food production, has also been damaged by industrial agriculture due to increasing farm specialization, ever-larger farms, and monoculture-cropping practices. With inorganic fertilizer use, the natural organic matter of the soil is not replenished, and despite chemical substitutes, many areas have seen a decline in soil productivity. Moreover, with the increased use of pesticides and herbicides, some insects, weeds, and fungi evade being controlled and eventually evolve resistance. Similarly, overuse of antibiotics in industrial livestock production is a global concern, given that agricultural pathogens are becoming resistant to antibiotics and are being passed to humans via the food chain, which causes health risks and reduces the efficacy of important antimicrobial drugs. These problems are further compounded by the overall loss of genetic diversity in agriculture.

Many people question the long-term ecological sustainability of this system and its contributions to environmental destruction, global climate change, and the worldwide depletion of fossil fuels. As a result, increasing attention has focused on **sustainable agriculture**, which promotes social, ecological, and economic health for food- and fiber-producing land and communities. Numerous locally based, low-input and knowledge-intensive farm systems exist that follow the principles of sustainable agriculture. Sustainable agricultural systems are often referred to as "alternative," yet this is a misnomer as these approaches are actually quite normal and are practiced by millions of farmers worldwide in the developing world.

Approaches that follow the principles of sustainable agriculture include organic farming, agroecology, holistic management, urban gardening, community-supported agriculture, and natural systems agriculture. The overwhelming similarity among sustainable agricultural systems is their reliance on local knowledge and their minimal ecological impacts. The knowledge that local farmers possess represents a rich and reliable source of information regarding agroecosystems and helps to replace some of the costly external inputs found in industrialized agriculture. In sustainable agricultural systems, farmers' local knowledge can improve nutrient cycling; crop productivity; profitability; and the conservation of soil, water, energy, and biological resources. Some of these savings are achieved in combination with the reduction or elimination of external, nonlocal, and nonrenewable inputs (e.g., oil, pesticides), which often have associated financial and environmental costs. Sustainable agricultural systems are often relatively labor-intensive. Almost any farmer worldwide can incorporate sustainable agricultural approaches into existing operations. Indeed, many people believe that sustainable agriculture will be necessary in future food production.

THE VON THÜNEN MODEL

One of the first and most influential models in economic geography concerned agricultural land use. It was proposed by Johann von Thünen, a German estate owner interested in economic theory and local agricultural conditions, in his book *The Isolated State,* published in 1826, one of the first models in economic geography. From his experiences as an estate manager, von Thünen observed that identical plots of land would be used for different purposes depending on their accessibility to the market. Von Thünen's aim was to uncover laws that govern the interactions of agricultural prices, transport costs, and land uses as landlords sought to maximize their profits.

The concept of economic rent, also called location rent, a relative measure of the advantage of one parcel of land over another, is central to **von Thünen's model** of agricultural land use. Differential rents may result from variances in productivity of different parcels of land and/or variances in the distance from the market. Von Thünen demonstrated that rent is the price of accessibility to the market. In other words, rents decline with the distance from a market center.

To explain agricultural land use, von Thünen described an idealized agricultural region about which he made certain assumptions. He envisioned an isolated state with a large city serving as the only market. A uniform plain surrounded the city. There were no extraneous disturbances in this idealized landscape; social classes and government intervention were absent. Transport to the central town increased at a rate proportional to distance. He concluded that near the town will be grown those products that are heavy or bulky in relation to their value and that are consequently so expensive to transport that the remoter districts are unable to supply them. With increasing distance from the town, the land will progressively be given over to products cheap to transport in relation to their value. Farmers near the central market pay lower transport costs than farmers at the margin of production. Farmers recognize this condition, and they know that it is in their best interest to bid up the amount that they will pay for agricultural land closer to the market. Bidding continues until bid rent equals location rent. At that price, farmers recover production and transport costs and landowners receive location rents as payments for their land. Competitive bidding for desirable locations cancels income differentials attributable to accessibility. The bid rent, or the trade-off of rent levels with transport costs, declines just far enough from the market to cover additional transport costs; hence, farmers are indifferent to their distances from the market. Location rent for any crop can be calculated by using the following formula:

$$R_k = e(p - a) - efk$$

where

R = location rent per unit of land at distance k

e = output per unit of land

k = distance to the market

p = market price per unit of output

a = production cost per unit of product (including labor)

f = transport rate per unit of distance per unit of output

Different bid rent curves for different crops (or land uses in general) reflect how competitive or profit maximizing they are at different distances from the market (Figure 29).

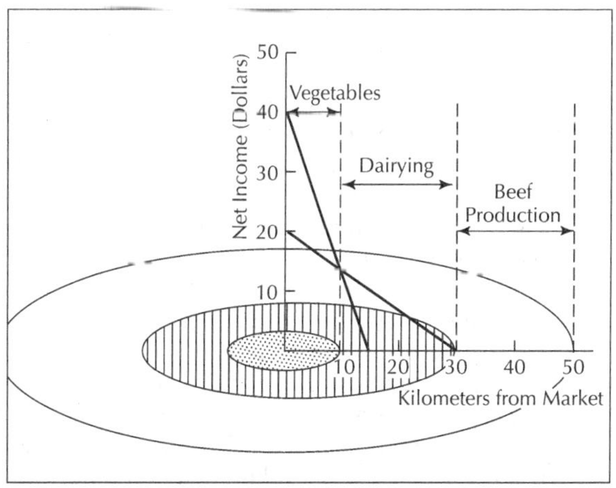

FIGURE 29 The von Thünen model illustrates why rents decline with accessibility and how competing land uses maximize rents.

When the rent gradient is located around the market town, it becomes a rent cone, the base of which indicates the extensive margin of cultivation for a single crop grown at a single intensity. Concentric rings or belts will form around the town, each with its own particular staple product. From ring to ring, the staple product will change.

Von Thünen was acutely aware that many conflicting factors—physical, technical, cultural, historical, and political—would modify the concentric patterns of agricultural land use. He modified some of his initial assumptions—equal transportation costs in all directions, for example—to approximate actual conditions more closely. Although he retracted the condition of a single-market town, he did not elaborate on the effects of several competing markets and a system of radiating highways. We can presume, however, that the tributary areas of competing markets would have a variety of crop zones enveloped by those of the principal market town and that a radiating highway system would have produced a "starfish" pattern.

Summary

The invention or discovery of agriculture, roughly around 10,000 B.C., in several parts of the world marked a momentous change in how human beings live, enabling settled communities, class-based societies, the state, and a more specialized division of labor than hunting and gathering, which had lasted 95% or more of the span of human existence.

All people depend on agriculture for their well-being, but in our agricultural pursuits, we necessarily modify the land. The production of agricultural landscapes thus involves the large-scale transformation of nature in many ways. When geographers speak of agricultural regions, they refer to the artificial division of the world among a variety of socially produced farming types. The most intriguing aspect of farming regions is not their number or extent but the similarity of land-use decisions farmers make within them. After reviewing agricultural origins and dispersals, we identified basic factors that influence agricultural land-use patterns: site characteristics, cultural preferences and perceptions, systems of production, and relative location.

Agriculture in developing countries is primarily of the subsistence variety. Subsistence agriculture includes shifting cultivation, intensive farming, and pastoral nomadism. Developing countries often have more than 50% of their labor force engaged in agriculture and use relatively little mechanized equipment.

By contrast, agriculture in developed countries employs a small percentage of the labor force and uses large-scale mechanized equipment and large inputs of energy and fertilizers. While outputs per hectare are comparable to intensive farming and even somewhat less, outputs per worker are as much as 50 times greater. The industrialization of agriculture—a process that is still ongoing in much of the world today—freed millions from the boring and arduous toil of the farm. Farmers who raise crops and livestock on huge farms are part of a vast agricultural system that includes machinery manufacturers; fertilizer, pesticide, and energy suppliers; grain mills and slaughterhouses; food processors; and wholesale and retail distributors. Although mixed crops and livestock farms are the most common in developed countries, other types of commercial agriculture also exist, including dairy farming; commercial grain farming, usually centered on wheat; cattle ranching; Mediterranean horticulture; and truck farms.

Like many other sectors under capitalism, agriculture exhibits the powerful role of the state in shaping markets and landscapes. Far from being a free market, agriculture in most societies, including the United States, is heavily subsidized by the government under programs that reflect the political power of the farm lobby. Price floors and subsidies contribute to the chronic overproduction and low prices that farmers typically face, and subsidies go disproportionately to large corporate agribusinesses that dominate the industry rather than to small family farms.

Because industrialized agriculture is so enormously energy-intensive, and because it has serious environmental consequences due to the energy and chemicals that it relies upon, the long-term ecological viability of commercial farming has come under question. In response, many farmers around the world have experimented with sustainable agricultural practices, which utilize fewer chemicals, less energy, more local inputs, more labor, and more local knowledge. There is a wide variety of types of sustainable agriculture, but many have been shown to be as productive as commercial agriculture with far lower environmental impacts.

One theory that helps us to understand the local organization of agriculture was formalized by Johann Heinrich von Thünen in the early nineteenth century. We described the model that von Thünen developed to explain patterns of local land use. We then presented some of the conclusions that can be drawn from this model: (1) There is an inverse relationship between location rent and transport costs; (2) there is a limit to commercial farming on a homogeneous plain with an isolated market town at its center; (3) land values and intensity of land use increase toward the market; and (4) crop types compete with one another and are ordered according to the principle of the highest economic rent. We saw how the basic von Thünen principles can be applied to agricultural land-use patterns at scales ranging from the village to the world. Contemporary von Thünen effects at the microscale are best observed in the developing world, where localized circulation systems provide a transport setting similar to that of early nineteenth-century Europe.

Key Terms

agribusiness
agricultural subsidies
capital-intensive production
cattle ranching
commercial agriculture
dairy farming
double cropping
feedlots
food stamp program
four-field rotation system
intensive subsistence agriculture
labor-intensive production
land degradation
Mediterranean agriculture
mixed crop and livestock farming
parity price
pastoral nomadism
peasant agriculture
plantation
price floor
shifting cultivation
slash-and-burn agriculture
spring wheat
stage model of agriculture
subsistence agriculture
sustainable agriculture
swidden agriculture
target pricing
Township and Range System
truck farming
vertical integration
von Thünen model
Wheat Belt
winter wheat

Study Questions

1. What proportion of the U.S. population works in agriculture? Why is it so low?
2. Describe the impacts of European colonialism on global patterns of agriculture.
3. What are plantations?
4. What is meant, exactly, by Boserup's agricultural intensification?
5. Summarize the peasant mode of production.
6. What is the main characteristic of commercial agriculture?
7. How have the number and size of U.S. farms changed over time?
8. Describe the geography of U.S. agriculture.
9. What are the causes and impacts of agricultural subsidies?
10. How does sustainable agriculture differ from commercial, industrialized forms of farming?
11. What is the von Thünen model and what does it explain?

Suggested Readings

Hart, J. 2003. *The Changing Scale of American Agriculture*. Charlottesville: University of Virginia Press.

Rumney, T. 2005. *The Study of Agricultural Geography*. Lanham, MD: Rowman and Littlefield.

Watts, M. 1996. "The Global Agrofood System and Late Twentieth-Century Development." *Progress in Human Geography* 20:230–45.

Web Resources

About.Com Geography of Agriculture
http://geography.about.com/od/urbaneconomicgeography/a/aggeography.htm

This site offers concise overviews of many geographical topics, including several aspects of agriculture.

Life on the Farm
http://web2.airmail.net/bealke

What is it like to live on a farm? Visit Chuck Bealke's virtual farm and get his biweekly remembrance or comment on farm life—much of it based on farming (soybeans, wheat, corn, hay, and polled Herefords) west of St. Louis. Links to many agricultural online resources.

National Agricultural Library (NAL)
http://www.nalusda.gov

The NAL is part of the Agricultural Research Service of the U.S. Department of Agriculture and is one of four national libraries in the United States. NAL is a major international source for agriculture and related information.

U.S. Department of Agriculture Quick Stats
http://quickstats.nass.usda.gov/

This Web page provides information on the U.S. agricultural sector.

Manufacturing

OBJECTIVES

- To acquaint you with the major manufacturing regions of the world
- To summarize deindustrialization in the developed world and the industrialization of parts of the developing world
- To reveal sector-specific dynamics through five industry analyses
- To show the trend toward flexible production and flexible labor
- To present the product life cycle

Chinese factory workers making garments illustrate both the combination of labor and capital that lies at the heart of the production process and the shift of manufacturing employment to East Asia.

Manufacturing

To manufacture is to make things—to transform raw materials into goods that satisfy needs and wants. Manufacturing is crucial because it produces goods that sustain human life, provides employment, and generates economic growth. It has played this role since the Industrial Revolution began in Britain in the late eighteenth century, when manufacturing generated the working classes of Europe, North America, and Japan.

Geographers who study manufacturing emphasize the locational behavior of firms and the structures of the places they create. First, we examine the nature of manufacturing, including the basic steps involved in the transformation of raw materials into final products. Second, we study the major regions in the world that produce goods—North America, Europe, and Japan. Manufacturing is highly unevenly located around the world. How did these clusters come to exist? Third, we explore three crucial industries in more depth: textiles, automobiles, and electronics. Fourth, we summarize the geography of U.S. manufacturing and the changes wrought by globalization. Fifth, the chapter introduces the notions of flexibility, post-Fordist production, and just-in-time systems, which have revolutionized the manufacturing production and delivery process.

Manufacturing involves deciding what is to be produced, gathering together the raw materials and semifinished inputs at a plant, reworking and combining the inputs to produce a finished product, and marketing the finished product. These phases are called *assembly*, *production*, and *distribution*. The assembly and distribution phases require transportation of raw materials and finished products, respectively. The production phase—changing the form of a raw material—involves combining land, labor, capital, and management, factors that vary widely in cost from place to place. Each of the steps of the manufacturing process has a spatial or a locational dimension.

Changing a raw material into a usable good increases its use or value. Flour milled from wheat is more valuable than raw grain. Bread, in turn, is worth more than flour. This process is termed **value added by manufacturing**. The value added by manufacturing is quite low in an industry engaged in the initial processing of a raw material. For example, turning sugar beets into sugar yields an added value of about 30%. In contrast, changing a few ounces of steel and glass into a watch yields a high added value—more than 60%. The cost and productivity of labor, and the availability of skills, plays an important role in high value-added manufacturing.

MAJOR CONCENTRATIONS OF WORLD MANUFACTURING

Manufacturing capacity and employment are highly unevenly distributed around the world and go far to explain the uneven spatial development that typifies the world economy. Four major areas account for approximately 80% of the world's manufacturing (Figure 1): North America; Europe; Western Russia and Ukraine; and parts of East Asia, notably Japan, South Korea, and China.

North America

North American manufacturing is largely centered in the northeastern and midwestern United States and southeastern Canada, the **North American Manufacturing Belt**, which accounts for one-third of the North American population, two-thirds of North American manufacturing employment, and one-half of its industrial output. The belt extends from the northeastern seaboard along the Great Lakes to Milwaukee, where it turns south to St. Louis, then extends eastward along the Ohio River valley to Washington, DC (Figure 2).

This area grew rapidly in the late nineteenth century, having access to raw materials from the continental interior as well as European markets. As capital flooded in, much of it foreign, the Manufacturing Belt developed a large labor

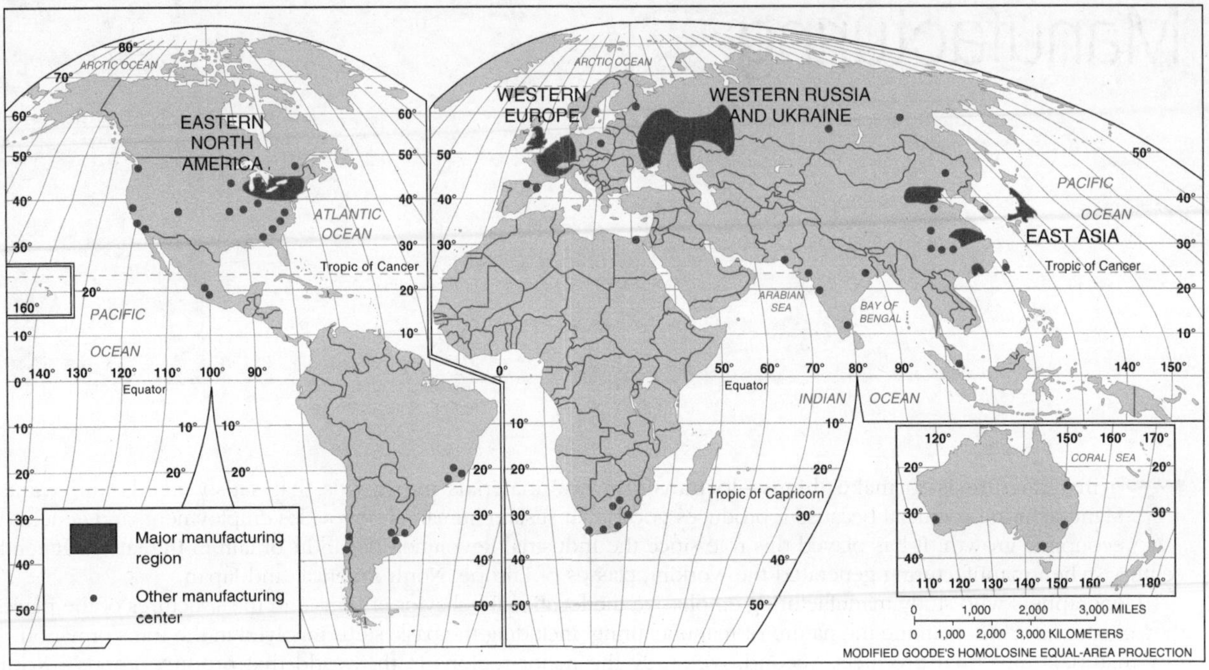

FIGURE 1 Worldwide distribution of manufacturing. The four main manufacturing regions include the northeastern United States and southern Great Lakes region, the northwestern European region, the eastern Soviet Union and Ukraine region, and the Japan/ South Korea region. See the text for a detailed elaboration of districts within each of these regions, as well as other manufacturing regions shown as dots.

pool, often comprised of immigrants from Southern and Eastern Europe. The transportation system included the St. Lawrence River and the Great Lakes, which were connected to the East Coast and the Atlantic Ocean by the Mohawk and Hudson rivers. This transportation system allowed the easy movement of bulky and heavy materials. Later, canals and railroads supplemented the river and lake system. The first major factories in the belt—the textile mills of the 1830s and 1840s—clustered along the rivers of southern New England. When coal replaced water as a power source between 1850 and 1870, and when railroads integrated the belt, factories were freed from the riverbanks.

Between 1850 and 1870, many urban areas enjoyed rapidly expanding industrial production. Manufacturing employment in New York City, Philadelphia, and Chicago soared more than 200% between 1870 and 1900. Factories concentrated in large cities for a combination of the following reasons: (1) They could be near large labor pools, including unskilled and semiskilled immigrants; (2) they could secure easy railroad and waterway access to major resource deposits, such as the Appalachian coalfields and the Lake Superior–area iron mines; (3) they could be near industrial suppliers of machines and other intermediate products, which lowered transport costs; and (4) they could be near major markets for finished goods. In other words, **agglomeration economies** accounted for the concentration of manufacturing activity in the Manufacturing Belt. The highly concentrated pattern of industrial production served the nation (and much of the world) well for about 100 years—roughly the century between 1870 and 1970.

Within the North American Manufacturing Belt, there are several districts. The oldest is southern New England, centered on the greater Boston metropolitan area. In Connecticut, Rhode Island, Massachusetts, and southern New Hampshire, textile and garment plants formed the basic industries of many towns and cities (Figure 3) since the early nineteenth century. Cotton was brought from the Southern states to be manufactured into garments, many of which were consumed locally and some of which were exported to Europe. As the region developed economically, wages became steadily higher, and the textile industry moved to the South in the early twentieth century in a classic example of a new pattern of uneven spatial development. Today, New England manufacturing is centered around electrical machinery, fabricated metals, and electronic products, including military goods. The region is noted for highly skilled labor, with nearby universities—including Boston College, Boston University, Massachusetts Institute of Technology, and Harvard University—providing the chief supply (see Figure 2).

The Middle Atlantic district includes the metropolitan areas of New York, Baltimore, Philadelphia, and Wilmington, Delaware. Industrial traffic from the Great Lakes terminates in New York City via the Mohawk and Hudson rivers, and from the city, foreign markets and sources of raw materials can be reached. The city is the largest

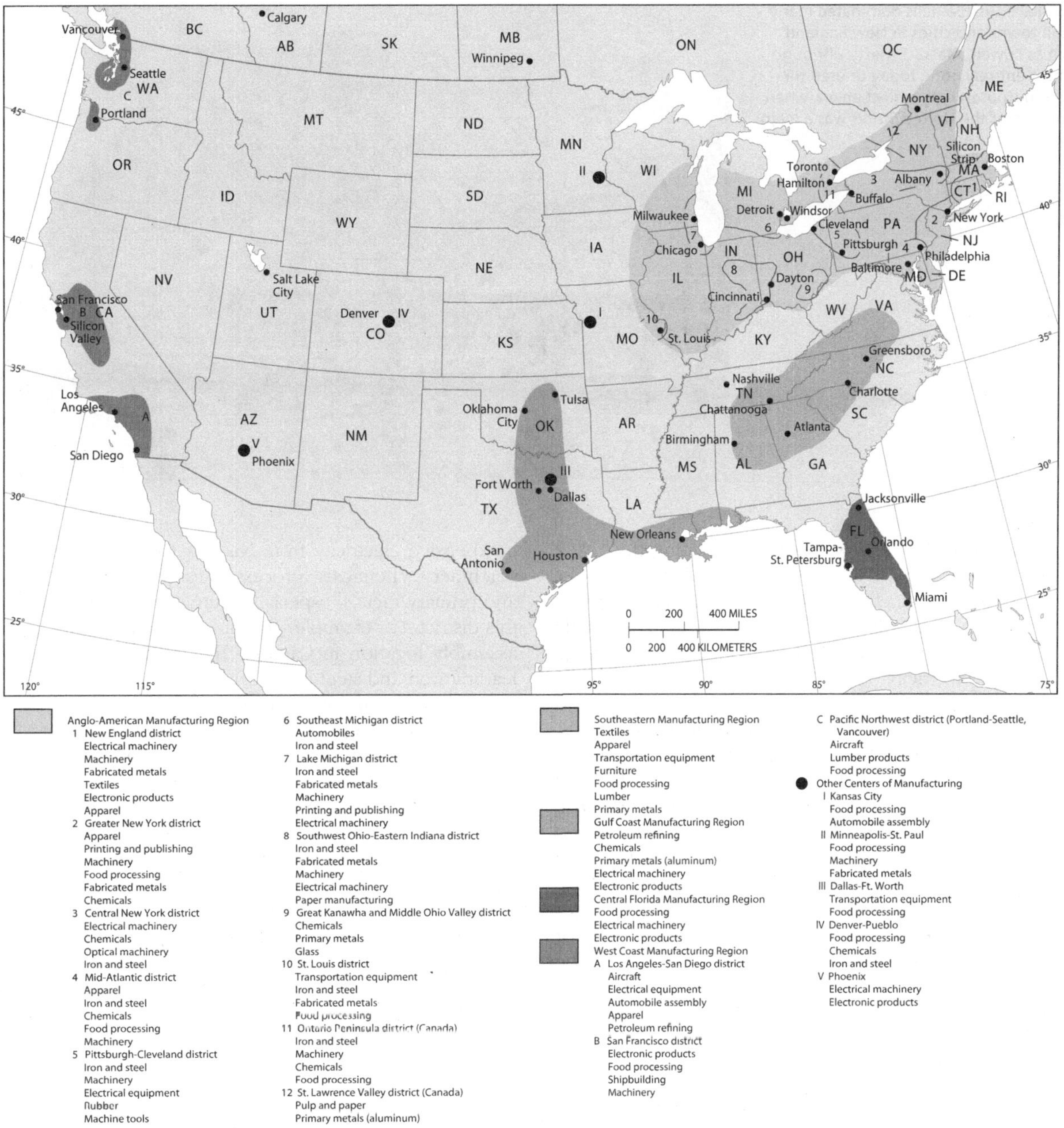

FIGURE 2 Manufacturing regions of the United States and Canada.

market and has the largest labor pool of the region. Because of the enormous agglomeration economies it offers, many corporate headquarters are located there. Not only do the region's infrastructure, financial clout, and role as center of global corporations give the New York district unparalleled proximity to trade with the rest of the world, it also has access to the population centers and manufacturing hubs of America (see Figure 2). It is also near financial, communications, and the news and media industries, which are important for advertising and distribution. In addition, it is the headquarters of the North American publishing industry: What Los Angeles is to film, New York is to the book and magazine world.

The central New York and Mohawk River valley district, including cities such as Buffalo, Rochester, Syracuse, Utica, Schenectady, and Albany, produces electrical machinery, chemicals, optical machinery, and iron and steel. These industries are concentrated along the Erie Canal and the

AP Photo/Julia Malakie

FIGURE 3 Textile mills dominated many small towns and cities in New England such as Lowell, Massachusetts; although almost entirely gone today or used for other purposes, they reflect an era when that part of the United States was a major textile producer.

Hudson River, the only waterway connecting the Great Lakes to the U.S. East Coast. Abundant electrical power produced by the kinetic energy of Niagara Falls provides inexpensive electricity to this district and explains its attraction for the aluminum industry, which requires large amounts of electricity.

The Pittsburgh–Cleveland–Lake Erie district, centered in western Pennsylvania and northeastern Ohio, is the oldest steel-producing region in North America. Pittsburgh was the original steel-producing center because of the coal supplies in the nearby Appalachian Mountains and iron ore deposits in northern Minnesota that could be accessed via the Great Lakes system. Besides iron and steel, electrical equipment, machinery, rubber, and machine tools are also produced in this region. Cleveland was long a major producer of steel and automobiles, and Akron was for decades the rubber-producing capital of the nation, the world's largest center of automobile tires and rubber parts.

The western Great Lakes industrial region is centered on Detroit in the east and Chicago in the west (see Figure 2). Toledo and Milwaukee were long known for the production of transportation equipment, iron and steel, automobiles, fabricated metals, and machinery. Detroit and surrounding cities, of course, were for many years the world's leaders in automobile manufacture, and Chicago has produced more railroad cars, farm tractors and implements, and food products than any other city in the United States. The convergence of railroad and highway transportation routes in this area makes it readily accessible to the rest of the country and a good distribution point to a national market.

Canada's most important industrial region stretches along southern Ontario near the St. Lawrence River Valley, on the north shore of the eastern Great Lakes (see Figure 2). This area has access to the St. Lawrence River–Great Lakes transportation system, is near the largest Canadian markets, has abundant skilled labor, and is supplied with inexpensive electricity from Niagara Falls. Iron and steel, machinery, chemicals, processed foods, pulp and paper, and primary metals, especially aluminum, are produced in this district. For example, Toronto is a leading automobile-assembly location in Canada, and Hamilton is Canada's leading iron and steel producer.

The 1960s marked the start of the steady gain of manufacturing employment in the Sunbelt, including the South and parts of the Southwest. Labor costs in the South were often lower because its labor force was less skilled and less unionized, in part because Southern states are "right-to-work" states, meaning that unions cannot force employees at an establishment to join. The Southeastern manufacturing region of the United States stretches from central Virginia through North Carolina, western South Carolina, northern Georgia, northeastern Alabama, and northeastern Tennessee (see Figure 2). It wraps around the southern flank of the Appalachian Mountains because of poor transportation connections across the mountains. Textiles are the main product, the industry having moved from the Northeast to the South to take advantage of less expensive, nonunion labor. Transportation equipment, furniture, processed foods, and lumber are also produced. Aluminum manufacturers moved to this region because of the inexpensive electricity produced by the more than 20 dams built by the Tennessee Valley Authority during the Great Depression. Birmingham was long the iron and steel center of the southeastern United States because of the plentiful iron ore and coal supplies nearby, although it never rivaled the large centers of the north such as Pittsburgh. The region is also the primary producer of American textiles.

The Gulf Coast manufacturing region stretches from southeastern Texas through southern Louisiana, Mississippi, and Alabama, to the tip of the Florida panhandle, including cities such as Houston, Baton Rouge, Mobile, and Pensacola. Because of nearby oil and gas fields, petroleum refining and chemical production are important here. The region also produces primary metals, such as steel in Birmingham, as well as aluminum, electrical machinery, and electronic products.

The expansion of manufacturing in the West is not easily explained by low labor costs. California is known for its relatively high labor costs, yet it has registered substantial increases in manufacturing employment since World War II. In California, labor costs were less important than the appeal of the physical environment and the role of the federal government (particularly defense spending). The West Coast manufacturing district does, however, represent an outstanding example of **industrial restructuring** in response to economic crisis and labor unrest. The Los Angeles–San Diego district has been successful in negotiating the transition from older forms of manufacturing to newer ones. Since the 1960s, the district has shed much of its traditional, highly unionized heavy industry, such as steel and rubber. At the same time, it has attracted a cluster of high-tech industries and associated services, centered on electronics and aerospace and tied strongly to enormous defense and military contracts from the U.S. government.

In the 1930s, the aerospace industry chose the greater Los Angeles–San Diego district in southern California because favorable weather much of the year meant unimpeded test flights and savings on heating and cooling the large aircraft plants. Federal government subsidies during World War II reinforced this choice. Because of the myriad electronic parts and equipment and the associated high-tech sensing and navigational devices required in aircraft manufacture, segments of the electronics industry were also attracted to this region and became anchored there 30 years later. Today, aircraft, apparel manufacture, and petroleum refining are important in Los Angeles; San Diego also specializes in pharmaceuticals, biotechnology, and military equipment.

The San Francisco Bay Area is another important West Coast manufacturing region, particularly computer hardware and software. *Silicon Valley*, the world's largest manufacturing area for semiconductors, microprocessors, software, and computer peripherals, is located just south of San Francisco in the Santa Clara Valley (Figure 4).

The Pacific Northwest district includes the cities of Seattle, Washington, and Portland, Oregon. Boeing Aircraft is the single largest employer, followed closely by the paper, lumber, and food-processing industries. More recent additions include Microsoft, the giant of the software industry.

Europe and Russia

Europe has a number of the world's most important industrial regions (Figure 5). They are located in a north–south belt starting in Scotland, extending through southern England, continuing from the mouth of the Rhine River in the Netherlands, through the Ruhr region in Germany, through northern France, to northern Italy. Good supplies

San Jose Convention & Visitors Bureau

FIGURE 4 San Jose, California. The growth of computer software has established San Jose, capital of Silicon Valley, as California's third- and the United States' eleventh-largest city.

of iron ore and coal, and highly skilled supplies of labor, fuel these industrial regions.

Although the Industrial Revolution started in Britain in the mid-eighteenth century based on textile and woolen manufacture, many other countries have since learned to produce their own iron, steel, and textiles. The world currently has an oversupply of these items and the market for British goods has decreased substantially. Britain's overall global competitiveness has been reduced by its outmoded factories, high labor costs, slow productivity growth, and deteriorating infrastructures. In contrast, when Germany and Japan rebuilt after World War II, they were able to modernize their plants and industrial processes that had been destroyed during the war with new capital stock. As a result, Germany and Japan became major industrial successes in late twentieth century, while Britain, more so than any other modern industrialized country, has suffered persistent industrial decline.

The largest European manufacturing region today is in the northern European lowland countries of Belgium and the Netherlands, northwestern Germany, and northeastern France, where the Rhine and Ruhr rivers meet. This region's economic backbone has been the iron and steel industry because of its proximity to coal and iron ore. Production of transportation equipment, machinery, and chemicals helped lead Western Europe into the industrial age. The Rhine River, which is the largest waterway of European commerce, empties into the North Sea in the Dutch city of Rotterdam. Because of its excellent location, Rotterdam became one of the world's largest ports.

The Upper Rhine–Alsace-Lorraine region is in southwestern Germany and eastern France. Because of its central location, this area is well situated to be a distribution hub to population centers throughout Western Europe. The main cities on the German side include Frankfurt, Stuttgart, and Mannheim. Frankfurt became the financial and commercial capital of Germany and the center of its railway, air, and road networks. Stuttgart is known for its precision goods and high value, high-volume,

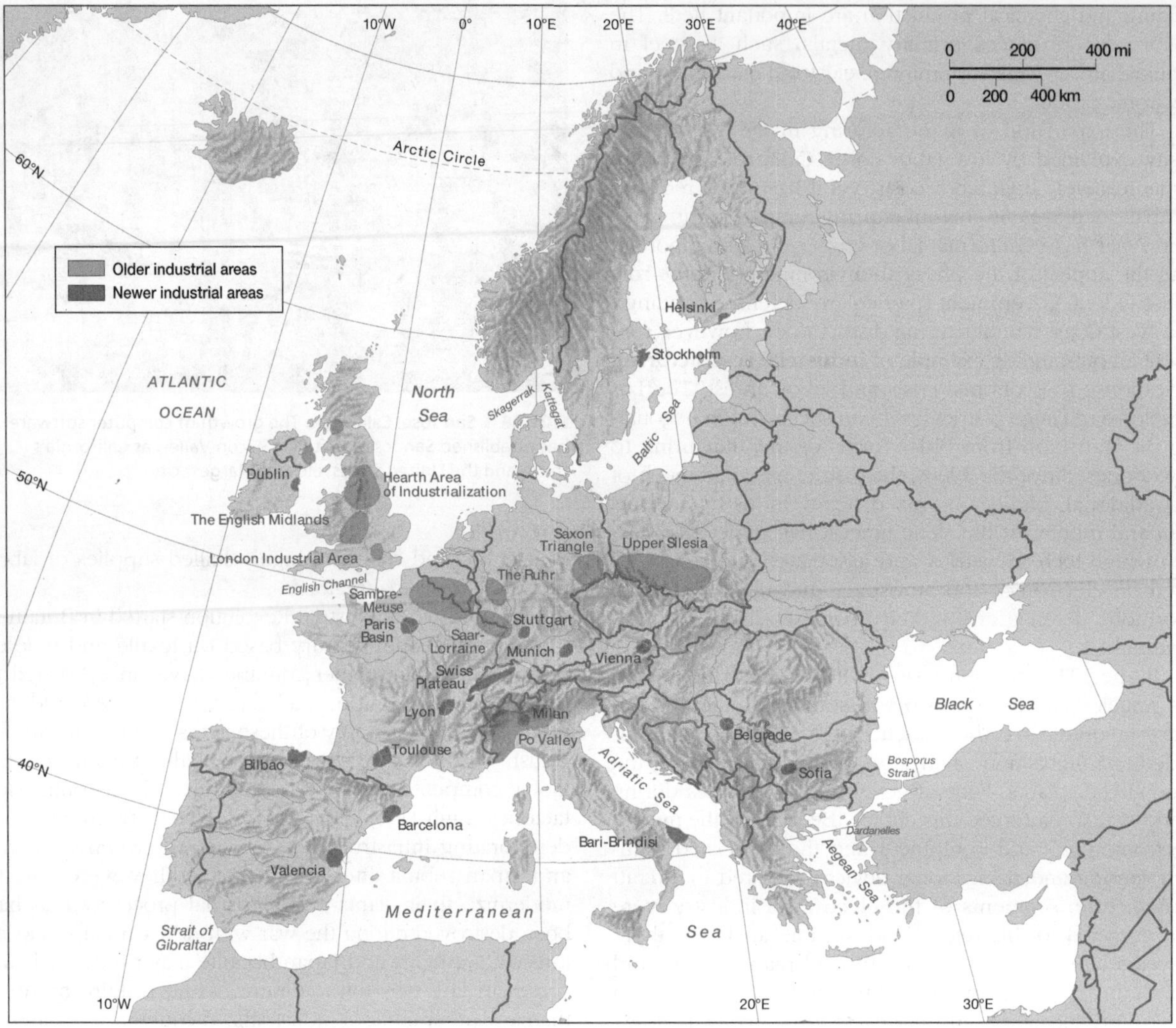

FIGURE 5 European manufacturing regions. Much European manufacturing takes place in a linear belt from Scotland to the Midlands of Britain, to the south, including the London area. This belt continues onward from the low countries of Belgium, Luxembourg, and the Netherlands, south along the Rhine River, including portions of France and Germany, and into northern Italy. These areas became major manufacturing regions not only because of the concentration of skilled laborers but also due to the availability of raw materials, principally coal and iron ore. In addition, good river transportation was available, as well as large consumer markets for finished products.

manufactured goods, including Mercedes Benz, Porsche, and Audi automobiles. Mannheim, located along the Rhine River, is noted for its chemicals, pharmaceuticals, and inland port facilities. The western side of this district, Alsace-Lorraine in France, produces a large portion of the district's iron and steel.

The Po River valley in northern Italy encompasses Turin, Milan, and Genoa; it encompasses only one-fourth of Italy's land but more than 70% of its industries and 50% of its population. This region specializes in iron and steel, automobiles, textile manufacture, and food processing. The Emilia-Romagna region is Italy's high-technology center (Figure 6), generating a number of high-value-added textiles, industrial ceramics, processed foods, and electronic equipment. The Alps, a barrier to the efficient distribution of German and British industrial products, give the Italian district a large share of the Southern European market as well as cheap hydroelectricity.

In Russia, the Moscow industrial region takes advantage of a large, skilled labor pool as well as a large market. It is known for its output of textiles: linen, cotton, wool, and silk fabrics. East of Moscow is the linear Volga region, extending northward from Volgograd (formerly Stalingrad) astride the Volga River. The Volga, a chief waterway of Russia, is linked via canal to the Don River and thereby to the Black Sea. This industrial region continued to develop

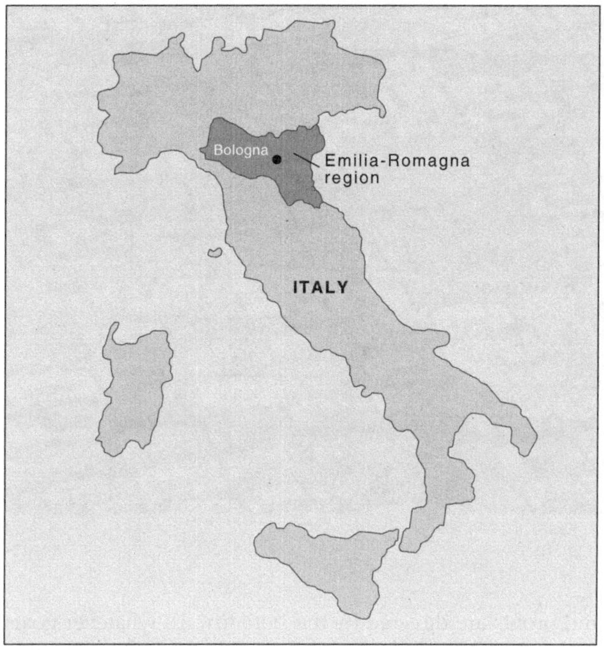

FIGURE 6 Italy's Emilia-Romagna area, Europe's largest conglomeration of high-technology firms, produces a variety of high-quality items ranging from shoes to electronics. It exemplifies the regional basis of competitive advantage and the critical role of agglomeration economies.

during World War II because it was just out of reach of the invading Nazi army that occupied Ukraine. It is the principal location of substantial oil and gas production and refining. Recently, a larger oil and gas field was discovered in West Siberia. Nonetheless, the Volga district is Russia's chief supplier of oil and gas, chemicals, and related products.

The Ukraine industrial region relies on the rich coalfield deposits of the Donets Basin. The iron and steel industry base is the city of Krivoy Rog, with nearby Odessa the principal Black Sea all-weather port (Figure 7). The area is collectively known as Donbass. Like the German Ruhr area, Ukraine's industrial district is near iron ore and coal mines, a dense population, and a large agricultural region and is served by good transportation facilities.

Just east of the Volga region are the Ural Mountains that separate the European and Siberian parts of Russia (see Figure 7). The Urals have large deposits of industrial minerals, including iron, copper, potassium, magnesium, salt, tungsten, and bauxite. Although coal must be imported from the nearby Volga district, the Urals district provides Russia with iron and steel, chemicals, machinery, and fabricated metal products. The Kuznetsk Basin centered on the trans-Siberian railroad, is the chief industrial region of Russia east of the Urals. Again, as in the case of Ukraine and the Urals districts, the Kuznetsk industrial district relies on an abundant supply of iron ore and the largest supply

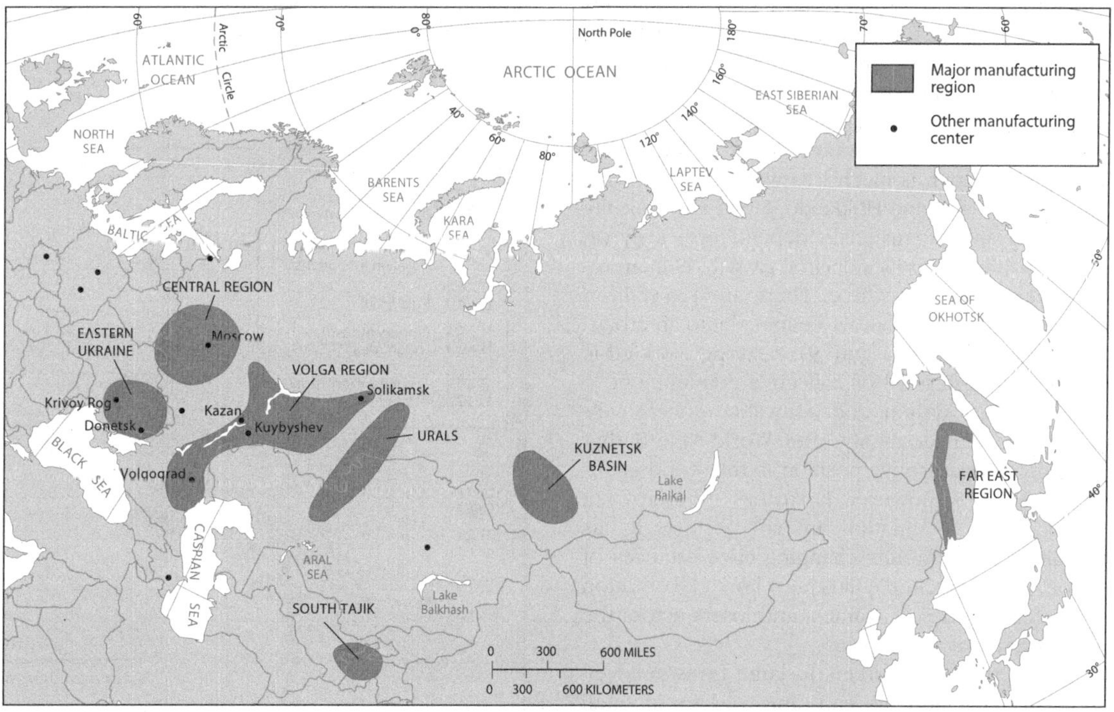

FIGURE 7 Manufacturing regions of Russia, Ukraine, and Central Asia.

Spencer Grant/Photo Researchers, Inc.

FIGURE 8 Firmly linked to the global economy by its reliance upon international trade, Japan's economy is heavily export-oriented, as demonstrated by these cars waiting on loading docks. But Japan's experience over the past two decades is a cautionary example, especially to fast-aging European economies. The country's financial crash in the early 1990s contributed to a slump in productivity growth. Soon afterward, the working-age population began to shrink. A series of policy mistakes caused the hangover from the financial crisis to linger. The economy failed to recover and deflation set in. The result has been a combination of weak demand and slowing supply.

of coal in the country. The Kuznetsk Basin is a result of the grand design of former Soviet city planners, who poured heavy investment into this region, hoping that it would become self-perpetuating and eventually the industrial supply region for Soviet Central Asia and Siberia.

East Asia

In the post–World War II period, Japan set about rebuilding itself into a potent economic power, and by the late twentieth century, Japan's output had increased dramatically in several manufacturing sectors. Today, it has the third-largest gross national product (GNP) in the world and is the world's leading producer of electronics, steel, commercial ships, and automobiles (Figure 8).

Compared to the natural resources found in the United States and Britain, Japan is much less well off. Except for coal deposits in Kyushu and Hokkaido, Japan is practically devoid of significant raw materials, depending heavily on imported raw materials for its industrial growth. Human resources, however, are not scarce. There are 126 million Japanese, one-quarter of whom are crammed into an urban-industrial core near Tokyo (Figure 9). A strong work ethic combined with a high level of collective commitment, a first-rate educational system, and innovative government policies produced an economy after World War II that achieved a highly competitive position in the world economy. Since the 1990s, however, Japan has suffered from persistent economic stagnation, in part because it has failed to adjust to the rapidly changing circumstances of globalization. Japan's example illustrates how globalization produces a steady series of winners and losers across the planet in different times and places.

Although permanent workers in large firms are well paid, especially in relation to part-time workers in small and medium-sized firms, the relative cost of labor is lower in Japan than in many other industrial countries. Savings and profits are directed by the state toward whatever goals are set forth by a unique collaborative partnership between Japan's famous Ministry of International Trade and Industry (MITI) and private enterprise. Under MITI's guidance, Japan has subsidized research and relocated industries such as steel making and shipbuilding "**offshore assembly**" in the newly industrializing countries (NICs), where labor costs are considerably lower.

Countries such as South Korea and Singapore have developed their own higher-value-added industries (e.g., consumer electronics). South Korea, with a manufacturing capacity centered on Seoul, has become the eighth-largest

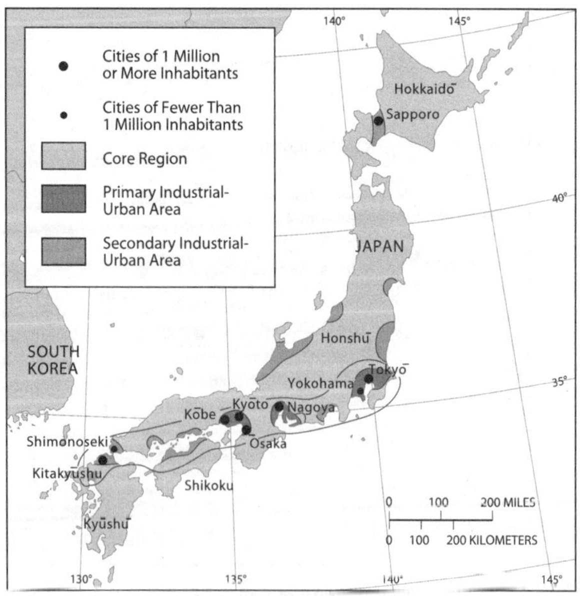

FIGURE 9 Japan's manufacturing core region.

economy in the world, a major producer of electronic goods, steel, autos, and ships. And, of course, China, which began to industrialize in the 1980s, has become a powerhouse in its own right, the world's second-largest economy. With light manufacturing centered on Hong Kong and Guangdong province in southern China and stretching along its Pacific Coast, China is the world's largest producer of steel, concrete, apparel, toys, and, increasingly, many other manufactured goods. To fuel its growth, China uses enormous amounts of coal and is now the world's largest producer of the gases linked to global climate change. Similarly, the island of Taiwan is an important center for apparel and electronics production.

DEINDUSTRIALIZATION

As capitalism underwent one of its periodic restructurings, a new **international division of labor** arose in the late twentieth century, the rate of world economic growth declined, and the long boom period following World War II drew to a close. This round of restructuring started with a deep recession in 1974–1975 following the first oil shock in 1973. One of its most visible effects in the advanced economies was **deindustrialization**, the decline in manufacturing capacity that is typically reflected in the loss of manufacturing jobs. As numerous corporations restructured or went out of business in a climate of intense international competition, millions of workers were laid off. Deindustrialization affected all of the world's developed economies to a greater or lesser extent (Figure 10). Britain, France, Germany, Canada, the United States, and, to a lesser extent, Japan, all lost tens of millions of well-paying jobs.

In the United States, in the late twentieth century all states in the Manufacturing Belt experienced job loss in manufacturing, and virtually all states in the South and the West registered manufacturing job gains (Figure 11). The deindustrialization in the Manufacturing Belt that resulted in increases in manufacturing employment in new industrial areas of the Sunbelt reflected another round of uneven spatial development, as some areas were abandoned by capital and others attracted it. As manufacturing plants closed, once-vibrant Manufacturing Belt cities such as Detroit and Youngstown, Ohio, were turned into economically stagnant regions, with devastating economic and social impacts on their communities. Many of its inner-city areas are littered with closed factories, bankrupt businesses, depressed real estate, and struggling blue-collar neighborhoods. Not surprisingly, poverty rates soared in such communities, as did problems in the housing market, retail trade, and local services as the multiplier effects of closed factories spread through interindustry linkages.

The effects of disinvestment on workers and their communities have been devastating. Victims of plant closings sometimes lose not only their current incomes but often their accumulated assets as well. When savings run out, people lose their ability to respond to life crises and often suffer depression and marital problems. Although job losses occurred in many occupations, some groups are more vulnerable than others. Unskilled workers are particularly likely to bear the costs of globalization, including job displacement. African American workers, many of them unskilled or semiskilled, were particularly hard hit, and by driving up unemployment, the deindustrialization of the inner city was largely responsible for the creation of impoverished ghetto communities. However, some old industrial cities—Pittsburgh, for example—have successfully built

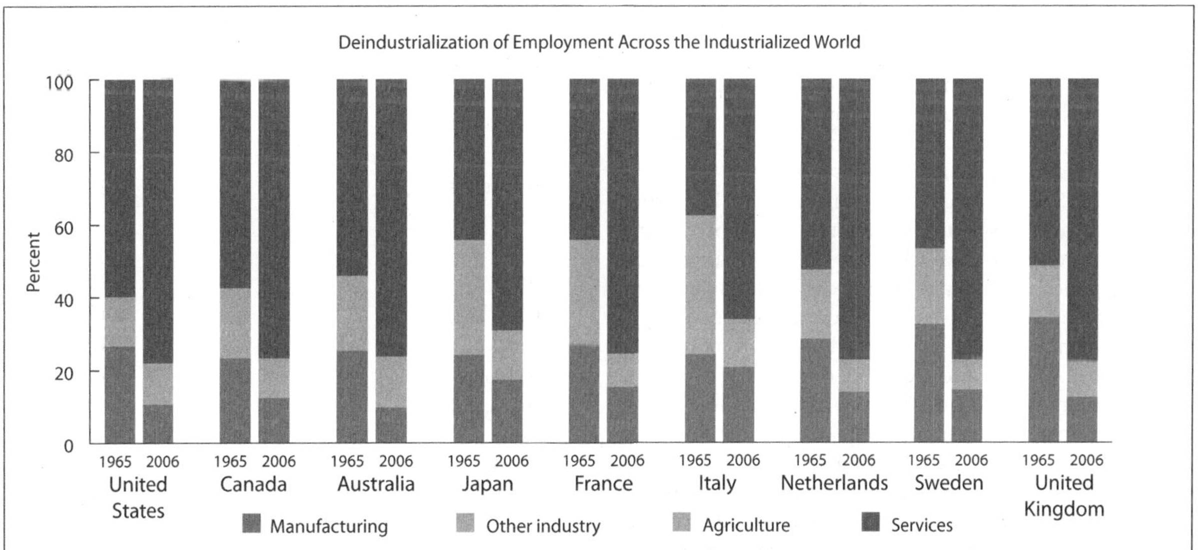

FIGURE 10 Deindustrialization of employment across the industrialized world since the 1960s has been part of a massive structural transformation in the nature of capitalism, reducing manufacturing to a small share of the total labor force.

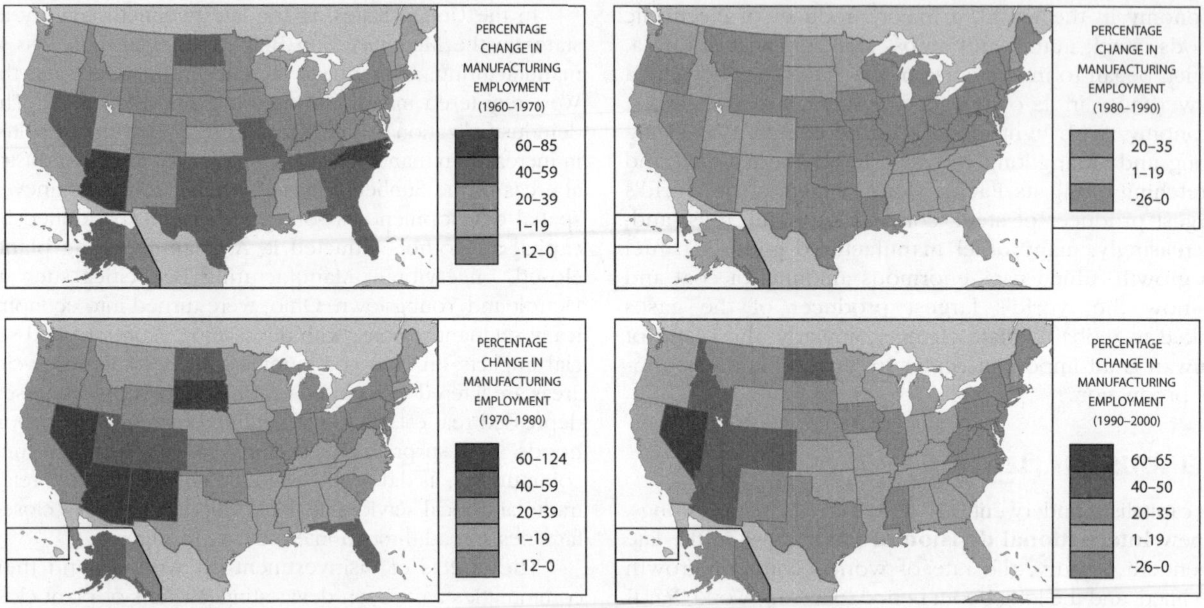

FIGURE 11 Changes in U.S. manufacturing employment by decade, 1960–2000. The steady losses in the northeastern Manufacturing Belt and the rise of Sunbelt producers reflected the sustained impacts of globalization, technological change, and the shift to lower-cost, less-unionized parts of the country. The microelectronics revolution and information technology allow firms much greater locational flexibility than in the past and contribute to the low-density, decentralized landscapes of industrial as well as urban areas.

new bases for employment in service industries, and others, such as Cleveland, show potential for doing so. Southern New England, which had suffered high unemployment rates throughout much of the post–World War II period, underwent a new round of industrial expansion based on electronics, which took advantage of its pool of highly skilled workers.

Several factors underlie the decline in manufacturing in the First World. First, the cost of wages in Europe and North America is much higher than in the developing world, creating an incentive for firms to save by relocating abroad (Table 1). Firms will, of course, pay high wages for skilled, productive labor; however, many developing countries that invested in education now have a labor force whose skills and productivity rival those in Europe, North America, and Japan. In addition to wages, many industrial firms faced high pension costs for their workers. In the United States, which lacks universal health insurance, significant costs are associated with providing health insurance and care. (Automobile companies, for example, spend, on average, more on health insurance per car than it does on steel.) Second, technological changes in manufacturing, which made it much more capital-intensive, steadily reduced the number of workers needed per unit of output. Thus, although the United States lost numerous manufacturing jobs, its manufacturing output (as percent of gross domestic product) stayed almost constant. Third, although industrial firms reinvested in new production techniques, they often failed to save and re-invest sufficiently in reseach and development to remain competitive on the world stage, thus diminishing their productivity in comparison to their competitors

TABLE 1 Hourly Wage Costs for Factory Workers, U.S. Dollars, 2008

Norway	34.64
Denmark	33.75
Germany	32.53
Netherlands	30.76
Finland	30.67
Switzerland	30.26
Belgium	29.98
Sweden	28.42
Austria	28.29
Luxembourg	26.57
United Kingdom	24.71
France	23.89
United States	23.17
Australia	23.09
Ireland	21.94
Japan	21.90
Spain	17.10
Israel	12.18
South Korea	11.52
New Zealand	9.14
Singapore	7.45
Portugal	7.02
Taiwan	5.97
Mexico	2.50
Sri Lanka	0.51

Source: U.S. Bureau of Labor Statistics.

abroad. Many firms instead spent their net revenues on high stock dividends. Finally, inadequate public investment in education and infrastructure, including in the United States, led to shortages of human capital and high transportation costs, which lowered productivity growth.

Conversely, manufacturing employment and output have increased sharply in lower-wage industrializing countries. Between 1974 and 2005, the advanced industrial countries lost 35 million jobs, whereas the newly industrializing countries gained 29 million. The most rapid growth of manufacturing output occurred in East and Southeast Asia, the world region with the fastest rate of economic growth since World War II. Japan, with high wages but low labor militancy, set the model for much of this region, a phenomenon sometimes called the "flying geese formation." This notion compares industrialization to a flock of birds flying in a V-formation, with Japan at the head, the NICs such as South Korea, Taiwan, Hong Kong, and Singapore closely behind, and other countries (Thailand, Indonesia, Malaysia) forming a third generation of "tigers." However, even Japan has witnessed stagnation in its manufacturing base as its economy has moved steadily into services. As the NICs industrialized, they acquired a capacity first in unskilled, light forms of manufacturing such as textiles and garments, then in successively more skilled, higher-value-added industries such as ships, steel, automobiles, and electronics.

The "sunrise" industries of the developing world thus corresponded to the "sunset" industries of the West. Deindustrialization in some places and industrialization in others are mirror images of each other, both reflecting the continual process of uneven development. It should be remembered that these changes are massive and reshape the lives of hundreds of millions of people, lifting many out of poverty and catapulting them from rural villages into industrial cities, often at great personal cost and suffering.

Susan Meiselas/Magnum Photos, Inc.

FIGURE 12 **Manufacturing Reeboks at the Lotus Plant, Philippines.**

THE DYNAMICS OF MAJOR MANUFACTURING SECTORS

Globalization occurs differently in different industries. Because various forms of manufacturing have their own specific technical, labor, and locational requirements, and because nation-states have approached different industries from a variety of regulatory perspectives, the production changes that have come about worldwide are unique to each industry. Five industries—textiles and garments, steel, automobiles, electronics, and biotechnology—dramatize the similarities and differences that exist among manufacturing sectors as they create their own geographies, reflecting and producing globalization in sector-specific ways.

Textiles and Garments

The textile and garment industries are dramatic examples of the globalization of manufacturing. Textile manufacture is the creation of cloth and fabric; clothing manufacture uses textiles to produce wearing apparel (Figure 12). About one-half of textile output is used in apparel; the rest goes into products such as carpets and industrial fabrics (e.g., upholstery in cars). Textiles and garment manufacture are the classic low-tech industry: unskilled, labor-intensive, relatively little technological sophistication, small firms, and few economies of scale. The industry is relatively easy to enter and exit and highly competitive, leading producers to seek savings by paying very low wages. Thus, textiles have long been among the most brutally exploitative industries (Figure 13).

Textiles were *the* leading sector of the Industrial Revolution, fueling the rise of a manufacturing base in cities like Manchester, England, and providing grist for the novels of Charles Dickens. Then, as today, the industry relied heavily on young women and child labor. By the early nineteenth century, it began to expand to the rest of Europe, including Lyon in France and northern Italy. In the late nineteenth century, textile production arose in Japan as part of its rapid modernization that began after 1868. In the United States, textile production generated the manufacturing cities of southern New England and later moved to the Piedmont states in the South, particularly North and South Carolina, which remain the core location of the rapidly shrinking American textile industry. Secondary concentrations of garment manufacturing are found in New York, a remnant of its heyday in the late nineteenth and early twentieth centuries, as well as in southern California, where sweatshops using immigrant labor are common.

Textile and garment production in the economically developed world declined steadily from the 1970s onward as the industry was confronted with waves of cheap imports from abroad. Lured by significantly lower wages in the developing world (Figure 14), especially in East Asia, the industry moved overseas (Figure 15). Today, major cotton fabric producers include China, the world's largest producer, where the garment industry has grown explosively, and

FIGURE 13 In this textile factory in Fortaleza, Brazil, women constitute the largest part of the workforce. Brazil is a major exporter of textiles to advanced industrial countries. Despite high rates of growth in many Third World countries, only a handful managed to narrow the gap with the northern, industrialized countries. Moreover, increased polarization between the richer and poorer countries has been accompanied by a rising trend in income inequality within countries such as in Brazil. The income share of the richest 20% has grown almost everywhere, while those in the lower ranks have experienced no rise in incomes to speak of. Even the middle classes have experienced little improvement.

India (Figure 16); in both, working conditions resemble those of Britain 200 years ago. A small industry has sprouted up in Central America, where sweatshops are found in Honduras, Guatemala, and El Salvador. Everywhere, the industry is notorious for its exploitation of young, predominantly female labor, many of whom work long hours for very low wages. By continually looking for new sources of cheap, easily exploitable labor, this industry has exhibited a very fluid geography over time, a pattern that continues in the twenty-first century.

Steel

The steel industry has played an enormously important role in the development of industrialized societies. Iron and steel production generate a wide variety of outputs essential to many other sectors as well, including, for example, parts for automobiles, ships, and aircraft; steel girders for buildings, dams, and other large projects; industrial and agricultural machinery; pipes, tubing, wire, and tools; furniture; and many others uses. The inputs into

FIGURE 14 Hourly labor costs in clothing manufacturing in various countries. China and other emerging nations are lifting the global economy, but their strength threatens to come at the expense of the United States and Europe. The world economy expanded at 5% in 2010, with China and several other developing countries growing at 9%, and the United States at 3%, with Europe at 2%. Emerging countries are benefiting from low-priced labor and low-priced exports, fueled by artificially low-valued currencies.

FIGURE 15 A man in India weaving silk illustrates how this process was done for centuries before the Industrial Revolution, a method still practiced in many developing countries today.

steel making are relatively simple, including iron ore, which is purified into pig iron; large amounts of energy, generally in the form of coal; and limestone, which is used in the ore purification process. Steel production is a very capital-intensive process, requiring huge sums of investment, and because the barriers to entry are high, it has generally been very **oligopolistic** (Figure 17). Involving highly skilled, almost entirely male jobs, it has high multiplier effects. Transport costs have traditionally been high in this sector, which has made it an ideal candidate for Weberian locational analysis.

The historical geography of steel production includes the important Midlands cities of Britain, such as Sheffield and Birmingham, which were critical in the early stages of the Industrial Revolution. A similar complex of steel production arose in the Ruhr region in Western Germany on the banks of the Rhine River. In the northeastern United States, the earliest iron and steel producers were very small and localized, using wood and charcoal as fuel and serving local markets. A number of changes in the third Kondratiev wave of the late nineteenth century, however, dramatically reshaped the industry. The Bessemer open hearth furnace made the production of steel relatively cheap, using huge amounts of energy in plants that were open 24 hours per day. Geographically, the industry came to center on the steel towns of the Manufacturing Belt, including, above all, Pittsburgh, which became the largest steel producing city in the world. Other centers within this complex included Hamilton, Ontario; Buffalo, New York; Youngstown and Cleveland, Ohio; Gary, Indiana; and Chicago. These locations allowed easy access to coal from Appalachia, iron ore from Minnesota and upper Michigan,

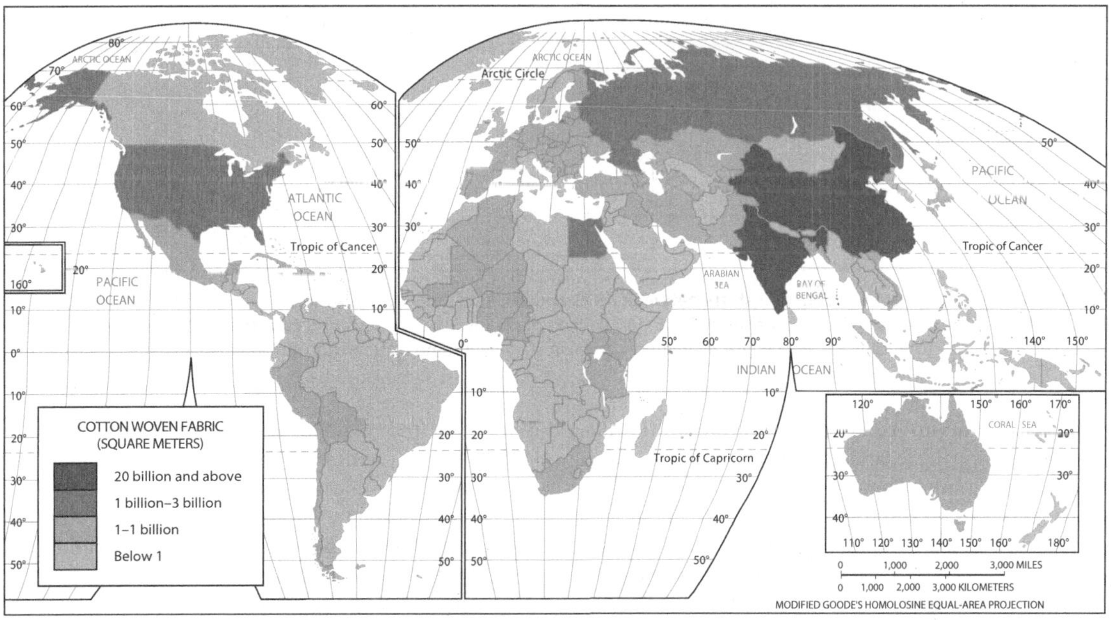

FIGURE 16 Worldwide distribution of textile manufacturing employment.

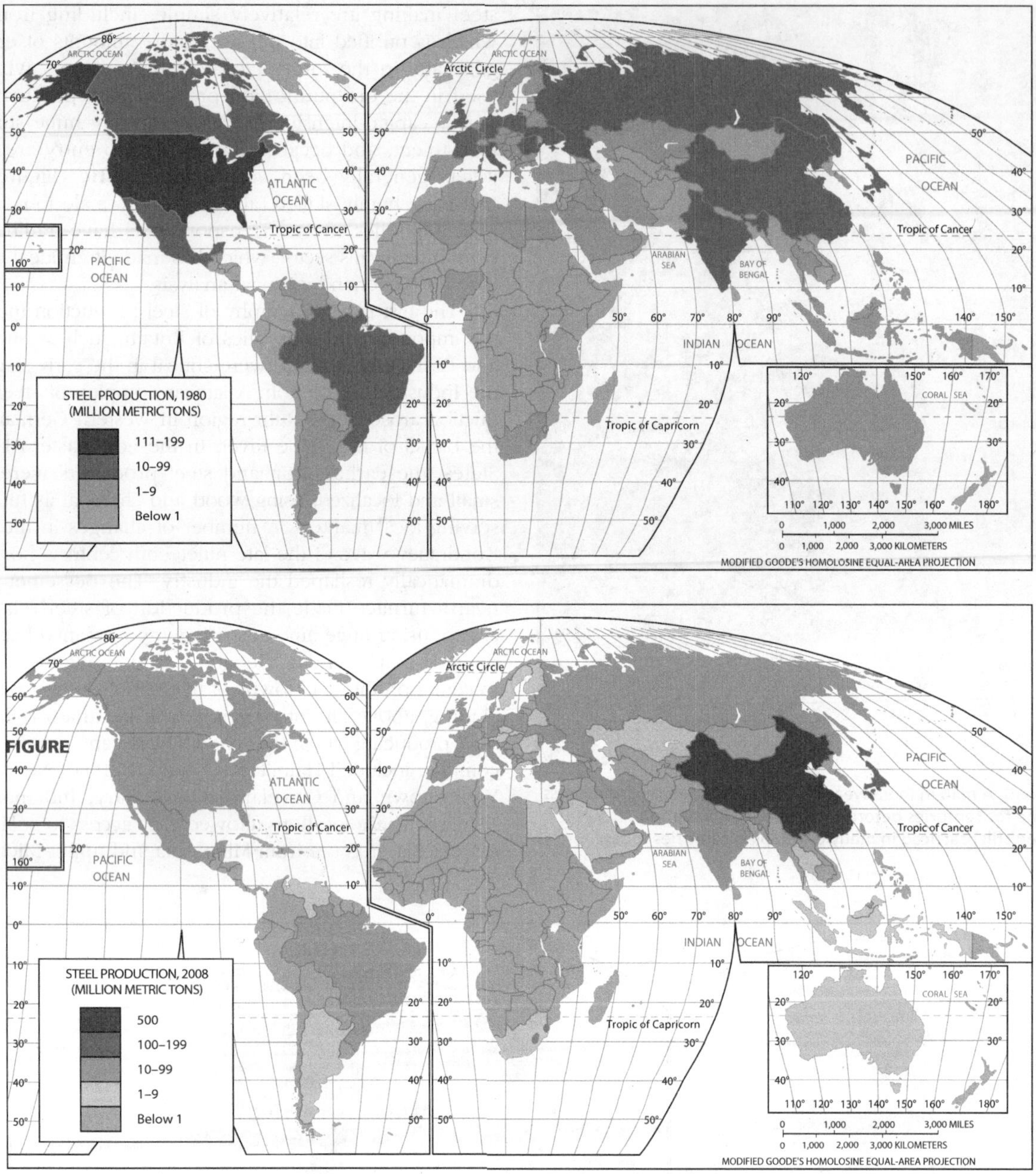

FIGURE 17 Worldwide steel production shifts between 1973 and 2008. In 1973, North America and Europe accounted for 90% of the world's steel output. Production by 2008 had shifted dramatically to developing countries in Latin America, South Asia, and East Asia. Total global steel output remained approximately the same during this period.

and cheap transportation via the Great Lakes and the railroads. By clustering there and taking advantage of the agglomeration economies that such locations offered, they had ready access to specialized pools of labor and suppliers of parts and services. The industry became increasingly oligopolized over time: For example, the rise of the U.S. Steel Company under Andrew Carnegie put 30% of the nation's steel output in the hands of one firm in 1900.

The United States dominated the world's steel industry in the early and mid-twentieth century, producing as much as 63% of the world's total output after World War II. However, new competitors that had recovered from the war entered the industry in the 1960s, first in the rebuilt factories of Europe (particularly Germany, France, and Spain) and Japan, and later in some developing countries (e.g., Brazil, South Korea) (Figure 18). The United States saw a gradual decline in the share of the world steel it produced. By 2005, that share equaled 8.3% of the world total (Figure 19). Conversely, the industry has flourished in East Asia (Figure 20), particularly China, now by far the

FIGURE 18 Steel exemplifies forms of manufacturing that are highly capital-intensive, relying on huge economies of scale to produce their output profitably.

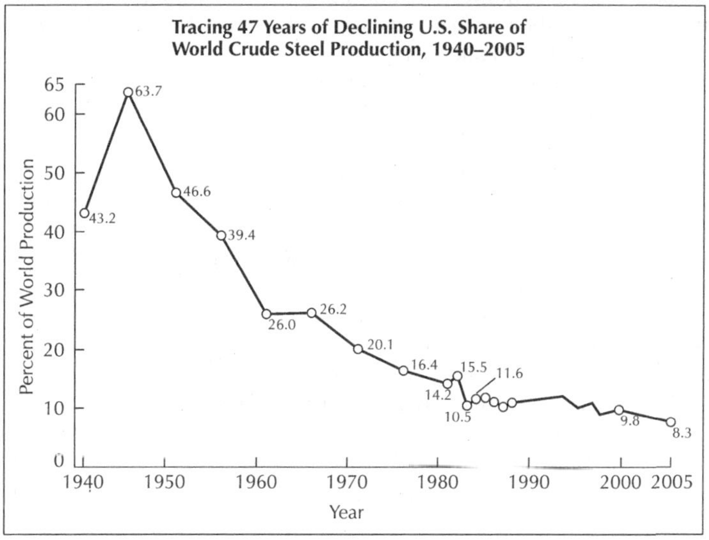

FIGURE 19 Decline in the U.S. share of world steel output after World War II. America has not been able to keep pace with China and other developing countries in steel making because of the cheap, nonunion labor available overseas.

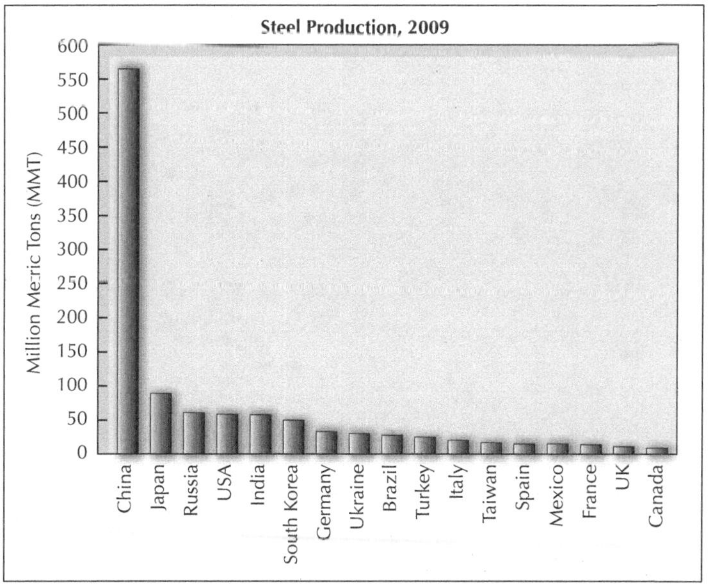

FIGURE 20 China, with its rapidly growing economy, has become by far the world's largest producer of steel, followed by Japan, Russia, and the United States. In a decade, China has gone from a huge, poor country to an economic colossus. Although its median per capita income, $6600 in 2010, is only one-seventh that of the United States ($46,499), the sheer size of its economy gives it a growing influence. Cheap labor; abundant natural resources; and a huge demand for new infrastructure of buildings, roads, airports, high-speed trains, and dams have made it the leader in world steel production.

world's largest producer of steel, with Japan in second place, Russia in third, and the United States in fourth. These shifts are all reflective of the continual reproduction of uneven development under capitalism.

The decline of the U.S. steel industry generated enormous problems for families living in the former steel towns. Waves of plant closures in the 1970s and 1980s generated high unemployment, rising poverty, depressed property values, and out-migration. Today, former giants of steel production in the Manufacturing Belt produce very little steel; Pittsburgh produces none at all. Steel offers a compelling example of the decline in industrial capacity. Between 1970 and 2007, the North American and western European proportions of global steel production declined from 67% to 41%, whereas developing countries' production levels increased from 10% to more than 33%. Many steel-manufacturing firms have gone out of business as the global steel-production capacity exceeds global demand. Because of government subsidies, steel mills in some countries, especially in Europe, have remained open in the face of dwindling quotas. The U.S. government, however, has been less willing to pay unemployment compensation to displaced workers and has allowed the U.S. steel industry to decline. Since the 1970s, U.S. production has decreased 33%, whereas employment in the steel industry has declined 66%.

The industry's response to crisis, other than plant closures, was both to call for protectionism from imports and to restructure. The introduction of computerized technology led to widespread changes in steel production, including the emergence of highly automated "minimills" that use scrap metal as inputs (Figure 21), are generally not unionized, and produce specialized outputs for niche markets, all of which are symptomatic of the growth of post-Fordist production (to be discussed later in this chapter).

Automobiles

From its inception at the dawn of the twentieth century, the automobile industry has unleashed major changes on cities and everyday life throughout the world. Originally, Henry Ford standardized the European invention of auto production, introducing the moving assembly line and a highly detailed division of labor to make automobiles affordable to the middle class. He also paid high wages to reduce turnover rates of his workers. So successful was his model of production that what became known as "Fordism" was widely imitated by other sectors of industry. Ford's success was centered in the Detroit region, adding another layer of investment to the Manufacturing Belt. Other than being Ford's hometown, Detroit enjoyed other locational advantages, including being the previous center of the wagon and buggy industry, numerous parts-producing firms, the rubber-producing center in nearby Akron, numerous rail lines and shipping routes that brought raw materials across the Great Lakes, and a large supply of skilled labor.

Throughout the twentieth century, the market expanded dramatically as incomes rose and as Western cities, particularly in the United States, increasingly developed around low-density suburban arrangements. Many auto companies purchased mass transit lines and destroyed them, as in Los Angeles, in order to force people to buy cars. In the 1950s and 1960s, the construction of the federal interstate highway system greatly enhanced the demand for cars. As it matured, the auto industry underwent a pronounced series of consolidations, becoming steadily concentrated in a handful of highly capital-intensive **transnational corporations (TNCs)**, a classic example of oligopolization (Figure 22). In no other industry do so few companies dominate the world scene. For example,

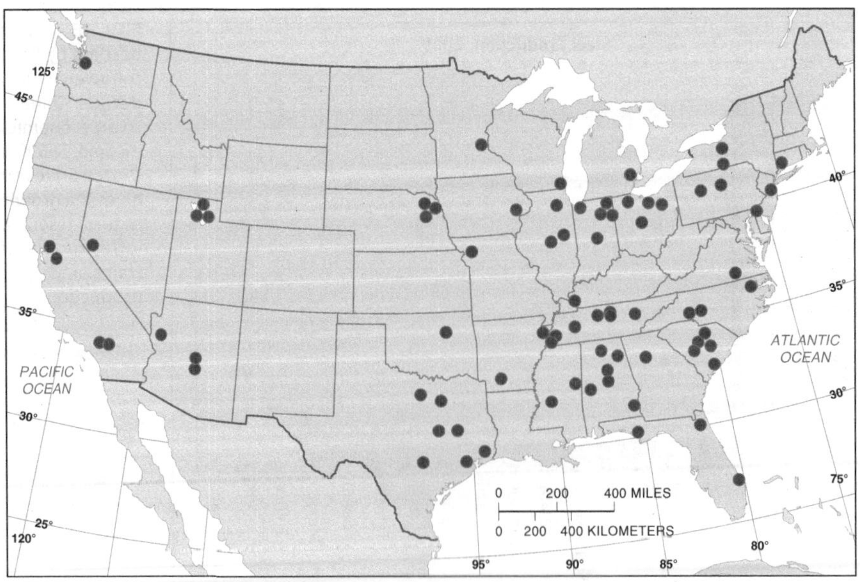

FIGURE 21 Minimills producing steel from scrap metal are more numerous than large, integrated steel mills. They are located near markets because their main input is abundantly available there.

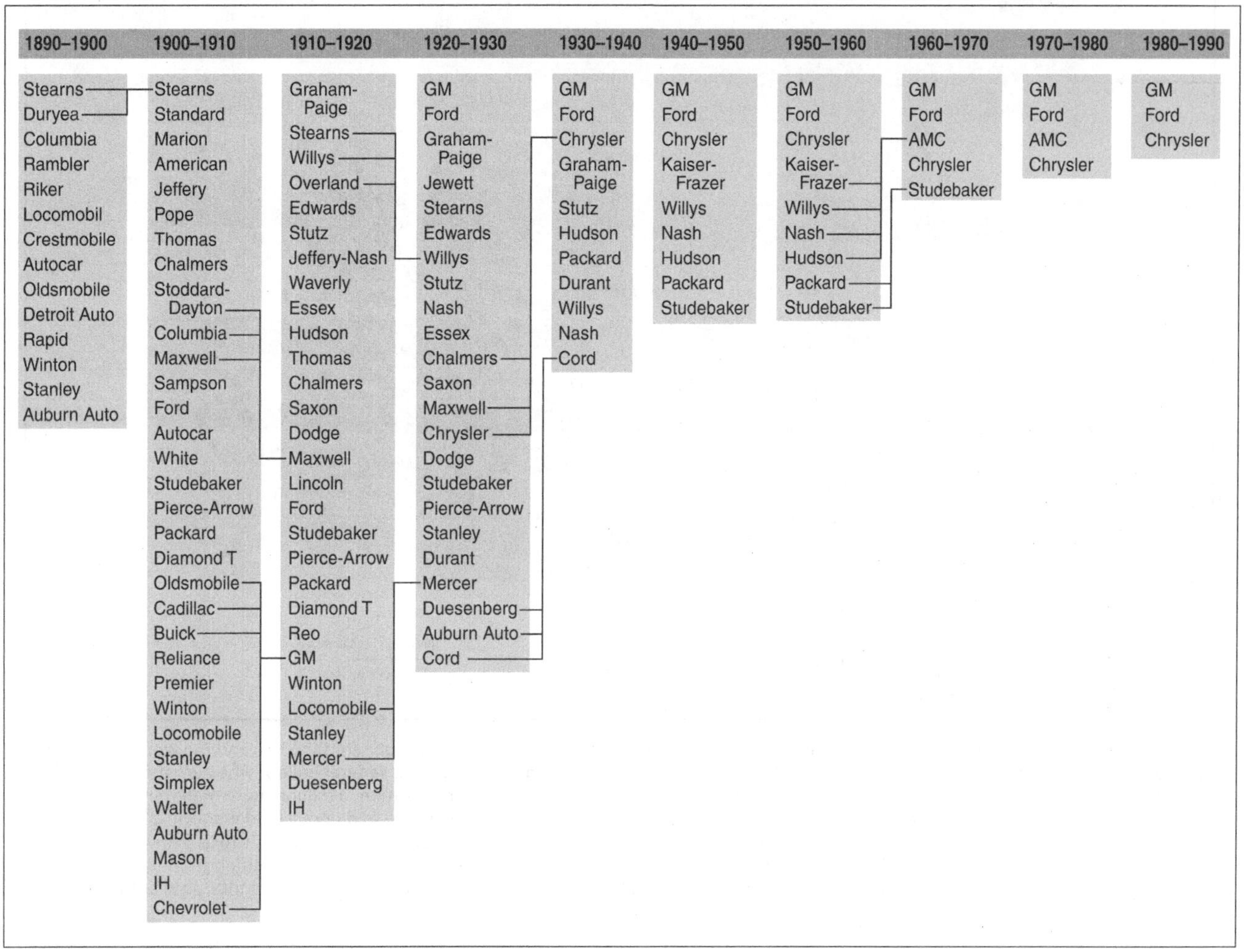

FIGURE 22 Oligopolization of the U.S. automobile industry. As the market expanded, larger firms could take advantage of economies of scale to eliminate or swallow up smaller ones, leading to corporate concentration in fewer hands. The process reveals the market as a battleground of firms, not simply a smooth mechanism coordinating supply and demand.

the world's 10 leading automobile manufacturers produce over 70% of the world's automobiles. In the United States, General Motors, Chrysler, and Ford account for 95% of national automobile output. However, the U.S. auto industry has been steadily besieged by competition from abroad and has suffered declining market share. Chrysler was purchased by the German firm Daimler-Benz, and in the wake of the 2007–2009 financial crisis, General Motors went bankrupt, saved from obliteration by massive federal government bailouts. In two short years, between 2005 and 2007, the industry laid off 40,000 workers, mostly in the greater Detroit area (Figure 23).

Worldwide automobile production extends across the globe and has expanded in capacity as the growing middle class in many developing countries seeks out cars. Figure 24 shows the world distribution of automobile production and assembly. Three major nodes of automobile production exist—Japan, the United States, and Western Europe. In 2008, Europe accounted for 25% of the world's automobiles; Japan, 20%; and the United States, 19%. The three developed regions of the world, East Asia, North America, and Europe, accounted for 72% of the automobiles produced. Japanese firms also invested heavily in the United States in the late twentieth century, setting up factories in much of the Midwest (Figure 25) that allowed them access to the American market, low transport costs, and freedom from fluctuating exchange rates and threats of U.S. protectionism. Smaller production centers exist in Brazil, Russia, South Korea, and India.

Electronics

Although its roots extend into the nineteenth century, the electronics industry underwent enormous changes during the microelectronics revolution of the late twentieth century. Microelectronic technology is the dominant technology of the present historical moment, transforming all branches of the economy and many aspects of society.

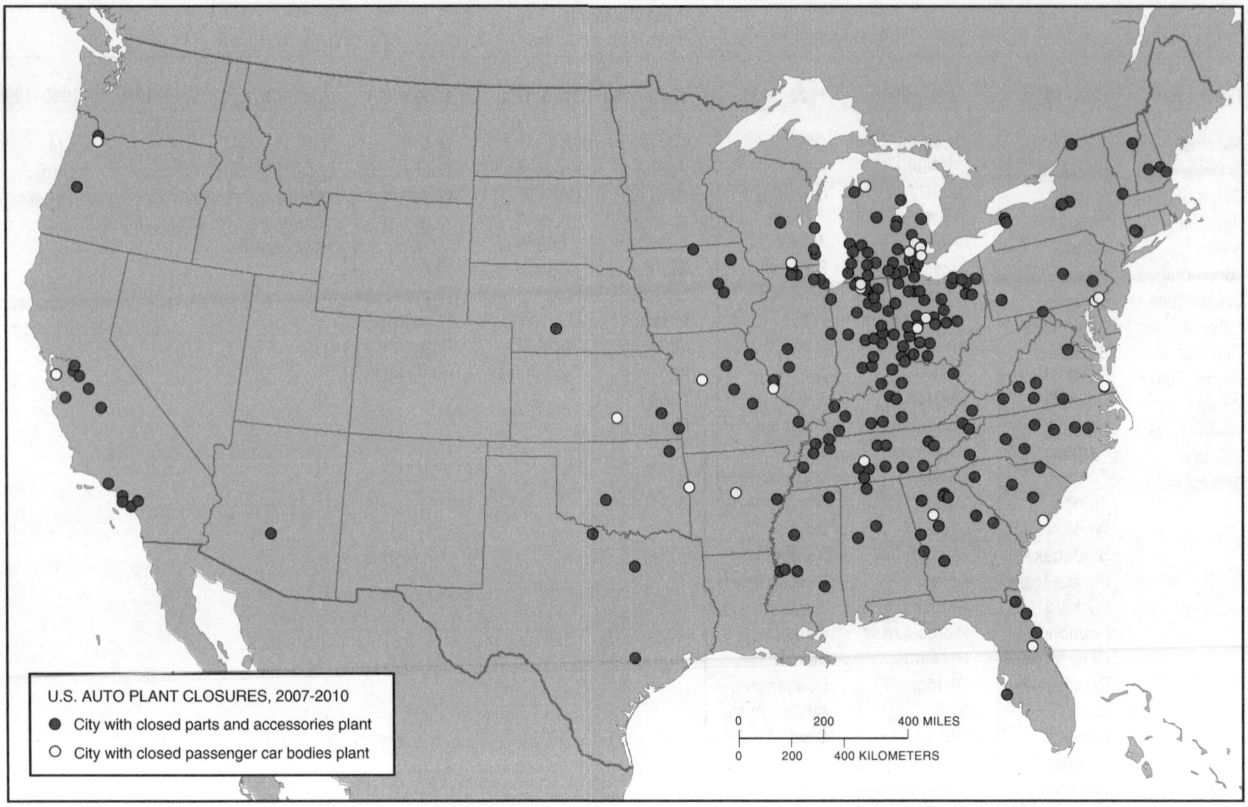

FIGURE 23 Job losses in automobiles in the United States, 2007–2010. The financial crisis that clobbered the U.S. and the world economy was particularly acute in manufacturing, symbolized by the bankruptcy of General Motors and its federal bailout. The auto industry shed more than 40,000 jobs in two years alone. In 2009, about 14 million cars were taken out of operation, 4 million more than rolled off the assembly lines and onto the roads of America. That was the first time more cars were scrapped than sold since World War II. In 2008, GM and Chrysler went through bankruptcy and were bailed out by the U.S. government. Sales fell over 20% in 2009 and 2010 as many consumers held off buying new cars, because of fears of losing their jobs. Sales went up somewhat in 2011, but heavy competition from Japan, Europe, and East Asia; high unemployment; a weak housing market; and toxic bank debt continue to hobble growth, and thus, new auto demand.

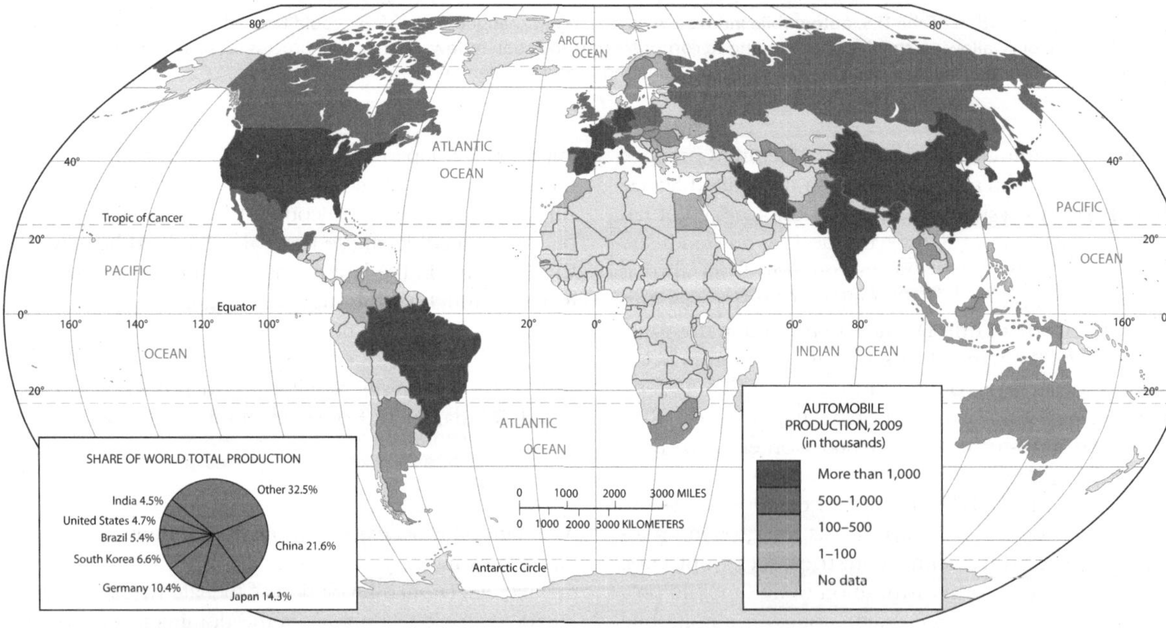

FIGURE 24 World distribution of automobile production and assembly. In 2009, Japan produced 11 million automobiles, which was 19% of the world's total output. In 1960, the United States produced half of the world's total automobiles, but by 2009, that proportion had dropped to 15%. In Europe, Germany, France, and Italy are the largest producers.

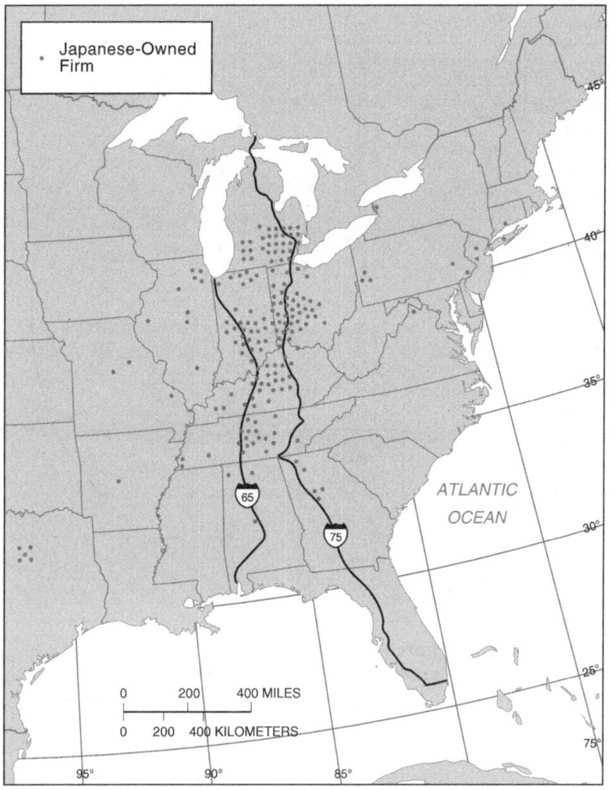

FIGURE 25 Distribution of Japanese-owned automobile parts manufacturing plants in the United States.

The radio was invented and produced as early as 1901, but the modern electronics industry was not born until Bell Telephone Laboratories built the transistor in the United States in 1948. The transistor supplanted the vacuum tube, which had been used in most radios, televisions, and other electronic instruments. The microelectronic transistor was a solid-state device made from silicon that acted as a semiconductor of electric current. By 1960, the *integrated circuit* appeared, which was a quantum improvement because transistors could be connected on a single small silicon chip. By the early 1970s, a computer so tiny that it could fit on a silicon chip the size of a fingernail came into production. Thus, the *microprocessor*, each of which could do the work of a roomful of vacuum tubes, was born. With these devices, information could be converted from an analog to a digital format in binary code, which made it much easier for computers to process. The microprocessor made possible the microcomputer, which in turn revolutionized the collection and analysis of information, particularly office work. As the industry generated wave upon wave of improvements, computing power and memory increased exponentially (Figure 26).

As capacity, speed, and miniaturization increased, the electronics industry discovered new applications, including calculators, electronic typewriters, personal computers, industrial robots, aircraft-guidance systems, and combat systems. New discoveries were applied in automobile construction to improve guidance, safety, speed, and fuel consumption. An entire new range of consumer electronics also became available for home and business use. The electronics industry, like textiles, steel, and automobiles before it, has come to be regarded as the modern touchstone of industrial success. Hence all governments in the developed market economies, as well as those in the more industrialized developing countries, provide substantial support programs for the electronics industry, particularly for microprocessors and computers.

For nearly two decades, from the 1960s through the 1970s, the United States dominated the field of semiconductor manufacture. However, by the 1990s, Japan took over this role. Figure 27 shows world production of electronic components, which includes semiconductors, integrated circuits, and microprocessors. The field is dominated by Japan and the United States, with other significant production in Western Europe and Southeast Asia. In 2007, Japan accounted for 40% of the world production of semiconductors; the United States, 21%; and Europe, 11%. In Southeast Asia, South Korea, Malaysia, Taiwan, Thailand, and China were significant manufacturers.

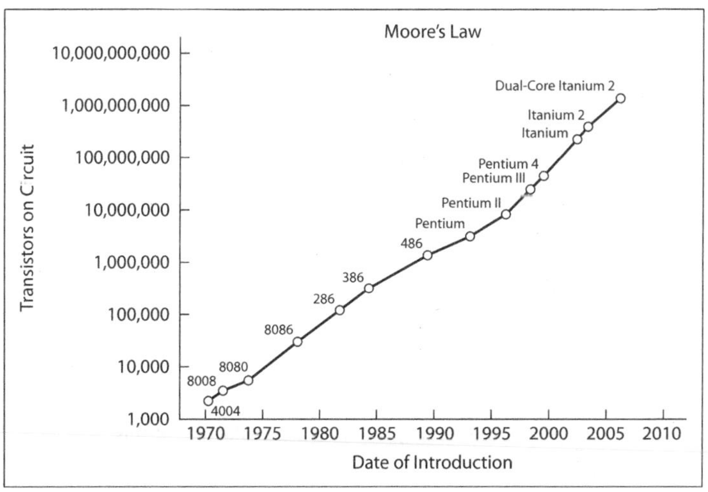

FIGURE 26 The spectacular growth in microelectronic computing power and memory has revolutionized contemporary capitalism, including manufacturing and information-processing activities.

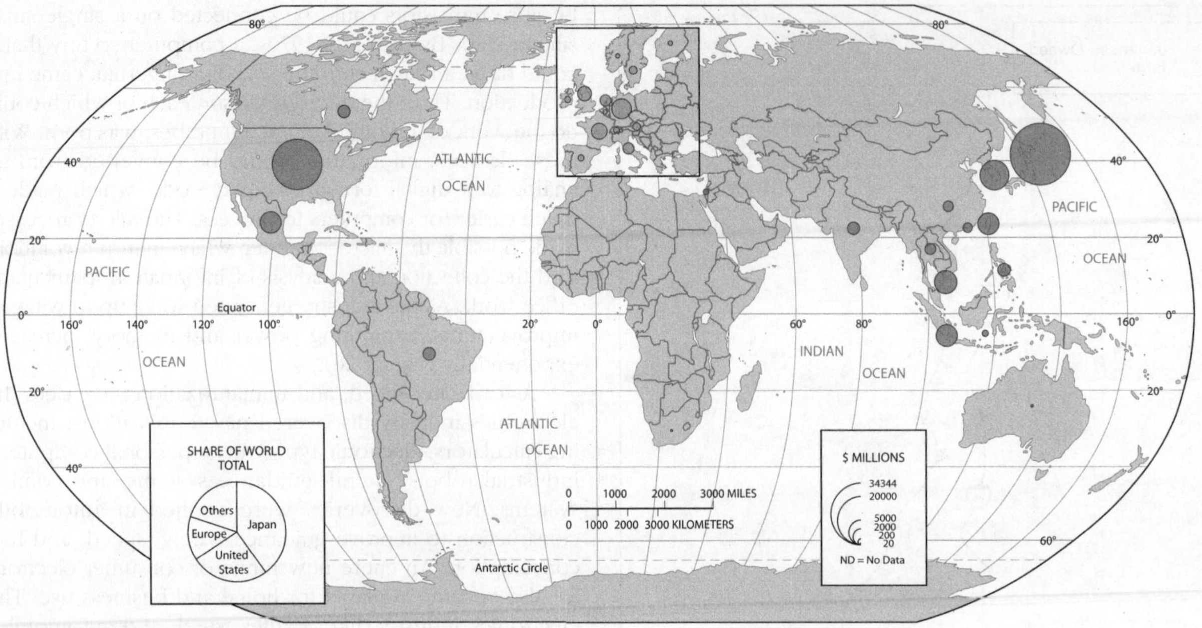

FIGURE 27 World production of electronic components, including semiconductors, integrated circuits, and microprocessors. Japan is the world's leading producer of electronic components, followed by the United States, Germany, France, and the United Kingdom. Growth in the electronics industry has been greatest in the Pacific Rim, with recent major centers of production developing in South Korea, Malaysia, Taiwan, and Thailand.

The world manufacture of consumer electronics such as televisions, computer peripherals, and microwave ovens has shifted steadily to developing countries, especially in East and Southeast Asia (Figure 28). Much of the television production that formerly occurred in the United States, Germany, and the United Kingdom now takes place in China and Malaysia. Outside Asia and North America, Brazil produces 87% of the televisions used in Latin America.

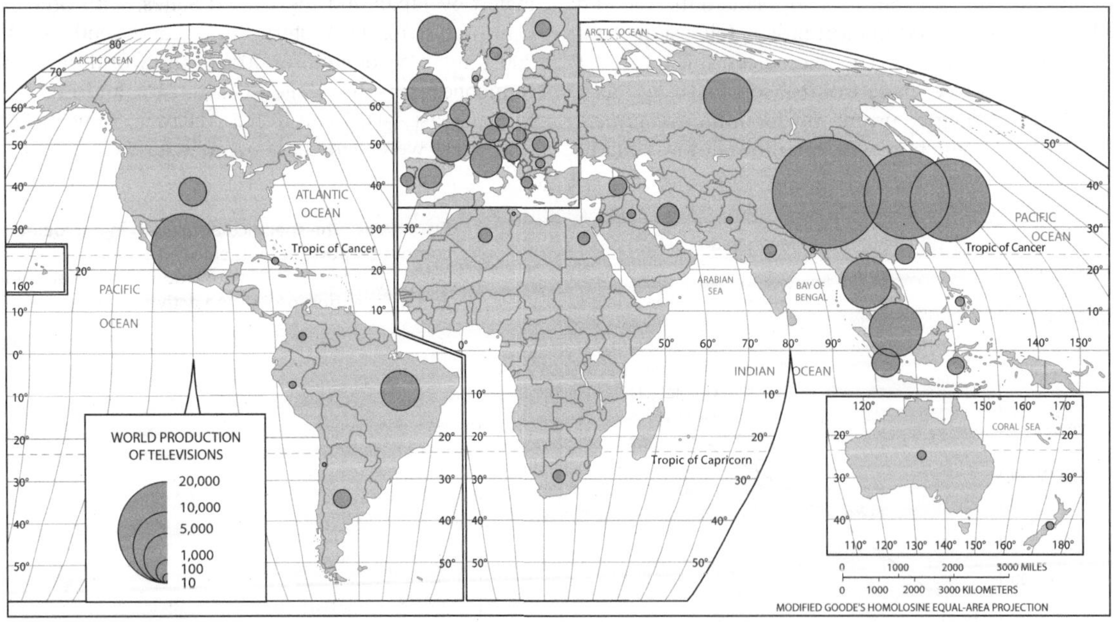

FIGURE 28 Worldwide television receiver production. East Asia has become the leading area in the manufacture of this commodity.

Case Study

Export Processing Zones

Export Processing Zones (EPZs), also known as Free Trade Zones (FTZs), Special Economic Zones (SEZs), Free Zones (FZs), Free Ports, and by other names, became a global development tool in low-income nations in the 1970s—in Latin America, the Caribbean, Asia, and, to a lesser degree, Africa—as innovations in international transportation and communication of the 1960s and 1970s allowed transnational corporations to lower production costs by outsourcing part or all of production to offshore locations. By the 1980s, almost all developing nations had established at least one EPZ.

EPZs are locations for foreign enterprises and nodes of international transportation. An EPZ is usually set up by a government to promote and capture foreign direct investment, international trade, technological transfer, and industrial deepening. These zones give preferential treatment and incentives to attract foreign enterprises and investors. The usual benefits provided to foreign firms are infrastructure, communication and financial services, elimination of trade barriers such as tariffs and quotas, exemptions from production and trade-related taxes and business regulations, friendly and simplified bureaucratic requirements and procedures, elimination of labor laws, and relaxation of environmental protections. Tax holidays, duty-free export and import, and free repatriation of profits are regularly offered.

EPZs are instruments for increasing foreign currency earnings, upgrading production technology, entering the world market, and providing employment. With foreign investment concentrating in these zones, the expectation is that there will be a multiplier effect to locally linked industries and a dispersal effect to the surrounding regions. Apart from being the locations for attracting foreign business and international exchange, EPZs are also demonstration centers for cultural borrowings from around the world.

The division of the industrial production process between corporate headquarters in home nations and production branches overseas is the basis for EPZs. These zones are designed to absorb international branch operations. Globalization gives rise to selective international economic integration, and EPZs are the transnational connection points in developing nations. For example, the Shenzhen SEZ in China was designated in 1979 when that country was transforming from a socialist to a market economy. It was the "window" to observe and learn from free market operations and the "bridge" for linking to the international economy. Shenzhen is an unusual EPZ in that it replicates the same set of development objectives and provisions, but it is as an experiment for a radical economic systematic change that gives the SEZ a special importance. Adjacent to Hong Kong, an international market economy free port, Shenzhen adopted trade regulations and industrial practices freely from its neighbor. In turn, Shenzhen's institutions, regulations, administration, and planning are used as models for the rest of China as the country opens up its economy to the world. Connecting to the global economy through Hong Kong initially, Shenzhen has grown with export manufacturing and spread the developmental effects to its northern region, the province of Guangdong. The SEZ has industrialized the province's economy and uplifted it from one of the poorest provinces in the Socialist era to one of the richest in the market era. Shenzhen has successfully fulfilled the promises of the EPZ, but what is more, it has become a transference point for foreign economic ideas and practices to the nation, playing a major role in marketing and internationalizing China's economy.

Even though there are sufficient successful examples of EPZs to attest to their effectiveness, there are nevertheless many counterarguments. In general, the multiplier and regional effects of many EPZs have been weak. The cost-effectiveness of direct government investment and the indirect costs in relation to revenue received have had poor results. Negative externalities have been reported, especially environmental pollution. One of the most frequently raised issues about EPZs as instruments of national development strategy is the competition among them. Since transnational firms seek economically advantageous sites for outsourcing and branch locations, and there are many suitable and available EPZs in the world, firms tend to seek out and bargain for the most advantageous zone. EPZs are forced to make their best offers to outbid their rivals. This competitive process is not only limited to international rivalry; it also generates intensive contests between EPZs within the same country.

The most controversial issue is labor practices. International firms relocating their labor-intensive production segments are often opposed to labor unions and minimum wage regulation. In order to be competitive, EPZs tend to avoid enforcing these institutional practices. Working conditions are substandard and wages are low. Most of the employment is for young women in "dead-end jobs." Often, EPZs do not offer new jobs, but replicate local jobs. Foreign firms, with better resources than the local firms, compete successfully at the expense of local industries for the existing "best" workers.

Biotechnology

Biotechnology (or simply "biotech") may be defined as the application of molecular and cellular processes to solve problems, develop products and services, or modify living organisms to carry desired traits. Arising after the discovery in 1973 of recombinant DNA, biotechnology has been a rapidly growing industry worldwide, with extensive linkages to agriculture, health care, energy, and environmental sciences. In 2008, the U.S. biotech industry (excluding medical equipment firms) consisted of roughly 1500 firms that employed 450,000 people and ranging in size from single proprietorships to firms of more than 500 employees. The mean national annual salary in the industry is $63,000, which is well above the national average.

Venture capital is critical to making basic research in biotechnology commercially viable. Most small biotech firms lose money, given the high costs and enormous amounts of research necessary to generate their output and the long lag between research and development and commercial deployment (generally on the order of 12–15 years of preclinical development). Only 1 in 1000 patented biotech innovations leads to a successful commercial product, and that may take 15 years. Venture capitalists may invest in many biotech firms, and one biotech firm may receive funding from several venture capitalists. Above all, venture capitalists look for an experienced management team when deciding in which companies they are willing to invest. Venture capitalists often provide advice and professional contacts and serve on the boards of directors of young biotech firms. As a biotech firm survives and prospers, its relations with investors often become spatially attenuated; that is, venture capitalists gradually withdraw from day-to-day direct management.

Since the industry began, various levels of government have been extensively involved in establishing biotechnology complexes. Because of the industry's rapid growth as well as its demonstrated and potential technological advances, it has been targeted by many national science policies as a national growth sector. The survival and success of biotechnology firms is heavily affected by federal research funds, primarily through institutions such as the National Science Foundation (NSF) and National Institute of Health (NIH). Other federal offices—the Small Business Technology Transfer (STTR), Small Business Innovation Research (SBIR), the Environmental Protection Agency (EPA), and the Food and Drug Administration (FDA)—also play a significant role. Federal policies regarding patents and intellectual property rights, subsidies for medical research, and national health care programs are all important to the industry. As well, roughly 83% of state and local economic development agencies have targeted biotech for development, and states have designed critical regulatory policies, educational systems, taxation, and subsidies to support it.

Biotech firms tend to cluster in distinct districts, and place-based characteristics are essential to the industry's success in innovation. Europe, for example, hosts the BioValley Network situated between France, Germany, and Switzerland. In Britain, Cambridge has assumed this role. Similarly, Denmark and Sweden have formed Medicon Valley. Geographically, a small handful of cities, particularly Boston, San Diego, Los Angeles, San Francisco, New York, Philadelphia, Seattle, Raleigh-Durham, and Washington, DC, currently dominate the U.S. biotech industry. Together, they account for three-fourths of the nation's biotech firms and employment (Figure 29). All of these cities have excellent universities with medical schools, state-of-the-art infrastructures (particularly fiber optics and airports), and offer a rich array of social and recreational environments.

Biotech firms tend to agglomerate for several reasons. In an industry so heavily research-intensive, the knowledge

FIGURE 29 Major clusters of the U.S. biotech industry (employment in thousands). The industry exhibits a pronounced tendency to cluster in major metropolitan areas, where it has essential access to universities and agglomeration economies.

base is complex and rapidly expanding, expertise is dispersed, and innovation is to be found in networks of learning rather than in individual firms. This observation is at odds with the popular misconception that such firms are the products of heroic individual entrepreneurs. In a highly competitive environment in which the key to success is the rate of new product formation, and in which patent protections lead to a "winner-take-all" scenario, the success of biotech firms is closely related to their strategic alliances with universities and pharmaceutical firms. Although many biotech firms engage in long-distance partnering, this tends to complement, not substitute for, co-location in clusters where knowledge is produced and circulated face-to-face, both on and off the job.

Because pools of specialized skills and a scientifically talented workforce are essential to the long process of biotech research and development, an essential element defining the locational needs of biotech firms is the placement of research universities and institutions and the associated supply of research scientists. Most founders of biotech companies are research scientists with university positions. Regional human capital may be measured by examining the prevalence of bachelor's degrees, which indicate a region's educational attainment, and the location and size of regional universities that grant PhD degrees in biology and related fields. However, to a large extent, local labor shortages can be mitigated through in-migration. Because knowledge is generated and shared most efficiently within close loops of contact, the creation of localized pools of technical knowledge is highly dependent on the detailed divisions of labor and constant interactions of colleagues in different and related firms. Successful biotech firms often revolve around the presence of highly accomplished academic or scientific "stars" with the requisite technical and scientific skills but also the vision and personality to market them. Often such individuals begin in academia and move into the private sector.

FLEXIBLE MANUFACTURING

In the aftermath of the turbulent 1970s—which brought on, among other things, the shift from fixed to floating exchange rates, the petrocrises, the rise of the NICs, massive deindustrialization in the West, and the microelectronics revolution—geographers and others began to recognize that capitalism in the late twentieth century was undertaking a new direction. The new world economy, characterized by mounting globalization and competition among nation-states, also included a profound shift in the nature of manufacturing, including changes in markets, technologies, and location. Now, as we have seen, capitalism is a highly dynamic economic system characterized by many such transformations in its history. In this sense, the creation of a new form of capitalism in the 1980s and 1990s was not particularly new. However, each age brings with it a new form of commodity production and consumption and often a new terminology to describe these changes. A common set of terms used to sum up these epochal changes is the shift from "Fordism" to "post-Fordism," also variously known as "flexibilism" or "the flexible economy" (Table 2).

TABLE 2 Differences between Fordism and Post-Fordism

Fordism	Post-Fordism
Vertically integrated	Vertically disintegrated
Long-run contracts	Short-run contracts, just-in-time inventory systems
Large firms	Small firms
Economies of scale	Specialized output
Competitive producers	Cooperative networks
Product price	Product quality
Mass consumption	Segmented markets
Unionized	Nonunionized
Unskilled labor	Skilled labor
Routinized work	Varied tasks
National linkages	International linkages

Fordism

Fordism, as the name indicates, is named after the American industrialist Henry Ford, who pioneered the mass production of automobiles in the early twentieth century by means of standardized job tasks, interchangeable parts (which date back to gun maker Eli Whitney), and the moving assembly line. Ford's methods, which were very successful; were widely imitated by other industries and soon became almost universal throughout the North American, European, and Japanese economies.

The precise moment when Fordism became the dominant form of production in the United States is open to debate. Some argue that it began as early as the late nineteenth century, when mass production first made its appearance, displacing the older, more labor-intensive (and less profitable) forms of artisanal production. For example, during the 1880s and 1890s, glass blowing, barrel making, and the production of rubber goods such as bicycle tires became steadily standardized, and the Bessemer process for fabricating steel was invented. Fordism, however, took mass production to a new level; it introduced highly refined divisions of labor within the factory, so that each worker engaged in highly repetitive tasks. To do this, Ford engaged the services of Frederick Taylor, the founder of industrial psychology, who applied time-and-motion studies to workers' jobs to organize them in the most efficient and cost-effective manner. By breaking down complex jobs into many small ones, Fordism made many tasks suitable for unskilled workers, including the waves of immigrants then arriving in the United States, and greatly increased productivity. Well paying, relatively unskilled jobs in manufacturing became the basis of social mobility for the growing working class.

Some see Fordism as a particular kind of social contract between capital and labor, one that tolerated labor unions (such as the Congress of Industrial Organizations [CIO] that came into being in the 1930s), and so regard Fordism as beginning in the crisis years of the 1930s. Yet others make the case that Fordism was the backbone of the great economic boom in the three decades following World War II, when the United States emerged as the undisputed superpower in the West, and so should be dated back only to the 1950s. Whenever its origins, Fordism is reflective of a historically specific form of capitalism that dominated most of the twentieth century.

Fordism came to stand for the mass production of homogeneous goods, in which capital-intensive companies relied heavily on economies of scale to keep production costs low and profits high. Thus, mass consumption and advertising, as the demand side of Fordism, would also come into being. It is no coincidence that starting in the late nineteenth century, the mass production of goods made them affordable, and that forms of consumption such as early shopping malls made their appearance.

Typically, firms working in this context were large and **vertically integrated**, controlling the chain of inputs from raw material to final product. In Ford's plants, for example, coal and iron ore entered one part of the factory and cars came out the other end, an example of extreme vertical integration. Well suited to large, capital-intensive production methods, this system of production and labor control was largely responsible for the great manufacturing complexes of the North American Manufacturing Belt, the British Midlands, the German Ruhr region, and the Inland Sea area of Japan.

While Fordism "worked" quite successfully for almost a century, ultimately it began to reach its social and technical limits. Productivity growth in the 1970s began to slow dramatically, and the petrocrises and rise of the NICs unleashed wave upon wave of plant closures in the United States. Because wages and salaries are often tied to the overall growth of productivity, these changes not only led to widespread layoffs but to the declining earning power of American workers, especially men, who formed the bulk of the manufacturing labor force. Rates of profit in manufacturing began to drop in the 1970s and 1980s, and many firms faced the choice of either closing down, moving overseas, or reconstructing themselves with a new set of production techniques. It is in this context that Fordism began to implode, giving way to post-Fordist, flexible production techniques, which have become widespread throughout the world economy today.

Post-Fordism/Flexible Production

Post-Fordism refers to a significantly different approach to the production of goods than that offered by Fordism. Flexible manufacturing allows goods to be manufactured cheaply, but in small volumes as well as large volumes. A flexible automation system can turn out a small batch, or even a single item, of a product as efficiently as a mass-assembled commodity. The new approach to production appeared, not accidentally, at the particular historical moment when microelectronics began to revolutionize manufacturing; indeed, the changes associated with the computerization of production in some respects may be seen as capitalists' response to the crisis of profitability that accompanied the petrocrises. Post-Fordism also reflected the imperative of American firms to increase their productivity in the face of rapidly accelerating, intense international competition (Figure 30).

The most important aspect of this new, or lean, system is the flexibility of the production process itself, including the organization and management within the factory, and the flexibility of relationships among customers, supplier firms, and the assembly plant. In contrast to the large, vertically integrated firms typical of the Fordist economy,

FIGURE 30 The flexible customized economy is replacing the standardized economy. Henry Ford's assembly-line approach led off the innovation phase of the standardized economy. The late twentieth century has been a comparable innovation period, launched by the baby boomers. Microcomputer technology, comparable to the automobile of the early part of the century in innovation, is multiplied in its potential by powerful, flexible software. This software and hardware can revolutionize any industry, just as the assembly line revolutionized industries throughout the Fordist period of the last century. New technologies put downward pressure on prices, but rising consumer demand from the spending wave exerts an upward pressure. A shakeout occurs, leading to survival of only the fittest companies.

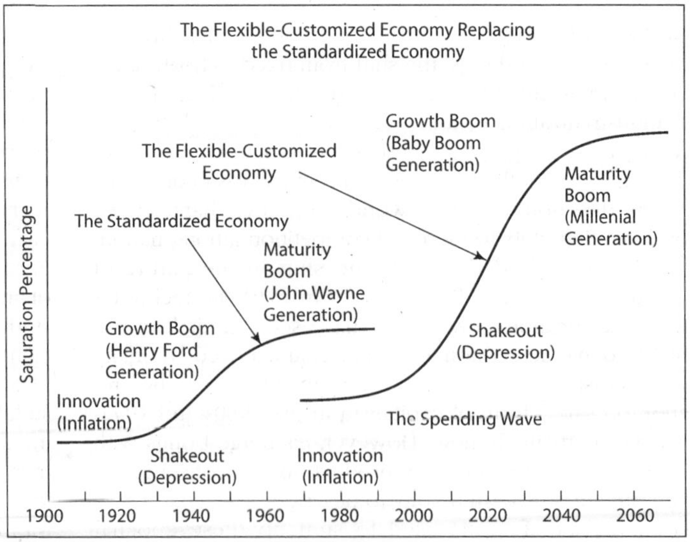

under **flexible production**, firms tend to be relatively small, relying on highly computerized production techniques to generate small quantities of goods sold in relatively specialized markets. Microelectronics, in essence, circumvented the need for economies of scale.

The classic technologies and organizational forms of post-Fordism include robots (Figure 31) and just-in-time inventory systems. The Japanese developed **just-in-time inventory systems** shortly after World War II to adapt U.S. practices to car manufacturing. The technique was pioneered by the Toyota Corporation (and hence is sometimes called "Toyotaism"), which obviated the need for large, expensive warehouses of parts (the "just-in-case" inventory system), saving rents in a country in which the price of land is high. Just-in-time refers to a method of organizing immediate manufacturing and supply relationships among companies to reduce inefficiency and delivery times. Stages of the manufacturing process are completed exactly when needed, according to the market, not before and not later, and parts required in the manufacturing process are supplied with little storage or warehousing time and cost. This system reduces idle capital and allows minimal investment so that capital can be used elsewhere.

Thus, many firms in the late twentieth century engaged in significant downsizing, ridding themselves of whole divisions of their companies to focus on their core competencies. Numerous middle-class jobs, such as middle managers, were eliminated in the process, contributing to the growing crisis of the American worker. Many companies reversed their old principles of hierarchical, bureaucratic assembly-line (Fordist) systems as they switched to customized, flexible, consumer-focused processes that can deliver personal service to niche markets at lower costs and faster speeds.

In this process, the use of subcontracting accelerated rapidly. Firms always face a "make-or-buy" decision (i.e., a choice of whether to purchase inputs such as semifinished parts from another firm or to produce those goods themselves). Under the relatively stable system of Fordism, most firms produced their own parts (i.e., decided to make rather than buy), justifying the cost with economies of scale, which lowered their long-run average cost curves. Large firms, for example, would have their own parts producers, trucks, or printing shops. Under post-Fordism, however, this strategy was no longer optimal: Given the uncertainty generated by the rapid technological and political changes, many firms opted to buy rather than make (i.e., to purchase inputs from specialized companies). This strategy reduces risk for the buyer by pushing it onto the subcontractor, who must invest in the capital and hire the necessary labor. Thus, the number of interindustry linkages grew rapidly as the economy shifted into a post-Fordist regime of production, a process that changed the locational requirements of many firms. Transport costs and proximity to suppliers and clients grew accordingly, raising the significance of agglomeration economies in many sectors.

A key to production flexibility lies in the use of information technologies in machines and operations, which permits more sophisticated control over the process. With the increasing sophistication of automated processes and, especially, the flexibility of the new, electronically controlled technology, far-reaching changes in the process need not be associated with an increased scale of production. Indeed, one of the major results of the electronic computer-aided production technology is that it permits rapid switching from one process to another and allows the tailoring of production to the requirements of individual customers. Traditional automation is geared to high-volume standardized production; the newer flexible manufacturing systems are quite different.

As interfirm linkages grew rapidly in the late twentieth century, many firms found themselves compelled to enter into cooperative agreements, such as strategic alliances, with one another. Quality control (i.e., minimizing defect

FIGURE 31 Japan utilizes one-half of all industrial robots in the world, an example of the increasing capital intensity of the production process, particularly in manufacturing.

rates) became very important. Many firms succeed in this environment by entering into dense urban networks of interactions, including many face-to-face linkages and interactions based on noneconomic factors—tacit knowledge, learning, reflexivity, conventions, expectations, trust, uncertainty, and reputation. Post-Fordism thus highlighted the culturally embedded nature of economic linkages.

Post-Fordist approaches to production, although varying from region to region, came to dominate much of the electronics and automobile industries, the minimills in the steel industry, and the new manufacturing spaces of California's Silicon Valley, Italy's Emilia-Romagna region, Germany's Baden-Wurtenburg, the Danish Jutland, and the British electronics region centered on Cambridge.

Summary

This chapter offered an overview of the changing empirical patterns of manufacturing and the reasons that underlay them and emphasized that manufacturing geographies are always fluid, changing over time. Four major regions of the world account for approximately 80% of the world's industrial production—northeastern North America, northwestern Europe, western Russia, and East Asia. The textile, clothing manufacture, and consumer electronics industries have steadily shifted from the developed world to the developing world. Automobile production and semiconductor manufacture have also experienced global shifts, from North America and Europe to Japan and mainland East Asia.

We explored the social relations that lead to industrial change, described worldwide manufacturing trends, and examined the recent history of industrial devolution in the developed world and of industrial revolution in the developing world. The processes of deindustrialization and industrialization are not temporary tendencies within the global system; rather, they constitute a zero-sum global game played by multinational nomadic capital. Multinational firms switched production from place to place because of varying relations between capital and labor and technological innovations in transportation and communications. With improved air freight, containerization, and telecommunications, multinational corporations can dispatch products faster, cheaper, and with fewer losses.

In the United States, millions of manufacturing jobs were lost between 1978 and 2008. To some extent, these losses were hidden by selective reindustrialization and the migration of manufacturing from the Manufacturing Belt to the South and the West, although even portions of the Sunbelt, such as Los Angeles, are steadily deindustrializing. Job loss at this scale entails enormous costs for local communities, many of which are plunged into permanent states of high unemployment, depressing family incomes, limiting purchasing power, and contributing to national economic decline.

The latter part of this chapter explored the ways in which manufacturing landscapes change over time. Capitalism is constantly reinventing itself through the process of "creative destruction," and one of its prime strengths is the capacity to adjust to rapid change. In the late twentieth century, as the Fordist regime of production began to collapse, the petrocrises, deindustrialization, the rise of the NICs, and the microelectronics revolution spawned the emergence of post-Fordist, flexible production techniques and organizational forms. Mounting international competition called for a restructuring of the production process in order to restore the conditions of profitability, and many firms downsized, subcontracted many functions, and adopted technologies like just-in-time inventory systems.

Key Terms

agglomeration economies
deindustrialization
flexible production
Fordism
industrial restructuring
international division of labor
just-in-time inventory systems
North American Manufacturing Belt
offshore assembly
oligopoly
post-Fordism
transnational corporation (TNC)
value added by manufacturing
vertical integration

Study Questions

1. What are the forces of production and the social relations of production?
2. Why are capital–labor relations a necessary starting point for studying economic geography?
3. What are four major world regions of manufacturing?
4. Summarize the historical development of the North American Manufacturing Belt.
5. When and why did the Manufacturing Belt begin to lose industries? Where, specifically, did they go, and why?
6. What are vertical integration and disintegration?
7. What is flexible production or post-Fordism?

Suggested Readings

Bluestone, B., and B. Harrison. 1982. *The Deindustrialization of America: Plant Closings, Community Abandonment, and the Dismantling of Basic Industry*. New York: Basic Books.

Dicken, P. 2010. *Global Shift: Mapping the Changing Contours of the Global Economy*. 6th ed. London: Guilford.

Knox, P., J. Agnew, and L. McCarthy. 2007. *The Geography of the World Economy*. 5th ed. London: Edward Arnold.

Linkon, S., and J. Russo. 2002. *Steel-Town U.S.A.: Work & Memory in Youngstown*. Lawrence: University Press of Kansas.

Web Resources

International Monetary Fund
http://www.imf.org/external/pubs/ft/issues10/

One perspective on the causes of global deindustrialization.

Econolink
http://www.progress.org/econolink

Econolink has selective descriptions of economic issues; many sites themselves have Web links.

Post-Fordism
http://www.generation-online.org/c/fcimmateriallabour.htm

Summary of the labor process under post-Fordism.

Services

OBJECTIVES

- To illustrate the difficulties in defining and measuring services
- To assess the diversity of services, including the range of industries and occupations
- To explore the reasons for the growth of services
- To describe the world of labor in services
- To provide case studies of financial and several producer services sectors
- To examine the globalization of services
- To sketch the nature of consumer services and tourism

Office workers exemplify the massive structural shift from manufacturing to knowledge-based economies and personify the growth of producer services, often the largest and most rapidly growing segments of postindustrial countries.

Services

Broadly defined, the tertiary sector of the economy consists of all those sectors engaged in the provision of services of various sorts, intangibles that include retailing, banking, real estate, finance, law, education, and government. "Services" encompass an enormous diversity of occupations and industries, ranging from professors to plumbers to prostitutes. Indeed, so great is the variation among firms and occupations within services that the term threatens to lose any coherence whatsoever. For this reason, it is simplistic to speak of "*the*" service sector because the phrase masks the enormous diversity among and within different service industries.

The traditional postindustrial perspective on services saw them as purely information-processing activities, including clerical work, executive and management decision making, legal services, telecommunications, and the media. Such a view regarded information processing as a qualitatively new form of economic activity; thus services were held to represent a historically new form of capitalism. Unfortunately, this view is mistaken. While many service jobs do involve the collection, processing, and transmission of large quantities of data, clearly others do not: The trash collector, restaurant chef, security guard, and janitor all work in the service sector, but the degree to which these activities center around information processing is minimal.

More recent perspectives stress services as another form of capitalist commodity production, involving the same dynamics and constraints of location, production, and consumption as other industries (i.e., manufacturing). In this light, services may be seen as an extension of market relations into new domains of output and activity. Such a view does not deny that services may indeed possess a logic somewhat different from that exhibited by manufacturing; however, it stresses the embeddedness of services within the broader social contours and relations of capitalism generally.

Throughout most of the economically developed world, various forms of services have replaced manufacturing employment. In large part, this trend reflects the persistent tendency of the international division of labor to shift manufacturing activities from economically developed countries to low-cost developing ones. The percent of national labor forces employed in various types of services varies considerably around the world (Figure 1). Generally, the more economically advanced countries have the highest proportions of service workers; in the United States, more than 80% of people work in these activities, whereas in poorer countries, notably Africa, where most people work in agriculture, the share of service workers is relatively low.

This chapter begins by pointing out problems in the definition of services and lays out some of the ways in which various service industries differ from one another. Next, it turns to the forces that underlay the growth of the service economy and then summarizes the reasons why many firms externalize services, or purchase them from subcontractors rather than produce services themselves. The chapter addresses the nature of labor markets in services, contrasting them with manufacturing and delves into the locational dynamics of services. Case studies of some key sectors are included, including finance and producer services such as accounting, design, and legal services. Finally, we explore the means by which services are traded among cities, regions, and countries, focusing on telecommunications and its impacts on electronic funds transfer, offshore banking, and the global back office.

DEFINING SERVICES

Services may be broadly understood as the production and consumption of intangible inputs and outputs and thus services stand in contrast to manufacturing, whose product can be "dropped on one's foot." What, for example, is the output of a lawyer? A teacher? A doctor? It is impossible to measure these outputs accurately and quantitatively, yet they are real nonetheless. The U.S. federal government estimates services output using revenues as a proxy;

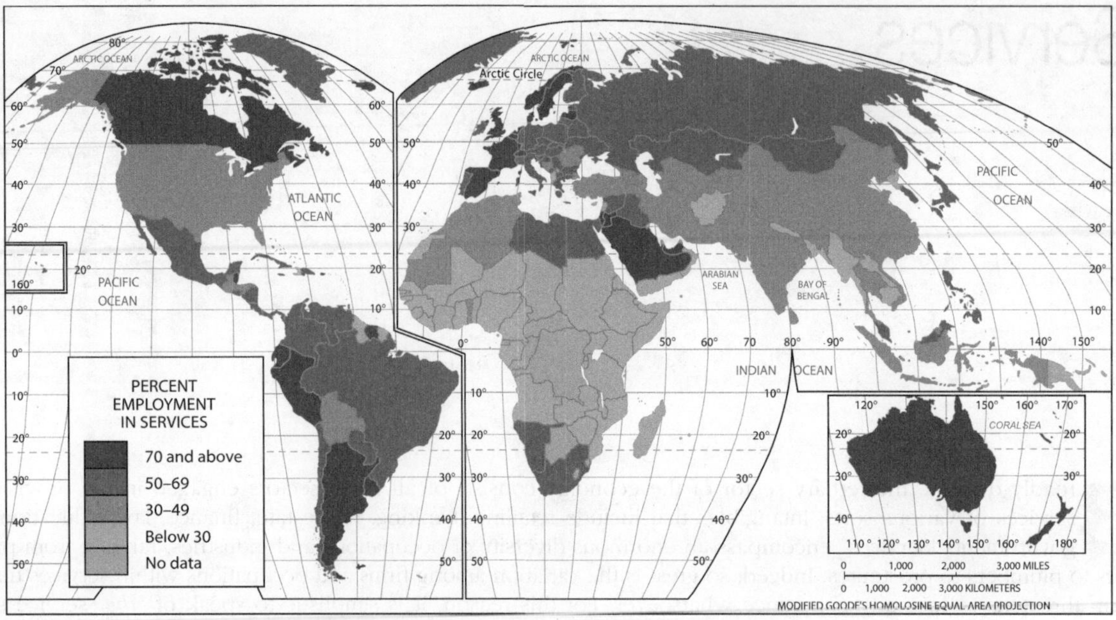

FIGURE 1 Proportion of workers in service industries by country. Note that developed countries have much higher proportions of service workers than less developed ones. (See the color insert for a more illustrative map.)

yet revenues are determined by output and prices, so this measure is distorted by changes in the relative prices of service outputs. To complicate matters, many services generate both tangible and intangible outputs. Consider a fast-food franchise: The output is assuredly tangible, yet it is generally considered a service; the same is true for a computer software firm, in which the output is stored on disks.

The fact that it is difficult, if not impossible, to measure output in services has enormous implications for analysis of them. For example, some critics of the service sector argue that the slowdown in U.S. productivity growth in the late twentieth century reflected the growth of services. Yet, if output in services cannot be adequately measured, how can one argue that output per employee in services is high or low, rising or falling? If wages reflect productivity increases, what is the relation between services output and salaries? Clearly, the lines between services and manufacturing are blurry and no simple definition will suffice.

Figure 2 shows the major categories commonly used to depict the changing nature of the U.S. employment structure over time. There is a broad consensus as to the major components of the service sector, including the following:

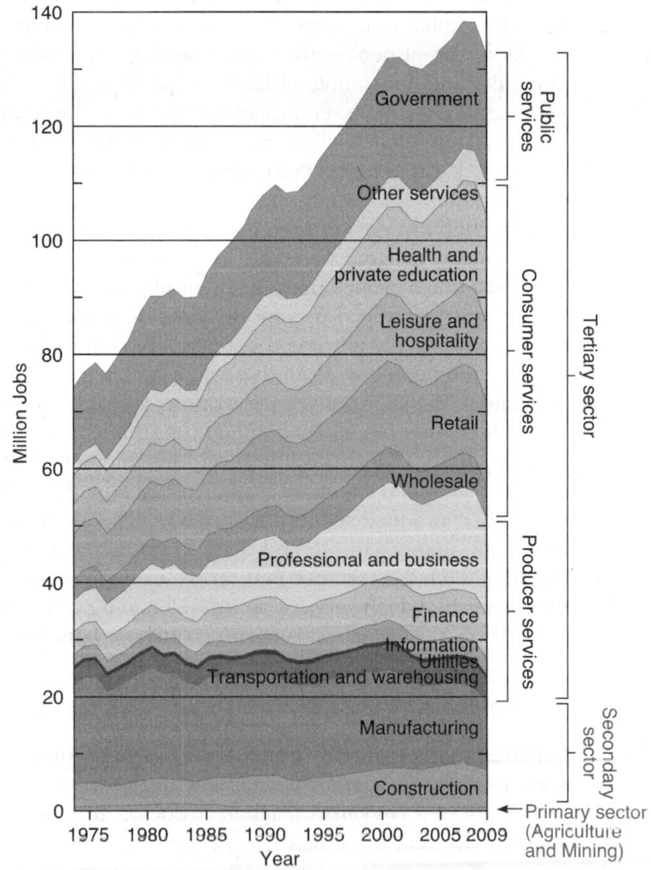

FIGURE 2 Employment change by economic sector in the United States, 1975–2009. Since 1975, the greatest employment gains have come in professional services, personal services, and social services. Manufacturing has remained relatively constant in absolute terms but declined as a proportion of the total, while agriculture and mining employment has declined in both relative and absolute terms.

a. Producer services are those primarily sold to and consumed by corporations rather than households. Many producer services also serve final demand, such as attorneys that cater to both commercial clients and individuals. Producer services are commonly divided into two major groups:
The **finance, insurance, and real estate (FIRE)** sector includes commercial and investment banking, insurance of all types (e.g., property, health, life, casualty), and the commercial and residential real estate industry.
Business services subsume legal services, advertising, computer services, engineering and architecture, public relations, accounting, research and development, and consulting.

b. Transportation and communications services include the electronic media, trucking, shipping, railroads, airlines, and local transportation (taxis, buses, etc.).

c. Wholesale and retail trade services firms are the intermediaries between producers and consumers.

d. Consumer services are those such as eating and drinking establishments, personal services, and repair and maintenance services, all of which have locational requirements closely associated with local demographics and transport structures. Entertainment, hotels, and motels comprise elements of tourism, the world's largest industry in terms of employment.

e. Government services at the national, state, and local levels include public servants, the armed forces, and all those involved in the provision of public services (e.g., public education, health care, police, fire departments, etc).

f. Nonprofit services include charities, churches, museums, and private, nonprofit health care agencies, many of which play influential roles in local economies.

Every definition of *services* is slippery, however. For example, does the term *services* refer to a set of industries or occupations? (The U.S. government uses industries.) Yet when measured on the basis of their daily activities, many workers in manufacturing are in fact service workers, including, for example, personnel in headquarters, administration, clerical functions, janitorial work, and research. Is the secretary who works for an automobile company part of manufacturing while the secretary who works for a bank part of services? The use of industrial versus occupational definitions is particularly critical given the growth of many **"nondirect" production workers** within many manufacturing firms, such as clerical, administrative, research, advertising, and maintenance workers. Clearly, different definitions of services have significantly varying implications for assessments of the size and composition of the service sector.

Using the standard definition of **intangible output**, services comprise the vast bulk of output and employment in most economically developed countries of the world. Indeed, more than 80% of the U.S. labor force is employed in services (Table 1); similar proportions hold in Europe, Canada, and Japan. Even as early as 1910, services exceeded manufacturing in the United States, indicating it was a **"postindustrial" economy** before becoming an industrial one! Further, services comprise the vast majority—often 90%—of all new jobs generated in these economies, indicating that they are not only predominantly service-oriented

TABLE 1 Composition of U.S. Labor Force (millions), 1950–2008

	1950	1960	1970	1980	1990	2000	2008
Agriculture	5.0	4.8	3.9	3.6	3.5	3.3	2.2
Mining	0.9	0.7	0.6	1.0	0.7	0.5	0.8
Construction	2.1	2.9	3.6	4.3	5.1	6.6	11.0
Manufacturing	16.4	16.8	19.4	20.3	19.1	18.5	15.9
TCPU[a]	3.9	4.0	4.5	5.2	5.8	7.0	7.7
Wholesale trade	2.6	3.1	4.0	5.3	6.2	7.0	4.1
Retail trade	6.6	8.2	11.0	15.1	19.6	23.3	16.5
FIRE[b]	1.8	2.6	3.6	5.1	6.7	7.6	10.2
Other services	5.3	7.3	11.4	17.5	27.3	39.9	53.9
Personal services		0.8	0.9	0.8	1.1	1.2	1.4
Business services		0.6	1.4	2.5	5.1	9.7	15.5
Health services		1.5	3.0	5.1	7.6	10.0	14.1
Education		0.6	0.9	1.1	1.6	2.3	2.4
Government	5.8	8.2	12.3	16.1	18.0	20.4	23.1
TOTAL	50.4	62.1	80.5	103.0	127.4	157.3	178.8

[a]Transportation, communications, and public utilities.
[b]Finance, insurance, and real estate.
Source: U.S. Bureau of Labor Statistics.

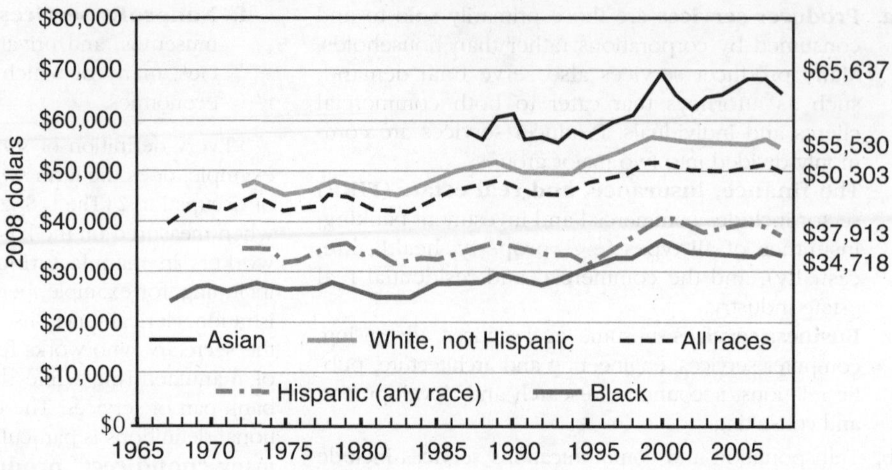

FIGURE 3 U.S. median household income, 1967–2008. Although productivity gains have kept prices low, the U.S. labor market has been besieged by deindustrialization and corporate restructuring; consequently, real income growth has been meager and mostly concentrated in the upper echelons. Asian American households have the highest median income, followed by whites, Hispanics, and African Americans.

but are becoming increasingly more so. Even in much of the developing world the services labor force comprises a large share of the labor force, including much of the "informal" (untaxed, unregulated) economy; this fact belies earlier, simplistic assertions that held that all economies inevitably move through a rigid series of stages (i.e., agricultural to industrial to postindustrial).

FORCES DRIVING THE GROWTH OF SERVICES

Why have services grown so rapidly? In economically developed countries, services employment has increased steadily even in the face of low rates of population growth, slowly rising rates of productivity and incomes, and significant manufacturing job loss. Services and manufacturing are intimately intertwined, but it is equally apparent that services exhibit growth and locational dynamics somewhat different from those of manufacturing, although both constitute commodity production in varying forms. Six reasons for the increase in services employment throughout the world are described next.

Rising Incomes

A standard explanation for the growth of services centers on gradually rising per capita incomes, particularly in the industrialized world (Figure 3). The demand for many services is relatively **income-elastic**, that is, increases in real personal income tend to generate proportionately larger increases in the demand, in contrast to the demand for most manufactured goods (Figure 4). Services with particularly high income elasticities include entertainment, health care, and transportation. Households in the United States spend more on services than they do on durable and nondurable goods combined (Figure 5).

An important reason contributing to this growth is the increasing value of time that accompanies rising incomes (especially with two income earners per family). As the value of time (measured by income) climbs relative to other commodities, consumers generally will attempt to minimize the time inputs needed for the accomplishments of many ordinary tasks. While this phenomenon also explains the demand for dishwashers and automobiles, it is especially important for the growth of services. The explosion of fast-food

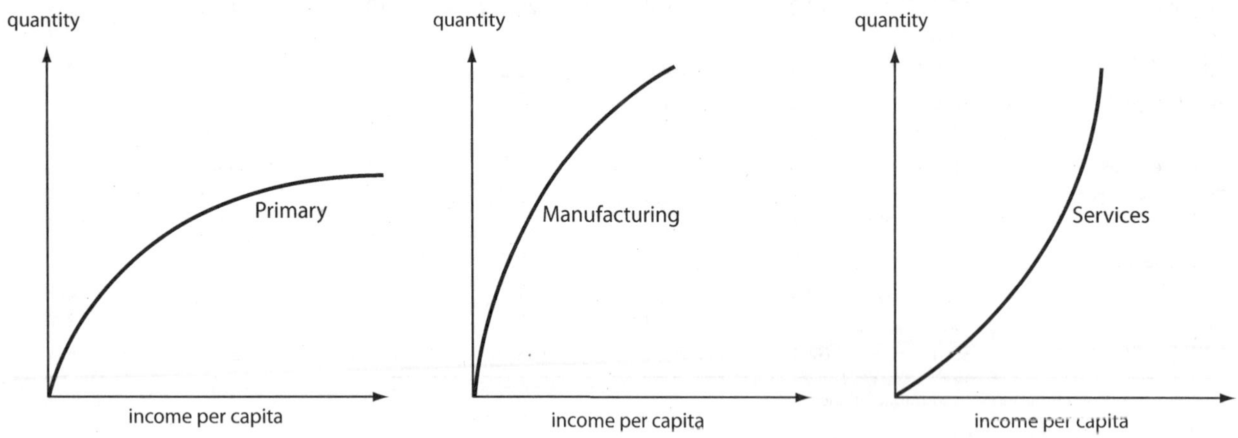

FIGURE 4 Income elasticities of demand for services are high compared to those for agricultural or manufactured goods.

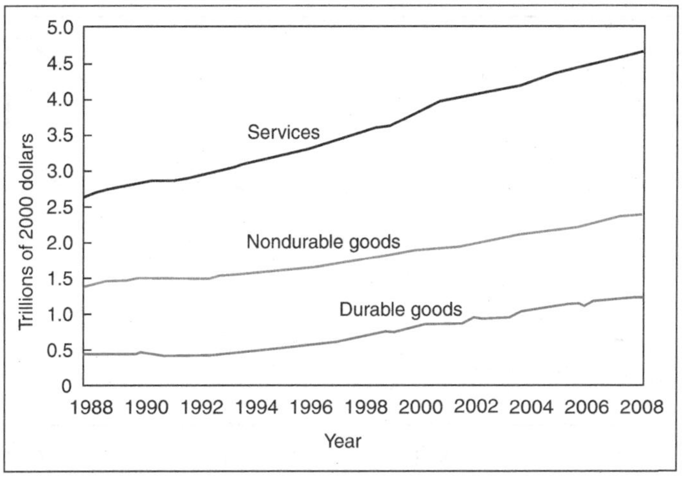

FIGURE 5 Personal spending on goods and services in the United States, 1988–2008. The income elasticity of most services compared to goods have led to their becoming more than half of the average household's expenditures, particularly for education, transportation, health care, and entertainment. As a country ages and develops, there is a natural economic transition as well as a demographic transition—from major spending on goods, like food, clothing, houses, refrigerators, cars, washers, and televisions, to major spending on services.

restaurants, for example, has little to do with the quality of food (or even the price) and much to do with attempts by consumers to minimize time spent cooking at home. Similarly, the growth of repair services reflects both increasingly sophisticated technologies (e.g., in automobiles or televisions) and a generalized unwillingness to spend limited recreation time doing such chores. Thus, the increasing value of time has led to a progressive externalization of household functions, so that which used to be done in-house becomes a for-profit commodity purchased through the market.

Demand for Health Care and Education

Rising levels of demand for health and educational services comprise an important part of the broader growth of the service economy. Health services employment and output have increased steadily throughout Europe, North America, and Japan (Figure 6), often leading to political conflicts about how to contain them, such as the debates in America about the rising costs of Medicaid and Medicare. In the United States, health care expenditures surpassed \$2.3 trillion in 2008 (Figure 7), roughly 17% of the gross domestic product and much higher than any other industrialized country (Figure 8). The only developed country to lack a national health insurance ("single-payer") system, the United States also has the most inefficient system, spending far more per capita on health care than do countries with single-payer systems (Figure 9) and devoting a larger share of its national output to health care than Europe or Japan. In many cities, health care is the largest and most important part of the local economy, including, for example, Pittsburgh, Pennsylvania, formerly the steel capital of the world, and Birmingham, Alabama.

The demand for health care has increased steadily in large part because of the changing demographic composition of industrialized countries. Throughout the developed world, the most rapidly growing age groups today are the

FIGURE 6 As populations age, health care has become a major part of the economies of economically developed countries. As health related expenses comprise a growing share of gross domestic product, debates and strategies to contain health-related costs have arisen as well. Medicare, Medicaid, Social Security, and guaranteed health care programs are under fire today in most of the developed world because of runaway costs. Putting in place reforms that slow down the rise in pension and health care spending is a priority, since the level of future governments' entitlements to the elderly dwarf today's debts.

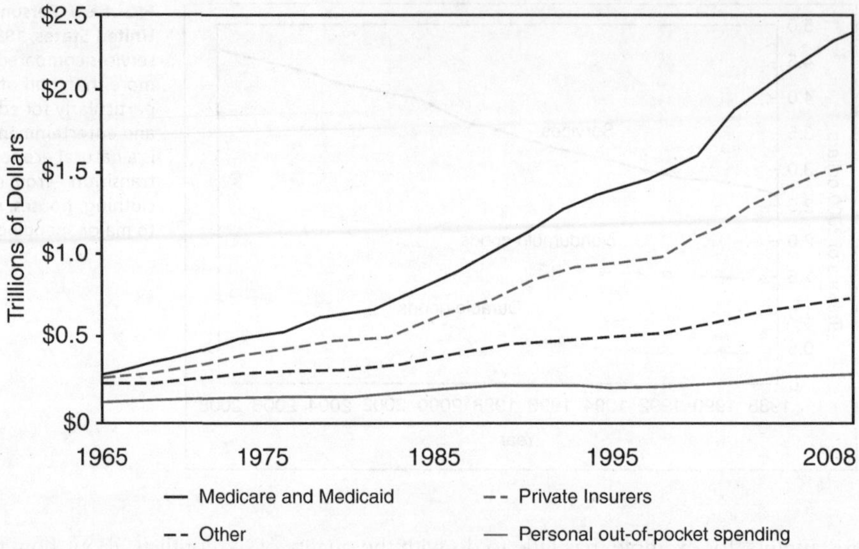

FIGURE 7 U.S. national health care expenditures, 1965–2008. One of the costs of an aging, rich population is runaway health care cost. But the age of austerity has already arrived in Europe and is about to land in America. Governments throughout Europe have already nationalized health care to keep costs down, and are now cutting health and social spending, and raising taxes, causing riots across Europe. Even rich countries find the runaway costs of health care too high, and the United States is now considering austerity measures like raising retirement age and raising taxes, although there is no movement to nationalize health care itself, like in many European countries.

middle-aged and the elderly, precisely those demographic segments that require relatively high per capita levels of medical care. As the baby boom enters its retirement years, the demand for health services will rise even higher. Higher life expectancies and soaring equipment, medical insurance and litigation, and pharmaceutical and research expenses have added to the costs of this sector.

Similarly, globalization, increasing technological sophistication, and increasing demand for more analytical skills (particularly numeracy and computer skills, as shown

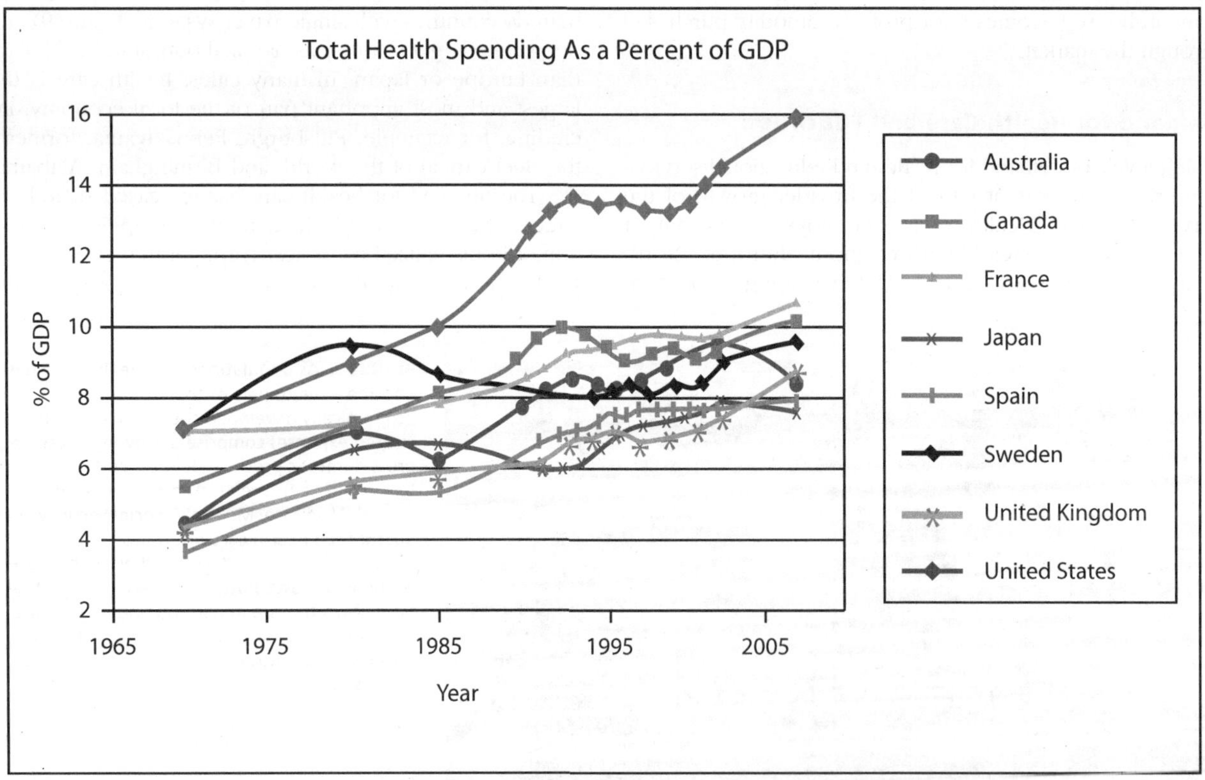

FIGURE 8 Percent of gross domestic product (GDP) dedicated to health care expenditures, 1970–2005, for major industrial countries. As the population has aged and the cost of health services has risen, this sector consumes one-seventh of all dollars spent in the United States. Although the United States has the least equitable system of health care access in the industrialized world—15% of adults have no health care insurance—its system is not the most efficient; European countries spend, on average, less on health care as a proportion of their total output.

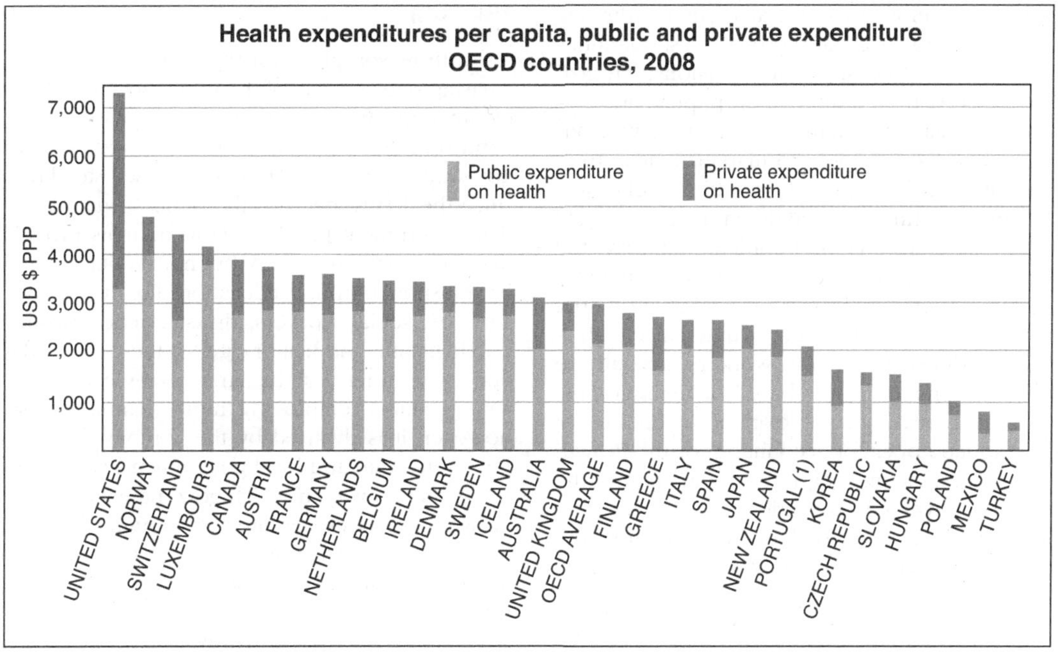

FIGURE 9 **Per capita health care expenditures, Organization of Economic Cooperation and Development (OECD) countries, 2008.**

in Figure 10) in the workplace have driven the increasing demand for educational services at all levels, a process reflected in higher enrollments in universities. A college education has become a prerequisite to middle-class jobs; thus, whereas the bulk of graduating high school students in the United States did not attend college in the 1960s, today almost 70% do so.

An Increasingly Complex Division of Labor

The growth of the service sector reflects the increasing complexity of the division of labor. A rising proportion of nondirect production workers are employed by firms in the manufacturing sector. All corporations today devote considerable resources to dealing with a complex marketplace and legal environment, including many specialized clients, complex tax codes, environmental and labor restrictions, international competition, sophisticated financial markets, and real estate purchases and sales. Deregulation—the lifting of state controls in many industries—increased the uncertainty faced by many firms and had significant impacts on the profitability, industrial organization, and spatial structure of numerous sectors.

FIGURE 10 The microelectronics revolution made computers cheap, powerful, and widely used in virtually all forms of production, including services, with powerful impacts on productivity and labor markets. First it was mainframes, then PCs, then laptops, and then the Internet innovation of networked computers and unlimited amounts of data and information. This has been followed by microelectronic revolutions in cell phone technology and broadband services, allowing for even mainstream users to leap into user-friendliness, affordability, and productivity increases, while inflation and interest rates fall, all the while making new friends on Facebook.

To negotiate this environment, firms require administrative bureaucracies to collect and process vast quantities of information and make strategic decisions; clerical workers to assist with mountains of paperwork; researchers to study market demand and create new products; advertisers and salespeople to market their output; and legions of people engaged in public relations, accounting, law, and finance to assist in an enormously complicated decision-making environment. Similarly, the introduction of sophisticated machinery requires maintenance and repair personnel, and offices or industrial plants require security and building maintenance staff—these are all nondirect production workers, and all are services workers. Unlike income-based or demographic explanations, the increasing complexity of the division of labor has the added appeal of accounting for the growth in producer services, the most rapidly growing part of the economy of most developed countries.

In line with the division of labor explanation, the growth of services reflects not the development of a new economy, but rather the revolutionary force of capitalism as it has generated dramatic shifts in the division of labor in society. Thus, the **outsourcing** of service functions by clients and the creation of new types of service occupations represent an extension of the division of labor. An increasingly specialized division of labor reflects both increasing specialization of activity with a resultant increase in the complexity of production and alterations in the way in which production is organized. Thus, research and development, design, market research, trial production, product testing, marketing, and sales are all essential parts of getting goods from producer to consumer. The fact that they can be separated in both time and space from the actual production process does not necessarily imply that they are not an integral part of the manufacturing sector. This means that the dramatic growth in business service employment reflects alterations in the way in which manufacturing production is organized rather than the development of a new type of service or knowledge economy.

The Public Sector: Growth and Complexity

A fourth reason given for the growth of services is the increasing size and role of the public sector and concomitant expansions in government employment. Governments contribute to the growth of services in two ways. First, government employment has increased steadily, especially since the 1930s, because the public demands the services that government provides. Governments provide services that are socially necessary but not profitable, such as public transportation and education, libraries, police and fire services, social workers, disaster relief, postal services, public health, and health care for the poor. Today the federal government is the largest single employer in the United States, employing more than 2 million people; however, it is dwarfed by the combined employment in numerous state and local governments (Figure 11).

A second way in which government contributes to the growth of services is indirectly, through a labyrinthine web of laws, rules, restrictions, and regulations, such as tax and antidiscrimination laws, contributing to the increase in the number of attorneys, accountants, consultants, and other specialists who assist firms (externally or internally) to negotiate the legal and regulatory environment. The degree of regulation that the government imposes creates changing levels of uncertainty in the market, and many service functions allow firms to deal with this phenomenon.

Service Exports

A fifth reason for the growth of services is the rising levels of service exports within and among countries. A widespread myth is that services always cater to local demand (i.e., they are nonbasic activities and are thus of secondary importance to manufacturing). This notion reflects the long-standing bias against services in economic thought. In countries where the vast bulk of workers are engaged in the production of services, it would be astonishing if services were *not* traded among places. The economies of many cities, regions, and countries derive a substantial portion of their aggregate

FIGURE 11 Employment in the federal, state, and local governments in the United States, 1950–2006. The vast bulk of public sector employment is at the local level.

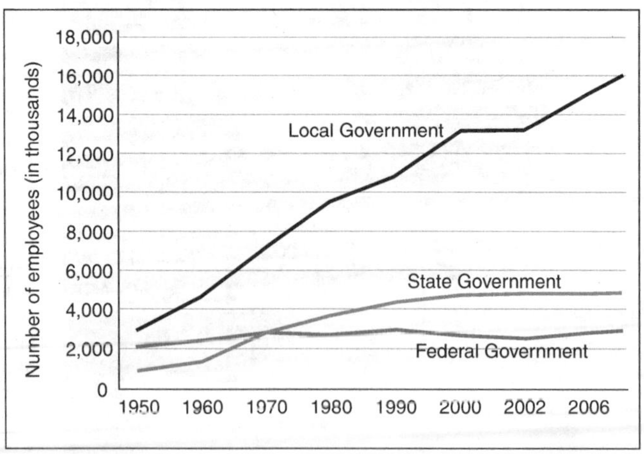

revenues from the sale of services to clients located elsewhere. Many urban areas export services to clients located in other parts of the same nation. Consider, for example, cities such as Las Vegas, which revolve exclusively around exports of tourism and entertainment services to visitors. Whenever a television company in one city sells advertising time to a client in another, that city exports services. New York City depends heavily on exports of financial, legal, and advertising services domestically as well as globally, and Washington, DC, exports government services to the rest of the country. "Export" in this context refers to the location of those who pays the bills: If the client is located elsewhere, then sales of a service are exported and form part of the local basic sector.

Services are also traded on a global basis and comprise roughly 20% of international trade. Internationally, the United States is a net exporter of services, with an annual trade surplus of roughly $144 billion in 2008, but runs major trade deficits of more than $840 billion in manufactured goods (Figure 12). Exports of services are one reason services employment has expanded domestically. Indeed, it could be said that as the United States has lost much of its comparative advantage in manufacturing, it has gained a new one in financial and business services. The data on global services trade are poor, but some estimates are that services comprise roughly one-third of total U.S. export revenues. These sales overseas take many forms, including tourism, fees and royalties, business services, and profits from bank loans. Unfortunately, the U.S. data in the Current Account (the government's system for collecting data on trade in goods and services) include repatriated profits from overseas manufacturing investments as a "service," while in reality these comprise returns from capital investments abroad, many of which are in manufacturing. Service exports do not generate jobs and revenues in all places equally: Cities such as New York, Los Angeles, and London, for example, which are critical to global capital markets, have benefited the most heavily.

The Externalization Debate

Yet another explanation for the growth of services lies in the **externalization** of many functions as large, previously vertically integrated firms downsized in response to mounting international competition. As a result, the amount and degree of subcontracting grew explosively. The externalization thesis holds that large firms used this strategy to minimize costs during economic downturns and periods of rapid restructuring. The alternative is that new business practices led to an increased need for external expertise not met in-house. In the face of strategic downsizing, cost-cutting, and the desire to be "lean and mean," many large companies no longer had sufficient staff to internalize producer service functions. The result was the creation of large numbers of new firms.

Firms turn to the market for services in cases where it is less expensive to acquire them externally than internally (i.e., when it is more efficient to buy rather than to "make" them). For example, companies may subcontract with legal, repair, trucking, maintenance, or advertising firms, or management consultants and software engineers. The decision to use external providers of producer services may be based on the lower costs of an outside supplier, or it may be based on a perception that external services are of higher quality if subcontractors are highly specialized in a given area. Firms also employ external producer services when they require types and quantities of expertise, information, and knowledge that are very different from their own. It makes limited sense to employ full-time staff if the demand is seasonal, unstable, or temporary in nature.

By employing external providers of expertise, a firm is not exposed to the risks associated with full-time employees. It does not have to pay Social Security costs, provide training, and invest in buildings to house these functions. It also means that during downturns in the economy, the firm can reduce its use of external expertise by not renewing contracts. Full-time workers are relatively difficult to remove from the firm; subcontracting spreads risks down the production chain.

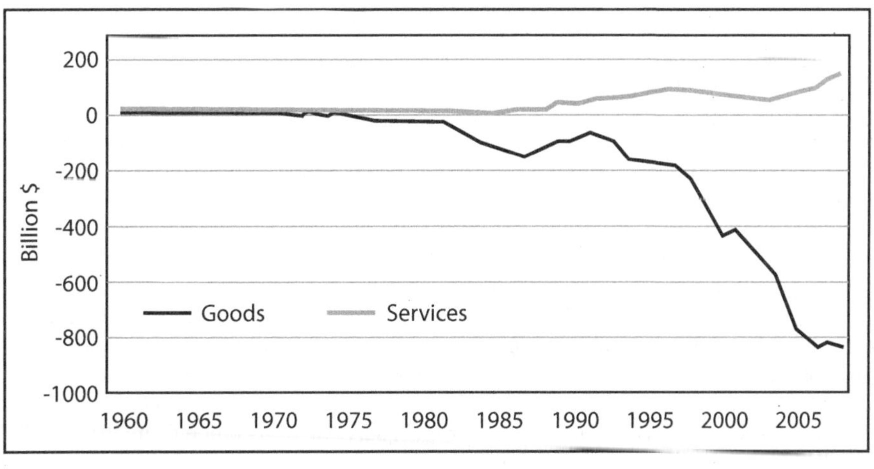

FIGURE 12 U.S. trade balance in goods and services, 1960–2008. Although the United States runs large trade deficits in manufactured goods, it usually has a small surplus in services. In 1920 employment in the primary, secondary, and tertiary sectors were each approximately 33% of the work force. Since then, the primary sector has plummeted to 3% and the secondary sector to less than 15% of the total workforce in America. The United States is a postindustrial society, whereby over 80% of the jobs are in services. Developing countries have taken over production of most manufactured goods in the world due to low factory wages and less access to the information economy. Two notable exceptions for the United States is the production and export of modern commercial aircraft, such as Boeing wide-bodied jets, and military equipment.

The growth of the service economy has been marked by the emergence of all sorts of niche markets, including those in which the supply of new skills and expertise is often very limited (for example, Web page designers in the early days of the Internet). As soon as the supply of expertise develops, purchasing companies may develop their own internal supply. In the meantime, external purchases are their only option. The externalization process created many producer service firms that primarily offer expertise, reflecting the growing technological and administrative complexity of the division of labor. As a result, many functions formerly performed in-house by manufacturing firms have been subcontracted to small suppliers, some of which are in the service sector. Subcontracting and externalization are often studied in the context of manufacturing firms but have also become increasingly common among large service firms, which face similar competitive pressures to reduce costs.

Firms face a choice as to when to externalize, and how much. This choice reflects both the external environment of the firm and the internal role of its key decision makers. Externalization is most likely when:

a. The firm faces severe in-house technical limitations (i.e., expertise)

b. The firm is an independently owned entity, not a branch plant; vertically integrated firms tend to provide more services in-house

c. The firm is sophisticated relative to its competitors (e.g., technologically, or operates in foreign markets)

d. Service inputs are diverse, shifting, and nonstandardized (e.g., installation, repair, or consulting)

Firms also externalize services when they face rising uncertainty—rapid change in products or technology—when the labor process resists easy automation, or when the optimal scales of operation of production processes are markedly different.

Essentially, externalization allows for external economies of scale to replace internal economies of scope. By externalizing, firms substitute variable costs for fixed ones and spread the risks of production over their subcontractors, a particularly vital concern during peak periods of demand and mounting levels of uncertainty. Among producer services that firms are likely to externalize, insurance ranks as the most common, followed by accounting, advertising, research, management consulting, advertising and public relations, engineering/architecture, market research, management/professional recruitment, computer installation/repair, commercial law, real estate, and temporary office help. Such services are ones that are not efficient or cost-effective to duplicate in-house and are only needed once or infrequently, on an unpredictable basis (e.g., troubleshooting). Typically, such services augment or substitute for the internal resources of firms; when the demand is regularized and predictable, they are likely to be internalized within the firm. Further, the degree of externalization depends on whether the firm is a single establishment or a branch plant, as the latter has the lowest degree of local externalization.

Statistically, the apparent growth of services has been inflated by externalization due to the limitations of the parameters used to measure economic activity. For example, a steel company may lay off a janitor during a period of contraction but then subcontract with a janitorial service; similarly, a ball bearing company may lay off a secretary and contract out to a clerical services company. Nothing has substantively changed in the process—the same jobs are performed—but the manufacturing sector in which the janitor or secretary used to work has registered a decrease in employment while the service sector has registered an increase. Because employment statistics rely on definitions based on industries and not occupations, this results in a reduction of manufacturing employment and an increase in services employment. In terms of services employment, therefore, the data are something of a statistical mirage; were definitions of services based on occupations and not industries, this circumstance would cease to be an explanation of services growth.

Nonetheless, given the structure of national employment accounting systems, contracting out represents one reason why services employment has increased, but the process serves as a reminder that the boundaries between services and manufacturing employment are often blurry. Some have argued that the vertical disintegration and externalization thesis has been exaggerated. One line of thought is that the demand for services is satisfied through the growth of many small firms that satisfy new needs, often consisting of a single individual filling a highly specialized niche providing customized services, with low overheads and few economies of scale.

LABOR MARKETS IN THE SERVICE ECONOMY

In the economically developed world, the vast majority—often more than 80%—of all jobs involve services of one form or another. Further, new jobs—on the order of 90%—are overwhelmingly concentrated in services. Of course, the production of intangibles includes an enormously diverse array of industries and occupations, but despite these variations, the diverse labor markets in which most workers find employment demonstrate several common characteristics.

Characteristics of Services Labor Markets

Services employment exhibits a number of properties that simultaneously resemble and differ from those in manufacturing. Among these are its relatively labor-intensive nature, income distribution, gender composition, relative lack of unionization, and educational requirements.

LABOR INTENSITY In contrast to labor markets dominated by manufacturing firms, services industries tend to be relatively labor-intensive, that is, they use relatively more labor per unit of output. Accordingly, the costs of wages and salaries for most services firms range from 70% to 90% of their total costs compared to 5% to 40% in most

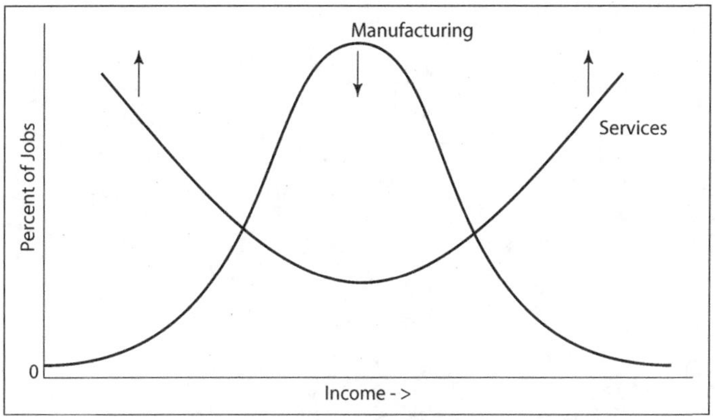

FIGURE 13 Depiction of the argument that manufacturing generates a middle-class set of incomes, whereas services incomes are polarized between the relatively high and low. Viewed in this way, deindustrialization and the shift to a service economy are responsible for the mounting income inequality in the United States.

manufacturing firms, depending on the degree of labor intensity or capital intensity. Of course, some services, such as finance, can be quite capital-intensive and generate huge outputs with minimal workers; others, however, such as education and medical care, require large numbers of employees. Whereas manufacturing output has been relatively easy to mechanize—witness the remarkable changes of the Industrial Revolution—in many services it is much more difficult or costly to replace workers with machines, particularly if the services involve variations in tasks, judgment, or dexterity. Obviously, some services sectors have seen enormous technological change—for example, personal computers in the office or automatic scanners in retail stores. For poorly paying jobs, however, firms' incentives to replace workers with machines may be relatively low.

INCOME DISTRIBUTION The distribution of incomes in services occupations has been a major source of concern to many social observers. The standard argument is that industrial economies generated a distribution of income that was relatively "normal" (in the statistical sense) (i.e., manufacturing created societies with a large middle class and relatively few rich or poor [Figure 13]). In contrast, services are frequently viewed as exhibiting a bifurcated wage distribution polarized between well-paying, white-collar managerial/professional jobs, on the one hand, which require a university education, and on the other, unskilled, low-paying jobs that require little to no higher education (e.g., retail trade, many medical services, security guards, etc.). Indeed, in contrast to early, optimistic postindustrial expectations that a service-based economy would eliminate poverty, a large share of new service jobs pay poorly, offer few benefits, and are part-time or temporary in duration, leading to widespread concerns about the "McDonaldization" or "K-Martization" of the economy. The most rapidly growing occupational groups in the United States, including over the next decade (Table 2), include professionals but also low-wage service workers and retail trade employees. The wage gap between skilled and unskilled workers contributed heavily to the rapidly growing income inequality in the United States (Figure 14).

The occupations with the greatest relative projected job growth are those having almost anything to do with computers as well as health care–related positions such as home personal care aides and medical assistants. In absolute terms, the greatest number of new opportunities will be in low-wage, unskilled positions such as food preparation, customer service, retail sales, cashiers, clerks, and security guards. Further, the United States has witnessed a steady growth in **contingent labor** (i.e., involuntarily part-time jobs typically filled by women and minorities), a process that has created a class of the "working poor" and helped to swell the ranks of the homeless caught between jobs that pay too little and housing that costs too much.

Concerns about the bifurcation of income in services are augmented by the fact that deindustrialization has eliminated many jobs in manufacturing, while services continue to grow. In general, services have tended to pay poorly compared to manufacturing: The average clerical position in the United States generates only 60% of the annual income of a blue-collar industrial worker, and the average retail trade job only 50% as much. Such fears are

TABLE 2 Projected Employment Change in U.S. Labor Force for Most Rapidly Growing Occupations, 2006 and 2016 (thousands)

	2006	2016	% Change
Home care aides	767	1156	50.7
Home health aides	787	1171	48.8
Customer service	2202	2747	24.8
Nurses	2505	3092	23.4
Teachers	1672	2054	22.8
Nurses aides	1447	1711	18.2
Food preparation	2503	2955	18.1
Janitors	2387	2732	14.5
Office clerks	3200	3604	12.6
Retail sales	4477	5034	12.4

Source: U.S. Bureau of Labor Statistics.

William Thomas Cain/Getty Images, Inc.

AP Wide World Photo

FIGURE 14 Retail trade is a major source of employment in economically developed countries and one of the largest forms of unskilled, low-wage services work. Wal-Mart, with 1 million American employees, is the largest private employer in the United States.

manifested in worries about the "declining middle class" and the polarization of postindustrial economies into distinct groups of "haves" and "have-nots." Statistically, however, there is little evidence to suggest that the distribution of incomes among services occupations differs significantly from that in manufacturing (Figure 15). Indeed, services generate the vast majority of employment that allows for millions of well-paid, middle-class suburbanites to live relatively comfortable lives. Others argue that the statistical evidence pointing to the mounting inequality in income distributions, particularly in the United States (Figure 16), reflects the increasingly regressive nature of taxation in the United States (where the affluent escape their share) as well as the growth of unearned income (e.g., rents, royalties, stock dividends), from which the rich derive the vast bulk of their earnings.

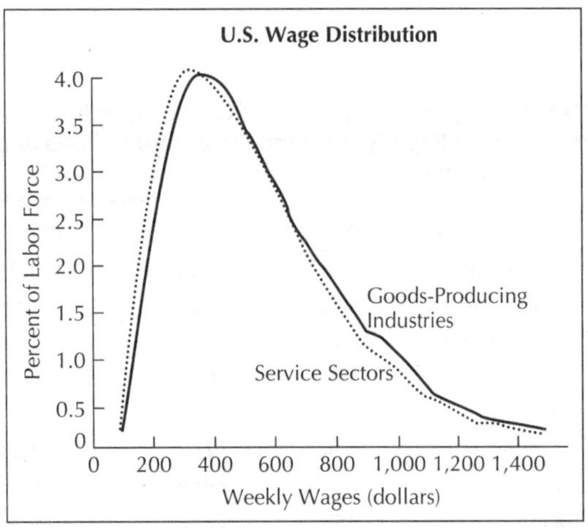

FIGURE 15 Distribution of wages in goods-producing and services industries in the United States. Contrary to the notion that services incomes are inherently polarized, in fact they closely resemble the distribution in manufacturing.

GENDER COMPOSITION A third widespread concern about the labor markets in services is gender composition. Manufacturing-based economies predominantly employed males; although some women worked in factories (particularly textiles and garment manufacturing), industrial economies characteristically saw relatively low rates of female labor force participation. Most women who worked outside the home did so either for brief periods prior to getting married or in a few specialized occupations such as teaching or nursing.

The growth of services since World War II has been accompanied by the steady increase in women in the paid labor force throughout Europe, North America, and Japan. The most rapid rise in women's labor force participation rates has been among married women with children (Table 3; Figure 17). In the United States today, women comprise 48% of all full-time employees. However, women's entry into services jobs has been in large part limited to so-called pink-collar jobs, including clerical and secretarial work, retail trade, health care (other than doctors), eating and drinking establishments, teaching, and child care (Figure 18). Most of these jobs pay relatively poorly, leading to widespread concerns about the feminization of poverty. Indeed, until very recently, women's presence in well-paying occupations such as physicians, attorneys, or corporate management has been relatively low, leading to complaints about the "glass ceiling" many women face in the workplace. Such a clustering in low-paying jobs is the major reason why women on average in the United States have incomes that are only 70% of those of men.

The rise of women's labor force participation in services carries several social implications. Some maintain that the increase in women's work participation serves as a strategic compensation to declining male incomes in the face of deindustrialization. The rise of the two-income family has significantly changed gender roles at home and led to (largely unsuccessful) pressures to redistribute

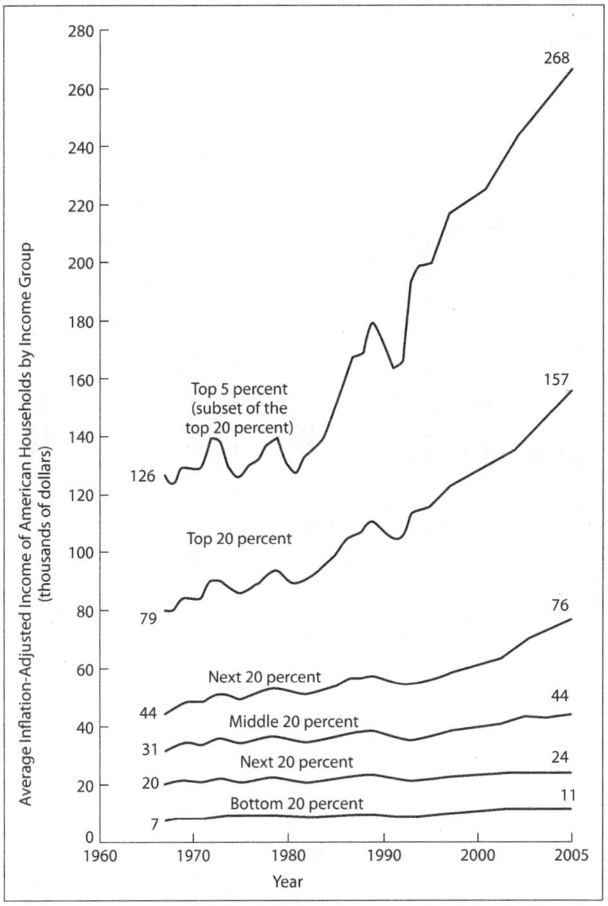

FIGURE 16 Income inequality in the United States, 1965–2005. Mounting income inequality reflects the impacts of globalization on the poor and the rapid rise in incomes of the top 5% who have benefited the most from the growth of unearned income and an increasingly regressive tax structure.

housework to men. In countries without national day care systems, such as the United States, childcare is an important constraint to women's job opportunities. For middle class families capable of paying for private childcare, this constraint can generally be overcome, although it represents a constant source of financial and emotional stress, complicating commuting patterns and changing the socialization patterns for the young: Most children today enjoy relatively little free time compared to earlier generations. For the poor, who cannot afford professional childcare, adequate or otherwise, the rise in women's paid work has created a childcare crisis. In many cases, unattended children, often called "latchkey kids" in the United States, spend long periods free from adult supervision, a phenomenon likely to lead to youths who engage in illicit or illegal activities.

LOW DEGREE OF UNIONIZATION In general, most services are nonunionized. Indeed, since World War II, the percentage of workers belonging to unions has declined steadily (from 45% in the United States in 1950 to 12% today), largely due to deindustrialization. Despite their rapid growth, jobs in services are rarely unionized; the vast majority of workers in retail trade and in many skilled professions (e.g., attorneys, accountants) do not belong to unions. Among American workers, some service employees are occasionally organized in medical unions (e.g., Union 1199), teachers' unions (e.g., American Federation of Teachers), or public-sector unions (e.g., American Federation of State, County, and Municipal Employees), but these are generally exceptions.

TABLE 3 20 Leading Occupations of Employed Women in the United States, 2005

Occupation	% Women
Secretaries	96.7
Elementary school teachers	96.7
Nurses	91.8
Home health aides	88.3
Cashiers	75.0
Administrative support	69.5
Retail sales	41.1
Customer service	70.1
Clerks	91.2
Accountants	60.8
Receptionists	93.9
Maids	89.7
Office clerks	83.8
Secondary school teachers	54.8
Waiters/waitresses	67.3
Financial managers	55.7
Teacher assistants	91.7
Preschool teachers	97.7
Social workers	76.1

Source: U.S. Department of Labor.

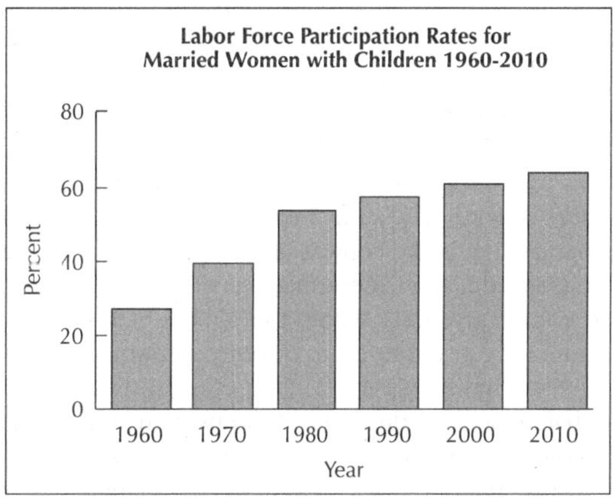

FIGURE 17 U.S. labor force participation rates for women with children under age 18, 1960–2010.

FIGURE 18 Women as percent of labor force in different industries. The gender composition of jobs varies widely among industries. Most women are concentrated in relatively low-paying pink-collar occupations.

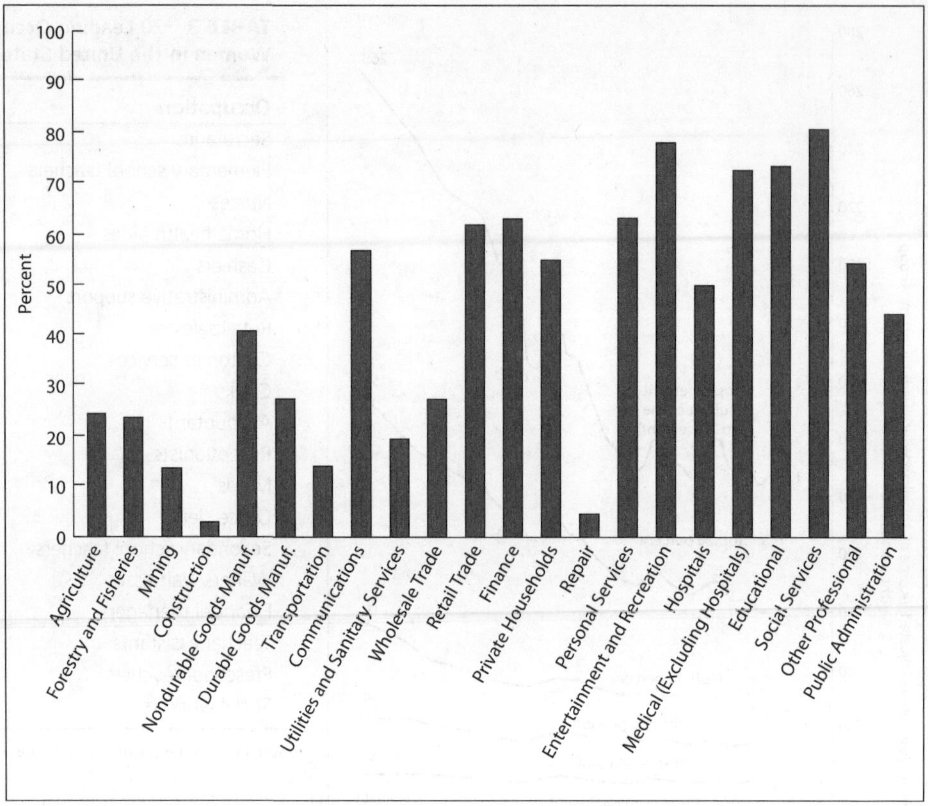

EDUCATIONAL INPUTS Finally, the relationship between the services sector and education is important. In general, skilled managerial and professional services require more education than do jobs in manufacturing, particularly literacy, numeracy, and computer skills. Unemployment rates are almost always lower for well-educated workers compared to those without a high school degree (Figure 19). Moreover, the average income of workers with advanced degrees is considerably higher than those without (Figure 20). College degrees pay for themselves many times over in terms of higher lifetime earnings.

Accordingly, the demand for higher-educational services has risen throughout the industrialized world as a university degree has become a prerequisite for obtaining middle-class jobs, leading to a surge in college enrollments. In the United States, for example, only 20% of high school graduates went on to attend universities after graduating in the 1950s, whereas today roughly 70% do so (although many do not graduate). Accordingly, many industrial countries have come to recognize that in the context of a knowledge-based economy, higher education is essential to national and local economic competitiveness. Within the United States, likewise, states that have spent funds to develop high-quality university systems have enjoyed a competitive advantage over those that neglected educational funding, particularly in high technology and producer services. People who are denied access to a university education—overwhelmingly the poor and minorities—are increasingly likely to be condemned to a lifetime of poverty. During the industrial economies of the early twentieth century, a strong back and arms ensured at least a lower-middle-class lifestyle; today, those who do not complete high school find few venues in which to compete successfully in the labor market.

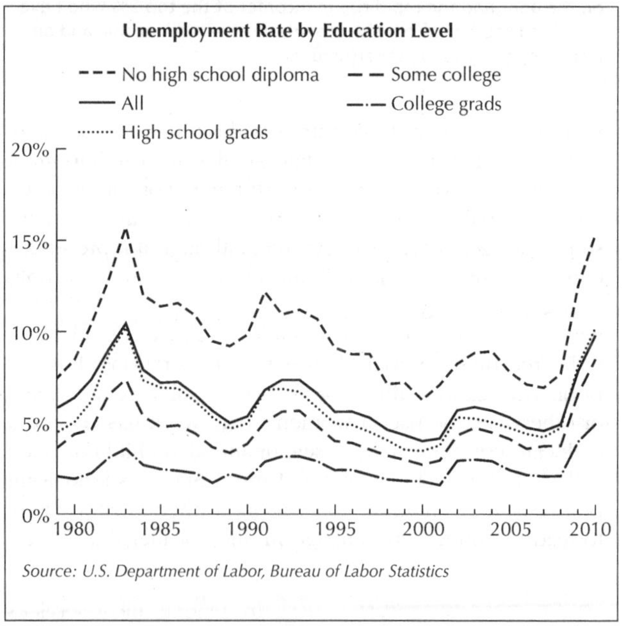

FIGURE 19 Unemployment rates by educational level in the United States, 1979–2010. The better educated workers are, the less likely they are to lose their jobs.

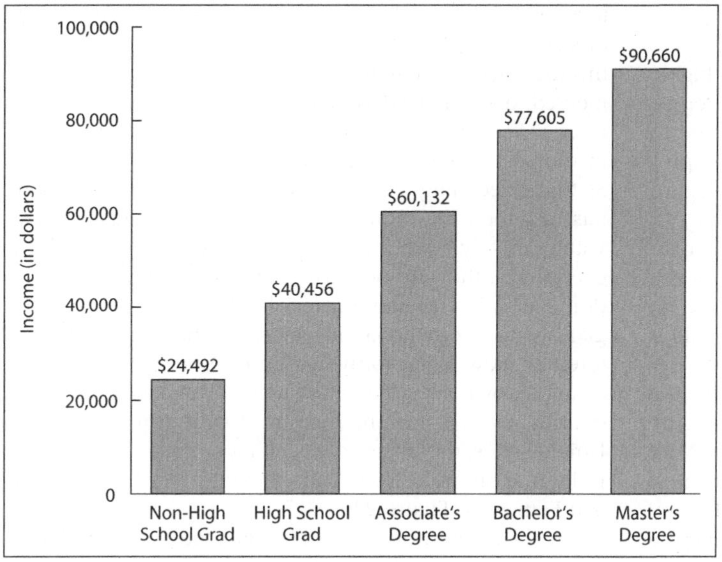

FIGURE 20 Median household income in the United States by educational level, 2007. Don't believe the claims about skyrocketing tuition costs and worthless education. The investment in higher education is the safest bet around. Not only does it correlate with higher lifetime earnings, but it prepares for long-term success in careers, while broadening minds and enriching lives. Now is a good time to go to school, as there are fewer jobs available and unemployment is high. The economy will eventually improve. The majority of American students attend public institutions where the average tuition is $7000 per year, for in-state students, and the costs are much less at junior colleges. Much of the tuition in Europe is free, but the entrance requirements are much stiffer.

FINANCIAL SERVICES

Finance is a critical part of contemporary economies and societies. The historical development of capitalism was accompanied by the formation of credit systems that led to the modern banking system. Today, economic geographers are often concerned with the geography of money in various forms, including the spatial structure of financial industries.

Finance is not a single industry but a set of closely related sectors that revolve around money in various forms. A central function of these institutions is to serve as intermediaries between borrowers and savers (i.e., individuals and institutions with excess capital to be lent and those who need capital for whatever purpose). Banking thus involves the exchange of money from savers to borrowers and assets (i.e., debt) from borrowers to savers. However, there are considerable differences in the markets involved, the size and behavior of firms, and the regulatory environment they face. The distinctions vary from country to country, as they are often based on government-mandated regulations that prohibit some types of firms from operating in some markets. The following distinctions are based on the American financial system and vary in other countries with different regulatory systems.

COMMERCIAL BANKING The largest component of financial services, and the one most people think of when they consider finance, is commercial banks. These are the institutions primarily involved in commercial loans (i.e., to firms, not individuals, including venture capital for start-ups and commercial real estate). Commercial banks provide most of the capital firms use for the building of office complexes, hotels, waterfront developments, and sports stadia. They also offer nonhousing personal loans, for example, to individuals who wish to purchase a boat or fund a college education and provide retail banking services such as checking and saving accounts and credit cards. However, retail banking is a relatively small part of their activities and often the segment that generates the least profit.

INVESTMENT BANKING Legally differentiated from commercial banks by the McFadden Act (1927) and the Glass-Steagall Act of 1933 (since repealed), investment banks are involved in buying and selling securities such as stocks, bonds, futures, and derivatives. They also provide specialized expertise and capital to facilitate corporate mergers, takeovers (friendly or hostile), and leveraged buyouts. Like commercial banks, investment banks buy and sell foreign exchange.

SAVINGS AND LOANS Also known as thrifts, savings and loans (S&Ls) have traditionally been involved in only one market, mortgages for houses. Thus they are the largest source of home loans and play a critical role in structuring the residential geography of cities (including such activities as redlining, or denials of mortgages to low-income minority communities). However, with the deregulation of the finance industry in the 1980s, savings banks began to diversify, penetrating the commercial real estate market, with horrific consequences that led to a large federal government bailout in the early 1990s, an example of state intervention.

INSURANCE Insurance is a sector largely centered on the commodification of risk. Insurance policies, paid for by premiums paid by households and firms, provide protection from unexpected loss. There are many types of insurance, including property (e.g., for houses), casualty, life, health, automobile, and even nursing home and pet care!

The Regulation of Finance

Compared to most industries, the finance industry is highly regulated. The reasons for the comparatively high degree of government control lie in the centrality of finance to current economies. The behavior of financial firms profoundly affects the national money supply, which in turn shapes national interest, inflation, and exchange rates.

Virtually all countries have a national bank of some sort to regulate their financial sector. In Britain it is the Bank of England; in Germany, the Deutschebank; in Japan, the Bank of Japan. In the United States, this role is played by the Federal Reserve, which was set up in 1913. The Federal Reserve, whose chair is appointed by the president, consists of 12 Federal Reserve Banks (Figure 21), the most important of which is located in New York City. It attempts to manage, among other things, interest rates and the money supply by regulating the federal funds rate, the reserve ratios of commercial banks, and the buying and selling of government securities.

In the 1930s, as the banking system was devastated by a wave of bankruptcies created by the Depression, the federal government introduced a variety of new regulations to stabilize the system. These included separating commercial and investment banking and a series of laws prohibiting interstate banking that were designed to protect small local banks from competition from larger out-of-state ones, particularly those in New York. The result of this regulatory system in the United States was a highly fragmented, decentralized system of banking. There are today more than 5000 banks in the United States, compared to 15 in Canada. However, the number of banks in the United States is declining as a wave of mergers and acquisitions has reshaped the industry. Most banks are small in size, with relatively few employees and assets, and serve local clients (i.e., they are nonbasic). Financially, the industry is dominated by "money-center" banks, large, commercial giants with international ties, most of which are headquartered in New York. Only 2% of all American banks, for example, own 50% of all bank assets. Thus, while financial employees are present in most towns throughout the United States, the largest concentrations are in large metropolitan areas near New York, Boston, Los Angeles, and San Francisco (Figure 22).

It should be noted that the regulated system of banking worked relatively well for roughly 50 years (i.e., the period between the Depression of the 1930s and the great changes that swept the world economy in the 1970s and 1980s). International banking was stabilized during this time by the Bretton-Woods Agreement of 1947, in which the United States and its allies set up the architecture for the global financial system after World War II.

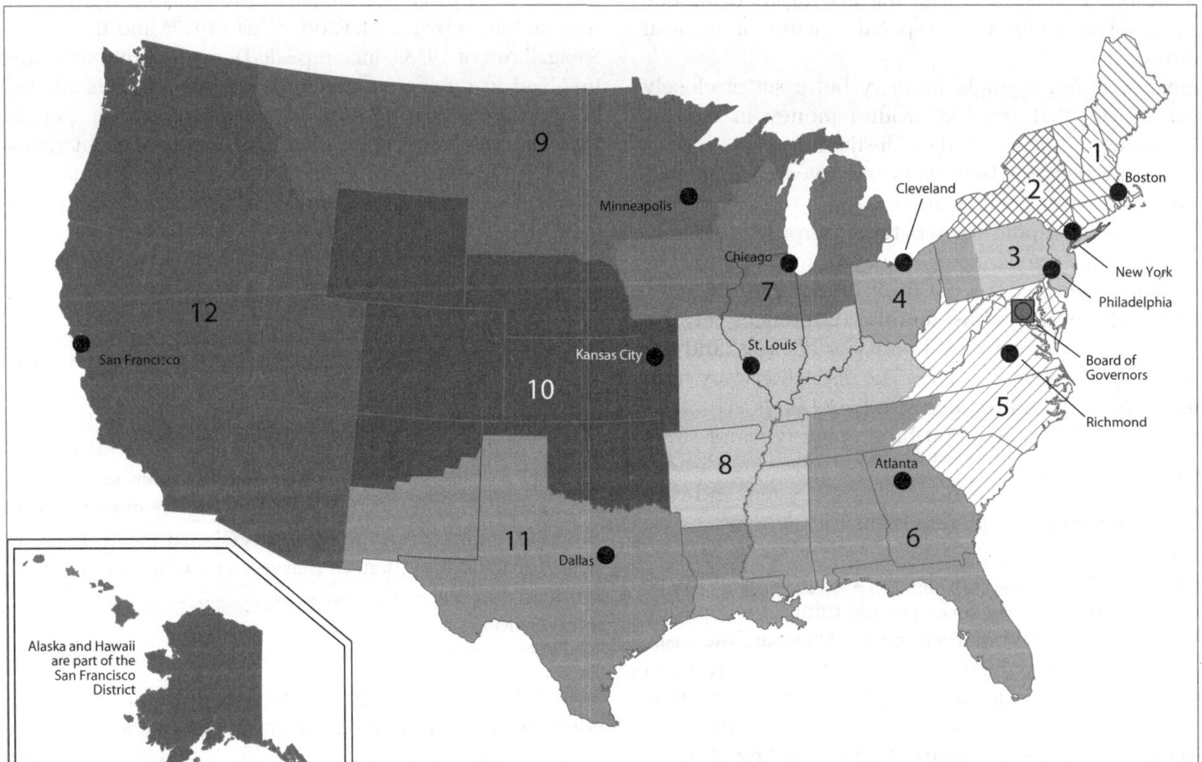

FIGURE 21 The 12 districts of the U.S. Federal Reserve, the agency responsible for controlling the nation's money supply The head of the Federal Reserve is appointed by the president for a 6-year term.

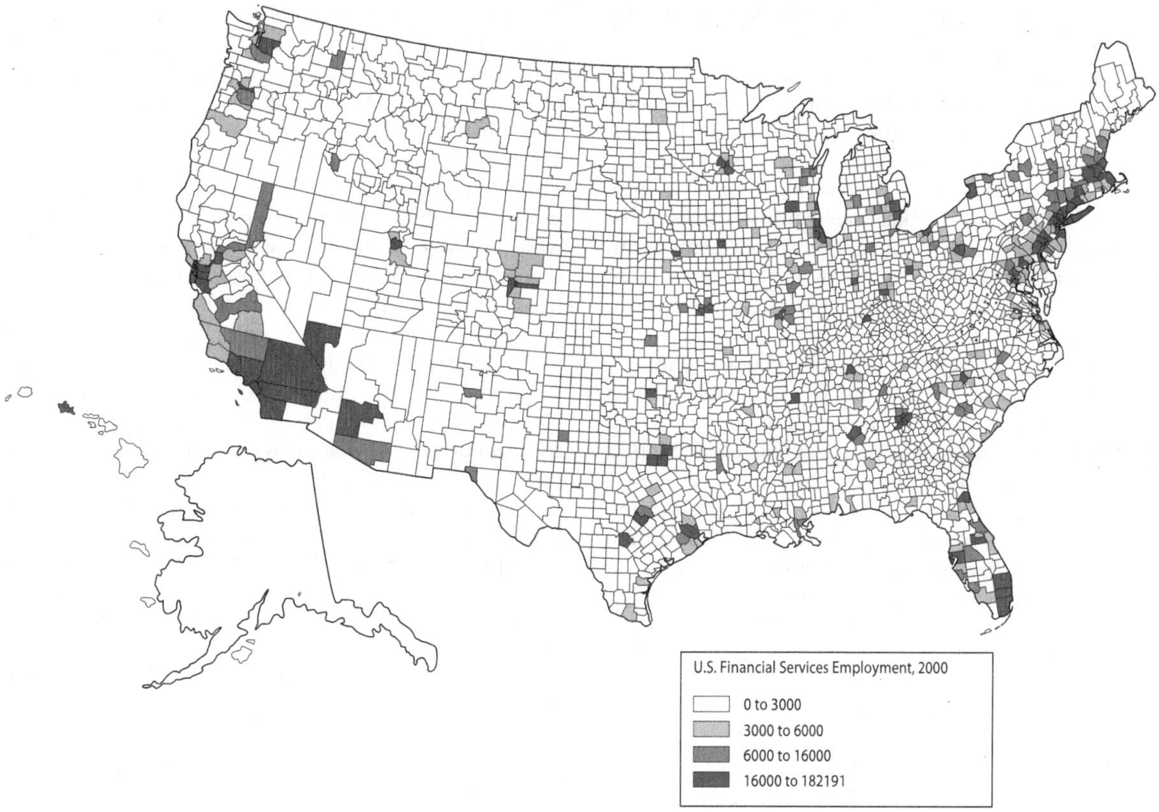

FIGURE 22 U.S. financial services employment, 2000.

The number of bank bankruptcies was very low. In the 1970s and 1980s, however, much as manufacturing was shaken by petrocrises and deindustrialization, the banking system was forced to confront a radically different environment. The collapse of the Bretton-Woods agreement and the shift to floating exchange rates opened up currency trading as a major source of revenues. Rising debt levels in the developing world were another factor that accentuated the globalization of banking as large banks extended loans to countries in Latin America and Asia. The microelectronics revolution introduced new ways of serving customers (e.g., automatic teller machines), and electronic funds transfer systems allowed money to be transmitted digitally. As the relatively safe period following World War II drew to a close under mounting international competition and domestic deregulation, commercial banks found themselves in an increasingly hostile and competitive environment, to which many succumbed. As finance became increasingly globalized, the pressures at home to deregulate mounted accordingly.

The Deregulation of Finance

Beginning in the 1970s, the U.S. government undertook a series of actions that had far-reaching consequences in many industries, particularly real estate. In 1974, money market mutual funds were introduced, which sharply increased competition for core banking deposits. In 1979, the Federal Reserve switched from a policy of stabilizing interest rates to a policy of slowing money growth in order to combat inflation. In 1980, Congress passed the Depository Institutions Deregulation and Monetary Control Act, and in 1982, the Garn-St. Germain Act, which permitted thrifts (savings and loans) to compete with commercial banks and eliminated geographic limitations on S&L lending. Deregulation included the removal of interstate banking regulations, which led to a national banking market dominated by a few giants. Finally, this process extended to the removal of restrictions governing pension and mutual fund portfolios, the abolition of fixed commissions on stock market transactions, the approval of foreign memberships on stock markets, and the repeal in 2001 of the Glass-Steagall Act, which had separated commercial from investment banking since the Great Depression.

These changes effectively liquidated the relatively stable division of markets and lending sources that had existed since the 1930s (commercial banks in commercial property lending, thrifts in residential lending), increased the liquidity and level of competitiveness of finance capital as new institutional players entered the commercial market, and markedly altered the role of the local state in attracting development. The two institutions most heavily affected by these developments were commercial banks and the savings and loans institutions.

An important consequence of deregulation was increased competition in banking, particularly as noncommercial banks invaded traditional commercial bank markets. S&Ls entered the commercial real estate market, and many large retailers (e.g., Sears) began to offer their own credit cards. For banks, these trends were most unwelcome because they reduced their profit margins. In response, many banks diverted funds in their investment and loan portfolios to higher-risk/higher-return opportunities or by becoming more leveraged in commercial real estate. Many banks, desperate for new avenues of investment, began to raise funds from foreign banks or on the increasingly globalized securities markets, or moved away from traditional lending practices into such higher-value-added practices as currency trading and cash.

Commercial real estate has always been the bread and butter of commercial banking; U.S. banks carry more than $800 billion in commercial real estate loans. Deregulation unleashed an enormous wave of investor-driven construction of commercial real estate in the 1980s and 1990s, as commercial banks and S&Ls, with funds augmented by junk bonds, joined mutual funds, insurance companies, and others in a massive surge of investment in these lucrative projects, particularly office towers and shopping malls. Many banks have the vast majority of their net worth tied up in offices, shopping centers, and hotels. Consequently, in the 1990s, a serious nationwide office glut appeared, vacancy rates climbed, and commercial rents fell, all of which precipitated a wave of commercial bank and S&L failures and a hugely expensive federal government bailout.

The Financial Crisis of 2007–2009

As we have seen, capitalism is regularly given to periodic downturns and crises. The latest example is the "Great Recession," the most severe and long-lasting economic crisis since the Great Depression of the 1930s. Although the crisis was not confined to financial services—indeed, manufacturing workers suffered the worst rounds of layoffs—it did originate there. The crisis should be seen within the context of the high levels of U.S. government, corporate, and household debt that had been accumulating for decades.

The causes of the crisis are complex and involve several forces. The deregulation of financial industries and very lax government oversight set the stage for an explosion of complicated activities, including derivatives (contracts whose value is derived from other values such as stocks, bonds, or commodities) and the "financialization" of mortgages, that is, the repackaging of individual home loans into bundles that were traded on financial markets. Meanwhile, the Federal Reserve kept interest rates very low, stimulating the demand for loans by individuals and corporations alike. Simultaneously, predatory lenders extended mortgages to numerous unqualified, often naïve individuals, generating a vast pool of "subprime" (often variable interest rate) loans susceptible to small disruptions in the economy. As a result, housing prices rose much faster than did household incomes in the 1990s and 2000s, a trend that is not sustainable in the long run.

As this house of cards began to collapse, a wave of mortgage defaults began to erupt, as many homeowners were not able to keep up with their monthly payments. Banks, in turn, initiated foreclosures on those who fell far behind in their payments; by mid-2009, almost 2% of mortgages were in this state. Accordingly, housing prices began to decline for the first time in years, and new construction came to a halt, crippling the construction industry. Many homeowners found themselves, in the lingo of the industry, "under water," that is, with mortgages that exceeded the value of their homes.

The housing crisis was most pronounced on the East and West coasts; the country's interior was less affected. In 2010, for example, 50% of mortgages in Arizona were under water. As the effects of the housing crisis rippled through the economy, consumer spending slowed down, depressing earnings in other sectors, many of which already faced mounting international competition. Banks became reluctant to extend new loans, leading to a "credit freeze" that deprived qualified borrowers and investors of much-needed capital. The Dow Jones Industrial Average plunged more than 60% between the fall of 2007 and early 2009, devastating investors as well as the lifetime savings of many individuals, some of whom had to postpone retirement as a result. Unemployment doubled, rising to 10.5% in June 2010, a level unseen since the 1930s, and millions more people simply dropped out of the labor market. During this period, the U.S. economy shed more than 8 million jobs, greatly depressing consumer spending, and the effects reverberated throughout the world. Manufacturing was particularly hard hit, and the automobile companies found themselves on the verge of bankruptcy (General Motors did indeed file). As income, sales, and property tax revenues declined, many state, county and municipal governments found themselves slashing social services just as ever more people needed public relief.

In response to the crisis, two presidential administrations—those of George W. Bush and especially of Barack Obama, who took office as the crisis was well underway—initiated rounds of federal interventions designed to aid the economy, a revival of the Keynesian economic policies common in the 1930s, which added to the federal government budget deficit. These measures included the Troubled Assets Relief Program (TARP), designed to assist troubled companies with $700 billion in funds. Central to this effort was a large "bailout" of the banks, including hundreds of billions of dollars to lenders and insurance companies, designed to free up the capital markets and get lending going again. For example, the government lent American Insurance Group more than $160 billion. The federal government came to own a large share of stocks in many firms, including General Motors and the American Insurance Group. Substantial aid was given to mortgage brokers, including the giants Freddie Mac and Fannie Mae. Additionally, a "stimulus package" of roughly $787 billion passed Congress, generating both assistance for beleaguered state and local governments and numerous federal projects designed to improve the infrastructure and the educational system. General agreement exists among economists that such measures likely saved the economy from another depression; however, as of mid 2010, the U.S. economic situation remains fragile and unemployment still hovers around 9.5%.

STUDIES OF MAJOR PRODUCER SERVICES BY SECTOR

Services are a very broad group of industries with widely varying attributes. To appreciate this diversity, we now turn to several important service industries, including accounting, design, and legal services.

Accounting

Accounting is one of the most important professions in advanced capitalist economies (Figure 23). Modern financial auditing began to emerge in the middle of the nineteenth century, when the ownership and control of firms became separated. With this separation, increasingly detailed financial statements were required so that investors would have sufficient information to make their decisions. For most companies, an annual audit undertaken by a professional accountant is a legal requirement, forcing them to enter into a short- to medium-term relationship with an accounting firm. This relationship allows the accountant to sell related services, such as tax advice, estate management, and consulting. The audit process provides an independent check for shareholders into the activities of publicly owned enterprises.

Design and Innovation

Management consultants support the production process by improving management systems and by increasing the productivity and profitability of their clients. They also encourage innovations, especially in administrative practices. Consultants can act as catalysts for introducing or encouraging change. There are, however, other producer services that play an important, direct role in supporting, encouraging, and developing innovation.

Traditionally, manufacturing rather than services has been regarded as the source of innovation. Today, services are an important part of innovation: Education provides a skilled workforce; universities conduct research and development; and research companies perform product testing, product naming, brand development, and product or service development. Some of the most important producer services directly involved in product innovation are provided by design companies. A manufacturer employs an industrial designer to assist in making changes that will increase the demand for a product. The task involves understanding public tastes and altering products to make them more attractive to the consumer. Design is an integral part of the production process but is separate from the actual manufacturing process; rather it informs and is informed by it. Small to medium-sized firms frequently employ private design companies; large companies (e.g., automobile manufacturers) have in-house design departments but can still employ external designers.

As happens with most providers of producer services, informal or formal networks of small design companies and sole practitioners develop. Networking with other companies allows information and experience to be shared and can also lead to collaborative partnerships, which allow small companies to provide a range of services that can compete with those provided by large companies. Small companies can be flexible as collaborations or temporary coalitions of different service suppliers, associations that can be re-formed on a client-by-client basis. This

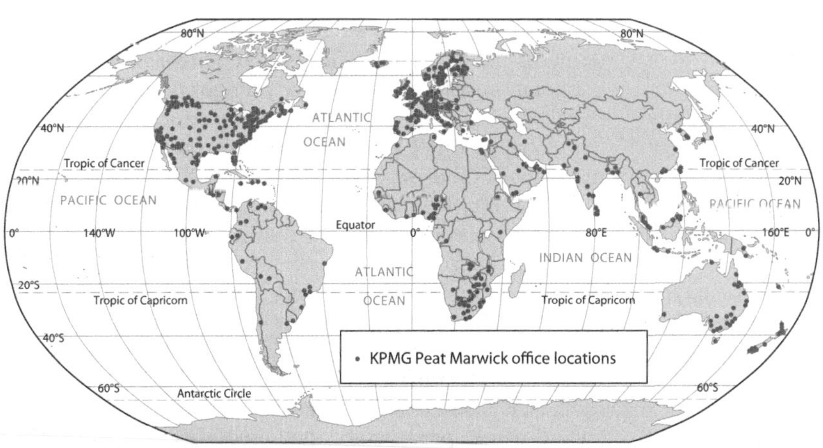

FIGURE 23 Offices of KPMG Peat Marwick, the world's largest accounting firm. Accountants, lawyers, and other professionals cluster in world cities to provide advise to major corporations and financial institutions. As centers of finance, world cities attract the headquarters of producer services-major banks, insurance companies, and specialized financial institutions, where corporations obtain and store funds for expansion and production.

strategy enables small companies to provide a variety of services but without having to employ full-time specialists. It also enables them to deliver services outside their local area by drawing on companies located in other places.

Legal Services

Legal services may be employed by individuals and households, on the one hand, as consumer services and by corporations, on the other hand, as producer services. The services that law firms provide to corporations, which may include accounting and investment advice, are highly specific to the needs of individual clients. These services include help in negotiating a highly complicated legal environment—torts, patents, and product liability; bankruptcy; employment law and antidiscrimination ordinances; antitrust restrictions; tax law; trade agreements; venture capital negotiations; joint venture agreements; technology transfer protocols; reinsurance; corporate mergers, buyouts, acquisitions, and takeovers; and intellectual property rights. In an increasingly uncertain production environment characterized by huge and unpredictable changes in markets, government regulations, technology, and output, the demand for legal services arises from the proliferation of management tasks, such as those made necessary by strategic planning and coordination.

In many sectors, deregulation has accentuated the uncertainty faced by many firms and increased the need for legal services, as did the wave of late-twentieth-century vertical disintegration of production, which led to a steady contracting out of many functions formerly performed in-house. More recently, many large law firms have diversified into advising, consulting, debt restructuring, and deal brokering, competing directly against other large services providers based in other industries.

In the United States, legal services are characterized by high degrees of spatial and functional agglomeration, primarily in large cities, and close interaction with clients and suppliers. The prime reason behind the geographical concentration of such firms in metropolitan regions is the ready access they offer to clients, suppliers, and ancillary services, most of which is accomplished through face-to-face interaction. Because personal relationships in which trust and reputation are of paramount significance figure prominently in the practice of law, legal services are heavily embedded in local cultural relations and require frequent personal contacts. For these reasons, and others, New York City remains the nation's capital of domestic corporate law, and Washington, DC, is the center of government-related legal practice.

The practice of law is dominated by large firms, a type of organization that began with the birth of corporate law in the late nineteenth century. These large firms currently influence disproportionately the structure of incentives, rewards, prestige, and employment in the sector, and they shape the nature of litigation in government, finance, and the business community. Large law firms commonly have several branch offices employing dozens, even hundreds, of attorneys.

Mounting international trade and investment has led to increasing demand for international legal services. After World War II, the outflow of U.S. capital across the world attracted numerous firms eager to serve multinational clients. Thus, the globalization of services often followed the path set by manufacturing firms. The foreign operations of international law firms were originally highly correlated with multinational corporate trade in goods, but increasingly it is the global markets in finance and real estate that dictate the geography of demand. During the first wave of globalization, in the 1960s and 1970s, most firms handled foreign work exclusively through their home office, but throughout the 1980s and 1990s, locating offices abroad came to be increasingly important. The principal markets for international law firms tend to be in highly developed countries with well-developed corporate economies, high degrees of international trade, and relatively high per capita levels of income (as the demand for legal services is income-elastic).

The practice of international law involves a crazy quilt of varying national legal systems, qualifications to practice, and barriers to entry for foreigners. What constitutes the profession of law varies around the world; there is considerable diversity as to the requirements and obligations of lawyers, including education, licensing requirements, scope of practice, standards of conduct, ethics, and forms of association. There are also significant national variations in the degree to which foreign attorneys may practice in different countries; virtually all countries constrain foreign attorneys from practicing law within their borders to some extent.

Law is a clubby work environment in which reputation is critical, and foreigners are often informally, if not formally, excluded as outsiders. The practice of law relies heavily on tacit local culture and networks, so international law firms must incorporate local knowledge with global expertise, a reflection of the limited substitutability that characterizes this sector. Local firms fearful of foreign competition may urge formal restrictions on foreign lawyers, such as establishing nontariff barriers like those the Japanese attempted to exclude American firms from their domestic market, difficult licensing requirements, mandated cooperation with local firms, or outright prohibitions from certain kinds of legal practice. Perhaps for these reasons, the globalization of legal services has occurred more slowly than that of other services.

Typically, it is large law firms with a long history of service to corporate clients that are best positioned to engage in international practice, which involves high start-up costs, short-term losses, and reliance on economies of scale. This fact favors a group of giants in a limited number of locations. Some of the largest law firms, such as Baker and McKenzie, are true multinationals in their own right. Size is a key asset in foreign markets because it offers "brand-name" predictability and credibility (i.e., the necessary critical mass of expertise); reputation is an important proxy for reliability and quality of service. International law firms may offer services either directly from their headquarters office or via overseas branch offices, subsidiaries, and affiliates with varying degrees of autonomy. Their clients may be either foreign corporations operating abroad or local companies drawing on the resources of the multinational firm, which often has large pools of attorneys with highly specialized expertise. Often, international legal services are provided through "co-counseling" agreements, strategic alliances between foreign and domestic firms, a common feature of post-Fordist capitalism. However, national variations in legal systems have made the formation of transnational law firms an exceedingly complicated matter.

THE LOCATION OF PRODUCER SERVICES

Where producer services are located and the functions that they perform are increasingly important given their rapid growth. Generally, highly skilled, high-value-added functions such as financial and business services concentrate in metropolitan regions (Figure 24). Two-thirds of business and professional services jobs in the United States are located in large urban areas. Agglomeration economies, such as the minimization of transaction costs between locally based suppliers and subcontractors, the lower costs of accessing a pool of labor with a wider range of skills, superior transportation infrastructure and facilities, access to specialized information, and a range of office accommodations and telecommunications facilities play a large part in explaining this tendency.

Interregional Trade in Producer Services

A view long held among many planners, academics, and development officials is that services are nonbasic: They function in the local or regional economy essentially as support for the basic goods-related activities that are engaged in interregional trade and that generate the revenues that create the demand for services. If one takes away the basic goods-producing activities, this view holds, the demand for services will decline and they will accordingly contract. However, these ideas are erroneous. In fact, producer services also play a basic role in regional development in their own right, although how they do so is less well understood.

Head offices are more likely to have an orientation toward exports; the extent to which independent, single-establishment firms have this orientation varies according to their size. It is generally the case that the more specialized the service, the more likely it is to derive revenues from exports. Advertising, architecture and engineering, design, computing, research and development, real estate, management consulting, or transport services are more export-oriented than legal, banking, or equipment rental services, which rely for a larger proportion of their sales revenues on local clients. Overall, interregional service trade within individual countries is considerable, around 35% of total sales and rising.

INTERNATIONAL TRADE IN SERVICES

Like agricultural and manufactured goods, services are traded internationally. It is difficult to measure international services transactions, however, because measurement is

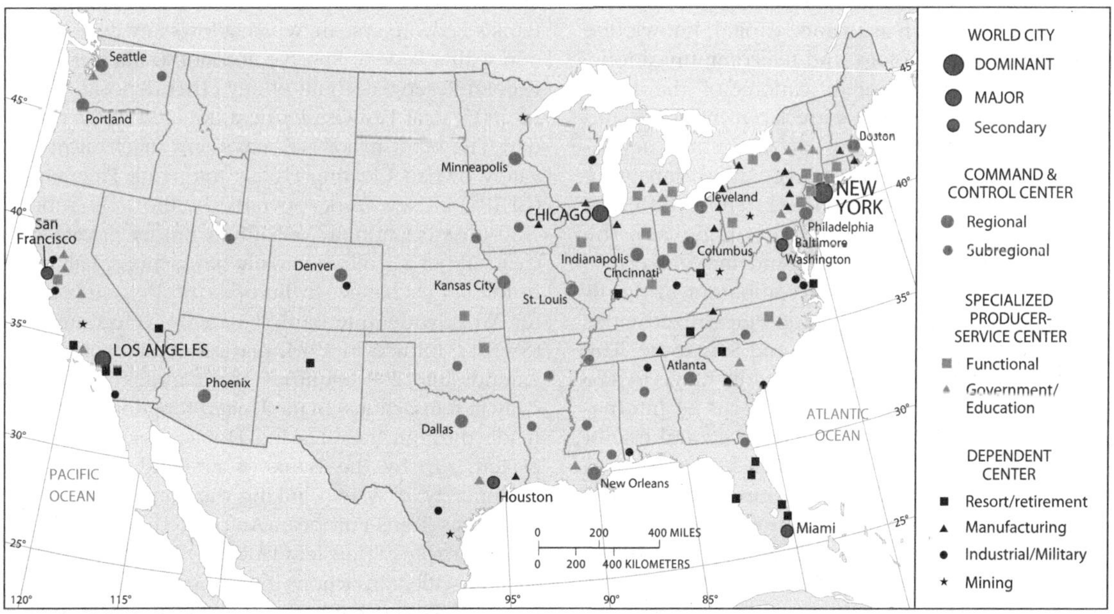

FIGURE 24 Business services cities in the United States.

complicated by developments in information technology. The Internet allows diffusion in geography and location; servers can be located in multiple locations and can be run by multiple companies. In addition, the Internet blurs the distinction between tangibles and intangibles. Goods that are ordered electronically and distributed internationally can be tracked by conventional taxation and tariff systems without too much difficulty. The same cannot be said for services that are supplied online, especially those that are digital or that can be rapidly digitized unlike the traditional across-the-border delivery of goods and services.

An example of this problem is electronic trade in packaged computer software; the sector also comprises services—advertising, computer services, financial services, entertainment, and other media, such as newspapers, periodicals, and music, radio, and television broadcasting—newspapers, software packages, CDs, and DVDs can all be converted into bits and delivered electronically. The United States has traditionally dominated electronic software distribution, but the packaged software market is becoming more global each year. The tracking of imports and exports is seriously hampered by measurement problems. An estimated 50% or more of packaged software is directly downloadable from the Internet, and electronic sales represent roughly 6% of the world software market. Border valuations of the trade are based on the medium (CD-ROMs) rather than the content (the software), which is significantly more valuable than the medium. Not only is the value of the software underestimated but the value of copyrighted works sold in foreign markets is also overlooked. In addition, computer software is often bundled with hardware (e.g., Microsoft software and Intel processors), and many other products, such as cameras, cars, and washing machines, also incorporate highly sophisticated electronic software.

Different services require a combination of different factors of production, such as finance capital, knowledge capital, information technology and telecommunications, labor supply, educational qualifications of the labor force, and so on. Some services are labor-intensive and involve simple tasks (call centers, data-processing services); others are knowledge-intensive and nonroutine (professional and business services). Thus, different types of services will locate in different regions in the ongoing attempt to minimize costs and maximize revenues. These issues are most clearly apparent in the financial and business quarters of large, globalized cities like New York, London, Hong Kong, and Singapore. The shift from comparative to competitive advantage as the basis for understanding trends and patterns of international trade in services has highlighted the central role of service firms rather than countries. The key players in this context are transnational service corporations (TNCs) that, in addition to shaping patterns of trade, are also associated with the rising importance of foreign direct investment (FDI). Noneconomic determinants of the growth in services TNCs include social and cultural differences between countries that also make it difficult to compete at arm's length (such as in law or advertising), thus encouraging direct representation in some form in order to match service supply with the nuances of local or regional characteristics.

Another perspective on services TNCs is the global value-added chain, which includes distribution, marketing, advertising, and similar functions. As business functions within TNCs have become more specialized, the links among different stages in the production process have spread out. In both services and manufacturing, high-value-added functions tend to agglomerate in large cities; conversely, support and clerical services have been widely dispersed. Clustering is now a global phenomenon; TNCs require co-location with information and knowledge providers, other service providers, and competitors.

Electronic Funds Transfer Systems

Financial markets all over the world were especially affected by the digital revolution, which essentially eliminated the costs of capital transactions and transmissions much in the same way that deregulation and the abolition of capital controls decreased regulatory barriers. Banks, insurance companies, and securities firms, which are very information-intensive, have led in the construction of an extensive network of leased and private telecommunications networks, particularly fiber-optic lines.

Electronic funds transfer systems (EFTS), in particular, which form the nervous system of the international financial economy, allow banks to move capital around at a moment's notice, taking advantage of geographic differences in interest rates and exchange rates and avoiding political unrest (Figure 25). One of the primary forms of EFTS is Real Time Gross Settlement (RTGS) systems, which handle money flows among financial institutions and governments. The largest of these is the U.S. Federal Reserve Bank's Fedwire system, which allows any depository institution with a Federal Reserve account to transfer funds to the Federal Reserve account of any other depository institution. In 2005, total Fedwire traffic amounted to $2.1 trillion per day. The other major U.S. payments mechanism is the privately owned Clearing House Interbank Payments System (CHIPS) in New York, operated by the New York Clearing House Association, a consortium run by private firms, that clears about $1 trillion in daily transactions, half of which is in foreign exchange. In Europe, the Belgian-based Society for Worldwide Interbank Financial Telecommunications (SWIFT), formed in 1973, plays a comparable role; SWIFT extends into 208 countries and handles 2.6 billion euros daily in transactions. In the United Kingdom, settlements are made through the Clearing House Association Payments System run by the Bank of England; in the European Community, a system linking the banks of member states known as Trans-European Automated Real-time Gross settlement Express Transfer (TARGET), which began in 1999, is used to settle transactions involving the euro. In Japan, starting in 1988, the Bank of Japan Financial Network System fulfills a comparable function.

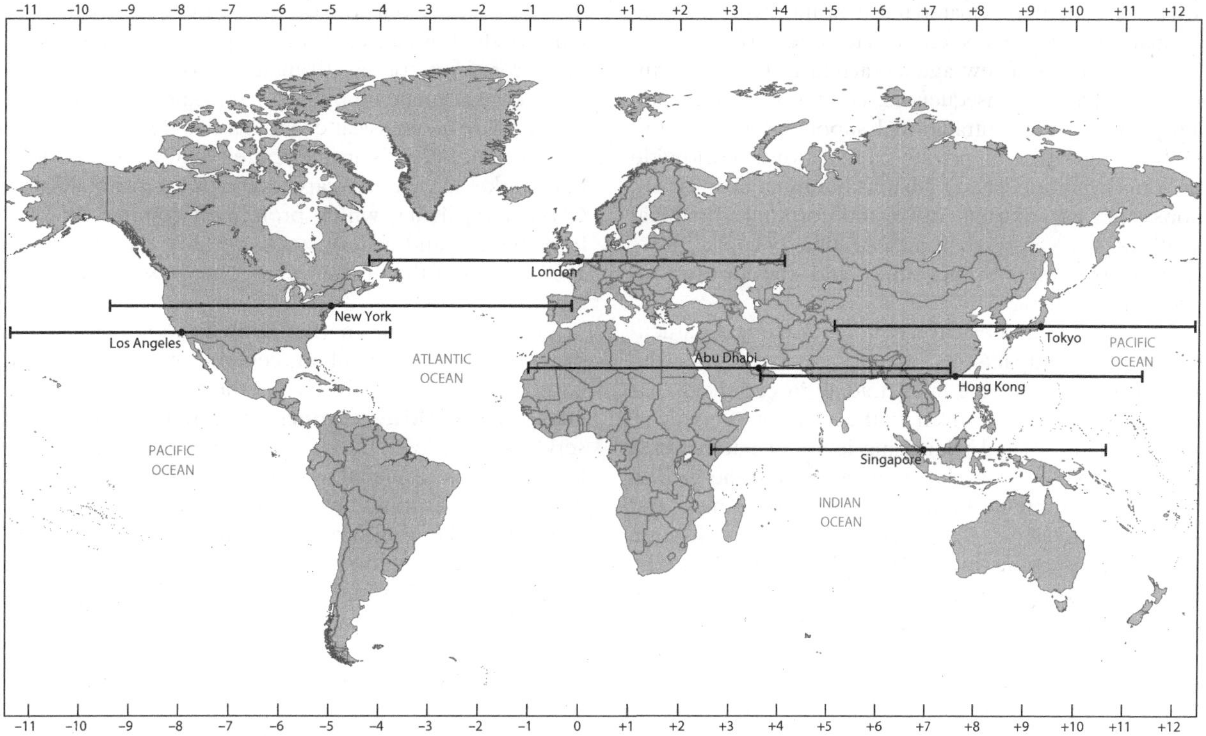

FIGURE 25 The world's major stock markets are linked by telecommunications networks, which greatly facilitates round-the-clock trading and the movement of capital across national borders. Horizontal lines indicate the hours when each exchange is open compared to other cities. The numbers at the bottom indicate time zones, with 0 representing Greenwich Mean Time.

Private firms have similar systems. Citicorp, for example, has a Global Information Network that allows it to trade $200 billion daily in foreign exchange markets around the world. MasterCard has its Banknet, which links all its users to a centralized database and payments clearing system. Reuters, with 200,000 interconnected terminals worldwide linked through systems such as Instinet and Globex, alone accounts for 40% of the world's financial trades each day. Other systems include the London Stock Exchange Automated Quotation System, the Swiss Options and Financial Futures Exchange, and the Computer Assisted Order Routing and Execution System at the Tokyo stock exchange. Such networks provide the ability to move money around the globe at stupendous rates (the average currency trade takes less than 25 seconds); supercomputers used for that purpose operate at teraflop speeds, or 1 trillion computations per second.

Similarly, in the securities markets, EFTS facilitated the emergence of 24-hour day trading by linking stock markets through computerized trading programs. Electronic trading frees stock analysts from the need for face-to-face interaction to gain information. Online trading also allows small investors to trawl the Internet for information, including real-time prices, and execute trades by pushing a few buttons, eroding the advantage once held by specialists. Trade on many exchanges has risen exponentially as a result. The National Association of Security Dealers Automated Quotations (NASDAQ), the first fully automated electronic marketplace, is now the world's largest stock market; lacking a trading floor, NASDAQ connects millions of traders worldwide through the over-the-counter market, processing 2000 transactions per second. EASDAQ, the European version of NASDAQ launched in 1996, operates similarly, albeit on a smaller scale. Facing the challenge of online trading head-on, Paris, Belgium, Spain, Vancouver, and Toronto all recently abolished their trading floors. The volatility of trading, particularly in stocks, also increased—hair-trigger computer trading programs allow fortunes to be made (and lost) by staying microseconds ahead of (or behind) other markets.

Liberated from gold, traveling at the speed of light as nothing but assemblages of zeros and ones, global money performs a syncopated electronic dance around the world's neural networks in astonishing volumes. The world's currency markets, for example, traded roughly 4 trillion in U.S. dollars every day in 2007, dwarfing the funds that change hands daily to cover global trade in goods and services. This ascendancy of electronic money shifted the function of the finance sector from investing to speculation, institutionalizing volatility in the process. Foreign investments have increasingly shifted from foreign direct investment (FDI) to intangible portfolio investments such as stocks and bonds, a process that reflects the securitization of global finance. Unlike FDI, which generates predictable levels of employment, facilitates technology transfer, and alters the material landscape over the long run, financial investments tend to create few jobs and are invisible to all but a few agents, acting in the short run with unpredictable consequences. Further, such funds are often provided by nontraditional suppliers: A large and rapidly rising share of private capital flows worldwide is no longer intermediated by banks but by nonbank institutions such as securities firms and corporate financial operations. Thus, not only has the volume of capital flows increased but their composition and the institutions involved have changed.

Globalization and electronic money have had particularly important impacts on currency markets. Since the shift to floating exchange rates, trading in currencies has become a big business, driven by the need for foreign currency that is associated with rising levels of international trade, the abolition of exchange controls, and the growth of pension and mutual funds, insurance companies, and institutional investors. The vast bulk (85%) of foreign exchange transactions involve the U.S. dollar. Typically, the moneys involved in these markets follow the sun. For example, the foreign exchange (FOREX) market opens each day in East Asia while it is still evening in North America; funds then travel west, bouncing from city to city over fiber-optic lines—from Tokyo to Hong Kong to Singapore to Bahrain to Frankfurt, Paris, or London, then to New York, Los Angeles, and back across the Pacific Ocean. (Given the continuous circularity of this movement, funds can originate anywhere and circle the globe within 24 hours.)

Electronic money can be exchanged an infinite number of times without leaving a trace, making it difficult for regulatory authorities to track transactions, both legal and illegal. The opportunities for money laundering are thus all the more numerous. As electronic money has become the norm, tax evasion has thus become an increasingly serious problem. Moreover, the jurisdictional question—who gets to tax what—is vastly more complicated. Digital counterfeiters can also take advantage of the system to use the Internet to create currencies anywhere.

Offshore Banking

One important effect of the telecommunications revolution on the world of finance is the emergence of offshore banking centers, which are loosely regulated or even unregulated. These centers reflect the growth of globalized "stateless money."

The origins of the contemporary, deregulated, globalized financial system lie in the Euromarket of the 1960s. The Euromarket arose in large part because of the large trade surpluses that western European countries enjoyed with the United States, which left them with large sums of excess dollars, known as Eurodollars. Thus, originally, the Euromarket comprised only trade in assets denominated in U.S. dollars but not located in the United States; today, it has spread far beyond Europe and includes all trade in financial assets outside of the country of issue (e.g., Eurobonds, Eurocurrencies), not just currencies. One of the Euromarket's prime advantages was its lack of national regulations: Unfettered by national restrictions, it has been regarded by neoclassical economists as the model of market efficiency. Banks in the United States invested in the Euromarket in part to escape domestic restrictions like the Glass-Steagall Act, which prohibited commercial banks from buying and selling stocks until it was repealed in 1997 as part of the broader deregulatory wave in financial markets.

Beyond the Euromarket, globalization spawned the growth of offshore banking centers, often in very small countries (and not always offshore, i.e., on islands). **Offshore banking centers** offer regular commercial services (e.g., loans), foreign currency trades and speculation, access to electronic funds transfer systems, asset protection (insurance), investment consulting, international tax planning, and trade finance (e.g., letters of credit). There are five major world clusters of offshore finance, including the Caribbean (e.g., the Cayman Islands, Bahamas, Panama); Europe (e.g., Isle of Man, Jersey, Luxembourg, Liechtenstein, Andorra, San Marino); the Middle East (Cyprus, Lebanon, Bahrain, Abu Dhabi); Southeast Asia (Hong Kong, Singapore); and the South Pacific (Nauru, Vanuatu) (Figure 26). The emergence of such places reflects the flows of electronic money across the global topography of financial regulation as large financial institutions shift funds to take advantage of lax regulations and low taxes found on the periphery of the global financial system. Many such places, although by no means all of them, have dubious reputations as centers of money laundering, that is, as places in which illicit funds garnered through international drug or weapons sales are converted into legitimate monies.

As the digital revolution allowed global capital to circulate more freely and rapidly, the technological barriers to moving money declined dramatically. Accordingly, spatial variations in the nature and degree of regulation rose in importance. Even small differences in regulations concerning taxes or repatriated profits may be sufficient to induce large quantities of capital to enter, or exit, particular places. Thus, many small states attempt to attract finance capital by deregulating as much as possible, lifting controls over currency exchanges, investment, and repatriated profits, and by eliminating taxes in the hope that global money will select their locale. By making it possible for funds to cross national borders at will, offshore banking reworks the notion of national sovereignty: Nation-states retain political sovereignty but sacrifice their financial sovereignty.

Back-Office Relocations

Large service corporations generally divide their labor force into two major segments: those associated with headquarters and those that carry out back-office func-

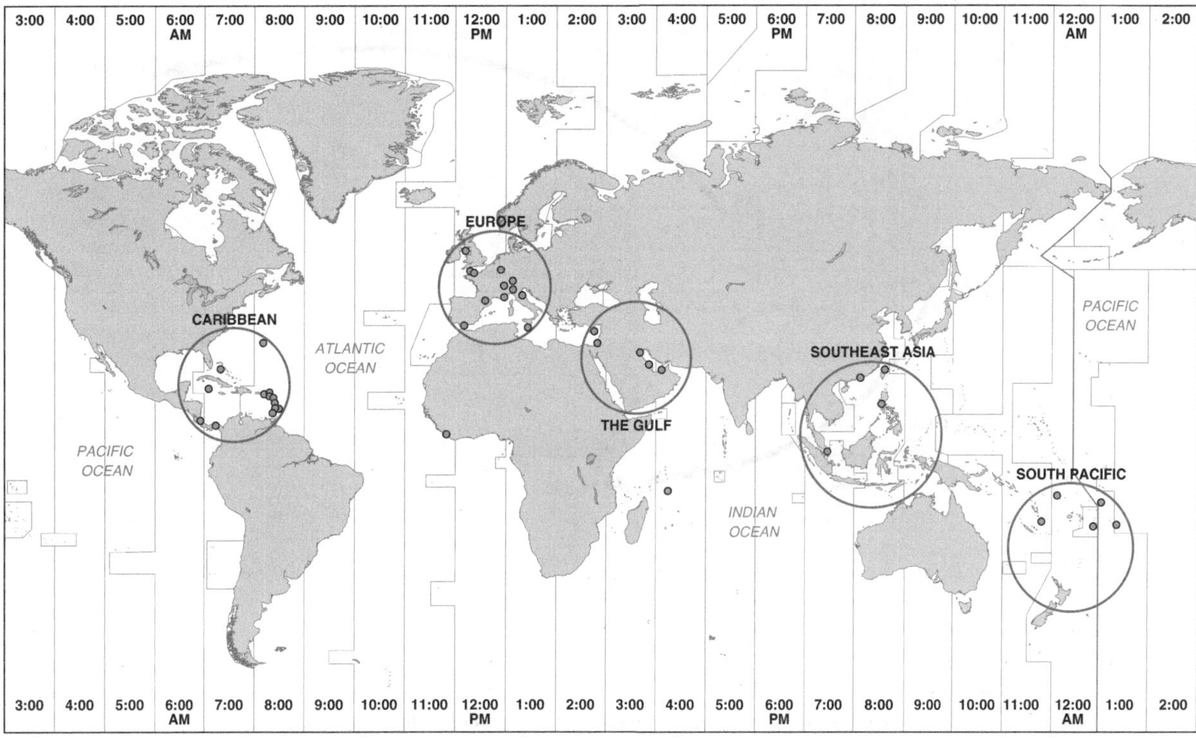

FIGURE 26 Five major areas of world offshore banking.

tions. Headquarters functions typically involve skilled professionals, including executives and managers as well as those who perform research, publicity, finance, legal work, accounting, engineering, marketing, and product development. **Back offices**, in contrast, perform data entry of office records, compile telephone books, catalog libraries, perform payroll and billing, and process bank checks, insurance claims, and magazine subscriptions. These tasks involve unskilled or semiskilled labor, primarily women, and frequently operate on a 24-hour-per-day basis. Unlike headquarters, back offices have few links to clients or suppliers. These facilities require extensive computer capacity, reliable sources of cheap electricity, and sophisticated telecommunications networks.

Historically, back offices were located next to headquarters activities (i.e., behind the front office) in downtown areas to ensure close management supervision and rapid turnaround of information. However, as central city rents rose in the 1980s and 1990s and companies faced shortages of qualified (i.e., computer literate) labor, many firms began to uncouple their headquarters and back-office functions, moving the latter out of the downtown to cheaper locations on the urban periphery. This uncoupling was made possible by the introduction of digital telecommunications, including microwave transmission. Back offices that moved to the suburbs often found pools of female labor willing to work part-time, generally consisting of married women with small children. Moreover, the suburbs offered ample parking, lower crime rates, and cheaper taxes and electricity. A considerable part of the suburban commercial boom of the late twentieth century consisted of back offices.

More recently, given the increasing locational flexibility provided by satellites and fiber-optic communications lines, back offices have also begun to relocate on a much broader, continental scale. Many financial and insurance firms and airlines moved their back offices from New York, San Francisco, and Los Angeles to lower-wage communities in the Midwest and South. Omaha, Nebraska, saw the creation of thousands of telegenerated jobs because of its location at the crossroads of the national fiber-optic infrastructure. Boulder, Colorado, Albuquerque, New Mexico, and Columbus, Ohio, have taken much the same direction.

Internationally, back-office dispersal began by moving to offshore locations. But this trend remained insignificant until transoceanic fiber-optic lines enabled relocation on an international scale. The capital investments in such operations are minimal and they possess great locational flexibility to maximize the ability to choose among places based on minor variations in labor costs or profitability. Offshore back offices are established not to serve foreign markets, but to generate cost savings for U.S. firms by tapping Third World labor pools where wages are low compared to those in the United States. Notably, many firms with offshore back offices are in industries facing

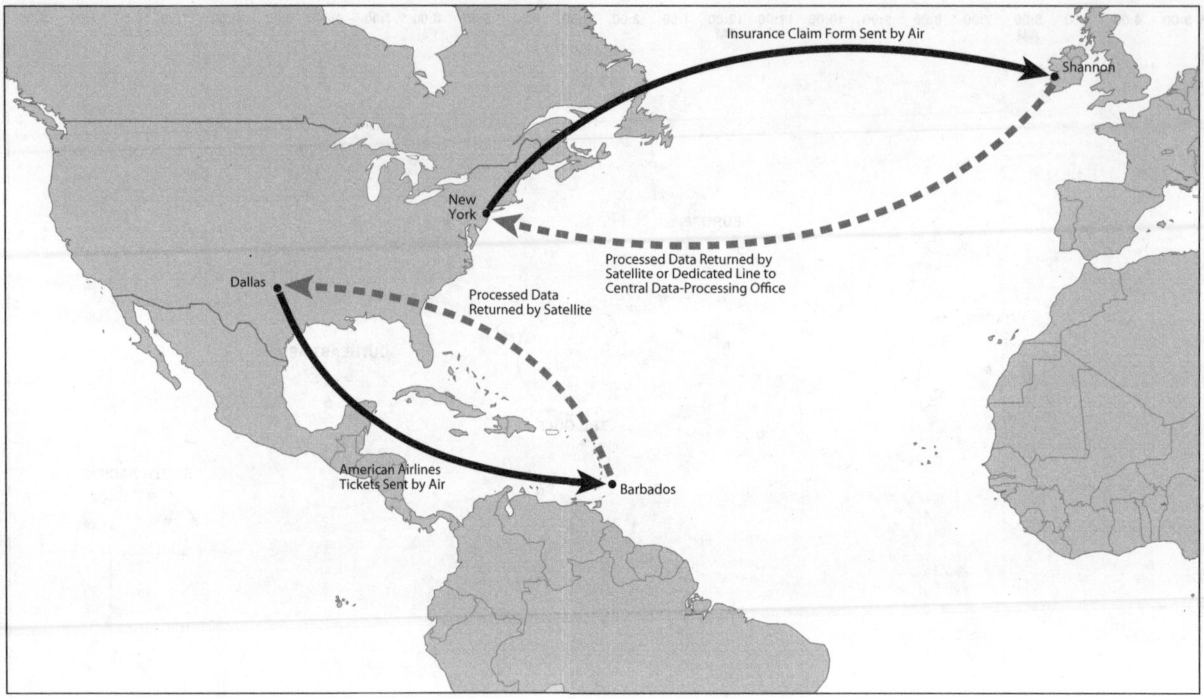

FIGURE 27 Mechanics of relocating American back offices to Ireland and to the Caribbean.

competitive pressures to enhance productivity, particularly insurance, publishing, and airlines. Several New York–based life insurance companies, for example, relocated back-office facilities to Ireland (Figure 27) with the active assistance of the Irish Development Authority. Often situated near Shannon Airport, the principal international transportation hub, documents are shipped in by Federal Express and digitized records are exported back via satellite or by one of the numerous fiber-optic lines that connect New York and London.

Likewise, the Caribbean, particularly Anglophone countries such as Jamaica and Barbados, has become a particularly important location for American back offices. American Airlines paved the way in the Caribbean when it moved its data-processing center from Tulsa to Barbados in 1981; through its subsidiary Caribbean Data Services (CDS), it expanded operations to Montego Bay, Jamaica, and Santo Domingo, Dominican Republic. Manila, in the Philippines, has emerged as a back-office center for British firms, with wages only 20% of those in the United Kingdom.

The microelectronics and telecommunications revolutions allowed for the dispersal not only of offshore banks and back offices but also telephone operators, call centers, graphic designers, Internet Service Providers (ISPs), and readers of magnetic resonance imaging (MRI) records, all of which can be transmitted easily through the Internet. Such trends indicate that globalization and telecommunications may accelerate the offshoring of many low-wage, low-value-added jobs from the United States, with dire consequences for unskilled workers.

Another form of low-wage, low-value-added services involves centers of telework, often labeled **call centers**, which are designed to handle high-volume sales, marketing, customer service, telemarketing, and technical support calls. Call center functions include telemarketing, customer assistance, and phone orders. They range greatly in size, from as few as five to as many as several thousand employees. Like back offices, call centers are primarily screen-based and do not require proximity to clients. The major cost consideration is labor, although the workforce consists primarily of low-skilled women, and high turnover rates are common. There are an estimated 80,000 to 100,000 call centers within the United States, employing between 3% and 5% of the national labor force, the majority of which are located in urban or suburban locations.

As with so many industries, call centers have become increasingly globalized. India, for example, has attracted a significant number of customer service centers near its software capital of Bangalore, where workers are trained to speak with the U.S. dialect of English and are able to gossip with customers about pop culture. Wages there, which average $2000 per year, are higher than average Indian salaries but only 10% of what equivalent jobs pay in the United States. The next time you call a company about your cell phone or hard drive or to make an airline reservation, for example, you are likely to be speaking to a worker in India.

CONSUMER SERVICES

The attention paid to services by economic geography has focused mostly on producer services, those that are sold primarily to other firms. However, a large part of the service sector concerns consumer services, those sold to individuals and households. Consumer services include a wide variety of industries and occupations—retail trade and personal services, eating and drinking establishments, tourism, and sports and entertainment facilities. To some extent, the health care and education fields can be considered consumer services.

Traditional geographic studies of consumer services focused on the travel costs incurred in their consumption. Many large retail firms today deploy marketing specialists to assess the market potential of store locations, including the possible number of consumers, land costs, and access to transportation routes, an approach that often employs geographical information systems (GISs). In economic base theory, consumer services are generally considered part of the nonbasic sector (i.e., the part of the economy that recycles revenues derived through the export base). Consumer services, therefore, have been seen as reliant on the wealth created in other sectors. Often, however, the line between basic and nonbasic industries is hard to draw. Financial and legal services, for example, may be consumed by both firms and households, and to this extent they, too, are part of the consumer services sector.

More recently, geographers have taken a fresh look at consumer services, as potentially basic sectors in their own right. In some cases, consumer services may form the basis of local economic development after all. In large cities with specialized department stores (e.g., New York City), retail trade lures nonlocal residents in and forms part of the economic base. Outlet malls in small towns have a similar function. Tourism and conventions also are consumer services that form part of the economic base.

The distribution of consumer services largely reflects the affluence of the local client base. When disposable incomes rise, consumer services tend to thrive (they are income-elastic), although we must remember that the expenditures on goods and services do not rise equally in all cases. The study of consumer services is thus intimately linked to local demographics (age, gender, and ethnic factors), the associated tastes and preferences of people, their inclination to save or spend, and the advertising and other sources of information that shape their preferences that affect spending and buying habits. The gender, age, and ethnic composition of the labor force is also important. Other forces that shape consumer services include changing household work patterns, the presence of chains and franchises, regulatory frameworks, and the relative prices of imports and other goods, which themselves reflect a galaxy of variables in the global economy. The ever-changing geography of consumer services is also heavily influenced by the location of different types of purchasing power. For example, as suburbanization drew much of the middle class and its incomes to the metropolitan periphery, the urban geography of retail trade was reshaped, leading to an explosion of suburban malls of various sizes.

Tourism

Tourism is defined as travel to a place outside of one's usual residential environment and involving a stay of at least one night. Tourism is an important consumer service with enormous international and local implications. It involves a large variety of visitors, including personal as well as business travelers and conventions, staying over a wide range of time. Domestic tourists generally outnumber international visitors by a substantial margin, although from a locality, both types of visitors involve basic sector activities. When tourists spend money at a destination, the destination exports a service.

The growth of tourism reflects the increase in disposable income among a considerable segment of the world's population. Although mass tourism began in the nineteenth century, it grew particularly rapidly after World War II, when it became an option for the working classes. Contributing to this trend was the automobile and the wide-body jet airplane, which greatly reduced transportation costs. Fundamentally, tourism reflects the highly income-elastic demand for leisure that accompanies economic development, a demand that takes various forms: the need to "get away from it all," to take a break from routine, and to seek out education and novelty.

An enormous industry, tourism forms part of the economic base of many countries, including, for example, Italy, France, and Spain, as well as smaller island states in the Caribbean, Hawaii, and Florida, and cities such as New Orleans, Las Vegas, Los Angeles, and New York City. Table 4 shows the distribution of the world's 797 million international tourist arrivals in 2008 (12% of the world's

TABLE 4 International Tourist Arrivals, 2008

Region	Visitors (millions)	% of World
Europe	441.5	55.4
Africa	37.3	4.7
North America	85.8	10.7
Caribbean	18.1	2.3
South and Central America	21.7	2.3
East Asia	87.6	11.0
Southeast Asia	48.3	6.1
South Asia	7.6	1.0
Oceania	10.1	1.3
Middle East	39.0	4.9
World	797.0	100.0

Source: World Tourism Organization.

Case Study

Medical Tourism

The globalization of economic sectors has expanded to include health care. Traditionally, health care has been relatively unaffected by international trade, and health care markets were relatively closed national entities. However, institutions such as the World Trade Organization and agreements such as the General Agreement on Trade in Services (GATS) have steadily opened up these markets to foreign competition, typically focusing only on privately provided health care. International trade in health care services includes medical tourism, telemedicine, temporary movements of medical professionals across borders, and affiliate offices of health care corporations abroad. Many countries worry about the impacts of this trend on national health care systems, particularly in terms of potential "brain-drain" disruptions in service to the poor and underserved places such as rural areas or from the public to the private sector and skewed national health care priorities that emphasize foreigners over domestic citizens. However, the GATS specifically exempts publicly provided health services that do not compete with privately provided ones and allows member countries to choose to what extent they may wish to liberalize health services, if at all. As with many services traded internationally, the medical services trade is often hampered by nontariff barriers, notably licensing requirements, and many states have policies that charge foreigners extra for treatment.

The United States, with an inefficient and inequitable health care insurance system (the only industrialized country without publicly provided national health insurance), has witnessed health care costs rise dramatically to roughly 17% of its gross domestic product (GDP), significantly higher than in Canada, Japan, or Western Europe. Rising employment opportunities in this sector have been complemented by imported labor from a variety of countries, particularly India and the Philippines. An aging baby boom of 80 million people (including 11,000 Americans who turn 50 daily, or one every 8 seconds) has led many to seek medical care abroad. The United States is also home to large numbers of uninsured people (47 million) as well as 108 million who lack dental insurance. Many American "working poor" earn too much to qualify for Medicaid but lack jobs that provide health care insurance. In addition, there are countless others who are underinsured, that is, who have health care insurance policies that cover only the bare minimum, leaving them to pay for many procedures on their own. Indeed, high medical costs are responsible for 62% of personal bankruptcies in the United States.

This population, faced with steadily rising medical costs, has created most of the demand for health care from foreign providers, generating a steady exodus of medical tourists to India, Thailand, Singapore, Mexico, Venezuela, Colombia, Malaysia, Cuba, the Philippines, Costa Rica, Turkey, the United Arab Emirates, and elsewhere. Medical tourism, in which the person physically travels to another country for the purpose of obtaining either medically necessary or optional health-related services, should be differentiated from the outsourcing of medical functions via the Web, such as teleradiology, or interpretations of U.S. computerized axial tomography (CAT) scans by doctors in India. Estimates of the number of Americans traveling abroad annually to seek medical care vary considerably, depending on the source, but typically range between 180,000 and 750,000, spending at least $2.2 billion.

Medical tourism has accelerated the evolution of health care in some developing countries toward specialized services oriented to the more lucrative parts of the health care market. Increasingly, hospitals in the developing world have built the necessary infrastructure to serve medical tourists, including specialized skills and equipment and support staff. As the opportunities for treatment abroad have multiplied, medical tourists have availed themselves of them. Initially synonymous with plastic surgery, medical tourism has long surpassed its early, purely cosmetic affiliation to include vital, even life-saving operations. For many countries, it is the most rapidly growing segment of the tourist market. Nonetheless, many potential patients are deeply fearful of the quality and reliability of overseas medical care, often associating it with poor-quality care or organs purchased from impoverished donors.

Concerns over medical tourism generally center on the quality of care received and liability for malpractice in the absence of an international body to regulate the industry. Courts are usually reluctant to assert jurisdiction over physicians in other countries, and medical tourism plaintiffs face a long, uphill battle in attempts to gain compensation for botched procedures overseas. In this context, debates over negligence, liability, fraud, misrepresentation of credentials or risks, or lack of informed consent can be the Achilles heel of medical tourism.

Unlike recreational tourists, who have sufficient disposable income to travel for leisure, many medical tourists are individuals and families of modest means, for whom out-of-pocket health-related costs are an onerous burden. Medical tourists are typically middle-income Americans evading impoverishment by expensive, medically necessary operations. Avoiding bankruptcy and protecting assets such as home equity are often prime motivations for medical tourists. Of course, not all medical tourists are on the brink of financial collapse: Many are financially comfortable and do have health insurance but prefer to avoid paying the steep copayments or higher premiums that they would still encounter in the United States, particularly in the case of expensive medical procedures.

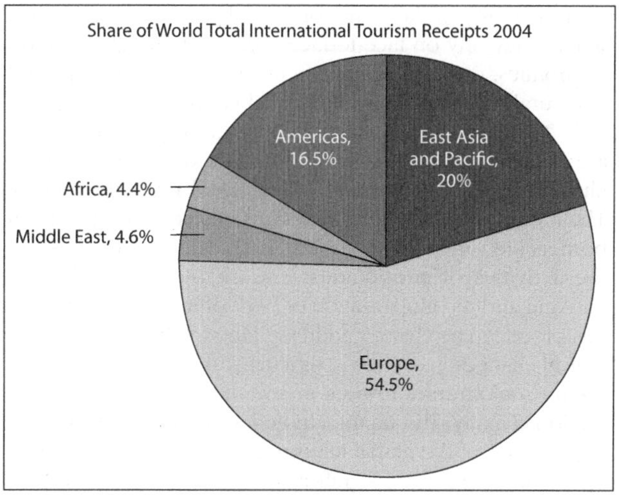

FIGURE 28 Distribution of global tourism receipts.

population): More than one-half of the world's tourists visited Europe (Figure 28). France, Italy, and Spain boast the largest tourist industries in the world, attracting large numbers of tourists from colder northern climates. To a lesser extent, East Asia and North America are also significant.

The types of tourism and tourist destinations are highly diverse, ranging from low-impact ecotourism to highly urbanized urban cores, from simple backpacking to luxury cruises, from safaris to writers' camps to honeymoon retreats, from individual explorers to package tours, from health resorts to Asian sex tourism, from dude ranches to gay night clubs, from museums and fine art institutes to tribal cultural events, from small ski resorts to the tropical playgrounds organized by Club Med. The facilities that accommodate tourists are similarly varied, ranging from large hotel chains to quaint bed-and-breakfast establishments. Tourism is often highly seasonal, fluctuating greatly over the course of a year, with corresponding changes in prices for hotels and travel.

The volume of tourists to a given destination will reflect, among other things, the information available to potential clients; their disposable incomes and willingness to travel; currency exchange rates; transportation and lodging supply and costs; the relative cultural familiarity or degree of exotic appeal the destination may have; concerns over crime; and unpredictable events such as terrorist attacks. Political restrictions, for example, prohibit U.S. tourists from visiting Cuba, which attracts visitors from Europe and Canada.

Economically, the tourism industry consists of a variety of sectors, including international and local transportation, hotels and motels, eating and drinking establishments, entertainment, and retail trade. To sustain these activities requires enormous investments in water, transportation, communications, and other infrastructures. The impacts of tourism are varied and often mostly confined to localized sectors that deal with visitors, such as hotels and retail trade. In cases where a large share of inputs is not purchased locally, tourism can generate relatively small positive impacts. "Tourist enclaves" in which visitors have few contacts with local residents are not uncommon. Often the jobs involved are unskilled (e.g., janitors) and pay low wages, with low multiplier effects. Indeed, tourist-dependent economies may be trapped in the "low road" to economic development. At times, tourist developments may generate resentment among locals, who can witness wealthy foreigners frolicking in the sun but who cannot enjoy the benefits themselves. In some cases, particularly in fragile ecosystems, tourism can have significant detrimental effects on local flora and fauna. Skiing in the Alps or backpacking in Nepal, for example, can damage local landscapes by generating garbage; similarly, high-density tourist developments (e.g., Hawaii) can create large quantities of waste.

The tourist area cycle model, which resembles the product cycle, holds that tourist areas, their attractions, and the number and type of tourists change in a predictable series of stages over time: (1) exploration, in which a small number of tourists are attracted to natural and cultural features and have considerable amounts of contact with local residents; (2) rising numbers of visitors; (3) development, where the tourist population during peak seasons is greater than the local population and foreign capital flows in; (4) consolidation, where marketing and advertising are used to offset potential declines in tourists; (5) stagnation, during which the peak number of visitors is reached, original attractions are displaced by newly created ones; and (6) decline, in which the number of visitors falls. Some places may experience rejuvenation by completely changing the tourist attractions or by capitalizing on previously untapped resources. This model has been much examined and its fit to real places debated, but it has important implications for planning because it confronts the assumption that tourist areas are unchanging and will always attract an increasing number of visitors.

Summary

Throughout the world, particularly in developed countries, the vast majority of people work in services firms—shops, offices, restaurants, hospitals, airlines, banks, universities, and hotels. A service is any economic activity that provides a benefit to the consumer that does not involve the creation of a physical good or product (i.e., services involve the production and consumption of intangibles).

This chapter explored the difficulties of defining services and emphasized the significant differences that exist among and within services industries. Services encompass both highly skilled, well-paying positions, such as international financier or brain surgeon, and low-paying, unskilled jobs, such as security guard, cashier, and janitor. It explored the reasons for the massive structural shift from a manufacturing economy to a services economy, noting the driving forces of changes in income and consumption patterns, demographics, international trade, and the increasingly complex production environment. The chapter paid particular attention to the externalization of services, by which many companies, in an age of globalized post-Fordism, shed the functions they previously performed in-house (vertical disintegration).

The chapter delved into the debates about the productivity of services, which is notoriously difficult to measure. Services have been blamed for the slowdown in productivity growth in the world over the past few decades, but the extent to which this claim is true is questionable. Labor markets in services received scrutiny, including the empirical patterns of growth in the United States as well as the features that differentiate services labor markets from those in manufacturing. It also offered several sectoral studies to demonstrate that the dynamics of services vary from sector to sector.

The geography of services is characterized by an ongoing tension between the pressures to agglomerate, typically in large metropolitan areas, and the decentralizing forces associated with standardization, routinization, and the availability of information technology. While highly skilled, white-collar functions tend to remain in cities, where they rely on face-to-face contact and agglomeration economies, unskilled functions have increasingly dispersed into rural areas and less developed countries.

Contrary to popular, erroneous stereotypes that services are always consumed where they are produced, this chapter illustrated interregional and international trade in services. The means by which companies sell services across national borders are varied and complex and include direct sales and the activities of subsidiaries. The chapter cited offshore banking and the globalization of back offices as examples of this process. The chapter addressed the multitude of technological changes in services, particularly those associated with the microelectronics revolution and the digitization of information. Finally, the chapter noted some of the economic implications of the global telecommunications infrastructure, including the role and significance of electronic funds transfer systems. The emergence of hypermobile, digitized money has been a defining feature of contemporary globalization and has brought with it new forms of global finance (offshore centers) as well as the internationalization of clerical work in the form of the offshore back office. These trends are likely to continue, if not accelerate, in the future.

Finally, the chapter noted the growing role of consumer services, not all of which are simply nonbasic in character. Tourism, arguably the world's largest industry in terms of employment, is a very powerful shaper of national and local economies in a variety of ways.

Key Terms

back offices
business services
call centers
consumer services
contingent labor
electronic funds transfer systems (EFTS)
externalization
FIRE (finance, insurance, and real estate)
income-elastic
intangible output
nondirect production workers
nonprofit services
offshore banking centers
outsourcing
postindustrial economy
producer services
services
wholesale and retail trade services

Study Questions

1. What proportion of the labor force in North America, Japan, and Europe works in services?
2. Why is it difficult to measure the productivity of services?
3. What are the major industries that comprise *the* service sector?
4. What are six reasons services have grown so quickly?
5. Why have services been externalized by many firms?
6. What are five ways in which labor markets in services differ from those in manufacturing?
7. Where do producer services tend to locate, and why?
8. What are the components of the FIRE sector?
9. What are the causes and consequences of the 2008–2009 "Great Recession"?
10. Why do U.S. law firms do business overseas?
11. What forms does international trade in services take?
12. What are electronic funds transfer systems?
13. Where are the major areas of offshore banking?
14. How have firms globalized their back offices?
15. Are all consumer services nonbasic? Explain.
16. Define *tourism*. What continent is the world's leader in this industry?

Suggested Readings

Bryson, J., P. Daniels, and B. Warf. 2004. *Service Worlds: People, Organizations, Technologies*. London: Routledge.

Foster, J., and F. Magdoff. 2009. *The Great Financial Crisis: Causes and Consequences*. New York: Monthly Review Press.

Krugman, P. 2008. *The Return of Depression Economics and the Crisis of 2008*. New York: W.W. Norton.

Schiller, D. 1999. *Digital Capitalism: Networking the Global Market System*. Cambridge, MA: MIT Press.

Shaw, G., and A. Williams. 2004. *Tourism and Tourism Spaces*. London: Sage Publications.

Solomon, R. 1999. *Money on the Move: The Revolution in International Finance since 1980*. Princeton, NJ: Princeton University Press.

Web Resources

Reser
http://www.reser.net

Home page of the European services research network.

World Tourism Organization
http://www.world-tourism.org

This site provides comprehensive data and analysis of international tourism flows and impacts.

Cities and Urban Economies

OBJECTIVES

- To explore the relationship between modern urban growth and the development of capitalism
- To analyze how cities are linked together through their economic bases and export sectors
- To describe how the supply and demand for housing is related to residential space
- To summarize the causes and consequences of suburbanization and urban sprawl
- To address the reasons, costs, and benefits of gentrification
- To illustrate the reasons for inner-city poverty and the multiple problems of the ghetto
- To discuss global cities in light of the current round of globalization
- To introduce the concept of urban sustainability

Large, congested metropolitan areas, such as downtown Los Angeles, California, testify to the power of agglomeration economies in clustering activity in dense urban areas and the important role such conurbations play in regional, national, and global economic linkages.

Mike Powell/Stone/Getty Images

Cities and Urban Economies

Cities lie at the heart of economic geography. For thousands of years, and particularly since the emergence of industrial capitalism, cities have played a uniquely important economic, political, and social role. A city is many things: It is a built environment—a tangible expression of religious, political, economic, and social forces that houses a host of activities in proximity to one another. Cities also consist of dense webs of social relationships that are fundamental to the formation and operation of economic ties and the division of labor. Cities are where the engines of capitalist growth tend to be concentrated, centers of innovation, corporate operations, and key nodal points that tie together vast flows of money, power, people, goods, and information. Cities are also depositories of cultural meaning, where the symbolic systems that people use to negotiate the world are produced and consumed. In short, cities, the foundation of modern life, represent humanity's largest and most durable artifact. They are living systems—made, transformed, and experienced by people.

This chapter summarizes some major issues pertaining to urbanization and economic geography. It opens with a historical overview of the role of cities in the development of capitalism and the urban division of labor. Next is a discussion of residential space, including the residential location decision, the filtering process of housing, and the supply and demand for housing. The dynamics and consequences of suburbanization and urban sprawl are explored, including the reasons that underlie it and its impacts on metropolitan cores and peripheries. Gentrification and the resurgence of certain inner-city cores are discussed, as well as the persistent problems of the inner city, including the crisis of the African American ghetto. Finally, the chapter concludes with some observations about global cities and the international urban hierarchy.

THE RISE OF THE MODERN CITY

From the sixteenth century onward, cities grew along with the emerging capitalist economy. A network of new towns spread across the European continent, fueled by population growth and rising productivity. A new merchant class came into being, and capitalization of the countryside drove many peasants into cities, where capitalist labor markets (i.e., the commodification of labor time) created the working class. The accumulation of capital, the growth of new social classes, new trade networks, inexpensive colonial labor and raw materials, and scientific and technological breakthroughs combined to destroy feudal barriers to production. The capitalist city provided lower transportation and communication costs for firms that needed to interact with one another; hence, most commercial and industrial enterprises concentrated in and around the most accessible parts of the city.

With the Industrial Revolution came the steady penetration of European societies by the new market relations. The industrialization of agriculture displaced thousands of rural workers, resulting in waves of rural-to-urban migration. In Britain, home of the Industrial Revolution, industrialization brought about the rapid growth of manufacturing centers in the Midlands as well as in London, Glasgow, and Belfast in Northern Ireland (Figure 1).

The rise of capitalism was accompanied by colonialism and the European conquest of much of the world. In the process, European colonization threatened the existing urban civilizations of Asia, Africa, and the Americas. Often, land-based cities in Asia, Africa, and the Americas were undermined by the growth of colonial ports (e.g., Lima, Peru; Rangoon, Burma; Jakarta, Indonesia). Centuries of European penetration and occupation also resulted in the growth of many cities such as Calcutta, Hong Kong, and Singapore that owe their origins to colonial foundations or to trading requirements. Thus, from the sixteenth to the nineteenth centuries, *colonial cities* dominated the urban patterns of Africa, Asia, and Latin America. After political independence and the development of the new international division

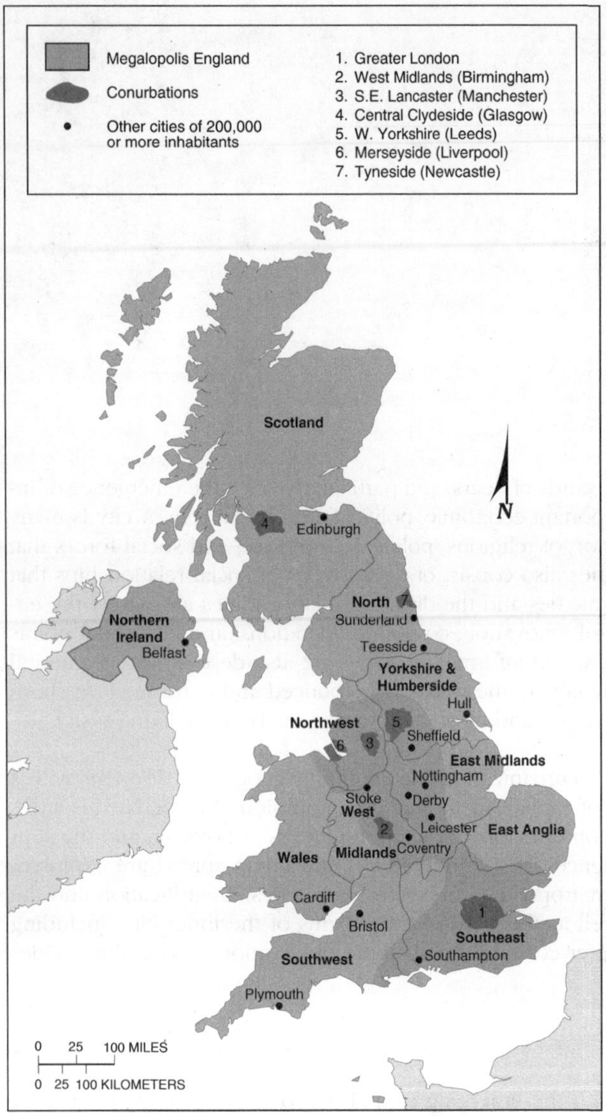

FIGURE 1 Urban Britain, home of the Industrial Revolution. Britain's early lead in industrialization created a network of cities that made it the most urbanized country in the world in the nineteenth century.

of labor, the transformation of the urban process in previously colonized countries was as profound as it was in nineteenth-century Europe and North America. The construction of a global, maritime-based world economy, mercantilism, and the growth of port cities in Latin America, Asia, and Africa can therefore be viewed as different facets of the same process.

Waves of rapid urban growth swept through Europe and North America in the nineteenth century (Figure 2). In the United States, the emerging Industrial Revolution caused explosive growth in eastern seaport cities like New York, Boston, and Philadelphia (very few cities existed west of the Appalachian Mountains until the latter half of the century). As the Manufacturing Belt came into being, cities like Chicago, Cleveland, Milwaukee, and St. Louis grew up around the consolidation of agricultural trade and the establishment of a national railroad system. The steel industry gave rise to Pittsburgh, Buffalo, and a host of other industrial centers. The Canadian urban core of southern Ontario and Montreal became established for similar reasons (Figure 3). Under the capitalist regime, cities were the location of much industrial innovation and prodigious increases in productivity, fueled by large numbers of immigrants. The standards of achievement were based on industry and technology. However, cities of this era were invariably ugly creations and horrifying environments for the laboring poor, who toiled under highly exploitative conditions. With few systems for managing urban waste, most cities were dirty, polluted, filled with garbage, and their inhabitants were persistently exposed to numerous disease agents.

The modern corporation has had a huge influence on the contemporary Western city. Corporate administrative buildings dominate skylines and take up extensive land areas. The geography of Western cities has also been affected by the repeated need of corporate enterprises to find ways to absorb their surpluses by creating new markets and commodifying ever larger spheres of social life. Since the late twentieth century, for example, large corporations and the federal government have poured resources into urban renewal projects in numerous central cities, reclaiming downtown space that once suffered from the impacts of suburbanization and deindustrialization. In short, the structure and form of the city are inseparable from the dynamics of capitalism, including state policy, as the logic of commodity production and consumption plays out unevenly in different urban areas.

URBAN ECONOMIC BASE ANALYSIS

Because they are always part of a broad capitalist division of labor, cities are integrated with a wider economic environment that both reflects and shapes their structure, location, and behavior in time and space. Linkages among cities are largely linkages among firms, which buy and sell parts, goods, and services to one another extensively. Firms thus simultaneously constitute and create geographies, that is, they are shaped by and in turn shape their local, regional, and global surroundings. **Economic base analysis** (sometimes also called export base theory) offers a means of understanding how cities are integrated with one another as well as a way to understand the essentials of how urban economies function.

An important part of the geography of capitalism is the tendency for cities and regions to specialize in the production of some outputs to the exclusion of others (i.e., develop a comparative advantage). Cities are, in short, part of the creation and reproduction of uneven development across space. American history provides numerous examples of urban specialization in the past and the present. For example, New York City has long been known as a garment-producing center and the capital of

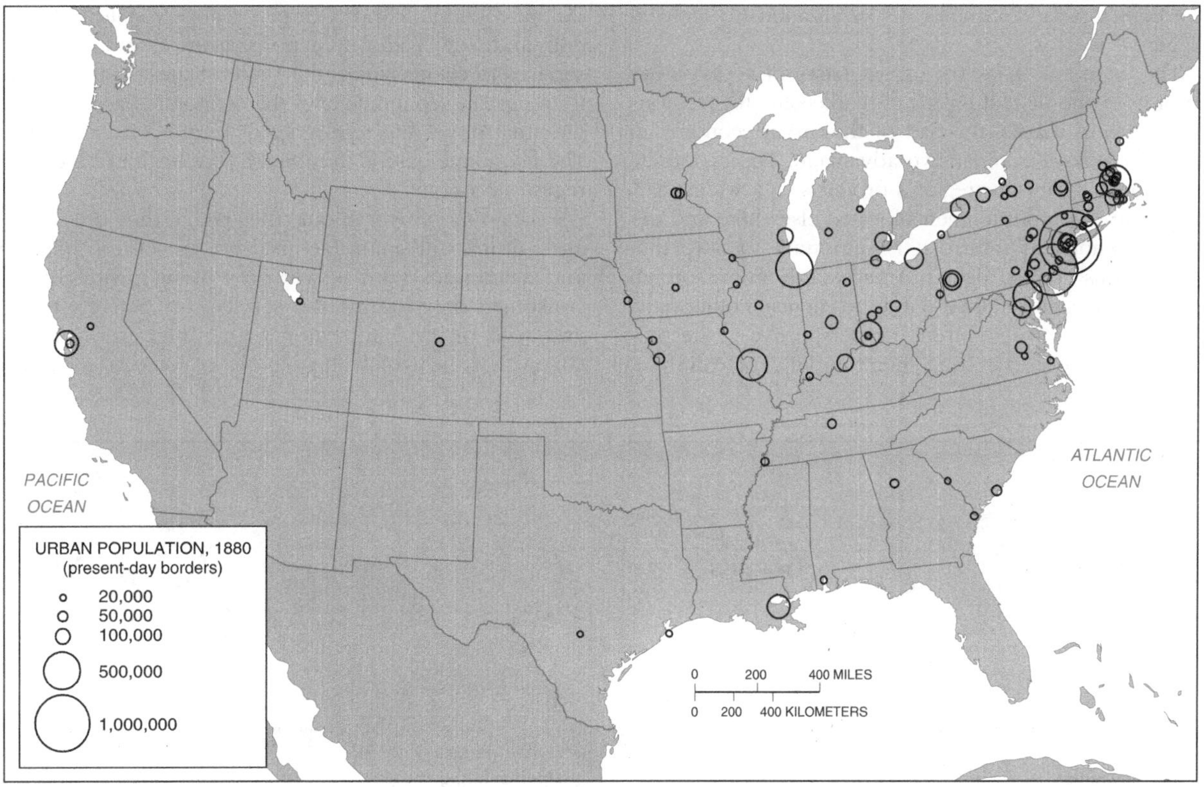

FIGURE 2 Urban population of the United States, 1880, just as the Industrial Revolution was creating the Manufacturing Belt.

finance, as well as other producer services; Detroit focused on automobile production; Akron, Ohio, was once the epicenter of rubber manufacturing; Pittsburgh, Buffalo, and Youngstown were renowned as centers of the steel industry; Minneapolis originated as a flour-milling center; Corning, New York, has long been an important glass-producing city; Cincinnati was once known as "porkopolis"; Memphis was the cotton seed oil capital of the Mississippi; Los Angeles is the leading film-producing center on the planet; and Seattle has long been closely tied to aerospace as the home of the Boeing Corporation. Often, urban specialization takes the form of small towns that arise in connection with agricultural production, timber and lumber, or mining centers.

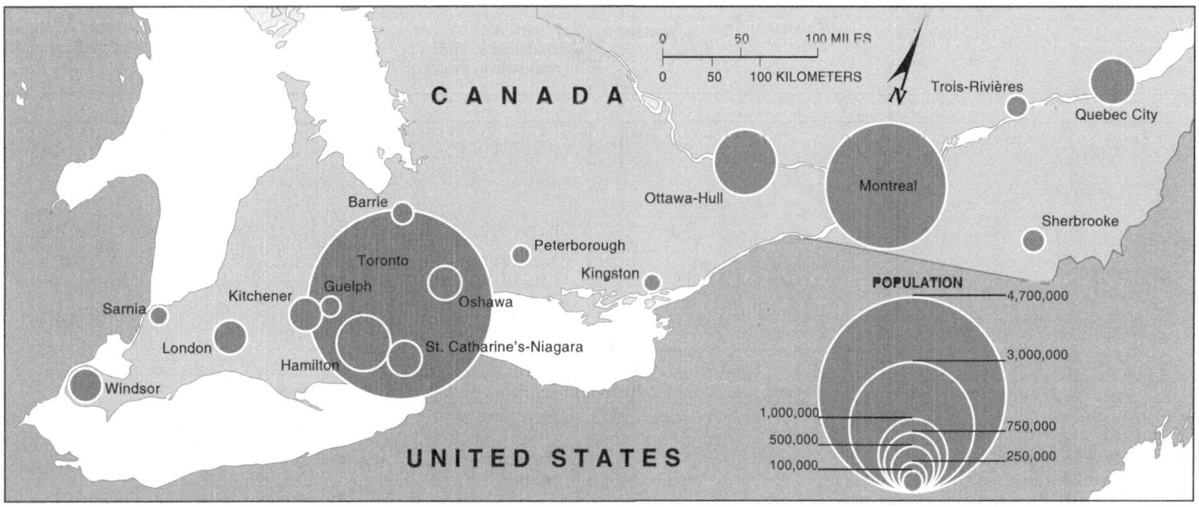

FIGURE 3 Canada's urban core lies in southern Ontario.

Some contemporary examples of specialization are depicted in Figure 4.

The economic base (or export base) of a city is the part of its economy that links it to markets in other regions and countries, that is, the part of the urban economy in which it enjoys a comparative advantage. The economic base exports a city's or region's products to a wider market, selling its output to clients located elsewhere and deriving revenues in return. It is important to note that *export* in this context does not necessarily mean foreign exports. Cities that sell their output to clients in other parts of the same country are, from the perspective of the producer, effectively exporting their output. For example, Dalton, Georgia, exports carpets to every state in the nation, and New York City exports financial and advertising services to clients across the United States. This approach to the urban economy views the economic base as the engine that drives the remainder of local economic activity. The economic base is thus vital to the health of a city's or region's economy.

A common myth about cities is that they only export agricultural and manufactured goods. Often, planners and academics view services as being produced and consumed only locally. But services are actively bought and sold on an interregional and international basis. Given that the vast bulk of the labor force is involved in

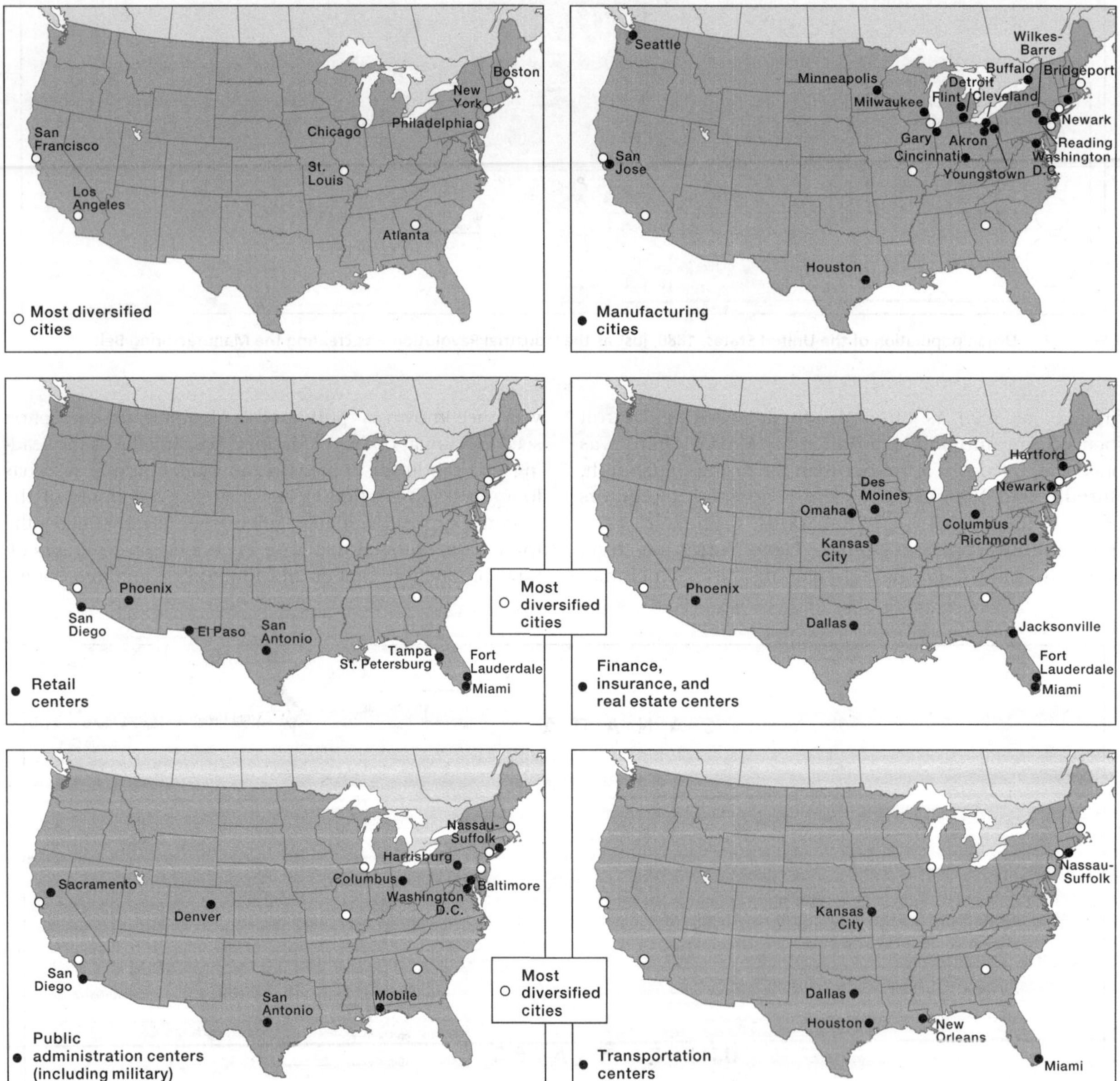

FIGURE 4 Economic specialization within the U.S. urban hierarchy.

service industries, it would be astonishing if services were *not* traded among cities. New York, for example, sells financial services; Pittsburgh, Birmingham, and Gainesville export health care services to clients in other cities and states; and Los Angeles exports television and film services. University towns export educational services to students who move there from other places to attend school; their tuition comprises a form of export earnings. Whenever a radio or television station in one city sells advertising time to a client located in another, it exports a service. When tourists descend on Disneyworld, New Orleans, Las Vegas, or San Francisco, those places export services. Even Washington, DC (as well as state capitals) exports government services in the sense that the costs of producing these (as well as the benefits) are paid for (and enjoyed by) people and firms located elsewhere. Thus, cities certainly can and do have service industries as an export base.

It is useful, therefore, to divide the urban economy into two broad segments, the economic base, or **basic sector** (B), and the **nonbasic sector** (NB). If the basic sector is export-oriented (selling agricultural or manufactured goods, or services), the nonbasic sector recycles corporate and personal incomes in ways that meet the demands of firms and households locally. Nonbasic economic sectors typically include retail trade, eating and drinking establishments, real estate, and personal services (e.g., beauty salons) (Figure 5). Essentially, these functions rely on local earnings and cater to local demand, and are not export-oriented. From the standpoint of economic base analysis, therefore, the nonbasic sector plays a less important role in shaping the urban economy than does the economic base. Of course, sometimes the division between these two sectors is not so simple. Some cities, such as New York, or outlet malls, for example, may export retail services, and some sectors, such as legal services, may be both basic and nonbasic (e.g., a law firm that handles corporate mergers in another city and local divorces). Firms typically sell to both local and nonlocal clients, which makes the boundary between the basic and nonbasic sectors fuzzy at times.

The relative sizes and relations between the economic base and the nonbasic sector are important to understand. Using a simple economic base model, total employment (T) equals the sum of the basic (B) and nonbasic (NB) sectors, or

$$T = B + NB.$$

Now let's introduce the concept of a **multiplier**, m. In its simplest form, the multiplier reflects the degree to which the basic sector "drives" the total economy, that is, the impacts of the export-oriented basic sector on the entire urban area. There are many types of multipliers and ways of estimating them, but for the sake of simplicity, let us define the multiplier (m) as the ratio of total to basic sector employment,

$$M = T/B.$$

Substituting ($B + NB$) for T in the preceding equation yields

$$m = (B + NB)/B = 1 + NB/B.$$

(Note this is the simplest possible definition of an *average* multiplier; more sophisticated approaches use *marginal* multipliers, which are calculated slightly differently.) This is a useful approach because it allows us to estimate the degree to which changes in the basic sector (ΔB) cause changes in the total economy (ΔT), that is,

$$\Delta T = m\ \Delta B.$$

Consider a very simple example. Say that a small lumber town has 1000 people in the labor force. (Total population will be larger than total labor force, depending on the labor force participation rate and demographic structure of the community.) Of that 1000, 400 work for a local paper mill. If the paper mill hires an additional 100 people, what will be the impact on the community's employment? We can answer this question simply enough by defining $B = 400$; therefore, $NB = 600$. Thus, the paper mill's multiplier is

$$m = 1 + 600/400 = 2.5.$$

Courtesy of The Taubman Company

FIGURE 5 Stamford Town Center, Connecticut. The shopping malls and retail stores here exemplify the urban nonbasic sector, which recycles incomes in the local economy to meet the needs of households and firms.

The change in total employment that an additional 100 in the basic sector generates is thus

$$\Delta T = m\ \Delta B = 2.5 \cdot 100 = 250.$$

The additional 100 jobs in the paper mill created 250 jobs in the community (including the 100 new jobs in the basic sector). Note that since the multiplier is always greater than zero, changes in the basic sector always create somewhat larger changes in the rest of the local economy; that is, since $m > 0$, $\Delta T > \Delta B$.

How did this process operate? The economic base model holds that when the basic sector expands (or contracts), the initial changes reverberate through the local economy by means of a series of interfirm linkages as well as changes in consumer spending. The total changes can be decomposed into three constituent parts. First, changes in total employment include changes in the basic sector itself, or **direct effects** (in this example, 100). Second, changes in total employment include changes in firms that sell goods and services to the export base through subcontracts, or **indirect effects**. For example, when the paper mill in the preceding example expands, it may increase its purchases of equipment, repair and maintenance services, office machinery, advertising services, legal services, and other inputs. Thus, the size of a multiplier—and thus the impact of the basic sector on the rest of the urban economy—is closely tied to the number and strength of linkages among firms. Every firm purchases inputs from other firms, that is, every purchase is a sale, depending on whether you take the perspective of the buyer or the seller. In the example above, firms that supply the paper mill with inputs will enjoy increases in sales as the paper mill buys more inputs from them. These are the indirect effects.

The size of the multiplier in many ways reflects the spatial distribution of backward linkages from the basic sector, that is, the geography of subcontracts between the exporting firms and their suppliers. To the degree that the basic sector firm subcontracts with suppliers located far away, many of the benefits of its growth will be exported. Firms that use nonlocal suppliers heavily, therefore, have much lower local impacts than do firms that rely on local suppliers of inputs. Conversely, the more the firms in the basic sector subcontract locally, the higher the local multiplier effect will be. For example, a law firm that uses local suppliers of paper or office equipment has greater impacts on its city than does a firm that purchases those inputs from suppliers in distant locations. Sometimes rules governing public subcontracts (e.g., for construction at universities) mandate the use of local employers for this reason.

Finally, the third set of changes resulting from multiplier effects are changes in consumer spending resulting from changes in the basic sector. When workers' incomes (in the exporting firm and its subcontractors) go up, they have more to spend. These expenditures, or **induced effects**, will occur primarily in the nonbasic part of the economy. Higher wages will lead to greater savings rates (depending on workers' propensity to save or spend) as well as higher expenditures on local real estate, food, entertainment, transportation, and other consumer goods and services. The size of induced effects will depend on the number of workers affected, their average salaries, and their spending habits and preferences.

It is important to note that multipliers are double-edged razors, that is, they may work against as well as for a region. Changes in the basic sector can be negative as well as positive, such as when a steel plant or auto assembly plant shuts down and devastates a local community. For example, Seattle was long dependent on the fortunes of the Boeing Company, giving rise to the saying "When Boeing sneezes, Seattle catches cold." In terms of economic base theory, if $\Delta B < 0$, then $m\Delta B$ will be < 0. For example, if a pulp mill closes down, or if a tool and die factory reduces its labor force (perhaps as part of a strategy of industrial reorganization and relocation overseas), then multiplier effects will amplify the initial losses. Subcontracts to supplying firms dry up, reducing the indirect effects. Lower workers' incomes in turn lead to negative induced effects, causing local businesses to suffer and real estate prices to decline (in long-suffering Detroit, the average housing price today is just $16,000). For this reason, the downtowns of many deindustrialized cities often have empty storefronts and housing that is relatively cheap.

Finally, diversity in the local economy is important. Typically, the basic sector declines and the nonbasic sector increases as city population grows, that is, the city moves from the lower to the upper tiers of the urban hierarchy (Figure 6). This pattern reflects the fact that large cities usually have a diversified economic base. Los Angeles, for example, has an economic base in the film industry as well as finance and producer services, port activities, and aerospace. Conversely, small towns often have very narrow economic bases centered on one industry, and often only

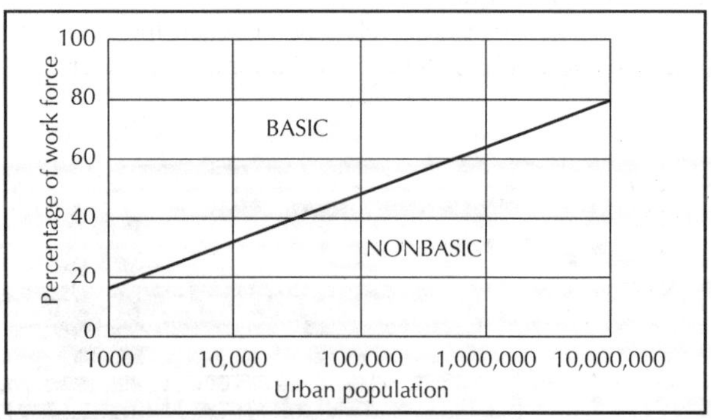

FIGURE 6 The relative sizes of the basic and nonbasic sectors vary greatly across the urban hierarchy. Generally, the economies of small towns such as those that rely on mining, agriculture, and lumbering specialize around the export of a single good; hence small-town basic sectors are relatively large. In large metropolitan areas, in contrast, diverse economies and large internal markets reduce the relative proportion of the basic sector in terms of the economy as a whole.

one firm, such as agricultural processing centers, mining, and lumber, which means their fortunes are very closely tied to a single firm or industry.

THE URBAN DIVISION OF LABOR

Within and among cities, there exists a division of labor, that is, social and spatial patterns of production that link individuals, firms, and urban areas together into complex networks. The division of labor is a fundamental aspect of capitalist society, and it has become steadily more specialized over time. Cities are central to this process: Indeed, from the perspective of firms, urban areas exist largely because they are advantageous to the production process and the creation of profits and value. The geography of the division of labor determines the spatial patterns of jobs and employment and therefore is central to understanding housing and commuting patterns as well.

At the core of the urban division of labor is the fact that different economic activities have different locational requirements, including the need for varying amounts of labor, land, and access to other firms. All firms, and therefore all urban areas, are enmeshed in production complexes—networks of linkages that bind them together. As the urban basic sector produces and exchanges goods and services with other cities, these linkages take a variety of forms, ranging from shipments of parts and raw materials to face-to-face meetings between executives and managers. Different sectors of production—garments, steel, electronics, business services—have different numbers and types of linkages; thus, different kinds of economic activity tend to locate in different parts of the city. All of this is a reflection of the broad imperative to minimize costs and maximize profits: Firms that do not find optimal locations tend to get driven out of business.

For many firms, agglomeration economies are critical, that is, access to other firms. By agglomerating, firms can produce what they cannot profitably do alone; they gain access to markets, corporate services (e.g., repair and maintenance, advertising, legal services), specialized labor and information, and the urban infrastructure. Some firms need to agglomerate more than others; generally, it is labor-intensive companies with extensive "forward" and "backward" linkages in the production process that must have the most access to clients and suppliers of inputs and services, and thus these tend to agglomerate in urban core areas, or **central business districts**. Because land in the central business district is very accessible, it is relatively expensive; firms that locate downtown therefore pay high rents, which is why downtown areas have tall buildings. Historically, firms that clustered downtown tended to be small manufacturing companies, such as clothing firms in the nineteenth and early twentieth centuries, but the same logic is true of many producer services today. As the economies of industrialized countries shifted from manufacturing to services, their central business districts witnessed changes in the landscape from smokestacks to skyscrapers. Conversely, larger, more capital-intensive firms, which often produce their own inputs rather than purchase them from other companies, tend to locate on the urban periphery, where they can obtain lower rents. Thus, manufacturing firms tend to seek out suburban areas rather than cluster in downtown regions, a process that accelerated rapidly after World War II with the interstate highway system. The relatively unskilled, lower-paying jobs like those in industrial branch plants and assembly jobs also tend to disperse to suburbs (e.g., industrial parks), smaller towns, and, increasingly, to the developing world.

Just as firms tend to seek out urban locations that are advantageous to them, collectively they create different types of labor markets. Generally, skilled, information-processing activities tend to cluster in downtown regions, where they can take advantage of the agglomeration economies available there. Thus, the clusters of steel and glass skyscrapers that dominate the heart of most large cities today tend to be filled with professional and technical workers with above-average levels of education who earn reasonable incomes. Today, many such employees are called members of the "creative class," that is, people whose jobs involve collecting, processing, and transmitting information in novel ways. Indeed, because creativity and innovation require extensive interactions among groups of people, urban core areas are central to the process of generating new ideas. However, urban core areas also have numerous low-paying jobs in sectors like retail trade and personal services.

In short, the spatial division of labor generates unequal geographical development within and among urban areas. But it is important to note that the urban division of labor is not static; it is always changing. The division of labor changes because the economy is in constant evolution, with new products, new production processes, changing forms of competition, technological changes, transformations in labor relations, and government intervention all altering the locational calculus of companies. For example, the Industrial Revolution had profound impacts on the structure of urban areas, generating large numbers of firms that increasingly became oligopolized over time; as such firms grew, they tended to disperse to the urban periphery, and more recently, to the global periphery. Similarly, the federal government's construction of the interstate highway system, with numerous roads that bypassed metropolitan areas, lowered transportation costs and altered the distribution of potential locations for firms, enhancing the locational value of suburban areas.

Typically, as firms became more capital-intensive, they tended to internalize the linkages that bound them to other firms; that is, they tend to make their inputs "in-house" rather than buy them from suppliers. Firms that undergo this process rely less heavily on agglomeration economies and have less need to be near other firms. As a result, they often seek out peripheral areas (e.g., suburbs or small towns) where labor and land costs are lower. In short, as firms go through the product cycle and become larger, they tend to vacate inner city areas. Under capitalism, urbanization has long tended toward industrial decentralization. In contrast, the rise of a service-based economy as well as the agglomerative pressures of post-Fordist production, in which subcontracts among firms have multiplied, have tended to foster renewed growth in centralized locations, including both gentrified downtowns and broader regions, such as Silicon Valley. The varying processes of corporate decentralization and centralization, which reflect the changing structures of capitalist production, go far to explaining the rise and fall (and sometime rise again) of different cities.

Finally, the division of labor within each city is also shaped by its interactions with other cities. Urban areas are embedded within urban hierarchies, forming city-systems within systems of cities of varying scales. The **urban hierarchy** consists of a series of cities, ranging from small towns, which often form around primary-sector, resource-extracting activities, to larger, more service-oriented conglomerations. In this way, urbanization generates a spatial division of labor that is national, and increasingly global, in scope. At the top of this system are **global cities** with linkages that stretch across the planet. Thus, cities can never be understood in isolation of one another: The key to understanding the economic health of cities lies in appreciating their ties with other places.

URBAN RESIDENTIAL SPACE

In addition to the spatial ordering of economic activities within cities, that is, spaces of production, there are also spaces of reproduction, which encompass the activities of the people who inhabit cities, including their residential locations. In short, people don't just work in cities, they also live in them. Three dimensions of the intraurban residential spatial structure are addressed here: residential location, the filtering process of housing, and the supply and demand for housing.

The Residential Location Decision

Typically, the most important criterion for people selecting a home is its accessibility to where they work. The **residential location decision** depends in part on how much money a family can afford to spend. A family's budget must cover living costs, housing costs, and transportation costs. Poor families, who have relatively little money to spend on commuting after living and housing expenses are deducted from their income, attach much importance to living close to where they work. The only way the inner-city poor can afford to live on high-rent land is to consume less space, that is, live in small, densely populated housing units such as apartment buildings. On the other hand, relatively wealthier families have plenty of money to spend on transportation; therefore, the proximity of residential sites to their places of employment is of little consequence to them. They can trade access to their jobs for larger housing lots farther away from the center of the city. Because most people prefer substantial quantities of space over short commuting times, the middle class tends to live in the urban periphery. This process of segregation generates the familiar, ironic pattern in which the relatively prosperous live on the cheaper land in the suburbs and the poor are concentrated in the inner city. Although cost, time, and mode of travel to work have important implications for residential location, many other dimensions of accessibility, such as nearness to personal services, shopping, and entertainment must be considered. In every situation, however, the rich can outbid the poor. The results in market societies are always relatively advantageous to the rich and relatively disadvantageous to the poor.

The Filtering Model of Housing

For most people, housing consists of units that have been previously owned or occupied by someone else. The **filtering model** describes the process by which houses pass from one household to another. The model is based on the principle that there is a decrease in housing quality through time through physical deterioration; deferred maintenance; technological obsolescence of fixtures, facilities, and design; and changes in housing construction standards and methods. The filtering model suggests that houses are occupied by families with progressively lower incomes as the quality of the house declines over time and it becomes lower in price. The dwelling becomes affordable to households that settle for progressively lower quality housing because of their more limited incomes (Figure 7), which is why low-income households occupy used rather than new housing. The filtering model also explains why poor households benefit from zoning and land-use policies that encourage the building of new, high-end housing throughout the city but especially in the suburbs. As medium-quality housing declines in value, that allows some of the lower-income households to move up. Housing filters down from the medium-quality submarket to the low-quality submarket, decreasing the price of low-quality housing. Naturally, the escalation of housing costs can reduce the positive benefits of filtering for medium- and low-income families.

Housing Demand and Supply

The geography of housing markets is fundamental to the ordering of residential space in urban areas. Key demand factors in the housing market are the number of people in a city and their average household size; the price of housing, which also reflects interest rates; inflation-adjusted or "real" income per household, which reflects the dynamics of the local labor market; and, in the case of owner-occupied housing, the home buyers' expectations of future changes in home prices (e.g., speculation). Important supply factors are the amount of construction of new housing units, construction standards, or building and zoning codes. Availability of financing may be viewed as a demand factor (long-term mortgages) or as a supply factor (construction loans), both of which are greatly shaped by interest rates.

As an example of the effect of a demand factor, consider an increase in the number of households in an area as a result of in-migration (say to an area whose export base is booming). If vacant housing of the desired kind is limited,

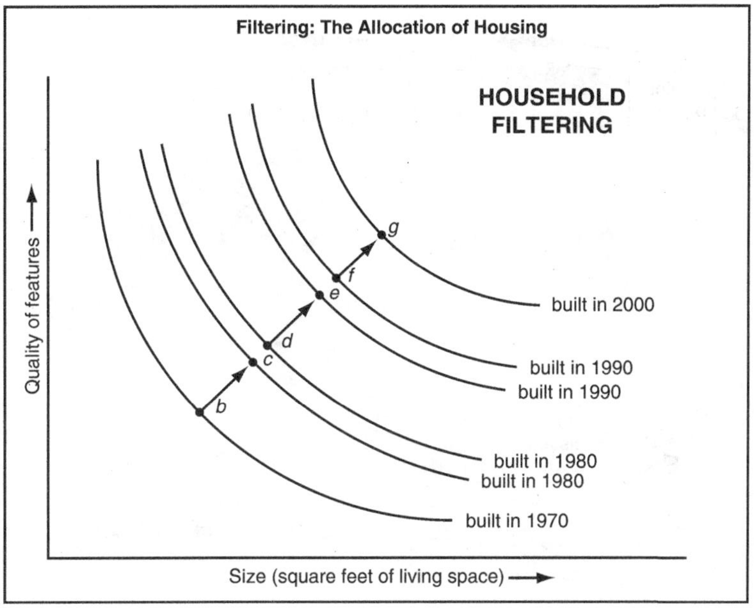

FIGURE 7 The filtering model of housing. As incomes rise, either over time or across socioeconomic groups, the ability to purchase more space of a higher quality also rises. Lower-income groups tend to rely on housing that has been previously occupied. Thus, the population filters up through the housing stock, or, conversely, the housing supply filters downward through the population. But massive bank lending to households that could not afford to repay the mortgage loans (the so-called sub-prime borrowers), has caused the housing bubble to burst in most of the world, starting in the United States in 2008. Americans have lost up to 40% of their equity in some regions, and 25% of American households are upside down, owing more on their mortgages than the house is worth. As of 2011, foreclosure sales comprised up to 30% of home sales in some parts of America, instead of the normal rate of less than 1%.

the price of existing housing will be bid up. This will induce new construction and create new demand for land, labor, and raw materials. If these items are abundant, their prices should remain constant. However, if supplies are limited, their prices will be bid up. This increase in building costs will lead to even further increases in housing prices, with the amount depending on how willing home buyers are to pay the higher prices. Thus, in places with rapidly growing populations, housing prices rise quickly, as happened in many Sunbelt cities in the 1990s and 2000s. Booming urban economies, typically those that are globally oriented with a concentration of producer services, will likewise generate explosive increases in rents and housing prices. Increases in income and a heightened expectation of future increases in home prices will have similar effects.

On the supply side, consider a case of vacant land where local authorities change the zoning from residential to nonresidential; that is, it is rezoned for commercial purposes. The reduction in available supply of land for housing will raise land prices and, hence, builder costs. The ability of builders to pass on the higher costs depends on the willingness of buyers to pay them. The more willing the buyers are to pay, the higher the demand pressures remain on land, labor, and so forth, and the more limited these items are relative to the demand, the greater the upward pressure on home prices. Similar effects occur in cases of few vacancies; higher labor, equipment, and material prices; and stricter building codes.

THE SPRAWLING METROPOLIS: PATTERNS AND PROBLEMS

As a result of the steady decentralization of jobs and successive innovations in urban transportation, the North American city has evolved from a mononucleated city to a multicentered metropolis. The roots of **suburbanization** extend deep into American history. As early as the late eighteenth century, the urban elite constructed country homes on the outskirts of cities on the East Coast. Brooklyn arose in much this fashion outside of Manhattan. Throughout the nineteenth century, those who could escape the filth, crowding, disease, and squalor of the inner city often did so, situating themselves along horse-drawn omnibus lines. By the 1880s and 1890s, with the introduction of the electric streetcar, the growing middle class increasingly relocated to the periphery. Precisely during this time, many inner cities were being filled by waves of impoverished immigrants. Thus, early suburbanization created a spatial bifurcation between American-born and foreign-born populations. Throughout the early and mid-twentieth century, the automobile played a key role in the growth of suburbs. In the culture of mass consumption that became preponderant during this period, movement *outward* from the inner city became deeply associated with movement *upward* socially (i.e., suburbs were linked to aspirations of escape from working-class life, the locus of the fabled "American dream"). It is important therefore to note that suburbanization was not simply produced by changes in transportation technology, influential as those were, but also involved transformations in the urban division of labor, class structure, housing markets, and the position of cities within the national and international urban hierarchies.

After World War II, suburbs exploded in size. In part, this growth reflected years of pent-up demand for housing that had accumulated during the Depression and the war, as well as the baby boom that began in the late 1940s and early to mid-1950s. With an economy enjoying low unemployment and rising real incomes, tens of millions of people made the choice to relocate to the urban periphery (as described in the discussion of residential location

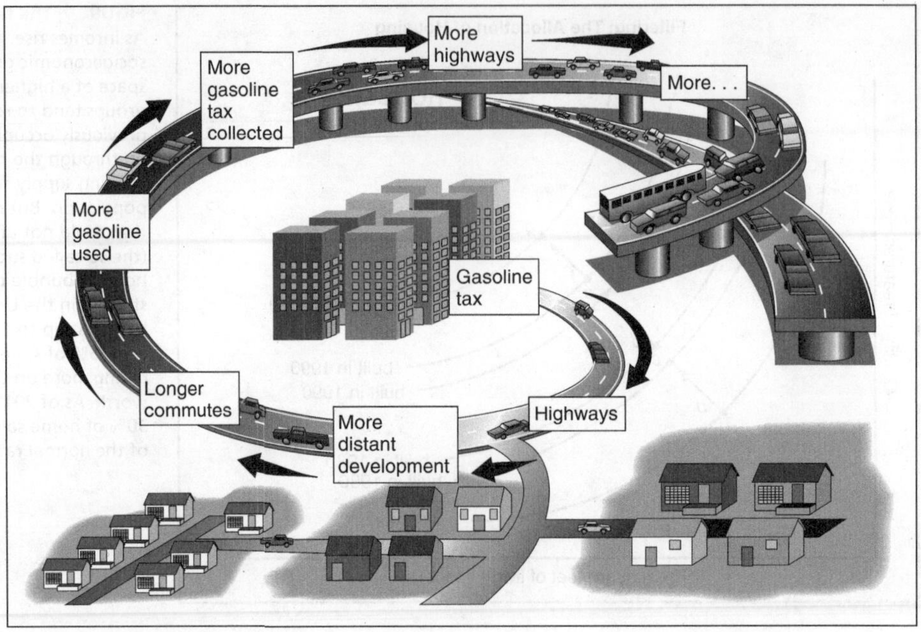

FIGURE 8 The development cycle ushered in by the federal Highway Trust Fund. Because federal funds are used so heavily to subsidize highways, the demand for automobiles grows ever larger, generating more funds through taxes on gasoline, which perpetuate the cycle of suburban growth and inner-city decline.

decisions above). Suburbanization, however, was not simply the product of an ostensibly "free market" but owes much to government policy, particularly the radial and circumferential highway network started by the federal government in 1956. In effect, the freeway system eroded the comparative advantage of the central business district, making most places along the expressway network just as accessible to the metropolis as the downtown used to be (Figure 8). Other factors that reinforced suburban sprawl were (1) low mortgage interest rates, (2) loan guarantees provided under federal housing and veterans' benefit programs, (3) property tax reductions for owner-occupied homes, (4) cheap transportation, (5) massive highway subsidies, and (6) most of all, cheap land.

The mass introduction of the automobile in the 1920s rapidly accelerated the process of suburbanization, leading to waves of people settling on the urban fringe as the single-family home became the most common type of housing in the nation. Automobile companies even began buying and closing down intraurban streetcar lines across the country to force the public to adopt the automobile. In Los Angeles in the 1930s, for example, a well-functioning public streetcar system was dismantled by General Motors. The freeway era has removed most restrictions on the intraurban mobility of the population (at least for the middle class) so that almost anywhere in a metropolitan area using land for residential purposes is feasible. As a result of industrial and demographic shifts to the periphery, downtown areas gave way to an ever-widening belt of suburban cities, including new neighborhoods, new business centers and office complexes, and new shopping malls. Far from simply being residential centers, suburbs have become commercial, financial, and retail centers in their own right. More Americans now live, work, play, shop, and dine within the confines of suburbs and **exurbs** (beyond the suburbs) than in city centers or rural areas (Figure 9).

The suburbanization of retailing was a response to the residential flight to the suburbs, new merchandising techniques, and the technical obsolescence of older retailing areas. The automobile provided customers with a convenient mode of transportation to shopping places, but downtown parking facilities were scarce and expensive. A need to improve the parking situation and to increase profits impelled retailers to the suburbs. Beginning in the 1920s, stores began spreading out from the downtown along main thoroughfares. It was not until the postwar years that

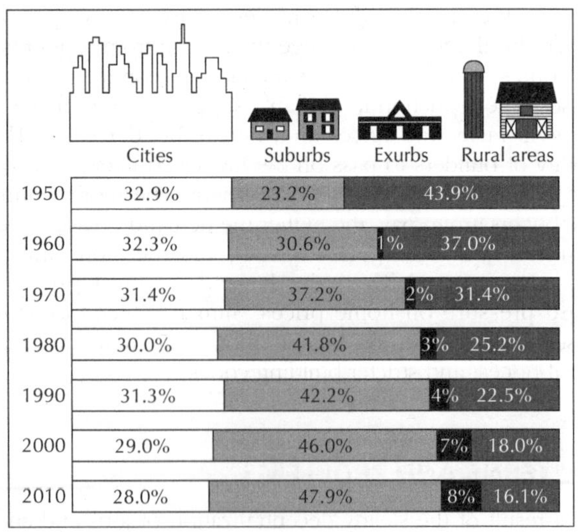

FIGURE 9 Proportions of Americans living in central cities, suburbs, exurbs, and rural areas, 1950–2010.

retailers moved to the suburbs in large numbers. First came the strip center, or neighborhood shopping center, consisting of a string of 10 to 30 shops, usually anchored by a supermarket. Then came the larger community center with a small department store or a variety store as the principal tenant. The success of these early centers, which catered to a limited area of trade, depended on location on a main road, free parking, and the persuasiveness of "drive-in" everything, self-service stores, and discount outlets.

These early neighborhood and, later, larger community shopping centers were in some cases supplanted by more attractive regional shopping centers that appeared after 1955. The newest and biggest of these centers in distant suburbs have several levels, three or more anchor department stores, and scores of specialty shops. Surrounded by huge parking lots, these shopping complexes are usually enclosed so that customers can shop in climate-controlled comfort. Unlike early suburban shopping centers, the giant regional shopping centers are catalysts that attract a variety of activities to an area.

Technical advances, such as the development of continuous-material flow systems, induced many manufacturers, especially those engaged in large-scale production of industrial goods, to spread out along suburban railway corridors where land was relatively cheap and abundant even before the turn of the century. Nonetheless, most manufacturers, despite the cost of truck transportation, decided to remain in or near the central city until the 1960s, when two important factors changed their minds: (1) the completion of the urban expressway system and (2) the scale economies to be realized in local trucking operations. The freeway network—built by the government, not the market—helped to neutralize the differential in transportation costs between inner city and suburb. As the locational pull of central-city water and rail terminals declined, most of the remaining urbanization economies of downtown were nullified.

The expansion of offices into the suburbs began in the early post–World War II years, when large corporations began looking for new headquarters on the urban periphery, where rents are lower and access to transportation much easier than in congested downtowns. The major factor in office-site selection typically is accessibility to an expressway. This trend led by the large corporations prompted an avalanche of similar moves by a host of smaller firms precisely when the growth of business services was generating considerable demand for new office space. Suburban office parks became a mainstay of employment growth on the metropolitan fringe.

Without doubt, the defining feature of the **edge city** is the huge regional shopping center. Super shopping malls are catalysts for other commercial, industrial, recreational, and cultural facilities. The result is the emergence of miniature downtowns. In many metropolitan areas, edge cities are unplanned, loosely organized, multifunctional nodes, and they are strongly shaping the geography of suburbia. Local and regional trends have combined with the general shift of the U.S. population and economic activity from the Northeastern and Midwestern Snowbelt to the Southern and Western Sunbelt, where U.S. metropolitan areas have been growing particularly rapidly.

Suburbanization had huge consequences for people and landscapes both on the urban periphery and in the center. At the edges of metropolitan areas, the steady conversion of agricultural land to residential and commercial property has eroded one of the nation's prime resources, farmland. Because many cities were established on river floodplains as agricultural centers, urban sprawl eats up the choicest farmland with the best soils. In the United States, rural-to-urban land conversion occurs at the rate of roughly 1 acre per minute. But the suburbs' success came at a price. Building a road and water infrastructure in relatively low-density areas such as suburbia is expensive. Today, many suburbs are faced with persistent problems with traffic, access to water, and housing costs increasingly out of the range of their residents.

Out to the Exurbs

Americans have been pulled in increasing numbers to the outermost edges of metropolitan areas, the exurbs. Housing affordable to most workers is not being built in adequate quantities near growing suburban employment centers of edge cities, commuting trips are becoming longer, and pollution levels from automobile emissions continue to increase. Rapidly growing communities, exurbs, lack urban amenities; they are "boomtowns" in remote areas for only one reason—access to affordable housing. The trade-off for people living in the exurbs is the stress of the long commute and the strain it places on family relations. This problem has now spread to most metropolitan areas of the country. Long work commutes have fueled the clamor for growth management in which the balance of jobs and housing and traffic congestion play important roles.

The information revolution and the customized flexible economy allow businesses, consumers, and entrepreneurs to move to the exurbs, thereby increasing their standard of living and quality of life. New technologies and infrastructures have opened up new low-cost land, factors that played a part in the move to the cities in the late 1800s and the 1920s and, after the Great Depression and World War II, also in the massive move to the suburbs. Automobiles, highways, telephones, and electrical energy made possible the move to the suburbs, where cheaper land meant larger properties and lower prices. Eventually, many suburbs matured into high-cost places to live. Families then began looking to small cities beyond the metropolitan area to escape the high costs of land, houses, and utilities and high taxes, congestion, and crime. Retirees as well as baby boomers have also flocked to the exurbs. Thus, exurban growth is the latest chapter in a long series of moves from the urban center to the periphery.

Typically, exurbs are 50 to 150 miles beyond the metropolitan fringe. They include small, resort-type, recreationally oriented towns that have lower costs and high appeal, perhaps with a university or with potential for business infrastructure. Provided that they can still reach their markets, businesses will relocate to the exurbs or small towns to reduce their costs and to become more competitive. At the outset, the businesses that move to smaller towns see much higher average growth and experience less competition.

The result of this move to the exurbs in the United States is an urban geography dominated by sprawling conurbations (continuous urban networks) that encompass thousands of square miles and tens of millions of people (Figures 10 and 11). The effect on central cities and suburbs can be disastrous, because it depletes the tax base and purchasing power of the metropolitan core.

Suburbanization and Inner-City Decline

For central cities, suburbanization has been catastrophic, producing uneven development on a metropolitan scale. The migration of jobs, particularly manufacturing jobs, led to rising unemployment and lower incomes in the inner city (Figure 12). Minorities were particularly affected; indeed, the deindustrialization of the inner city was a prime factor in the creation of the inner-city ghetto. Many once-stable working class African American neighborhoods, for example, disintegrated in the face of persistent joblessness and mounting poverty. Because the white population was generally much better positioned to flee inner cities than were minorities, suburbanization contributed to a marked increase in racial segregation within metropolitan areas. Declining property, income, and sales tax revenues made it increasingly difficult for central city governments to offer adequate public services (particularly education) precisely when the demand for them rose steadily. Suburbs typically are incorporated as independent municipalities to avoid paying property taxes to central cities, a process that created a metropolitan political geography in which large inner cites are surrounded by archipelagoes of small suburbs on the fringes. Declining populations also eroded the political power of inner-city governments at the state and federal level and their capacity to acquire funds for housing and transportation. In short, suburban expansion and inner-city decline may be seen as two sides of the same coin.

Gentrification

During the past 30 years, just as the collapse of the inner city was widely perceived to be irreversible, a resurgence of growth began in many downtown areas, a process known as **gentrification**. Explanations of gentrification point to the growth of producer services that accompanied the late twentieth-century globalization (Figure 13). Producer services often rely heavily on agglomeration

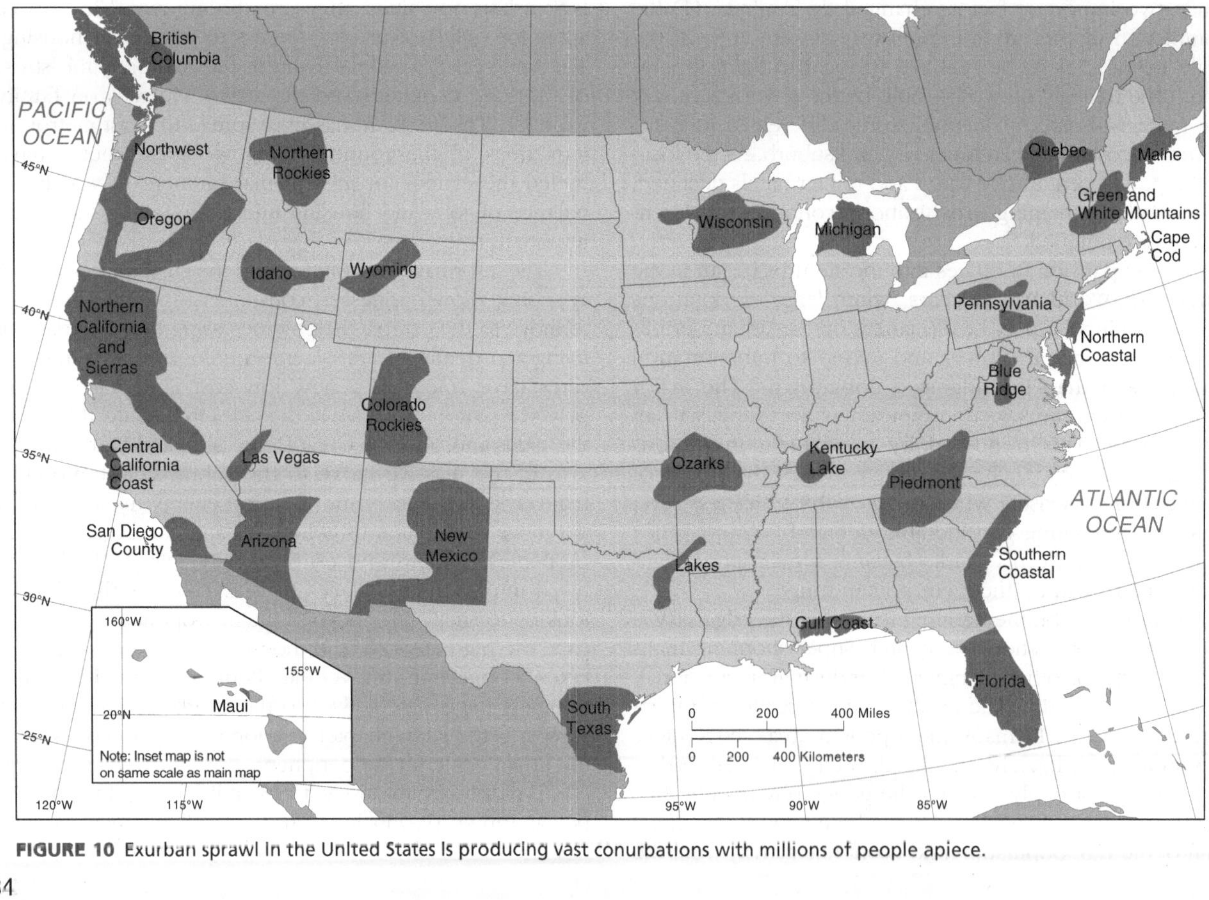

FIGURE 10 Exurban sprawl in the United States is producing vast conurbations with millions of people apiece.

FIGURE 11 Cheap transportation, particularly via the automobile, has fostered waves of decentralization of jobs and people to the urban periphery. American urban form is thus locked into access to low-cost gasoline. Urban sprawl contributes to long commuting times and rapid rural-to-urban land conversion as local farmland is converted into housing developments and shopping centers.

economies and cluster in centralized locations—downtowns. Just as suburbanization was based on the changing urban division of labor after World War II (if not before), so too did gentrification reflect the reworking of cities in the face of a globalized, service-based economy. The growth of producer services—finance, legal services, accounting, advertising, engineering, sales, and so forth—created large numbers of well-paid positions that demanded university educations. These jobs were typically filled by young urban professionals—"yuppies"—who brought with them considerable disposable income and purchasing power. Often the initial wave of gentrification, typically consisting of gays, artists, and "urban homesteaders" in relatively inexpensive lofts, was succeeded by couples employed in professional services and living in luxury condominiums or co-ops. Finally, commercial gentrification followed in the form of publicly subsidized office and hotel complexes, waterfront developments, convention centers, and sports stadia. Gentrification is thus not just a market-driven phenomenon but involves the state. This series of stages is evident, for example, in downtown Boston, parts of Manhattan, central Philadelphia, Baltimore, Pittsburgh, Chicago, and numerous other large cities (Figure 14).

Other explanations focus on the changing demographic structure of families, particularly the increase in two-income families, and residential preferences, such as attraction to the diversity and historic architecture of inner-city areas. Many baby boomers with small families and two-incomes (including "DINKS," or "double-income, no-kids" couples) strongly preferred the recreational opportunities offered by the inner city. Some of the gentrification trend involves a "back-to-the-city" movement from the suburb, but the number of families involved was relatively small. For every family that moved back to the inner city, eight moved to the suburbs, so residential revitalization in and around the central business district clearly did not spawn a return-to-the-city movement by suburbanites. In fact, most of this reinvestment was undertaken by those already living in the central city.

From the perspective of many long-time residents of inner-city communities, often made up of working-class or relatively low-income minorities, gentrification represents a horrific tidal wave of change. Rents and housing prices go up, displacing those who cannot afford them and contributing to urban homelessness. Local community-oriented stores, frequently in place for decades, may disappear, unable to compete with expensive trendy boutiques catering to the elite. Often gentrification involves the displacement of ethnic minorities by higher-income whites. For these reasons, gentrification is often despised and resisted by those who have little to gain and much to lose. Gentrification, then, reflects the reworking of once-dilapidated areas of the inner city by the types of capital associated with globalized services. Because it involves the in-migration of people who are very different from local residents in terms of social class and ethnicity, gentrification is simultaneously an economic, political, cultural, and geographic phenomenon.

PROBLEMS OF THE U.S. CITY

As we have seen, the socially and geographically uneven development of the contemporary metropolis has produced a pattern of suburban sprawl and inner-city decline. From an infrequent social entity at the beginning of the twentieth century, suburbs have evolved into major growth centers for industrial and commercial investment, and a suburban way of life has been adopted by millions of people. The decentralized metropolis fostered large-scale consumption and prosperity in the past, but it is causing real problems now. In some instances, the urban fringe has pushed out farther than workers are willing or able to commute, and urban sprawl has generated externalities such as uneven development, traffic congestion, pollution, and the irrational use of space, which increasingly impinge on the life of urban residents. Furthermore, recurrent fiscal crises threaten to bankrupt central cities.

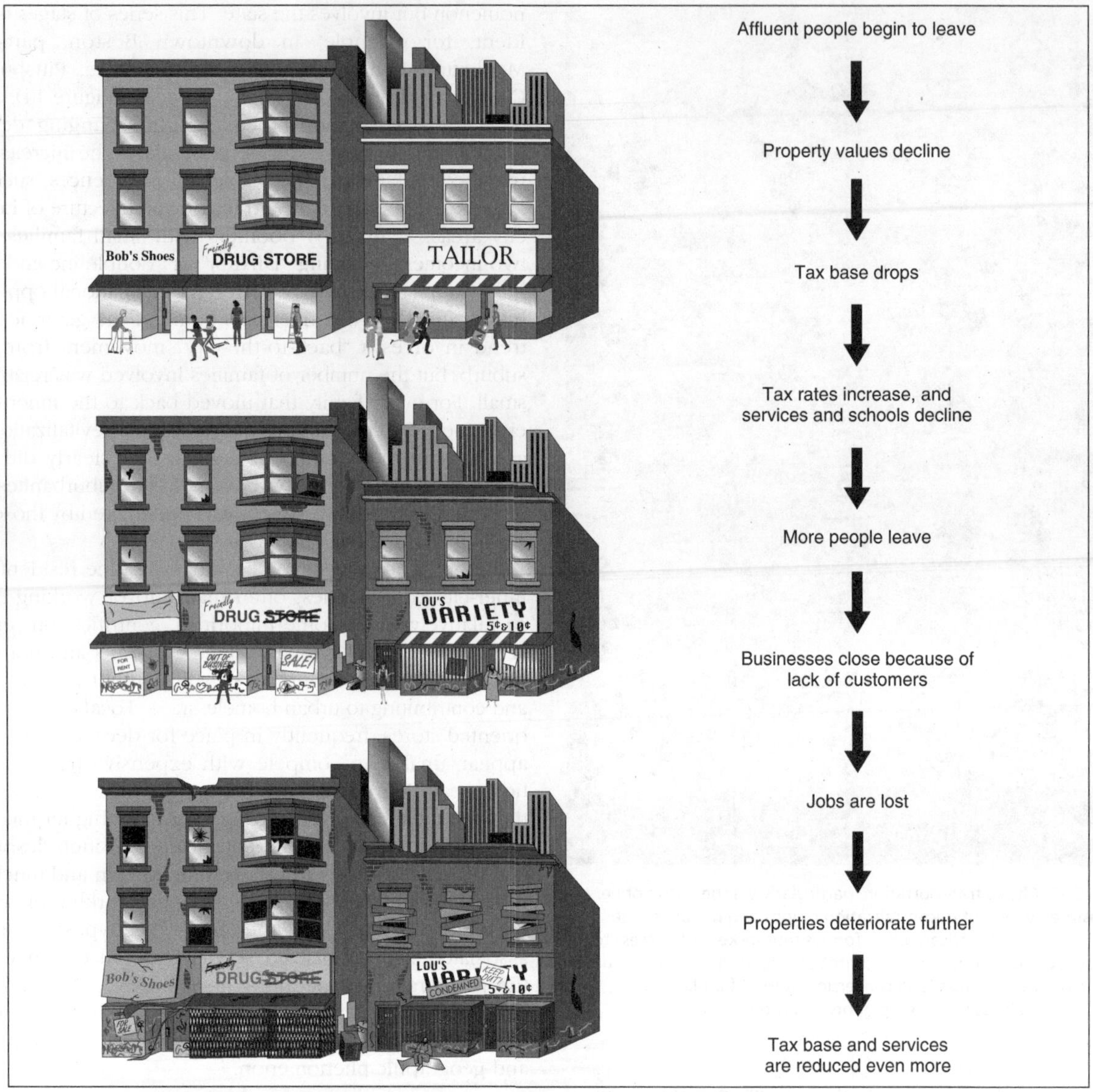

FIGURE 12 The cycle of suburban migration and urban decay. People moving to the suburbs from the city create the cycle of urban blight, which continues in most U.S. inner cities to this day. Cities are responsible for providing public schools, maintaining local roads and facilities, police and fire protection, waste collection, public water and sewer, welfare services, libraries, and local parks. The major source of revenue to pay for these services is property taxes, the amount of which declines in the face of sustained out-migration to the suburbs, although gentrification occasionally stimulates central-city tax revenues.

In the United States, the development of metropolitan-wide governance systems to tackle metropolitan-wide problems (e.g., congestion, air pollution) is blocked by the political independence of the suburbs. Some groups (i.e., the predominantly white suburban middle class) have been able to organize to protect their advantages. These groups are not willing to give up clear-cut, short-run benefits for less certain long-run gains. Thus, metropolis-wide planning in the face of a bewildering multitude of rigid and outdated municipal boundaries—1200 of them in the New York metropolitan area alone—is extremely difficult, if not impossible, to implement. Yet without planning, without redrawing areas of municipal authority, the continued profitability and stability of the metropolis and capitalist society are threatened.

For decades, most municipal governments in Western Europe and North America have concentrated on relieving congestion and the social pressures produced by poverty by demolishing block upon block of old housing, factories, and other buildings. A major problem for local governments has been where to rehouse the people affected by urban clearance. One solution has been to replace crowded terraced streets with blocks of tall apartments. However, simply building new

© David Forbert/SuperStock

FIGURE 13 Propelled by the surge in producer services over the past three decades, gentrification occurs in many forms, including residential construction of luxury co-ops and commercial construction of hotels, sports arenas, and waterfront developments. Corporate America reclaimed the inner cities of the United States in the 1990s, generating both new levels of wealth and inequality simultaneously.

housing does not address the fundamental causes of urban poverty and decline.

Urban Decay

Affluent residents moving to the suburbs begin the process of exurban migration and urban blight. Local governments' major source of revenue to pay for municipal services comes from local property taxes, the yearly proportional assessment based on real estate market value. If property values increase or decrease, property taxes are adjusted accordingly. Therefore, exurban migration causes, through the laws of supply and demand, suburban property values to escalate because of the influx of affluent people, and so suburban governments enjoy increasing tax revenue with which they can supply local services. As part of the induced effects of corporate decentralization, property values in the central city decline, however, because of declining demand, leading to an declining tax base. Those who are unable to move to the suburbs are usually lower-income and disadvantaged people who are not property owners and require public assistance; thus, cities must carry a disproportionate burden of welfare obligations, as well as the burden of increasing crime rates and the deterioration of their infrastructure that is due to declining tax revenues (see Figure 12).

An eroding tax base forces local governments to cut local services and raise the property tax rate. Therefore, property taxes on a home in the central city are often several times higher, proportional to assessed valuation, as those on a home in the suburbs. The difference could be as much as $5000 to $8000 per year in many cities based on an average-size home. At the same time, parks, schools, streets, libraries, and garbage collection services deteriorate from neglect, leading to the vicious cycle of exurban migration and urban decay (Table 1).

The Crisis of the Inner-City Ghetto

The cores of American metropolitan areas have been centers of poverty for a long time. Many immigrant neighborhoods of the late nineteenth century were such centers of poverty. The formation of the African American ghetto can be traced back to the 1920s, when several circumstances combined to send millions of workers to the North. The mechanization of agriculture reduced job opportunities in the South; in the North, a steady growth in manufacturing jobs and congressional quotas imposed on immigration restricted the supply of labor, creating relatively high wages.

Photograph courtesy of the Greater Minneapolis Convention & Visitors Association

FIGURE 14 Downtown Minneapolis–St. Paul reflects not only the agglomeration of producer services in the central business district but also the city's role as the premier center of services in the northern Midwest.

TABLE 1 Proportion of Population in Urban Areas, 1950–2010

Region	1950	1970	1980	1990	2000	2010
World	29	37	41	45	50	56
Developed countries	54	67	72	74	75	76
Underdeveloped countries	17	25	31	35	39	45
Africa	16	23	30	31	33	37
Latin America	41	57	69	71	73	79
East Asia	17	27	29	36	38	50
South Asia	16	21	28	30	31	32
North America	64	74	74	75	75	79
Russia and Central Asia	39	57	66	68	70	73
Europe	56	67	72	74	75	73
Oceania	61	71	71	71	72	70

As a result, large numbers of African Americans moved north into cities such as Boston, New York City, Baltimore, Philadelphia, Pittsburgh, Cleveland, Gary (Indiana), Chicago, and Milwaukee, taking jobs in railroads, steel mills, meat packing, shipyards, stockyards, machine tools, auto assembly, glass and rubber plants, and other industrial enterprises (Figure 15). These jobs, while still paying less than those filled by whites, nonetheless created relatively stable working-class communities of homeowners and low crime rates.

This window of opportunity, however, was very limited in time. Following World War II, inner cities were besieged by waves of economic dislocation. The flight of manufacturing to the suburbs, accompanied by the exodus of the white middle class, created a sustained downturn in the economic fortunes of inner cities as the multiplier effects of industrial decentralization unfolded. Trapped in unskilled or semiskilled jobs, many minority inhabitants were unable to move to the suburbs, constrained by their lack of skills in a deteriorating labor

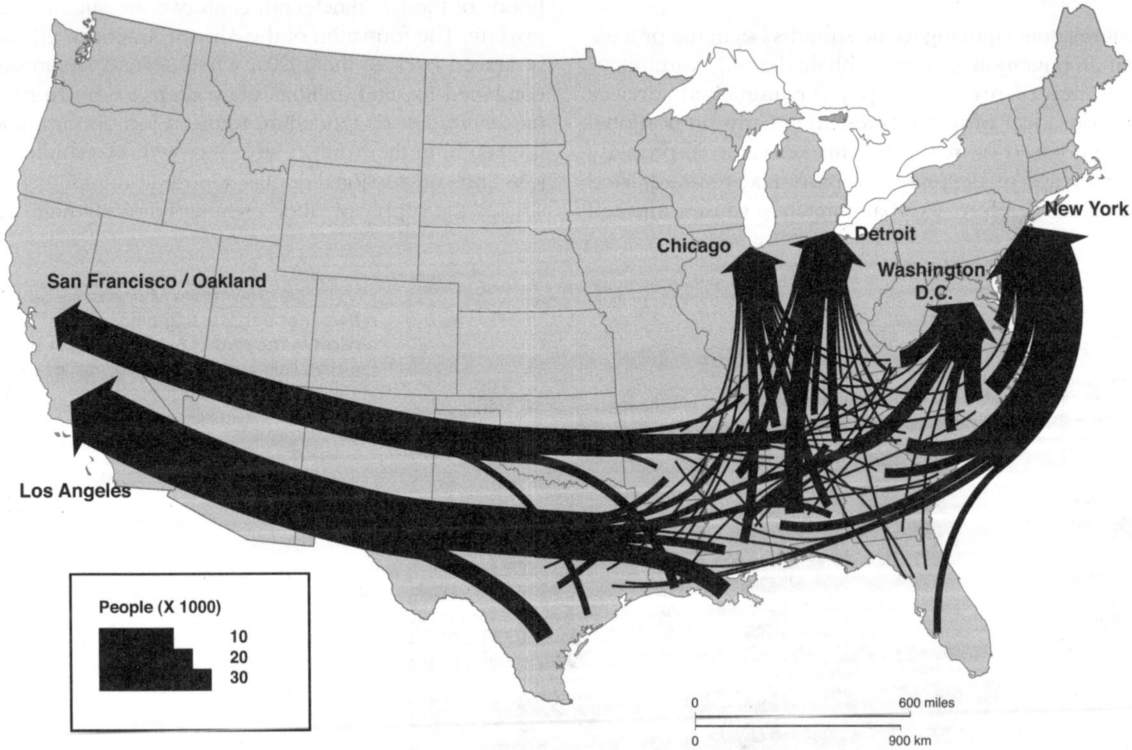

FIGURE 15 African American migration streams to the North and the West, 1920s–1940s. Lured by relatively well-paying industrial jobs and fleeing discrimination in the South, black migrants in the early twentieth century formed the nuclei of the many African American communities in large U.S. cities today.

market. Many suburbs actively resisted attempts by the poor to locate there, using, for example, zoning laws that specified a minimum lot size that put houses out of reach of the poor. In the central city, neither the low-paying jobs of consumer services (e.g., retail trade and personal services) nor the well-paying jobs of producer services, which required advanced educations, offered alternatives for the poorly educated. Low-wage service jobs often paid the minimum wage and carried few or no health benefits, pensions, or paid holidays; nonetheless, each advertisement often drew (and still draws) large numbers of desperate applicants. Inner city poverty, in short, is not the fault of the poor, but the outcome of deep structural forces that have changed the suite of opportunities that inner city inhabitants face.

Unsurprisingly, unemployment rates rose steadily in inner cities from the 1960s onward, accompanied by deepening poverty and a degeneration of the social fabric. Working-class neighborhoods were steadily transformed into impoverished ghettos. In 2009, the official definition of the poverty line in the United States was $21,756 for a family of four (the number varies depending on the number of children involved), a level that severely underestimates the number of people living in financially strapped circumstances. More realistic measures of poverty would reflect a far larger pool of poor people than do official statistics. Whatever the threshold used, inner-city poverty rates, especially among minorities, are often twice or three times the levels found among whites and suburbanites. Today, African American incomes are, on average, only two-thirds those of white families. The overall U.S. poverty rate in 2009 was about 13%—10.5% for whites, 25.3% for African Americans, and 21.5% for Latinos/Hispanics (Figure 16), which reveals how class and racial inequality are deeply intertwined.

One result of white exodus was that minorities became an increasing proportion of the population of inner-city areas. In 2008, for example, African Americans comprised 78% of the population of Detroit, 38% of Chicago, and 24% of New York City. Including Latinos and Asian Americans, minorities frequently make up over one-half of the populations of large cities. One result has been an increasing number of minority politicians, including numerous African American mayors.

Many observers point to deindustrialization as the primary cause of the inner-city ghetto; in contrast, conservative explanations focus on the role of welfare and public assistance. From the latter perspective, transfer payments to low-income families popularly known as "welfare" (e.g., Temporary Assistance to Needy Families) allegedly create a disincentive to work, a view that tends to blame the victim, attributing poverty to the behavior of poor people. But this perspective ignores the macroeconomic context of urban poverty, the decline in available jobs, and the considerable drop in earning power of low-income workers brought on by globalization and technological change. It also overlooks the fact that the burden of public assistance has declined substantially in postinflationary dollars over the past four decades.

Also noteworthy is the fact that in the face of high, and rising, unemployment in the late twentieth century, the African American family underwent a severe contraction in its scope and viability. Long a strong institution during the centuries of slavery, in which women played a crucial role, the African American family was profoundly reshaped by urban deindustrialization. As low skilled jobs evaporated, the supply of marriageable, employed males declined steadily, leaving relatively few marriage partners for women of the same age cohort. As a result, teenage pregnancy and out-of-wedlock births steadily increased; the poorest, most economically vulnerable families are typically headed by women alone. Many mothers struggle heroically to raise their children in the face of low wages and persistent poverty, but two incomes are always better than one in this context. Children in homes led only by women are often deprived of the role models provided by two working parents that have become the norm among white Americans.

As a result of the changes in inner-city labor markets, including poverty, unemployment, low wages, hopelessness, and despair, many (but by no means all) residents in low-income inner-city communities may be regarded as belonging to an urban "underclass." The term has been criticized as dehumanizing to its members, but the concept nonetheless retains validity; it adequately describes unskilled, poorly educated, chronically unemployed, and often unemployable people without access to schooling, skills, or adequate jobs. The persistence of the underclass remains one of urban America's greatest challenges today as it is linked to high inner-city rates of crime, unemployment, poverty, drug abuse, and other problems.

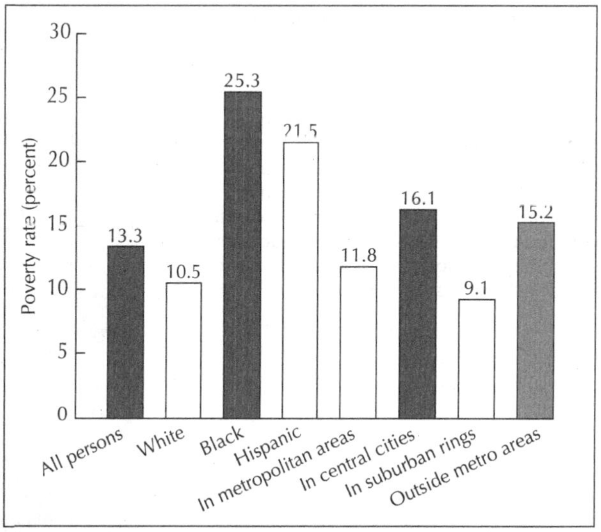

FIGURE 16 Poverty rates in the United States, 2009, by ethnicity and location. Black and Hispanic poverty rates are three times that of whites, while central-city rates are twice that of suburban areas.

© SuperStock/SuperStock

FIGURE 17 Abandoned by capital, with high unemployment rates, the inner cores of American metropolises are increasingly divided between a high-wage sector centered around globalization and producer services and a low-wage sector that keeps the unskilled trapped in poverty. These economic conditions are mirrored in the landscapes and lives of the people who inhabit them.

In addition to poverty, inner cities suffered other, related problems that lowered the quality of life for their inhabitants. Major problems include the following:

- A crisis in *housing*: Inner-city housing markets are unable to provide sufficient housing for the poor. In the 1950s, urban renewal, a federal program to clear space in inner cities for freeways and corporate capital, destroyed millions of housing units, contributing to an ongoing housing shortage. Many private landlords disinvested in their communities, applying the profits from their rents to other parts of the city and on occasion even burning some units for the insurance funds. Disinterested landlords, and tenants who rent and can ill afford to invest in upkeep, led to persistent lack of investment in the housing stock, a process matched by lack of corporate and public investment in such communities (Figure 17). A long series of cutbacks in federal housing assistance magnified the problem. As a result, the physical landscape in many inner-city areas became increasingly decrepit. Gentrification also exacerbated the housing problem as developers constructed homes for the wealthy elite and no one else.

  The most vulnerable people in the face of this process are the homeless—the 1 to 2 million Americans without homes (Figure 18), including those caught between jobs that pay too little and housing that costs too much, the long-term unemployed, children recently released from the foster care system, those with drug problems, military veterans, and rising numbers of homeless women. Homelessness, in short, is not the result of choice—an argument that conveniently and inaccurately blames the victim—but the product of political and economic processes that have systematically worked against the poor, the unskilled, and the uneducated.
- The *inadequate funding of inner-city public schools*, which often suffer from large classes, are plagued by violence, are taught by underpaid and overworked teachers, and have buildings in serious need of repair, leads many students to drop out, perpetuating the employment mismatch (the supply and demand for skills; see below). By any measure, inner-city schools in the United States compare negatively to their suburban counterparts, with lower scores on standardized tests and lower graduation rates. Because education is the major vehicle for social mobility, poor quality schools offer little opportunity for impoverished students to rise above their circumstances.
- *Health care* for low-income residents in inner cities is typically very poor. Many residents do not have

FIGURE 18 The homeless, such as this panhandler in New York City, reflect the myriad ways in which deep structural political and economic forces leave some social groups desperately poor. Far from being a voluntary condition, homelessness is the product, among other things, of jobs that pay too little and housing that costs too much.

© Viviane Moos/CORBIS

jobs that carry health insurance as a benefit. Roughly 37 million Americans currently lack health insurance, a population that is disproportionately concentrated in inner cities. Often, residents must use overcrowded hospitals as their primary source of care, or go without, a factor that contributes to relatively high infant mortality rates and low life expectancy among African Americans. Illegal drug use has compounded this problem, as has the epidemic of AIDS.

- *Crime rates* in inner-city areas are substantially higher than in suburban areas and disproportionately involve young minority males. However, the white fear of crime tends to be exaggerated—the victims overwhelmingly are also minorities. Easy access to handguns contributes to high rates of violent crime and homicide rates. Crime rates are largely determined by the status of the labor market: When jobs are plentiful, crime goes down. Nonetheless, harsh incarceration policies have led to the imprisonment of more than 2 million Americans, and 10% of all African American males are either in prison or on probation.

Employment Mismatch

Private and public discrimination (e.g., redlining by lending institutions, discriminatory actions of realtors, screening devices adopted by subdivision developers) and exclusionary zoning practices give the poor, especially minorities, little recourse but to locate in inner-city areas. Meanwhile, most of the new employment opportunities matched to the work skills of these people are out in the suburbs. This disjunction is known as the **spatial mismatch**. The poor are faced with the problem of either finding work in stagnating industrial areas of the inner city or commuting longer distances to keep up with the dispersing job market. Although **reverse commuting** has increased, the barriers are many, including transportation constraints—such as the increased time and cost of the daily journey to work, and inadequate public transportation for those without cars—and communication constraints—such as the difficulty in obtaining timely information concerning new job opportunities. Other serious obstacles to suburban employment faced by the inner-city poor include few and substandard work skills and biased hiring practices. In the face of these problems, many otherwise employable persons give up job hunting altogether and become instead part of the growing number of unemployed in inner-city neighborhoods. The spatial mismatch is an important factor in the high unemployment levels of black youths. Overall, unemployment levels for both black and white youths rise the greater the distance to job opportunities. The principal reason for higher black unemployment rates is that the commuting time is higher for blacks than for whites. However, the mismatch hypothesis accounts for approximately only half of inner-city unemployment, leaving another 50% to be explained by labor-market discrimination, differences in education, and so forth.

GLOBAL CITIES

To this point, we have considered cities at the national and regional levels, that is, within the context of the national division of labor. Now let us shift our attention to the international scale. Around the world, the proportion of people living in urban areas (Figure 19) reflects the economic

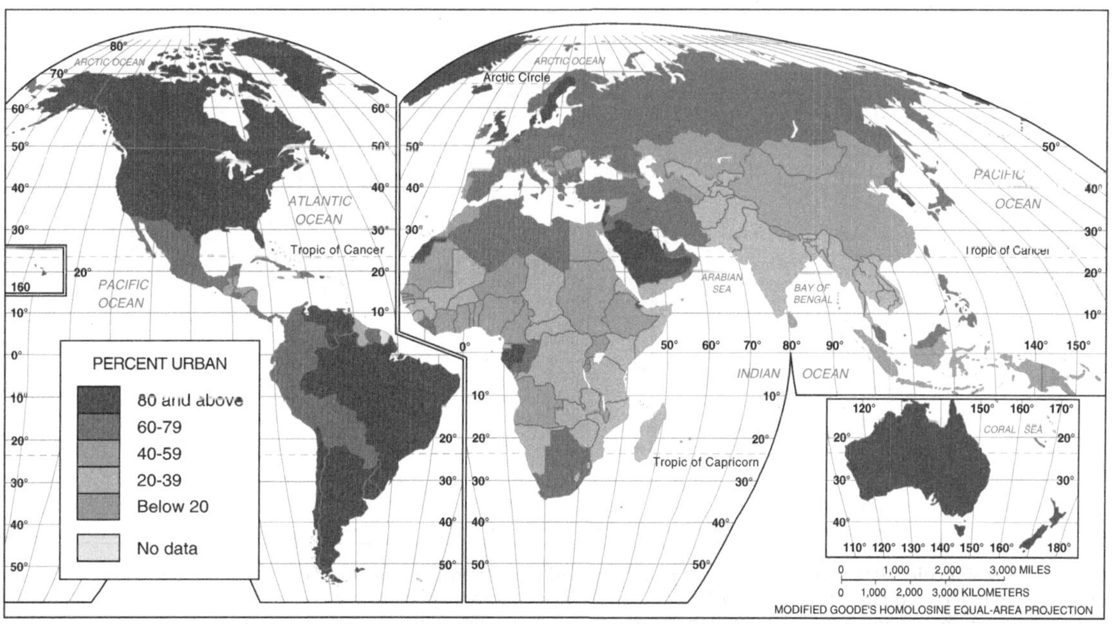

FIGURE 19 Urbanization rates around the world. Usually, more economically advanced countries have a higher proportion of people living in cities.

wealth and power of countries. Industrial and postindustrial societies tend to be highly urbanized, whereas large shares of the labor force in most developing countries are in agricultural and rural areas. However, rapid population growth in the developing world has led to the rapid growth of cities there.

The international division of labor is forging many cities of the world into an integrated, composite system. Figure 20 depicts the location of the largest cities in the world, many of which are located in developing countries. However, while cities in Asia, Africa, and Latin America are important as centers of people and shape the everyday lives of billions, by and large they are not the major centers of control in the global economy, which lie in the economically developed world. Position in the worldwide urban hierarchy, therefore, depends less on population size than it does on a city's role in the global division of labor, corporate hierarchies, and access to the political and financial wealth that controls the world economy.

The wave of intense **industrial restructuring** that started in the 1970s is changing the global urban hierarchy. The process of restructuring involves the movement of industrial plants from developed to developing areas within or between countries; the closure of plants in older, industrialized centers, as in the American Rust Belt; and the technological improvement of industry to increase productivity. The forces behind restructuring include the need for multinationals to develop strategies for locating new markets and to organize world-scale production more profitably, the national policies of developed countries for improving their future international competitive position, and the national policies of developing countries to attract subsidiaries of multinationals. These strategies of multinationals and governments have contributed to major shifts in employment and trade. The greatest effects have been felt in the urban centers of developed countries and in the larger cities of underdeveloped countries.

One of the most significant repercussions of the internationalization of economic activity has been the growth of "global cities," particularly London, New York City, and Tokyo (Figures 21 and 22). Global cities act as the "command and control" centers of the world system, serving as home to massive complexes of financial firms, producer services, and corporate headquarters of multinational corporations. In this capacity, they are arenas of interaction, allowing face-to-face contact, political connections, artistic and cultural activities, and elites to rub shoulders easily. Other cities (e.g., Paris, Toronto, Berlin, Los Angeles, Osaka, Hong Kong, Singapore, and Shanghai) (Figure 23) certainly can lay claim to being national cities in a global economy, but since the late twentieth century the trio of New York, London, and Tokyo has played a disproportionate role in the production and transformation of international economic relations.

At the top of the international urban hierarchy, global cities are simultaneously (1) centers of creative innovation, news, fashion, and culture industries; (2) metropolises for raising and managing investment capital; (3) centers of specialized expertise in advertising and marketing, legal

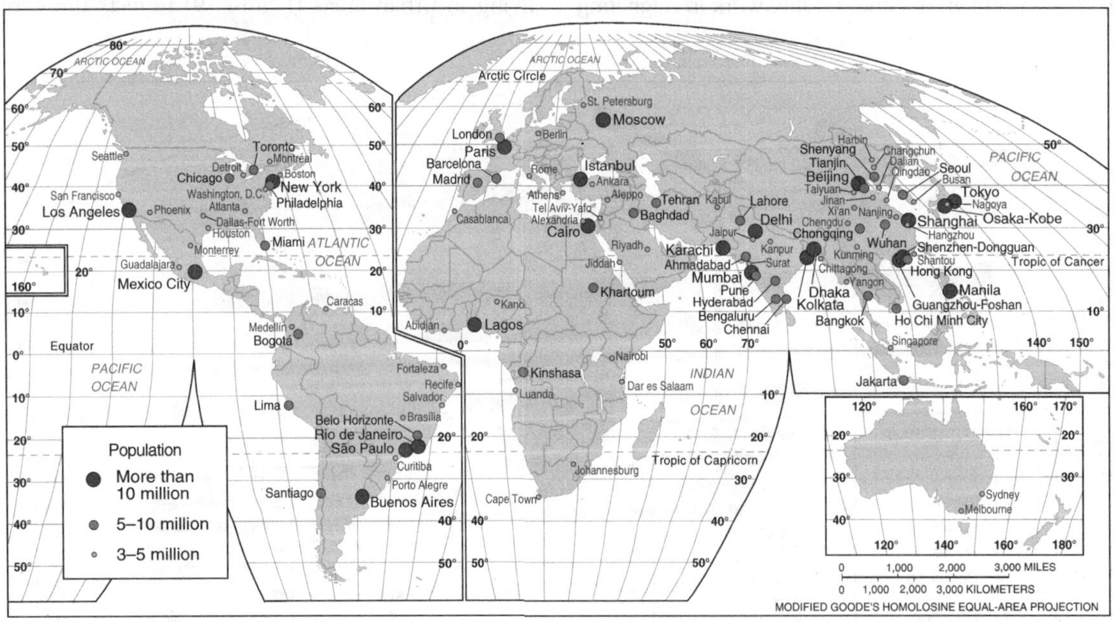

FIGURE 20 Major cities of the world. Cities with populations of more than 2 million are shown on this map. While urbanization of a country usually increases with per capita income, more than 10 million cities are located in the developing world, especially in East and South Asia. The growth of large cities in developing countries suggests the overall rapid growth of populations and the migration from rural areas to the city as more jobs are offered in the manufacturing sector and fewer laborers are required in the agricultural sector.

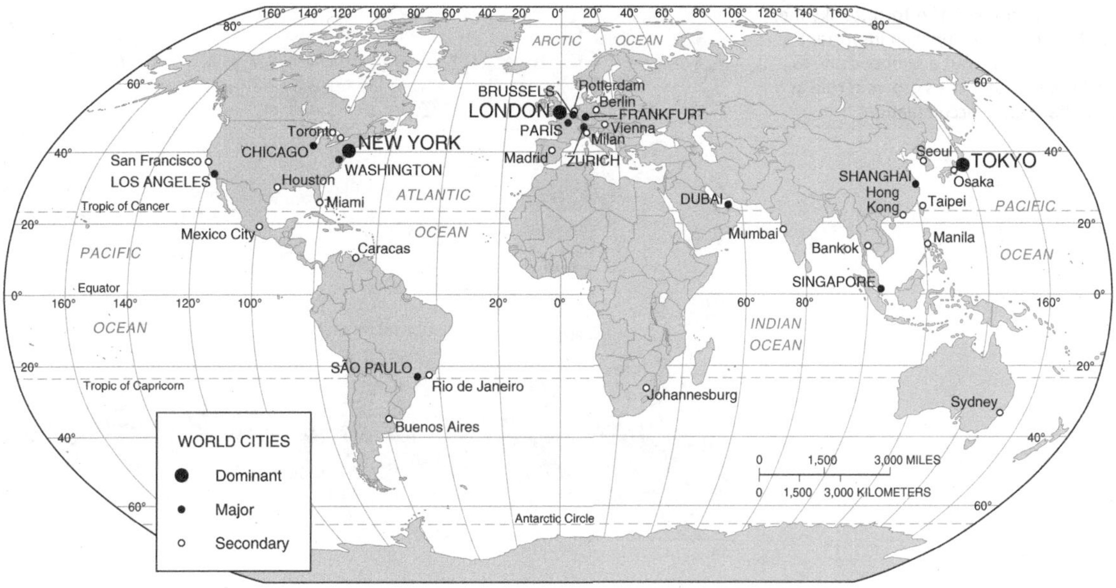

FIGURE 21 Global cities, such as London, New York, and Tokyo, dominate the world system of cities. They are not the largest in terms of population, but they are the primary command and control centers of the world economy.

services, accounting, computer services, and so on; and (4) the management, planning, and control centers for corporations and nongovernmental organizations (NGOs) that can operate, with increasing ease, all over the planet. Global cities make possible the specialized expertise on which so much of the current global economy depends. Each is tied by vast tentacles of investment, trade, migration, and telecommunications to clients and markets, suppliers and competitors, scattered around the world. All three metropolises are endowed with enormous telecommunications infrastructures that allow corporate headquarters to stay in touch with global networks of branch plants, back offices, customers, subcontractors, subsidiaries, and competitors. This phenomenon again illustrates how geographic centralization can be facilitated by telecommunications.

Singular and economically preeminent world cities are points into and from which global capital flows. These *world cities* are usually the largest and most important in terms of international corporate headquarters, financial markets, population, employment, and output. They are control points in the world economy where the critical mass of capital and decisions concerning production and marketing create dramatic world economic development. Because high levels of economic development have been primarily located in developed countries and the

FIGURE 22 Manhattan, the economic heart of the New York metropolitan area, is the product of global patterns of trade, investment, and immigration, as well as the powerful agglomeration economies so critical to producer services such as finance, banking, and business services.

FIGURE 23 Looming over the Toronto skyline, Canada's CN Tower is not only an important communications center, but a symbolic one as well, a testimony to that city's growing role as a center of global capital accumulation.

globalization of economic activity has little touched developing countries, the world cities are few among a hierarchy of dominant, major, and secondary cities. London, New York City, and Tokyo, the largest cities of Europe, North America, and Japan, respectively, have the most important stock exchanges, major corporate headquarters, international banks and financial institutions, and the largest grouping of communications and business services. An important second tier of world cities includes Paris, Berlin, Madrid, Milan, Rotterdam, Zurich, and Vienna in Western Europe; Los Angeles, Houston, Miami, San Francisco, and Toronto in North America; Bangkok, Bombay, Hong Kong, Osaka, Seoul, and Taipei in East Asia; Buenos Aires, Caracas, Mexico City, Sao Paolo, and Rio de Janeiro in Latin America; Johannesburg in South Africa; and Sydney in the South Pacific.

The transformation of the Pacific Rim city of Los Angeles from a regional metropolis in the 1960s to a global center of capital, corporate headquarters, financial management, and trade today has been remarkable. Through selective deindustrialization and reindustrialization, a cluster of technologically skilled and specialized occupations has been complemented by a rapid recycling of low-skilled workers from declining heavy industry to low-wage industries and by a massive influx of Third World immigrants and part-time workers. Sprawling, low-density Los Angeles is symbolic of urban restructuring: It combines elements of Sunbelt expansion, globalization, an economy centered on finance and the Hollywood film complex, and exploitation of immigrants. The traditional Chicago School of urban studies focused on a manufacturing-oriented city embedded in the national division of labor, but many contemporary scholars point to the incipient "Los Angeles" model as suggestive of the shape and form of modern, globalized, service-oriented urbanization.

Also having a place in the urban hierarchy are cities that are centers of a more specialized variety of producer services to specific industries, for example, research and development (R&D) services in Detroit that cluster around the motor vehicle industry; steel agglomerations in Chicago, Pittsburgh, and Baltimore; office equipment in Rochester, Atlanta, and Los Angeles; and semiconductors in San Jose, Phoenix, and Boston. Included in these specialized centers are the cities that provide government and education services, including college towns and most state capitals.

URBAN SUSTAINABILITY

The environmental impacts of cities are enormous, including air, soil, and water pollution (some of which is global in scope); their requirements for food, water, and energy; and their effects on ecosystems near and far. Some have labeled cities ecological black holes due to their replacement of natural surfaces with paved ones, biologically unproductive land uses, and consumption of natural resources far beyond their physical size. The quest for balance among economic, social, and environmental conditions may be defined as sustainable development and as the achievement of a way of living that can be continued long into the future without deterioration in any of these conditions. **Urban sustainability** is the ideology and practices behind "green cities," the attempt to reduce energy use and urban environmental impacts, such as carbon emissions.

Efficiencies in energy consumption can be accomplished in many ways: through the shared walls in apartment buildings compared to single-family homes; multisectoral industrial clustering to reduce transport; and adoption of more sustainable values as social norms—for example, recycling, reduced meat eating, smaller cars, and use of public transportation. The large urban cores with their high population densities can generate the greatest advantages; the resource use and lifestyles of the suburbs are unsustainable. Through strategies such as smart

Case Study

Environmental Impacts of Cities

All cities impact the environment. At the turn of the twenty-first century, cities occupied only 3% of earth's land surface; however, their environmental influence extends across the globe. Cities' voracious appetites for resources and their dispersal of pollutants into the air and water and on land result in local, regional, national, and global environmental degradation. For example, the greater Los Angeles area, a megacity, covers much of five counties and includes over 17 million people. The direct effects of Los Angeles include smog, displacement of native habitats, contamination of surface and unconfined groundwaters, and the deposition of litter on city streets and beaches. The megacity's direct environmental impacts on the western United States include air pollutants that diminish visibility at national parks of the Intermountain west, mountains of urban waste in landfills, and the diminished flow and increased salinity of the Colorado River.

However, the direct effects on the region are only the beginning of the megacity's environmental impacts. The coal-fired plants in the West and Midwest that provide electricity to Los Angeles do so by creating the associated environmental effects of surface and underground mining, coal combustion, and waste disposal. Power plant emissions send sulfur dioxide across national boundaries, affecting Canada's forests. The cars driven in Los Angeles not only produce emissions materials for the cars come from mines, chemical plants, manufacturing centers, and cities that also dispose of wastes on land and into the air and water. Beyond these influences on local, regional, and global physical and biological environments are those on local, regional, and global populaces who sustain the city's appetite for goods and services.

In less than a few hundred years, the pernicious environmental effects of cities have spread beyond their immediate locales, which had largely absorbed them, to become globally pervasive. Three hundred years ago, the world's population of approximately 600 million lived in what today would be considered rural settings with scatterings of villages and towns and about two dozen cities with over 100,000 inhabitants. Even at their worst, the environmental effects of most cities were largely confined within city limits; they were local sinkholes of misery, disease, stench, and smoke-filled interiors and outdoors were scenes of rotting refuse and human feces. They were ecosystems that supported rats, vermin, and disease, and social structures vulnerable to recurrent famine. Cities changed irrevocably in response to the widespread consumption of fossil fuels introduced during the Industrial Revolution and the increased crop yields of the Green Revolution. Commerce and trade provided jobs, and people migrated to cities. With more people, cities consumed more resources and attracted more people. In 1800, only 3% of the world's 900 million people lived in towns of more than 5000. Fewer than 45 cities had populations over 100,000. In 2008, over half of the world's 6.5 billion people were urban. Cities have evolved from local and regional centers of trade into today's drivers of national and global economies.

Urban appetites for goods and services grew at a faster rate than the increase in urban population due to (1) higher per capita consumption, (2) socioeconomic factors such as cultural migrations, and (3) political and economic policies that treated the environmental costs as economic externalities. The metabolism of cities varies among developed versus developing nations and even within nations. An estimate of the daily intake and output of energy, air, and water for a city of a million people in the United States in the 1980s estimated intakes of over 500 million kilograms of water, almost 10 million kilograms of energy, and almost 2 million kilograms of food. The daily output included somewhat less than 500 million gallons of sewage, almost a million kilograms of air pollutants, and 10 million kilograms of solid waste.

It is a challenge to geographers to agree on a methodology for calculating and analyzing urban metabolisms and their ecological footprint as well as a model for the flows of materials into and from cities. Every city on every continent impacts the earth. The cumulative environmental effects are innumerable, with feedback loops that make a full assessment of the total environmental impact of a city almost impossible. Assessing the effects of cities on even their local urban environment includes assumptions about costs and benefits, intangibles like the quality of life, and issues of social justice.

The impacts within the city's borders include surface disturbances of the land that result in changes in albedo and hydrologic capacities and the consequences of waste disposal. It is estimated that globally, one-third to two-thirds of the solid waste of cities is not collected, resulting in contaminated land and water as well as stench. But the effects of cities on the land within their borders are minor compared to environmental repercussions beyond their borders. On regional and global scales, the earth's geosphere is impacted by the needs for food, construction materials, and resources to sustain city economies. The need for food results in global losses of topsoil from wind erosion of cultivated fields and water erosion following deforestation. The addition of nutrients, such as phosphate, to increase crop production relies on the mining and processing of chemicals with attendant environmental effects. The need for construction materials results in mining and resource extraction of sand, gravel, limestone, and road materials, including the petroleum extracted for asphalt. The need for raw materials to sustain urban economies requires mining, petroleum extraction, and animal products to provide, among other things coal for electricity and petroleum for plastics and transportation.

Cities must have water and most are located near water supplies. As cities grew, the local and regional water sources within the city limits proved inadequate to the need. Cities in arid and semiarid regions disrupt natural water regimes by importing water for municipal and industrial uses and withdrawing groundwater from aquifers more rapidly than it is replenished. The demands cities place on agriculture justify construction of dams, diversions, and drainage systems. Cites also use water to dispose of wastes. Until the economic and social consequences of disease outweighed convenience and local economics, cities disposed of their sewage and liquid industrial wastes in local surface waters. It is estimated that even in the first decade of the twenty-first century, 220 million urban dwellers did not have clean drinking water and 420 million did not have simple latrines. Shallow, unconfined groundwaters in most urban areas are biochemically degraded with by-products of human waste, fertilizers, trace metals such as mercury, and newer contaminants such as pharmaceuticals and antibiotic-resistant pathogens. The contamination of waterways associated with cities includes contamination of oceans and waterways used to transport materials to and from cities and involves petroleum spills, invasive species from flushed bilge waters, and ubiquitous plastic flotsam.

Cities depend on many kinds of fuels for heat, transportation, and industry. The combustion of fossil fuels emits gases that affect local, regional, and global air quality. Effects within city boundaries differ according to topography, culture, and national regulations. It is estimated that in 1988 over 1 billion city dwellers breathed unhealthy air. Breathing the air pollutants for a day in Calcutta equates to smoking a couple of packs of cigarettes. The impact of cities on the atmosphere is rarely constrained, however, by city borders. The combustion of fossil fuels for transportation within and between cities discharges carbon dioxide and nitrous oxides into the atmosphere. The need for electricity and heat, largely met by burning coal and natural gas, releases sulfur dioxide into the atmosphere, which has the downwind consequences of acid rain and increased greenhouse gases such as carbon dioxide and water vapor. The need for food results in emissions of methane and particulate pollution from agriculture. The need for raw materials for industry result in a potpourri of gaseous pollutants emitted locally and at their places of refining or manufacturing. The cumulative effects of local and global discharges of pollutants into the atmosphere, largely to sustain urban economies, are the inconvenient truths of global warming and climate change.

Cities affect the biota within their borders and, to an even greater extent, the biosphere of regions that supply their food and raw materials. Urban ecosystems favor generalists: omnivores such as rats, animals that adapt to buildings such as pigeons, and biota appreciated by humans such as cats, dogs, and grass for lawns. Urban land uses displace natural biota, and urban corridors fragment habitats. But compared to the effects of the agriculture required to sustain growing urban appetites, those impacts are minor. The oceans, fished for millennia as an apparently inexhaustible source of nutrition, have been so harvested of some species of mammals and fish as to bring them to the brink of extinction. Aquaculture brings environmental challenges analogous to those of feedlots for cattle. Meanwhile, globalization and the increased concentrations of urban populations have accelerated the spread of invasive species and enhanced the spread of diseases to animals as well as to humans. Even more extensive than a city's effects on its urban biota or its effects on regional and global ecosystems will be the as-yet-unknown effects of atmospheric emissions on global climate and, consequently, on earth's biosphere.

growth, mass transit, carbon neutrality, and green buildings, urban planners can promote sustainability.

Urban sustainability presents new challenges for the development, redevelopment, and management of cities. A small but growing number of cities are applying urban sustainability policies in the form of improved environmental management, regional planning to combat sprawl, and downtown revitalization. Often, these strategies include a "brown agenda" of urban sanitation, a "green agenda" of environmental services, resource and energy conservation, and a social agenda that pursues the development of just and fair institutions and governance and economic, educational, and health opportunities.

In line with the sustainable development mantra to "think globally, act locally," different local conditions make suitable different approaches. Cities in the developing world might try to find better, long-term solutions to their slums and the living conditions of slum dwellers; those in the economically developed world might pursue a shift to a new, less environmentally destructive economic base. From an ethical perspective, one needs to ask: what is it we are interested in sustaining? City managers, planners, and elected officials, need to think about sustaining their particular cities over the long term and adopt the preferred strategy of a comprehensive approach to maintaining or enhancing economic, social, and environmental conditions. A sustainability agenda may be difficult to reconcile with the attempts in most, if not all, cities to enhance economic competitiveness, and the scale at which sustainability is implemented is important. It is disingenuous to talk about sustainability on a local scale if it is achieved by transferring costs, inequities, or waste products elsewhere that may perpetuate global environmental injustice. The pursuit and the study of urban sustainability entail a perspective that moves from the local scale of a single city, down to the individual neighborhoods, businesses, parks, and natural areas within it and up to the global scale of the world system of cities.

Summary

Cities exist in societies that create the conditions necessary for the appropriation of *surplus product*, that is, the excess that a society generates above the costs of production. In the nineteenth century, the stratified societies of Europe and North America experienced a massive urban transformation during the Industrial Revolution. During this period, cities, especially large manufacturing ones, were ugly creations and horrifying environments for the poor. Denied access to the fruits of rapid economic growth, the worker bore the social costs of urban industrialization. The early nineteenth-century industrial city was characterized by many small, relatively powerless enterprises. Toward the end of the nineteenth century, however, the market mode of economic integration shifted from individual to monopoly capitalism. The mechanization of agriculture and lure of factories concentrated the majority of the populations of Europe, Japan, and North America in urban areas by the 1920s.

Cities are organized into urban hierarchies, webs of interaction that stratify them based on their economic significance. On the intraurban scale (i.e., within the metropolitan area) spatial ordering exists because firms and people sort themselves in response to the differential distribution of opportunities and constraints. We examined the residential selection process, the dynamics of how people choose houses that lead to the creation of neighborhoods, and nonrandom patterns of residential space. We examined how the filtering of housing reflects changes in the supply and demand, and how land and labor markets generate dense concentrations of people in urban cores and relatively few on the periphery.

Suburbanization and urban sprawl reflect the pronounced, prolonged tendency toward decentralization characteristic of most cities in the industrial world. This trend reflects many combined forces, including changes in incomes and residential preferences; the search by manufacturing and office firms for cheaper land, taxes, and labor on the urban periphery; and the role of the state. The flip side of urban sprawl, although much smaller in scope, is gentrification, which in part reflects the rapid growth of producer services agglomerated in inner cities.

Urban areas, especially inner cities, are beset by a multitude of ills and problems. We traced the massive impacts that suburbanization and deindustrialization had on metropolitan cores, which were drained of many well-paying jobs and suffered rising unemployment and poverty. Thus, suburban growth and inner-city decline are two sides of one coin. These issues have particularly affected minority-dominated areas, including African American and Latino neighborhoods, which, in addition to a low number of jobs, also suffer from an insufficient low-income housing, inadequate schools and health care, and high crime rates.

The world's urban hierarchy is being reworked in light of the huge wave of contemporary globalization. Global cities, the command and control centers of the world economy, are sites where corporate headquarters are concentrated along with specialized producer services. They act as major centers of information processing, finance, tourism, media, and fashion. These cities, particularly New York, London, and Tokyo, are places whose decisions affect most people on the planet. A secondary tier of important local command centers complements these major nodes.

Finally, we considered the matter of urban sustainability. Cities are enormous consumers of resources and generate a large share of the world's environmental problems. Sustainable development aims to minimize the long-term costs of cities and promote lifestyles and forms of production that consume less energy and generate less waste. The steps toward such goals are clear, including greater use of mass transit, green buildings, enhanced recycling, and multifamily dwelling units—even if the political will to achieve them generally is lacking.

Key Terms

basic sector
central business district
direct effects
economic base analysis
edge city
exurbs
filtering model
gentrification
global city
indirect effects
induced effects
industrial restructuring
multiplier
nonbasic sector
residential location decision
reverse commuting
spatial mismatch
suburbanization
urban hierarchy
urban sustainability

Study Questions

1. How did the Industrial Revolution change cities?
2. How is the urban division of labor organized?
3. Why does the nonbasic sector increase with urban size?
4. Summarize the household location decision process.
5. How does the housing filtering model work?
6. What are the major causes of suburbanization and urban sprawl?
7. What are some of the major U.S. inner-city urban problems today and what causes them?
8. What is gentrification and what causes it?
9. What are world cities? What and where are the largest ones?
10. What steps can be taken to make cities more sustainable?

Suggested Readings

Florida, R. 2004. *The Rise of the Creative Class.* New York: Basic Books.

Knox, P., and L. McCarthy. 2005. *Urbanization: An Introduction to Urban Geography*, 2nd ed. Englewood Cliffs, NJ: Prentice Hall.

Knox, P., and S. Pinch. 2005. *Urban Social Geography: An Introduction*, 5th ed. Englewood Cliffs, NJ: Prentice Hall.

Pacione, M. 2001. *Urban Geography: A Global Perspective.* London: Routledge.

Sassen, S. 1991. *The Global City: New York, London, Tokyo.* Princeton, NJ: Princeton University Press.

Taylor, P. 2003. *World City Network: A Global Urban Analysis.* London: Routledge.

Web Resources

Center for Urban Policy Research
http://www.policy.rutgers.edu/cupr/

A well-known center for the analysis of urban space, with links to publications and other resources.

Globalization and World City Research Network
http://www.lboro.ac.uk/gawc/

Data and reports by world's leading authorities on the global urban system.

USA CityLink Project
http://usacitylink.com

Comprehensive listing of Web pages featuring American cities.

Transportation and Communications

OBJECTIVES

- To place modern transportation systems in a historical perspective
- To illustrate the nature of cost-space and time-space convergence or compression
- To demonstrate the relationship between transport and economic development
- To emphasize the critical role of transportation policy
- To examine communications innovations and online computer networks
- To summarize the social and economic impacts of the Internet

© JS Callahan/tropicalpix/Alamy

Although the United States has not invested in a high-speed rail system, many countries rely upon them heavily to facilitate interactions across the urban hierarchy. This is a high-speed train at Taipei Main Station, Taiwan.

Transportation and Communications

Improvements in transportation promote spatial interaction; consequently, they spur specialization of location. Thus, local comparative advantages—the ability of places to compete with one another—are facilitated, in part, by declines in **transport costs**. By stimulating specialization, better transportation leads to increased land and labor productivity as well as to more efficient use of capital.

Transport networks are constructed to facilitate **spatial interaction**, the movement of goods, people, and information among and within countries and cities. With increasing distance, there is attenuation or reduction in the flow or movement among places—the distance-decay effect. The underlying principle of distance decay is the friction of distance, that is, the costs of overcoming space. For individuals, these costs include the out-of-pocket expense of operating a vehicle or truck combined with the cost of a person's time. With longer distances, it is more expensive to ship commodities because of the labor costs of drivers and operators as well as the over-the-road costs of vehicle operation, such as fuel and maintenance.

In today's world, relatively little is consumed close to where it is produced; therefore, without transport services, most goods would be worthless. Part of their value derives from transport to market. Transport costs, then, are not a constraint on productivity; rather, efficient transportation increases the productivity of an economy because it promotes specialization of location. As societies abandon self-sufficiency for dependency on trade, transport makes it generally possible for their wealth and incomes to rise, although not equally for everyone.

Transportation and communications are key to understanding economic geography. How does the geographic pattern of transport routes affect development? How do changing transport networks shape and structure space? What are the impacts of transport costs and transit time on the location of facilities? How is information technology (IT) changing the way we live, work, and conduct business? This chapter provides answers to these questions in its discussions of transport costs and networks, transport development, transportation and communications innovation, and metropolitan transportation policy. It explores the historical development of modern transport systems, some general properties of transport costs, and the central role played by the state. Later, the discussion turns to telecommunications and its impacts.

TRANSPORTATION NETWORKS IN HISTORICAL PERSPECTIVE

For most of human existence, people have occupied narrowly circumscribed areas mostly isolated from other areas of settlement. Gradually, improvements in transportation systems changed patterns of human life. Control and exchange became possible over wider and wider areas and facilitated the development of more elaborate social structures such as far-flung empires. Prior to the development of railroads, overland transportation of heavy goods was slow and costly. Movement of heavy raw materials by water was much cheaper than by land. For this reason, most of the world's commerce was conducted by water transportation, and the important cities were maritime or riverine cities.

Human history, and people's relations in space and time, changed dramatically when capitalism spread over the globe. From the sixteenth century onward, there were great revolutions in science and trade, voyages of discovery and conquest, and a consequent increase in the amount of productive, commodity, and financial capital. During the Industrial Revolution of the late eighteenth and nineteenth centuries, the speed with which people, goods, and information crossed the globe accelerated exponentially when inanimate energy was applied to transportation in the form of the railroad and the steamship. Capitalism required an expanding world market for its goods; hence, it broke down the isolation of preindustrial economies.

The engine that drove this economic expansion was capital accumulation, that is, production for the sake of profit. In an effort to increase the rate of accumulation, all forms of capital had to be moved as quickly and cheaply as possible between places of production and consumption. To annihilate space by time, some of the resulting profits of commerce were devoted to developing the means of transportation and communication. "Annihilation of space by time" does not simply imply that better transportation and communication systems diminish the importance of geographic space; instead, the concept refers to how and by what means space is used, organized, created, and dominated to facilitate the circulation of capital. Time and space appear to us as "natural," that is, as somehow existing outside of society, but a historical perspective on how capitalism has changed our experience of them reveals time and space to be social constructs. Different societies experience and give meaning to time and space in different ways, and the changes unleashed by capitalism reconfigured these experiences worldwide.

The steady integration of production systems around the globe does not change their absolute location (*site*), but it does dramatically alter their relative location (*situation*). If we measure the distances between places in terms of the time or cost needed to overcome them (the friction of distance), then those distances have steadily shrunk over the past 500 years, particularly over the last 100. Transport improvements thus increase the importance of relative space. The progressive displacement of absolute space by relative space means that economic development becomes less dependent on relations with nature (e.g., resources and environmental constraints) and more dependent on social relations across space.

Beginning in the sixteenth century, canals were constructed in Europe to bring stretches of water into locations that needed them (Figure 1). The height of technology involved in the canal system was the pound lock, developed in the Low Countries and northern Italy. Until the nineteenth century, canals were the most advanced form of transportation and were built wherever capital was available. Road building was the cheap alternative where canals were physically or financially impractical. In the eighteenth century, the most active period of canal building coincided with the early industrial period, and the vast increase in manufacturing and trade fostered by the canals paved the way for the full Industrial Revolution. The canals were financed by central governments on the Continent and by business interests in England, where a complex network was built during the past 40 years of the eighteenth century and the first quarter of the nineteenth century. Somewhat later, artificial waterways were constructed in North America (e.g., the Erie Canal). They supplemented the rivers and the Great Lakes, the principal arteries for moving the staples of timber, grain, preserved meat, tobacco, cotton, coal, and ores.

Technological developments in sea-going transport before the Industrial Revolution concentrated on improving ships (e.g., better hulls and sails) to improve the economics of shipping goods over increasing distances, but such improvements had reached their limits by the mid-1800s. However, it was clear even as early as 1800 that traditional means of transport had become inadequate. The rapid expansion of commerce and industry overtaxed existing facilities. The canals were crowded and ran short of water in dry periods, and the roads were clogged with traffic in wet periods. These problems contributed to a general crisis of profitability by the late eighteenth century. The result was an effort to develop mechanical energy as the motivating power for transport.

Marvullo/Belgian National Tourist Office

FIGURE 1 A canal in the merchant city of Bruges, Belgium. This Hanseatic town was the north European counterpart to Venice during the mercantile period.

What paved the way for technical advances in transportation was the invention of the steam engine by James Watt in 1769. Its application to water transportation in 1807 and to land transportation in 1829, through the development, respectively, of the steamship and locomotive began an era of cheap transportation. In Europe, an expanding network of railways helped to create markets and provided urban populations with an excellent system of freight and passenger transportation (Figure 2). In the United States, the railroad was an instrument of national development; it preceded virtually all settlement west of the Mississippi, helped to establish urban centers such as Kansas City and Atlanta, and integrated regional markets. Today the Amtrak system (Figure 3) is the passenger rail system in the United States. However, relatively low demand and minimal public subsidies have kept the U.S. rail network very poorly developed compared to the

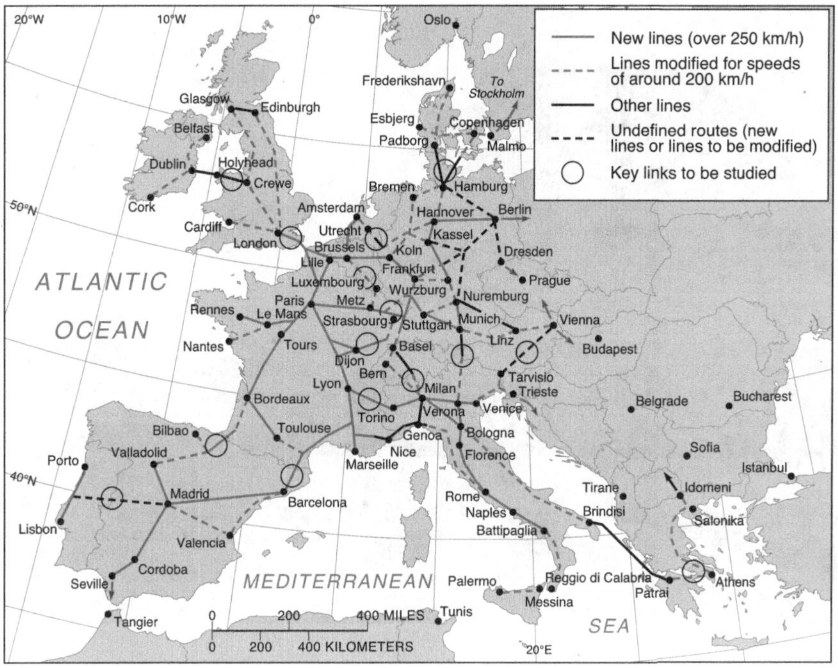

FIGURE 2 Western Europe's rail networks, which are extensive and generally high-speed, link its numerous cities in an efficient and productive transportation system. The system reflects the continent's population density, the relatively high cost of automobile ownership (which makes mass transit attractive), and government policies and subsidies.

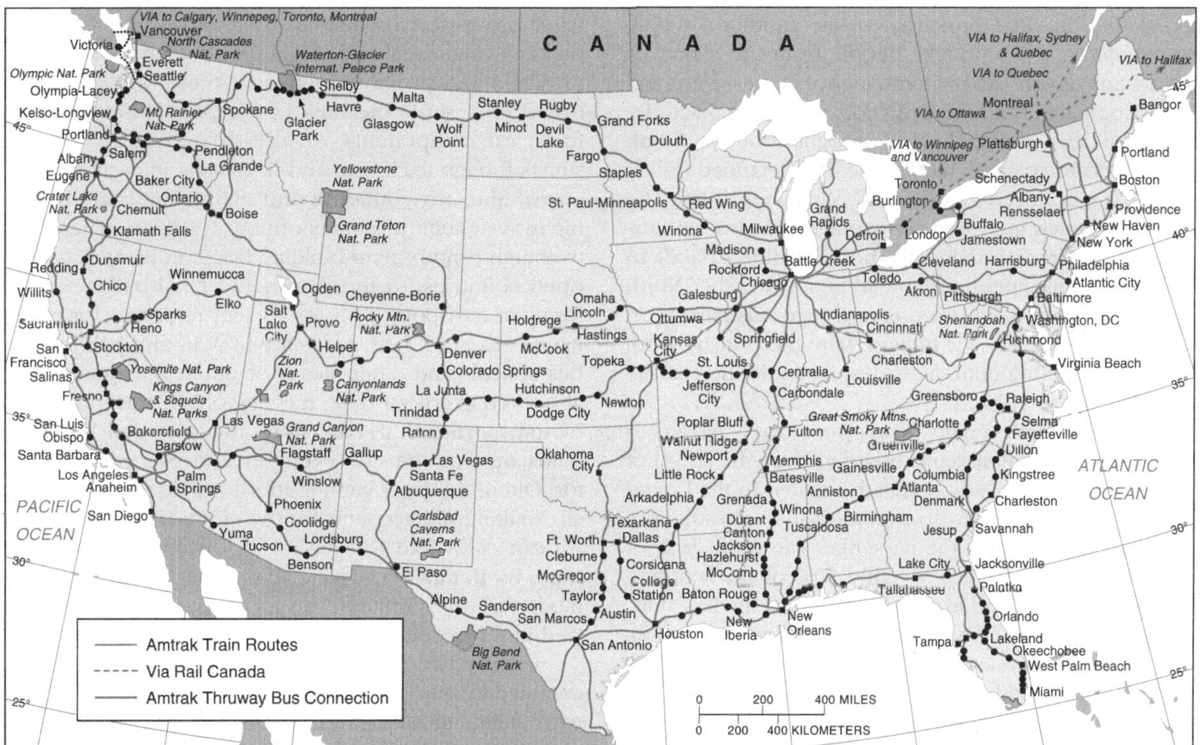

FIGURE 3 Amtrak's network. In contrast to Europe, the United States lacks a high-speed rail network, although trains are important to the movement of goods, if not people.

FIGURE 4 Amtrak, the dominant passenger train system in the United States, is relatively underfunded, slow, and inefficient compared to high-speed inter-urban rail systems in place in Europe and Japan.

high-speed systems of Europe (Figure 4). In developing countries, railroads link export centers to the economies of Europe and North America.

Until the 1880s, cities were mainly formed around the pedestrian and business establishments necessarily agglomerated in close proximity to one another. Usually, a 30-minute walk would take people from the center of town to any given urban point; hence, cities were extremely compact. The transformation of the compact city into the modern metropolis was brought about by Frank Sprague's electric traction motor. The first electrified trolley system was introduced in Richmond, Virginia, in 1888. The innovation, which increased the average speed of intraurban transport from 5 miles per hour (walking pace) to more than 15 miles per hour, spread rapidly to other North American and European cities, as well as to Australia, Latin America, and Asia. Electric trolleys were the primary form of urban commuting until the widespread adoption of the automobile in the 1920s.

In the nineteenth century, roads were merely feeders to the railroads. Road improvements had to await the arrival of the automobile in the early twentieth century. In the United States, heavy reliance on the automobile is a result of a national love affair with the passenger car and a lack of alternatives. In most cities, roughly 90% of the working population travels to and from work by car; in the less auto-dependent cities like New York, cars still account for two-thirds of all work-related trips. Public transportation is relatively scarce, and confined to a few large cities. By comparison, in Europe, where cities are less extensively suburbanized and average commuting distances are half those of North America, only 40% of urban residents use their cars. In Tokyo, a mere 15% of the population drives to work. In these cases, commuting trains and buses are the norm.

In the developing world, insufficient capital investments in transportation have created a crisis—the result of a mismatch between inadequate budgets for the transportation infrastructure and services and the need of the majority of the population for mobility. Governments that favor private car ownership by a small but affluent and politically influential elite distort their country's development priorities and promote inefficient transportation systems that do not serve the bulk of the population. Importing fuels, car components, or already assembled cars consumes foreign exchange and negatively impacts trade balances. Similarly, building and maintaining an elaborate highway system devours enormous resources. There was a twentieth-century road-building boom in many less developed countries, to the detriment of railroads and other forms of transport. With insufficient resources for maintenance, many of these roads are now in disrepair. In cities, bus systems and other means of public transportation are also often in a poor state, meeting only a small proportion of transportation needs; slow and crowded buses and trains are symptoms of severe underinvestment in this sector. Often, the poor cannot afford public transportation at all. Walking still accounts for two-thirds of all trips in large African cities like Kinshasa and for almost one-half the trips in Bangalore, India. Pedestrians and traditional modes of transportation are increasingly being marginalized in the developing countries.

In the past 175 years, transport changes have not been confined to railroads and roads. At sea, ships equipped first with steam turbines and then diesel engines facilitated the rapid expansion of international trade. In addition, the opening of the Suez Canal in 1869 and the Panama Canal in 1914 dramatically reduced the distance of many shipping routes (Figure 5), reconfiguring trade networks and

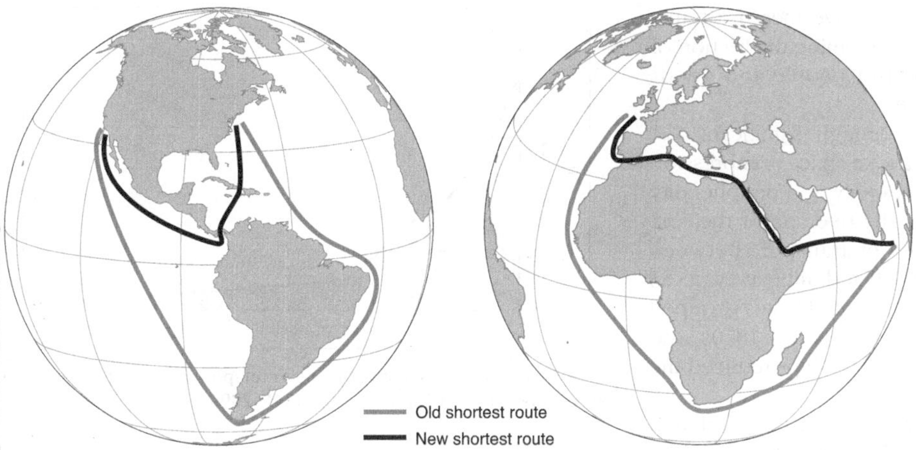

FIGURE 5 The Panama and Suez canals greatly shortened world shipping routes, reflecting and contributing to the time-space compression of the industrial revolution and the new geographies it brought into being.

changing the geography of port cities. In ocean shipping, container ships, which use standardized containers that can be efficiently stacked and easily switched between ships and trucks, have become the basic transoceanic carrier. Planes have ousted passenger liners and trains as the standard travel mode for long-distance passengers. The shipment of cargo by air, however, is still in its infancy. Heavy, bulky, and low-valued goods are always shipped by water (e.g., oil, flour), such as through the Suez Canal (Figure 6); only perishable, high-value, or urgently needed shipments are sent by air freight (e.g., pharmaceuticals).

TIME-SPACE CONVERGENCE OR COMPRESSION

Transport networks are constructed to facilitate the movement of goods, people, and information among countries and cities as well as within cities over networks that range greatly in their degree of complexity and connectivity. These flows represent the exchange of supplies and demands at different locations. The term **distance decay** describes the attenuation or reduction in the flow or movement among places with increasing distance between them, although most food shipments, passenger trips, natural resource flows, and commodity movements occur within regions and within countries, rather than between them. The underlying principle of distance decay is the **friction of distance**, that is, the costs of overcoming space.

If geography is the study of how human beings are stretched over the earth's surface, a vital part of that process is how we know and feel about space and time. As we have mentioned, although space and time appear as "natural" and outside of society, they are in fact social constructions; every society develops different ways of dealing with and perceiving them. Time and space are thus socially created, plastic, mutable institutions that profoundly shape individual perceptions and social relations. Transport improvements have resulted in what geographers call **time-space convergence** or **time-space compression**—that is, the progressive reduction in the cost of travel and

FIGURE 6 The Suez Canal, constructed in 1869 to connect the Mediterranean with the Red Sea, dramatically shortened shipping routes between Europe and Asia, with important impacts on both sides of the network. Part of the wave of globalization in the nineteenth century, it continues to exert significant impacts on trade patterns and port activities throughout the Mediterranean and Indian Ocean.

travel time among places. Similarly, if we measure transport costs in terms of the cost of overcoming the friction of distance, ever-cheaper movement of people and goods leads to **cost-space convergence**.

Transport improvements have brought significant cost reductions to shippers, creating a cost-space compression that altered the geographies of centrality and peripherality of different places. For example, the opening of the Erie Canal in 1825 reduced the cost of transport between Buffalo and Albany from $100 to $10 and, ultimately, to $3 per ton. Railroad freight rates in the United States dropped 41% between 1882 and 1900. Between the 1870s and 1950s, improvements in the efficiency of ships reduced the real cost of ocean transport by about 60%.

Cheaper, more efficient modes of transport widened the range of distances over which goods could be shipped economically, contributing to the growth of cities. They enabled cities to obtain food products from distant places and facilitated urban concentration by stimulating large-scale production and geographic division of labor. Furthermore, transportation improvements changed patterns of urban accessibility. North American cities have grown from compact walking and horse-car cities (pre-1800–1890), to electric streetcar cities (1890–1920), and, finally, to dispersed automobile cities in the recreational automobile era (1920–1945), the freeway era (1945–1970), the edge city era (1970–1990), and the exurban era (1990–present).

Developments in transportation have also cut travel times extensively. For example, the travel time between Edinburgh and London, a distance of 640 kilometers, decreased from 20,000 minutes by stagecoach in 1658 to less than 60 minutes by airplane today (Figure 7). During the period of rapid transport development, time-space convergence was especially marked. In the 1840s,

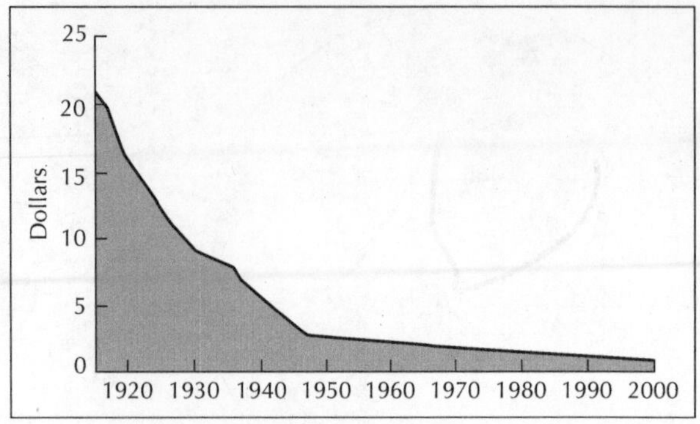

FIGURE 8 Cost-space convergence in telephone calls between New York City and San Francisco, 1915–2000.

Cost-space convergence in telephone calls between New York City and San Francisco from *Human Geography: Landscapes of Human Activities* by J. A. Getis and J. Getis, McGraw-Hill, 1992

travel time between Edinburgh and London was longer than 2000 minutes by stagecoach, but by the 1850s, with the arrival of the steam locomotive, the travel time had been reduced by two-thirds, to 800 minutes. By 1988, the rail journey between Edinburgh and London took 275 minutes. When the line was electrified in 1995, travel time was reduced to less than 180 minutes. Similarly, in communications, the steadily declining cost of long-distance telephone calls increased the interactions among cities, such as between New York City and San Francisco (Figure 8).

Air transportation provides spectacular examples of time-space convergence. In the late 1930s, it took a DC-3 between 15 and 17 hours to fly the United States from coast to coast. Modern jets now cross the continent in about 5 hours. In 1934, planes took 12 days to fly between London and Brisbane. Today, the Boeing 747 SP is capable of flying any commercially practicable route nonstop. The result is that any place on earth is within less than 24 hours of any other place, using the most direct route.

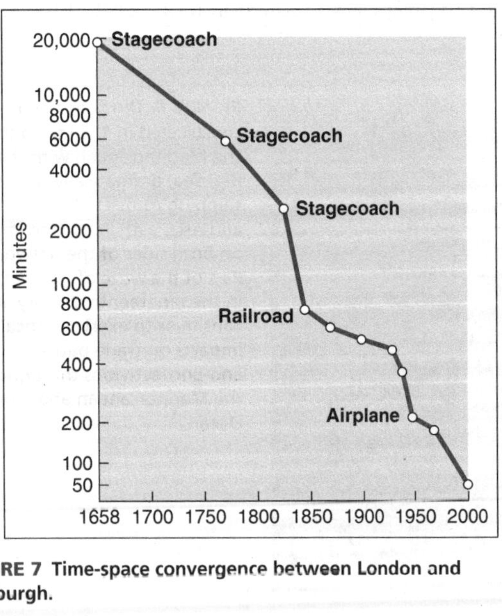

FIGURE 7 Time-space convergence between London and Edinburgh.

Time-space convergence between London and Edinburgh from "Central Place Development in a Time-Space Framework" by Donald G. Janelle, *The Professional Geographer*, Vol. 20, No. 1. (1968). Reprinted by permission of the publisher, Taylor & Francis Ltd, www.informaworld.com

TRANSPORTATION INFRASTRUCTURE

Transportation infrastructures allow countries to specialize in production and trade and form integrated spatial divisions of labor. Figure 9 depicts the world's major roads and highways and indicates the unevenness that both reflects and contributes to uneven spatial development. In some countries, transportation is slow, expensive, and difficult. Much of the developing world has poorly developed infrastructures, which inhibits economic development: Farmers cannot get their goods to market, and people cannot move easily to jobs. By contrast, the well-developed infrastructures of North America and Europe reflect their level of economic development.

Fast and efficient transportation systems allow the development of natural resources, regional specialization of production, and internal trade among regions. For example, even though India appears to be well connected, it is a country whose economic growth is hampered by an inadequate transportation and communications infrastructure. Passenger and commodity traffic on India's roads has

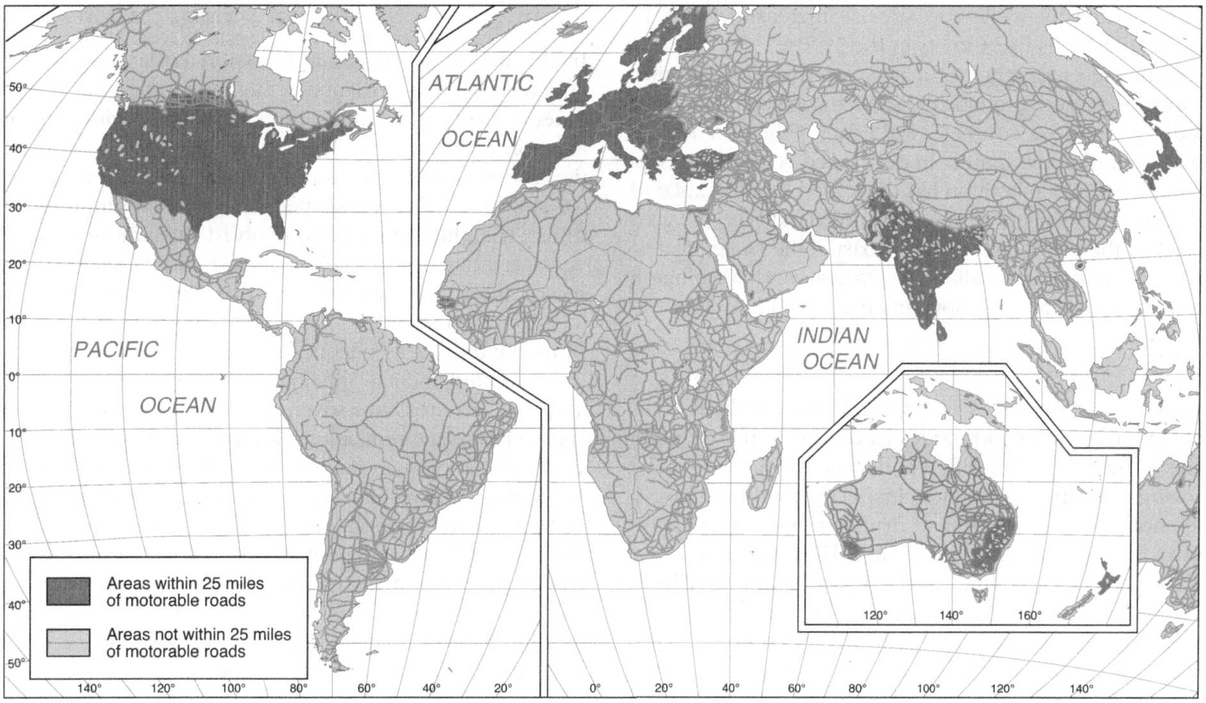

FIGURE 9 Major world roads and highways. Note areas within the United States, Europe, India, Southeastern Australia, and New Zealand are virtually all within 25 miles of roads and highways. Only a few exceptions exist, in the Western mountain regions of the United States.

increased 30-fold since independence in 1947. Roads are overcrowded and 80% of villages lack all-weather roads. Improved accessibility would allow regional specialization of production and cash-crop farming, especially of fruits and vegetables that spoil quickly. Most countries of the world have simple transport networks penetrating the interior from ports along the ocean. These railroads and highways are called tap routes. Tap routes are the legacy of colonialism or the product of neocolonialism. Such routes facilitate getting into and out of a country, but they do not allow for internal circulation or circulation between countries in the same region.

GENERAL PROPERTIES OF TRANSPORT COSTS

Transportation costs appear deceptively simple but are far more complex upon further scrutiny. They can be categorized as either **terminal costs** or **line-haul costs** (Figure 10). Terminal costs must be paid regardless of the distance involved. They include the cost of loading and unloading, capital investment, and line maintenance. Line-haul costs, in contrast, are strictly a function of distance. For example, fuel costs are proportional to the distance a load must be moved.

Recently, container-handling facilities have tied trucks and ships together. The first facility was developed in 1956. By the early 1970s, numerous carriers entered into the container ship business. At first, the greatest appeal of the container ship was its speed and economy in port and its facilitation of the multimodal transport of goods. For example, commodities from Japan and other Pacific Rim countries could be transported economically to Europe via North America. Later, container operations sped up the

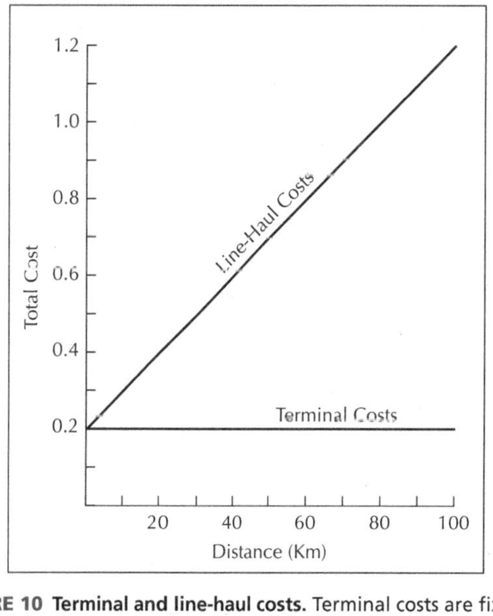

FIGURE 10 Terminal and line-haul costs. Terminal costs are fixed costs that are invariant with respect to distance. Line-haul costs incurred "over the road" are variable costs (e.g., fuel, labor time) that rise proportionate to distance.

ocean voyage as well—top speeds increased from 15 knots in the 1950s to 33 knots in the 1970s.

With the emergence of a new international division of labor, ports have continued to modernize their methods of handling cargo to compete with each other for shares of global commodity traffic. In developed countries served by many ports, competition has decreased the relevance of the traditional concept of the port hinterland (i.e., the area served by the port). On the West Coast of the United States, for example, ports in California, Oregon, and Washington compete fiercely for the mounting trade with East Asia.

Carrier Competition

Competitive differences in terminal and line-haul costs among various transport modes lead firms to use different forms of transport over different distances (Figure 11). The trucking industry has low terminal costs partly because it does not have to provide and maintain its own highways and partly because of its flexibility. However, trucks are not as efficient in moving freight on a ton-kilometer basis as are railroad and water carriers. Of the three competing forms of transport, trucks involve the least cost only out to distance D_1. Railroad carriers have higher terminal costs than truck carriers, but lower than water carriers, and a competitive advantage through the distance $D_1 - D_2$. Water carriers, such as barges, have the highest terminal costs, but they achieve the lowest line-haul costs, giving them an advantage over longer distances.

Freight Rate Variations and Traffic Characteristics

When there is an absence of competition among transport modes, a carrier can set rates between points to cover costs, and in the absence of government intervention, a carrier may set unjustifiably high rates. Intermodal competition or government regulation reduces the likelihood of such practices. Competition among carriers reduces rate differences among them. For example, the opening of the St. Lawrence Seaway in 1959 resulted in lower rail freight rates on commodities affected by low water-transport rates.

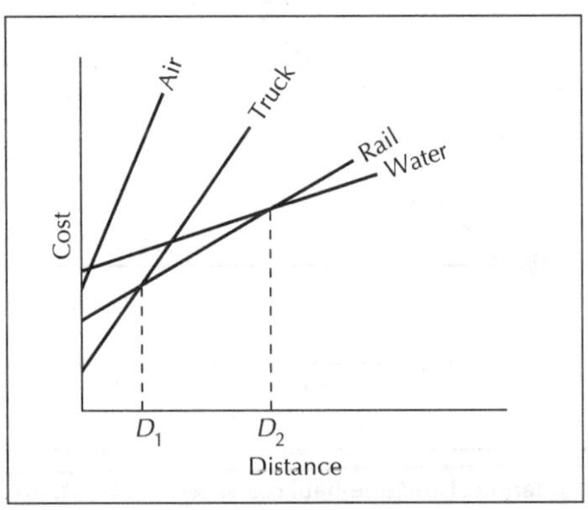

FIGURE 11 Variations in terminal and line-haul costs for air, truck, rail, and water.

Regimes for International Transportation

In the international arena, transport rates and costs are affected by the governance of the transport mode. To illustrate, consider the contrasting governance regimes of civil aviation and shipping. The international regime for aviation is dominated by nation-states; the international regime for shipping has been shaped by large shipping corporations. The shipping regime evolved over more than 500 years and has been more concerned with facilitating commerce than with national security. The regime for civil aviation developed in the early twentieth century and primarily reflects a concern for national security.

The fundamental principle governing international aviation is that states have sovereign control over their own air space. From this principle, rules and procedures have developed that permit countries to regulate routes, fares, and schedules. As a result, many countries, developed and developing, have secured a market share that is more or less proportional to their share of world airline traffic. Developing countries have been able to compete with companies based in the industrialized world on an equal footing; for example, Air India, Avianca, and Korean Air Lines can challenge Delta, Air France, and British Airways.

The international shipping regime has left many developing countries in a weak position with regard to nurturing their own merchant fleets. In a world of global markets, few underdeveloped countries have much influence when it comes to setting commodity rate structures. Lack of control over international shipping is an important area of concern in the Third World's quest for development. Although the shipping regime is dominated by shipping firms, the market is inherently unfair in that it favors developed countries over developing countries. Hence, less developed countries are faced with rate structures that work against them, inadequate transport services, a perpetuation of center-periphery trade routes, and a lack of access to decision-making bodies. Those less developed countries that generate cargoes such as petroleum, iron ore, phosphates, bauxite/alumina, and grains cannot penetrate the bulk-shipping market, which is dominated by the vertically integrated MNCs (multinational corporations) based in developed countries. Cartels of ship owners called liner conferences set the rates and schedules for liners (freighters that ply regularly scheduled routes).

The global pattern of container ports reflects the geography of production and trade. In the past, the largest ports were located in Europe and North America. For many years, Rotterdam, at the mouth of the Rhine River in the Netherlands, was Europe's primary port and the largest in the world. Up until the 1980s, the largest ports in the United States were located on the Atlantic coast (e.g., New York City), as most American trade was with Europe. The rapid economic growth of East Asia, however, has changed these patterns. Today, the world's largest ports are located in Hong Kong and Singapore (Figure 12). Because most U.S. trade is across the Pacific Ocean with East Asia, West Coast ports such as Los Angeles, Oakland, and Seattle have

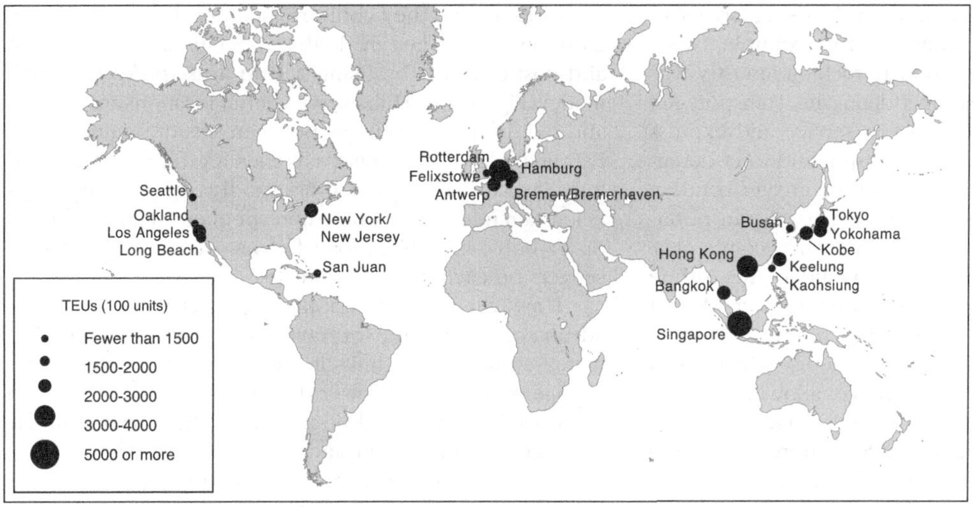

FIGURE 12 Most of the world's major container ports are located in East and Southeast Asia and in Europe. New York City, Los Angeles, and Vancouver represent major container ports in North America. Such ports are "hub ports," acting as major centers where container traffic splits into feeder flows to and from the centers within the hub's respective hinterland.

surpassed East Coast ports such as New York City in trade volume. These changing patterns reflect the ways in which globalization is changing the domestic economic geography of the United States and other countries.

Transport innovations have reduced circulation costs and fostered the new international division of labor (Figure 13). They have encouraged the decentralization of manufacturing processes in industrialized countries, both from major cities toward suburbs and smaller towns and from central regions to those more peripheral. They have also encouraged the decentralization of manufacturing processes to those less developed countries with an abundance of low-wage labor.

Port Authority of New York & New Jersey

FIGURE 13 Container cargo handling at the Maersk Line Terminal, Port Newark, New Jersey. Containerization has greatly improved the operation, management, and logistics of oceangoing freight. The container revolution has impacted more than conventional shipping and international trade. Newly designed cellular vessels have much faster turnaround time in port and have improved cargo-handling productivity at ports. Moreover, the interface between water and land transportation has expanded. The new container trains have enhanced the economy and scale of rail transportation.

The container revolution and bulk-air cargo carriers enabled MNCs based in the United States, Japan, and Europe to locate low-value-added manufacturing and high-pollution manufacturing processes "offshore" in more than 80 Third World free-trade zones. Almost one-half of these zones are in Asia, including Hong Kong, Taiwan, Malaysia, and South Korea. Free-trade zones are areas where goods may be imported for packaging, assembling, or manufacturing free of duties and then exported. These global workshops are geared to export markets and often have few links to the national economy or the needs of local consumers. They tend to be located near ports and international airports and in areas virtually integrated into global centers of business.

TRANSPORTATION, DEREGULATION AND PRIVATIZATION

Long-established national transportation policies and regulation were the global norm until the 1970s. The purpose of regulating airlines and rail carriers was to ensure quality, protect companies and customers, and establish these quality and safety standards throughout the industry. During this period, providers not only provided basic transportation services but also met a social obligation, such as providing service to low-income, unprofitable areas.

By the late 1970s, with the rise of conservative neoliberalism, international trade regulators required that new transportation operators be allowed to enter the market in order to ensure efficiency and maximize benefits. The move toward **privatization** and **deregulation** had begun. Market advocates criticized regulation for creating inefficiency, limiting competition, and raising prices to consumers. The Swedish railways, for example, were deregulated by 1968, and the British trucking industry was also deregulated in that year. In Britain, the Transport Act of 1980 removed all controls on bus service and express service between cities; the 1985 Transportation Act deregulated local bus service inside and outside greater London; and the British government sold nationalized transportation companies and many municipally owned companies. In the United States, deregulation included the Airline Deregulation Act of 1978 and the Motor Carriers Act of 1980, which loosened rules on trucking. In the developing world, Sri Lanka deregulated all bus routes, while China deregulated long-distance coach service, allowing fares to vary. Nigeria followed suit, by privatizing Nigeria Airways and its National Shipping Line; Singapore privatized Singapore Airlines and its mass transit corporation.

The most important result of U.S. airline deregulation has been more competitive fares and the survival of the most efficient companies (as well as bankruptcies of others). Domestic U.S. airline routes are now open to any carrier, and the development of a hub-and-spoke network has reduced costs. The number of direct flights has been reduced, and air service now requires at least one stop in an airline hub city, unless the service is between very large cities. Flights from smaller cities are directed into larger city airports or hubs and then linked to final destinations by direct flights. Privatization and deregulation have kept fares down. In 1976, only 15% of passengers on domestic air routes used discount fares; today, 90% of passengers use discounted tickets. However, as average fares have fallen on long-haul routes, fares on short routes have risen.

Hub-and-Spoke Networks

In order to remain competitive, the airlines that survived the shake-out following deregulation restructured their networks so that they could reduce direct flights between most city pairs. They made their operations more efficient and cost-effective by using a **hub-and-spoke network** model (Figure 14). Hubs serve central locations that collect and redistribute passengers between sets of original cities. Extremely large passenger volumes are funneled through hubs, and this allows the airlines to fly larger and more efficient aircraft and to offer more frequent flights between major hubs, increasing load factors—the number of passengers on each plane (Figure 15).

However, hub-and-spoke networks have some disadvantages, especially for travelers who must negotiate an increased number of links in their trips, frequently with a change of planes and fewer direct flights. Also, the system creates congestion at the main hub cities, which affects efficiency both in the air and on the ground. Airlines must make careful decisions about the location and exact number of hubs so that their operation is competitive with other airlines. Not all cities have fared equally well. Some airports had a precipitous decline in traffic after the new system prompted by deregulation began. Others, benefiting from the mad scramble to reduce fares and elevate efficiency, became megahubs: Atlanta, Chicago, Dallas, and Denver became major hubs for two or more airlines; and Salt Lake City, Minneapolis–St. Paul, Memphis, and Detroit have also become major centers.

PERSONAL MOBILITY IN THE UNITED STATES

Automobiles

An important dimension of changes in urban transportation concerns personal mobility in the United States, which is at its highest level in history—individuals are making more and longer trips and own more vehicles. Three factors account for this. The first is the overall performance of

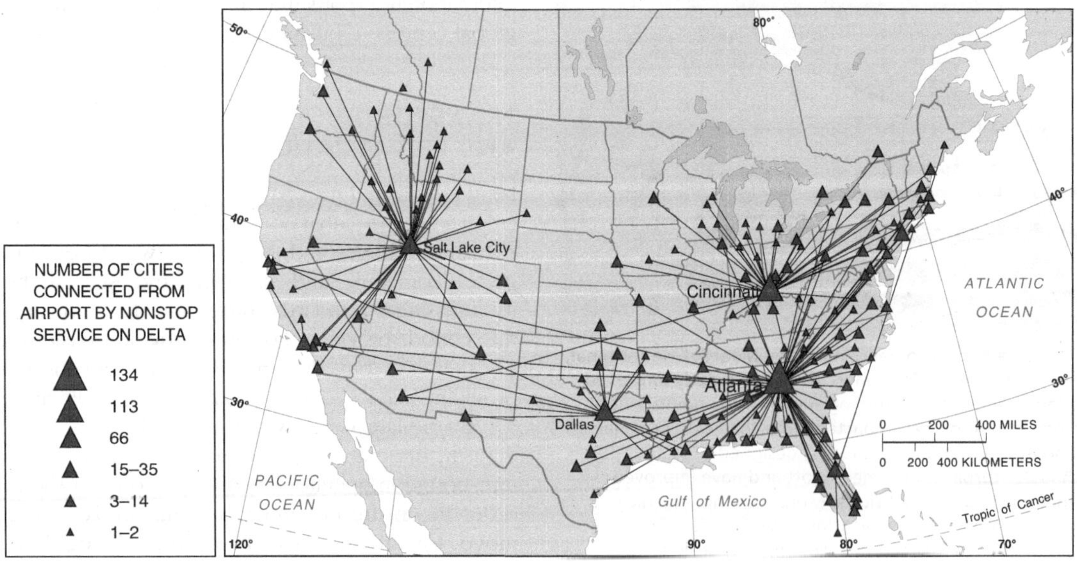

FIGURE 14 Hub-and-spoke networks for Delta Airlines.

Frederick P. Stutz

Jeff Greenberg/PhotoEdit

FIGURE 15 Delta Airlines' network has hubs in Atlanta and Cincinnati.

the national economy: When more people have more money to spend on transportation, greater automobile ownership and greater travel distances result. The second factor is the spread of low-density exurbs. On average, distances between home and job have increased, leading to longer commutes. A third reason for increased mobility is participation of women in the workforce. Over the past 40 years, many more women have come to own vehicles and have entered the workforce full-time, and so their need to travel has increased.

In the late twentieth century, the number of households, drivers, vehicles, vehicle trips, vehicle miles traveled (VMT), person trips, and person miles of travel all increased at a much faster rate than the population. Both vehicle miles traveled and annual number of vehicle trips per household have increased annually. The number and percentage of households that did not own a vehicle has dropped, and the number of households with three or more vehicles has increased dramatically. These trends have led to significant congestion in many cities, particularly at peak usage times or rush hours (Figure 16).

The **journey-to-work** trip has continued to account for the largest proportion of travel by U.S. households, in terms of both total miles of travel and number of trips. In 2008, the average home-to-work trip was 12 miles, and social and recreational trips averaged just over 11 miles per trip. Other family or personal business trip lengths averaged 7 miles, and shopping trips averaged 5.6 miles. Vehicle occupancy has declined but is explained partially by the greater number of vehicles per household and the decrease in average household size during this period.

As average vehicle prices continue to increase, Americans are retaining their vehicles for a longer period. For example, in 1969, 42% of household automobiles were 2 years old or less; by 2008, only 14% were 2 years old or less. The number of automobiles that were 10 or more

Image copyright egd, 2010. Used under license from Shutterstock.com

FIGURE 16 Rush-hour congestion reflects the highly uneven distribution of demand for transportation that occurs throughout the day, as well as the enormous and unsustainable reliance on the automobile found in Western societies. Rush hours waste countless hours annually and generate many of the gases that contribute to global climate change.

years older increased from 6% in 1969 to 37% by 2008. Older vehicles burn energy less efficiently and cleanly than newer vehicles and contribute to energy and air pollution problems.

High-Speed Trains and Magnetic Levitation

Magnetic levitation (maglev) technology eliminates mechanical contact between a vehicle and the roadbed, thus eliminating wear, noise, and alignment problems (Figure 17). The vehicle floats on a cushion of air one-half foot above the guideway, supported by magnetic forces. The Japanese and German governments have spent approximately $1 billion each on magnetic levitation research. In the future, **maglevs** may transfer passengers between U.S. cities separated by up to 300 miles, at over 300 miles per hour, using far less energy and time than automobiles, Amtrak, or even air carriers. One could shuttle between Boston and Washington, or Chicago and Minneapolis, or Los Angeles and San Francisco in less than an hour. Maglevs presently are twice as fuel efficient as automobiles and four times as efficient as airliners, producing little or no air pollution. In the future, maglevs may be built alongside highways; they will occupy far less room than airports (the Dallas–Fort Worth airport consumes as much land as a 65-foot-wide right-of-way coast to coast).

In the meantime, high-speed conventional rail systems have been improved to include tilting train technology. The passenger car carriage tilts inward on curves, allowing increased speeds on existing track curvatures. For example, the Swedish X2000 train can travel up to 150 miles per hour; it could be used in the U.S. northeast rail corridor. Amtrak, the national passenger railroad service, is currently making heavy investments in the tilting train technology. ISTEA, the Intermodal Surface Transportation Efficiency Act of 1991, identified five existing rail corridors that could be suitable for the development of high-speed trains: San Diego–Los Angeles–San Francisco, Dallas–Houston–San Antonio, Miami–Orlando–Jacksonville, Pittsburgh–Chicago–Minneapolis, and Washington–New York City–Boston.

TELECOMMUNICATIONS

The transmission of information is every bit as important as the movement of people and goods. An abundance of information and its availability facilitate and accompany economic development and political liberty. Because the circulation of information is critical to the operation and success of large, complex economies, the history of capitalism has been accompanied by succeeding waves of innovation in communications (Figure 18).

Telecommunications are not a new phenomenon. It took its first form in 1844 with Samuel Morse's invention of the telegraph, which allowed communications to become detached from its dependence on transportation. The telegraph made possible the transmission of information concerning commodity needs, supplies, prices, and shipments worldwide—information essential to the efficient conduct of international commerce. Telegraphy grew rapidly in the United States, from 40 miles of cable in 1844 to 23,000 miles by 1852. The first transcontinental telegraph wire was installed in 1861. Thus the telegraph was important in the American colonization of the West, where it displaced the Pony Express, and helped to form a national market by allowing long-distance circulation of news, prices, stock market, and other information. In 1868, the first successful trans-Atlantic telegraph line was laid, part of the international time-space compression that accompanied the Industrial Revolution.

For decades after the invention of the telephone in 1876, telecommunications was synonymous with simple telephone service. Just as the telegraph was instrumental to the colonization of the American West, in the late

FIGURE 17 The French Train à Grande Vitesse (TGV) is one of a series of high-speed intercity rail systems found in Western Europe and Japan that link producers and consumers across their countries (and between them) in an effective skein of interactions. Fast and energy-efficient train systems offer an opportunity to provide affordable inter-urban transportation, as long as societies and governments dedicate the necessary resources to build and maintain them.

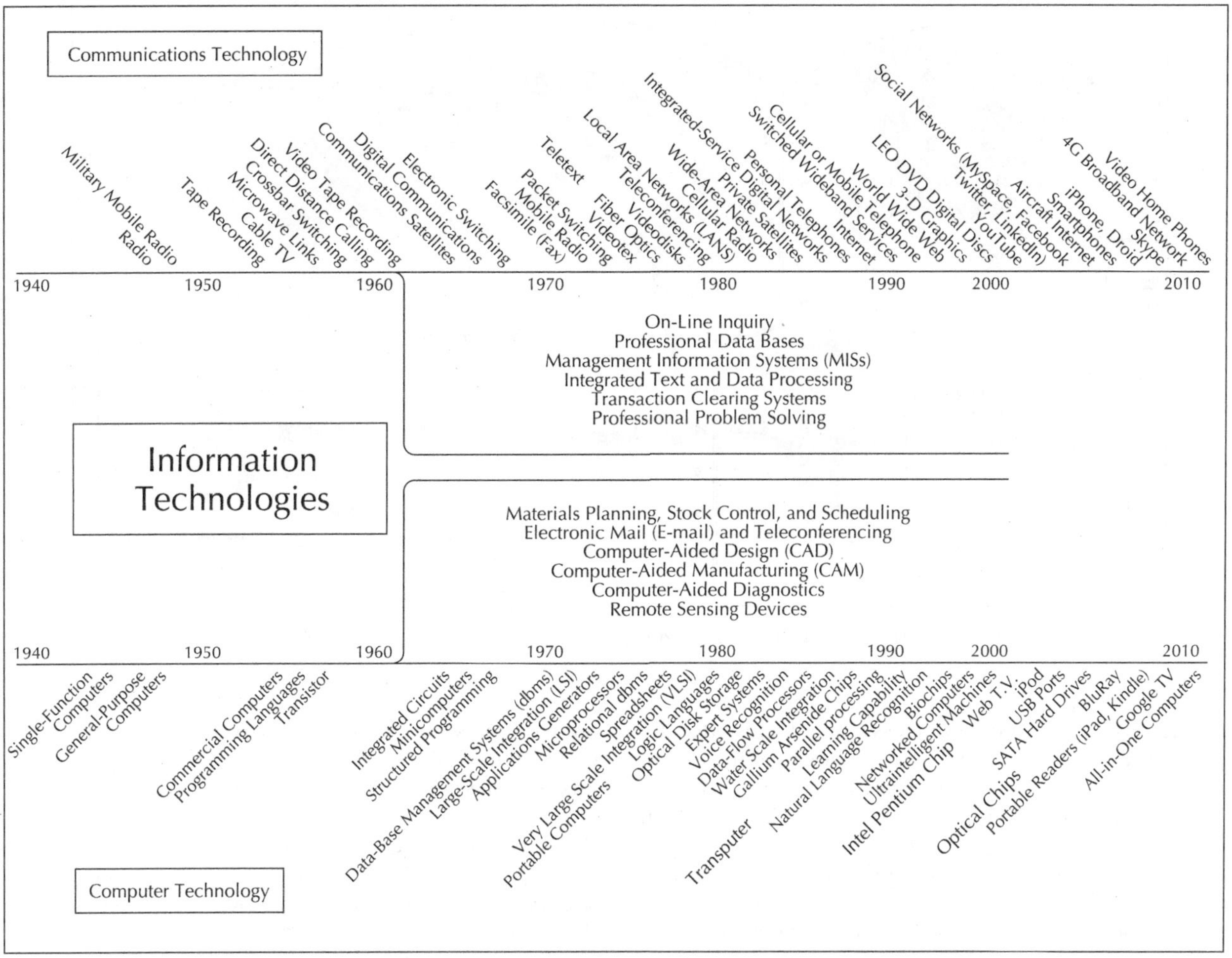

FIGURE 18 The history of innovations in computer and communication technologies.

nineteenth century the telephone became critical to the growth of the American city system, allowing firms to centralize their headquarters functions and spin off branch plants to smaller towns. These growing multi-establishment corporations utilized the telephone to coordinate production and shipments. In the 1920s, the telephone, like the automobile and the single-family home, became a staple of the increasing middle class and had significant social effects on friendship networks, dating, and other ties. In the 1950s, direct dialing eliminated the need for shared party lines, and the first international phone line was laid across the Atlantic Ocean. Even today, despite the proliferation of several new forms of telecommunications, the telephone remains by far the form most commonly used by businesses and households.

From 1933 to 1984, the American Telegraph and Telephone Company (AT&T) enjoyed a monopoly over the U.S. telephone industry. Congress exempted AT&T from antitrust laws in return for its commitment to guarantee universal access, which eventually resulted in a 98% penetration rate—the availability per 100 people—among U.S. households. When the deregulation of industry became widespread in the 1970s, telecommunications were not exempt. In 1984, AT&T was broken up into one long-distance and several local service providers ("Baby Bells"), and new firms such as MCI and Sprint entered the field. Faced with mounting competition, telephone companies have steadily upgraded their copper cable systems to fiber-optic lines, which allow large quantities of data to be transmitted rapidly, securely, and virtually error free.

Telephones are a common measure of a nation's communications infrastructure (Figure 19). Telephone penetration rates are highest in the economically developed world. Africa has less than 1% of the world's telephones although it contains 15% of the world's population, and one-half of the world's population has never made a telephone call. But landlines are rapidly being supplemented or replaced by wireless technologies; indeed, there are more cell phones today than landline phones.

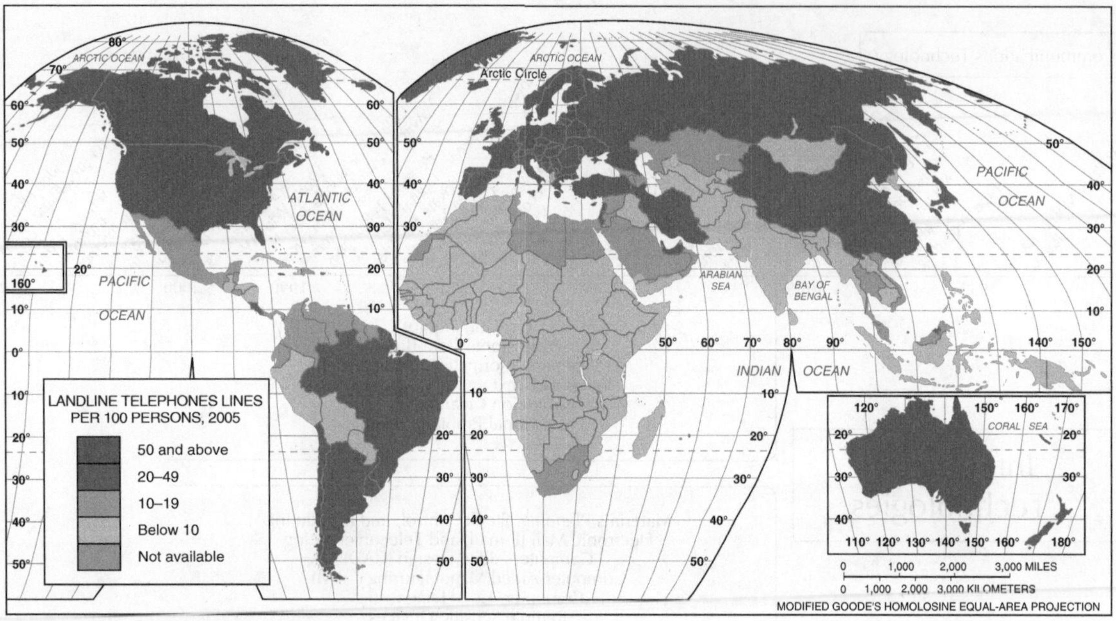

FIGURE 19 Landline telephones per 100 people worldwide by country, 2005.

For developing countries, wireless technologies offer the possibility of "leapfrogging," that is, moving directly into newer, lower-cost forms of technology.

The advances in microelectronics of the late twentieth century was particularly important in the telecommunications industry, which is quite possibly the most rapidly expanding and dynamic economic sector in the world today. Innovations in processing power have led to exponential increases in the ability of computerized systems to analyze and transmit data. The ability to transmit vast quantities of information in real time over the planetary surface is crucial to "digital capitalism." No large corporation can operate in multiple national markets simultaneously, coordinating the activities of thousands of employees within highly specialized corporate divisions of labor, without access to sophisticated channels of communications. Thus, telecommunications are important to understanding broader issues pertaining to globalization and the world economy, including the complex relations between firms and nation-states.

Fiber-optic and Satellite Systems

Today, two technologies—satellites and fiber-optic lines—are the basis of the global telecommunications industry. The transmission capacities of both technologies grew rapidly in the late twentieth century as the microelectronics revolution unfolded. Multinational corporations, banks, and media conglomerates typically employ both technologies, often simultaneously, either in the form of privately owned facilities or circuits leased from shared corporate networks. Roughly 1000 fiber-optic and two dozen public and private satellite firms provide international telecommunications services—a network of fiber lines, the nervous system of the global financial and service economy, linking cities, markets, suppliers, and clients around the world (Figure 20).

Although to a great extent they overlap, satellite and fiber-optic transmission providers form separate market segments. Fiber is heavily favored by large corporations for data transmission and by financial institutions for electronic funds transfer systems. Satellites tend to be used more often by international television carriers. Telephone and Internet service providers use both. The two types of telecommunications carriers are differentiated geographically as well: Because their transmission costs are unrelated to distance, satellites are optimal for low-density areas (e.g., rural regions and remote islands), where the relatively high marginal costs of fiber lines are not competitive. Fiber-optic carriers prefer large metropolitan regions, where dense concentrations of clients allow them to realize significant economies of scale in cities where frequency transmission congestion often plagues satellite transmissions. Satellites are ideal for point-to-area distribution networks; fiber-optic lines are preferable for point-to-point communications, especially when security is of great concern.

Historically, the primacy of each technology has varied over time. From 1959 to 1980 (i.e., before the invention of fiber optics), satellites enjoyed only limited competition from transoceanic copper cable lines, which had low capacity. From the 1970s onward, innovations in microelectronics allowed fiber-optic lines to erode the market share of traffic held by satellites (Figure 21). And since then, new data transmission techniques, such as the so-called frame delay format, have raised transmission speeds nearly 30-fold over that of the 1990s technology.

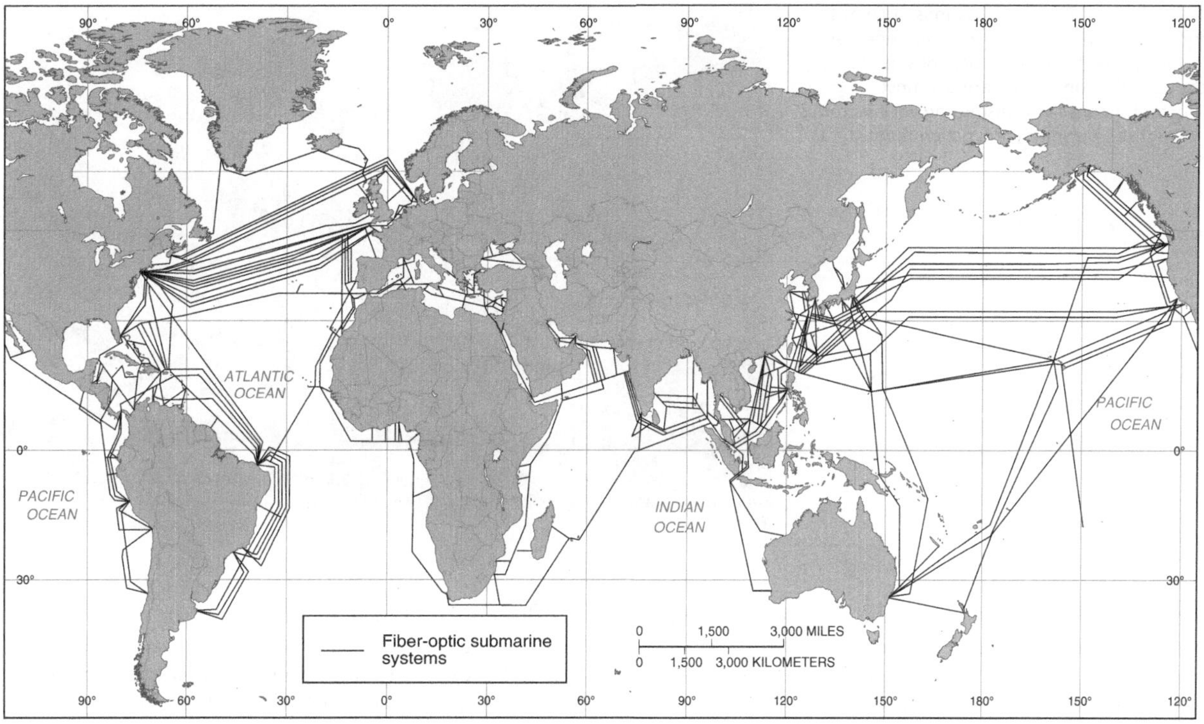

FIGURE 20 The global network of fiber-optic cable. Beginning in the 1980s with the laying of cable across the Atlantic, many telecommunications giants have expanded the grid, increasing exponentially the ability to transmit enormous quantities of data.

More recently, the satellite industry has seen a resurgence due to the growth of wireless and cellular phone traffic. Low-orbiting satellites—large-scale, *low earth orbit* (LEO) satellite systems—provide telephone communications to and from anyplace on earth and are ideal for the rapidly expanding cell phone market. These private, global satellite constellations transmit television, radio, and voice signals and fax and computer data (Figures 22 and 23).

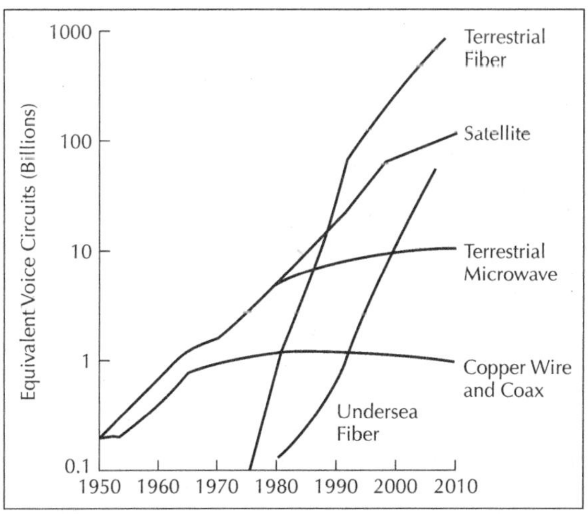

FIGURE 21 Global telecommunications capacity. Fiber-optic and satellite transmission have surpassed copper wire and the coaxial cable. The increase has been exponential regarding voice circuitry.

Telecommunications and Geography

There exists considerable popular confusion about the real and potential impacts of telecommunications on spatial relations, in part due to the exaggerated claims made in the past. We often read, for example, that telecommunications mean "the end of geography." Often such views hinge on a simplistic, utopian technological determinism that ignores the complex relations between telecommunications and local economic, social, and political circumstances. For example, repeated predictions that telecommunications would allow everyone to work at home via telecommuting, dispersing all functions and making cities obsolete, have fallen flat in the face of the persistent growth of densely inhabited urbanized places and global cities. In fact, telecommunications are usually a poor substitute for face-to-face meetings, the medium through which most sensitive corporate interactions occur, particularly when the information involved is irregular, proprietary, and unstandardized in nature. Most managers spend the bulk of their working time engaged in face-to-face contact, and no electronic technology can yet transmit the subtlety and nuances critical to such encounters. It is true that networks like the Internet allow some professionals to move into rural areas where they can conduct most of their business online, gradually permitting

Image copyright Caitlin Mirra, 2010. Used under license from Shutterstock.com

FIGURE 22 Satellite earth stations, such as the Very Large Array, New Mexico, form one part of the vast global telecommunications network, although fiber-optic lines are the most important means for transmitting various forms of corporate, personal, and government data.

them to escape from their long-time reliance on large cities and face-to-face contact. Yet the extent to which these systems facilitate decentralization is often countered by other forces that promote the centralization of activity.

For this reason, even after a century of advances in telecommunications, from the telephone to fiber optics, most high-wage, white-collar, administrative command and control functions are still clustered in downtown areas. Telecommunications are ideally suited for the transmission of routine, standardized forms of data, facilitating the dispersal of functions involved with their processing to low-wage regions. In short, there is no particular reason to believe that telecommunications inevitably lead to the dispersal or deconcentration of functions; by allowing the decentralization of routine functions, information technology actually enhances the comparative advantage of inner cities for nonroutine, high-value-added functions that are performed in person. Thus, telecommunications facilitate the simultaneous concentration and deconcentration of economic activities. And so popular notions that "telecommunications will render geography meaningless" are simply naïve. While the costs of communications have decreased, as they did in transportation, other factors have risen in importance, including local regulations, the cost and skills of the local labor force, government policies, and infrastructure investments. Economic space, in short, will not evaporate because of the telecommunications revolution. Exactly how telecommunications are deployed is a matter contingent on local circumstances, public policy, and the local niche within the national and world economy.

NASA

FIGURE 23 One of the satellites in the tracking and data relay satellite system in orbit around the earth is the TDRSS satellite. The system provides advanced tracking and telemetry services for a number of other satellites, as well as commercial telecommunications services.

Within cities, digital networks have contributed to an ongoing reconstruction of urban space, but telecommunications networks tend to be largely invisible to policy makers and planners and receive little attention. Many city governments are willing to invest in new roads or water control projects; urban planners and economic development officials, however, have often overlooked or ignored altogether the role that telecommunications can play in stimulating economic growth. Although the telecommunications industry per se is relatively small and capital-intensive, generating few jobs, and does not guarantee economic development, such systems have become necessities for many firms. In short, telecommunications are necessary, but not sufficient, to induce economic growth.

Although large cities typically have much better developed telecommunications infrastructures, the technology has rapidly diffused through the urban hierarchy into smaller towns and is becoming increasingly equalized among regions. In the future, therefore, the marginal returns from investment in this infrastructure are bound to diminish, minimizing competitive advantages based on information systems infrastructure alone and forcing competition among localities to occur on other bases, such as the cost and quality of labor, taxes, and the regulatory framework. Regions with an advantage in telecommunications generally succeed because they have attracted firms for other reasons.

However, information systems have steadily impacted urban settings in the form of transportation informatics, which include a variety of improvements in surface transportation such as smart metering, electronic road pricing, synchronized traffic lights, automated toll payments and turnpikes, automated road maps, trip planning information and navigation, travel advisory systems, electronic tourist guides, remote traffic monitoring and displays, and computerized traffic management and control systems—all of which are designed to minimize congestion and optimize traffic flow (particularly at peak hours), enhancing the efficiency, reliability, and attractiveness of travel. Wireless technologies such as cellular phones allow more productive use of time otherwise lost to congestion. Such systems do not so much comprise new technologies as enhance existing ones.

Teleworking is often touted as the answer to reducing transportation costs, easing demands on energy, and reducing environmental impacts. The trend to wireless terminals is growing because the hardwiring of computers and peripherals to networks is costly. Wireless terminals allow computers and other devices to communicate via infrared or electromagnetic signals and make it possible for computers to function within company or international networks. The popular desktop and laptop PCs are the first generation of wireless terminals. Mobile, hand-held units can now send large data files or e-mail using satellite communications technology and will eventually replace pagers and cellular phones. This trend toward wireless terminals is significant because it allows more portability and eliminates the need to be connected and disconnected from local area networks.

With the digitization of information, telecommunications have joined with computers to form integrated networks (Figure 24). New technologies such as fiber optics have complemented and at times substituted for telephone lines. Fax services and 800 number toll-free calls are now standard for virtually all companies, and even newer technologies such as electronic data interchange and wireless services are becoming increasingly popular. Like the railroad system of the nineteenth century and the interstate highway system of the twentieth century, the information highway of fiber-optic cables, satellites, and wireless grids links billions of computers, telephones, faxes, and other electronic products all over the world (Figure 25).

GEOGRAPHIES OF THE INTERNET

Among the various networks that comprise the world's telecommunications infrastructure, the largest and most famous is the **Internet**. This system delivers many services to homes, offices, and factories, including e-mail, telephone calls, TV programs and other video images, text, and music. By now, digital reality and everyday life for hundreds of millions of people have become so thoroughly

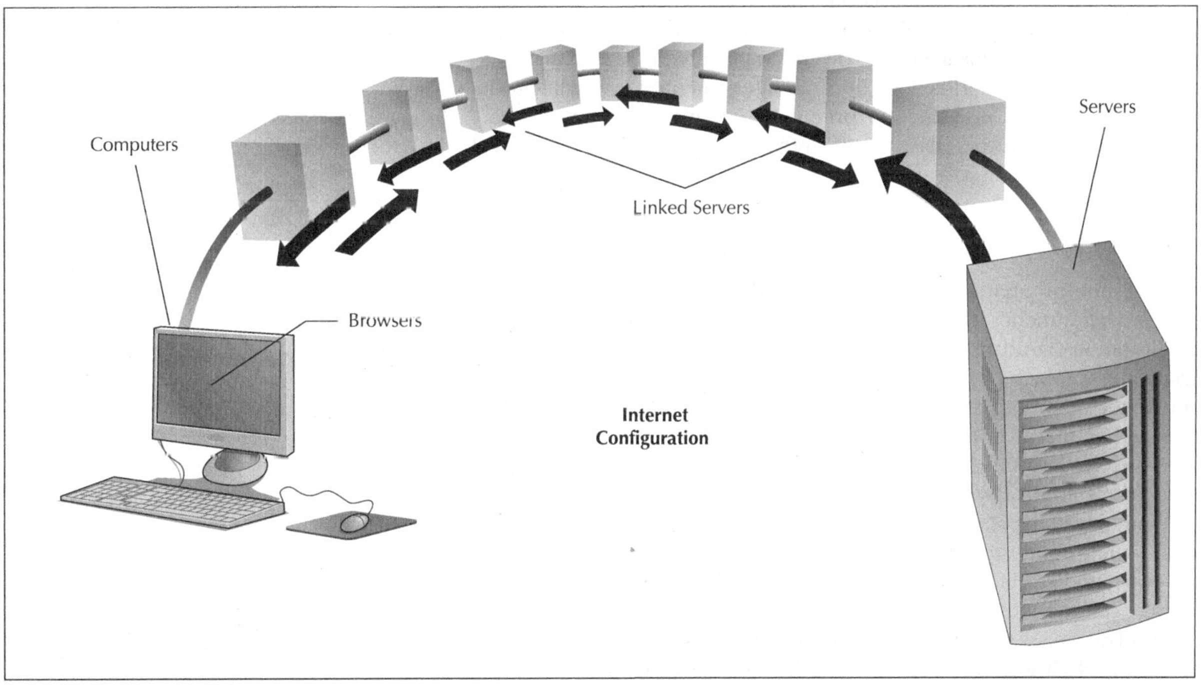

FIGURE 24 The convergence of computer technology and communications technology yields information technologies (IT). Two initially distinct technologies have now combined to impact the world economy. Communications technology is concerned with the transmission of information, while computer technology is concerned with the processing of information.

AP Photo/ Clement Ntaye

FIGURE 25 Small shopkeepers, such as this Sudanese refugee running a telephone repair shop in Cairo, Egypt, represent the gradual growth of digital information technologies in the developing world.

fused that it is difficult to disentangle them. The Internet is used for so many purposes that life without it is simply inconceivable for vast numbers of people. From e-mail to online shopping and banking to airline and hotel reservations to playing multiplayer video games to chat rooms to Voice over Internet Protocol telephony to distance education to downloadable music and television shows to blogs to YouTube to simply "Googling" information, the Internet has become much more than a luxury—it is a necessity for huge numbers of people in the economically developed world. In this context, simple dichotomies such as "offline" and "online" fail to do justice to the diverse ways in which the "real" and virtual worlds have become interpenetrated for hundreds of millions. Like the printing press and the automobile, the Internet has significantly changed the ways we work, interact, consume, and live.

Origins and Growth of the Internet

The Internet originated in the 1960s under the U.S. Defense Department's Defense Agency Research Projects Administration (DARPA), which designed it to withstand a nuclear attack. Much of the durability of the current system is due to the enormous amounts of federal dollars dedicated to research in this area. In the 1980s, control of the Internet was transferred to the National Science Foundation, and in the 1990s control was privatized via a consortium of telecommunications corporations. The Internet took on a global scale through the integration of existing telephone, fiber-optic, and satellite systems, which was made possible by the technological innovation of packet switching, TCP/IP (Transmission Control Protocol/Internet Protocol), and the **ISDN (Integrated Services Digital Network)**, by which individual messages may be decomposed and the constituent parts transmitted by various channels and then reassembled, virtually instantaneously, at the destination. In the 1990s, graphical interfaces developed in Europe greatly simplified the use of the Internet, leading to the creation of the World Wide Web.

The growth of the Internet has been phenomenal (Figure 26); indeed, it is the most rapidly diffusing technology in world history. Technological changes will further increase the utility and popularity of the Internet in the future. Mobile phones make it possible for consumers to access the Internet from any location, not just in the home or at the office. Broadband connectivity is becoming increasingly mainstream, allowing for innovations such as on-demand television. This next phase of Internet expansion will produce the real information revolution for everyday consumers: TV-quality video and voice-activated commands will allow them to enjoy the Internet as a practical

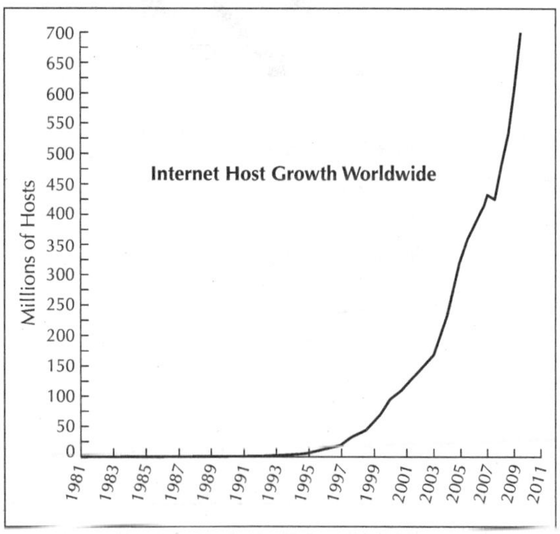

FIGURE 26 The growth of the Internet, 1981–2010. Roughly 1.9 billion people, or 28% of the planet, use the Internet today.

TABLE 1 Internet Usage by World Region, June 2010

Region	Internet Users (millions)	Penetration Rate (%)
Africa	110.9	10.9
Asia	825.1	21.5
Europe	475.1	58.4
Middle East	63.2	29.8
North America	266.2	77.4
Latin America	204.7	34.5
Oceania	21.2	61.3
World	1966.5	28.7

Source: Internet World Statistics, http://www.internetworldstats.com.

home appliance, useful for entertainment, communication, and shopping.

In mid-2010, an estimated 1.9 billion people, or roughly 29% of the world's population in more than 200 countries, were connected to the Internet. However, there are large variations in Internet penetration rates (percentage of people with access) among the world's major regions (Table 1), ranging from as little as 0.2% in parts of Africa to as high as 86% in Scandinavia. Penetration in the United States was 65% of households and 77% of the population. Inequalities in access to the Internet among nations reflect the long-standing bifurcation between the economically developed and developing worlds (Figure 27). While virtually no country is utterly without Internet access, the variations in accessibility among and within nations are huge. Given its large size, the United States—with more than 185 million users—dominates when measured in terms of absolute number of Internet hosts. The world's highest penetration rate is in Norway (85.7%). In Europe, the greatest connectivity is in relatively wealthy nations such as the Netherlands (85%), Denmark (84%), and Finland (83%); Eastern Europe lags considerably behind, and in Russia only 27% of the population uses the Internet. In Asia, access is greatest in South Korea (77%) and Japan (74%); about 25% of China is hooked up, although the numbers there are growing rapidly and already amount to more than 338 million users. In Latin America, the largest numbers of users are found in Brazil (34%) and Mexico (25%). On the African continent, the Internet is largely confined to South Africa.

In all cases, per capita incomes are the key; the Internet can only be used by people with resources sufficient to own computers and to learn the essential software. In many developing countries, where most people cannot afford their own computers, Internet cafes are popular. Variations in the number of users are also reflected in the geography of Internet flows (although flow data are much harder to come by than are place-specific attribute data): 80% of all international traffic on the Internet is either to or from the United States (Figure 28), fueling fears among some people that the Internet is largely a tool for the propagation of American culture.

Social and Spatial Discrepancies in Internet Access

The significant discrepancies in access to the Internet largely occur along the lines of wealth, gender, and race. Access to computers linked to the Internet, either at home or at work, is highly correlated with income; wealthier households are

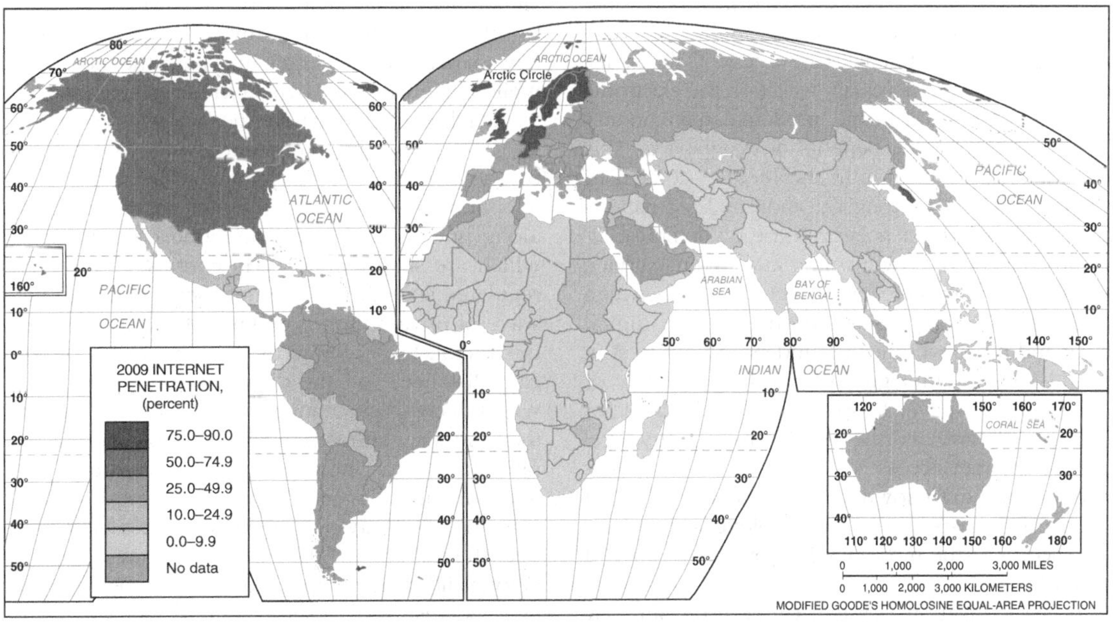

FIGURE 27 Internet penetration rates worldwide by country (proportion of population with access). Although cyberspace appears "placeless," there are in fact profound geographic disparities in who gains entry that reflect and contribute to the unevenness of modern capitalism.

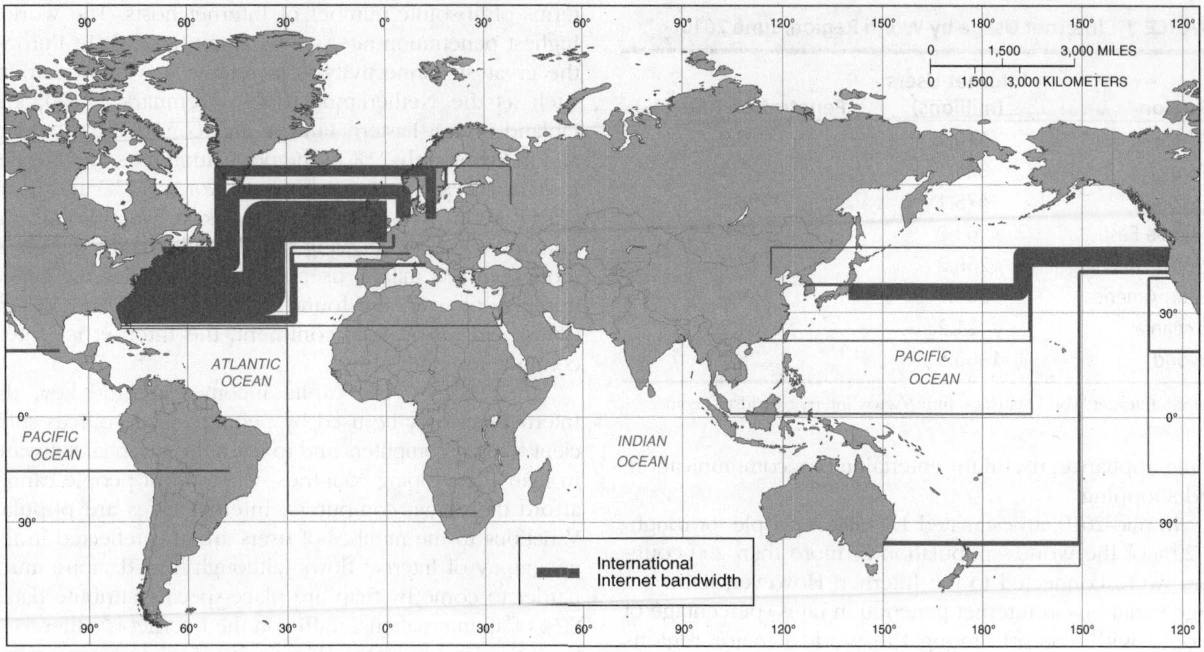

FIGURE 28 International information flows on the Internet. Eighty percent of all traffic is to or from the United States.

far more likely to have a personal computer at home with a modem than are the poor. In the United States, white households use networked computers more frequently than do African American or Latino households. The elderly likewise often find access to the Internet intimidating and unaffordable, although they do make up the fastest-growing demographic group of users. American Internet users thus tend to be white and middle class, relatively well educated, younger than the national average, and employed in professional occupations demanding college degrees.

Social and spatial differentials in accessing the skills, equipment, and software necessary to get on the electronic highway threaten to create a large, predominantly minority underclass deprived of the benefits of cyberspace. This phenomenon must be viewed in the broader context of the growing inequalities throughout industrialized nations generated by labor-market polarization (i.e., deindustrialization and growth of low-income, contingent service jobs). Modern economies are increasingly divided between those who are comfortable and proficient with digital technology and those who neither understand nor trust it, disenfranchising the latter group from the possibility of citizenship in cyberspace. Despite the falling prices for hardware and software and the fact that basic, entry-level machines for Internet access cost less than $1000, this is still an exorbitant sum for low-income households. Internet access at work is also limited for employees in poorly paying service jobs (the most rapidly growing category of employment).

The public educational system cannot offer an easy remedy to the problem of the digital divide. Even in the United States, wide discrepancies in funding and in the quality of education among school districts, particularly between suburban and central city schools, may reproduce this inequality rather than reduce it. Most public libraries offer free access to the Internet; however, in many municipalities mounting financial constraints have limited access to these public systems. Even within the most digitized of cities there remain large pockets of "offline" poverty. Those who need the Internet the least, who already live in information-rich environments with access to information through many non-Internet venues (e.g., newspapers and cable TV), may have the most access to it, while those who may benefit the most (e.g., through electronic job banks) may have the least chance to log on.

Internationally, Internet access is deeply conditioned by the density, reliability, and affordability of national telephone systems. Most Internet communications occur along lines leased from telephone companies, some of which are state regulated (in contrast to the largely unregulated state of the Internet itself), although the global wave of privatization is ending government ownership in this sector. Prices for access vary by length of the phone call, distance, and the degree of monopoly: In nations with telecommunications monopolies, prices are higher than in those with deregulated systems, and hence usage rates are lower. The global move toward deregulation in telecommunications will lead to more use-based pricing (the so-called pay-per revolution), in which users must bear the full costs of their calls, and fewer cross-subsidies among different groups of users (e.g., between commercial and residential users), a trend that will likely make access to cyberspace even less affordable to low-income users.

Case Study

Chinese Internet Censorship

With more than 1.9 billion users, the Internet has become a major vehicle for political action and public speech. Not surprisingly, many governments seek to restrict access to this medium. Typically, governments that seek to impose censorship do so using the excuse of protecting public morality from ostensible sins, such as pornography or gambling, although more recently combating terrorism has emerged as a favorite rationale. Deliberately vague notions of national security and social stability are typically invoked as well.

In a country with more than 380 million Internet users in 2010, Chinese Internet censorship is arguably the world's most severe, although cyberspace in China remains relatively free compared to the traditional media. The state has encouraged Internet usage, but only within an environment that it controls. In the early phases of Internet development, China did little to regulate cyberspace, but as chat rooms and blogs pushed the boundaries of allowable dissent with a steady stream of criticism of government officials, it began to tighten control significantly after 2000. In many respects, China's state-led program of Internet development serves as a model for other authoritarian governments elsewhere.

International Internet connections to China are squeezed through a selected group of state-controlled backbone networks. The government deploys a vast array of monitoring devices, collectively but informally known as the "Great Firewall," which include publicly employed monitors and citizen volunteers, to screen blogs and e-mail messages for potential threats to the established political order. There are numerous components to the Great Firewall that operate with varying degrees of effectiveness. Popular access in China to many common Web services, such as Google and Yahoo!, is often heavily restricted. The national government hires armies of low-paid commentators to monitor blogs and chat rooms, inserting comments that "spin" issues in a light favorable to the Chinese state. However, a large share of censorship occurs via Internet companies themselves, which monitor chat rooms, blogs, networking services, search engines, and video sites for politically sensitive material in order to conform to government restrictions. Web sites that help users circumvent censorship are prohibited. Instant messaging and mobile phone text messaging services are also heavily filtered, and a program called QQ is automatically installed on users' computers to monitor communications. In June 2009, the government attempted to require manufacturers to install filtering software known as Green Dam Youth Escort on all new computers but retreated in the face of a massive popular and corporate outcry.

The Great Firewall system began under an initiative known as the "Golden Shield," which China developed with the aid of U.S. companies Nortel and Cisco Systems and extended beyond the Internet to include digital identification cards with microchips containing personal data that allow the state to recognize faces and voices of its 1.3 billion plus inhabitants. The envy of authoritarian governments worldwide, the Golden Shield has been exported to Cuba, Iran, and Belarus.

The Chinese government has periodically initiated shutdowns of data centers housing servers for Web sites and online bulletin boards, disrupting use for millions. Blogs critical of the government are frequently dismantled, although for the most part the government outsources this function to blog-hosting companies. E-mail services like Gmail and Hotmail are frequently jammed; before the 2008 Olympics, the Facebook sites of critics were blocked. Police frequently patrol Internet cafes, where users must supply personal information in order to log on, while Web site administrators are legally required to hire censors popularly known as "cleaning ladies" or "big mamas."

Google, the world's largest single provider of free Internet services, famously established a separate, politically correct (by China's government standards) Web site, Google.cn, which censors itself to comply with restrictions demanded by the Chinese state, arguing that the provision of incomplete, censored information was better than none at all. In early 2010, responding to the ensuing international criticism, Google announced it would no longer cooperate with Chinese Internet authorities and withdrew from China, although later in 2010 the two parties negotiated a deal that allowed Google to re-enter this market.

Social Implications of the Internet

In an age in which personal social life is increasingly mediated through computer networks, the reconstruction of interpersonal relations in the digitized spaces of cyberspace is of the utmost significance. The fact that cybercontacts differ from face-to-face ones, however, serves as a useful reminder that telecommunications change not only what we know about the world but also how we know and experience it.

Much Internet usage revolves around entertainment, personal communication, research, downloading files, and online games. However, the Internet can also be used to challenge established systems of domination and legitimate and publicize the political claims of the relatively powerless and marginalized. The Internet has given voice to countless groups with a multiplicity of political interests and agendas, including civil and human rights advocates, sustainable development activists, antiracist and antisexist organizations, gay and lesbian rights groups, religious movements, those espousing ethnic identities and causes, youth movements, peace and disarmament parties, pacifists and those who espouse nonviolent action, animal rights groups, and gays living in homophobic local environments. By facilitating the expression of political positions that otherwise may be difficult or impossible to broadcast, the Internet allows for a dramatic expansion of the range of voices heard about many issues. In this sense, it permits the local to become global. Within the Internet itself one finds all the diversity and contradictions of human experience: Cyberpolitics mirror those of its nonelectronic counterparts, although the boundaries between the two realms are increasingly fuzzy. Indeed, in a sociopsychological sense, cyberspace may allow for the reconstruction of "communities without propinquity," groups of users who share common interests but not physical proximity.

Finally, there is also what may be called the "dark side" of the Internet, in which it is deployed for illegal or immoral purposes. Hackers, for example, have often wreaked havoc on computer security systems. Such individuals are typically young men playing pranks, although others may unleash dangerous computer viruses and worms. Most hacks—by some estimates as much as 95%—go unreported, but their presence has driven up the cost of computer firewalls. The dark side also includes unsavory activities such as identity theft; counterfeit drivers' licenses, passports, and Social Security cards; securities swindles; and adoption scams. Credit card fraud is a mounting problem; 0.25% of Internet credit card transactions are fraudulent, compared to 0.08% for non-Internet transactions. Some Internet sites even offer credit card "marketplaces," where people who hack into merchant accounts and steal large numbers of card numbers offer to sell them wholesale.

E-Commerce

The impacts that telecommunications have had on businesses are often lumped together under the term **e-commerce**, which may refer to both business-to-business transactions as well as links between firms and their customers. In general terms, information technology lowers the transaction costs among corporations, which helps to spur productivity. Some claim that such systems were instrumental in the restructuring of many corporations in response to mounting global competition, as they downsized in favor of flatter corporate hierarchies. Many firms sought improved productivity by accelerating information flows within the firm and lower costs by reducing intermediaries and distribution costs.

One important version of e-commerce concerns electronic data interchange (EDI) systems, which are generally used in business-to-business (B2B) contacts. **EDI, electronic data interchange**, can be defined as the electronic movement of standard business documents between and within firms. EDI uses a structured, machine-retrievable data format that permits data to be transferred between networked computers without rekeying. Like e-mail, EDI enables the sending and receiving of messages between computers connected by a communication link such as a telephone line. Common uses of EDI include updated advertising, online product catalogues, the sharing of sales and inventory data, submissions of purchase orders, contracts, invoices, payments, delivery schedules, product updates, and labor recruitment. E-commerce reduces delays and marketing and delivery costs and has led to a greater emphasis on connectivity, ideas, creativity, speed, and customer service.

Data on "e-tailing" or electronic retailing, reveals the growing commercialization of the Internet: In 1993, 2% of all Web sites were commercial (i.e. "dot-com") sites; by 2008, 70% were so categorized. Shopping by Internet requires only access (e.g., a modem), a credit card, and a parcel delivery service and it allows effortless comparison shopping. The most successful example perhaps is Amazon.com, started by Seattle entrepreneur Jeff Bezos, which now accounts for 60% of all books sold online. Other e-tailing examples include online auctions (e.g., eBay), Internet-based telephony (e.g., Skype), and Internet music (e.g., the downloading of MP3 music files), which has provoked a firestorm of opposition from music companies concerned about infringement of their intellectual property rights and declining over-the-counter music sales. Internet sales have also provoked worries about tax evasion and sales of illegal goods (e.g., pharmaceuticals from abroad).

However, despite predictions that "click-and-order" shopping would eliminate "brick-and-mortar" stores, e-tailing has been slow to catch on; it accounts for only 3% of total U.S. retail sales, perhaps because it lacks the emotional content of shopping. Shoppers using this mode tend to be above-average in income and relatively well educated. Web-based banking has also experienced slow growth, even though it is considerably cheaper for banks than automatic teller machines, as have Internet-based bill payments, mortgages, and insurance. Internet-based sales of stocks (e.g., E*Trade) now comprise 15% of all trades. One particularly successful e-tailing application has been the travel reservation and ticketing business, where Web-based purchases of hotel rooms and airline seats (through services such as Travelocity and Priceline.com) have caused a steady decline in the number of travel agents. Electronic publishing, including more than 700 newspapers worldwide, has been extended to e-books and e-magazines, which, unlike printed text, can be complemented with sound and graphics. Other services offered are the searching of Internet databases and classified ads. Webcasting, or broadcasts over the Internet (typically of sports or entertainment events), demands high-bandwidth capacity but comprises a significant share of Internet traffic today. Web-logs, or "blogs," have become increasingly important sources of personal, social, and political commentary and are alternatives to the mainstream media and a voice for independent views.

Internet advertising has proven to be a difficult Internet business, in part because the Internet reaches numerous specialized markets rather than mass audiences, but cyberspace does allow specialized companies to reach niche markets all over the globe. E-advertising makes up only 1% of the total ad revenues in the United States and it overwhelmingly focuses on computer and software offerings. Indeed, many users have become wary of "spam" e-mail (unwanted commercial messages), which constitutes an ever-larger, and increasingly annoying, share of e-mail traffic (by some estimates as high as 75%).

Another form of e-commerce involves universities, many of which have invested heavily in Web-based **distance-learning** courses. Although such programs are designed to attract nonlocal and nontraditional students, many of whom may not be able to take lecture-based courses in the traditional manner, they also reflect the mounting financial constraints and declining public subsidies that many institutions face. These institutions may see distance learning as a means of attracting additional students, and tuition, at relatively low marginal costs. The largest example of Web-based instruction is the University of Phoenix, based in Arizona but with students located around the world; with more than 400,000 students, it is now the largest university in the world. Distance learning has provoked fears that it will open the door to the corporatization of academia and the domination of university education by the profit motive; some have questioned whether the chat rooms that form an important part of its delivery system are an effective substitute for the face-to-face teaching and learning that classrooms offer. It remains unclear whether Web-based learning is an effective complement or substitute for traditional forms of instruction. Distance-learning programs, some suggest, may be better suited to professional programs in business or engineering than to the liberal arts.

More moral ambiguity is attached to Internet-based gambling systems, which include a variety of betting services, especially those concerning sports events, and even online slot machines in which gamblers may use their credit cards. (Some complain that online gambling doesn't adequately substitute for the heady experience of a gaudy casino in Las Vegas or Atlantic City.) Because the geography of legal gambling is highly uneven, the existence of such systems may challenge the laws of communities in which gambling is illegal. Offshore gambling centers have grown quickly, particularly in the Caribbean, which started when Antigua licensed its first Internet casino in 1994. By 2008, an estimated 1000 online casinos, mostly in the Caribbean, attracted roughly 12 million users.

E-Government

The placing of government and social service information online (office locations, departmental telephone numbers, city council minutes, government documents, and so on) is a valuable first step toward building an Internet-savvy community—but only a first step. Increasing people's access to such information serves the important function of creating a more well-informed citizenry.

Information technology's greatest potential in this area may lie in transforming the very nature of local government, making it possible to reconfigure traditionally monolithic, downtown city halls into a network of small, neighborhood-based "branches" linked electronically to a slimmed-down city "headquarters." Under this scenario, almost all government and social services would be dispensed in the neighborhood—either from kiosks or in small, multifunction neighborhood service centers. Such structural reengineering could further reduce the staffing, operational, and office costs of government; minimize traffic congestion and pollution; and increase people's access to officials and services while creating a government that is more institutionally sensitive to neighborhood concerns.

E-government can take a variety of forms, ranging from the simple broadcasting of information to community integration (i.e., allowing user input), in which the network minimizes duplication of effort. E-government, for example, makes possible the digital collection of taxes, voting, and provision of some public services, particularly information. Such activities boost the efficiency and effectiveness of public services; for example, it could allow online registration of companies and automobiles; electronic banking; utility bill payments; applications for government programs, university enrollment, and licenses; access to census data; and reductions in the waiting time involved when paperwork filters through government bureaucracies.

E-Business

Hundreds of thousands of companies, from small start-ups to the *Fortune* 500, use the Internet to promote their businesses. Web sites are a passive form of advertising, and potential customers often do not encounter a particular company's site unless they happen to be looking for it—turning the World Wide Web into something of a high-tech Yellow Pages. In an effort to reach a broader Web audience, therefore, many companies have taken a cue from the television advertising model and have begun to sponsor high-profile, Web-based news, information, and entertainment sites where they can gain exposure to customers that they otherwise might never reach.

For all their promotional potential, however, information networks are more than effective advertising and marketing vehicles. They can, in fact, change the very nature of work. In much the same way that government telecommunications networks may allow government agencies to distribute their workforces among communities, large businesses can establish dedicated neighborhood **telework centers** that give them access to potential employees, such as rural residents, homemakers, people with disabilities, or individuals without transportation, whom they otherwise might not be able to attract. By moving operations out of congested and high-priced central cities, telework centers simultaneously can increase workers' productivity while reducing companies' rent, transportation, and labor costs.

Telecommuting, of course, is an idea that has been around for decades, and the practice has not yet taken off in the way many of its advocates forecast. This has been due in part to business's lack of interest in telecommuting and to its reliance on traditional, fixed-site management

© Alamy Images

FIGURE 29 The Internet and telework allow some parents to work from home, but create difficulties in balancing professional and personal obligations.

practices. But the growing number of home-office, temporary, and contract workers makes telework not only feasible but increasingly necessary. With workers demanding more flexible schedules, shorter commutes, and relief from traffic congestion, companies may find that telecenters are a powerful tool for retaining or attracting high-quality workers. What's more, with videoconferencing, e-mail, and high-speed computer networks, telecommuting finally has become technologically practical in a wide range of job categories and work situations (see Figure 29).

Health Care

Escalating health care costs have severely strained the U.S. health care system. Much of the cost increase is due to the enormous volume of recordkeeping and data transfer involved—a problem well suited to a technological solution. In the United States alone, billions of dollars per year could be saved by using advanced communications technology for the routine transfer of laboratory test data and more orderly collection, storage, and retrieval of patient information. Thus telemedicine, in which physicians at one location use remote viewing techniques to diagnose and provide advice to patients located somewhere else, has grown in popularity, particularly in rural areas with inadequate access to health care. In some cases, such technologies facilitate the training of physicians—in virtual surgeries, for example. The Internet has also become an important source of medical information, dramatically changing the traditional doctor–patient relationship: Two-thirds of those who have gone online have searched for health-related information there, although much of it is inaccurate or misleading.

Summary

This chapter examined two major systems of spatial interaction—the transportation of people and goods, and the communication of information—that are critical to the ever-changing structure of global capitalism. We considered some of the factors other than distance that play a role in determining transport costs—the nature of commodities, carrier and route variations, and the regimes governing transportation. Transport costs remain critical for material-oriented and market-oriented firms, but they are of less importance for firms that produce items for which transport costs are but a small proportion of total costs. For these firms, transit time is more crucial than cost. Modernized means of transport and reduced costs of shipping commodities have also made it possible for economic activities to decentralize. Multinationals have taken full advantage of transport developments to establish "offshore" branch-plant operations.

Movements of goods and people take place over and through transport networks. Our account of the historical development of transportation explained how improvements over the centuries have resulted in time-space and cost-space convergence. Improved transport and communications systems integrated isolated points of production into a national or a world economy. And although the friction of distance has diminished over time, transport remains an important locational factor. Only if transportation were instantaneous and free would economic activities respond solely to aspatial forces such as economies of scale. We also considered the role of the state in transportation policy, which has varied around the world and among different sectors (rail, airlines, trucking, etc.). Because of the tremendous increase in travel demand in large cities of the developed world, innovation in urban transportation systems is necessary. For example, in the

United States, vehicle miles traveled, automobile ownership, and total vehicle trips have increased rapidly.

Communications and information technology are transforming the world economy at rates never before thought possible. Profound implications, even many that the world cannot yet measure, accompany this IT explosion, at the center of which are the microprocessor, networked computers, and the Internet. We explored the origins, growth, and size of the Internet, which connected about 1.9 billion people worldwide in 2010, about 29% of the planet; noted the uneven access to it experienced by people around the world and within the United States; and touched on some of its many consequences, including the growth of cybercommunities. We pointed out that the social divisions that exist offline are replicated online. We also explained the nature and impacts of e-commerce. The impacts of the telecommunications revolution may only be just at their beginning. We noted a number of ways the IT revolution may carry forward into the future, including in the areas of employment, health care, and government services.

Key Terms

cost-space convergence
deregulation
distance decay
distance learning
e-commerce
electronic data interchange (EDI)
friction of distance
hub-and-spoke networks
Internet
ISDN (Integrated Services Digital Network)
journey-to-work
line-haul costs
maglev
privatization
spatial interaction
telework centers
teleworking
terminal costs
time-space compression or convergence
transport costs

Study Questions

1. Under capitalism, what force drove improvements in transportation velocities?
2. What are terminal and line-haul costs?
3. How do transport costs enter into location theory?
4. What is time-space convergence/compression?
5. How did deregulation affect the structure of airline networks?
6. How does the global distribution of telephones mirror colonialism?
7. What role has fiber optics played in global telecommunications?
8. What factors shape access to the Internet?
9. What is e-commerce? E-government?
10. What is electronic data interchange?

Suggested Readings

Brunn, S., and T. Leinbach, eds. 2001. *Wired Worlds of Electronic Commerce*. London: John Wiley.

Dodge, M., and R. Kitchin. 2001. *Mapping Cyberspace*. London: Routledge.

Graham, S., and S. Marvin. 2001. *Splintering Urbanism: Networked Infrastructures, Technological Mobilities and the Urban Condition*. London. Routledge.

Kellerman, A. 2002. *The Internet on Earth: A Geography of Information*. Hoboken, NJ: John Wiley.

Schiller, D. 1999. *Digital Capitalism: Networking the Global Market System*. Cambridge, MA: MIT Press.

Standage, T. 1998. *The Victorian Internet*. New York: Walker and Company.

Warf, B. 2008. *Time-Space Compression: Historical Geographies*. London: Routledge.

Web Resources

Atlas of Cyberspace
http://www.cybergeography.org/atlas/

Most comprehensive collection of maps on the Internet. Type in a starting location and an ending location. The program provides door-to-door or city-to-city directions.

E*Trade
http://www.etrade.com

This flat-rate broker allows buying or selling of 5000 shares or less of stock for a fee of $14.95; a penny a share more above that fee for more than 5000 shares.

Internet World Stats
http://www.internetworldstats.com/stats.htm

Up-to-date data on Internet users internationally.

Consumption

OBJECTIVES

- To offer a historical overview of consumption and consumerism
- To summarize sociological, neoclassical, and Marxist views of consumption
- To analyze the geographies of consumption at multiple spatial scales
- To note the environmental impacts of mass consumption

Behind the façade of consumerist fantasylands, shopping malls, such as this mall in Moscow, Russia, are important places in which the output of capitalism is consumed, people are socialized to become consumers, and the culture of the commodity is reproduced.

Consumption

Economic geography has studied production in exhaustive detail, but it has long ignored its equally important counterpart, consumption. Consumption did have a role in traditional models of space and location theory; only recently, however, have economic geographers come to view the issue in *social* terms. Yet consumption is critically important to understanding how economies are stretched over the earth's surface, their impacts, and how they change. Consumption and production cannot be neatly separated and are closely intertwined: Most people work in order to consume, and consume in order to live. Thus, we can think of consumption and production as two facets of one circular process (Figure 1).

As long as there have been people there has been consumption. The term **consumption** means a variety of things: The Latin word *consumere* meant "to use up," and for centuries consumption was the name for the disease tuberculosis. Today, consumption is typically taken to mean the use of goods and services to satisfy individual and collective wants and needs. Consumption is a complex topic because it is where the various spheres of social life, such as production, class, gender, the relations between government and market, and the individual and society, intersect. Far more than simply an "economic" phenomenon, consumption is also a cultural and psychological construction that reflects and affects people at both the social and personal levels. Consumption is a major interface between the individual and society.

Of course, consumption is enormously important economically; it constitutes the bulk of the economic activity of most countries. For example, roughly 55% of the U.S. gross domestic product (GDP) consists of consumer or household purchases of different types (Figure 2); in comparison, business spending (investment) is only 15%, foreign trade is 18%, and government spending is 12%. Not surprisingly, indices such as the Consumer Confidence Index are important to economists, and the ways in which consumers spend their incomes are enormously important to different industries. In the United States, for example, spending on food and clothing has declined as a share of disposable income while spending on medical care has soared (Figure 3). Consumption also tends to be seasonally sensitive, often at its greatest around the Christmas shopping season, when crowded stores earn a large share of their annual revenues.

American consumers are not only vital to the U.S. economy but to many other economies around the world as well. By purchasing vast volumes of imports, U.S. consumers help increase the export revenues of countries that export goods and services to the United States. Thus, cyclical swings in American consumer spending can have global repercussions. The same is true, although to a lesser extent, of consumers in other countries with large GDPs and consumption of imports, such as several European states, although no other country rivals the United States in both size of consumption and enormous trade deficits.

THE HISTORICAL CONTEXT OF CONSUMPTION

Although it may seem "natural," consumption as a social process has a history and a geography that reflect the extent and ways in which people have consumed over time. Most of what we know of premodern consumption focuses on elites, who had the greatest purchasing power. In feudal Europe, for example, in which there was relatively little trade and most goods were produced and consumed locally, trade was largely confined to luxury goods such as silks, spices, porcelain, and gems that were affordable only to the wealthy. Levels of consumption were hobbled by low average spending power. In medieval Europe, there was no single word for consumption; only with the advent of taxation in the sixteenth and seventeenth centuries did diverse commodities like clothing and food come to be seen as having something in common. We may thus posit that consumption as a social category is essentially an invention

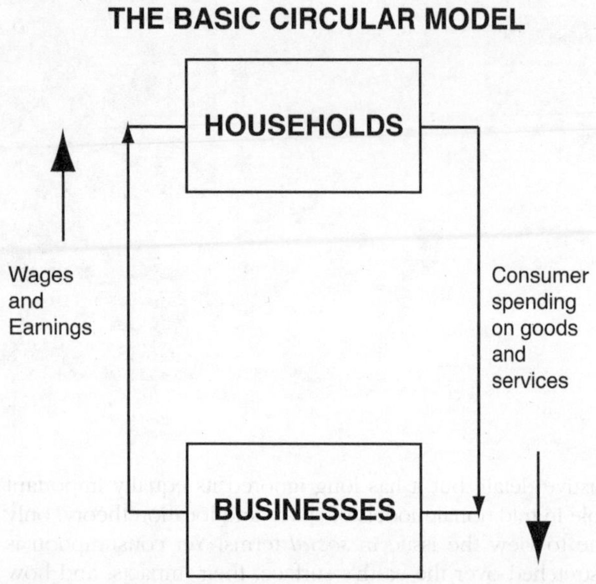

FIGURE 1 Production and consumption are two intertwined dimensions of capitalism. Without demand, there will not be a supply. Because both production and consumption are shaped by complex geographies, their interrelations across space and time are complicated, contingent, and ever-changing.

of capitalism. The rise of the colonial world economy in the sixteenth century simultaneously created a class of people with significantly large incomes to spend and made available products from a wide array of places around the globe, such as sugar from the Caribbean, cod from New England, and spices from Asia.

The modern form of consumption—**consumerism**—came about largely because of the Industrial Revolution of the late eighteenth and nineteenth centuries. Industrialization, inanimate energy sources, and mechanized production generated a sudden, dramatic lowering of the prices of many goods and lifted millions of people above the level of subsistence. Food and clothing, for example, became more affordable to the growing middle class. Simultaneously, gradually increasing incomes began to transform workers into consumers, leading to the birth of a consumer society. This process had enormous implications in terms of class consciousness, essentially allowing consumption to displace the politics of work; that is, buying gradually became more important than the struggles over workplace hours, pay, and conditions. In the 1830s, the first department stores appeared, in Paris, which were the retail trade version of the factory. The growth of mass production in the nineteenth century was accompanied by mass consumption. Because capitalism thrives on newness, advertising became an integral part of this process, forever generating new needs and desires and converting luxuries into necessities. In the late twentieth century, changes in the world economy, including deindustrialization and the explosive growth of producer services, induced concomitant changes in consumption. For example, niche markets became increasingly specialized (e.g., specialty beers, shoes, and other goods) and consumers more sophisticated, demanding higher quality goods.

It should be noted that originally this process met some resistance; consumerism threatened much older, religious ways of looking at the world. In particular, ideas of thrift, embodied in such slogans as "waste not, want not," gradually gave way to a morality that regarded money as the measure of social status and consumption as the means to obtain it. Objections often had religious overtones, often expressed as scorn for the material world and the moral superiority of asceticism. Ultimately, however, the ethic of the rising bourgeois class triumphed. The growth of capitalism, the ideas of the Enlightenment, and the growing dominance of individualism encouraged the **commodification** of consciousness, in which the self was defined through its wants, not social obligations. Thus consumerism was an integral part of the ascendancy of modernity; if tradition meant the regulation of desire, modernity tended to unleash it as rising incomes gave consumers ever more purchasing power and freedom to indulge in whims.

Consumerism encompassed the entire experience of shopping in many forms. The purchase of goods through early catalogues and mail orders was made possible by the growth of postal systems and through illustrated advertisements (which were useful in societies with low literacy rates). As workers fought for higher wages and lower working hours, including the 8-hour work day and the 5-day work week, consumption increasingly came to be equated with leisure. In the process, everyday life changed dramatically. Child rearing, for example, increasingly included the purchase of goods to comfort children. Clean clothes became the norm expected of an increasingly larger fraction

FIGURE 2 More than half of the U.S. economy consists of consumption by households and individuals. Small changes in consumer spending therefore have enormous repercussions on the economy as a whole. For this reason, consumer confidence levels, shopping habits, demographics, changing preferences, and inclinations to spend versus save are closely monitored by social scientists and policy makers.

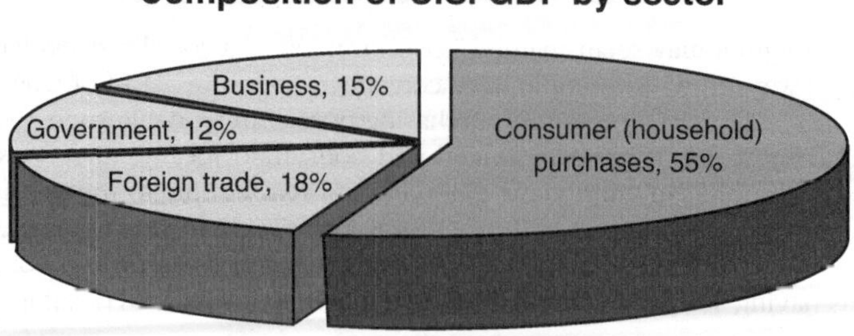

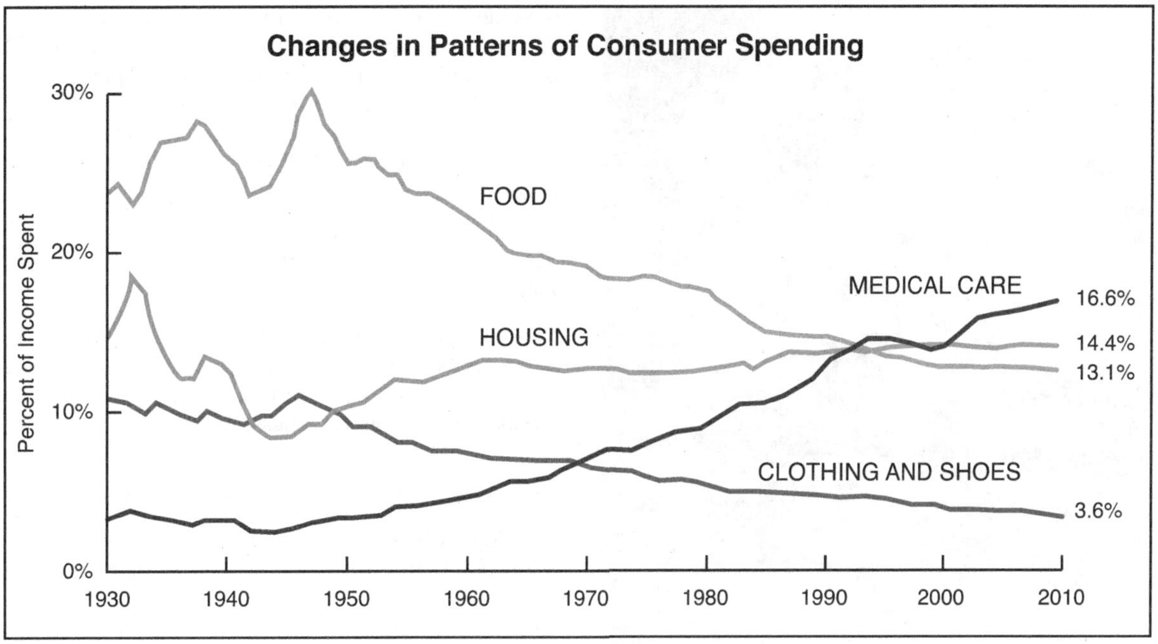

FIGURE 3 Changes in patterns of consumer spending reflect several influences, including the costs of goods, demographic trends, and shifts in preferences. In the United States, consumers spend relatively less on food and clothing and more on medical care.

of society. By the early twentieth century, the association of consumption and romance led to the invention of dating. Consumption increasingly came to be regarded as the only means to happiness, status, success, relief from pain and boredom, and a positive self-image.

Beginning primarily in Britain and the United States, consumerism engulfed Western Europe in the nineteenth century. After World War II, it spread to large parts of the world, where it displaced many traditional, noncapitalist value systems, although its influence was uneven geographically, often resisted, and yet almost always successful in the end. Typically, the growth of consumerism was slowest in societies in which entrenched religious objections were difficult to overcome. But as ever larger portions of the world's population embraced this lifestyle, consumerism came to be the model for many kinds of social relations within such institutions as education, religion, and family life. Everyone, it seems, has become essentially a buyer or a seller, for better or for worse. In many ways, the spread of consumerism reflects the enormous influence of the United States in the world economy and the diffusion of capitalism. Multinational corporations, increased international trade, rising literacy levels, expanding middle classes, and television and ubiquitous advertising have also played key roles.

After World War II, the expansion of consumerism was closely linked to the growth of credit. Prior to this period, people saved for expensive items or bought them on "layaway." During the Great Depression, General Motors introduced financing for cars, an early form of credit. It was the introduction of the credit card, however, that greatly accelerated the capacity to consume by essentially undermining the notion of **delayed gratification**. In consumerist societies, credit cards can enhance the quality of life for many and are useful in emergencies or for the purchase of major luxuries (Figure 4). Psychologically, credit cards are often regarded as synonymous with freedom or as a rite of passage for adolescents. In the United States, the size of the "credit card nation" is staggering; 181 million Americans held 1.5 billion cards in 2010 and had a total of $852 billion in debt. Average household consumer debt levels have exploded (Figure 5). This debt, including debt incurred by small businesses, has generated a wave of personal bankruptcies; indeed, more people go bankrupt each year than go to college.

The traditional view of credit was grounded in the Puritan work ethic, which emphasized frugality, delayed gratification, and saving; debt was often regarded as sinful. The moral view of consumption was that it was supposed to be held in line with one's income. The late twentieth-century marketing onslaught by banks and credit card companies annihilated this worldview—their marketing campaigns centered on the commodification of fun and they sold credit cards as a marker of social sophistication and a carefree lifestyle. The deregulation of banking produced a more competitive environment, and credit cards became the golden goose for many banks, earning higher rates of return than other banking activities. Banks and credit card companies make a profit by charging merchants processing fees and by charging consumers relatively high interest rates. Banks aggressively targeted many social groups that they had previously ignored, such as college students and

FIGURE 4 Consumer using credit card.

the elderly. Consumers were facing problems arising from the end of the postwar boom: deindustrialization, globalization, and stagnant incomes. But their spending continued to rise even though incomes did not, the average U.S. savings rate declined steadily (Figure 6), and many households found themselves drowning in debt they could not afford.

THEORETICAL PERSPECTIVES ON CONSUMPTION

Consumption may be thought of from several conceptual angles, three of which are discussed here: the sociological, neoclassical economic, and Marxist notions. Each has its merits as well as its weaknesses.

Sociological Views of Consumption

Sociologists have long examined the role of consumption in relation to individual and household standards of living, class, and status. In the early twentieth century, Max Weber argued that delayed gratification lay at the heart of the Protestant ethic and that consumerism had mutated this ethic into the unmitigated pursuit of pleasure. American sociologist Thorsten Veblen, studying the profligate consumption of the rich during the late nineteenth-century era of the robber barons, coined the famous term **conspicuous consumption** to denote consumption as a social statement to others, that is, as an expression of wealth and status—for example, large, elaborate mansions. This line of thought led to a long series of works on the relations between consumption and identity—shopping as something much more than simply the purchase of goods but as symbolic of who we are and what

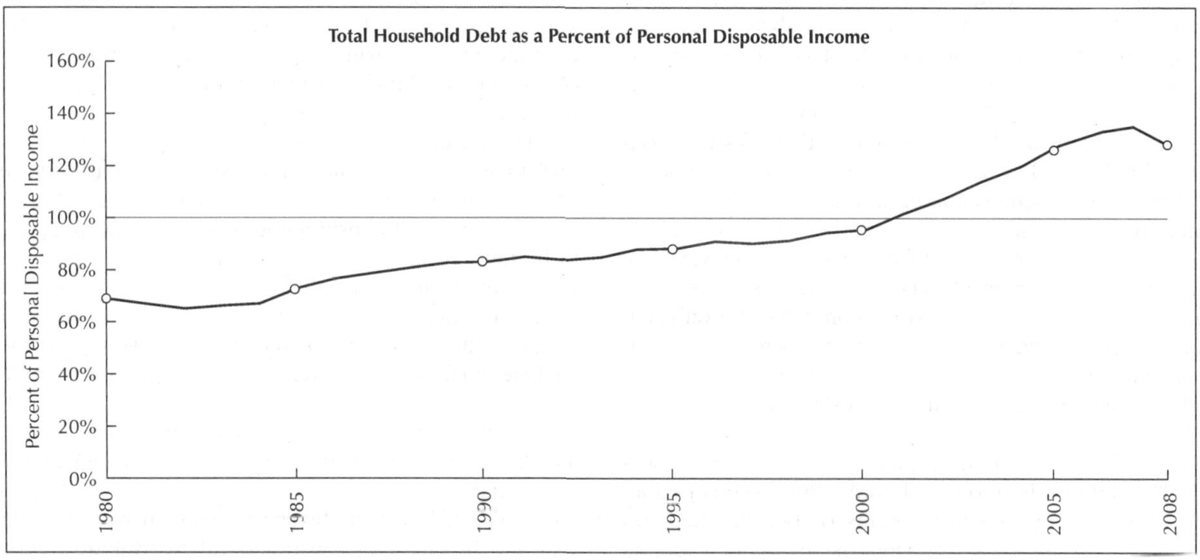

FIGURE 5 Faced with stagnant incomes, rising costs, and ferocious advertising, American households have increasingly turned to debt to maintain their standard of living.

FIGURE 6 Mounting international competition, unchanging disposable incomes, and a love of consumption have combined to generate a long-term decline in the U.S. savings rate.

we mean. From this point of view, it is ironic that in societies like the United States many people find their "unique" individual identities by purchasing mass-produced commodities.

Sociologists have noted how consumption varies greatly by social category, a phenomenon of vital importance to marketers. Consumption is stratified by class and income. For example, low-income families are apt to consume more of cheaper foods like potatoes and less of expensive ones like steak. Consumption also varies by age. Typically, people's levels of consumption remain relatively constant (although the types of goods and services they buy vary), but because their incomes fluctuate over the life cycle, they display different patterns of spending and saving with age (Figure 7). As people age, they tend to spend more on medical care, which is a major cause for the growth in health care services. Consumption is also highly gendered. Men and women buy very different items and shop in different ways. For example, men are more likely to purchase alcohol, motorcycles, and cigars; women are the primary buyers of lipstick and clothes. Markets for magazines, movies, shoes, and cars are similarly gendered. Because women are typically responsible for buying household necessities, they comprise the bulk of consumers; as a result, many advertisers appeal to them directly. Other critical sociological categories are race and ethnicity, which are intertwined with the class and gender dimensions of consumption, as different ethnic groups have different preferences for food, clothes, magazines, television shows, automobiles, and other goods and services. Finally, consumption habits change often due to a variety of factors, including age and ethnic demographics, tastes and preferences (e.g., concerns over health), and income levels.

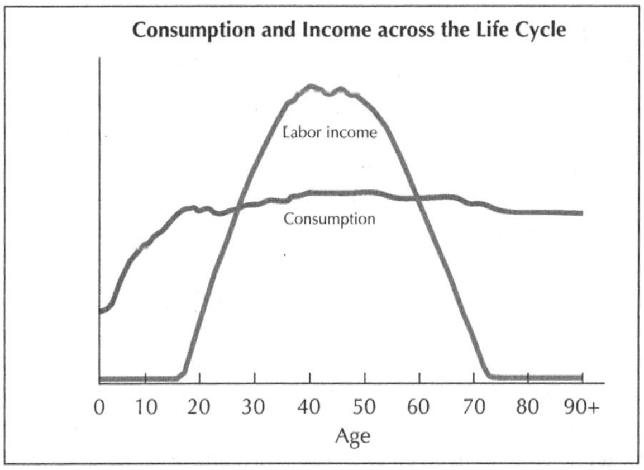

FIGURE 7 While consumption remains relatively constant across the life cycle, income levels vary markedly, generating different patterns of saving or spending by age.

Advertising is a critical factor in the sociological understanding of consumption. It is essentially a mechanism for funneling people's consciousness to the commodity to generate new wants and desires, and it plays a major role in the construction of a society's roles and role models. Consumerist societies are saturated by advertising; for example, two-thirds of the space in newspapers is taken up by advertisements, and the average American child has seen 1 million television advertisements by age 20. Advertising encourages us to meet nonmaterial needs through material means; it generates a belief that commodities will impart a desirable status, sexuality, physical prowess, happiness, and so forth.

So extensive has consumerism become that it has generated an epidemic of "**affluenza**," which may be variously described as the dogged pursuit of more, an obsessive quest for pleasure, materialism run amuck. Many people find themselves on an endless treadmill of permanent discontent. Ironically, never has so much meant so little to so many. To pay for this lifestyle, Americans often suffer a "time famine" in which they work more hours than they did in the 1960s and get 20% less sleep than they did in 1900. The average American parent spends 6 hours per week shopping and only 40 minutes with their children. Surveys of average satisfaction levels, which peaked in the United States in the 1950s, indicate that money does not necessarily buy happiness. Even as consumption levels have risen, so have psychological markers of unhappiness, such as depression and suicide. In their need to be constantly entertained by the new, many people find themselves bored as the exotic quickly becomes commonplace. Moreover, consumerism often carries into interpersonal relationships, turning citizens into consumers and collapsing personal and community social obligations into the self. Rather than be citizens, everyone becomes simply a consumer whose sense of community extends no further than the shopping cart. The steady, long-term declines in voting and volunteering and disinvestment in public areas are manifestations of this phenomenon.

Neoclassical Economic Views

Neoclassical economics has been the traditional and dominant viewpoint from which to understand consumption. Its analysis places the importance of demand over that of production. Emerging in the 1870s, neoclassical economics detached economics from political economy and drew upon the long tradition of utilitarian thought initiated by the philosopher Jeremy Bentham (1748–1832), a tradition long dominant in the United States, Canada, and much of Europe and one that emphasizes the individual. Bentham devised what he called a "hedonic calculus" based on mythical units called "utils" that were standards of happiness. Arguing that people are inherently self-interested, Bentham suggested the motivating principle of greatest happiness; that is, people will maximize pleasure and minimize pain. In the nineteenth century, neoclassical economists began to model this idea in highly simplified but powerful ways, often invoking complex mathematics.

Individual consumers, according to this perspective, are modeled after the pattern of the mythical creature ***Homo economicus***, who is an all-knowing being in terms of his or her opportunities in the world and their respective costs, benefits, and consequences. The level of satisfaction that individuals experience, the "utility" of a particular combination of goods, is reflected in indifference curves that map the trade-offs each person is willing to make to achieve equal satisfaction (Figure 8). Thus, along one utility curve the person is equally happy; the higher the utility curve is, the higher the person's happiness is.

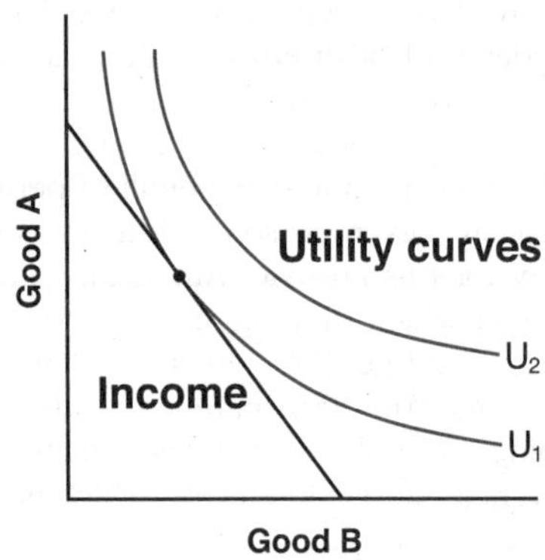

FIGURE 8 The neoclassical economic view of consumption begins with **utility maximization.** Utility, or satisfaction, is equal along any utility curve (e.g., U_1 or U_2). With a limited income, a consumer will attempt to maximize his or her utility by consuming the combination of goods or services in which the slope of the utility curve just touches the income constraint (point Z). While elegant, and popular among economists, this view ignores the social origins of consumer desires and demand.

The law of diminishing marginal utility holds that as a person consumes increasing quantities of a given good, the pleasure or use they derive from it—their utility—increases, but at a decreasing rate. Because consumption is always limited by income, each person must maximize his or her total utility or happiness by allocating income among different goods. Since we cannot spend more than we earn (at least in the long run), the model does not allow the consumer to rise from Utility curve 1 to Utility curve 2. The slope of the income constraint reflects the relative prices of Good A and Good B, that is, the amount one must give up of one in order to purchase a unit of another. The optimal amount consumed (point Z) occurs when the consumer maximizes utility by equating the marginal satisfaction derived from the consumption of each good with the marginal cost as reflected in the income line. Essentially, utility curves reflect what consumers would *like* to do, income constraints depict what they *can* do, and the point of tangency explains what they *actually* do.

Aggregate consumption patterns in this model reflect the multitude of individual choices about how to spend income. Because consumer demand is held to be the engine that drives the economy, this view is often equated with the notion of consumer sovereignty. Spending levels, in turn, reflect the dynamics of the labor market, unemployment rates, wage levels, prices, attitudes toward saving and spending, and the various factors that determine the shape of utility curves, such as the demographic composition of a society. Consumer tastes and preferences are constantly remolded by a number of social issues, including fads and fashions and concerns over health and safety (e.g., in the United States beer and wine are displacing hard liquor, chicken has surpassed beef consumption). Consumption patterns also reflect the income- and price-elasticity of the demand for different commodities, that is, the degree to which demand responds to changes in incomes and prices. For example, increased consumption of services is partly driven by the relatively high income-elasticity of many intangibles relative to goods.

The neoclassical economic view of consumption holds that markets are always optimally efficient economically (and hence morally optimal as well). Although the neoclassical view is internally consistent within its own terms of reference, it suffers from several problems, including the lack of any historical or social context and its inability to do justice to the rich semiotics and social dimensions of consumption. In part, this failure arises because neoclassical economics does not represent the consumer, or consumption, as a *social* act—that is, one embedded within broader relations of class, gender, ethnicity, and power—but as a purely individual one. For example, neoclassical economics offers no account of the origins of utility curves or why they assume their particular form; they are simply taken as given. "Individual" choices are always part of web of public policy choices: The simple act of buying butter reflects federal subsidies to dairy farmers, health laws, the use of artificial chemicals and hormones, and so on. Social categories in neoclassical economics, if they arise at all, are defined largely by their relations to consumption: Class in conventional social analysis, for example, refers to income and socioeconomic status. Adam Smith's invisible hand assumes choices are independent, not socially produced, and that they don't affect each other. In the real world, consumer choices are strongly affected by one another (e.g., fads), as well as by social institutions such as the media.

Finally, even when individuals act rationally, that is, maximize their utility, that can lead to collective irrationalities or market failures. These occur when the market price does not reflect all of the social costs of a good. If each person at a concert stands up for a better view, for example, everyone is forced to do so and is therefore worse off. Each person's decision to drive is individually rational, but collectively these decisions create irrational traffic jams and gridlock. Such examples point to the difficulty in reducing *social* problems to *individual* causes and solutions.

Marxist Views of Consumption

A third interpretation of consumption comes from Marxism, which argues that social science must penetrate the veneer of the capitalist system to reveal the social relations that lie beneath. The expansion of capitalism historically has been predicated upon a widening and deepening of commodity relations, that is, the development of new markets and the transformation of goods and services that were formerly outside of the market into commodities (e.g., housing, child care, transportation, education). Although Marxism has generally placed more importance on production and labor over consumption, Marx did argue that unless products are entirely consumed in the market, firms and employers cannot realize the profits generated at the workplace.

Marx argued that commodities are not simply *things*, but embodiments of social relations. To view commodities separately from their social origins is to commit the error of commodity fetishism; for capitalist producers the functionality of market relations obscures the social relations. Marx argues that the social character of labor appears objective, given the nature of products:

> The relations connecting the labour of one individual with that of the rest appear, not as direct social relations between individuals at work, but as what they really are, material relations between persons and social relations between things. . . . To [producers], their own social action takes the form of the action of objects, which rule the producers instead of being ruled by them.

When seen in this way, commodities are not simply items on the shelf or advertised on television, they become complex combinations of labor, nature, and ideology. If one learns to view the simple act of shopping in social, not simply individual terms, then every voyage to the grocery store becomes an exercise in deciphering how class, production, ideology, and other forces manage to bring so many goods together from around the world.

Marxism drew upon classical economics to differentiate the **use value** of commodities—the qualitative, subjective dimensions—from their **exchange value**, the quantitative price they command on the market. For example, the use value of an apple is its taste and the relief from hunger it offers; its exchange value is what it sells for. Critically, for Marxists, labor too is a commodity whose use value to employers is less than its exchange value in wages. Class is thus defined by relations to production, not consumption.

Marxism suggests that the extraction of surplus value by employers inevitably leads to **underconsumption** by the working class and the tendency toward crisis. Employers, in this view, cannot by definition pay their workers the value of their output, or no profit would be generated. Thus, capitalism is perpetually faced with the problem of producing too much, which drives down prices and, ultimately, profits. For Marxists, this line of thought is the most severe of the internal contradictions that capitalism faces and lies at the core of capitalists' incessant need to find new markets.

GEOGRAPHIES OF CONSUMPTION

Drawing upon the work of sociologists, historians, philosophers, and anthropologists, geographers have engaged in numerous lines of inquiry with which to connect commodities to their social and spatial origins. This body of work has tended to divide into three major categories.

First, in the tradition of humanistic geography, some geographers have examined the relations between consumption, the body, and individual experience. The body is the most intimate of geographies, the site of intentionality, where mind and a basis of identity reside. Bodies appear natural but are actually social constructions, inscribed with social meanings. The geography of the body locates it within social relations, as a place within a network of places. A considerable literature, for example, has looked at food, its origins and cultural meanings in different geographic contexts, and its role in the unfolding of daily life. Eating is the most intimate relation between a body and the environment, and food consumption plays a major role in shaping bodies (witness the epidemic of obesity plaguing the United States today, in part due to the widespread consumption of fast food [Figure 9]).

Similarly, geographers have examined the shopping mall not simply as an economic phenomenon, but as a cultural site full of meanings. For many Americans, time spent in shopping centers ranks third in importance only to time at home or at work or school; many prefer shopping to sex. Shopping is the nation's dominant form of public life, and the shopping mall the only remaining pedestrian space in which to congregate. Thus, the mall is not just a place to consume, but a metaphor for public life in general. For

Nancy Siesel/NYT Pictures

FIGURE 9 Fast-food preparation exemplifies standardization of production that accompanied the standardization of consumption. The "McDonaldization" of many forms of social activity is widespread.

example, the West Edmonton mall in Canada (Figure 10), the first and largest megamall in the world, has 600 stores, 18,000 employees, and generates 1% of all of Canada's retail trade. Inside it contains a golf course, skating rink, fantasyland hotel, and four submarines. Similarly, the Mall of America in Bloomington, Minnesota, has 520 stores, chapels, a roller coaster, aquarium, and rain forest. In this environment, fantasy, fun, and the commodity are merged into a seamless whole. Malls are carefully engineered to maximize throughput of people and turnover of goods; anchor stores are carefully located and exits are minimally visible. Symbolically, this environment is designed to transport shoppers to a mythologized looking-glass world where the only thing that matters is commodities. All "backstage" functions involving the production, transportation, and storage of goods are carefully hidden.

Third, geographers have focused on consumption in the context of the global economy, particularly the manner in which commodities are produced, distributed, and consumed through the use of **commodity chains** (also called value chains). A commodity chain is a network of labor and production processes that produce a commodity; it extends from the raw material to various stages in processing and delivery and ends in consumption. For example, the coffee commodity chain begins with the grower, typically an impoverished farmer, and extends through the processing plant, exporters, traders, roasting companies, retailers, and, finally, the consumer (Figure 11). Similarly, the meat commodity chain extends from ranchers

© WALTER BIBIKOW/DanitaDelimont.com

FIGURE 10 West Edmonton Mall in Canada exemplifies the ways in which spaces of consumption lie at the boundary of reality and fantasy, enticing consumers into a world in which status and happiness are allegedly guaranteed by purchasing commodities.

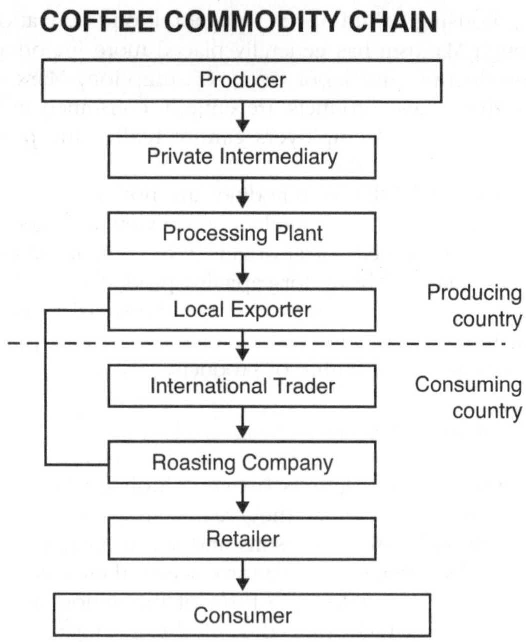

FIGURE 11 The coffee commodity chain represents the multiple steps involved between producer and consumer, including the growing, harvesting, and roasting of the beans; intermediaries involved in transporting, distributing, and marketing them; and, finally, the consumer. Different goods have different commodity chains that stretch unevenly over time and space, connecting producers, consumers, and those in between in complex networks of causality. In contrast to the utility-maximizing point of view, commodity chains are explicitly social and spatial in outlook.

CASE STUDY

Commodity Chains

A commodity chain encompasses a set of interrelated activities associated with the production of a good or service, such as design, manufacturing, retailing, marketing, advertising, and consumption. In commodity chain analyses, the economy is conceptualized as an assemblage of individual chains, each possessing a unique logic, organization, and spatiality. Using the lens of the commodity chain, geographers have explored the dynamics of production and consumption for a range of commodities, including fashion, furniture, footwear, gold, diamonds, fruit, vegetables, coffee, chocolate, and cut flowers. The focus is on the relationships between buyers and suppliers of a particular product, tracing the flow of material resources, value, finance, and knowledge between sites along a chain. This approach marks a break with horizontal approaches to the economy, which focus on one sector such as retailing or on one stage in the life of a commodity such as manufacturing.

The interest in commodity chains can be traced to consumers' growing political concerns about the origins, quality, and ethical implications of products, and with the vast distances and inequalities that often differentiate producers and consumers. A key problem is that the images and meanings of consumer goods often bear few traces of the production processes and environments that created them. The aim of a commodity chain analysis is to uncover these connections.

Many commodity chains are buyer-driven, that is, they are dominated by large retailers and brand-name marketers who establish decentralized production networks in a variety of exporting countries. These dynamics are common in labor-intensive, consumer goods industries like clothing, footwear, toys, and electronics. Production is generally executed by tiered networks of contractors that make finished goods according to the design specifications of retailers or marketers. In buyer-driven chains, profits derive from innovations in design, sales, and branding, rather than from advantages in scale, volume, or technology.

Some believe buyer-driven chains have recently become dominant over producer-driven chains. One example is the case of fresh fruit and vegetable chains, where supermarkets exercise heightened control in determining what products will be offered to consumers and when, as well as dictating the quality, appearance, and packaging of produce. In order to achieve this control, large grocery chains have been shifting from sourcing products through wholesale markets to establishing their own tightly managed supply chains.

Commodity chains may affect whether firms in developing countries can upgrade their industrial operations, including reorganizing production systems or implementing new technology, shifting into the production of more sophisticated lines, or expanding into new areas such as research, design, or marketing. A central issue is the role of buyers in transferring knowledge to their suppliers.

Commodity chain analysis provides a helpful framework for understanding some of the central characteristics and geographies of contemporary capitalism. Labor organizations are increasingly using commodity chain analysis to map the connections between workers around the world, as well as between workers and consumers at a variety of sites and on at many scales. Unions have begun organizing across commodity chains and developing transnational networks (e.g., alliances with other unions). The commodity chain approach also affords the political potential to challenge relations of exploitation, inequality, and injustice across space.

to feedlots to packinghouses to cold storage to grocers and finally to consumers. The gold commodity chain begins with miners working under horrendous conditions in such countries as South Africa, through diamond cutters (usually in Europe), to retailers and consumers in the developed world, who celebrate gold's romantic allure without considering the conditions of its production. At each stage, the commodity is transformed in some way and value is added. The same company may control one or more stages in a commodity chain, depending on how vertically integrated or disintegrated the production process is. Because different nodes where these activities are carried out are spatially separated, commodity chains are geographical as well as economic and cultural phenomena.

Commodity chains are thus a means of understanding the ways economic activity reverberates through the production process, the linkages among different economic sectors, and the flows of value over time and space and help us bridge the artificial separation between consumption and production. They allow us to see the commodity as more than just a thing and as an embodiment of processes at different spatial scales. Essentially, commodity chains are mechanisms that allow us to trace the impacts of consumption decisions backward through the production and distribution process, broadening our scale of analysis from the local to the global. They trace the commodity through the complex, contingent lines of causality, linking sellers and buyers across multiple spatial scales.

Over time, with the expansion of capitalism globally, commodity chains have become longer and longer. They allow us to understand how globalization has unleashed a tidal wave of cheap imports that has propelled the high rates of consumer spending in societies like ours. By letting us understand the link between consumption and production, the concept allows us to recognize the sacrifices made by low-wage laborers trapped in sweatshops in the developing world in order to provide American consumers with cheap goods. Such a perspective reveals consumption as being simultaneously an economic, cultural, psychological, and environmental act that simultaneously encompasses both the world's most abstract space, the global economy, and the most intimate, the individual person and body.

ENVIRONMENTAL DIMENSIONS OF CONSUMPTION

In addition to being an economic and cultural phenomenon, consumption is also an important environmental one. Because consumption is linked to other domains—transportation and production, for example—every act of consumption, both direct consumption by households and individuals and indirect consumption of resources in the production process, imposes changes on the environment. Often, because commodity chains are long, the environmental impacts may be felt thousands of miles away or on the other side of the world.

Traditionally, environmental destruction has been blamed on the poor. This view, taking its cue from Malthusianism and the high birth rates found in many developing countries, maintains that the large numbers of the world's impoverished peoples have been the cause of the clearing forests for farms, soil erosion, and other environmental predicaments. There is no question that rapid population growth contributes to environmental destruction, but overconsumption in the economically developed world is even more destructive. It is overconsumption that generates most of the planet's environmental problems.

Inevitably, as standards of living rise, and thus consumption, a society puts increasing stress on the environment, using more resources and producing more waste (Figure 12). For example, in 2006, the United States consumed 15 times more raw materials than it did in 1900, but its population had only tripled. In many respects, mass consumption is disastrous to the world's ecologies. Whole ecosystems are sacrificed to support our lifestyle. Tropical regions are deforested to produce plywood and newspaper. Entire landscapes are ravaged by the tremendous hunger of both the developed and developing world for minerals, water, and energy to produce houses, hamburgers, and clothes. Our demand for seafood is emptying the world's oceans: 75% of the fish stocks are overfished, the size of the catch not limited by fishing technology, but by the reduction of stocks. Pollution, deforestation, and habitat destruction are generating the greatest loss of biodiversity in the planet's history. For economically developed countries, the effects are often not experienced directly because globalization allows us to export problems.

Of the many forms that consumption takes, private transportation via the automobile is perhaps the most wasteful. Urban sprawl and long commutes have locked us into an energy-intensive lifestyle. In the United States, for example, there are more cars than drivers. Transportation consumes 40% of the total U.S. energy budget. With 5% of the world's people, we consume 25%

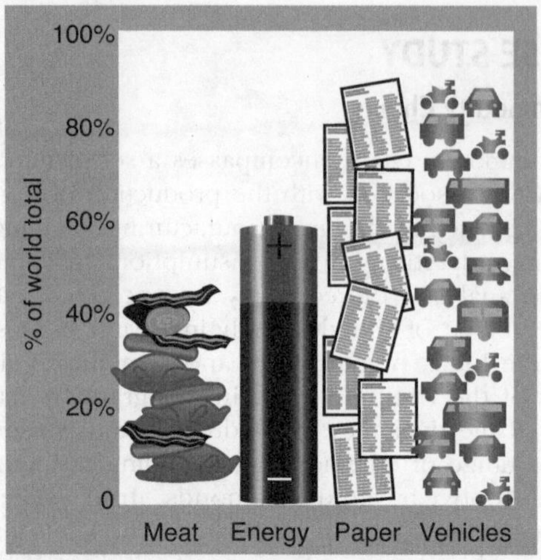

FIGURE 12 The economically developed world, which constitutes only one-fifth of the planet's population, consumes the bulk of the world's resources due to its higher incomes and privileged position internationally. This unevenness in consumption levels reflects the historical dynamics of capitalism, which simultaneously produces wealthy and impoverished societies around the globe.

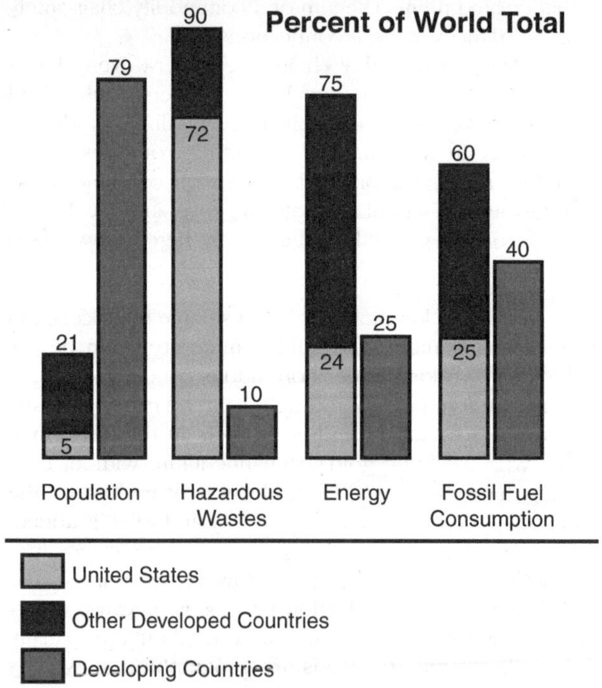

FIGURE 13 The result of mass consumption is mass waste; many countries produce enormous amounts of garbage, much of it nonbiodegradable, which has horrendous effects on ecosystems, water supplies, air quality, and standards of living.

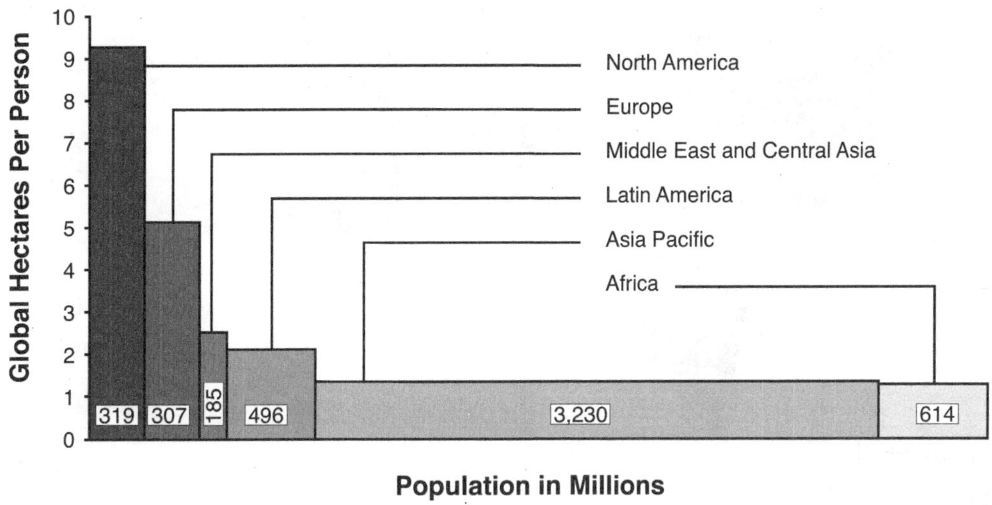

FIGURE 14 Ecological footprints measure the amount of the world's resources used and the biophysical impacts on the planet. Their size is directly proportionate to the income and consumption levels in each society. Consumption is thus simultaneously an economic, social, psychological, geographic, and environmental phenomenon.

of its energy. Petroleum consumption is a primary generator of the greenhouse gases that are a major factor in anthropogenic global climate change.

A remarkable feature of the world economy is the uneven distribution of purchasing power and resource consumption, a pattern that reflects the legacy of colonialism and the unevenness of capitalist development. Today, roughly 20% of the world consumes about 80% of its vital resources, including 40% of the world's meat, 60% of its energy, and 80% of its paper and vehicles (Figure 13). As a consequence, the First World generates disproportionate amounts of the world's hazardous wastes (Figure 14). As much of the developing world embraces the consumerist lifestyle, it, too, has demanded a larger share of energy, food, and other materials. Many ecologists are highly doubtful that the world can generate the resources necessary to support the entire planet's population at a level approaching that of the United States today.

An **ecological footprint** is an estimate of the amount of resources necessary to support a person's lifestyle. During his or her lifetime, the environmental impact of one U.S. citizen is 200 times more than that of a child in Mozambique. (Go to **www.footprintnetwork.org** to estimate your own footprint, which is a function of the size of your house, your diet, mode of transportation, etc.). The economically developed world has much larger footprints than do societies in Asia, Africa, or Latin America (Figure 15). Taken together, the footprints generated by each country reflect the per capita environmental impact of the total global population (Figure 16).

FIGURE 15 Fresh Kills landfill, Staten Island. The flip side of consumption is the production of waste. Given the large degrees of packaging and high levels of consumption found in Western societies, inevitably enormous quantities of garbage are generated, often with serious environmental consequences.

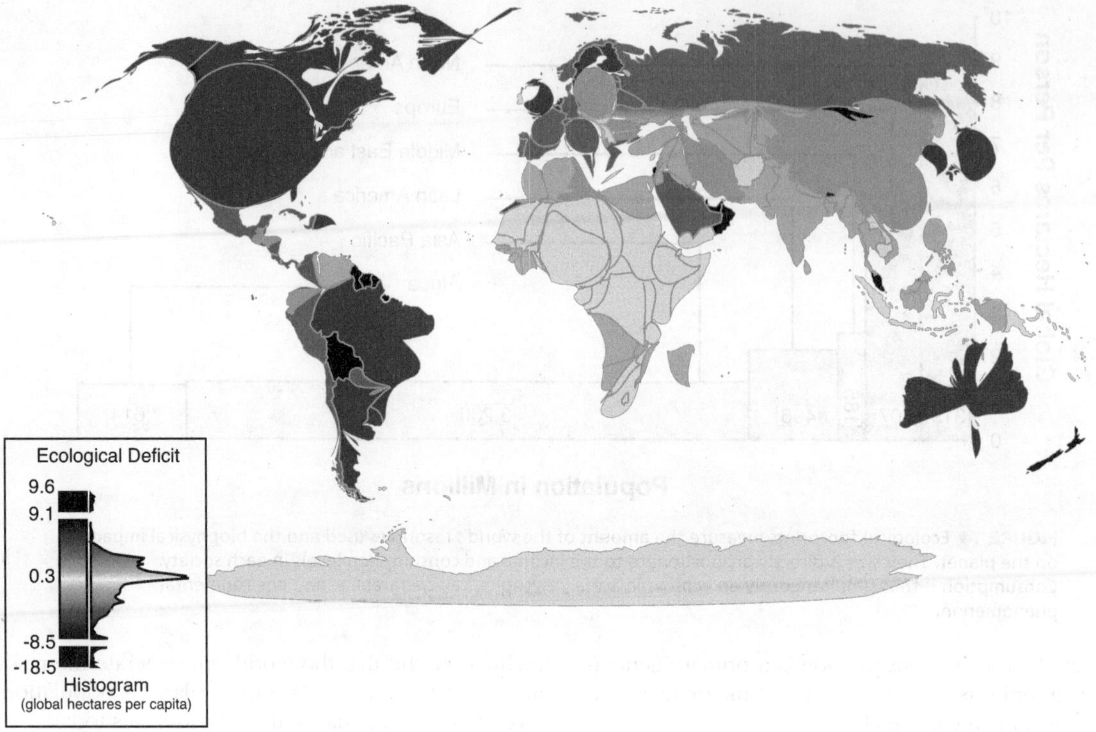

FIGURE 16 Because the developed countries consume so much, they have much larger ecological footprints than do the developing countries. Economic growth in some developing countries, however, such as India and China, has enlarged their ecological footprints.

Summary

Consumption, a complex phenomenon, has economic, cultural, political, and environmental dimensions. Historically, consumerism arose on the heels of capitalism, and the Industrial Revolution, which created a working class with disposable income and sufficient free time, brought on mass consumption. In the United States and similar societies, consumerism has reached its apex, and it has spread rapidly to much of the rest of the world. However, surveys reveal that despite higher levels of income and consumption, the populations of these societies are not necessarily happier than they were a half-century ago.

Sociology views consumption as a social act, in which individual choices are made and constrained by one's relative social status. Demand does not simply appear spontaneously in the individual consumer, but is generated by outside pressures, such as advertising, which is exceptionally adept at making people believe that luxuries are necessities. Neoclassical economics models demand using utility curves and the desolate figure of *Homo economicus*. The Marxist labor theory of value concludes that employers always extract more value from workers than workers can consume, leading to the chronic problem of capitalism—overproduction and underconsumption.

Geographically, consumption can be studied on multiple spatial scales. The individual body, for example, reflects one network of consumption. But the body is also part of a social construction. Groups of shoppers at a mall engaging in consumption are enclosed in many webs of symbolic meaning, and they are in a spatial environment that is often deliberately constructed to entice them to spend as much as possible. Commodity chains that link consumption with production and describe the various stages through which commodities pass from raw material to final product may extend thousands of miles or even globally.

Consumption inevitably entails some degree of environmental transformation. Popular opinion often holds the developing world responsible for most of the world's environmental ills, but statistically it is the mass-consumption lifestyle of the First World that generates most of problems, such as global warming. A relatively small proportion of the world—around 1.7 billion people, or 20%—consumes up to 80% of the world's resources. As consumerism has spread to much of the world, the impacts on ecosystems have multiplied accordingly. Ecological footprints are one means of analyzing the environmental costs of consumption.

Key Terms

affluenza
commodification
commodity chains
conspicuous consumption
consumerism
consumption
delayed gratification
ecological footprint
exchange value
Homo economicus
underconsumption
use value
utility maximization

Study Questions

1. When and where did consumerism originate?
2. Why don't rising levels of consumption make everyone happier?
3. What are the major theoretical perspectives on consumption?
4. How is the body a geographical locus of consumption?
5. Are most of the world's environmental problems generated in the developing world?
6. What is an ecological footprint?

Suggested Readings

Bell, D., and G. Valentine. 1997. *Consuming Geographies: We Are Where We Eat*. London: Routledge.

De Graaf, J., D. Wann, and T. Naylor. 2001. *Affluenza: The All-Consuming Epidemic*. San Francisco: Berrett-Koehler Publication.

Goss, J. 1999. "Once Upon a Time in the Commodity World: An Unofficial Guide to the Mall of America." *Annals of the Association of American Geographers* 98:45–75

Hartwick, E. 2000. "Towards a Geographical Politics of Consumption." *Environment and Planning A* 32:1177–1192.

Mansvelt, J. 2005. *Geographies of Consumption*. Thousand Oaks, CA: Sage.

Stearns, P. 2001. *Consumerism in World History: The Global Transformation of Desire*. London: Routledge.

Whybrow, P. 2006. *American Mania: When More Is Not Enough*. New York: Norton.

Web Resources

Global Footprint Network
http://www.footprintnetwork.org/en/index.php/GFN/page/footprint_basics_overview/

Calculate your ecological footprint or one for any part of the world.

International Trade and Investment

OBJECTIVES

- To explain the theoretical bases of international trade and factor flows, including comparative and competitive advantage
- To examine the effects of trade barriers such as tariffs, quotas, and nontariff barriers
- To present the dynamics of foreign direct investment
- To understand the financing of international trade, including the impacts of exchange rates
- To state the role of trade organizations such as cartels, the World Trade Organization (WTO), and regional trade agreements

A Japanese-owned automobile plant in Bellefontaine, Ohio, United States, reflects global patterns of foreign direct investment, particularly how international capital combines with labor to create changing geographies of wealth and poverty.

Jay LaPrete/Bloomberg via Getty Images

International Trade and Investment

Since World War II, most national economies around the world have become more integrated than ever as they have become more specialized within a global division of labor. Technological breakthroughs in transportation and communications, the decline in protectionism, and the growth of transnational corporations have contributed to this process. International business includes the international transmission of goods, services, information, and capital. Increasingly large numbers of companies invest in foreign countries to acquire raw materials, to penetrate markets, and to exploit cheap labor. The expansion of production overseas has been matched by a parallel, symbiotic expansion of services. International trade is expanding, and its composition and patterns are changing. But in many respects, it is now less significant than the international movement of capital.

The late twentieth century marked a watershed in the world economy. First, industrialized countries experienced a slowdown in their economic growth rates, in part due to the petroshocks of the 1970s and the ensuing deindustrialization. Increases in oil prices reduced real income in the advanced countries and dealt a particularly harsh blow to the oil-importing Third World countries. These oil shocks left a permanent imprint on the structure of global finance, trade, and investment. Second, competitive rivalry among industrialized countries increased significantly as they developed their productive capacities in different sectors and sought foreign revenues via exports. Third, global financial markets underwent a profound series of alterations. In 1973, the old Bretton-Woods monetary arrangement, which involved fixed monetary exchange rates and the convertibility of the U.S. dollar into gold, collapsed and was replaced by a system of fluctuating exchange rates, in which supply and demand dictated relative values of currencies. This change had serious effects on the relative prices of imports and exports worldwide. A global electronic network allowed vast sums of money to be traded internationally, creating an almost seamless financial market around the planet. The World Trade Organization became a permanent body for the regulation of barriers to trade. The fourth structural change was massive geopolitical realignments. Japan and other East Asian newly industrialized countries (NICs) enjoyed rapid industrial growth. China grew rapidly to become the world's second-largest economy. The Soviet bloc collapsed, sending waves of turmoil throughout central Asia and Eastern Europe. And Europe pursued a relentless strategy of economic unity through the European Union, a path followed to some extent in North America under the North American Free Trade Agreement.

This chapter examines the concepts and patterns that underlie the expanding world of international business. First, it reviews the major theories of trade, including classical comparative advantage and competitive advantage. Second, it discusses international capital markets, including exchange rates and foreign direct investment (FDI). It focuses on the role played by multinational corporations. Then the chapter considers major obstacles to trade, particularly tariffs, quotas, and nontariff barriers. Finally, it examines ways in which protectionism has declined, including the roles of trade organizations such as the General Agreement on Trade and Tariffs (now the World Trade Organization) and regional trade agreements such as the European Union and the North American Free Trade Agreement.

INTERNATIONAL TRADE

Trade among countries has long been central to capitalism and a major factor in linking various parts of the world together. World trade has jumped from $2 trillion annually in 1980 to over $11 trillion today. Why are so many countries, large and small, rich and poor, deeply involved in international trade (Figure 1)? One answer lies in the unequal distribution of productive resources among countries (i.e., uneven spatial development), which can be offset to some extent by trade. However, whether a country can export successfully depends not only on its resources but also on the

FIGURE 1 World growth in exports by value, 1965–2005. Even though world exports have grown steadily over the time period represented here, the growth of world trade is under some question now. The global financial crisis of 2008–2010 has left an unprecedented degree of unemployed workers and underused factories in its aftermath. At the same time, across the rich world, the supply of workers is about to slow, while the number of pensioners rises. In Western Europe the working age population will shrink by 0.5% a year going forward, while in Japan the rate of shrinkage will be about 1.0% per year. America's aging population demography is more favorable, but growth in its working-age population will eventually slow down to a third of its post–World War II average. The world is expanding at 4% in the second decade of the twenty-first century, but growth of the United States is less than 3% per annum, and growth of Europe is less than 2%—so low that there will be little, if any, reduction in the 40 million unemployed in these two major economic blocs. Clearly, most European countries waited too long to overhaul their welfare states. The added costs of global recession have now forced them to do the politically undesirable: chop social spending and raise taxes, leading to further production slowdowns. The United States must now deal with the same issues.

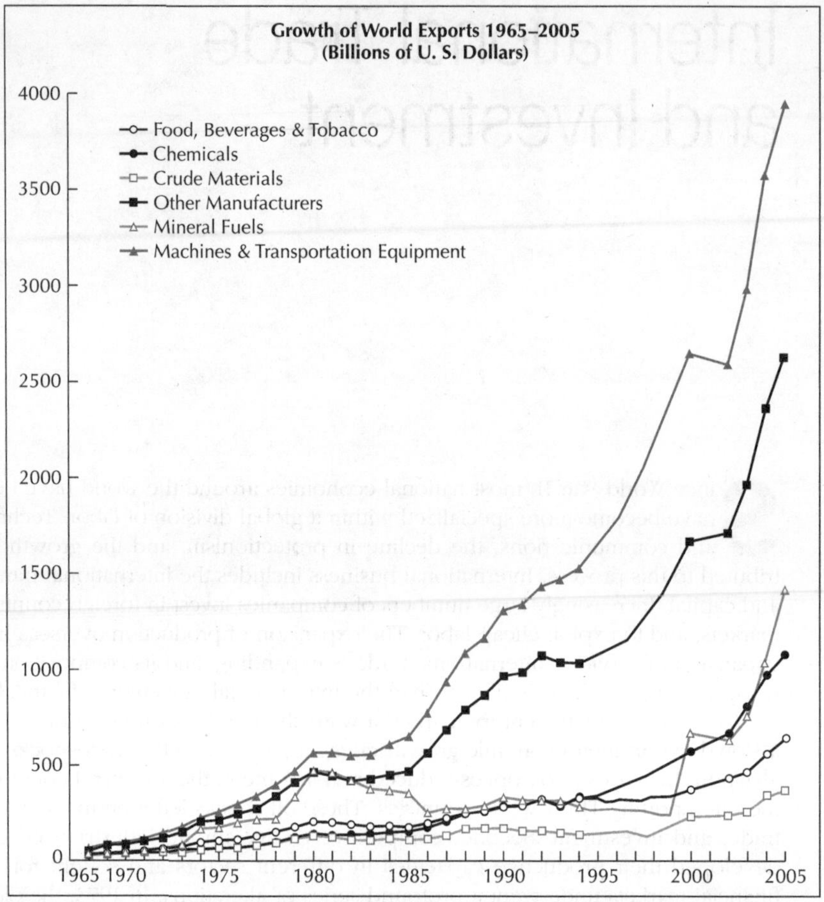

opportunity, ability, and effort of producers to trade and the capacity of local producers to compete abroad.

Production factors—labor, capital, technology, human capital and skills, entrepreneurship, and land containing raw materials—vary enormously from country to country. Some countries have populations large enough to support industrial complexes and domestic markets; others do not. One country is home to workers adept at running modern machinery; another abounds with scientists and engineers specializing in research-intensive products; yet a third has huge pools of unskilled workers. The imbalance in natural and human-made resources accounts for much of the international interchange of production factors and the products and services that the factors can be used to produce. There may be numerous reasons why countries cannot use their productive factors to best advantage, including inflation, exchange rates, labor conditions, governmental policies, and laws. Other countries are hobbled by the legacy of colonialism, drought, or political violence.

Inflation, which is an increase in the general level of the prices of goods and services, can be detrimental to a country's ability to compete domestically or internationally. Exchange rates, the prices of currencies in foreign exchange markets, can also influence competitiveness. For example, if a currency is overvalued in relation to other currencies, local producers may find it difficult not only to compete with foreign imports but also to export successfully. In addition, recurring labor disputes that interrupt production can create obstacles for exporters. Governments can encourage or discourage their export sectors through trade policies, currency manipulation, and the allocation of public resources to infrastructure or education. Finally, the competitiveness of exporters is affected by labor laws, tax laws, and patent laws.

Trade by Barter and Money

For much of human existence, trade was primarily conducted on a barter basis, the direct exchange of goods or services for other goods or services. Barter still occurs within some traditional markets in underdeveloped countries and is of importance to some countries internationally. Russia and eastern European countries use barter to trade among themselves and with underdeveloped countries. Major oil-exporting countries such as Iran and Nigeria barter oil and gas for manufactured goods. Despite its widespread use, particularly by governments that have turned toward economic protectionism, barter is a cumbersome way of conducting international exchange. Even within a country, consumers would find it difficult to barter goods or services to satisfy all their daily needs. In fact, more time would be devoted to exchange than to production. Money greatly simplifies exchange and trade within and among countries, although it does present problems such as those associated with exchange rates.

COMPARATIVE ADVANTAGE

One of the hallmarks of capitalism is its tendency to generate uneven economic landscapes, that is, great differences in the types of economic activity from place to place as well as in the standards of living and life chances that those activities create. Different regions have long specialized in the production of different types of goods and services. In Europe during the Industrial Revolution, for example, Britain became a major producer of textiles, ships, and iron; France produced silks and wine; Spain, Portugal, and Greece generated citrus, wine, and olive oil; Germany, by the end of the nineteenth century, was a major exporter of heavy manufactured goods and chemicals; Czechs were selling glass and linens; Scandinavia sold furs and timber; and Iceland and Canada exported cod to the growing middle classes. Similarly, within the United States different places acquired advantages in some goods and not others: The Northeast dominated light industry, particularly textiles; the Manufacturing Belt became the center of heavy industry; Appalachia developed a large coal industry to feed the furnaces of the industrial core; the South grew crops such as cotton and tobacco; the Midwest became the agricultural behemoth of the world; the Rocky Mountains sold coal and copper; and the Pacific Northwest was incorporated into the national division of labor based on the expanding timber and lumber industry.

When regions or countries specialize in the production and export of some goods or services, they enjoy a **comparative advantage**. This notion was first introduced by the famous nineteenth-century economist David Ricardo (Figure 2), a contemporary of Thomas Malthus and one of most famous figures in the history of economics. Like all classical political economists, he held to the labor theory of value (the value of goods reflects the amount of socially necessary labor time that goes into their production) and thus ignored demand, and so labor productivity was the central element of his model of trade. Ricardo concluded that nations will specialize in the production of a commodity that they can produce using the least labor, compared to other nations.

Ricardo's classic example of this concept is demonstrated in Table 1, which illustrates the allocation of labor time in England and Portugal, two long-time allies and trading partners, before and after they specialized and thus traded. In the first table, which depicts the labor hours per unit of wine or cloth that England and Portugal must each dedicate to the production of one unit of each good, it is evident that Portugal has an **absolute advantage** in both goods (i.e., it can produce both of them with fewer labor hours

Library of Congress LC-USZ62-99849

FIGURE 2 David Ricardo (1772–1823) was one of the most influential figures in the history of economic thought. By spatializing Adam Smith's notion of comparative advantage, he laid the theoretical foundations for the analysis of international trade and regional development.

than can England). If Portugal is more efficient in both goods, does it make sense for Portugal to trade? The answer is yes, because even the most efficient producer benefits from trade. Ricardo's analysis examined what happens when each country allocates its resources to the good it can produce most efficiently compared to its trading partners (i.e., when it acquires a comparative advantage). Thus, in the second table, England produces only cloth (two units at 100 hours each) and Portugal produces only wine (two units at 80 hours each). In the process of specializing, that

TABLE 1 Ricardian Example of Comparative Advantage

Before Specialization (labor hours/unit)				
	Wine	**Cloth**	**Total**	
England	120	100	220	
Portugal	80	90	170	
units	2	2	390	
After Specialization (labor hours/unit)				
	Wine	**Cloth**	**Total**	**Savings**
England	0	200	200	20
Portugal	160	0	160	10
units	2	2	360	30

is, of producing for a market that consists of both economies together rather than either alone, each country frees up some resources that would otherwise have been dedicated to the inefficient production of a good in which it did not have a comparative advantage. England saves 20 labor hours, Portugal saves 10, and the combined trading system thus saves 30, which can be reallocated toward investment (the original model is static and says nothing about change over time).

The Ricardian model—the simplest of many, more complex notions of comparative advantage—has important implications for economic geography. First, it shows how powerfully trade and exchange shape local production systems. It demonstrates that trade allocates resources to the most efficient (i.e., profitable) ends at minimum cost. The only costs of free trade are borne by inefficient producers, in this case, English wine growers and Portuguese textile producers (or today, U.S. auto and textile producers). Second, Ricardian notions of comparative advantage reveal that specialization reduces the total costs of production; thus, trade improves efficiency even without reallocating resources. Productivity arises from many sources, including technological change (which is absent in this model), but an increasingly specialized division of labor has long been central to capitalism's growth and success. For this reason, the vast majority of economists favor free trade as beneficial to all parties concerned. Third, the Ricardian model points out that large markets allow more specialization than do small ones; thus, the combined market of England and Portugal allows more specialization than either could achieve alone. Adam Smith, the great political economist of the eighteenth century, noted the same thing when he stated that the "division of labor is governed by size of the market." In this case, when the market expanded from one country to two, it allowed firms to specialize and become more efficient in the process. Large markets allow firms to develop economies of scale and become more efficient. Small economies, with limited domestic markets, thus tend to be inefficient ones.

Transport Costs and Comparative Advantage

Clearly, just as there is no specialization without trade, there can be no trade without transportation. Goods must be moved across space from producer to consumer, and these transport costs must ultimately be borne by those who consume the goods. To the degree that transport costs affect the delivered price of commodities, they also influence consumers' willingness to buy them and thus the competitiveness of the regions that export them. If transport costs are low, their impacts on the division of labor will be low.

However, sometimes, particularly for heavy and bulky goods, transportation costs may increase the market prices of exports/imports prohibitively, as demonstrated in Figure 3. In the two regions here, A and B, the supply for the good is identical but the demand differs greatly. Producers in A can sell their good in region B for a high price, and thus earn a higher rate of profit. Unfortunately, the market price plus transport costs (MP + TC) of the imported goods in region B exceed the domestic production price in region A; in other words, the transport costs make the exports too expensive to ship across regions.

Throughout the history of capitalism, declines in transport costs have made it progressively easier for regions to realize their comparative advantage; thus, lower transport costs have contributed to lower production costs and higher standards of living. For example, New Zealand became a major producer of lamb following the introduction of refrigerated shipping in the late nineteenth century. Similarly, the Pacific Northwest began to export vast quantities of wood and paper to the cities of the Midwest and East Coast following the completion of the transcontinental rail lines in the 1890s.

Heckscher-Ohlin Trade Theory

David Ricardo's two-country, two-product theory of comparative advantage can be expanded by considering several countries and commodities and by taking into account several production factors. The multifactor approach to

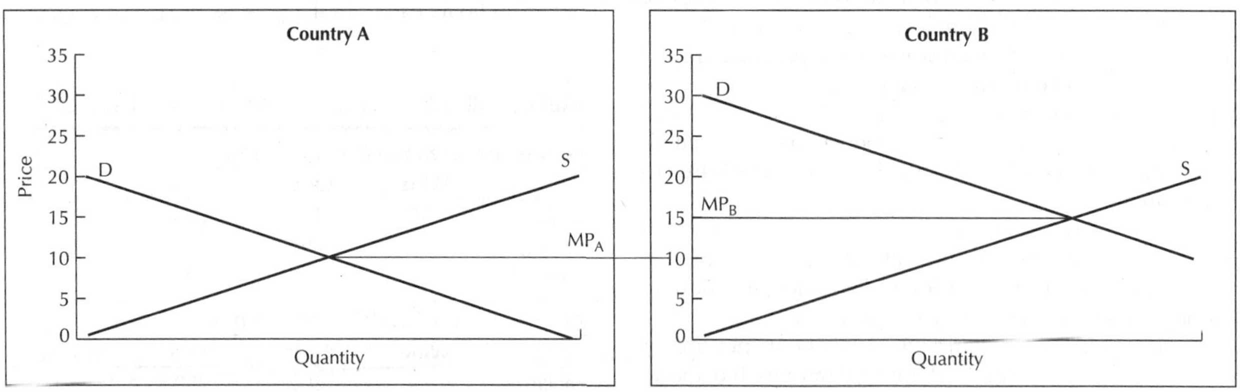

FIGURE 3 The influence of transport costs on comparative advantage. When the costs of shipping exports become greater than the differential in market prices between two countries, trade is inhibited. However, declining transport costs have steadily allowed for comparative advantages to flourish worldwide.

trade theory derives from work by two Swedish economists, Eli Heckscher and Bertil Ohlin. The Heckscher-Ohlin theory takes the view that a country should specialize in producing those goods that demand the least from its scarce production factors and that it should export its specialties in order to obtain the goods that it is ill equipped to make. Unlike the original Ricardian model, the theory includes demand and allows for the production of more than one good. Typically, in this formulation, specialization of production will be incomplete, that is, countries may continue to produce some of a good even if they do not enjoy complete superiority in the costs of production.

The Heckscher-Ohlin theory argues not only that trade results in gains but also that wage rates will tend to equalize as the trade pattern develops. The reasoning behind this factor-price equalization, as it came to be called, is as follows: If a country specializes in a labor-intensive good, its abundance of labor diminishes, the marginal productivity of labor rises, and wages increase. Conversely, in a different country specializing in capital-intensive goods, labor becomes less scarce, the marginal productivity of labor falls, and wages also fall.

Inadequacies of Trade Theories

Trade theories are based on restrictive assumptions that limit their validity. They generally ignore such considerations as scale economies and transportation costs. Scale economies improve the ability of a country to compete even in the face of higher factor costs. Trade theories assume perfect knowledge of international trading opportunities, an active interest in trading, and a rapid response by managers when opportunities arise. However, corporate executives are often ignorant of trading opportunities. Even if they are aware, they may fear the complexities of international trade.

Other inadequacies of trade theories include the assumptions of homogeneous products, perfect competition, the immobility of production factors, and freedom from governmental interference. But products are not homogeneous; oligopolies exist in many industries; and production factors such as capital, technology, management, and labor are mobile. Governments interfere with trade, they can raise formidable barriers to the movement of goods and services, as well as labor, and affect interest, inflation, and exchange rates, which in turn affect the prices of imports and exports.

The most important shortcoming of trade theories, however, is their failure to adequately account for the role of multinational corporations. In these corporations, trading decisions are made on the microeconomic level by managers, not by governments. Multinational corporations also operate from a multinational perspective rather than from a national perspective. When international trade occurs between different affiliates of the same company, it is referred to as **intracorporate trade**. Special considerations, such as tax incentives or no competition from other affiliates of the same company, can often play a pivotal role in a company's international decisions.

Fairness of Free Trade

Free trade is best from the standpoint of efficiency, but is it fair given the existence of **unequal exchange** between developed and developing countries? This question is raised by world-systems theorists such as Wallerstein, for whom free trade was a weapon wielded by rich countries to pry open the markets of poor ones. Their argument is that an artificial division of labor has made it difficult for developing countries to earn export revenues from free trade.

The British were instrumental in creating an unfair global division of labor in the eighteenth and nineteenth centuries. Implicit in the argument in favor of free trade was the notion that what was good for Britain was good for the world. But free trade was established within a framework of inequality among countries. Britain found free trade and competition agreeable only after becoming established as the world's most technically advanced industrial nation; prior to that period, Britain (and later, the United States) used protectionism extensively. For example, Britain levied high tariffs against textile imports from its colony India. Having gained a large initial advantage over other countries, Britain then threw open its markets to the rest of the world in 1849. Other countries were pressured to do the same. The pattern of specialization that resulted was obvious. Britain concentrated on producing manufactured goods, such as vehicles, engines, machine tools, paper, and textile yarns and fabrics, and exporting them in exchange for a variety of primary products such as furs, wines, silks, and bulk imports such as timber, grains, fruit, and meat. In this way, uneven spatial development was fostered and perpetuated. Although many countries gained from the application of this artificial division of labor, none gained more than Britain. The only way other countries could break out of this division of labor was by interfering with free trade. The United States was highly protectionist in the nineteenth century, and tariffs on imports were a major source of federal government revenues prior to the adoption of the income tax. Germany, France, Japan and other countries with embryonic industries did the same.

The original division of labor changed little until after World War II, when a new global structure began to evolve. The basic trend was export-led industrialization, concentrated in a few countries. For the best-off poor countries, industrial growth is geared to the needs of the old imperial powers. Thus, the growth of manufacturing in the Third World, under multinational corporate auspices, is not a portent of its emancipation from an unfair division of labor.

Worsening Terms of Trade

A deterioration in the **terms of trade**—the prices received for exports relative to the prices paid for imports—exemplifies the problem for less developed countries of unequal exchange. By and large, less developed countries export raw materials and semiprocessed goods—agricultural commodities, lumber, fish, and minerals. Primary commodities account for about 70% and 47%, respectively, of the total exports of low- and middle-income countries (excluding China and India). The proceeds from these exports are needed to pay for imports of manufactured goods, which are vital for continuing industrialization and technological progress. Shifts in the relative prices of commodities and goods can therefore change the purchasing power of the exports of less developed countries dramatically. The situation is exacerbated because many of these low- and middle-income countries are dependent on a single commodity for much of their revenues, which leaves them vulnerable to oscillations in the price of that good (Table 2).

In the late twentieth century, less developed countries experienced a worsening in their terms of trade caused by a decline in the prices of primary commodities and an increase in the prices of manufactured goods. Globally, there are numerous producers of primary sector goods in competition with one another, and productivity gains in

TABLE 2 Single-Commodity-Dependent Countries

Product's Percentage of Total Export Earnings		
40% to 59%	**60% to 80%**	**More Than 80%**
Agriculture and Fishing		
Benin (cotton)	Bhutan (spices)	Dominica (bananas)
Burkina Faso (cotton)	Burundi (coffee)	Rwanda (coffee)
Burma (lumber, opium[a])	Ethiopia (coffee)	Uganda (coffee)
Chad (cotton)	French Guiana (seafood)	
Cocos (Keeling) Islands (copra)	Guadeloupe (bananas)	
Comoros (spices)	Malawi (tobacco)	
El Salvador (coffee)	Maldive (seafood)	
Equatorial Guinea (cocoa, lumber)	Martinique (bananas)	
Finland (wood products)	St. Lucia (bananas)	
Ghana (cocoa)	Seychelles (seafood)	
Grenada (spices)		
Honduras (bananas)		
Iceland (seafood)		
Kiribati (copra, seafood)		
Mali (cotton)		
Mauritania (seafood)		
Nicaragua (seafood)		
Sudan (cotton)		
Crude Oil and Petroleum Products		
Congo	Bahrain	Algeria
Ecuador	Gabon	Angola
Syria	Libya	Brunei
Yemen	Venezuela	Iran
		Iraq, Kuwait
		Nigeria
		Oman
		Qatar
		Saudi Arabia
		Trinidad & Tobago
		United Arab Emirates
Metals and Minerals		
Central African Republic (diamond)	Botswana (diamonds)	Nauru (phosphates)
Chile (copper)	Guinea (aluminum)	Zambia (copper)
Jamaica (aluminum)	Niger (uranium)	
Liberia (iron ore)	Papua New Guinea (copper)	
Mauritania (iron ore)	Suriname (aluminum)	
Togo (phosphates)		

[a]Although impossible to quantify, Myanmar's opium exports may exceed 40% of total exports.

this sector have increased the supply much more than the demand, leading to falling prices, which is good for consumers but bad for exporters. Maintenance of trade deficits was possible only because the less developed countries had access to external sources of finance. The economies of many less developed countries are characterized by structural rigidity. They cannot alter the composition of exports rapidly in response to changing relative prices. Thus, if their commodity export prices decrease, they have no alternative but to accept declines in their terms of trade (Figure 4).

Another factor that may lead to worsening terms of trade is technological advances in developed countries. Advanced technology enables industrial economies to (1) reduce the primary content of final products; (2) produce high-quality finished products from less valuable or lower-quality primary products; and (3) produce substitutes for existing primary products (e.g., synthetic rubber for naturally grown rubber). These developments are irreversible. The demand for many primary products may be inelastic for price decreases, but in the long run it may be very elastic for price increases. A rise in the price of a raw material is an incentive for industrial research geared to economizing on the commodity, or substituting something else for it, or producing it in the importing country.

COMPETITIVE ADVANTAGE

The traditional theory of comparative advantage is very simplistic and unrealistic. Ricardo never gave an adequate account of why regions specialize in some goods and not others, instead offering a picture that is static with respect to time, overemphasizes labor productivity and does not explain its variations, ignores consumption as well as the role of economies of scale and agglomeration, says nothing about the nature of competition, and is silent concerning the impacts of public policy. Some of these issues have been addressed by neoclassical models in successively more sophisticated and complex models of comparative advantage.

A different model, advocated by Michael Porter, is called the theory of **competitive advantage**. Unlike the Ricardian model, which was useful for understanding the simpler economies of the early Industrial Revolution, Porter's model focuses on the social creation of innovation in a knowledge-based economy. Porter begins by dismissing two commonly held myths about the source of national competitiveness, cheap labor and abundant natural resources. Is cheap labor central to economic success? Cheap labor is, on a worldwide scale, virtually ubiquitous. Countries that have succeeded in competing internationally, such as Germany or Japan, have done so with labor costs well above those of their competitors. Why? Because their labor is productive and well educated and because their economies endow workers with sufficient capital. In contrast, countries with the cheapest labor, say most of Africa, have done poorly in the global economy. Nor are abundant natural resources necessary for economic success. Japan, for example, has done well despite having virtually no resources, and many developing countries with resources are trapped in low-wage economies that export raw materials. In a global economy, flows of oil, minerals, and foodstuffs are available anywhere.

What, then, does determine economic success? The key, in Porter's formulation, is productivity growth: Over the long run, rising productivity creates wealth for everyone, if not equally. Productivity growth in turn is a reflection of many factors, including the education and skills of the labor force, available capital and technology, government policies and infrastructure, and the presence of scale economies (as discussed earlier). In the context of global markets, all firms can maximize scale economies. Porter emphasizes that competitive advantage, unlike the Ricardian view, is dynamic and changes over time. The goal of

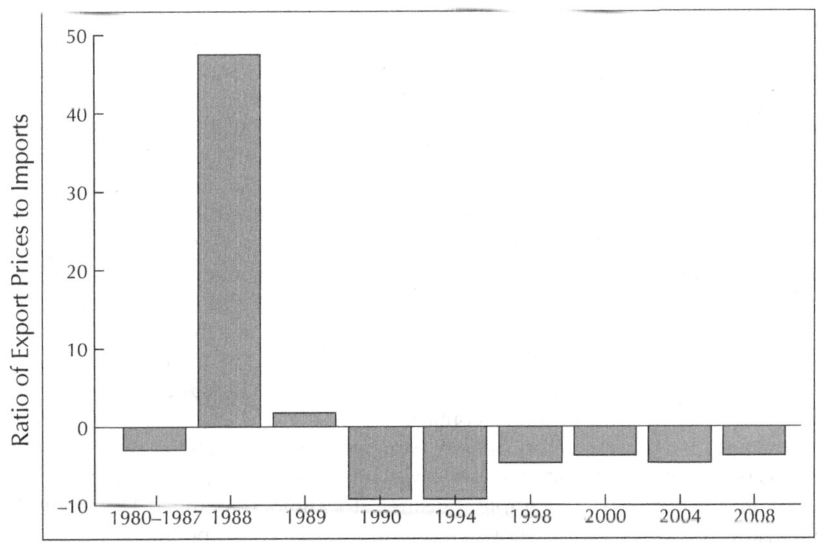

FIGURE 4 The worsening terms of trade for ores, minerals, and nonferrous metals, 1980–2008. Many underdeveloped countries cannot alter the composition of exports rapidly in response to changing relative prices.

national development strategies is to move into high-value-added, high-profit, high-wage industries as rapidly as possible. Such goods have high multiplier effects and do the most to trigger rounds of growth. To accomplish this goal, firms and countries should seek to sell high-quality goods at premium prices in differentiated markets. Quality is a key variable here; countries often acquire reputations for producing high- or low-quality goods, earning (or not earning) brand loyalty as a result. Finland is well known for its production of cell phones, for example, just as Japan is well regarded for its automobiles. When moving into high-value-added goods, nations should seek to automate low-wage, low-skill functions and retain knowledge-intensive ones.

Porter argues that although the global economy is increasingly seamless, competitive advantage is created in highly localized contexts (i.e., within individual metropolitan areas). Globalization does not eliminate the importance of a home base. Thus, countries that succeed internationally do so because a few regions within them move into cutting-edge products and processes. For example, the propulsive regions of the U.S. economy include Silicon Valley, Boston's Route 128, and New York City's financial and producer services firms; in Europe, they are Italy's Emilia-Romagna, that continent's largest high-technology region, as well as Germany's Baden-Wurtenburg, Denmark's Jutland peninsula, and the Cambridge region of the United Kingdom; in Japan, the government has actively constructed a series of technopolises toward this end.

The overall determinants of competitive advantage include the following:

1. Skilled labor, good educational systems, and adequate technical training.
2. Agglomeration economies, including pools of expertise, webs of formal and informal interactions, trust, linkages, strategic alliances, trade associations, integrated networks of suppliers and ancillary services.
3. A culture that rewards innovation: adaptation, experimentation, risk tolerance, and entrepreneurship; this includes heavy levels of corporate and public research and development and the continual upgrading of capital and skills. Corporations must engage in ongoing and organizational learning, anticipating changes in markets and demand; rigid corporate bureaucracies, like public ones, lead to complacency and short planning horizons.
4. Competitive markets at home; uncompetitive markets (i.e., private or public monopolies) exhibit little innovation. In the world economy today, increasingly sophisticated buyers spur a constant upgrading in the quality of output.
5. Adequate financing and venture capital.
6. Public policies that encourage productivity growth, including subsidized research, export promotion, improved educational systems; and an up-to-date infrastructure (i.e., airports, telecommunications).

The theory of competitive advantage concludes that four attributes of a nation combine to increase or decrease its global competitive advantage and world trade: (1) factor conditions; (2) demand conditions; (3) supporting industries; and (4) firm strategy, structure, and competition.

Factor conditions include land, labor, capital, technology, and entrepreneurial skill:

1. ***Human Resources*** The quantity, skill, educational level, productivity, and cost of labor.
2. ***Physical Resources*** Raw materials and their costs, location, access, and transport costs.
3. ***Capital Resources*** All aspects of the money supply and availability to finance the industry and trade from a particular country, including the amount of investment capital available; the savings rate; the health of money markets and banking in the host country; government policies that affect interest and exchange rates and the money supply; levels of indebtedness; trade deficits; public and international debt; and so forth.
4. ***Knowledge-Based Resources*** Research, development, the scientific and technical community within the country, its achievements and levels of understanding, and the likelihood for future technological support and innovation.
5. ***Infrastructure*** All public services available to develop the conditions necessary for producing the goods and services that provide a country with a competitive advantage. Included are transportation systems, communications and information systems, housing, cultural and social institutions, education, welfare, retirement, pensions, and national policies on health care and child care.

These five factors are identified in current international and economic circles as the keys to the competitive advantage of a nation in the foreseeable future.

Demand conditions are the market conditions in a country that aid the production processes in achieving better products, cheaper products, scale economies, and higher standards in terms of quality, service, and durability. Demand conditions cause firms to become innovative and therefore to produce products that will sell not only in the domestic market but also in the world market.

To be competitive internationally, firms require access to networks of other firms specialized in different tasks in the economy. For example, large financial institutions require law firms, marketers, and advertisers. Often large companies use management consultants or similar business services, subcontracting tasks that require heavy investments in human capital. Access to these industries, which generally provide expertise, is often done through face-to-face contact.

Firm strategy, structure, and competition relates to the conditions under which firms originate, grow, and mature. For example, because stockholders demand U.S. companies to show short-term profits, U.S. corporate performance may be less successful than it would be if it were judged over a much longer time period, as is Japanese and German corporate performance.

State support of corporate strategy and performance is important. For example, a country can regulate taxes and incentives so that investment by a firm is high or low. In addition, competition within a country can impose demands on company performance; new business formations often pressure existing firms to improve products and lower prices and thus increase competitiveness.

INTERNATIONAL MONEY AND CAPITAL MARKETS

In addition to trade, **capital markets**, or long-term financial markets, form another component of the international financial system. Stock exchanges, futures exchanges, and tax havens have proliferated. American, European, and Asian multinational corporations take advantage of **tax-haven countries**, countries where taxes on foreign-source income or capital gains are low or nonexistent.

The global expansion of the financial system has three components: the internationalization of (1) currencies, (2) banking, and (3) capital markets. **International currency markets** developed with the establishment of floating exchange rates in 1973 and with the growth in private international liquidity, mostly in the form of eurocurrencies.

Capital movements take two major forms. The first type involves lending and borrowing money. Lenders and borrowers may be in either the private or the public sector. The public sector includes governments or international institutions such as the World Bank and agencies of the United Nations. The second type of capital movement involves investments in the equity of companies. If a long-term investment does not involve managerial control of a foreign company, it is called *portfolio investment*. If the investment is sufficient to obtain managerial control, it is called *direct investment*. Multinational corporations are the epitome of direct investors.

Monetary capital is the result of historical development. Unlike a natural resource like iron ore, it must be accumulated with time as a result of the willingness of a society to defer consumption. Low-income countries have low capacities to generate investment capital; all the capital that they do generate is usually employed domestically. Developed countries have much greater capacities for generating investment capital. They provide most of the world's private-sector capital, although a few fast-growing countries such as the NICs are also capital exporters.

Optimally, financial markets should produce an efficient distribution of money and capital throughout the world. However, there are many barriers to optimal distribution. Personal preferences of investors, practices of investment banking houses, and governmental intervention and controls confine money and capital movements to well-worn paths. They flow to some areas and not to others, even though the need in neglected areas may be greater.

International Banking

Paralleling the internationalization of domestic currency is the internationalization of banking. International banks have existed for centuries; for example, banking houses founded by the Medici and the Rothschilds helped to finance companies, governments, voyages of discovery, and colonial operations. The banks of the great colonial powers—Britain and France—have long been established overseas. American, Japanese, and other European banks went international much later. Major American banks moved into international banking in the 1960s, and the Japanese banks and their European counterparts in the 1970s. Banks were enticed into international banking because of the explosion in foreign investment by industrial corporations in the 1950s and 1960s. The banks of different countries "followed the flag" of their domestic customers abroad. Once established overseas, many found international banking highly profitable. From their original focus on serving their domestic customers' international activities, banks evolved to service foreign customers as well, including foreign governments.

Euromarkets

Eurocurrencies are bank deposits that are not subject to domestic banking legislation. With relatively few exceptions, they are held in outside countries, "offshore" from the country in which they serve as legal tender. They have accommodated a large part of the growth of world trade since the late 1960s. The eurocurrency market is attractive because it provides funds to borrowers with few conditions; it also offers investors higher interest rates than can be found in comparable domestic markets.

At first, euromarkets involved U.S. dollars deposited in Europe; hence, they were called *eurodollar* markets. Although the dollar still represents about 80% of all eurocurrencies, other currencies, such as the deutsche mark and yen, are also vehicles of international transactions. Therefore, *eurocurrencies* is preferred to the less accurate term *eurodollar*. However, even *eurocurrencies* is a misnomer. Only 50% of the market is in Europe, the major center of which is London. Other eurocenters have developed in the Bahamas, Panama, Singapore, and Bahrain.

EXCHANGE RATES AND INTERNATIONAL TRADE

In international trade, the buyer country must swap its currency for the currency of the exporting country. If a retail chain in the United States wants to buy televisions, video camcorders, or DVD players from Japanese firms, the buyer must convert U.S. dollars to Japanese yen in order to satisfy the terms of the purchase. As is the case with most international transactions—exports and imports—the seller receives payment in the currency of its own country, not in the currency of the purchasing country. The value of a currency compared with that of another country is called the **exchange rate**, that is, the amount of one currency required to purchase one unit of foreign money. Exchange rates are never stable but fluctuate constantly over time, with important impacts on the prices of imports and exports. For example, the U.S. dollar strengthened abroad during the 1990s as a result of economic stagnation abroad and currency devaluation by European governments. However, the dollar declined during the 2000s as the U.S. government hoped it would improve its trade deficits.

To illustrate this notion in more detail, Figure 5 shows the changing relationship between the Mexican peso and the U.S. dollar as the peso was devalued. The demand curve, *D*, shows dollars sloping downward to the right. U.S. citizens will demand more Mexican pesos if they can be purchased with fewer dollars. Point P_0 suggests that fewer pesos can be purchased with a dollar, whereas point P_1 suggests that many more pesos are available per dollar. The demand for Mexican pesos in the United States is based on the amount of goods and services that a U.S. citizen wants to purchase in Mexico. A lower exchange rate for the peso makes Mexican goods less expensive to Americans.

In Figure 5, the supply of pesos is upward-sloping to the right, which means that as the number of dollars increases per 10,000 pesos, more pesos are offered in the marketplace. Mexican residents desire more goods from the United States when the dollar exchange rate (price) for the peso is high. The more dollars per 10,000 pesos, the relatively cheaper American products are for Mexicans. Therefore, Mexican residents will demand more dollars with which to purchase American goods and will consequently supply more pesos to foreign exchange markets when the exchange rate for the peso increases. The equilibrium position is reached when supply and demand conditions for foreign exchange is based on supply and demand of international goods produced in Mexico demanded by Americans and American goods demanded by Mexicans.

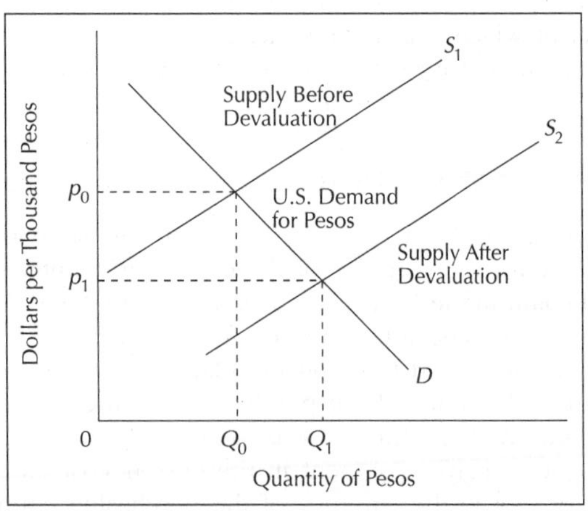

FIGURE 5 Determining the exchange rate of the U.S. dollar and the Mexican peso.

Line S_2 represents devaluation of the peso because of economic restructuring in Mexico. This restructuring effectively reduced the number of dollars per 10,000 pesos on the international market. The result was that fewer American products could be purchased for the same amount of pesos because American goods and services became relatively more expensive. Cross-border purchases by Mexican border residents decreased dramatically, as did the international flow of goods and services from America to Mexico. At the same time, the quantity of pesos available to Americans increased from Q_0 to Q_1. American purchasers poured into border communities and increased their travel to the main tourist destinations within Mexico. The international flow of goods from Mexico to the United States increased because Mexican goods and services were relatively less expensive.

When the dollar appreciates in foreign exchange markets relative to other currencies, it can buy more foreign currency and, therefore, more goods and services from other countries (Figure 5). As a result, American retailers import more goods to the United States. At the same time, the appreciated U.S. dollar means more costly American goods and therefore less demand for them. Exports from America decline under these circumstances. Thus, a strong dollar, which means that the dollar can buy more units of foreign currency, is not always desirable for U.S. trade. Conversely, when the dollar depreciates and is "weak," it can buy fewer units of foreign currency and therefore fewer goods and services. Imports usually decline under these circumstances.

Why Exchange Rates Fluctuate

Exchange rates fluctuate for five reasons. First, as countries become wealthier, they typically import more goods from abroad. The result is that the increased demand for foreign currency raises the exchange rate of their currencies and decreases the value of the home country's internationally.

A second factor is the inflation rate of a nation. If the inflation rate of one nation increases faster than that of its trading partners, the currency of the nation with high inflation will depreciate compared with the currency of its trading-partner nations. Consequently, the products of the trading-partner nations will be more attractive to consumers in the country with high inflation. For example, when the U.S. inflation rate increases, the demand for Canadian dollars increases as investors seek a currency that will not lose its value so quickly. This demand raises the U.S. dollar price of both the Canadian dollar and Canada's exports to the United States. At the same time, Canadians demand fewer of the comparatively higher-priced U.S. goods.

Third, domestic demand is a factor in determining exchange rates. Real income growth and the relative price levels between countries affect domestic demand. However, domestic demand also depends on consumer tastes and preferences. Americans will pay higher prices for specialty items and technologically advanced foreign

TABLE 3 Trade in the U.S. Economy, 1960–2008

	1960		1975		2000		2008	
	$ Billions	*% of GDP*	*$ Billions*	*% of GDP*	*$ Billions*	*% of of GDP*	*$ Billions*	*% of of GDP*
Exports	23.3	4.9	136.3	8.6	937	11.6	1826	12.0
Imports	22.3	4.4	122.7	7.7	1,048	12.9	2522	17.0
Net exports	4.7	.9	13.6	0.8	−111	1.3	−696	4.7

goods such as electronic consumer products from Asia, French wines and perfumes, German automobiles, and Italian shoes than for comparable domestic products. An increase in the demand for foreign goods decreases domestic demand and causes the dollar to depreciate.

Fourth, a country's currency may appreciate on foreign exchange markets if its domestic interest rates rise and provide a higher yield to foreign investors. Foreign investors increase their demand for dollars in order to purchase companies overseas and thus supply more of their currency in exchange for the target country's currency. When domestic interest rates are low compared to rates of return in other nations, they do not attract foreign investors. Conversely, as interest rates rise, a country's currency becomes more attractive to investors elsewhere.

Fifth, currency speculation affects exchange rates worldwide. Speculation is a prime cause of the "bubbles" that have affected finance-dominated capitalism for the past several decades. Speculation is often driven by expectations of future events, such as hopes that prices in a market will rise or fears of political disasters, which encourage individuals to sell the currency of that country to buy foreign currencies. If all people reacted in the same manner to such events, the market would be driven down for that currency against the dollar, and the anticipated depreciation would actually occur.

U.S. TRADE DEFICITS

Because the United States is by far the world's largest national economy, its trade situation deserves a closer look. The United States enjoyed a trade surplus throughout most of its history. Starting in the 1970s, however, the volume of its imports began to exceed the volume of exports (Table 3). The merchandise trade deficit was $25 billion in 1980, but by 2009, it had jumped to $850 billion. There are several causes of this **trade deficit**: an overvalued dollar, which makes imports cheap and exports expensive, and American consumers' voracious demand for imported goods, often fueled by consumer debt.

As Figure 6 shows, the international value of the U.S. dollar has changed widely over time. Increases in the value of the dollar mean that U.S. currency and products are relatively expensive to foreign nations, whereas foreign goods are less expensive to Americans. The U.S. dollar peaked in value internationally in 1985, which increased the amount of foreign imports and decreased the amount of its exports. Since 1985, the value of the dollar has fallen steadily. While the U.S. demand for imported goods remains strong, its exports have not kept pace, leading to severe trade deficits of $800 billion annually (Figure 7). In short, Americans import significantly more products from other countries than they export to them. Many households and

FIGURE 6 International value of the dollar, 1973–2009. The value of the dollar against other currencies exerts a significant influence over the prices of U.S. exports and imports. The dollar's value is determined by many things, including rates of inflation and interest between trading partners, speculative flows in currency markets, and government policy. The cheaper the dollar, the more foreign countries can afford to buy U.S. imports, thus improving job growth in America.

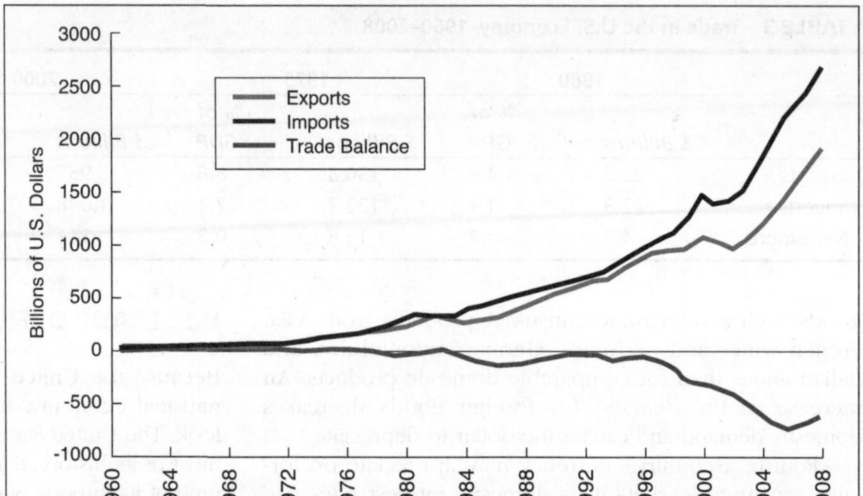

FIGURE 7 The U.S. merchandise trade deficit, 1960–2008. Because imports have exceeded exports, the United States regularly runs a deficit of more than $800 billion annually. Several factors account for this trade deficit. The first is that the United States is a rich country and can afford to import more than it exports. But the relatively high cost of labor in America makes American products more expensive outside its borders. Another factor is that some countries, such as China, undervalue their currency, leading to export-led economic growth. China's underpricing of exports and overpricing of imports hurts trading countries, from Brazil to India.

corporations finance these purchases through debt, as do many corporations and the federal government.

Results of the U.S. Trade Deficit

The U.S. deficit has had several results: First, an increase in the volume of imports pushes prices of domestically produced goods down as they compete with cheap imports. This is good for consumers but not good for producers, who see their profit margins fall. For example, U.S. automobile manufacturers, steel manufacturers, and textile producers have been seriously affected by the trade deficit. The United States is a large net debtor instead of a creditor nation, owing foreign governments more than they owe it. Foreign debt includes federal government debt held by foreigners (largely Asians and West Europeans), primarily in the form of Treasury Department securities, as well as private debt issued by corporations. The total U.S. foreign debt was roughly $14 trillion in 2009 (about 90% of gross domestic product [GDP]), making the United States by far the largest debtor nation in the world. In other words, American consumers have been subsidized because more goods and services flowed into the country than flowed out of the country. The reverse is true for its lenders, including particularly Japan and China. The United States has been living above its means, and its consumers have received an economic boost, but this situation is not likely to be tenable in the long run. Large federal budget deficits and huge balance of trade deficits have led to the so-called selling of America to foreign investors. Foreign investors now own 26% of America's total domestic assets.

CAPITAL FLOWS AND FOREIGN DIRECT INVESTMENT

Closely related to trade flows is the flow of capital, including **foreign direct investment (FDI)**. Given the massive scale of FDI today, understanding the rationale for such investments is important.

Three strategic profit motives drive a firm's decision to operate abroad. One motive for many direct investments is to obtain natural or human resources. Resource seekers look for raw materials or low-cost labor that is also sufficiently productive. A second motive is to penetrate markets. The third goal is to increase operating efficiency. These three motives are not mutually exclusive. Some segments of a transnational corporation's operations may be aimed at obtaining raw materials, whereas other segments may be aimed at penetrating markets for the products made from the raw materials. Automobile producers such as Ford, for example, rely on a global network of parts suppliers (Figure 8). These operations may also result in some production and market efficiencies.

There may be strong motivations for a firm to internationalize, but there are also compelling constraints. Prominent among these are the uncertainties of investing or operating in a foreign environment. Consumers' incomes, tastes, and preferences vary from country to country. Japanese consumers, for example, are wary of foreign products, at least those that are not name brands. Cultural differences in business ethics and protocol complicate the task of conducting business in two or more languages. Added to these barriers are differences in laws, taxation, and governmental restrictions.

World Investment by Transnational Corporations

Transnational corporations (TNCs) are the leading sources of foreign investment in the world and one of the most prominent dimensions of globalization. TNCs have a long history stretching to the dawn of capitalism in the sixteenth century, when their roots were laid down in the chartered monopolies like the British East India Company that played a crucial role in European colonialism. In the twentieth century, the numbers of TNCs grew exponentially (Figure 9), as did their economic and political

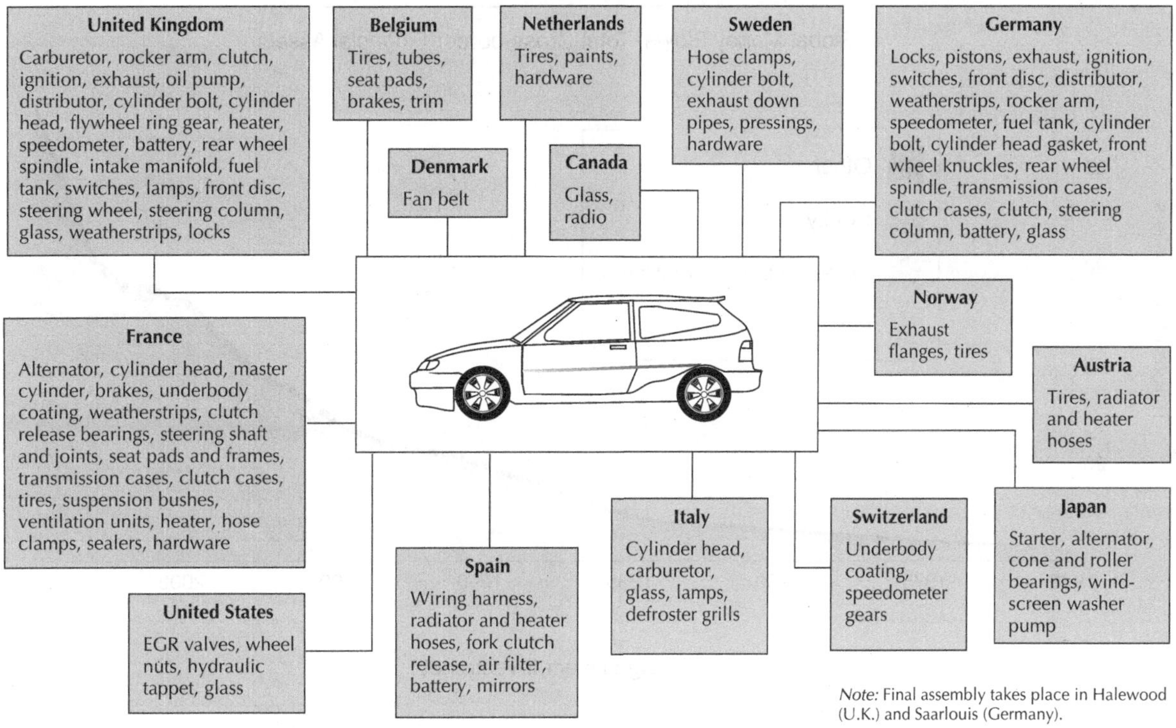

FIGURE 8 The international car: the component network for the Ford Focus (Europe).

significance. Today, they generate trillions of dollars in flows of cross-border financial assets and FDI (Figure 10). So rapid has their growth been in the late twentieth and early twenty-first century that the total values of their assets, sales, and exports is an order of magnitude larger than it was in the early 1980s (Figure 11).

Most TNCs and FDI originate in economically developed countries, which have the surplus capital to invest abroad. Despite common impressions that TNCs always invest in developing countries where labor costs are lower, the reality is that most FDI is concentrated in the developed world (Figure 12). American firms lead the world in FDI, but their share of the total is slipping. The rate of increase has been most rapid for companies from Western Europe; however, some developing countries have also increased their outflow of FDI, such as Brazil, Singapore, South Korea, and Taiwan. Investment in the developing world has focused mainly on a handful of countries—particularly China (Figure 13). Availability of natural resources, recent economic growth, and political and economic stability were among the factors that attracted foreign investment to less developed countries.

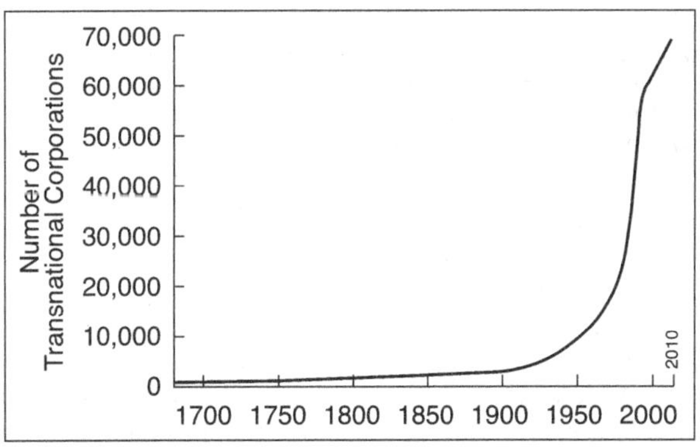

FIGURE 9 The history of transnational corporations extends to the dawn of the age of colonial expansion, but their numbers multiplied rapidly throughout the twentieth century.

Investment by Foreign Multinationals in the United States

Foreign firms invest in the United States, and in other developed economies, to gain access to its domestic markets and to avoid protectionist measures (Figure 14). For example, cars produced by Japanese transplants in the United States are exempt from quotas. FDI in the United States grew rapidly from 1970, when it was a skimpy $13 billion, to 2008, when it amounted to $1.6 trillion. In fact, it increased more rapidly than did U.S. foreign investment overseas. The popularity of investing in the United States was a result of the economic and political stability of the country and the relatively inexpensive dollar in world markets.

Trade deficits and FDI are closely linked. The U.S. trade deficit led to the outflow of American dollars into

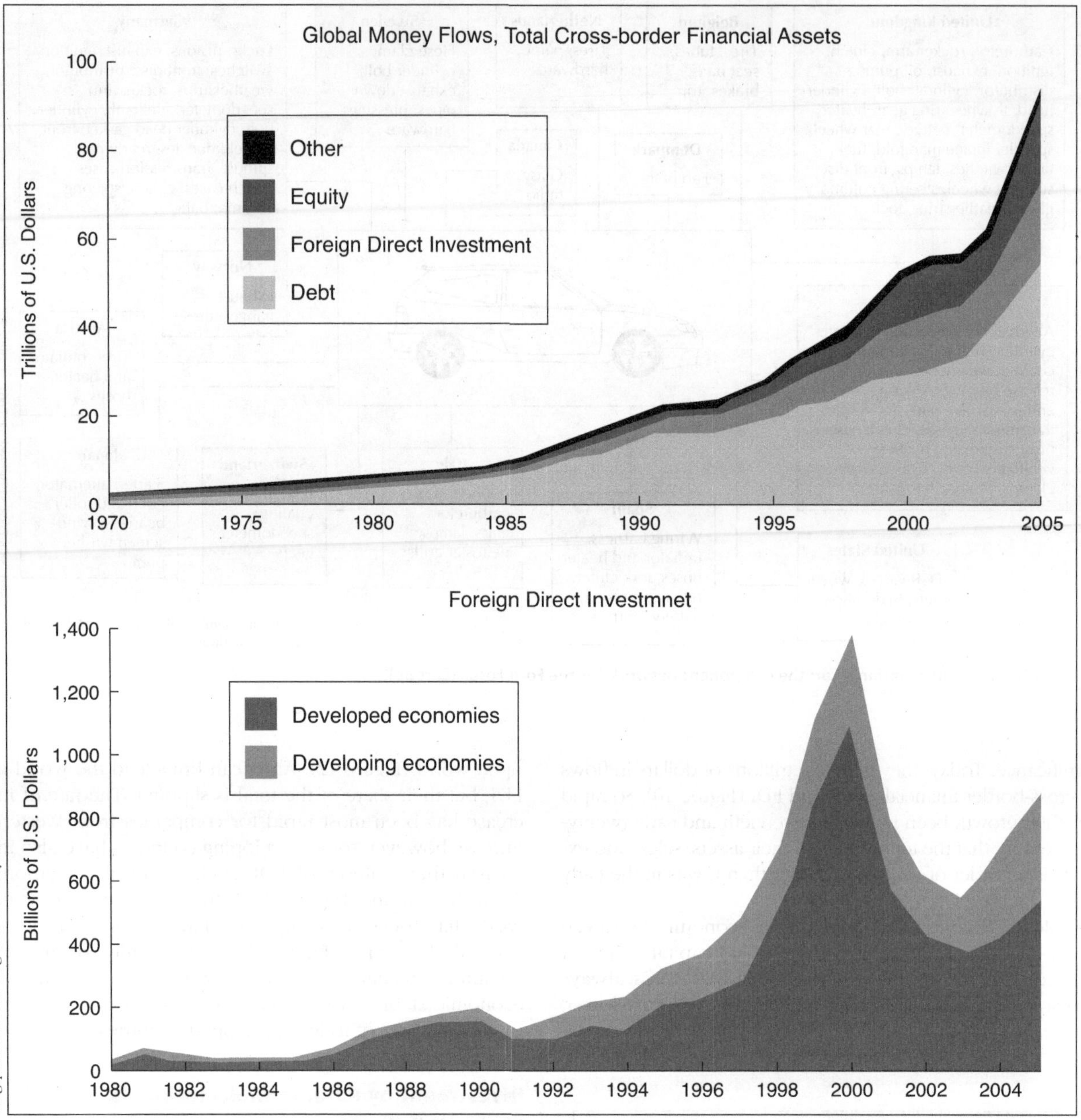

From *Our World in the 21st Century: The Global Economy*, http://afrikanisation.blogspot.com/2007/10/global-economy.html

FIGURE 10 Global money flows and foreign direct investment by transnational corporations have grown exponentially. To remain competitive in the global economy, transnational corporations carefully review their production costs to identify fabrication stages that can be performed by low-paid, low-skilled workers in developing countries. Hence, the substitute transportation cost for lower wage costs. Operations that require more skilled workers or better quality control remain in the host country.

foreign hands. This money allowed foreign governments and corporations to buy American real estate and factories. Ironically, many foreign firms that sell to U.S. markets found it cheaper to produce goods from plants that they own and operate *in* the United States. As a result, foreign investment in the United States increased sharply. Because of cultural affinities and the lack of language barriers, the United Kingdom is still the lead FDI investor in America (but not if one includes purchases of federal government debt, in which case China and Japan are the leaders). FDI in the United States is mainly in manufacturing, chemicals, electrical machinery, electronics, pharmaceuticals, and services.

The geography of FDI in the United States varies by investor country. Originally, European countries invested in the Middle Atlantic and Great Lakes states of the north central United States. Canadian investment has been the strongest, not unexpectedly, along the border states from the western north central region through New England, including the South Atlantic states. But the Pacific region

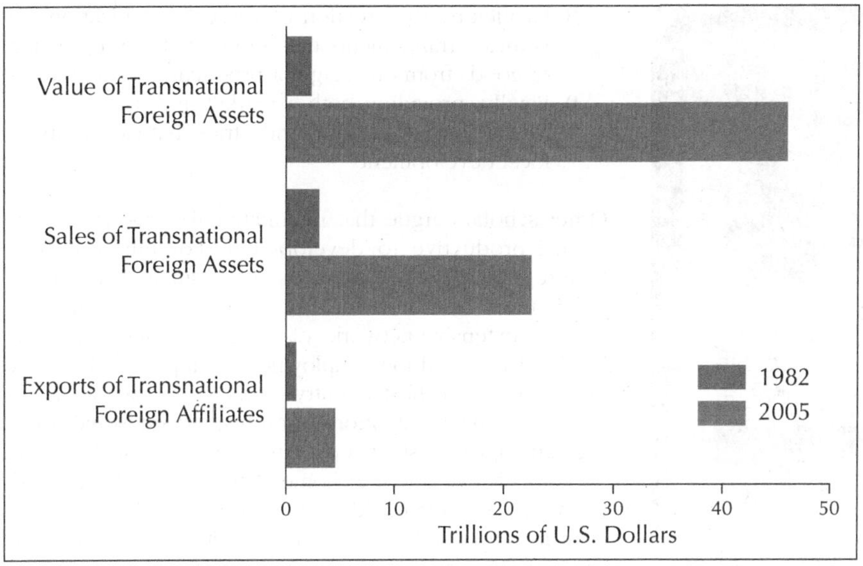

FIGURE 11 Compared to the early 1980s, the foreign assets, sales, and exports of the world's transnational corporations have jumped significantly. Transnational corporations select low-wage countries for work through the process known as outsourcing. Here, the host company either runs a subsidiary in the developing country or turns over much of the production to independent manufacturers that operate in the low-cost country. Outsourcing contrasts with the more traditional approach of manufacturing in which a company controls all phases of the manufacturing process, usually in the original country. This latter approach, used by Henry Ford and many others later, is called *vertical integration*.

of the United States leads overall in FDI, primarily because of recent heavy investment by the Japanese and other East Asian countries (Figure 15).

Effects of Foreign Direct Investment

Is widespread FDI desirable? Should the operations of multinationals be controlled? There is no unanimity of opinion, particularly when investment in less developed countries is the issue. There are two opposing attitudes regarding the presence of multinationals in less developed countries. Some argue that multinational firms offer great potential for aiding the economic development process (Figures 16 and 17). In this view, the TNC is an efficient social, economic, and political institution that accomplishes the following tasks for the less developed nations:

1. Raising, investing, and reallocating capital
2. Creating and managing organizations
3. Innovating, adopting, perfecting, and transferring technology
4. Distributing product, performing maintenance, marketing, and selling

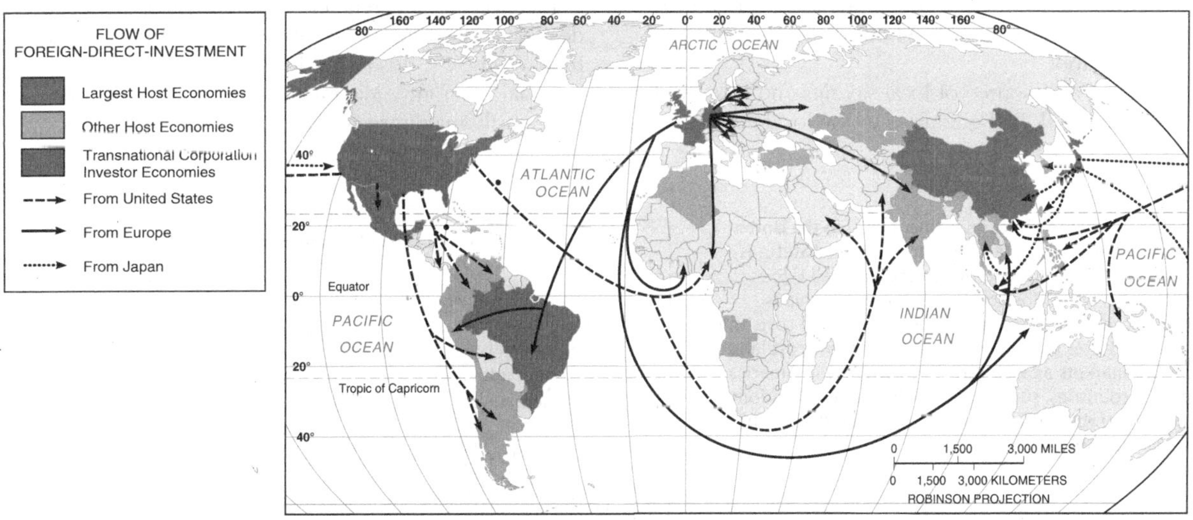

FIGURE 12 Global flows of foreign direct investment. Multinationals' foreign direct investment originates, for the most part, in the United States, Japan, the United Kingdom, Germany, and France. These transnational corporations have invested most of their resources in other developed countries. In addition, U.S. multinational corporations are more likely than the Japanese or Europeans to invest in Latin America. European multinational corporations are more likely to invest in Eastern Europe and the Middle East, while Japanese transnationals are more likely to invest resources in East Asia.

© Alex Segre/Alamy

FIGURE 13 Foreign direct investment by U.S. firms in the Pacific includes the ubiquitous McDonald's.

5. Furnishing local elites with suitable career choices
6. Educating and upgrading both blue-collar and white-collar labor
7. Serving as a source of local savings and taxes and in supplying skilled graduates to the local economy
8. Facilitating the creation of vertical organizations or vertical arrangements that allow for the progression of goods from one stage of production to another
9. Finally, providing both a market and a mechanism for satellite services and industries that can stimulate local development

Other scholars argue that multinational corporations are counterproductive to development. Foreign firms can bankrupt local producers, who may be small or undercapitalized, and establish local monopolies. Transnationals that use extensive networks of foreign suppliers have few local linkages and low employment multipliers. They may not improve the host country's balance of payments because of heavy repatriations of profits; indeed, some countries attempt to restrict repatriation of profits to keep the benefits of investment at home. Although the balance-of-payments problem could be avoided in part if multinational firms reinvested more of their profits in the host country, it is uncertain that the national interest would be served. Reinvestment increases foreign control of the economy and the denationalization of local industry. Some argue that the multinational firm is an assault on political sovereignty, with its demands for government subsidies, training programs, and tax breaks. This latter issue is particularly important given the long history of TNC involvement in the politics of developing countries, ranging from bribes to helping military officers overthrow democratically elected governments. Moreover, the transnational system internationalizes the tendency to unequal development and to unequal income.

To be sure, multinationals are imperfect organs of development in developing countries, and their potential for the exploitation of poor countries is tremendous. There is, therefore, an inherent tension between the multinational's desire to integrate its activities on a global basis and the host country's desire to integrate an affiliate with its national economy. Maximizing corporate profits does not necessarily maximize national economic objectives.

FIGURE 14 Foreign direct investment (FDI) in the United States by source area, 2007. FDI in the United States is defined by the U.S. Bureau of the Census as all U.S. companies in which foreign interest or ownership is 10% or more. Countries with higher production costs than the United States look at the attraction of low-cost labor and huge markets as an incentive to operate in America. High-cost countries, such as Japan, the United Kingdom, the Netherlands, and Germany, account for 70% of the FDI in the United States.

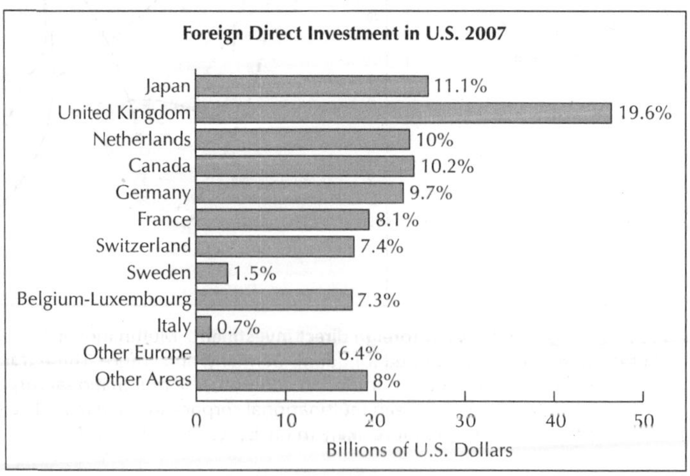

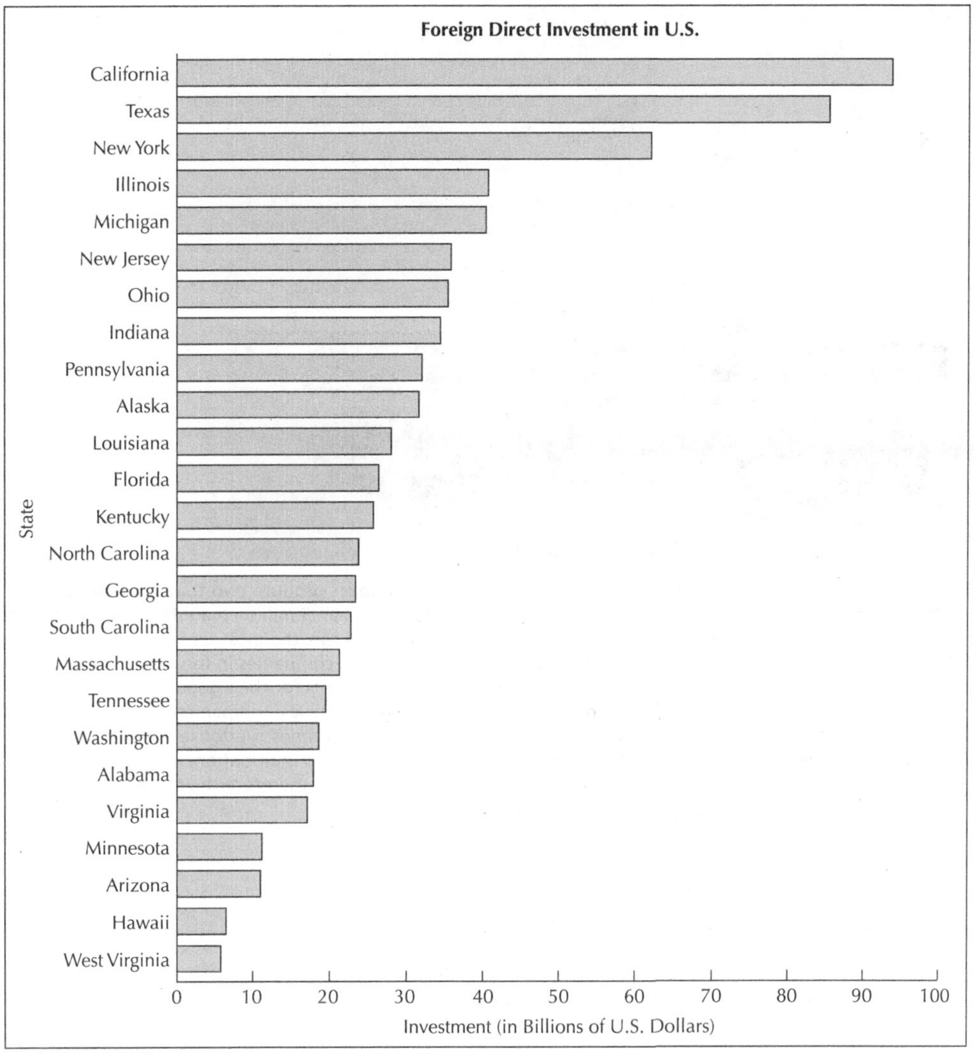

FIGURE 15 Foreign direct investment (FDI) in the United States, 2005. The value of FDI during this period has quadrupled. The top four states accounted for more than a third of the total FDI, and California, Texas, and New York led all states. A large influx of foreign direct investment has brought prosperity to the Rust Belt states of Michigan, Illinois, Indiana, and Ohio, with extensions into Kentucky and Tennessee.

Ford Motor Company

FIGURE 16 Foreign direct investment. Ford Motor Company headquarters in Britain. Ford Is a true multinational corporation, with 330,000 employees scattered in 30 countries throughout the world. Although accused of exploiting local labor in developing countries, Ford and other multinationals have helped to stimulate economies and, therefore, provide foreign sources of revenue.

Ford Motor Company

FIGURE 17 U.S. Ford plant in Europe. General Motors and the Japanese manufacturers of automobiles produce two-thirds to nine-tenths of their products in their home countries. The Ford Motor Company produces two-thirds of its more than 5 million manufactured vehicles abroad. Ford has been one of the most prolific of the American multinational corporations. The 1960s saw the most rapid expansion of American multinational investment, with over 300 new subsidiaries being set up annually by American companies in foreign countries. The advantage American firms saw in multinational expansion was the rapid development of postwar markets for their goods in Europe, Japan, and throughout the world. By establishing manufacturing plants in those foreign countries, companies could reduce expensive transportation costs for finished products as well as import duties. Another major factor, sometimes overlooked, is transfer pricing. Japan and European countries have higher rates of corporate taxation than does America. Shrewd multinationals can invoice their subsidiaries in these foreign countries in such a way as to show low profits in high-tax countries and therefore shelter their income.

The impacts of TNCs depend on both the characteristics and bargaining power of the corporation and the host country. Both the larger and wealthier of the less developed countries and of the TNCs have more bargaining leverage. A consumer-products manufacturing corporation will accept more controls to gain access to a country with a large market. Manufacturing industries with advanced and dynamic technologies are more difficult to control than firms involved in raw materials. Similarly, the degree of host-country control varies across industries and states. Host governments' policies can range from open hostility to open encouragement. Corporations prefer to invest in countries that follow an outward-oriented, export-led development strategy, impose few controls, offer incentives, and appreciate the employment, skills, exports, and import substitutes that foreign investment can bring, an outlook exemplified by the NICs.

BARRIERS TO INTERNATIONAL TRADE AND INVESTMENT

Like trade, international flows of the production factors can help to reduce imbalances in the distribution of natural resources. Whereas trade *offsets* differences in factor endowments, factor movements *reduce* these differences. International trade and flows of factors would occur more commonly if barriers did not exist. The main barriers relate to management, distance, and government.

Management Barriers

Trade and investment expansion can be reduced by a number of managerial characteristics. These include lack of awareness of opportunities, lack of skills, fear, and inertia.

Firms may have the potential to expand but fail to do so because they are satisficers—they settle for less than the optimal outcome. Until the economic crisis of the 1970s, many U.S. firms paid little attention to foreign markets; they were satisfied with the large domestic market. Firms may also have the will to go international but may lack knowledge of potential markets. The burden of recognizing export opportunities rarely falls solely on the managers of individual companies, however. Most national and local governments are actively involved in increasing international awareness and in promoting exports.

Firms may have the potential and will to go international *and* an awareness of the opportunities, but they may be thwarted by the complexity of international business and ignorance of foreign cultures. Governments and universities can aid companies by providing education. Knowledge of intermodal rate structures, freight forwarders, shipping conferences, and customs brokers (firms that contract to bring other companies' imported goods through local customs) is vital for the conduct of international business. Just as necessary is knowledge of foreign cultures. In the past, for example, U.S. firms could not penetrate the Japanese market, partly because of a failure to appreciate Japanese culture.

Government Barriers to Trade

No country permits a completely free flow of trade across its borders. Governments have erected barriers to achieve objectives regarding trade relationships and indigenous economic development. Trade barriers include **tariffs**—schedules of taxes or duties levied on products as they cross national borders—and **nontariff barriers**—quotas, subsidies, licenses, and other restrictions on imports and exports. These kinds of obstacles (apart from political bloc prohibitions) are the most pervasive barriers to trade.

Free-market enthusiasts advocate free trade because it promotes the increased economic efficiency and productivity that results from international specialization. They argue that trade, a substitute for factor flows, benefits each participating nation and that deviation from free trade will inhibit production. It follows, then, that **protectionism** adversely affects the welfare of the majority. Protectionism drives up the costs of imports, and products that compete with them, thus adding to inflation, and invites retaliation on the part of trading partners.

What are some of the major arguments in favor of protectionism? One of the most common is that it saves jobs by keeping cheap foreign imports out. This argument is often put forward by industries having difficulty competing internationally and contradicts the principle of comparative advantage: By purchasing cheap imports, consumers have more money to spend on other things. Conversely, protectionism takes the money away from consumers that generates jobs in consumer-oriented sectors and transfers it to inefficient industries that can't compete internationally. Each job "saved" in a declining, protected sector comes at enormous expense to the public in the form of more expensive goods. Thus, protectionism does not generate jobs, it *redistributes* them, imposing costs on large numbers of people for the benefit of a few. Typically, beleaguered producers are well organized politically while consumer groups are not.

A second argument is that protectionism is necessary to reduce dependence on foreign supplies of critical goods necessary for national defense. This line of thought led, for example, to the formation of the Strategic Petroleum Reserve, large salt domes in Louisiana and Texas where the federal government stores significant amounts of oil in case of a national emergency. However, deeming what goods are, and are not, essential for national defense is often arbitrary: cardboard boxes? Styrofoam? Moreover, it is more efficient to store such goods than to use tariffs to keep foreign suppliers at bay.

Still another argument is that tariffs can be used to protect an **infant industry** that is less efficient than a well-established industry in another country. The infant-industry argument was invoked to justify protectionist policies in nineteenth- and twentieth-century America and nineteenth-century Germany. It was also used to justify the protectionism of the less developed countries in the 1960s. Although these arguments have some merit, free marketers recommend other approaches to attain desired goals. For example, they suggest that if grounds exist for protecting an infant industry until it has grown large enough to take advantage of scale economies, this could be accomplished through providing a subsidy rather than through a protective tariff. In each case, protectionism provides a politically appealing argument that caters to troubled industries and regions, but it is a poor economic argument.

Despite long-standing pleas for free trade and low tariffs, throughout the history of the United States tariff rates have been unquestionably high (Figure 18). In the nineteenth century, tariffs were the federal government's primary source of income, until the adoption of the federal income tax in 1913. The Compromise Tariff of 1833 and the Smoot-Hawley Tariff of 1930 made 70% of imports subject to tariffs. The reasons for tariffs are clear. High-powered, politically savvy lobbyists hired by special-interest groups that stand to gain economically from tariffs and quotas press the government for protection. The public, which must then absorb these tariffs and quotas as a surcharge on all imported products, is politically uninformed and not well represented in Washington, DC. In 1947, however, the United States and 25 other nations signed the

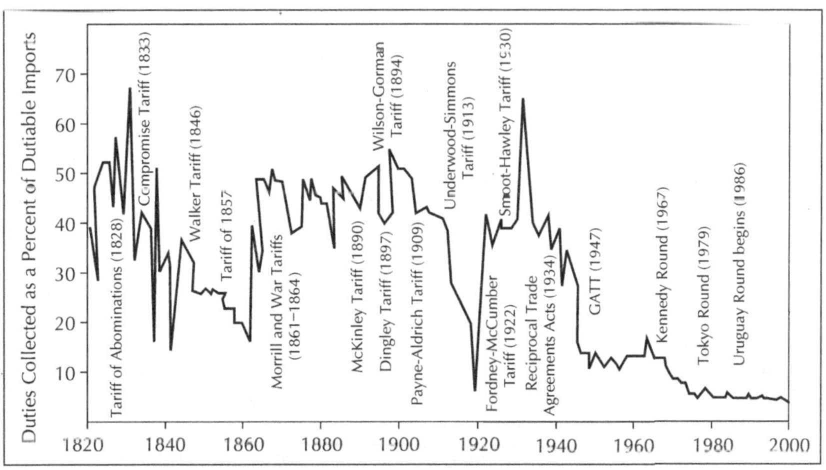

FIGURE 18 U.S. average tariff rates, 1820–2000. Historically, prior to the inception of the General Agreement on Tariffs and Trades (GATT) in 1947, U.S. tariff rates had been high. They reached a peak in the mid-1930s, with the Smoot-Hawley tariff, designed to protect U.S. markets from foreign goods, precipitating a trade war with Europe. Today, the United States has the lowest barriers to trade in the world, although on occasion American administrations will impose tariffs for political gain.

General Agreement on Tariffs and Trade (GATT). GATT established multinational reductions of tariffs and import quotas and now has more than 110 signatories. The Uruguay Round of tariff negotiations (1986–1994) once again reduced tariffs, so U.S. tariff rates currently average approximately 5% on dutiable imports.

Tariffs, Quotas, and Nontariff Barriers

Tariffs are the most visible of all trade barriers, and they can be levied on a product when it is exported, imported, or in transit through a country. The tariff structure established by the developed countries in the post–World War II period works to the detriment of underdeveloped countries. The underdeveloped countries encounter low tariffs on traditional primary commodities, higher tariffs on semimanufactured products, and still higher tariffs on manufactures. These higher rates are, of course, intended to encourage firms in industrial countries to import raw materials and process them at home. They also discourage the development of processing industries in the developing world.

In recent years, the relative importance of tariff barriers has decreased, whereas nontariff barriers have gained significance. The simplest form of nontariff barrier is the **quota**—a quantitative limit on the volume of trade permitted. A prominent example of a product group subject to import quotas in developed countries is textiles and clothing. Since the early 1970s, these have been subject to quotas under successive Multifibre Arrangements (MFAs), which have created a worldwide system of managed trade in textiles and clothing in which the quotas severely curtail underdeveloped-country exports. Another common nontariff barrier is the **export-restraint agreement**. Governments increasingly coerce other governments to accept "voluntary" export-restraint agreements by which the government of an exporting country is induced to limit the volume or value of exports to the importing country.

Other nontariff barriers include discretionary licensing standards; labeling and certificate-of-origin regulations; overly detailed health and safety regulations, especially on foodstuffs, aimed at discouraging imports; and packaging requirements. Increasingly, loose, or break-bulk, cargo is unacceptable to mechanized transportation handlers in developed countries. These are only a few of the hundreds of nontariff barriers devised by governments. The evidence indicates that these barriers in developed countries are higher for exports from developing countries than they are for exports from rich, developed countries.

Effects of Tariffs and Quotas

The economic effect of tariffs and quotas in the host country is the development and expansion of inefficient industries that do not have comparative advantages. At the world level, tariffs and quotas penalize industries that are relatively efficient and that do have comparative advantages. The result is less international trade and higher costs for consumers.

Figure 19 shows the economic effects of a protective tariff and an import quota. Let's first deal with the case of a protective tariff. Line D_d represents domestic demand in, for example, the United States for cassette players, whereas line S_d is the domestic supply. (Disregard the S_d+ Quota line for now.) The domestic equilibrium position is at price P_3 and quantity Q_3. Now assume that the U.S. market for CD players is open to world trade. The world price is lower than the domestic price because, compared with Japan, Malaysia, or Taiwan, the United States has the comparative disadvantage of high labor costs. The world equilibrium price is P_1. At this price, Americans will consume the quantity Q_5 at point *N*. With the low price, the domestic supply is only Q_1 at point *P*, with the quantity $Q_1 - Q_5$ supplied by foreign imports. Next, let's say that the United States imposes a tariff on the import of CD players. This tariff raises the price from P_1 to P_2. The equilibrium price and quantity are now P_2, Q_4, respectively, at point *Q*.

The first reaction will be a decline in quantity demanded by American consumers, from Q_5 to Q_4, as they back up their demand curve toward the higher price. American consumers are hurt by the tariff because they can buy fewer goods at a higher price. While the consumers move back up the demand curve to point *Q*, domestic producers, with a higher price opportunity, increase their production and move up their supply curve from point *P* to point *R*. Domestic production has increased from Q_1 to Q_2. Consequently, we can understand why domestic producers send lobbyists to Washington, DC, to secure tariffs that give the producers a relative advantage in the market.

The increased tariff reduces the number of East Asian imports from $Q_1 - Q_5$ to $Q_2 - Q_4$. The U.S. government, not the East Asian supplier, receives the tariff monies, $P_1 - P_2$. At the same time, the market shrinks because of reduced demand and domestic supply increases. The shaded area represents the amount of tariff revenue paid to the U.S. government. This revenue is an economic transfer from the consumers of the country to the government.

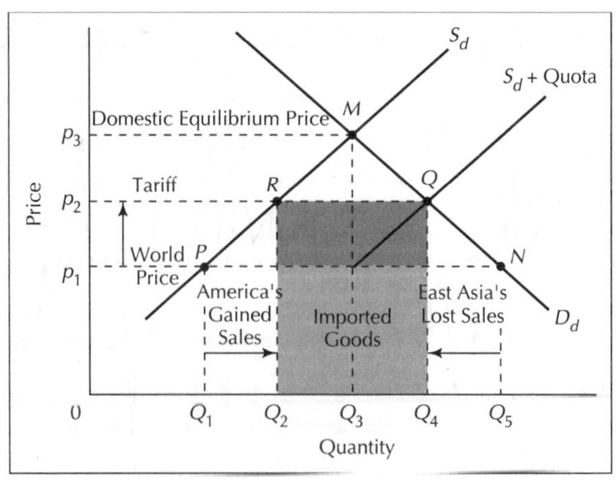

FIGURE 19 The economic impacts of tariffs and quotas.

The result of levying a tariff is reduced world trade and reduced efficiency in the international economic system, which hurts foreign suppliers, aids domestic producers, and costs the consumer. The indirect effect is that the supplying countries have a smaller market in America and thus earn fewer dollars with which to exchange or invest in American resources. As FDI decreases, trade deficits may increase.

Next, let us consider the effects of levying an import quota. The difference between a tariff and an import quota is that a tariff yields extra revenue to the host government, whereas a quota produces revenue for foreign suppliers. Imagine that the United States subjects a foreign nation to an import quota, rather than imposing a tariff (see Figure 19). The import quota in this case is $Q_2 - Q_4$. The quantity $Q_2 - Q_4$ is the number of CD players that foreign producers are allowed to supply. Note that for easy comparison this example limits the quota to the exact amount of imported goods in our tariff example. The quota establishes a new supply curve, S_d + Quota, with an equilibrium position at point Q. The new supply of CD players is the result of domestic supply, plus a constant amount, $Q_2 - Q_4$, which is supplied by importers.

The chief economic outcomes are the same as with the tariff example. The price is P_2 rather than P_1, and domestic consumption is reduced from Q_5 to Q_4. The American manufacturers enjoy a higher price for their goods, P_2 rather than P_1, and increased sales, Q_2 rather than Q_1. But the main difference is that the shaded box, paid by the domestic consumer on imports of $Q_2 - Q_4$, is not paid to the U.S. government. Instead, the extra revenue in the shaded box is paid to the foreign supplier. That is, no tariff exists, and consequently, the foreign supplier keeps all the revenue in the box $Q_2 - Q_4 - Q - R$.

The result is that for local consumers, and their government, a tariff produces a better revenue situation than a quota does. Tariff money can be used to lower the overall tax rate and provide social services and infrastructure for the population as a whole. However, either case is detrimental to international trade and economic efficiency and takes away from the comparative advantages of supplying nations.

Government Stimulants to Trade

Not only do governments attempt to control trade, they also attempt to stimulate trade. Examples of governmental assistance in promoting exports include market research, provision of information about export opportunities to exporters, international trade shows, trade-promotion offices in foreign countries, and free-trade zones, areas where imported goods can be processed for reexport without payment of duties. Advocates of free trade believe that government intervention to promote trade is yet another obstacle to free trade and is a form of subsidy to politically influential corporations. In their view, gains from trade should result from economic efficiencies, not from government support.

REDUCTIONS OF TRADE BARRIERS

There have been several efforts to eliminate some of the trade barriers that were erected in the past. GATT and the Uruguay Round were two of these efforts. In addition, the United States continues to try to penetrate Japan's closed markets.

General Agreement on Tariffs and Trade

The most notable reduction of trade barriers was a multilateral effort known as GATT, which was put into operation in 1947. When 23 countries signed the agreement, they thought they were putting in place one part of a future World Trade Organization (WTO), an organization that would have wide powers to police the trading charter and regulate international competition in such areas as restrictive business practices, investments, commodities, and employment. It was to be the third in the triad of Bretton-Woods institutions charged with overseeing the postwar economic order—along with the **International Monetary Fund (IMF)** and the **World Bank**. But the draft charter of the WTO was never ratified by the U.S. Congress and GATT remained a treaty without an organization.

More than 100 member countries administer GATT through a process of negotiation and consensus. This has resulted in a substantial reduction of tariffs. However, GATT's rules proved inadequate to cope with new forms of nontariff barriers, such as export-restraint agreements. Also, services, which now account for about 30% of world trade, were not covered by GATT at all. GATT was also of little help to developing countries with limited trading power. Since the 1970s, the liberal trading order that GATT helped to uphold has been steadily challenged by renewed protectionist threats, especially in the guise of nontariff barriers. Between 1981 and 1990, the proportion of imports to North America and the European Union (EU) that were affected by nontariff barriers increased by more than 20%. Trade between developed and underdeveloped countries is also increasingly affected by nontariff barriers; roughly 20% of less developed countries' exports are covered by such measures. In the coming years, pressure on governments in developed countries to protect domestic jobs through trade barriers is likely to mount.

The problem with GATT agreements arose from European and U.S. opposition to reducing and phasing out agricultural subsidies (a problem that has reoccurred under the World Trade Organization that was finally established in 1995). The EU wanted to maintain **export subsidies**, which are government payments that reduce the prices of goods to buyers abroad. Another type of subsidy, the domestic farm subsidy, constitutes direct payments to farmers according to their production levels in order to subsidize their output, a massive form of state intervention in agriculture. The result is increased domestic food output, which unfairly competes with the agricultural products of less developed countries on the world market, and often bankrupts small farmers throughout the world. Both types of subsidies are artificial barriers that reduce prices on the world market and provide advantages to local farmers.

In 1993, the Uruguay Round of GATT negotiations reconvened in an attempt to further resolve trade disagreements. Because most of the industrialized world was in the midst of an economic slowdown or an outright recession in 1992, there was much interest in the international trade measures of GATT to be discussed at Uruguay. The United States was attempting to pry open foreign markets, especially those considered to be unfair traders. Finally, in Geneva in 1993, 117 nations agreed to reduce worldwide tariffs, lower subsidies, and eliminate other barriers to trade. In reducing worldwide protectionism, the GATT nations have chosen to improve resource allocation, which will eventually increase trade and employment and raise wages and standards of living. The United States obtained most of what it sought: the opening of agricultural markets, cuts in industrial tariffs, intellectual property rights, and the opening of markets to the world's service industries. There were both losers and gainers. America's heavy equipment manufacturers, toy makers, and beer brewers were joyful, but the pharmaceutical industry, Hollywood film makers, and textile manufacturers were not pleased.

World Trade Organization

In 1995, the last round of GATT established a permanent **World Trade Organization (WTO)**, to which most of the world's countries now belong (Figure 20). Until the WTO was enacted, countries that had trade conflicts had to resolve their own problems. The WTO provides third-party arbitration to settle disagreements between nations over trade. The judgments are enforced through retaliatory trade sanctions by all members of the WTO, meaning that a country gives up part of its sovereignty to this multilateral world organization. Moreover, a country's power to control flows across its borders is to some extent lost. For some countries that do not abide by international trade agreements, this is a scary proposition.

Under the WTO, quantitative limits on imports have become illegal. For example, Japanese and Korean import restrictions on U.S. rice, peanuts, dairy products, and sugar are now banned. Perhaps the most important aspect of the WTO to the United States is **intellectual property rights**, because it has a strong competitive advantage in this area. All signatories of the WTO are required to protect patents, copyrights, trade secrets, and trademarks. This measure is designed to end the wholesale pirating of computer programs, videocassettes, musical recordings, books, and prescription drugs widely practiced in some developing countries (e.g., China). The WTO calls for free trade in financial services, shipping, and audiovisual products—movies, television programs, and musical recordings. The WTO also prohibits members from requiring in products a certain proportion of content manufactured within their borders. This practice was widely employed as a device to limit the use of imported parts and components and, thus, to bolster local employment.

The WTO has been criticized for its failure to establish production standards that address environmental concerns. Goods can be part of international trade without meeting the production standards of other countries, but many signatories have argued that the environmental standards under which a good has been produced should be a consideration. Others argue that globalization, as represented by the WTO, is dominated by large corporations, which are secretive and undemocratic. Labor unions are

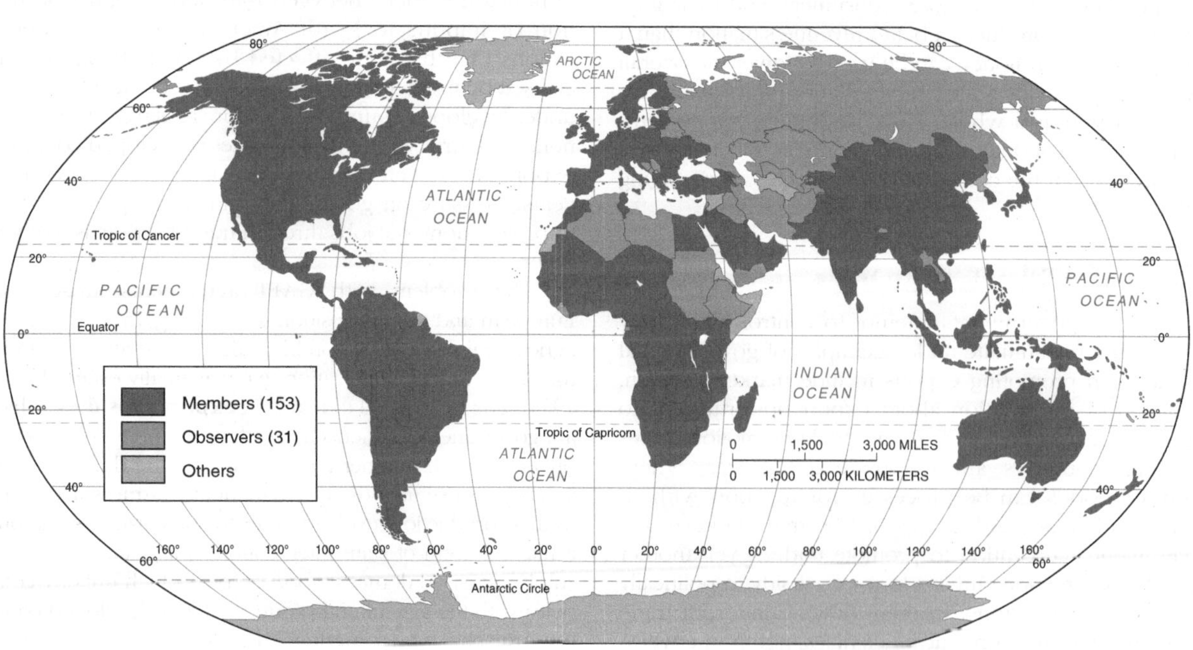

FIGURE 20 WTO members, WTO observers, and non-WTO members, 2009.

concerned about the exploitative conditions in sweatshops and the downward pressure on wages that globalization has created. Others protest the austerity programs of the IMF. For these reasons, and others, WTO and similar meetings have attracted globalization protestors (Figure 21), including the notable "Battle in Seattle" that occurred in 1999 and subsequent demonstrations at meetings of ministers of the Group of Eight (G-8).

Government Barriers to Flows of Production Factors

Although not as complex as trade barriers, obstacles to the free flow of capital, labor, and technology constrain international managerial freedom. Exchange controls and capital controls are the main barriers that interfere with the movement of money and capital across national borders. Exchange controls, which restrict free dealings in foreign exchange, include multiple exchange rates and rationing. In multiple-exchange-rate systems, rates vary for different kinds of transactions. For example, a particular commodity may be granted an unfavorable rate. Foreign exchange may also be rationed on a priority basis or on a first-come, first-served basis. Thus, exchange rates are political tools bearing little relationship to economic reality. Capital controls are restrictions on the movement of money or capital across national borders. They are typically designed to discourage the outflow of funds.

All countries regulate immigration, but the movement of workers from poorer to richer countries was the dominant pattern during the long postwar boom. When the boom ended, jobs moved to the workers. One reason for this change was tighter immigration laws in the advanced industrial countries. These laws strengthened the position of domestic labor and resulted in a growth in managed trade and a decline in managed migration.

Technology, which is highly mobile, can be transferred in many ways: equipment export, scientific and managerial training, books and journals, personal visits, and the licensing of patents. Political and military considerations regulate the export of technology. Although these controls are not yet terribly onerous, demands for more stringent controls are on the increase. Labor unions in advanced industrial countries are one source of demand for control. The unions attribute domestic job loss to the export of high technology.

Multinational Economic Organizations

As nations turn inward to concentrate on problems of economic growth and stability, we are witnessing a resurgence of protectionism. But also in evidence is a strong, simultaneous countermovement toward international interdependence. Scores of multinational organizations have sprung up that for the most part are loosely connected leagues entailing little or no surrender of sovereignty on the part of member nations. Some of these international organizations are global in scale, the most inclusive being the United Nations (UN), with 191 member nations accounting for more than 99% of humankind. Much of the UN's work is accomplished through approximately two dozen specialized agencies, such as the World Health Organization (WHO) and the International Labor Organization (ILO). Other international organizations have a regional character; for example, the Association of South-East Asian Nations (ASEAN) and the Asian Development Bank (ADB). Many international organizations are relatively narrow in focus—mostly military, such as the North Atlantic Treaty Organization (NATO), or economic, such as the Organization of Petroleum Exporting Countries (OPEC).

Some international organizations are discussion forums with little authority to operate either independently or on behalf of member states, for example, GATT and the Organization of Economic Cooperation and Development (OECD). Others, such as the IMF and the World Bank, have independent, multinational authority and power, performing functions that individual states cannot or will not perform on their own. Some international organizations integrate a portion of the economic or political activities of member countries—as, for example, the EU. International

© Steven Rubin/The Image Works

FIGURE 21 Protests against the World Trade Organization. The WTO was formed in 1995 to reduce barriers to world trade by negotiating reductions in international tariffs and trade restrictions among the world countries, freeing up the movement of products, money by banks, corporations, and wealthy individuals. The WTO enforces agreements and attempts to stop violations, for example, of intellectual property rights in the age of the Internet, and infringement of patents. However, many people feel that the WTO is a secretive organization supportive of corporations rather than working class and poor people.

organizations that promote regional integration are the most ambitious of all. Some observers believe that regional federations are necessary in the process of weakening nationalism and developing wider communities of interest. However, if a rigid, inward-looking regionalism is substituted for nationalism, the ultimate form of international integration—world federation—will be difficult to achieve.

This section examines international economic organizations that affect the environment in which firms operate and thus influence developing countries. We look at international financial institutions, groups that foster regional economic integration, and groups such as commodity cartels that deliberately manipulate international commodity markets.

International Financial Institutions

International financial institutions are largely a phenomenon of the post–World War II period. The IMF and the International Bank for Reconstruction and Development (IBRD), or World Bank, were established in 1945 as part of the Bretton-Woods agreement. These institutions are significant sources of multilateral capital, especially aid for developing countries, which is particularly important for those countries that do not have access to private capital markets. The IMF is an international central bank that provides short- to medium-term loans to member countries (Figure 22), and the IBRD is an international development bank that provides longer-term loans for particular projects. Both institutions are made up of clusters of governments, each paying a subscription or quota based on the size of its economy. Because quotas determine a member's voting power, the banks are dominated by the most powerful economies—particularly, by the United States.

International Monetary Fund

FIGURE 22 The entrance of the International Monetary Fund (IMF) headquarters in Washington, DC. The IMF attempts to maintain foreign exchange balances and to promote economic modernization and growth in the Third World. Adjustment programs generally include measures to manage demand, improve the incentive system, increase market efficiency, and promote investment.

The IMF and the World Bank were originally established to prevent a recurrence of the crisis of the 1930s. At first, they embodied Keynesian principles, which offered a rationale for state intervention in markets. Starting in the 1980s, however, the IMF and the World Bank adopted firm neoliberal, as opposed to dependency, interpretations of development. Under pressure from the United States, they took on an adamantly market-oriented stance and imposed it on the governments of less developed countries that needed their assistance, often at the cost of enormous human suffering. Loans from the IMF and the World Bank, therefore, tend to reflect U.S. economic and foreign policy.

This neoliberal "Washington Consensus" revolves around requiring less developed countries to follow tight monetary policies to combat inflation. This serves investors and bondholders well but raises interest rates for consumers. They are required as well to liberalize their financial markets by adopting a variety of deregulatory programs. In the name of budget balancing, governments are often forced to reduce subsidies for the poor for public transportation, kerosene, or cooking oil. Privatization policies encourage or require governments to sell off public assets to private investors, often at reduced prices; all over the world, formerly state-owned or state-operated power plants, hydroelectric facilities, bus routes, airlines, telecommunications firms, and other assets are rapidly being sold to the private sector on the assumption that it is more efficient than the public sector. Yet public services often exist precisely to overcome market failures, particularly the inability of markets to provide adequate services to the poor.

Similarly, IMF policies require countries to liberalize trade policies, including ending tariffs, quotas, nontariff barriers, and subsidies. Critics note that such liberalization is often just a smokescreen for increased penetration of markets in less developed countries by firms from the United States, so the IMF is accused of doing the dirty work of American capital. Even as less developed countries are forced to give up subsidies, the U.S. government lavishes them on its farmers, giving them an unfair advantage in selling low-priced crops to foreign markets. Thus, the IMF imposes requirements on less developed countries that governments of developed countries would never accept.

Critics deride these policies as neoliberal "market fundamentalism" (see Stiglitz, 2002). By the IMF's own admission, its policies have often exacerbated the problems of less developed countries, such as during the Asian financial crisis of the late 1990s. The macroeconomic models employed to buttress these policies are often highly oversimplified and underestimate the complexity of local political and social contexts. Tight monetary policies can generate recessions and lead to high unemployment. Many less developed countries lack a proper institutional environment for privatization to work, including bankruptcy procedures, protection of property rights, and debt repayment programs. Moreover, the Washington Consensus increases inequality in less developed countries by protecting and rewarding investors at the expense of the poor.

Alex Wong/Getty Images, Inc.

FIGURE 23 Headquarters of the World Bank, Washington, DC. The World Bank doles out loans to countries to reform governments and legal institutions; develop banks and financial institutions, which may give business loans to local firms; and develop infrastructure, transportation, and the social services sector of a deadlocking country.

In pursuing policies that dramatically emphasize economic stabilization over job creation, the IMF is eager to bail out bankers but never the impoverished masses. Finally, market fundamentalism tends to be overly optimistic about the private sector and overly pessimistic about the public sector.

Some developing countries have turned to two subsidiary World Bank organizations: the International Finance Corporation (IFC), founded in 1956, and the International Development Association (IDA), founded in 1960. These organizations provide loans with stipulations less stringent than those of the IBRD. For example, the IDA may charge no interest on loans, grant 10-year grace periods (no repayment of principal for the first 10 years), or allow 50-year repayment schedules for poorer developing countries. Because the IDA is much less creditworthy than the IBRD, all its resources must come from member-government contributions. These banks reflect the desire of developing countries to exert control over their financing needs. Of the three banks, the Asian Development Bank is most under the control of developed countries, in particular Japan. The AFDB has been the most independent, but it is also the smallest.

International financial institutions are important to international business. The IMF, the World Bank (Figure 23), and regional development banks annually finance billions of dollars of the import portion of development projects. This can be valuable business for foreign companies involved in the projects, either occasionally as part owners or, more commonly, as contractors or suppliers.

REGIONAL ECONOMIC INTEGRATION

Regional economic integration occurs when sovereign nations form a single economic region. It is a form of selective discrimination in which both free-trade and protectionist policies are operative: free trade among members and restrictions on trade with nonmembers. Four levels of economic integration are possible. At progressively higher levels, members must make more concessions and surrender more sovereignty (Figure 24). The lowest level of economic integration is the **free-trade area**, in which members agree to remove trade barriers among themselves but continue to follow their own trade practices with nonmembers. A **customs union** is the next higher level of integration. Members agree not only to eliminate trade barriers among themselves but also to impose a common set of trade barriers on nonmembers. The third type is the **common market**, which, like the customs union, eliminates internal trade barriers and imposes common external trade barriers; this regional grouping, however, permits free mobility to factors of production. At a still higher level, an **economic union** has the characteristics of the common market plus a common currency and a common international economic policy.

There are a variety of regional trade organizations in the world (Figure 25). They range from loosely integrated free-trade areas such as the Latin American Free Trade Association (LAFTA) to common markets such as the EU. There are North-South ties between the EU and LAFTA,

	Abolish tariffs and intra-trade restrictions	Common tariff to offshore trade	Liberalized endowment factor movement	Harmonization of economic policies	Unification of policies by common organization
Free Trade Area	*				
Customs Union	*	*			
Common Market	*	*	*		
Economic Union	*	*	*	*	
Political Union	*	*	*	*	*

FIGURE 24 Types of regional economic integration.

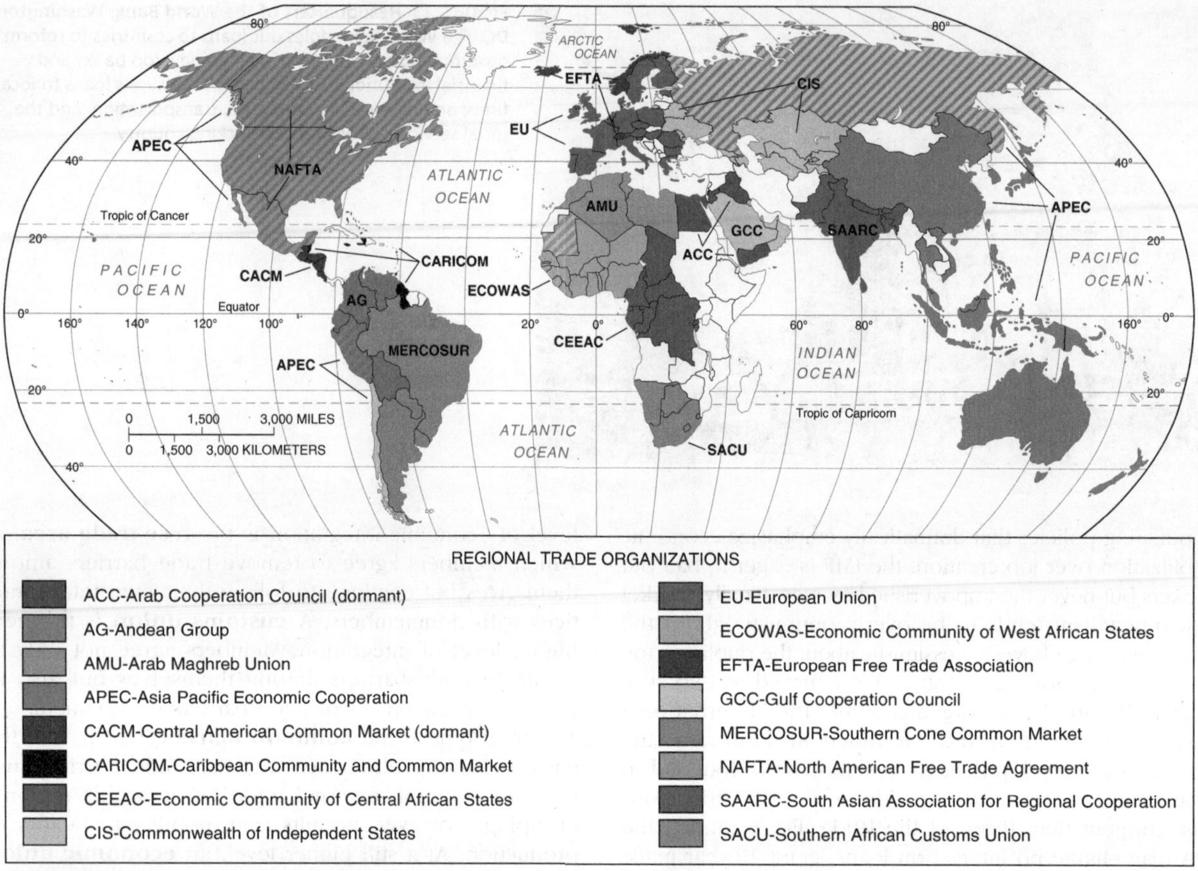

FIGURE 25 Regional trade organizations around the world.

and South-South ties between LAFTA and ASEAN. Most of these links are bilateral; that is, agreements between nations within different regions. Fully fledged interregional integration has yet to be achieved. Indeed, regional groups are more concerned with closer economic integration *within* regions than *among* regions.

Barriers to successful regional integration are stronger in developing countries than in developed countries. The most significant barriers are political—an unwillingness to make concessions. Without concessions to the weaker partners of a regional group, the benefits from cooperation pile up in the economically more prosperous and powerful countries. Another difficulty is that developing countries have not historically traded extensively among themselves. Still another obstacle to integration is the lack of sufficient transportation and power networks. Nonetheless, much potential for integration exists in developing countries, particularly because many are too small and too poor to grow rapidly as individual units.

Many developing countries turned to regional integration schemes in the 1960s and 1970s because they needed to gain access to larger markets, to obtain more bargaining power with the developed countries than they could if they adopted a "go-it-alone" policy, to create an identity for themselves, to strengthen their base for controlling multinational corporations, and to promote cohesive solidarity. How companies feel about regional groupings differs from one company to another. Companies that enjoy a secure and highly profitable position behind national tariff walls are unlikely to favor removal of these barriers. Conversely, companies that see the removal of trade barriers as an opportunity to expand their markets see integration as a favorable development. Similarly, companies that traditionally exported to markets absorbed by a regional grouping have a strong interest in integration. They perceive these enlarged markets to be more attractive than they were in the past. But as outsiders, these foreign companies will be subject to trade controls, whereas barriers for internal competitors will decrease. Thus, they may lose their traditional markets because they are outside the integrated group of countries. As a result, there is an incentive to invest inside the regional grouping. This is why many U.S. firms invested directly in the European Economic Community (now EU) countries.

The European Union

The most successful example of economic integration is the **European Union (EU)**. It began in 1957 with six nations: France, West Germany, Italy, Belgium, the

Netherlands, and Luxembourg, and since then has added most European countries, including a steady expansion into Eastern Europe. By 2009, the EU had 27 members with a total population of more than 500 million (Figure 26). The EU today is the largest single trade bloc in the world and accounts for 40% of international trade, which is three times its world share of population.

The intent of the EU was to give its members freer trade advantages while limiting the importation of goods from outside Europe. It called for (1) the establishment of a common system of tariffs applicable to imports from outside nations; (2) the removal of tariffs and import quotas on all products traded among the participating nations; (3) the establishment of common policies with regard to major economic matters such as agriculture, transportation, and so forth; (4) free movement of and access to capital, labor, and currency within the market countries; (5) transportation of goods, commodities, and people across borders with no inspection or passport examination; and (6) a common currency.

The EU has made tremendous progress toward its stated goals. The member nations have enjoyed more efficient, large-scale production because of potentially larger markets within the EU, permitting them to achieve scale economies and lower costs per manufactured unit, something that pre-EU economic conditions had denied them. However, the current stumbling block seems to be the lack of a common economic unit of currency. In addition, some northern countries scoffed at open borders that would allow southern Europeans to immigrate and thereby take advantage of social welfare programs of cradle-to-grave economic assistance.

Americans are concerned about the economic power of the EU. Present tariffs have been reduced to zero among EU nations, whereas tariffs for U.S.-made products have been maintained. Consequently, importing goods to EU nations is relatively more difficult. At the same time, increased prosperity among EU nations makes them potential customers for more American exports.

THE EU'S SINGLE CURRENCY On New Year's Day 1999, many (but not all) members of the EU adopted a unified currency, the **euro** (Figure 27); Britain, Denmark, and Sweden opted to retain their own currencies. The currency exchange rates of participating countries are locked in to one another and to the euro. However, the euro floats against non-EU currencies such as the U.S. dollar (Figure 28). When members of the European Union adopted the single currency, they were aware that the sacrifice would be great. Each country effectively surrendered the right to independently balance its own budget and manage its own debt. Each country relinquished its individual monetary identity. There is no question that when the economic efforts of 25 countries and 500 million people are combined, European goods and services will be better represented in the world economy by a currency capable of maintaining price stability in member countries (Figure 29)

North American Free Trade Agreement

The economic pressure placed on the United States by the EU led it to promote freer trade through GATT and to develop the U.S.–Canadian Free Trade Agreement, which

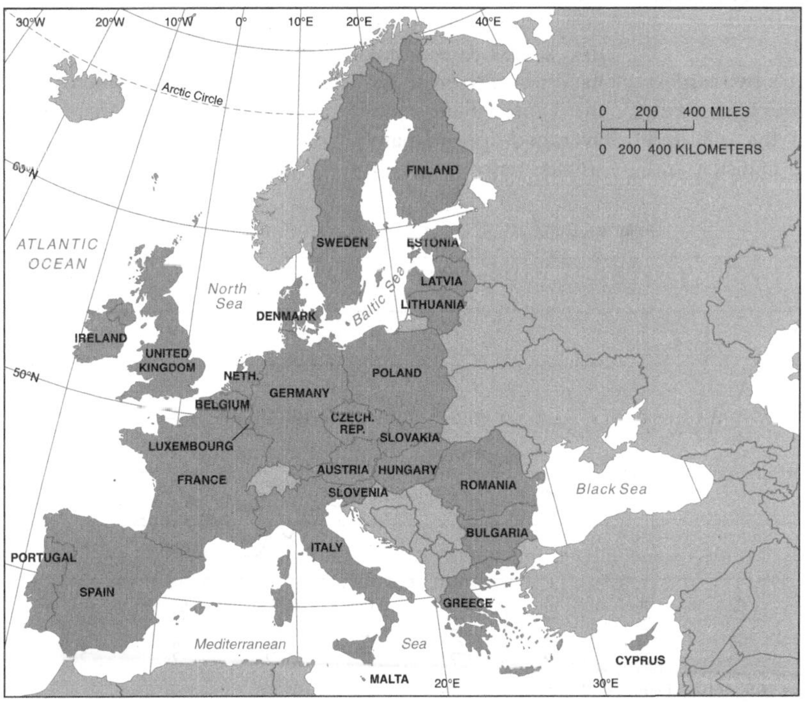

FIGURE 26 The European Union, 2009. With half a billion people and generally prosperous economies, it is one of the world's major players in the global economy. The European Union was founded in 1958 as a trade block to reduce the cost of imports between six countries—Belgium, Luxembourg, Netherlands, France, Germany, and Italy—in the aftermath of World War II. With the emergence of the economic powerhouses of the United States and East Asia, the EU has expanded to 27 countries as of 2010, with more applications coming in from Turkey, Croatia, Macedonia, Albania, Bosnia, and a host of others.

© Tom Grill/Corbis

FIGURE 27 The euro. A single currency has replaced national currencies in most, but not all, of the member states of the European Union.

phased out trade barriers between the world's two largest trading partners. In 1992, the **North American Free Trade Agreement (NAFTA)** was signed by the U.S. and Mexican presidents and the Canadian prime minister, and in 1994 the U.S. Congress approved it. NAFTA encompasses 310 million Americans, 33 million Canadians, and 100 million Mexicans. Unlike the EU, NAFTA includes a developing country. Access to the North American market is coveted by EU countries and Japan. Proponents of NAFTA argued that EU nations would negotiate a free-trade agreement between the two blocs.

NAFTA was not well received by all parties in North America. The critics' main argument was that it removes lower-skilled assembly and manufacturing jobs from the United States and transplants them to Mexico, where labor costs are one-fifth to one-eighth as much. In addition, companies might be tempted to flee the more stringent climate in the United States regarding environmental pollution and workplace safety controls. Critics of NAFTA also suggest that Japan, Korea, Taiwan, and other East Asian countries might build plants in Mexico and export goods duty-free to the American and Canadian markets, which would hurt U.S. firms and workers.

The principal argument in favor of NAFTA was that free trade would enhance U.S., Canadian, and Mexican comparative advantages: raise per capita income in Mexico and increase Mexican demand for goods from the United States and Canada. Another argument was that higher living standards in Mexico would help control the flow of undocumented aliens crossing the U.S. border, now estimated to be 1 million per year. With free trade, wages would rise in Mexico; therefore, undocumented aliens could stay home and work in their native country.

By reducing restrictions on the establishment of U.S. and Canadian financial service subsidiaries, NAFTA also frees up trade in financial and investment services. Thus, NAFTA deregulated Canadian and U.S. banking, securities brokering, and insurance operations in Mexico. In addition, Canadian and American trucking companies have free access to the Mexican market and Mexican truckers have free access to markets in the north. Before NAFTA, the United States' tariff rates on imports from Mexico averaged approximately 5%, too low for NAFTA to have substantially raised the incentive for Canadians and

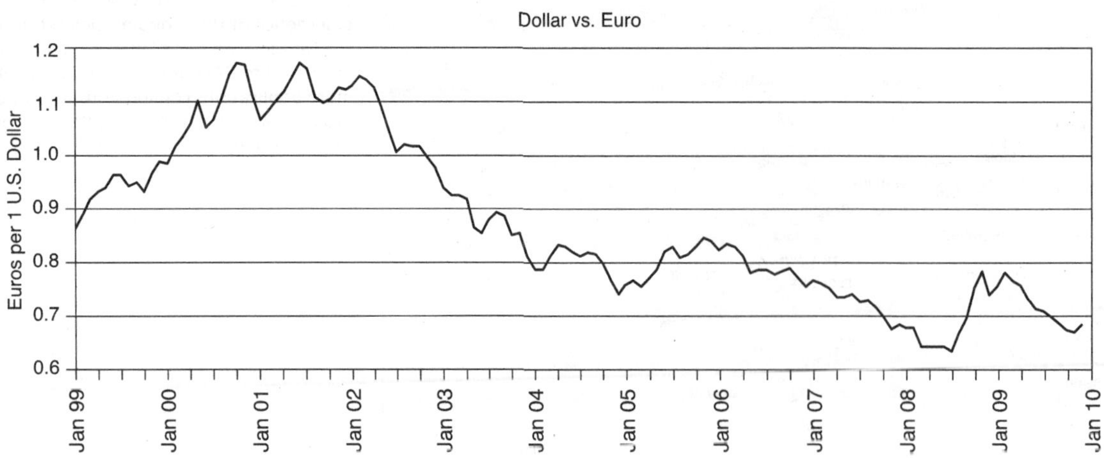

FIGURE 28 Value of the euro in terms of the U.S. dollar, 1999–2009.

FIGURE 29 London. The largest city in Europe, London is one of the world's premier financial centers, the largest center of currency exchange, and a major hub of producer services. London, New York City, and Tokyo are the greatest centers of world trade and banking in the world today, with Shanghai coming on fast. London developed as the preeminent center in Europe because of its large market; good port for overseas trade; its relatively remote location from warring Europe; its large supply of skilled labor; and a merchant class with access to capital and proximity to banks, insurance firms, investors, and government support needed for business growth.

United States firms to invest in Mexico. A free-trade zone, extending 100 miles south of the border, had been in operation before NAFTA; NAFTA has in effect extended the maquiladora export assembly zone to all of Mexico, which means that duties are imposed only on the value added to imports by manufacture in the maquiladora plants.

After 17 years, NAFTA's lower tariffs have expanded trade and investment with Mexico, but American imports have grown more quickly than its exports, turning what was once a U.S. trade surplus with Mexico into a deficit. Mexico saw its U.S.-bound exports jump in the years since 1994 by 241%, while imports from the United States climbed by 170%. While U.S. tariffs disappeared overnight on many Mexican and Canadian exports, powerful lobbies have slowed the trend in some sectors, particularly agriculture. So, Mexican imports of cheaper, subsidized U.S. corn has flooded the country and bankrupted 2 million farmers there. Moreover, two-thirds of the increased shipments between the U.S. and Mexico are "revolving-door exports," that is, products that stay in Mexico just long enough to be assembled into a product that is then sold in the U.S. market at a cheaper price than if produced in the United States itself. Mexico's burgeoning maquiladora industry has grown by 150% since 1994 but has skimmed manufacturing jobs and depressed wages in U.S. border states. The U.S. Department of Labor has certified a loss of more that 600,000 U.S. jobs to Mexico as a result of NAFTA, many of them in manufacturing—a small fraction of the 150 million jobs that make up the U.S. labor force, about 15 million of which are in manufacturing. About an equal number of jobs have been created on the U.S. side of the border but these are concentrated in lower-paying service occupations. Meanwhile, the United States continues to lure Mexican workers, many of whom were uprooted from rural communities when Mexico opened its markets to subsidized U.S. agricultural goods.

NAFTA was grossly oversold on both sides of the border as an instrument for reducing unwanted immigration. Illegal border crossings have increased steadily since 1994, mainly because of Mexico's disappointing economic performance and a large and growing wage disparity between the two countries. A very low minimum wage, 20% unemployment, and the lure of higher wages in the United States continue to draw millions of Mexicans. Full of undocumented immigrants, many communities along the U.S. side of the border remain as mired in poverty as they were a decade ago. The net impacts of these immigrants, who are largely unskilled and often illiterate, has been hotly debated. Many note that they take jobs typically unwanted by U.S. citizens, particularly picking fruits and vegetables and working in the garment industry, construction, and retail trade. Many concede that they increase the supply of unskilled labor and force wages down in those labor markets. Others note that they pay income and Social Security taxes but are ineligible to receive benefits such as Social Security. Many live in desperate destitution, vulnerable to employers who take advantage of their precarious situation.

Critics alleged that NAFTA would create a "giant sucking sound" of industries moving to Mexico to reap high profits and low wages. But when comparing Canada and Mexico in terms of their auto industries, Canada has done a better job of increasing its auto exports. The largest exporter of cars and trucks to the United States is no longer Japan, but Canada. Since 1989, Canada has doubled the number of cars and trucks it exports to the United States. A decade after the NAFTA agreement was signed, which many feared would send the U.S. auto industry to Mexico, free trade is having the opposite effect. The United States actually gained jobs in the auto industry, though not higher-paying ones. Meanwhile, Canada has also been gaining jobs. Canada employs almost 60,000 workers installing mostly U.S. parts into motor vehicles—about 85% of which were exported to the United States.

By 1995, Mexico had lost 13% of its 137,000 auto manufacturing jobs when foreign competition (particularly from Asia) wiped out Mexican parts suppliers. One year after NAFTA began, the peso collapsed and auto sales plummeted 70%. But due to the devaluation of the peso and reduced

Case Study

North American Free Trade Agreement (NAFTA)

The North American Free Trade Agreement (NAFTA) was ratified in 1994, linking Canada, Mexico, and the United States under a regime of liberalized trilateral commerce that harmonized procedures for defining rules of origin, expedited customs clearance for cross-border trade, and provided for a wide variety of institutional reforms to safeguard environmental interests and the rights of workers (sidebar agreements). The accord also included provisions to maintain stable flows of energy, easier business travel, and reduced restrictions on foreign direct investment (FDI). Despite many side agreements covering spheres such as labor conditions and environmental protection, a central goal of the accord was to achieve a phased elimination of import duties for most products and services by 2004. This goal has been achieved, though many non-tariff trade barriers are still in place. A further goal was to dilute long-standing restrictions on capital mobility (FDI), and pave the way for new investment across the trilateral region. The net result over the past 17 years has been a dramatic expansion of intra-industry trade among the three nations.

Much of this trade is conducted on an intracorporate basis. For example, Ford USA can import components or final assemblies on a duty-free basis from its subsidiaries located in Canada or Mexico—provided that such inputs qualify as 50% North American (i.e., "local content"). Automotive companies from outside NAFTA can also compete on the same basis, provided that 50% NAFTA content can be demonstrated. Transnational corporations headquartered outside the NAFTA zone have played an important role, too, especially in sectors such as automotive products, aerospace, and electronics.

From the perspective of traditional trade theory, the NAFTA concept made good sense in 1994. With respect to comparative advantage, Mexico was cast as labor abundant with low skills, Canada had lots of natural resources, while the United States was capital rich with high labor skills. Linking the three economies together via free trade would theoretically allow for factor specialization, expanded output, and an aggregate rise in consumer welfare. Specialization would also promote economies of scale. Critics of the accord, in contrast, argued that a free trade agreement between two advanced industrial economies and a developing nation within a geographically contiguous region would spur a mass exodus of capital from north to south, as well as downward pressure on northern wages. Neither perspective appears to be entirely correct.

Thus far, there is no evidence that NAFTA has caused trade diversion to any significant degree. Trade diversion takes place when a regional free trade agreement artificially displaces a lower cost producer from outside the club as a result of policy-driven cost advantages that accrue to in-club producers. NAFTA does not seem to be a factor in this regard, as witnessed by massive imports from China, India, and other low-cost producers located in the newly industrializing countries of Asia and South America.

Economic models have failed to uncover significant evidence of trade diversion, and points to either a zero or negligible impact of NAFTA on U.S. industrial employment. Some studies, in fact, suggest that NAFTA's tariff reductions have decelerated the pre-NAFTA trend toward employment decay in the manufacturing sectors of both Canada and the United States. Between 1994 and 2000, for example, total employment in the U.S. manufacturing sector actually increased (albeit by a modest 1%). On a less optimistic note, there is at least some evidence that trade liberalization has contributed to rising income inequality among regions for all three NAFTA members. This said, most economists and economic geographers would agree that it is hard to disentangle the trade, income, or employment effects of free trade agreements from broader factors such as capital intensification (process automation), exchange rate movements, import competition from outside the bloc, and declining industrial competitiveness (e.g., lack of investment in new technology). For example, some studies have found a negative relationship between NAFTA-assisted export growth and local employment levels. Capital intensification appears to be the key driver in this rather odd relationship.

Although NAFTA has been a topic of considerable debate, the accord is here to stay. Looking to the future, at least five areas of contention are likely to persist and/or emerge. These areas of contention include:

1. The increasingly adversarial relationship between the United States and Mexico regarding illegal immigration
2. The persistence of non-tariff trade barriers based on technical or environmental standards
3. The hardening of U.S. ports of entry in light of post-9/11 security threats
4. The management of NAFTA expansion into the Free Trade Area of the Americas
5. The development of more efficient border management systems

Some of these concerns might be addressed by moving toward a secure external perimeter, where the boundaries of all three countries are monitored and controlled by common standards to facilitate security compliance. This will probably not happen soon, as such a scenario would push NAFTA closer toward a customs union (which is not a politically popular option). Other concerns might be addressed by further policy harmonization, notably with regard to border management (e.g., fast-track entry for people and merchandise). Still other concerns might be dealt with by broadening the mandate of NAFTA's dispute resolution system to include issues such as illegal immigration, antiterrorism measures, and environmental compliance.

labor costs, Mexican production of Nissan, Volkswagen, GM, Ford, and Chrysler vehicles hit 1.4 million in 2003, approximately 85% of which were exported to the United States. Yet new equipment and more efficient production have meant fewer jobs for Mexican auto workers than expected.

NAFTA has been beneficial to the auto industry's competitiveness because the trade agreement allowed U.S. automakers to win back a share of the domestic market that they had lost to cheaper Asian and European cars. In order to do this, they had to send some jobs to neighboring countries to keep costs down; unfortunately, large numbers of U.S. workers have been laid off. The pressure to keep reducing the cost of labor to produce a car is continuing.

In the 1990s, the United States proposed an extension of NAFTA that would include 34 nations across the Western Hemisphere. The Free Trade Area of the Americas (FTAA) is sometimes referred to as "NAFTA on steroids." Given the long U.S. hostility toward Cuba, that country would be excluded. Early negotiations to design the FTAA were met with hostile demonstrations. If the FTAA is successful, it would create the largest free-trade zone in the world, exceeding even the EU.

OPEC

A **cartel** is an agreement among producers that seeks to artificially increase prices by arbitrarily raising them, by reducing supplies or by allocating markets. The most successful commodity cartel is the **Organization of the Petroleum Exporting Countries (OPEC)**. Founded in 1960, membership in OPEC has fluctuated over time as members joined and dropped out; in 2010, it consists of 12 countries—Venezuela, Ecuador, Saudi Arabia, Iran, Kuwait, Libya, Nigeria, Angola, Iraq, Algeria, Qatar, and the United Arab Emirates (Figure 30); Indonesia and Gabon are former members. The success of OPEC at raising oil prices encouraged other underdeveloped countries to create new nonoil cartels.

The first oil shock OPEC created occurred in 1973, followed by another in 1979. The price for a barrel of crude petroleum oil peaked in 1981 at $36. American oil companies commenced major new exploration and billions of dollars worth of infrastructure was erected in territories of North America that were rich in oil, low-grade oil shales, and tar sands, notably Colorado, Wyoming, Alberta, and Montana. With the subsequent decline in world oil prices, these new oil operations and oil explorations likewise declined, which sent shock waves through the oil industry and depreciated home and business values throughout Houston, Texas, and other oil-dependent cities.

In the late 1980s, the price of a barrel of crude petroleum on the world market sank to $15, but it rose again gradually in the 1990s to $21 because of the first Persian Gulf war and the destruction of wells in Iraq and Kuwait, where output has not returned to prewar production levels because of the devastation. The lower output, especially after the beginning of the Iraq war in 2003, required higher levels of production or refining capacity by other OPEC nations. By 2004, prices exceeded $120 per barrel, only to fall to roughly $70 in 2010.

World oil prices depend on the resumption of economic growth in not only the developed nations but also the less developed nations, and the resulting growth in demand for crude petroleum. The resumption of production and refining in Russia and Central Asia will also affect world petroleum prices. At present, OPEC produces less

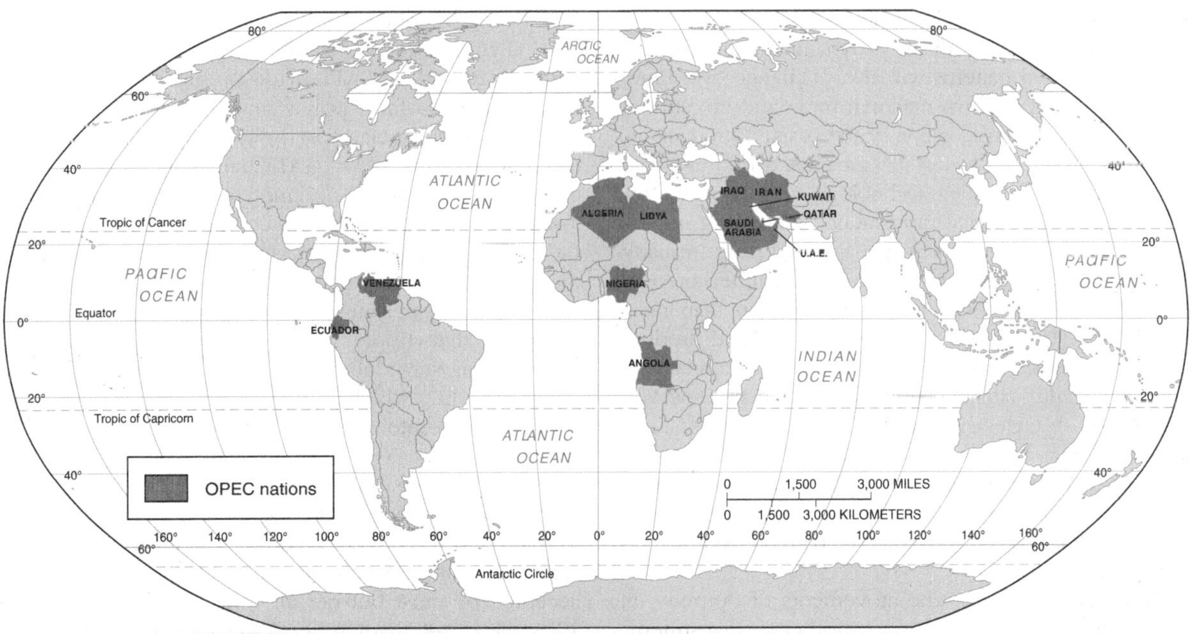

FIGURE 30 The Organization of Petroleum Exporting Countries (OPEC). Although dominated by Arab and Middle Eastern states, OPEC includes Venezuela and Ecuador in South America and Nigeria and Angola in Sub-Saharan Africa.

than 50% of the world's output but controls 80% of the proven reserves. Surprisingly, despite their tremendous oil reserves of several trillion barrels, OPEC nations will have a problem generating enough money either to maintain their production or to increase capacity because most have serious problems with debt. Much of their petrodollar accumulation is spent on military equipment to keep the wealthy sheiks in power.

Summary

This chapter examined some aspects of the international sphere of economic geography. International business is any form of business activity that crosses a national border. It includes the international movement of almost any type of economic resource—merchandise, capital, or services.

Our discussion of traditional trade theory introduced comparative advantage as the underlying explanation for international trade. The original, Ricardian conception argued that specialization within a larger international division of labor led to higher standards of living for all parties involved. Beyond predicting that everyone gains something from trade, including the most and least efficient producers, classical trade theory neglects too many issues to be realistic, including the roles of innovation, government, and economies of scale. Free trade was established in the nineteenth century within a colonial framework of inequality among countries. As primary goods producers, developing countries became dependent on foreign demand and, therefore, vulnerable to the business cycle of expansion and contraction in developed countries. Thus, some critics argue that free trade is a patina that disguises the ability of rich countries to take advantage of less developed ones. One country's free trade may appear like another's exploitation.

A major extension of the original theory of comparative advantage, Porter's notion of competitive advantage, includes the roles of labor skills, knowledge and innovation, agglomeration economies, financing, and government policy. Unlike comparative advantage, which is static, the theory of competitive advantage is dynamic, noting that regional and national patterns will always change over time.

The discussion of international trade and investment was extended to a consideration of the basis of production-factor flows. Production factors that are most readily movable among countries are capital, technology, and labor. We focused primarily on capital movements, enhancing understanding of FDI by using a managerial perspective. In many respects, the international movement of capital, technology, and managerial know-how is now more important than international trade. Theories of international trade and production-factor movements emphasize the benefits of a relatively deregulated, market-oriented business environment. However, a number of obstacles significantly impede flows of merchandise, capital, technology, and labor. These obstacles include distance, managerial behavior, and governmental barriers. Much progress was made in reducing governmental barriers—tariffs, quotas, and nontariff barriers—during the long postwar boom.

International trade—the movements of outputs—is only one facet of globalization. Another is investment. There are many types of investment, including, for example, purchases of government debt, but most attention has focused on FDI. Most FDI is organized through TNCs and originates in the economically developed countries. However, most TNCs invest in developed, not developing, countries. For example, various west European, Canadian, and Japanese firms invest heavily in the United States, where they generate jobs, wages, output, and profits.

Money must be exchanged on international markets for goods and services, and so exchange rates play a critical role in influencing the prices of imports and exports among countries. Determination of exchange rates allows world trade to function. Explaining why exchange rates fluctuate is no easy matter but is related to levels of real output, inflation rates, demand factors, and currency speculation in trading-partner countries.

Despite pressures for protectionism, countries continue to participate in myriad multinational operations. Major organizations that can be important to international business are international financial institutions and groups that promote regional integration. A common way of shaping global output and trade in a given sector is commodity cartels. Leaders in developed countries view commodity cartels as an unfortunate departure from market-oriented principles. In contrast, most leaders of less developed countries view commodity cartels as a means to reduce their vulnerability in a world of unequal exchange. The most famous cartel, OPEC, has played a fundamental role in affecting the price of petroleum worldwide.

GATT reduced trade barriers worldwide. The Uruguay Round of GATT made progress on a number of difficult issues—farm policy, intellectual property rights, and trade barriers related to the growing international trade in services. In the 1990s, GATT transitioned into the WTO, a permanent institution designed to minimize trade barriers and arbitrate disputes among countries. A current issue at the center of WTO policies is the huge subsidies paid by the governments of countries such as the United States and western European states to their agribusinesses, resulting in low prices that cripple farmers in less developed countries.

The most widely acclaimed regional integration to date is the EU, which at present involves 27 states in a powerful trade bloc of more than 500 million people that manages almost 50% of worldwide trade. The EU is the most complete economic integration among countries the world has ever seen, including the elimination of barriers to trade and investment among its member states as well as flows of labor. A common currency, the euro, has been accepted by most but not all of its members. Starting in 1994, the U.S.–Canadian free trade agreement was extended to include Mexico in the form of NAFTA.

Key Terms

absolute advantage
capital markets
cartel
common market
comparative advantage
competitive advantage
customs unions
economic union
euro
European Union (EU)
exchange rate
export-restraint agreement
export subsidies
foreign direct investment (FDI)
free-trade area
General Agreement on Tariffs and Trade (GATT)
infant industry
intellectual property rights
international currency markets
International Monetary Fund (IMF)
intracorporate trade
nontariff barriers
North American Free Trade Agreement (NAFTA)
Organization of the Petroleum Exporting Countries (OPEC)
protectionism
quota
regional economic integration
tariff
tax-haven country
terms of trade
trade deficit
transnational corporations (TNCs)
unequal exchange
World Bank
World Trade Organization (WTO)

Study Questions

1. Why does international trade occur?
2. What are the inadequacies of existing trade theory?
3. What is meant by the terms of trade? Why have they declined for many Third World nations?
4. What forces have driven the internationalization of banking?
5. How do exchange rates affect trade?
6. What caused the huge U.S. trade deficits?
7. What is foreign direct investment? What are its impacts?
8. How has the U.S. FDI balance changed over time? Why?
9. What are the principal barriers to international business?
10. What are the major arguments in favor of protectionism?
11. What are tariffs, quotas, and nontariff barriers?
12. What was the GATT and what is the WTO?
13. Why were the IMF and World Bank established?
14. What is regional economic integration and why does it occur?
15. Compare and contrast the EU and the NAFTA.

Suggested Readings

Dicken, P. 2010. *Global Shift: Mapping the Changing Contours of the World Economy,* 6th ed. New York: Guilford Press.

Friedman, T. 2005. *The World Is Flat: A Brief History of the Twenty-First Century*. New York: Farrar, Straus and Giroux.

Hugill, P. 1993. *World Trade since 1431: Geography, Technology, and Capitalism*. Baltimore: Johns Hopkins University Press.

Peet, R. 2003. *Unholy Trinity: The IMF, World Bank, and WTO*. New York: Zed Books.

Perkins, J. 2005. *Confessions of an Economic Hit Man*. New York: Plume.

Porter, M. 1990. *The Competitive Advantage of Nations*. New York: Free Press.

Stiglitz, J. 2002. *Globalization and Its Discontents*. New York: W. W. Norton.

Web Resources

The World Trade Organization
http://www.wto.org

The principal agency of the world's multilateral trading system. Its home page includes access to documents discussing international conferences and agreements, reviews of its publications, and summaries of the current state of world trade.

The World Bank
http://www.worldbank.org

A leading source for country studies, research, and statistics covering all aspects of economic development and world trade. Its home page provides access to the contents of its publications, to its research areas, and to related Web sites.

U.S. Department of Commerce
http://www.doc.gov

This government department is charged with promoting American business, manufacturing, and trade. Its home page connects with the Web sites of its constituent agencies.

International Trade Patterns

OBJECTIVES

- To describe the evolving pattern of international commerce
- To document the emerging markets for global exports
- To examine global trade flows of six major commodities groups

Container ships at a port in Melbourne, Australia, reveal the enormous quantities of goods that are traded internationally, mostly by sea. The economies of scale that such ships derive and mechanized loading and unloading systems help to keep the costs of imports and exports affordable, linking regions around the world through commodity flows.

James Lauritz/Photodisc/Getty Images

International Trade Patterns

Capitalism is an economic and political system forever in flux. Incessant change is the norm in market-based societies, and this pattern continued in the late twentieth century.

The past three turbulent decades witnessed major changes in the volume and composition of international trade. World trade in goods and services jumped from $2 trillion in 1980 to over $6.7 trillion in 2007, or more than 36% of gross world product, a clear sign of the increasing integration of national economies (Figure 1), although this number declined slightly in the face of the global economic crisis of 2008. Of these exports, agricultural goods comprised 7.6% by value; mining ores, fuels and minerals another 10.8%; all manufactured goods 61.3%; and all services 20.2%. Because a small group of economically developed countries produce the bulk of the world's tradable goods, and have the disposable incomes to afford imports, global trade is largely confined to a "triad" consisting of Europe, North America, and East Asia (Figure 2).

The changing structure of trade has affected various world regions differently, leading to fluid export geographies (Figure 3). Changes in international supply and demand, prices, production and transportation technology, production techniques, and government policies all play out differently in unique local contexts. For example, OPEC (Organization of the Petroleum Exporting Countries) recorded a meteoric rise in the value of their exports in the 1970s and a precipitous decline in the 1980s and 1990s; in the 2000s, oil prices climbed again as economies around the world, including China in particular, increased their demand for energy, then fell as recession reduced demand. North America, Europe, and East Asia experienced a drop in their export earnings after the oil crisis, but as a group they recovered and now account for 80% of the value of world trade. East Asian newly industrializing countries (NICs) led the world in export growth, with China surging to the forefront. With the exception of the major oil exporters, less developed countries that depend heavily on the export of a few primary commodities have fared badly. For many of them, the growth in primary commodity exports has been negative since 1980.

In addition to the expanding volume of trade, increased diversification of trade ties represents one of the most significant developments in the contemporary world economy. Industrialized countries still trade primarily among themselves, but the proportion has declined from more than 75% in 1970 to around 66% today. They have increased their share of exports to less developed countries and their imports from less developed countries have increased still more. Another major development has been the growth of manufacturing exports from less developed countries to developed countries and, to a lesser extent, to other less developed countries. Manufacturing exports now account for about 40% of total nonfuel exports from these countries, compared with 20% in 1963, and less developed countries now supply 13% of the imports of manufactures by developed countries, compared with only 7% in 1973. Yet only a handful of Asian and Western Hemisphere countries are involved in this development.

WORLD PATTERNS OF TRADE

Trade simultaneously reflects economic and spatial differences in production and consumption among nations and in turn helps to generate those differences. International trade can be understood on the basis of the theory of comparative and competitive advantage and has grown more rapidly than the output of individual countries, reflecting the ways in which they are tied together by globalization. However, the volume, growth, and composition of trade vary widely among the world's major trading countries and regions. Table 1 summarizes the major commodities that dominate world trade by value.

FIGURE 1 Postwar growth in world output and exports. Because of a global recession, world output is slowing down. Faster productivity growth could help mitigate the slowdown, but this does not seem to be forthcoming soon. Before the financial crisis hit the world in 2008, the trend in productivity growth was flat or slowing in the rich countries even as it soared in the emerging world. Growth in output per worker in America had risen sharply in the late 1990s, thanks to the increased output of information technology (IT), and again in the early part of the first decade in the twenty-first century, as the gains from IT spread throughout the economy. By 2005 this growth began to wane. Japan's productivity slumped in the early 1990s; Western Europe's, overall, has weakened since the mid-1990s as well. There seems to be a relentless shift of economic gravity toward the developing world. Since developing countries are more populous than rich ones, and since rich countries' populations are aging, the developing countries will inevitably come to dominate the world economy.

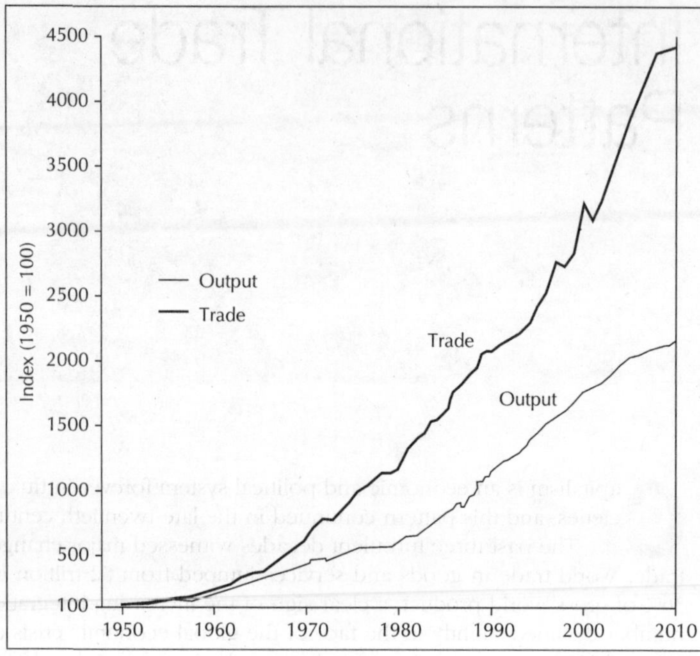

The United States

The United States is by far the world's largest trading nation, accounting for more than $4 trillion worth of combined exports and imports in 2008. During the 1950s, it accounted for 25% of total world trade but now accounts for only 19%, a reflection of the growing competitiveness of other countries, particularly in Europe and East Asia. From 1960 to 1970, the United States enjoyed a net trade surplus as a result of its strength in manufacturing, low oil prices, and the weak value of the dollar. However, following the petroshocks of the 1970s and deindustrialization, this surplus turned into growing trade deficits. In 2008, the

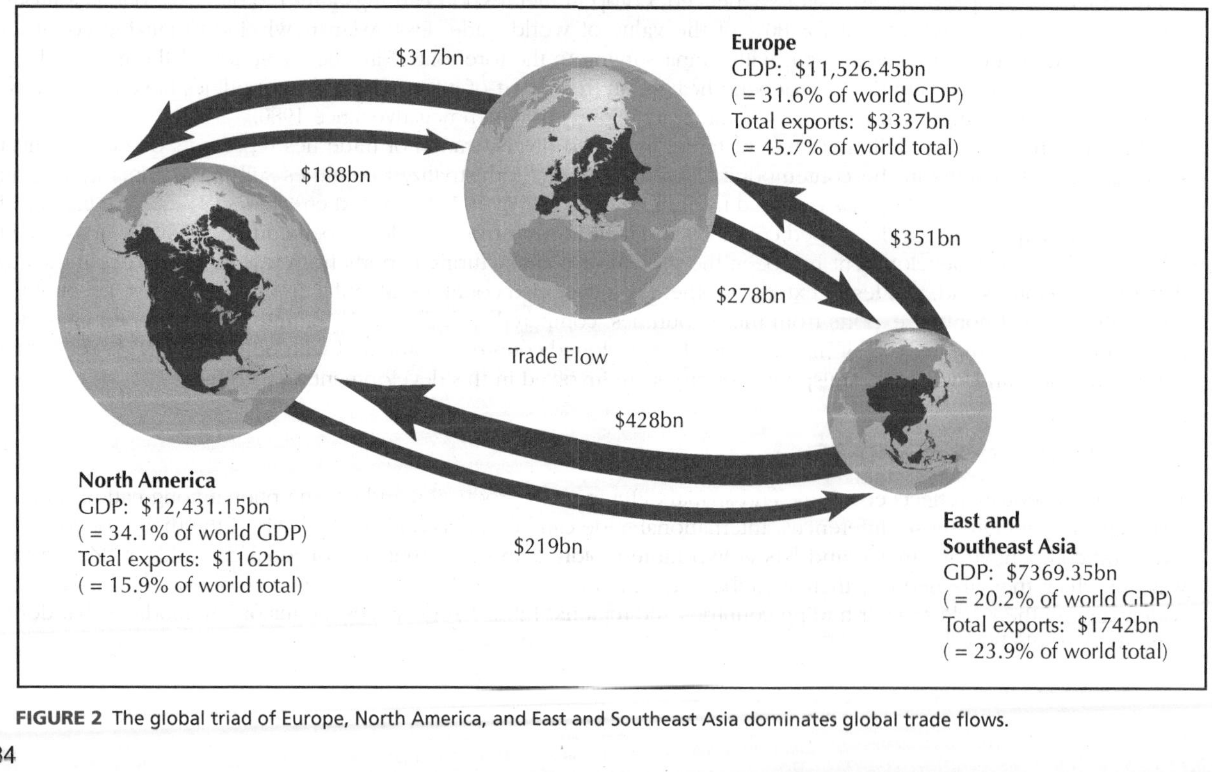

FIGURE 2 The global triad of Europe, North America, and East and Southeast Asia dominates global trade flows.

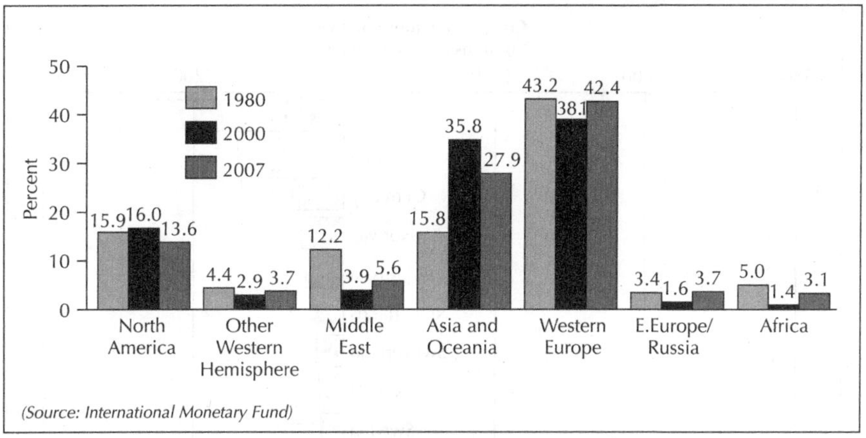

FIGURE 3 Distribution of world exports, 1980, 2000, and 2007. Western Europe leads the world in the distribution of world exports, a reflection of the large size of the European Union.

TABLE 1 Total World Exports of Goods and Services, 2007

	Value ($ billions)	%	% Change from 2006
Agricultural products	1128	6.8	19
Minerals and fuels	2659	16.0	15
Manufactured goods:	9500	57.3	15
Iron and steel	474	2.9	27
Chemicals	1115	6.7	19
Automobiles and parts	1183	7.1	16
Textiles and clothing	583	3.5	10
Office and telecom equipment	1514	9.1	4
All commercial services	3290	19.8	18
Transportation	750	4.5	19
Travel	855	5.2	14
Other services	1685	10.2	20
TOTAL	16,577	100.0	8

trade deficit was about $780 billion, compared with $75 billion in 1993. When trade in services and returns to capital investments are included (i.e., in the **current account**), the enormous size of the U.S. deficit puts it in a unique position internationally (Figure 4). The major U.S. trading partners include Canada, China, Mexico, and Japan (Table 2).

U.S. MERCHANDISE TRADE Figure 5 shows the composition of U.S. merchandise trade with the world in 2007. More than 70% of U.S. exports are manufactured items. Machinery and transportation equipment accounted for the largest single proportion of exports (43%), whereas agricultural products amounted to just 8%. Despite decades of deindustrialization, the United States remains the world's largest manufacturer today, although much of its output is consumed domestically. In part this status reflects the rounds of investment that accompanied the microelectronics revolution, in which computerization

TABLE 2 Major U.S. Trading Partners, 2008

	Exports ($ billions)	Imports ($ billions)	Total ($ billions)	% of Total Trade
Canada	261.4	335.6	596.9	17.6%
China	71.5	337.8	409.2	12.0%
Mexico	151.5	215.9	367.5	10.8%
Japan	66.6	139.2	205.8	6.1%
Germany	54.7	97.6	152.3	4.5%
United Kingdom	53.8	58.6	112.4	3.3%
Korea, South	34.8	48.1	82.9	2.4%
France	29.2	44.0	73.2	2.2%
Saudi Arabia	12.5	54.8	67.3	2.0%
Venezuela	12.6	51.4	64.0	1.9%
Brazil	32.9	30.5	63.4	1.9%
Taiwan	25.3	36.3	61.6	1.8%
Netherlands	40.2	21.1	61.4	1.8%
Italy	15.5	36.1	51.6	1.5%
Belgium	29.0	17.4	46.4	1.4%

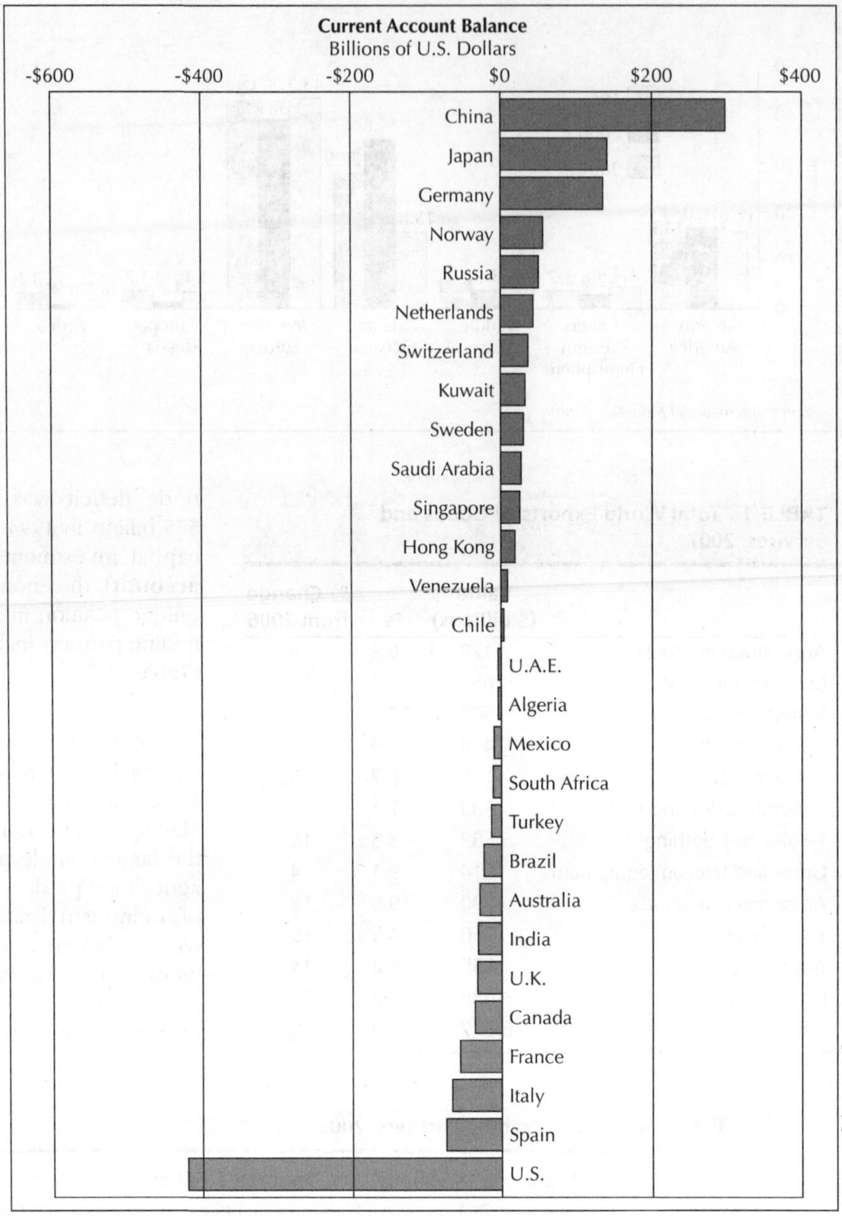

from www.DollarDaze.org. Used by permission

FIGURE 4 The current account balances of the world's major economies reveal the substantial size of the U.S. deficit in comparison to other countries. China has the largest positive current account balance, and the United States has the largest negative balance. China has never accepted the rules of the World Trade Organization, only following them when they suit its interests. If China were to revaluate its currency, to some 20% higher, this revaluation would immediately create fewer imports to the United States and more exports from the United States, creating 1 million new jobs in the United States. As the main architect and guardian of the world economy's old order, the United States faces a difficult choice: resist Chinese world trade ambitions and risk a trade war, in which everyone loses; or do nothing and let China remake the world trading system.

dramatically raised productivity levels in steel, automobiles, and other sectors.

Manufactured goods also account for approximately 75% of the traded goods flowing into the United States. The country has essentially farmed out much of its labor-intensive manufacturing to developing countries, shipping semifinished goods to countries such as Mexico for manufacturing and reimport, a production system that underpins Mexico's maquiladora system. At the same time, the United States exhibits an expensive taste for foreign-made items such as automobiles from Germany and Japan; shoes from Italy and Brazil; electronic items, apparel, and toys from the Far East, and perfume and wine from France. Other imports include one-half of the country's fossil fuels supply.

The destination of U.S. exports has gradually undergone an important shift—away from the traditional European markets and toward East Asia, Mexico, and Canada. The reasons behind this change include the rapidly growing economies of Asia, so that trans-Pacific trade now exceeds that across the Atlantic. Another reason is the progress made under the World Trade Organization in liberalizing tariffs in many countries. The North American Free Trade Agreement (NAFTA) was such a liberalization agreement and has increased trade among Mexico, the United States, and Canada, although

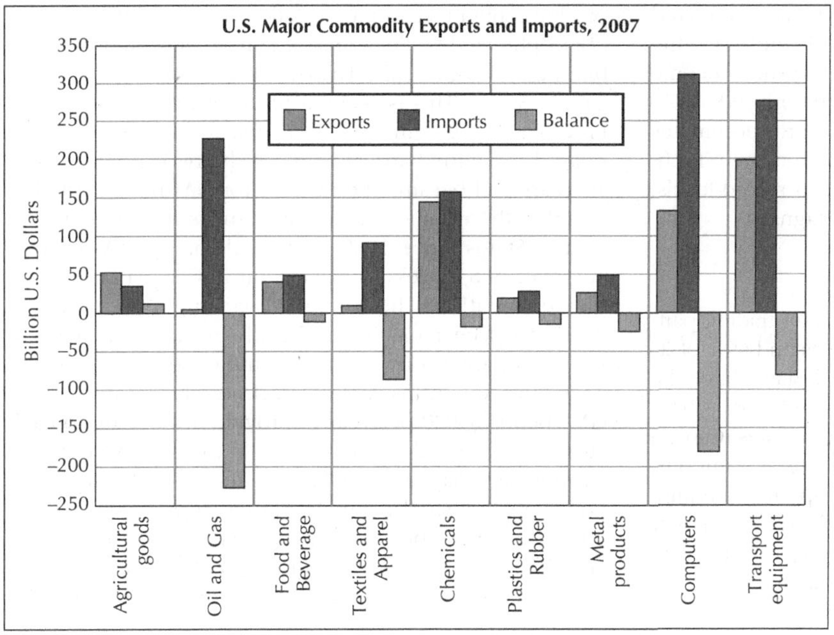

FIGURE 5 U.S. major commodity exports and imports, 2007. The trade deficit has exploded because imports of manufactured goods, especially automobiles, exceeded exports.

the process has generated both winners and losers. In addition, there have been increasing amounts of U.S. and foreign investment in the Asian economies, and this has produced rapid increases of trade in capital and intermediate goods. Investment has also come from Chinese and Japanese multinational corporations.

There are several manufacturing sectors in which the United States is competitive internationally, including medical equipment, transportation equipment, and computer hardware. The medical equipment sector is one of the most competitive sectors of the U.S. economy; its export growth has averaged 15% per year over the past 5 years. Two major U.S. exports are aircraft and motor vehicles. In 2009, the aircraft sector had a $20 billion surplus, but the U.S. auto sector showed a trade deficit of $120 billion, which wreaked havoc with companies like Ford and General Motors. World sales of motor vehicles grew just 1.2% annually over the past 10 years in the developed markets because of the prolific manufacturers and fell during the recession of 2007–2009. The U.S. computer equipment industry commands more than 75% of the world's computer sales, but the United States has had a trade deficit in this sector since 1992, as semiconductor production and assembly of consumer electronics moved to East Asia and Mexico.

U.S. SERVICES TRADE The United States is a world leader in computer software, supplying 49% of the $153 billion world market for packaged software per year. The world market for software grew 35% annually between 1995 and 2009. Asia and Latin America are the fastest-growing markets for computer software. Implementation of **intellectual property rights** through the World Trade Organization will reduce piracy, which has been a major problem for the industry.

The United States is also the world leader in the production, use, and export of information services and commands one-half of the world market. American firms produce and use the most advanced software, and exports have grown rapidly. The information industry is especially affected by government policy regarding market access, intellectual property rights, privacy protection, data security, and telecommunications services. The largest markets for U.S. information services will be the industrialized countries, especially the United Kingdom, Japan, and Canada. The most important emerging markets will be China and Korea, followed by Mexico, Brazil, and Argentina.

Services do not just consist of fast food and travel, but also information software, telecommunications, advertising, and entertainment. In the United States, services account for about one-fourth of gross domestic product (GDP) and 30% of total exports. Finance is another important service area. The U.S. stock market's gains rallied most equities worldwide, spurring companies to set records for mergers, acquisitions, and initial public offerings. Simultaneously, U.S. investment bankers were heavily involved in cross-border mergers and acquisitions.

The U.S. economy has reemerged as a magnet for the world's surplus savings because of its superior growth compared to other industrial nations. One reason for the better U.S. economic performance is that labor is not demanding its piece of the pie as it is in Europe and Japan. Europe's higher minimum wage, greater degree of union power, and government support for the disadvantaged have all put a floor under the wages of the lowest paid. This means that income polarity is far more pronounced in the United States than elsewhere. In the United States, wage earners in the top 10% income bracket make

4.4 times what those in the bottom 10% make, compared with 2.5 in Europe and Japan. Between 1979 and 2008, the poorest one-fifth of American families saw their income drop by 14%, while the wealthiest one-fifth enjoyed a 39% increase. In the United States, capital clearly dominates labor; in Europe, that is not true to the same degree. Profit ratios of publicly held U.S. companies are at record levels, while average family incomes remained stagnant.

Canada

The United States is Canada's most significant trading partner, accounting for 82% of Canada's exports and 69% of its imports (Table 3). Canada and the United States have the largest bilateral trade relationship in the world. For a country roughly one-tenth the size of the United States, Canada accounts for 21% of U.S. exports and 20% of U.S. imports.

Canada exports automobiles and transportation equipment, industrial supplies, and industrial and machinery parts, which combined account for 60% of Canada's total exports. Canada also has vast supplies of natural resources, including forest products, iron ores, metals, oil, natural gas, and coal. On the other hand, Canada imports industrial plant and machinery parts, transportation equipment, and industrial supplies from the United States. High-tech manufactured goods are also a chief import from the United States. Other American exports to Canada include automobiles, trucks, vehicle parts, paper and paperboard, computers, and software. The best prospects for U.S. exports to Canada include computers and peripherals, automotive parts, telecommunications equipment and automobiles. Canada's second most important regional partner is Asia, followed by Western Europe and Japan.

Canada's economy is markedly more internationalized than that of the United States. Whereas the United States exports approximately 13% of its output, Canada exports approximately 21% of its GDP. Because of its relatively small population, Canada cannot attain the economies of scale necessary for highly efficient plants and, therefore, imports many of its goods. As with many small countries, especially in Europe, the smaller the country, the more dependent it is on foreign markets for imports and exports.

Because of its vast amounts of hydroelectric power and its ability to manufacture hydraulic turbines and electric generators, Canada exports energy resources. It also produces high-tech communications equipment, including fiber-optic cables. For Canada, automobiles and automobile parts represent the largest category of exports to the United States. This is partly a result of the U.S.–Canadian Free Trade Agreement (FTA), enacted in 1989, which phased out trade barriers between the two countries over 10 years and became the nucleus for NAFTA. Canada has doubled the number of cars and trucks it exports to the United States since 1989 as a result of NAFTA. It costs automakers $300 less to make a car in Canada because of Canada's national health care financing system, which offers for free most of the health insurance that some U.S. companies buy for their employees. (In contrast, 47 million Americans currently lack health care insurance.) GM spends $5000 per year to provide health care to each current and retired U.S. employee but less than $1000 per current Canadian worker. Free trade has come at a cost, however: Industrial restructuring has improved Canada's competitiveness but many manufacturing workers were laid off in the process. Unemployment, while declining, still remains high, making job creation the government's chief objective.

A fast-growing sector of U.S.–Canadian trade is U.S. service exports, which jumped from $10 billion in 1988 to $17 billion in 2009. Canada's financial services market continues to expand as a result of the 1987 accord between the Toronto Securities Commission and the U.S. Department of Finance. This agreement allows the deregulation and integration of the financial securities industry, removing the distinctions among banks, trusts, insurance companies, and brokerages.

Like the United States, Canada has seen more recent trade growth with Asia and the Pacific Rim than with its traditional trading partner, Western Europe. Since 1980, the Pacific has eclipsed the Atlantic as the leading arena of North American commerce.

The European Union

Europe's trade, as a proportion of total world trade, is disproportionately large compared with its population of one-half billion people (Table 4). The European Union (EU) is the largest trading block in the world, with exports and imports each totaling about $1.9 trillion in 2008. The EU ranks second as an export market to the

TABLE 3 Canada's Major Export and Import Sources, 2008 ($ billions)

Region	Exports	Imports
Europe	36.3	51.4
United States	332.1	273.7
Asia	21.3	56.8
Russia	3.2	4.3
Middle East	3.7	4.4
Latin America	10.4	27.9
Africa	2.7	8.4

TABLE 4 European Union's Major Export and Import Sources, 2008 ($ billions)

Region	Exports	Imports
Europe	3926.1	3903.5
United States	354.5	247.2
Asia	395.7	709.7
Russia	120.8	196.9
Middle East	132.3	91.4
Latin America	74.4	110.0
Africa	138.1	162.2

United States, after East Asia. Although it possesses only one-fifteenth of the world's population, it accounts for 50% of world trade because of (1) the strength of the EU, (2) short distances and well-developed transport systems among member countries, and (3) complementary trade flows among its smaller states. Some European countries are comparable in population and size to individual U.S. states. The proximities and complementarities of Europe, with many relatively small countries close to one another, make intraregional trade ideal: 75% of all exports from European nations go to other European nations. Italy, France, and the United Kingdom have populations of 60 million each, while Germany has roughly 80 million. Other countries are much smaller. Some have food resources, such as Denmark and France; some have energy resources, such as Norway, the Netherlands, and the United Kingdom; some produce iron, steel, and heavy equipment, such as Italy, Germany, France, and Spain; and others produce high-value consumer goods.

The EU faces structural problems, such as high labor costs, economic rigidity that stalls the growth of smaller companies, industrial obsolescence, and costly social welfare programs, which tend to create relatively high levels of unemployment. However, the EU is the fastest-growing market for U.S. high-technology exports and remains the principal destination for U.S. foreign direct investment (FDI). The EU is as well the largest source area for FDI in the United States. The leading U.S. exports to the EU are aircraft; data processing and office equipment; engines and motors; measuring, checking, and analyzing equipment; and other electronic equipment. The most promising sectors for U.S. exports to the EU include telecommunications equipment, computer peripherals, software, electronic items, pollution control devices, machinery, medical equipment supplies, and aircraft.

The leading economies of the EU include Germany, a major exporter as well as importer of automobiles. Britain, which has not accepted the euro, is a significant exporter and importer of manufactured goods. The economy of France, which is more agricultural and less productive than Germany's, is nonetheless a significant producer and consumer of industrial and consumer goods. Italy, the smallest of these four, exhibits a growing strength in the export of engineering products.

Most international trade between the United States and Europe takes the form of *intra-industry trade*—investment in foreign affiliates that produce abroad, rather than shipments of U.S.-produced goods to target export markets. These sales accounted for more than half of U.S. affiliate sales worldwide and were four times as large as U.S. affiliate sales in Canada or Asia. Consequently, Europe's overall importance for U.S. companies and the U.S. economy is much greater than trade statistics indicate. These U.S. companies' overseas affiliates are major importers of products manufactured in the United States. In 2009, sales by U.S. parent companies to their European affiliates made up 40% of U.S. exports to Europe.

Latin America

Latin America comprises a group of countries with different levels of population, income, and economic development. Latin America has some of the poorest countries in the world (e.g., Bolivia and Paraguay in South America, Haiti in the Caribbean, and Nicaragua and Guatemala in Central America). For centuries, under Spanish colonialism, their traditional role in the world economy was to provide primary materials, namely agricultural exports and mineral resources, to Europe and North America. This pattern has been typical of less developed nations. Latin American countries today are diverse not only with regard to population and size but also with respect to development and natural resources. Some, such as Argentina, have grain surpluses; others, such as Venezuela and Mexico, are rich in iron ore and oil. Brazil has a wealth of minerals and is a strong producer of manufactured goods. In 2008, 62% of Latin America's exports, mainly food, minerals, and fuels, went to the United States. For a long time, Latin America had an import-substitution policy for industrialized products. However, Latin America's new hope to achieve wealth and a prominent place in the world economy is centered on export-led industrialization, led by Mexico, Brazil, and Chile.

MEXICO The chief trading partner of most Latin American states is the United States. Mexico's balance of merchandise trade is mainly in labor-intensive manufactured products (52%), many of which flow back to the United States from plants along the border (**maquiladoras**) that are owned by U.S., European, and Japanese multinational corporations.

Petroleum and its by-products, as well as agricultural products, account for 45% of Mexico's exports. Although Mexico is one of the world's largest exporters of energy, oil provided only slightly more than 35% of its export revenue in 2008. Sixty percent of Mexican imports are semifinished industrial input materials for final production; another 23% is manufacturing and plant equipment. These types of imports are necessary for Mexico to maintain its status as a newly industrializing nation.

Following the enactment of NAFTA on January 1, 1994, Mexican markets continued to be open to foreign competition. Mexico's modernizing economy is no longer protected from foreign competition and an improved investment environment. Mexico today is the third-largest market for U.S. exports, after Canada and Japan. Foreign investment opportunities are particularly strong in Mexico, where a neoliberal government has welcomed foreign capital and privatized airlines, highway construction, railroad services, and water and energy projects.

NAFTA generated costs and benefits for Mexico, as it did in the United States. Inflation dropped, but more than 1 million farmers were driven out of business by subsidized U.S. grain exports. Under NAFTA, half of U.S. exports to Mexico have not been subject to Mexican tariffs—semiconductors, computers, machine tools, aerospace equipment, telecommunication and electronic equipment, and medical supplies. Since 1994, when Mexican import barriers were sharply reduced or eliminated under NAFTA, autos, auto parts, semiconductors, machine tools, and certain fruits and vegetables, including apples, realized export increases of between 100% and 10,000%.

In short, NAFTA has neither lived up to the fervent expectations of its proponents nor fulfilled the most dire warnings of its opponents. It has enhanced Mexico's ability to supply U.S. manufacturing firms with low-cost parts but has not made Mexico economically independent. To some extent, Mexico's growth has been limited by American barriers to Mexican exports, including tomatoes, avocadoes, and trucking services. Finally, it must be remembered that compared to the EU, NAFTA is much more modest; for example, it has failed to lift regulations on cross-border movements of labor.

SOUTH AMERICA South America represents a large and diverse picture of economic growth, change, and stagnation. The southern countries of Argentina, Uruguay, Chile, and Bolivia have had strong ties to Western Europe. Each country is more tied economically to Europe, the United States, and East Asia than to one another. However, trade among many South American countries is facilitated by the *Mercosur Trade Agreement*, which eliminated tariffs on most goods traded among the five member states (Argentina, Brazil, Paraguay, Uruguay, Venezuela).

As a result of high indebtedness and high interest rates, and low prices for many export goods, some Latin American countries can barely keep up with debt service on their international loans. Mexico, Brazil, and Argentina each owe over $100 billion to the developed world, and several other Latin American countries owe close to this amount. Most of the loan money was put into urban **infrastructures**, but high world interest rates, oil prices, tariffs and agricultural subsidies in the developed world, and international recessions have minimized exports and thus foreign revenues, which are necessary to repay the debt. Brazil is a case in point. Exports are a little more than one-tenth those of the United States, but the population is approximately two-thirds that of the United States. Latin American governments have been forced by the International Monetary Fund (IMF) to devalue their currencies and to invoke austerity programs by restructuring their economies, raising taxes, decreasing public expenditures, and selling government-owned business enterprises, such as state banks, power companies, metal refineries, and transportation and airline companies.

Argentina, which after World War II enjoyed a standard of living comparable to Europe, suffered a steady decline in the late twentieth century. To find favor with working-class voters, President General Juan Perón engaged in **import substitution**, outlawed outsourcing, developed legislation preventing employers from dismissing employees, and strengthened unions' ability to strike, a pattern that was followed in South America during the 1940s through the 1960s. The Peronist populist actions, intended to reduce the gaping inequalities in Argentine incomes, did not achieve their stated purposes: reduction of poverty and real-income disparities. Instead, they resulted in hyperinflation lasting for decades. The Argentine government has faced a floundering economy hit by hyperinflation, high budget deficits, and an overcapitalized public sector.

In Brazil, erratic domestic policies and high inflation persist, accentuating its high degrees of poverty and inequality. Brazil continues to be pressured by the need to maintain high interest rates to finance domestic government debt and borrowing and to prevent capital outflows from the country. Domestically produced ethanol and recent discoveries of offshore oil have allowed the country to become self-sufficient in energy. Brazil largely exports primary-sector products, such as soybeans, iron ore, and coffee, but also exports transportation equipment and metallurgical products. The largest single-country market for Brazilian exports is the United States, accounting for 20% of its total exports (Table 5). Brazil remains the third-largest market in the Western Hemisphere after Canada and Mexico.

East Asia

The fastest-growing world trade region is East Asia and the Pacific. After Europe, this region has the largest amount of internal world trade, amounting to 23% of the world's total. In this region, Japan traditionally took the lead role. Since the 1960s, Japan has been joined by the **Four Tigers** of Taiwan, South Korea, Singapore, and Hong Kong. More recently, however, new emerging tigers followed suit with rapidly growing economies: Thailand, Malaysia, Indonesia, and China.

While the rest of the world reeled from two major oil-price hikes in the 1970s, East Asia forged ahead with unprecedented growth. Several factors contributed to this success, including U.S. economic and military subsidies, the Confucian culture dedicated to learning, and governments that actively promoted a shift into export promotion. In addition, Japan and other countries protected home markets with high import duties. Unlike America, where short-term profits were important to satisfy stockholders, banks, and financial institutions, Japan encouraged reinvestment and long-term growth cycles. These long cycles

TABLE 5 Brazil's Major Export and Import Sources, 2008 ($ billions)

Region	Exports	Imports
Europe	40.2	35.8
United States	29.2	32.9
Asia	30.5	27.4
Russia	4.2	2.2
Middle East	9.2	3.4
Latin America	48.6	24.4
Africa	8.0	13.0

allowed firms time to develop products and to reinvest in the highest-quality production systems before the owners or employees could reap any of the profits. Further, many of the Asian/Pacific countries acted as resource supply centers for the United States from 1965 to 1975, during the Vietnam War, which allowed them to collect a large volume of U.S. dollars.

The East Asian/Pacific governments switched from import substitution to export promotion, with a new emphasis on electronics, automobiles, steel, textiles, and consumer goods, whereas the other developing countries in Latin America, Africa, and South Asia did not have such policies. From 1970 to 2000, foreign investment in the region, especially in Japan and the Four Tigers, grew tenfold. This investment was led not only by U.S., British, German, Canadian, and Australian firms but also by the Japanese.

The United States plays an important role in East Asian trade. Until 1970, Western Europe was North America's chief trading partner; after that, however, East Asia surpassed Western Europe. North American trade with Asia and the Pacific outpaced trade with Western Europe by 30%, nearly hitting the $450 billion mark.

Japan

The growth of the Japanese economy after World War II is nothing short of an economic miracle. Japan developed into the world's third-largest economy, after the United States and China. Japan is much wealthier than China, which has nearly 9 times the population, although China's economy is growing more quickly.

Japan is the world's third leading international trading nation (Table 6). Japan exports manufactured goods and imports raw materials, food, and industrial components. Japan's economy essentially involves converting raw materials into high-value-added products, with high inputs of technology and labor. Unlike Western Europe and North America, Japan enjoys a huge export trade surplus. Although the United States is by far its largest trading partner, both for exports and for imports, Japan has a diversified trading base. In 2008, 30% of Japanese exports went to the United States and Canada, and 25% of its imports came from Canada and the United States. After Canada, Japan has been the second-largest market for U.S. exports for many years, but the U.S.–Japan trade deficit widened to $130 billion. This constituted more than one-third of the U.S. merchandise trade deficit (the other largest U.S. trade deficit was with China).

TABLE 6 Japan's Major Export and Import Sources, 2008 ($ billions)

Region	Exports	Imports
Europe	82.2	56.2
United States	139.0	78.9
Asia	214.4	250.8
Russia	16.5	13.3
Middle East	34.1	167.4
Latin America	39.3	26.1
Africa	11.1	19.3

The United States sells fewer exports to Japan than it imports, adding to large U.S. trade deficits. The United States exports primary-sector goods such as grains (Figure 6), feed, fruit, lumber, and nonoil commodities to Japan in return for its imports of high-value-added manufactured items—electronics and automobiles, primarily. The United States does sell to Japan some high-value-added goods such as Boeing aircraft, and U.S. products still account for the largest single proportion of goods imported by Japan. Indonesia, Australia, China, South Korea, and Germany follow. The United States and Indonesia fill a large need for energy that Japan cannot meet domestically. In addition, because of Japan's mountainous terrain, agricultural and food

Scott Olson/Getty Images

FIGURE 6 Grains that are mass produced on large, efficient farms, such as this farm in Carmi, Illinois, can be transported quickly and cheaply due to economy of scale. Although agricultural products comprise a small share of international trade by value, because they are relatively cheap, they nonetheless form an important part of global commodities flows by weight and volume.

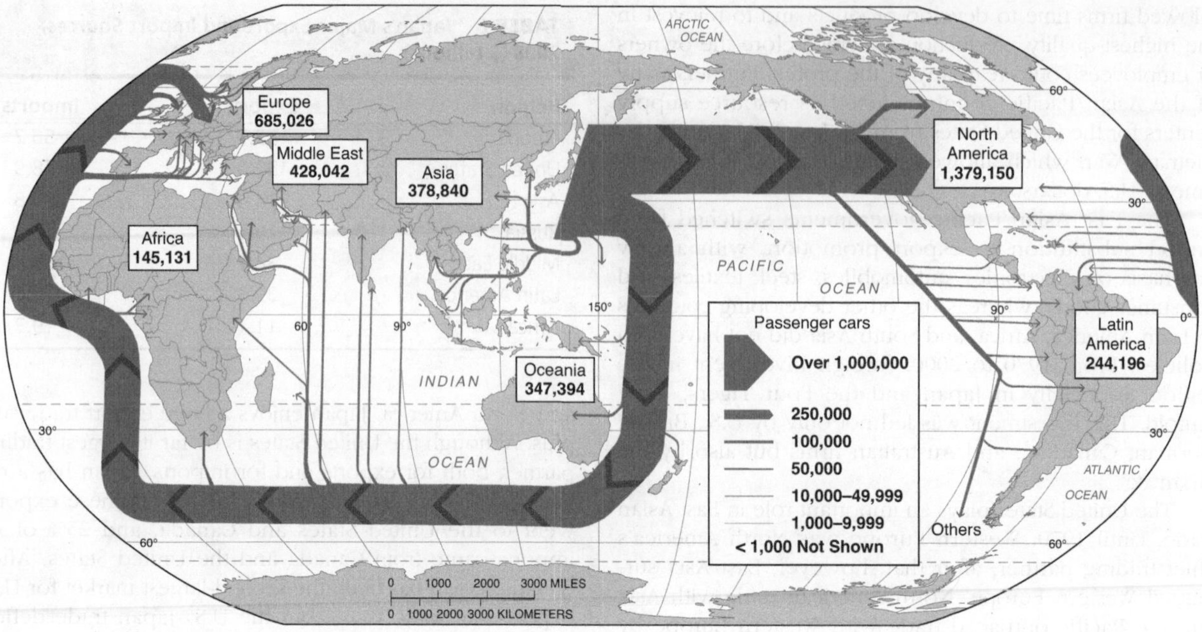

FIGURE 7 Geography of Japanese car exports. As the world's largest producer of automobiles, Japan controls a substantial share of the market in North America, Australia, and Europe.

products compose 15% of total imports. Chemicals, textiles, and metals are also imported.

Japan's trade surplus with Southeast Asian countries exceeded its surplus with the United States. As a group, Southeast Asian countries and the NICs are now a more important export market for Japan than is the United States. To counterbalance the impact of the high-priced yen, Japanese companies have been moving labor-intensive assembly manufacturing operations to Southeast Asian countries—mostly to Thailand and Malaysia, but more recently to China.

The world dominance of Japan in the manufacture of motor vehicles is phenomenal (Figure 7). Fully 22% of its exports are motor vehicles, followed by high-tech office machinery, chemicals, electronic components, iron and steel products, and scientific and optical equipment (Figure 8). The Japanese have been accused of dumping goods on local markets to weaken rivals and force

FIGURE 8 Forklifts and other technologies for moving and storing commodities reflect how automated such flows have become. The movement of commodities is essential to the functioning of markets, the realization of profits, and the reproduction of capitalism in general. When seen in this light, even everyday scenes acquire a new set of meanings.

them to sell their market shares to Japanese firms. At the same time, the Japanese have restricted foreign firms from selling in Japan by a huge amalgam of import duties, tariffs, and regulations.

Japan in the late twentieth century became a major player on the world financial scene. In 2008, Japanese banks were 2 of the 10 largest in the world (a decrease from the height of the "bubble economy," when they made up 6 out of 10), although Tokyo is still one of the world's premier financial centers, alongside London and New York. Japan has managed all this with only minimal food, mineral, and energy resources. It has shown the way for other East Asian and Pacific countries to follow suit.

In the 1990s, the Japanese economy endured a prolonged, serious recession prompted by the collapse of the so-called bubble economy. Equity values plunged and investment in North American properties tapered off to almost zero. The declining asset values left Japanese banks with nonperforming loans and precarious balance sheets. With a stable or declining population, Japan experiences low rates of growth in its labor force. Moreover, the Japanese government, plagued by corporate scandals, has been unable to deregulate and privatize extensively. The economy is facing economic restructuring similar to that of North America (i.e., away from manufacturing and into services). Although Japan has liberalized its trade relations considerably since 1990, removing many tariffs and quotas, it still has numerous nontariff barriers in place that discourage imports.

China

Following a long period of isolation lasting into the 1950s and 1960s, China opened up to international trade and investment after the death of Mao Zedong in 1976. During the 1980s, under the policies of Deng Xiaoping, China allowed foreign companies to set up joint ventures there. **Special Economic Zones (SEZs)** were created along the coast to produce goods for world markets. These economic zones received tax incentives but were subject to much legal red tape. Today, China has become a major actor in world trade (Table 7). It has an enormous pool of workers, low wages, and relatively high levels of worker productivity. The result has been a dramatic increase in foreign trade, the primary trading partners being Japan and the United States. Over the last two decades, China recorded an annual growth in GDP of 8% to 14%, continuing the surge in investment-led growth that began in the 1980s. Indeed, if current rates of economic growth continue, China may well be the world's largest economy by 2040 (Figure 9). Already, China's rapidly growing economy has led it to import vast quantities of raw materials (Figure 10): Middle Eastern oil, iron ore from Australia, soybeans from Brazil, and numerous minerals and oil from Africa.

By all measures, however, China is still a poor country. It struggles to provide its many people with sufficient food and housing. Relatively little capital is available for start-up programs; therefore, China has opened doors to foreign companies, especially in the industrial sector. Foreign investment has flooded in to establish factories, to

TABLE 7 China's Major Export and Import Sources, 2008 ($ billions)

Region	Exports	Imports
Europe	223.3	108.1
United States	273.1	80.7
Asia	133.6	127.9
Russia	32.6	25.4
Middle East	60.8	82.8
Latin America	70.1	71.8
Africa	42.4	4.7

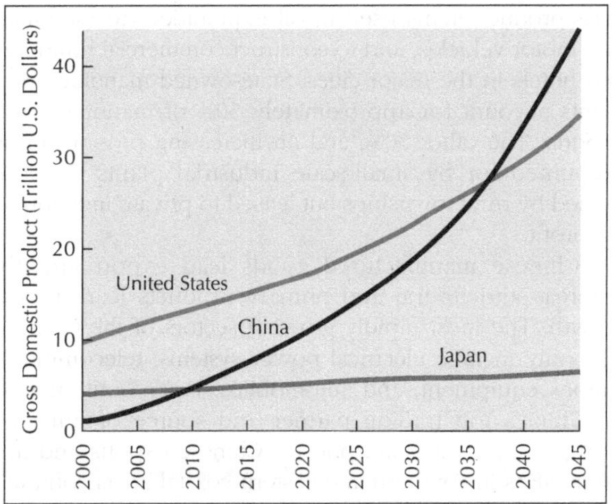

FIGURE 9 China's rapid economic growth, which has made it the world's second-largest economy, may see it overtake the United States by the year 2040 to become the world's largest economy.

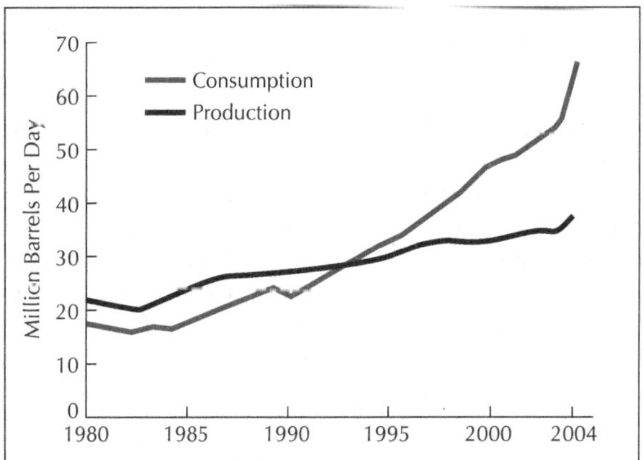

FIGURE 10 China's oil demand, 1980–2004. As China industrializes, its demand for raw materials, including petroleum, has become voracious. Rising demand for oil will inevitably lead to a long-term increase in prices paid at the pump.

mass-produce items used in oil exploration, to manufacture motor vehicles, and to construct commercial buildings and hotels in the major cities. State-owned manufacturing plants account for approximately 50% of manufacturing exports. The other 50%, and an increasing proportion, is accounted for by small-scale industrial plants that are owned by rural townships but leased to private individuals for profit.

Chinese manufactured goods lead export growth, whereas agricultural and primary products lead import growth. The most rapidly growing sectors of the Chinese economy include electrical power systems, telecommunications equipment, and automobiles. Japan continues to be China's top trading partner and source of imports. Hong Kong is the main port for Chinese exports and imports. SEZs in southern and eastern coastal areas continue to outpace the rest of the nation in economic growth and trade. Imports from the United States—cotton, fats and oils, manufactured fibers, fertilizer, aircraft, wood pulp, and leather—have been growing rapidly, but the United States ranks fourth after Japan, the EU, and Taiwan. The U.S. trade deficit with China has increased steadily and stood at about $226 billion in 2009, more than one-third of the American total. In the near future, China is expected to dislodge Japan as the country with which America has its biggest trade deficit.

Imports to the United States from China have grown 30% per year and exports by 15%. Part of this imbalance is due to China's intricate system of import controls. Most products are subject to quotas, licensing requirements, or other restrictive measures. Three out of four toys sold in America are foreign-made, and 60% of those imports come from China. Sixty percent of all shoes sold in America come from China. Other major goods imported into the United States from China include clothing, telecommunications equipment, household appliances, televisions and computer chips, computers, and office equipment. China's higher trade barriers protect the inefficient state-run companies that still employ two-thirds of the urban workers.

China's transition to economic superpower status continues unabated. The economy has grown at an annual clip of about 9%. In 2003, China began to institute private property laws allowing personal ownership of land and buildings. As a result, real estate tycoons abound in China today, especially in coastal cities like Shanghai. China is becoming the world's factory, with the growing trade surpluses to prove it. Ford and General Motors plants churn out autos for sale in East Asia. Most Chinese cannot yet afford autos, but televisions, washers, and stoves are selling well. The web of interconnections between foreign companies headed by overseas Chinese investors and a nominally communist China is growing.

Taiwan

Taiwan, with a population of 23 million and a GDP of $700 billion, sustains a 6% to 7% economic growth rate. Long-term growth and prosperity have led to increasing land and labor costs in Taiwan and a gradual restructuring of the economy as low-wage sectors have fled. Manufacturers of labor-intensive products such as toys, apparel, and shoes and circuit boards have moved offshore, mainly to Southeast Asia and to China. Large state-run enterprises still account for one-third of Taiwan's GDP, but major efforts are underway to privatize power generation.

Exports account for over 35% of GDP in Taiwan's export-oriented economy—one of the four newly developing NICs, or Four Tigers. Manufacturing growth is now concentrated in technology-intensive industries, including petrochemicals, computers, and electronic components. Historical trade patterns on the Pacific Rim are likely to keep Taiwan dependent on capital goods imported from Japan. Over the past 10 years, Taiwan has made great progress in lowering trade barriers and improving market access to foreign goods.

South Korea

Another miracle of the Pacific Rim is South Korea, a country of 48 million, with a GDP of $1.3 trillion in 2008. Over the past 10 years, its growth in GDP has averaged over 8%, and its GDP per capita is $26,000 (which compares favorably with the $1700 GDP per capita in North Korea), putting South Korea's standard of living on a par with parts of southern Europe (e.g., Portugal). South Korea's rapid advancement has turned this nation into one of the most economically powerful in the world (Table 8).

South Korea's largest export sectors include transportation equipment (e.g., Hyundai), computers and peripherals, and ships (Figure 11). It is implementing an ambitious transportation infrastructure development program that includes major high-speed rail and transit programs, airport development, and highway construction. Once the world's largest producer of tennis shoes, South Korea's success has driven up labor costs to the point where companies such as Nike have fled to lower-cost countries such as China, Vietnam, and Indonesia.

Australia

Australia's main trading partners are the United States, the EU, and Japan. Exports go primarily to Japan, which accounts for 20% of GDP. The next largest share, 10%,

TABLE 8 South Korea's Major Export and Import Sources, 2008 ($ billions)

Region	Exports	Imports
Europe	34.9	33.2
United States	45.3	38.3
Asia	151.1	120.4
Russia	10.2	9.2
Middle East	20.4	82.0
Latin America	28.5	12.6
Africa	11.1	6.5

The World Bank

FIGURE 11 Pusan, South Korea, is a major shipbuilding port that facilitates that country's exports of steel. The term "East Asian miracle" describes the rapid growth and development of South Korea and other countries after World War II. It was a remarkable achievement, bringing with it unprecedented improvements in living standards for East Asia and for the majority of people on the planet. The rate of global shift speaks to these countries' achievements. Because of globalization, the Confucian work ethic and good policies, many developing countries are catching up with their more developed neighbors. But there is another explanation for the rapid shift in the global center of economic gravity toward the Far East: the comparative lack of growth in the rich economies of America, Western Europe, and Japan.

goes to South Korea, followed by New Zealand, the United States, China, and Singapore. The EU accounts for approximately another 11%. Because of its small population, 19 million, industrial supplies, automobiles, and industrial equipment account for more than 60% of imports. Japan has made its greatest market penetration into Australia and accounts for 50% of all vehicles purchased there. The United States leads the list of importers, accounting for 23%, followed by EU countries with 35%, and Japan with 13%.

Australia is one of the leading raw-material suppliers in the world. It exports primary-sector products ("rocks and crops"), mainly ores and minerals, coal and coke, gold, wool, and cereals. It is the world's largest exporter of iron ore and aluminum and the second- or third-largest exporter of nickel, coal, zinc, lead, gold, tin, tungsten, and uranium. Consequently, Australia's current problem is to withstand the declining world prices of raw materials. To cushion against fluctuations in these prices, Australia needs to industrialize so that it can transform its raw materials into finished products and become an exporter of higher-value items. Doing so is nearly impossible with a small industrial base that demands consumer products before industrial products. It has, however, become a significant exporter of wine.

India

In South Asia, India has the world's second-largest population, with 1 billion people, but a relatively small economy, a reflection of the huge pools of poverty found there. Its world trade is minuscule but growing (Table 9). As a result of the Green Revolution, India is self-sufficient in food production. Today, it is an exporter of primary products, gems and jewelry, textiles, clothing, and engineering goods. In order for its factories to operate, it must import industrial equipment and machinery and crude oil and by-products, as well as chemicals, iron, and steel. Twelve percent of its imports include uncut gems for its expanding jewelry trade.

In international trade, India is no longer dominated by its former colonial overseer, Britain. It exports primarily to the United States and Japan. Britain and other EU countries account for 21% of total exports and 27% of total imports. Because India represents such a large pool of demand, most manufactured goods and consumer goods are consumed locally, not exported. Since 1990, India has also become an exporter of cereals and grains, and textiles and clothing are now the chief exports.

India's GDP has continued to increase at a rate of 5% to 6% annually. Record levels of foreign capital investment have stimulated a capital market over the last decade. But Indian public investment and infrastructure continue to be insufficient for a country with developmental goals. The Indian government began economic reforms in the 1990s to liberalize and deregulate the economy, privatize government-owned industry, and open India to international competition. As part of the global trend toward neoliberalism, the IMF persuaded India to turn its back on a policy of trade protection and import substitution that had been in place since the country became independent in 1949. Import tariffs were slashed, and the government

TABLE 9 India's Major Export and Import Sources, 2008 ($ billions)

Region	Exports	Imports
Europe	28.8	37.9
United States	24.5	20.5
Asia	39.1	57.9
Russia	1.5	6.2
Middle East	28.8	18.5
Latin America	6.2	6.5
Africa	10.7	4.6

loosened its hold on business by dismantling the licensing system that governed all economic activity, moving strongly toward deregulation of the private sector.

Most economic growth has been concentrated in western India, a region with a long history of trade ties. Mumbai (formerly Bombay) has become a significant financial and media center (e.g., Bollywood movies). One of the results is the accelerating development of information technology centers and industrial parks around the city of Bangalore, where companies produce software for international markets. Bangalore is India's Silicon Valley, and foreign multinationals have set up operations in a huge science park; it is also a major region for call centers.

The biggest export opportunities to India from developed regions of the world include electronic components. Because of increasingly stringent environmental regulations and growing industry awareness, markets for pollution control equipment are increasing at an annual rate of 40%. Food processing and packaging equipment is in demand as India's agricultural sector employs 70% of the country's workforce.

South Africa

The Republic of South Africa, with a population of 52 million and a GDP of $200 billion, is the most productive economy in all of Africa and accounts for almost 50% of the entire continent's output. Manufacturing now accounts for 19% of GDP, which indicates a diversification from the traditional African dependence on gold, copper, and diamond exports.

From World War II to 1995, South Africa practiced the policy of apartheid, or racial separation, that kept economic power concentrated among a few large economic enterprises. The remarkable peaceful transfer of power that occurred when apartheid collapsed produced an upswing in business confidence and a relaxation of foreign embargoes against South Africa. Job growth has been sluggish, however, and unemployment is 40% among blacks. Apartheid also produced widespread illiteracy, unemployment, and social problems that will be expensive to remedy and will take years to overcome. In addition, one-third of the South African adult population is infected with HIV. Nevertheless, the developed countries find South Africa to be an attractive market because of the pent-up demand for goods and services. Leading U.S. exports to South Africa include aircraft and aircraft parts, industrial chemicals, computer software, pharmaceuticals, medical equipment, telecommunications equipment, and building and housing products. The United States is South Africa's largest trading partner.

Russia

The disintegration of the Soviet Union in 1991 was one of the most momentous occurrences in recent history. It was then that Russia and its former client states in Eastern Europe began a long, slow, and painful transition to market economies. While Eastern Europe, to one degree or another, managed this transition relatively well, in Russia, transition led to economic collapse. The old central-planning systems have broken up, but the new systems that will replace them are not yet fully developed.

For many years following the disintegration of the USSR, Russia's average national income was less than it was before. For the majority of the population, this situation produced rising unemployment, crime, and poverty. Although rich in resources such as gold, natural gas, petroleum, and timber, Russia's government and economy have been so corrupt that the benefits of the sale of these assets have been concentrated in the hands of a tiny, wealthy oligarchy. Russia and its neighboring republics have been savaged by inflation. However, in the 2000s, the economy picked up, growing by 7% annually, with a 2008 GDP of $2.1 trillion. Unemployment has leveled off, but difficult inflationary battles continue, and the infrastructure is decaying. One sign of Russia's integration into the world economy is its oil industry, which has attracted the interest of transnational corporations.

What does the future hold for this region with regard to international trade? There is a certain complementarity in Eurasia. Western Europe needs the minerals, oil, natural gas, and other raw materials that are in vast supply in Russia and the republics of the former Soviet Union. At the same time, the eastern bloc nations need foodstuffs and industrial equipment and machinery to resume their economic production. Multinationals from every developed country are investigating the potential for investment in the former Soviet Union. Aeroflot, the National Electric Utility, and Gazprom (Russia's gas company) have become investment targets. Automobile manufacturers from Western Europe, Japan, and the United States are also investigating opportunities, as are consumer electronics producers. Although much business and property have been privatized, a third of all manufacturing plants remain largely in state hands as does agriculture.

The Middle East

The Middle East contains approximately two-thirds of the world's oil reserves, with Saudi Arabia accounting for more than one-third of the world's total. Other oil producers and exporters are Iraq, Iran, Bahrain, Kuwait, the United Arab Emirates, Oman, and Qatar (Figure 12). Countries in the region without significant oil supplies include Egypt, Israel, Jordan, Lebanon, Turkey, Yemen, Morocco, and Tunisia.

Inexpensive oil from the Middle East fueled the world (literally) for a long time. In fact, the United States and the Western European nations have enjoyed a large supply. Cheap Middle Eastern oil helped rebuild Europe and Japan after World War II. However, in 1973 and 1979, oil supplies were interrupted by Arab boycotts of the West for its support of Israel in the 1973 war, by the overthrow of the Shah of Iran in 1979, and by the Iraq–Iran war, and prices rose dramatically. As a result, OPEC's revenues soared and a worldwide recession was triggered, which accelerated deindustrialization in the West and essentially ended the post-WWII boom. However, because of squabbling among

FIGURE 12 Iranian oil field and peasant herder. Iran remains an important source of world oil supplies, but squabbles among OPEC members have kept world prices low.

OPEC members, particularly over the Iraq–Iran war, oil prices decreased to less than $10 per barrel in 1986. Oil exporters' revenues plummeted. In the 1990s, prices declined, and OPEC nations had lost much of their stranglehold on the world market, dropping to approximately to OPEC's pre-1973 levels. By 2008, however, oil climbed to $40 per barrel, only to decline again as recession gripped the West and demand fell.

While all these fluctuations in oil prices were occurring, other sources of oil, synthetic fuels, and solar, geothermal, and nuclear power sources were being explored. OPEC's largest market was Western Europe, which in the past had few supplies of fossil fuels. Before the oil crises, the Middle Eastern nations fulfilled 45% of Western Europe's energy needs. However, that proportion dropped as a result not only of the exploitation of the North Sea oil fields but also increased coal production in central Europe.

Outside of the oil-exporting countries, Israel exports cut and polished diamonds, machinery, computer software, and telecommunications equipment, and it is the only high-tech economy in the Middle East. Throughout the region, continued conflicts, such as in Iraq and that between Israel and the Palestinians, has depressed the tourist industry.

MAJOR GLOBAL TRADE FLOWS

Six major commodity groups merit further attention as fundamental to understanding international patterns of trade. microelectronics, automobiles, steel, textiles and clothing, grains and feed, and nonoil commodities.

Microelectronics

Microelectronics includes semiconductors, integrated circuits and parts for integrated circuits, and electronic components and parts. Japan and the East Asian countries, especially the Four Tigers of South Korea, Taiwan, Hong Kong, and Singapore, together dominate the world production and export of microelectronics (Figure 13). Although the United States no longer leads the world in the manufacture of semiconductors, it is still a major player in the global trade flow of microelectronics; the single largest flow of these products is from the United States to developing countries. Canada and the western European nations of the EU and the European Free Trade Area (EFTA) account for a much smaller proportion of overseas trade in this category. However, intra-European trade in microelectronics accounts for $15 billion.

Automobiles

Automobiles account for the largest single flow of trade within the EU. Exports of automobiles within and from Europe have been heavy. Germany is Europe's largest producer of cars, followed by France, Spain, and the United Kingdom. The United States imports large volumes of European automobiles as well, including Mercedes, Audis, Porsches, BMWs, Volkswagens, Peugeots, Fiats, and Renaults. Between 1960 and 2007, Japanese automobile manufacturers made major penetrations in the world automobile market (Figure 14), including the European automobile market. The largest volume of flow of trade in automobiles, however, is from Japan to the United States; automobiles are the commodity responsible for the largest share of the U.S. trade deficit.

The Big Three U.S. automakers scaled down operations substantially during the time that Japan was increasing its market share, although General Motors, which went bankrupt in 2009, and Ford remain among the world's largest automobile manufacturers. These U.S. companies have diversified, so both Ford and General Motors are now involved in substantial production for the international market. But Japan is continuing to capture an ever-larger share of the American automobile and light truck market. Much of the U.S. demand for cars is dictated by the price of gasoline: When oil is cheap, American consumers bypass passenger

FIGURE 13 An electronics plant in Guangdong, southern China, exemplifies the shift of that industry to the developing world, the commodification of labor that ensues, and capitalism's conquest of the Asian continent. During the world recession of 2008 and the sputtering recovery of 2009–2011, East Asia's expansion has eclipsed that of other parts of the world, attracting an influx of foreign capital. Within developing East Asia, China is growing the fastest, with its economy expanding at an annual clip of between 8% and 11% yearly. South Korea, Singapore, and Taiwan are now famous for their rapid post–World War II growth and development. But Thailand, Malaysia, Laos, and Mongolia are now also growing at almost double-digit rates.

cars for light trucks, minivans, four-wheel drives, and sports utility vehicles; when oil is expensive, they seek out fuel-efficient cars like those that Japanese manufacturers produce.

Japanese MNCs such as Toyota, Nissan, Mitsubishi, Mazda, and Honda have also invested heavily in factories in the United States, by which they can escape import quotas and have access to skilled, compliant labor. Starting with the Honda plant in 1982 in Marysville, Ohio, Japanese-made vehicles that are assembled and built in America account for nearly half the U.S. Japanese car sales. Locally made Japanese cars are cheaper than those imported from Japan because they escape the high prices of parts and labor associated with the yen. These American-made Japanese cars are called *Japanese transplants.*

Steel

America has lost as much as two-thirds of its steel employment in the past 30 years and now is a net importer of steel, but Western Europe continues to lead the world in steel production and trade. In addition, the EU sends billions of dollars worth of steel to developing countries, although the single largest flow of steel is from Japan to developing countries.

To be produced profitably, steel requires large and highly efficient plants, which are possible only with tremendous capital investments and large economies of scale. In the post–World War II period, steel made by traditional producers in Europe and North America became uncompetitive as new production centers began to emerge in Brazil, South Korea, Taiwan, and Japan. The migration of steel production to the Third World reflected the growing importance of labor costs, government subsidies, and taxes to the delivered cost of steel. There are a few problems of the British and U.S. steel industries: insufficient reinvestment, reluctant unions, narrow-minded management, and lack of government support of an ailing industry. However, under the impetus of the microelectronics revolution, the development of mini-mills has restored some efficiency to U.S. steel producers.

FIGURE 14 The rise of Japanese automobile manufacturers. As of 2010, Toyota was still the number one auto producer in the world, with 7.2 million cars produced, followed by GM with 6.5 million, Volkswagen with 6.1 million, Ford with 4.7 million, and Hyundai with 4.6 million cars produced.

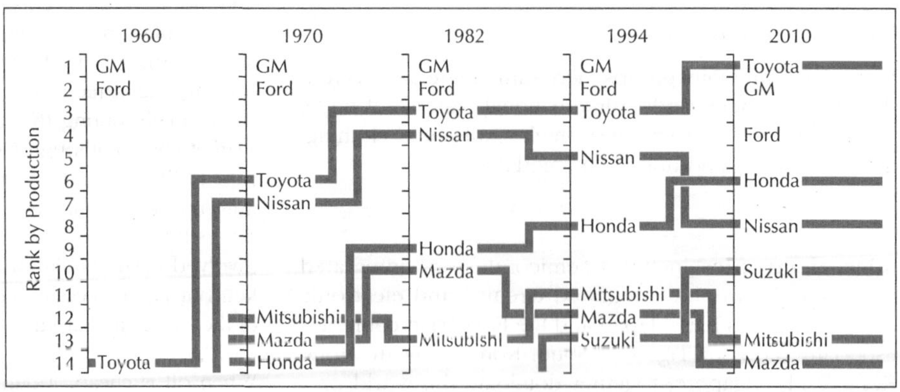

Textiles and Clothing

Labor-intensive textile and clothing manufacture has largely shifted to developing countries, including Central America, South Asia, and parts of East and Southeast Asia (e.g., China, Thailand, Indonesia), where labor costs are much lower (Figure 15). Correspondingly, textile production in the past 40 years declined in the United States and Western Europe. International trade in textiles reflects these shifts in production. Developing countries account for a growing share of global textile exports. Major gainers include the East Asian countries of China, Hong Kong, South Korea, and Taiwan. Eastern Europe, Russia, Japan, and Canada are relatively small players in the world textile and clothing trade.

Grains and Feed

Wheat, corn, rice, other cereals, feed grains, and soybeans are included in the category of grains and feed. The United States is the world leader in this category, although Canada is also a major exporter. Among the developing countries, India, Egypt, and Argentina are some of the largest net exporters. Japan, with its small base of agriculture and arable land, is a net importer, as is Eastern Europe and Russia. Trade within the EU is large.

Grains, feeds, and food products have become a steadily declining share of world trade. Some of this reduction is because Western seeds, grains, and fertilizers are now commonplace in Third World nations, and the Green Revolution has made it possible for some developing countries to provide for themselves. Another reason for the reduction is the worsening terms of trade for primary goods; prices of manufactures and energy have risen rapidly, which gives producers of feed, grains, and agricultural products less leverage in world international commerce.

Nonoil Commodities

The term *commodities* has different meanings, but in the discourse of international trade it is often taken to mean raw materials. Nonoil commodities include copper, aluminum, nickel, zinc, tin, iron ore, pig iron, uranium ore, crude rubber, wood and pulp, hides, cotton fiber, and animal and vegetable minerals and oils. The United States is a large exporter of this broad group of goods, primarily to the economies of Europe and Japan, but the developing countries of the world lead in the export of raw materials. The largest single flow of raw materials is from developing countries to Japan, which lacks significant natural resources and arable land, and from developing countries to Western Europe. Since the early days of capitalism, the international division of labor was based on international trade, which led the world's periphery—colonies and today's less developed countries—to trade their raw materials for industrialized goods, primarily from Europe and other developed nations in North America and Japan.

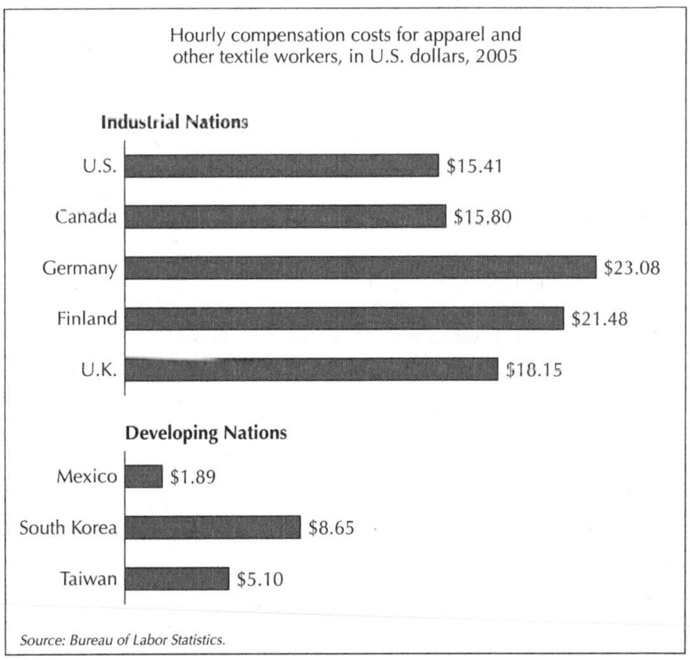

FIGURE 15 Hourly labor costs for apparel and textile workers, 2005.

Summary

International trade has grown rapidly over the past 40 years and now comprises roughly one-quarter of the world's total output of goods and services. Trade patterns simultaneously reflect and shape the specialization of production among and within different countries. Thus, trade must be viewed within the context of national changes in the level and composition of GDP. The growth of trade has occurred more quickly than the growth in national output, meaning that countries have become increasingly interconnected. However, important shifts in production internationally have contributed to changes in trade patterns, particularly the offshoring of many low-wage, low-value-added manufacturing sectors from developed countries to less developed countries. More generally, recent changes in global trade must be seen in light of the broader dynamics of globalization, including trade deregulation under the World Trade Organization, the rise of the NICs, and changes in global finance, including currency exchange rate fluctuations.

This chapter charted several major dimensions of international trade. It focused on the shifting nature and composition of U.S. exports and imports, which have become increasingly tied to East Asia, particularly Japan and China, although Canada remains the largest trading partner. NAFTA has expanded U.S. trade greatly with Mexico, as well as foreign investment in the form of maquiladores. The rising U.S. trade imbalance in merchandise has been offset to a small extent by a surplus in the trade balance in services. The chapter described the major role of trade on the other side of the Pacific in spurring the growth of the original "tigers," South Korea and Taiwan, as well as new tigers—Thailand, Malaysia, and Indonesia, but, above all, China. Russia, a potential giant in trade, has remained mired by corruption and inefficiency in the difficult transition from Communism to a market-based economy. Finally, the chapter summarized the world trade patterns in six major commodity groups, including microelectronics, automobiles, steel, textiles, grains, and nonoil commodities.

Key Terms

current account
Four Tigers
import substitution
infrastructure
intellectual property rights
maquiladoras
microelectronics
Special Economic Zones (SEZs)

Study Questions

1. What has happened to the level of world trade since World War II? Why?
2. Why does the United States trade more with East Asia than with Europe?
3. What are Mexican maquiladoras?
4. Describe the geography of Japan's automobile exports.

Suggested Readings

Dicken, P. 2010. *Global Shift: Mapping the Changing Contours of the World Economy,* 6th ed. New York: Guilford Press.

Grant, R. 2000. "The Economic Geography of Global Trade." In *A Companion to Economic Geography*, E. S. Sheppard and T. J. Barnes. Oxford: Blackwell.

Hiscox, M. 2001. *International Trade and Political Conflict: Commerce, Coalitions, and Mobility*. Princeton, NJ: Princeton University Press.

Porter, M. 1990. *The Competitive Advantage of Nations*. New York: Free Press.

Van Marrewijk, C. 2002. *International Trade and the World Economy*. Oxford: Oxford University Press.

Web Resources

International Trade Administration
http://www.ita.doc.gov/td/industry/otea/

Up-to-date reports and data about international trade and disputes.

World Bank
http://www.worldbank.org/

The Bank's view of the world and its various activities across the globe.

Trade Statistics
http://www.census.gov/indicator/www/ustrade.html

The U.S. Census Bureau's page for current and historical data on U.S. trade highlights, regulations, and partners.

Development and Underdevelopment in the Developing World

From Chapter 14 of *The World Economy Geography, Business, Development,* Sixth Edition, Frederick P. Stutz, Barney Warf.

OBJECTIVES

- To outline the multiple definitions of development
- To acquaint you with the major economic problems inhibiting development in the vast parts of the world that are economically underdeveloped
- To describe major theories and perspectives on development
- To examine the causes of poverty in the world today
- To explore the role of women in the world economy and gender roles in the workplace
- To shed light on successful development strategies, such as in East Asia
- To introduce sustainable development as an important strategy in light of the world's limits on growth

For a large section of the human race, such as these street dwellers in Mumbai, India, poverty is the norm; while the type and severity of poverty in the developing world vary over space and time, it remains the driving force behind many social ills, including disease, social unrest, and massive unhappiness.

Development and Underdevelopment in the Developing World

The modern world had its origin in Europe in the late fifteenth and early sixteenth centuries, when capitalism began in earnest and eventually displaced feudalism throughout the continent. True "modernity," however, arrived largely on the heels of the Industrial Revolution; the Enlightenment; and the massive political, social, economic, cultural, and technological changes of the nineteenth and twentieth centuries.

One of the most enduring and striking characteristics of the modern world is the division between rich and poor countries. By the nineteenth century, an international system had developed in which a wealthy minority of countries industrialized (in Western Europe, North America, and Japan) using primary products produced by the impoverished majority of their colonies. More recently, this original global division of labor has given way to a new one: The wealthy minority is increasingly engaged in office work and producer services while parts of the developing world have turned to hands-on manufacturing jobs on the global assembly line as well as in agriculture and raw-material production. The creation of today's world, with its rich core and poor periphery, was not the result of a conspiracy among developed countries, but was the outcome of a systemic process—that process by which the world's political economy functions and, in so doing, produces uneven spatial development. In short, the birth of the modern world system and the schism between rich and poor countries are both part of the process that lies at the heart of global political economy.

This chapter deals with how this world of unequal development came about—how present structures are the result of the past. We begin by examining the word *development* and the idea it represents, noting that it embodies many different concepts and measures. Next we survey the major regions of the developing world and then discuss the characteristics of less developed countries and some of the barriers to their development. We compare and contrast three major schools of thought on this issue: modernization theory, dependency theory, and world-systems theory. Finally, we turn to development strategies, including the central role of trade, and conclude with an examination of the potentials and pitfalls of Third World industrialization.

WHAT'S IN A WORD? "DEVELOPING"

If Europe, the United States, Australia, and Japan—all of which enjoy relatively high standards of living and material consumption—are described as developed countries, then what adjective should we use to describe the poorer countries of the world? Each of the following terms has flourished in succession: *primitive*, *backward*, *undeveloped*, *underdeveloped*, *less developed*, *emerging*, and *developing*. Today, many people and policy makers use the word *developing* and, increasingly, the phrase **less developed countries**, but social scientists favor the term *underdeveloped*.

Underdeveloped was formerly used to describe situations in which resources were not yet developed. People and resources were seen as existing in, respectively, a traditional and "natural" state of poverty. The problem with this view is that it regards poverty as natural and inevitable, rather than as a social product, and masks the origins and historical contexts and processes that *make* people poor. Many scholars argue that poverty is produced, not just a given, much like a building or a shirt or a TV show. **Underdevelopment** describes not an initial state, but rather a condition arrived at through the agency of imperialism, which set up the inequality of political and economic dependence of poor countries on rich countries. Thus, instead of viewing underdevelopment as an initial or *passive state*, we should view it as an *active process*.

While economic development may seem like a straightforward concept, in fact it involves several complex, even contradictory, goals and concepts. Broadly speaking, **development** entails the growth of per capita income and the reduction of poverty. However, some countries that have experienced rapid growth of per capita income have seen simultaneous increases in poverty, unemployment, and inequality. Beyond average levels of income, other measures of development include income equality, nutrition, health, infant mortality, access to education, and civil liberties. In some cases, development may actually accelerate income inequality. For example, in India, the much-heralded Green Revolution, which depends on fertilizer and water inputs, mainly benefited farmers who were already wealthy and who owned large tracts of land.

Other measures of development include capital inflows, the capacity to produce capital goods, balances in trade, and several major trading partners. When the amounts of food, adequate shelter, and health and educational services are sufficient to meet basic human needs, a minimal state of development exists; when they are insufficient, a degree of underdevelopment prevails. Even more intangible measures include longevity, self-esteem, and personal freedom.

More down-to-earth development goals include the following:

1. An adequate, healthful diet and clean drinking water
2. Sufficient health care
3. Environmental sanitation and disease control
4. Employment opportunities commensurate with individual talents
5. Sufficient educational opportunities
6. Individual and collective freedom of expression and freedom from fear
7. Decent housing
8. Economic activities that do not impose undue costs on the natural environment
9. Social and political milieus promoting equality

In conventional usage, *development* is a synonym for economic growth. But growth is not development, except insofar as it enables a country to achieve the nine goals. If these goals are not the objectives of development, if modernization is merely a process of technological diffusion, and if the spatial integration of world power and world economy is devoid of human referents, that is, insensitive to people's psychological and political needs, then development should be redefined.

How Economic Development Is Measured

Geographers and other social scientists measure economic development by a number of social, economic, and demographic indexes. The principal ones are gross domestic product (GDP) per capita, the distribution of the labor force by economic sector, ability to produce consumer goods, educational and literacy levels, the health status of a population, and its level of urbanization.

GDP PER CAPITA By far the most common measure of wealth and poverty internationally is GDP per capita, that is, the sum total of the value of goods and services produced by a national economy divided by its population. As shown in Figure 1, GDP per capita is more than $20,000 in most developed nations. At the same time, the United Nations

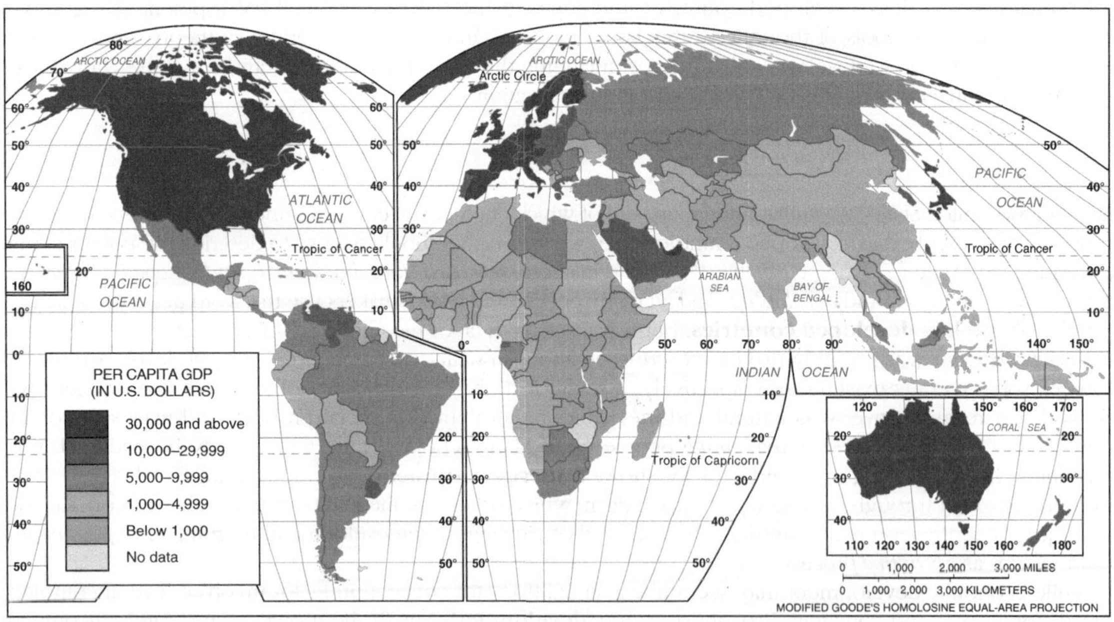

FIGURE 1 Annual gross domestic product (GDP) per capita. GDP per capita—the sum of a country's output divided by its population—is the most common, if flawed, measure of standards of living. GDP per capita varies from less than $250 per year in the most impoverished nations to more than $30,000 annually in much of Europe, Japan, and North America.

estimates that 3 billion people live on less than $2 a day, or $750 per capita per year, and that 1 billion people live on $1 per day or less. Japan, North America, Western Europe, Australia, and New Zealand have the highest per capita incomes in the world. The Middle East, Latin America, South Asia, East Asia, Southeast Asia, and sub-Saharan Africa have the lowest. However, GDP is a flawed measure in several respects: It does not capture nonmarket, noncommodified economic activity (e.g., barter and subsistence production or household domestic labor) and is vulnerable to fluctuations in exchange rates and the costs of living.

Per capita purchasing power is a more meaningful measure of actual income per person (Figure 2). The relative purchasing power in developed nations is more than $20,000 per capita per year, whereas in Africa it is much less than $1000 per capita per year. Per capita purchasing power includes not only income but the price of goods in a country. The United States is surpassed by Japan, Scandinavia, Switzerland, and Germany in per capita income. However, it surpasses almost all countries in per capita purchasing power because goods and services, particularly housing, are relatively inexpensive in the United States compared with those in other industrialized nations. In other respects, however, including poverty rates and income inequality, the United States lags behind much of Europe.

ECONOMIC STRUCTURE OF THE LABOR FORCE The distribution of jobs by sector also bespeaks a country's economic development. Economists and economic geographers divide employment into three major categories:

1. The **primary sector** involves the extraction of materials from the earth—mining, lumbering, agriculture, and fishing.
2. The **secondary sector** includes assembling raw materials and manufacturing (i.e., the transformation of raw materials into finished products).
3. The **tertiary sector** is devoted to the provision of services—producer services (finance and business services) wholesaling and retailing, personal services, health care and entertainment, transportation and communications.

In less developed countries, a large share of the labor force works in the primary sector, primarily as peasants and farmers (Figures 3 and 4). In the United States, only 2% of the labor force is engaged in agriculture, whereas in certain African nations, India, and China, more than 75% of the laborers work in fields or farms.

EDUCATION AND LITERACY OF A POPULATION Economic development can also be measured by the extent and quality of education in a country, including the proportion of children who attend school. The *literacy rate* of a country is the proportion of people in the society who can read and write (Figure 5). The number of students per teacher is another measure of access to education; small classes allow more student-teacher interaction and greatly

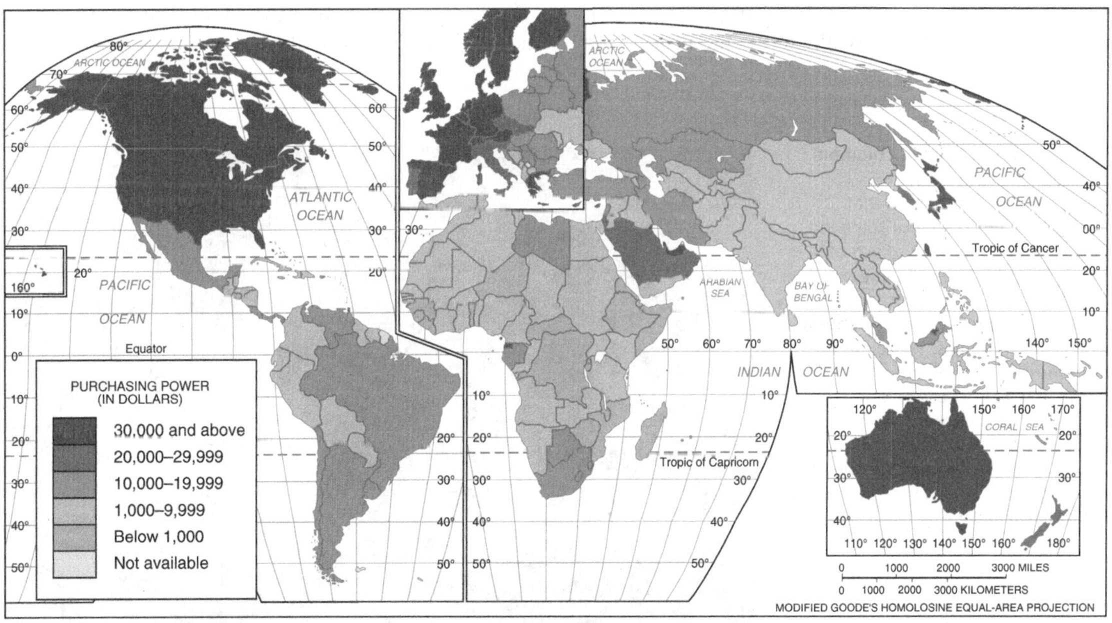

FIGURE 2 Per capita purchasing power. Per capita purchasing power is a better measure of a country's relative wealth than is GDP per capita because it includes the relative prices of products. For example, Switzerland, Sweden, and Denmark have higher per capita GDPs than the United States. However, the United States has the world's highest per capita purchasing power because of relatively low prices for food, housing, fuel, and services.

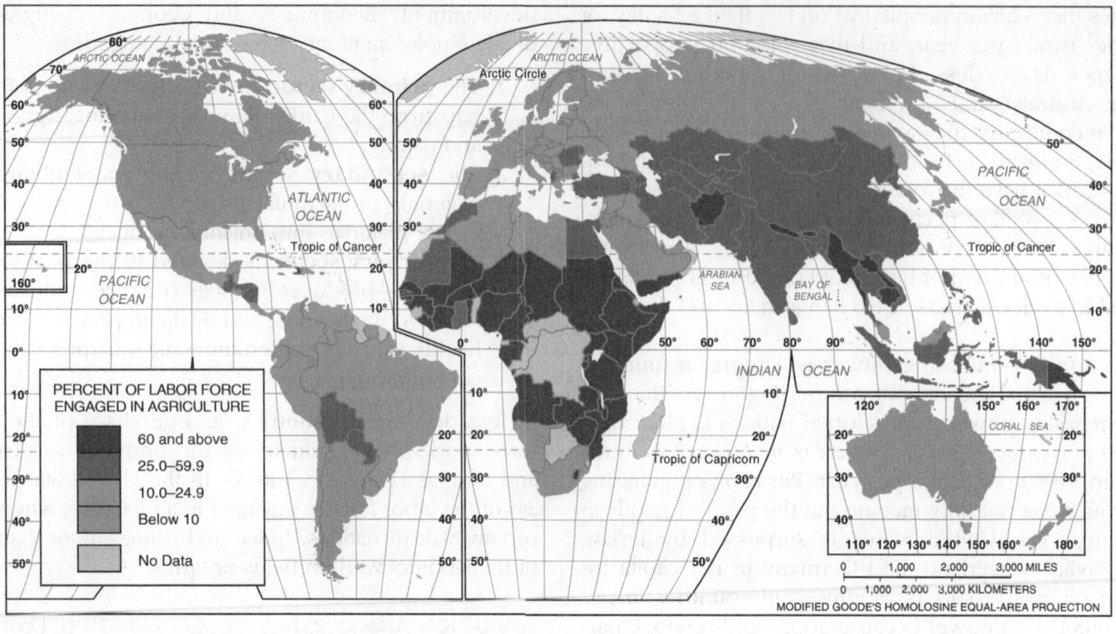

FIGURE 3 Percent of labor force in agriculture. Typically, large shares of workers in poor countries are in agriculture, often working in preindustrial conditions, whereas in economically advanced countries a small fraction of the labor force is so employed.

facilitate learning. Richer, First World societies generally have low student-teacher ratios, whereas poor countries have high ones (Figure 6). Moreover, in many developing countries, teachers' salaries are low, the buildings are decrepit, and there are insufficient funds for textbooks, uniforms, or equipment such as computers. Notably, despite these shortages, many countries have ample funds for their militaries, which often swallow up far more than what is spent on development, indicating that insufficient investment in education is a policy choice, not a "natural" limitation.

There are typically vast gender differences in literacy within developing countries. In many impoverished societies, desperately poor rural families can often only afford to send one child to school (the others are working the fields), and that child is very likely to be male, given

FIGURE 4 Harvesting coffee in Colombia. This photograph illustrates the preindustrial types of agriculture that employ large shares of the labor force in the primary economic sector in much of the developing world. Often, landless farmers (*campesinos* in Latin America) must sell their labor power at low wages in order to feed their families. These are the populations most vulnerable to globalization, low export prices, and exploitative local landowners. Their poverty reflects the ways in which global, national, and local forces are telescoped into places to generate both wealth and suffering simultaneously.

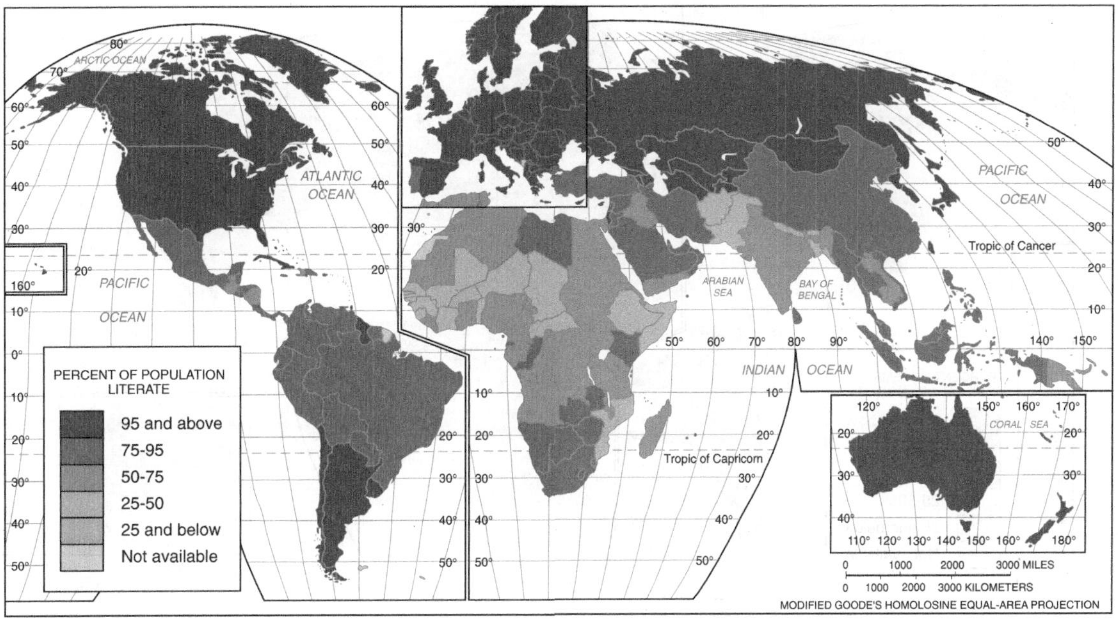

FIGURE 5 Literacy rates. A great disparity in literacy rates exists between inhabitants of developed countries and less developed countries. In the United States and other highly developed countries, the literacy rate is more than 98%. Notice the large number of countries in Africa and South Asia where the literacy rate is less than 50%.

the sexism entrenched in much of the world. Unfortunately, worldwide, only 75 girls attend school for every 100 boys. In areas particularly disadvantageous to women, their literacy rates tend to be much lower than men's (Figure 7). In many countries, the literacy rate of women is less than 25%, whereas that of men is between 25% and 75%. Schools catering only for girls help in this regard (Figure 8), as do courses taught on television (Figure 9), which can access rural areas quickly and cheaply. The Middle East and South Asia, where women

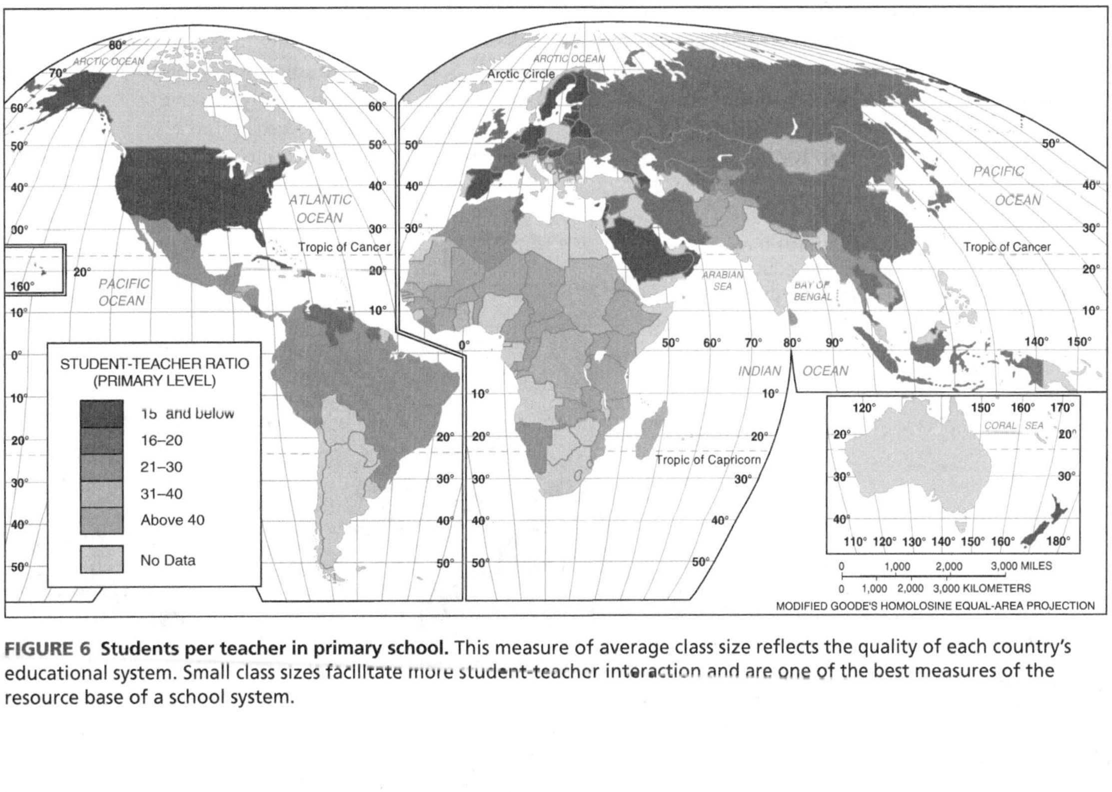

FIGURE 6 Students per teacher in primary school. This measure of average class size reflects the quality of each country's educational system. Small class sizes facilitate more student-teacher interaction and are one of the best measures of the resource base of a school system.

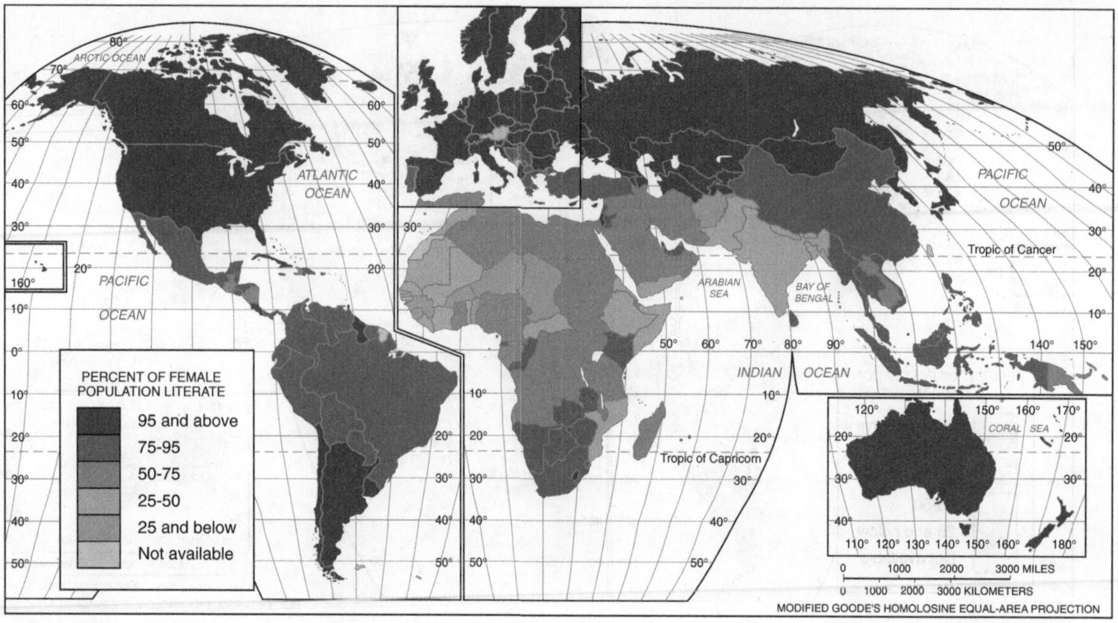

FIGURE 7 Literacy rate of women. The gender gap in literacy is most pronounced in the developing world, especially Africa, South Asia, and the Middle East. In most developing countries, the female literacy rate is much below the rate for men. Raising women's literacy rates is one of the most efficient means of stimulating economic and social development.

are frequently regarded as subservient to men, show the greatest disparities. However, in the highly developed world, the literacy rates of men and women are almost identical. The regions that have low percentages of women attending secondary school also generally have poor social and economic conditions for women, including job opportunities. Indeed, raising women's literacy rates has been shown to lower fertility rates and to empower women economically and politically. In addition, because more people can read and write, newspapers, magazines, and scholarly journals proliferate and foster communication and exchange, which leads to further development by informing people of opportunities, best practices, and so forth.

HEALTH OF A POPULATION Measures of health and welfare, in general, are much higher in developed nations than in less developed countries. One measure of health and welfare is diet, typically measured as caloric consumption per capita (Figure 10). In developed nations, the population consumes approximately one-third more than the minimum daily requirement and is therefore able to maintain a higher level of health. But in some areas of every country, calories and food supplies are insufficient, even in the United States, where significant pockets of hunger and malnutrition exist. Conversely, an overabundance of cheap food and inadequate exercise have generated an obesity epidemic in the United States, and to some extent, many

FIGURE 8 An all-girls school in Kenya reflects that country's investment in human capital. The education of girls is an important means of raising standards of living and reducing birthrates, as well as providing economic opportunities for women.

UN/DPI PHOTO/Raccah

FIGURE 9 African children watching television. Although literacy levels have increased throughout much of the developing world, they are still frequently lower for women than for men, often by a considerable margin, a reflection of patriarchal family and school systems that systematically deny girls the opportunity for an education equal to that of boys.

other countries, where lack of exercise and high-calorie diets (e.g., junk food, government-subsidized corn sweeteners) have caused waistlines to balloon. Obesity has become an epidemic worldwide; there are today more obese people than malnourished ones.

People in developed nations also have better access to doctors, hospitals, and health care providers. Figure 11 shows worldwide access to physicians as measured by persons per physician. For relatively developed nations, there is 1 doctor per 1000 people, but in developing countries, each person shares a doctor with many thousands of others. Africans by far have the worst access to health care (in some countries there are more than 15,000 people per physician, effectively meaning that most people *never* see one), followed by Southeast Asians and East Asians. Everywhere, wealthier societies have better access to health care, although there are huge discrepancies within them as well, such as in United States, with its large pool of uninsured people.

Infants and children are the most vulnerable members of any society, in part because their immunological systems are not as well developed and partly because they lack effective political power to shape public policy. In developed nations, on the average, fewer than 10 babies in 1000 die within the first 100 days; in many less developed nations, more than 100 babies die per 1000 live births, the result of poor prenatal care, malnutrition, and infectious diseases. When there are economic downturns, droughts, or disruptions of the food supply brought on by war, they are generally the first to die. The geography of infant mortality (Figure 12)—the proportion of babies

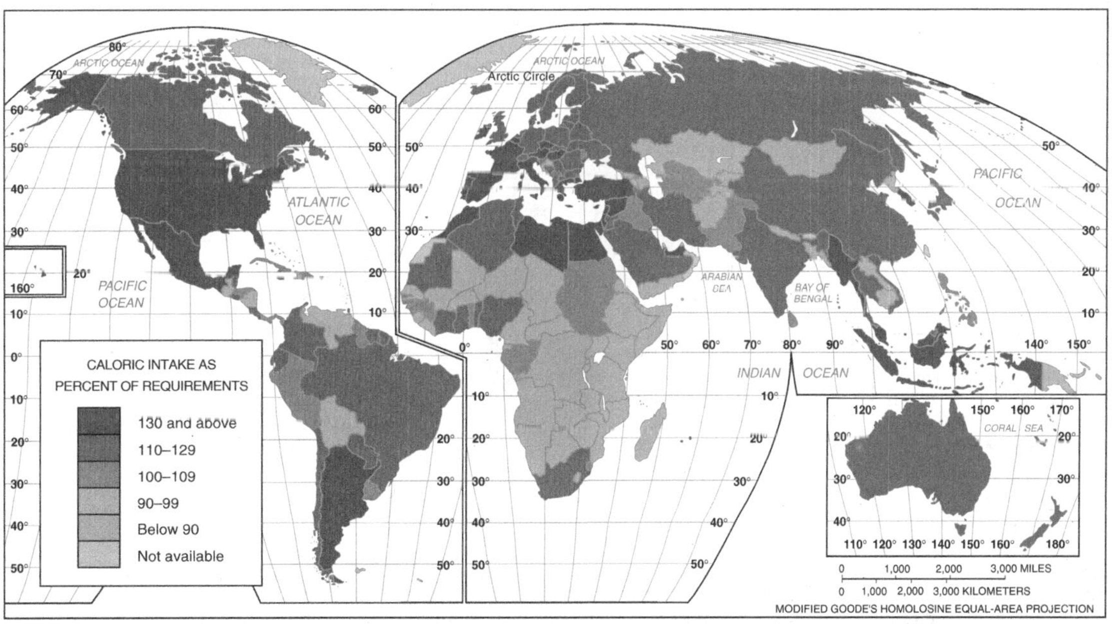

FIGURE 10 Daily caloric consumption per capita as a percentage of minimum nutritional requirements. Access to adequate nutrition is another measure of economic development. In many impoverished nations, malnutrition may be chronic. Conversely, many economically advanced countries, particularly the United States, suffer from an epidemic of obesity.

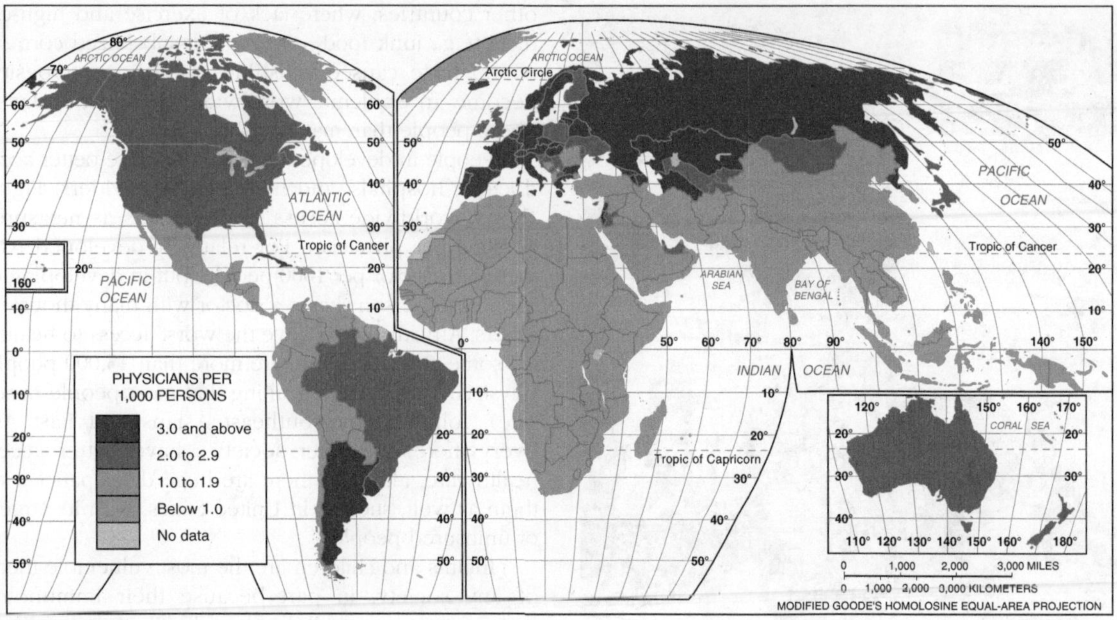

FIGURE 11 Persons per physician. An important measure of economic development is the number of persons per physician in a country. This measure is a surrogate for health care access, which includes hospital beds, medicine, and nurses and doctors. In most of Africa, there are 10,000 people per physician, or more, while Europe and the United States average 1 doctor for every 200 people.

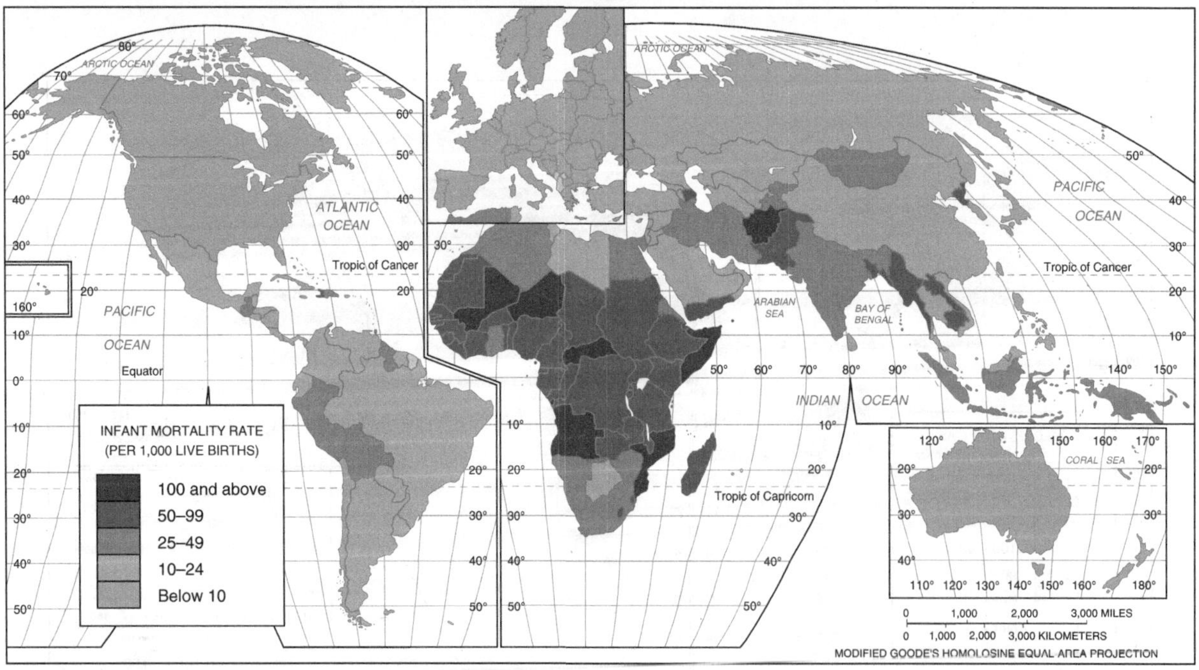

FIGURE 12 Infant mortality rates. Babies are the most vulnerable members of any society, and the percentage who die before their first birthday is another measure of economic development and well-being. In the poorest nations, particularly in Africa, more than 10% of babies die before their first birthday.

who die before their first birthday—is thus perhaps the best measure of economic development, or the lack of it. In much of Africa, more than 10% of infants do not live through their first year. In the developed First World, by contrast, infant mortality rates are very low. Notably, Cuba's infant mortality rate—6.0—is lower than that of the United States, which is 7.0, a discrepancy that reflects Cuba's investment in health care and the widespread unavailability of health insurance for the U.S. poor.

AIDS has emerged as a significant threat worldwide. More than 25 million people have died of this disease, and an additional 33 to 45 million are infected with the HIV virus. The epicenter of the AIDS epidemic is sub-Saharan Africa (Figure 13), where in some countries 40% of the adult population is infected. The sub-Saharan region accounts for more than 60% of the people living with HIV worldwide, or some 25 million men, women, and children. In India, researchers estimate that by the year 2020, 50 million people could be HIV positive. Half the prostitutes in Bombay are already infected, and doctors report that the disease is spreading along major truck routes and into rural areas, as migrant workers bring the virus home. In China, AIDS is spreading rapidly.

The social consequences of this epidemic are catastrophic, including tens of millions of children who have lost their parents to the disease. The labor forces of many developing countries have been affected by the loss of adults in their prime working years, and public health systems have been stretched beyond the breaking point. Because AIDS has a long lead time in which infected people do not show symptoms, and because sexual behavior is very difficult to change, many fear that AIDS could lead to a depopulation of large parts of the world in the future comparable to the Black Death of the fourteenth century.

The reliability and quality of the food supply, access to clean drinking water, public health measures, ability to control infectious diseases, and access to health services all shape how long people live. People in economically advanced countries have the luxury of living a relatively long time—often more than 75 years, on average—while those in poorer countries live considerably shorter lives (Figure 14). In parts of Africa, few adults can realistically expect to live past their fiftieth birthday. Thus, the geography of life expectancy is another measure of the health and welfare of a population.

CONSUMER GOODS PRODUCED The quantity and quality of consumer goods purchased and distributed in a society is another measure of the level of economic development in that society. Easy availability of consumer goods means that a country's economic resources have fulfilled the basic human needs of shelter, clothing, and food, and more resources are left over to provide nonessential household goods and services. Automobiles, textiles, home electronics, jewelry, watches, refrigerators, and washing machines are some of the major consumer goods produced worldwide on varying scales. In industrialized countries, more

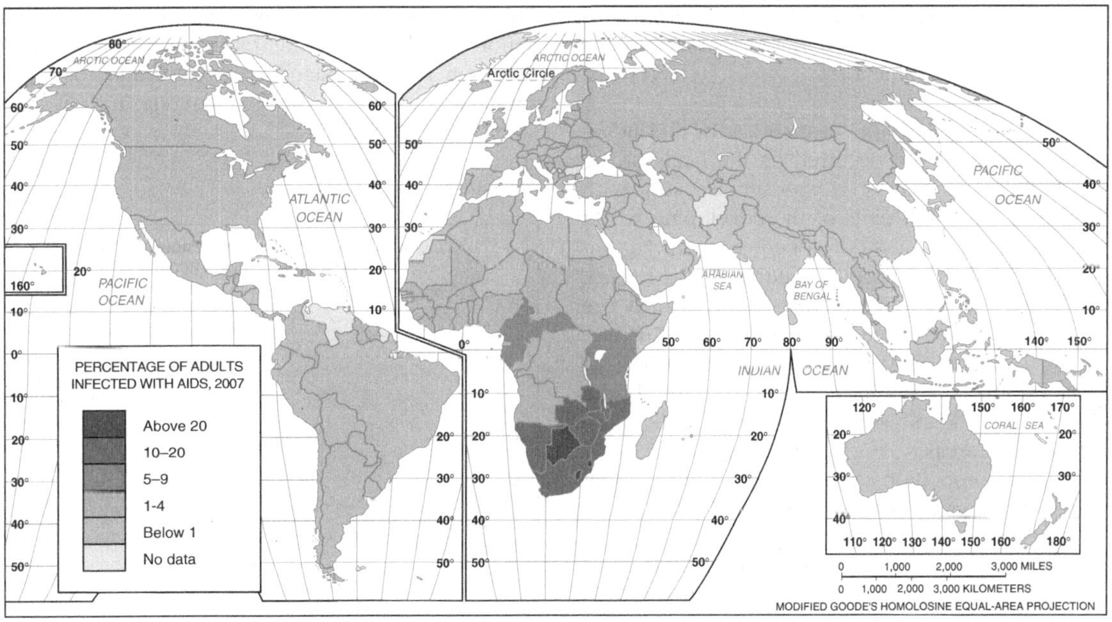

FIGURE 13 The geography of AIDS, 2005. Although AIDS does not kill as many people as malnutrition or heart disease, it ranks among the world's leading killers today, and the most rapidly growing. More than 25 million people have died of AIDS, and another 45 million are infected. The epicenter of the epidemic is Africa, where in some countries 40% of adults are HIV infected. The disease is only now making inroads into the huge populations of India and China and has the potential for creating a catastrophic depopulation in the future.

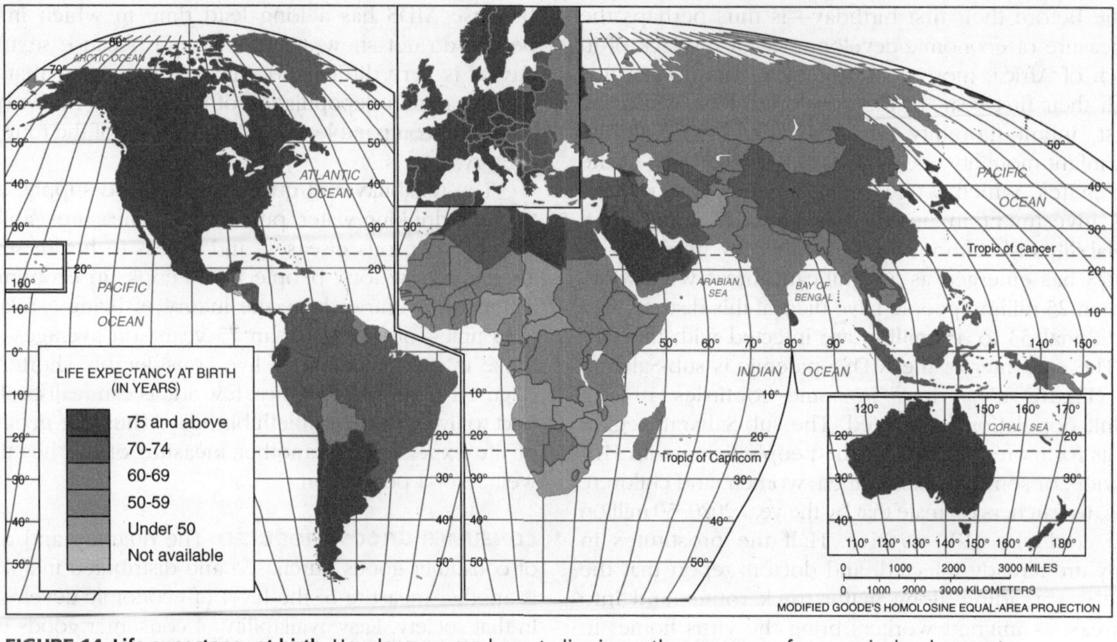

FIGURE 14 Life expectancy at birth. How long we can hope to live is another measure of economic development or the lack of it. In societies where people enjoy sufficient access to food, public health measures, and medical care, life expectancies are often over 75. In the poorest countries, in contrast, most people cannot expect to live beyond 50.

than one television, telephone, or automobile exists for every two people. In developing nations, only a few of these products exist for a thousand people. For instance, the ratio of persons to television sets in developing countries is 150 to 1, and population to automobiles is 400 to 1. The number of consumer goods such as telephones and televisions per capita is a good indicator of a country's level of economic development.

URBANIZATION IN DEVELOPING COUNTRIES In the industrialized West, urbanization occurred on the heels of the Industrial Revolution, that is, it was synonymous with industrialization. Although parts of the developing world are urbanized, in general less developed countries lag behind the economically advanced countries in this regard. However, cities throughout the developing world are growing quickly, much more so than those in Europe, North America, Japan, or Australia. Today, about 56% of the world's people live in cities (Figure 15), an all-time high—for the first time in history, the majority of the planet resides in urban areas.

However, the proportion of each country's people that lives in cities varies widely around the globe (Figure 16). In parts of Latin America, urbanization rates resemble those of North America or Australia, where 75% or more of the people live in cities. In Asia and Africa, however, the proportions are generally lower. Only 45% of China's people live in cities, and in wide swaths of Africa less than 20% do so.

Urbanization in the developing world differs significantly from that in the West. Because the historical contexts of less developed countries differ from that of the West, particularly due to the impacts of colonialism and their very different mode of incorporation into the global division of labor, the patterns of urban growth are also qualitatively different. Some countries in the developing world had well-established traditions of urbanization before the Europeans came, such as the Arab world and China. In others, colonial powers such as Britain played a major role, constructing cities like Calcutta in India, Rangoon in Myanmar (formerly Burma), or Singapore; similarly, the Dutch founded Batavia, Indonesia, which later became Jakarta.

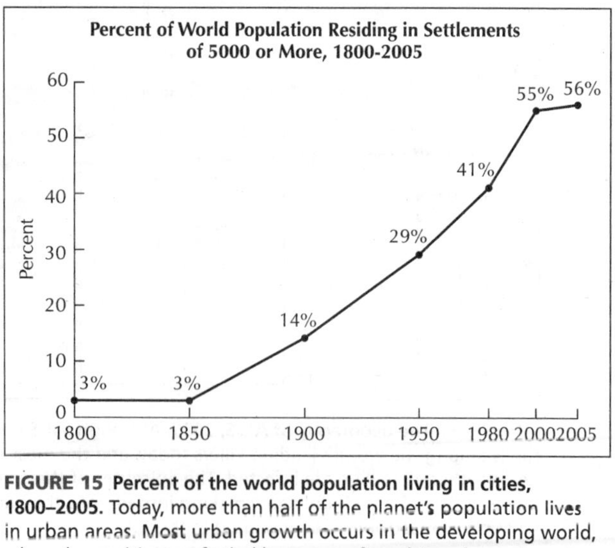

FIGURE 15 Percent of the world population living in cities, 1800–2005. Today, more than half of the planet's population lives in urban areas. Most urban growth occurs in the developing world, where large cities are fueled by waves of rural-to-urban migrants.

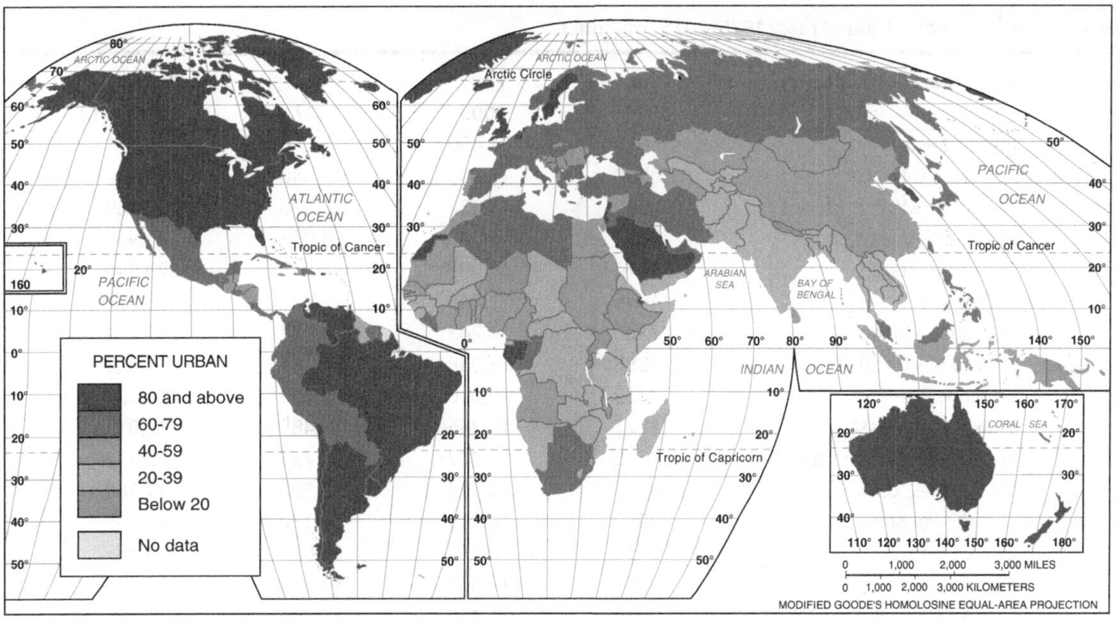

FIGURE 16 Urbanization rates. The proportion of people living in cities is another measure of economic development, although not a very accurate one. Generally, poor countries are less urbanized than wealthy ones. However, in Latin America even relatively poor countries such as Brazil have high urbanization rates, where much of the poverty is clustered in large urban conurbations such as São Paolo and Rio de Janeiro.

Today, the vast bulk of the world's urban growth is in the developing world. Moreover, cities in the developing world are growing much more rapidly than their counterparts in the developed countries. Most of the world's largest cities, for example, are found in Latin America and Asia, not Europe or North America (Table 1). In 2008, the greater Tokyo metropolitan region, with 36 million people, was the world's most populous, while New York City, Mexico City, Mumbai, and São Paolo, Brazil, vied for second place, with roughly 20 million inhabitants each. Many of the others, with populations over 5 million, are located in China and India (Figure 17).

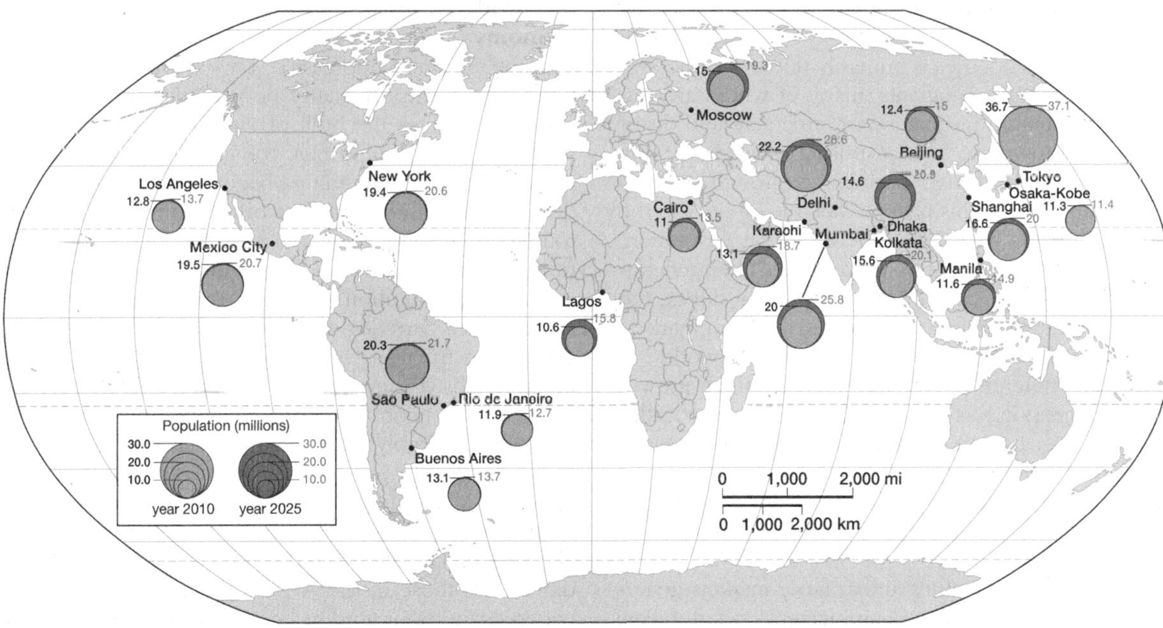

FIGURE 17 The location of the world's largest cities. The majority of the world's biggest cities are located not in the developed but in the developing world, particularly in China and India.

TABLE 1 World's Largest Urban Areas, 1975–2025 (millions)

Rank	Agglomeration, Country	Pop. 1975	Pop. 2000	Pop. 2007	Pop. 2025 (est.)
1	Tokyo, Japan	26,615	34,450	35,676	36,371
2	New York City, USA	15,880	17,846	19,040	19,974
3	Ciudad de México, Mexico	10,690	18,022	19,028	20,189
4	Mumbai (Bombay), India	7082	16,086	18,978	21,946
5	São Paulo, Brazil	9614	17,099	18,845	20,544
6	Delhi, India	4426	12,441	15,926	18,669
7	Shanghai, China	7326	13,243	14,987	17,214
8	Kolkata (Calcutta), India	7888	13,058	14,787	17,039
9	Moscow, Russia	7623	10,016	13,752	19,324
10	Dhaka, Bangladesh	2221	10,285	13,485	17,015
11	Buenos Aires, Argentina	8745	11,847	12,795	13,432
12	Los Angeles, USA	8926	11,814	12,500	13,160
13	Karachi, Pakistan	3989	10,019	12,130	14,855
14	Cairo, Egypt	6450	10,534	11,893	13,465
15	Rio de Janeiro, Brazil	7557	10,803	11,748	12,775
16	Osaka-Kobe, Japan	9844	11,165	11,294	11,365
17	Beijing, China	6034	9782	11,106	12,842
18	Manila, Philippines	4999	9958	11,100	12,786
19	Istanbul, Turkey	3600	8744	10,061	11,177
20	Paris, France	8558	9692	9904	10,007
21	Seoul, South Korea	6808	9917	9796	9740
22	Lagos, Nigeria	1890	7233	9466	12,403
23	Jakarta, Indonesia	4813	8390	9125	10,792
24	Chicago, USA	7160	8333	8990	9516
25	Guangzhou, China	2673	7388	8829	10,414

While natural growth rates in cities in the developing world are a little higher than that in the West, urbanization in less developed countries is primarily due to the massive influx of rural-to-urban migrants, many of whom are displaced by agricultural mechanization, unequal land distribution, unemployment, lack of public investment, low crop prices, war, and high population growth in rural areas. To some extent, this pattern was true of cities in the developing West as well during the Industrial Revolution. However, because few developing countries have generated the industrial job growth that the West did during the nineteenth and early twentieth centuries, the labor markets and employment conditions in the developing world are fundamentally different. Thus, we must see urban growth and the rural crisis in less developed countries as two sides of one coin. Because many migrants move to the cities believing that there are greater opportunities to be found there (perceptions that may be erroneous due to imperfect information), they often find themselves plunged into desperate circumstances.

Less developed countries' urban labor markets generally do not generate sufficient employment opportunities, leading to high unemployment rates, or underemployment, in which migrants do not utilize their skills. Many find work in low-paying service jobs in the **informal economy** (Figure 18), which includes a wide array of jobs that are generally oppressive, low-paying, and offer little financial security. In many poor countries, the informal sector constitutes the bulk of employment opportunities. In Manila, the Philippines, for example, 30,000 people earn their living recycling garbage on the Payatas garbage dump; adults and children work together, stepping over rotting debris to scavenge bits of plastic or tin cans. Others find marginal incomes selling trinkets and food on the streets, in prostitution, selling illegal drugs, in the black market, illicit currency exchange, or as casual day laborers doing construction. In contrast, formal-sector, export-oriented jobs (Figure 19) that tie developing country cities to the global economy, such as the garment industry or electronic assembly plants—which still offer higher wages than most local opportunities, however exploitative they appear to Western eyes—tend to be the exception, not the norm.

Residential patterns in developing countries' cities also differ from those in the West. Whereas in the developed countries the poor tend to be a minority, often consigned to the city center; in the developing world, the poor may be a majority of the city's inhabitants, depending on the

Case Study

Remittances

Remittances are the transfer of a portion of migrant workers' wages back to their families, although a broader definition includes the sending of products (foodstuffs, electronics, automobiles, etc.) and the transfer of ideas formed and beliefs adopted and from migrant destination workplaces to native communities, or social remittances. Remittances are an important geographic phenomenon due to the sheer volume and extent of monetary flow from more affluent to poorer nations and the fact that they are disproportionately received by poorer members of these developing countries. Such wealth transfer allows households to make important economic decisions concerning health, education, occupation, and daily living conditions. In many poorer nations, remittances have been shown to effect broad-scale reduction in poverty and mortality through increases in household income and better access to health care.

Prior to the 1970s, relatively little remittance income flowed among countries. Since that time, remittance transfers have increased dramatically in volume. Between 2000 and 2007, global remittance flows more than doubled, from $132 to $337 billion. The developing world alone received $251 billion in 2007. Remittances represent substantial contributions to overall GDP for many developing world economies. In 2006, remittances constituted more than 10% of the GDPs of 24 poorer nations.

Most world regions have received substantial remittance income. In 2007, Latin America and the Caribbean received the highest amount of remittance income ($61 billion), closely followed by East Asia, Europe, and South Asia. On a per country basis, India, China, and Mexico each received over $25 billion in remittances in 2007. The area most in need of international assistance, sub-Saharan Africa, received only $11.7 billion in remittance transfers in 2007.

How remittances are used is highly dependent on the economic status of the recipient household. Poorer households, especially those that practice subsistence agriculture, often use remittances for family maintenance, including diet diversification through the purchase of food grown elsewhere. They are more apt to purchase medicines, improve the condition of their homes, and allow their children to spend more time in school. At times, when a household's primary breadwinner is away, remittances are used to contract agricultural workers to tend a household's land.

More economically stable remittance-receiving households have been criticized for the disproportionate use of remittances to increase conspicuous consumption. It is common to encounter multilevel "trophy" homes in communities that receive large volumes of remittances. Other nonessential items purchased with remittances include packaged food, electronics, and high-end clothes and automobiles. There is also a strong expectation in many villages that migrants will host large celebrations upon their return.

However, more affluent remittance-receiving households do not spend all their remittance income on conspicuous consumption. Occasionally, remittance receipts are used to establish small businesses or to expand and improve the condition of farming operations. Additionally, they are used to pay for the construction of a parent's home or to fund the migration—often illegal—of other family members. And, just as with lower-income remittance recipients, a large volume of remittances is used to improve the education of children through the purchase of school supplies, placing children in private schools, and the financing of higher education.

Throughout much of the developing world, many households use remittance transfers as an economic diversification strategy. The most common type of remittance transfer is from migrant workers earning and sending money from the developed world to a household in the developing world. Within-country urban-to-rural remittance transfers are also common. For families that have fallen on hard times, it is not unusual to send one or more members to an economically more robust area to earn sufficient income to ride out the tough times. In many agrarian communities, the sending of migrant workers during slow periods of agricultural productivity, following crop failures, or to make ends meet between harvests is not uncommon.

It has long been argued that the remittance system has not fulfilled its potential as a catalyst for development in poorer countries. Many early studies on the topic concluded that the bulk of remittances are used for family maintenance and consumption rather than as seed money to start businesses. While it is indisputable that remittances fuel consumption, some scholars believe that they do spur development as they cycle their way through local and regional economies. Even when remittances are used to fuel conspicuous consumption, they argue, secondary and tertiary recipients (local vendors and suppliers) of remittance income will use this infusion of money to create or expand businesses, which will have a positive impact on local and regional development. Finally, remittance income is also used to further the education of recipients' children and to reduce household poverty—both of which are certainly long-term economic investments.

AP Photo/Charles Dharapak

AP Photo/Khalid Mohammed

FIGURE 18 The informal economy—unregulated, untaxed, and generally consisting of low-paying service jobs—comprises a variety of occupations that employ large numbers of the urban poor in both the developing and industrialized countries. In less developed countries, opportunities in the formal sector, including export-oriented multinational firms and the government, are often relatively scarce.

overall level of economic development of that nation. Often, the relatively wealthy command the city centers; the poor live in the urban periphery. In many Latin American cities, for example, housing near the old colonial plaza in the center is relatively expensive.

Billions of people today live in a "planet of slums." In dilapidated houses, often self-made from cheap materials like cinder blocks or sheets of corrugated tin, many urbanites inhabit squalid neighborhoods that go by a variety of names: slums, Brazilian *favelas*, Indonesian *kampong*, Turkish *gacekondu*, South African townships, and West African *bidonvilles* (or "tin can cities") (Figures 20, 21, and 22). Countless numbers of people are forced into conditions that lack adequate housing, electricity, roads, transportation systems, schools, clean water, or sewers. High unemployment, lack of access to health care, inadequate education and road systems all conspire to produce vast pools of suffering. Often densities in such places are far higher than in any Western city. When migrants seize buildings or land that belongs to wealthy landowners, the government may deem such "**squatter settlements**" to be illegal and tear them down with bulldozers. Such communities can become more than just cesspools of misery; they can turn into breeding grounds of resentment and political activism against corrupt and uncaring governments.

GEOGRAPHIES OF UNDERDEVELOPMENT

A map of the planet shows a clear separation between the developed and less developed countries of the world. A line drawn at 30° north latitude would put most of the developed countries to the north and the underdeveloped to the south, a division often known as the *North-South split*. Note that this dichotomy is not synonymous with the Northern and Southern hemispheres of the globe; most of the world's landmass is in the Northern Hemisphere, including much of the developing world (e.g., India and China). Australia and New Zealand, which are in the Southern Hemisphere, nonetheless belong economically to the North. *North* and *South* are thus shorthand terms to describe the First and Third worlds, respectively, that is,

Susana Gonzalez/Getty Images, Inc.

FIGURE 19 Assembly plants such as these in Mexico are an excellent example of a symbiotic relationship between less and more developed countries. The latter get cheap labor, while the former benefit through a boost in the local economy. Today, austerity is transforming world economic development and politics. The age of entitlements was about giveaways; the age of austerity will be about takebacks. Governments in the rich world are caught in a vice. Without unpopular spending cuts and tax increases, unmanageable deficits may choke their economies. But those same spending cuts and tax increases also threaten growth. The problem is that slowdowns in the developed world—the United States, Europe, and Japan—will affect emerging economies that rely heavily on exports to them, like Mexico, pictured here.

The World Bank

FIGURE 20 Most of the urban poor in the developing world live in decrepit, makeshift shelters, a reflection of the low incomes, inadequate employment opportunities, the lack of profit to lure builders of commodified housing, and their generally marginalized position with the world economy.

they are social products, not referents to the world's physical geography.

The less developed countries of the world—which contain the vast majority of the planet's population—include diverse societies in Latin America, East Asia (except Japan), Southeast Asia, South Asia, the predominantly Muslim world of Southwest Asia and North Africa, and sub-Saharan Africa.

Latin America

Most inhabitants of this region are descendants of either Spanish or Portuguese colonists, slaves brought from Africa, and various indigenous peoples. Many people are *mestizos,* or mixed race. Even though many different native languages are still spoken by the native American peoples, Spanish and Portuguese are the dominant languages and Catholicism is the most widespread form of religion. This region shows a higher level of urbanization compared to other less developed nations, as reflected in the massive conurbations of Mexico City and São Paulo.

Latin America's population is largely clustered along the coast, mainly the Atlantic Ocean, with the interior scarcely populated. Most of the region's economic activity is located along the south Atlantic coast, including Argentina, Uruguay, Brazil, Venezuela, and Mexico. One of the most striking characteristics of Latin America is its highly unequal distribution of wealth. A large share of the arable land is controlled by a few very wealthy families, who rent out parcels to landless tenant farmers.

Several factors combine to restrict development in Latin America. Many of the region's economies rely on

FIGURE 21 The favelas of Brazil are name among many for regions of informal housing inhabited by the poor found throughout the developing world. Roughly 1 billion people live in such communities, typically with minimal access to clean water, transportation, education, or adequate health care.

© Paul Conklin/PhotoEdit
© Dave G. Houser/Corbis
© Michael Brennan/CORBIS
© Catherine Karnow/CORBIS
© Reuters/ CORBIS
© Howard Davies/CORBIS

FIGURE 22 Most of the world's people live in economically underdeveloped countries, and the world's largest cities are usually found there. Often the poor comprise the bulk of the population and, driven from agricultural regions by unequal land distribution and low commodity prices, flock to urban areas in chains of rural-to-urban migration. With inadequate employment opportunities, low wages, and insufficient access to capital and public services, billions of people live in decrepit shantytowns without adequate water or sanitation.

low-priced primary-sector goods for export, such as coffee, bananas, orange juice, beef, soybeans, and mineral ores. Government policies that regulate development, but do not encourage it, are partly to blame. Repeated American military and political interventions have caused significant dislocations. Foreign investment has been relatively small, although it is growing.

Southeast Asia

This region, consisting of Indochina and the island countries of the Philippines, Malaysia, and Indonesia, is a potpourri of peoples. Culturally, it is a mix of Buddhism and Islam, with several major languages. Indonesia is the most populous country in the region and comprises over 13,000 islands and 240 million people.

Indochina was long plagued by horrific warfare, which included French, Japanese, American, and Chinese invasions, that shattered the economies of Vietnam, Laos, and Cambodia. Despite modest growth, these countries remain poor. Myanmar remains an isolated country trapped in poverty. The area is abundant in tin and contains substantial petroleum reserves, especially in Indonesia and the tiny oil-rich sultanate of Brunei. The region's cheap labor has made it a leading manufacturer of textiles and clothing, but there are widespread discrepancies among states in terms of their economic development. Malaysia and Singapore are now major producers of electronic goods, and Singapore is a center of finance and telecommunications of worldwide significance. Thailand and Indonesia have recently joined the ranks of the **newly industrializing countries (NICs)**, attracting foreign capital and developing a more diverse export capacity. Cities such as Bangkok have become congested with automobiles and expensive to live in. The Philippines, in contrast, remains mired in severe poverty, a reflection of the Spanish land grant system and endemic government corruption.

East Asia (Excluding Japan)

East Asia is a vast, heavily populated region that has enjoyed the most rapid rate of economic growth in the world since World War II. Culturally, it is dominated by Buddhism and Confucianism.

China is the region's giant; it has 1.3 billion people and has become a worldwide economic powerhouse. Roughly half of Chinese live in rural areas in about 700,000 agricultural villages. From the ascendancy of communism in 1949 to the death of Mao Zedong in 1976, China was an isolated country ruled by a Communist Party pursuing unadulterated socialism. Major events of this period include the Great Leap Forward, in which 30 million starved to death, and the Cultural Revolution of 1966–1976, a period of massive social disruption.

Since 1979, China has undergone an enormous economic and social transformation. Most of the communist regulations have been loosened, and in an ostensibly socialist country market imperatives dominate. Farmers now own land, control their own production, and sell their products on the market. The government still has a firm grasp on the everyday lives of the citizens, but a majority of Chinese feel the country is better off today than it was several decades ago. China is the largest recipient of foreign direct investment (FDI) in the developing world, and its coastal regions have experienced rapid economic growth, including the miraculous transformation of Guangdong province in the south, which has close ties to Hong Kong. Rapid rates of construction in Shanghai and Beijing reflect the new Chinese economy and a growing middle class. In the interior of the country, however, economic progress has been much slower, and great poverty remains.

Outside of China, Taiwan and South Korea have grown rapidly since the 1960s, making them text examples of NICs. They benefited from being front-line states during the Cold War and received massive U.S. military and economic assistance. With little arable land, they relied heavily on skilled labor and export promotion policies to advance into high-value-added goods, including steel, ships, and electronic goods, and have sizeable middle classes as a result. South Korea has become one of the most successful countries of the developing world; North Korea, in contrast, suffers deepening poverty and famine in part due to its notoriously dictatorial government.

South Asia

This area boasts the world's second-largest population and some of the world's poorest people. Formerly a British colony, it split into India and Pakistan in 1947; Bangladesh seceded from Pakistan in 1971. India and Pakistan have fought several wars in the past and continue to dispute possession of Kashmir; both now possess nuclear weapons.

Hinduism predominates in India; Pakistan and Bangladesh, with roughly 173 million and 147 million people, respectively, are overwhelmingly Muslim. Population densities are high from the natural rate of population increase, and the region has huge pools of poor people. India, the largest country of the region with over 1.1 billion people, overshadows its neighbors demographically and economically. It is the leading producer of certain world crops such as cotton. However, its government long practiced policies of import substitution and protectionism that greatly slowed economic growth. Recently, as the government has turned toward more export-oriented policies, western India has enjoyed modest economic growth from the production of automobiles, movies, call centers, and software. In contrast, eastern India lags behind, as has Bangladesh.

Most of South Asia's people live in villages and subsist directly off of the land, growing rice under quasi-feudal social systems. Many cities in the region—Calcutta, Madras, Delhi, Mumbai, Dacca, and Karachi, for example—contain large numbers of urban poor. Economic development progresses slowly as most of the achievements are erased by the rapidly growing population. Women tend to suffer harsh social and economic circumstances, and their opportunities for advancement are limited.

Middle East and North Africa

Physically dominated by deserts, this region lacks a substantial agricultural base; due to chronic shortages of water, many food products must be imported. Islam is the overriding religion, with the exception of Israel. The region's major asset is its immense petroleum reserves, the revenues from which are used to finance economic development. Culturally, this domain is dominated by the Arab states, which exhibit considerable variations in economies and standards of living. Not all countries in the Middle East have large oil reserves, which are concentrated mainly in the Persian Gulf region. In general, Arab countries with large populations (e.g., Egypt) lack oil reserves, while the major petroleum exporters (e.g., Saudi

Arabia, Kuwait, United Arab Emirates) have relatively small populations. Population growth in the Arab states continues at very high levels, creating societies with large numbers of unemployed young people.

The reasons for the region's low level of development despite the presence of petroleum reserves include repressive, corrupt governments that do not favor growth. Most Arab governments are at best only quasi-democratic. Some analysts argue that the revenues from petroleum inhibit the motivation to diversify the economies; in this sense, resources can be a curse rather than a blessing. Islamic traditionalism also challenges economic development in some areas; for example, the role of women in business and public life is sharply restricted, and financial markets are hindered by the Koranic prohibition against interest, although in some countries such as Bahrain and the United Arab Emirates this obstacle has been circumvented. The region has also been severely affected by political instability. Fundamentalist Shiite Muslims took over Iran in 1979 and their attempts to cleanse the land of Westernized values and institutions elsewhere have promoted social instability. Several wars hampered economic growth in the region as well, including the 1979–1989 war between Iraq and Iran, in which 1 million people died; the two American wars against Iraq (1991 and 2003), the latter of which is ongoing, if diminished; civil conflicts in Lebanon, which decimated its banking industry; several wars between Israel and its Arab neighbors (1948, 1956, 1967, and 1973); and constant, ongoing strife between Israel and the Palestinians.

Israel is the region's only country not dominated by Muslims and is its sole democracy. Israel is also the Middle East's most economically advanced state, in part because of large American subsidies, and has become a significant exporter of high-technology products and computer software. However, Israel has long been mired in conflict with the Palestinians and, at times, others of its Arab neighbors, which has depressed the political stability of the region, including tourism and foreign investment.

Sub-Saharan Africa

Sub-Saharan Africa is by far the poorest region in the world. Despite the fact that it has a relatively low average population density and an abundance of natural resources, most people live in poverty and suffer from poor health and a lack of education. The legacy of the long European colonial era still lives on in African countries; the borders established by their colonial masters often take no account of ethnic groups and tribal areas, which has led to violent tribal conflicts. Over the past 30 years, wars in Angola, Mozambique, Rwanda, Congo, Liberia, Sierra Leone, Somalia, Ethiopia, Eritrea, and Sudan have killed over 30 million people. Famine claims the lives of millions more annually. The governments of most of Africa have proved to be too corrupt and indifferent to the needs of their populations, spending most of their limited funds on the military rather than on their civilian populations. Foreign investment is minimal, and manufacturing uncommon. Africa's economies are largely centered on the production of raw materials, including petroleum, copper, uranium, diamonds, and gold. Low prices on the world commodities market have contributed to the continent's economic malaise, and often the profits of such enterprises are not reinvested in the country from which the resources were extracted. Africa also suffers from high rates of population growth, as its fertility rates are the highest in the world. The demand for farmland and overgrazing have stripped many agricultural areas of their potential to grow crops. Deforestation has decimated the continent's rich ecosystems and contributed to the worldwide crisis in biodiversity loss. AIDS has claimed the lives of millions, and in many countries more than one-third of all adults are infected with HIV. The epidemic has produced millions of orphans and shows little sign of slowing down. Malaria and other diseases are endemic in much of the continent.

As a result of numerous wars, corrupt governments, rapid population growth, economic mismanagement, disease, drought, and lack of foreign investment, Africa is the only region in the world that has become poorer since World War II. By any measure, Africans are the poorest people in the world and the ones most in need of economic growth.

CHARACTERISTIC PROBLEMS OF LESS DEVELOPED COUNTRIES

The developing world encompasses a vast array of societies with enormous cultural and economic differences. Nonetheless, several characteristics tend to be found throughout these nations to one degree or another. Obviously, the more economically developed a country is, the less likely it is to exhibit these qualities.

Rapid Population Growth

Can we ascribe the problems of less developed countries to unsustainable gains in their populations? We must be careful to avoid the simplistic Malthusian error of attributing all of the world's problems to population growth. Nonetheless, in many countries, the rate of growth does exceed productivity gains in agriculture and other sectors, diminishing the average standard of living. Many scholars argue that rapid population growth must be controlled if development is to succeed.

In many less developed countries, particularly in Africa, rapid rates of population growth tax the available food supplies, generating food insecurity and starvation. Rapid population growth also reduces the ability of households to save, inhibiting the accumulation of domestic investment capital and helping to perpetuate a cycle of poverty (Figure 23). More investment is required to maintain a level of real capital growth per person. If public or private investment fails to keep pace with the population, each worker will be less productive, having fewer tools and less equipment with which to produce

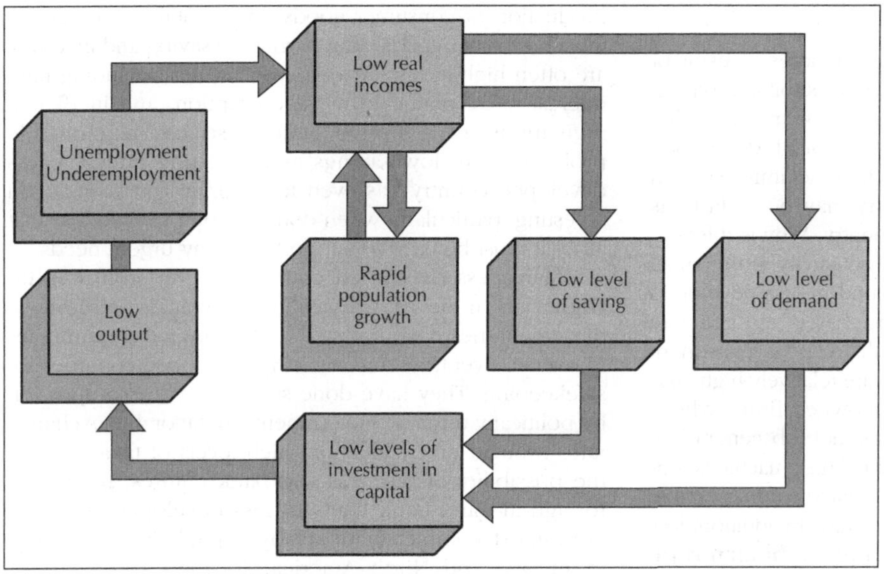

FIGURE 23 The cycle of poverty. Economic development means intervention in a host of intertwined variables that conspire to keep people poor. These include the historical context of a developing country and its linkages to the world system as well as internal factors such as population growth, investment levels, and government policies.

goods. Declining productivity results in reduced per capita incomes and economic stagnation. In agriculturally dependent countries, rapid population growth means that the land must be further subdivided and used more heavily than ever. Smaller plots inevitably lead to overgrazing and overplanting to meet the pressing need for more food from a limited amount of space. Rapid population growth also generally entails large flows of rural residents to urban areas and greater urban problems in many less developed countries. Housing, congestion, pollution, crime, and lack of medical attention are all seriously worsened by the growth in urban populations.

Assuredly, rapid population increases—especially the number and proportion of young dependents—creates serious problems in terms of food supply, public education, and health and social services; it also intensifies the employment problem. However, a high rate of population growth was once a characteristic of present-day developed countries and it did not prevent their development. This observation makes it difficult to argue that population growth necessarily leads to underdevelopment or that population control necessarily aids development. Thus it is erroneous and simplistic to blame all the less developed countries' problems on population growth, which is only one of several factors that contribute to poverty. Focusing on rapid population growth detracts from other, more important but politically controversial factors such as the legacy of colonialism, the dynamics of the world system, and governments' indifference to investing in their populations and infrastructure.

Unemployment and Underemployment

Unemployment and underemployment are major problems in less developed countries. Their economies rarely generate enough jobs for all, for a variety of reasons. **Unemployment** is a condition in which people who want to work cannot find jobs; **underemployment** means that working people usually work much less than 8 hours per day, not as much as they would like, and that their skills and talents are underutilized. Reliable statistics on unemployment and underemployment in less developed countries are difficult to obtain, but unemployment in many developing countries often exceeds 20%, a level comparable to that of the Great Depression in the United States.

Many cities in less developed countries have recently experienced rapid flows of migrants from rural areas as a result of low agricultural prices, inadequate public investment, rural unemployment, and lack of land reform. Large numbers of migrants are attracted to cities by the expectation of jobs and higher salaries, expectations that may not be met in reality. Thus, these flows exemplify the influence of imperfect information upon spatial decision making. Once in the cities, many migrants cannot find work and thus contribute to unemployment. Other migrants find limited employment as shop clerks, handicraft peddlers, or street vendors, often forming part of the unregulated, low-wage, informal economy.

Unemployment and underemployment are not the sole reasons for the problems of less developed countries. Jobs are generated only when there is sufficient investment, which is lacking in most less developed countries. Because domestic pools of investment capital are limited, poor countries rely on **foreign direct investment (FDI)** supplied by transnational corporations, which can be fickle, and can involve social, economic, and environmental costs. Thus, unemployment is much more complex than simply the willingness of people to work; it involves a complex, globalized institutional framework in which public and private resources are allocated, jobs are generated, occupations and skills created and maintained, and investment capital is channeled.

Low Labor Productivity

Are the problems of less developed countries a result of low labor productivity? It is true that a day's toil in a typical developing country produces very little commercial value compared with a day's work in a typical developed country. This is particularly evident in agriculture. Human productivity in a developing country may be as little as one-fiftieth of that in a developed country. Why is this?

One reason for low levels of productivity is the small scale of farming operations often found in less developed countries, which are often subsistence-based and generate few *economies of scale*. Another is that capital investment rates are often very low, interest rates are relatively high, and most capital is generated by foreign-owned firms, whose major incentive is exports and profits, not job generation. Most developing countries lack the buildings, machines, engines, electrical power networks, and factories that enable people and resources to produce efficiently. In addition, less developed countries are less able to invest in **human capital**. Investments in human capital—such as education, health, and other social services—make unskilled workers into productive workers. Schools are often inadequate, literacy rates may be comparatively low, and technical and management skills are in short supply. Many highly skilled professionals, such as engineers and doctors, emigrate to economically advanced countries, where they are more highly paid. The United States and Europe, for example, are replete with some of the best-trained labor from less developed countries, including doctors, engineers, mathematicians, and scientists who have come looking for better salaries from business or the government. This immigration has contributed to a so-called **brain drain**, whereby less developed countries lose talented people to the developed world.

Although it must be acknowledged that low labor productivity is a universal attribute of less developed countries, it is not a *causative* factor. The important question to consider is this: What prevents labor productivity from improving in developing countries? To answer this question, we must examine other dimensions of their poverty.

Lack of Capital and Investment

Most less developed countries suffer from a lack of capital in the form of machinery, equipment, factories, public utilities, and infrastructure in general. The more capital, the more tools available for each worker; thus there is a close relationship between output per worker and per capita investment and, in turn, between investment and income. If a nation hopes to increase its output, it must find ways to increase per capita capital investment.

In most cases, increasing the amount of arable land is no longer a possibility in most less developed countries. Most cultivable land in the world is already in use. Therefore, capital accumulation must come from savings and investment. If a less developed country can generate savings and export revenues, and invest some of its earnings, resources will be released from the production of consumer goods and capital goods (goods that make other goods). But barriers to saving and investing are often high in less developed countries, including high interest rates, political unrest, corruption, and inefficient regulations. (The United States also has had notable problems with low savings and investing rates.) A less developed country has even less margin for savings and investing, particularly when domestic output is so low that all of it must be used to support its many urgent needs.

Many less developed countries suffer from **capital flight**, which means that wealthy individuals and firms in these countries have invested and deposited their monies in overseas ventures and banks in developed countries for safekeeping. They have done so for fear of expropriation by politically unstable governments, unfavorable exchange rates brought on by inflation, high levels of taxation, and the possibility of business and bank failures. Inflows of foreign aid and bank loans to less developed countries are almost completely offset by capital flight to banks in Europe and North America. In 2008, for example, Mexicans were estimated to have held about $130 billion in assets abroad, which is roughly equal to Mexico's international debt.

Finally, obstacles to investment in less developed countries have impeded capital accumulation. The two main problems with investment are (1) lack of investment opportunities comparable to those available in developed countries and (2) lack of incentives to invest locally. Usually, less developed countries have low levels of domestic spending per person, so their markets are weak compared with those of advanced nations. Without domestic industries, consumers must turn to imports to satisfy their needs, especially for high-value-added products. There is also a lack of infrastructure to provide transportation, management, energy production, and community services—housing, education, public health—which are needed to improve the environment for investment activity.

Inadequate and Insufficient Technology

Typically, less developed countries lack the technologies necessary to increase productivity and accumulate wealth. Some acquire new production techniques through technology transfer that may accompany investment by transnational corporations, as happened in the NICs of East Asia. Similarly, OPEC nations benefited from foreign technology in oil exploration, drilling, and refining. Unfortunately, for less developed countries to put this technology to use, they must have a minimal level of capital goods and human capital for the installation, repair, and maintenance of machinery. The need is to channel the flow of technologically superior capital goods that have high levels of reliability to the developing nations so that they can improve their output.

In developed countries, technology has been developed primarily to save labor and to increase output, resulting in capital-intensive production techniques. However, in less developed countries, capital intensification tends to displace workers, eliminating critically needed jobs. Thus less developed countries need capital-saving technology that is labor-intensive. In agriculture, much of the midlatitude technology of the developed countries (e.g., wheat harvesting combines) is unsuited for low-latitude agricultural systems with tropical or subtropical climates and soils. In short, to understand the impacts of technology, we must examine them in the social and economic context in which they occur.

Unequal Land Distribution

In most developing countries, in which a large part of the population lives in rural areas, land is a critical resource essential to survival. Unfortunately, land is often highly unequally distributed, and a small minority of wealthy landowners controls the vast bulk of arable (farmable) land. In Brazil, for example, 2% of the population owns 60% of the arable land; in Colombia, 4% owns 56%; in El Salvador, 1% owns 41%; in Guatemala, 1% owns 36% of the land; and in Paraguay, 1% owns an astounding 80% of all land suitable for farming. These numbers reflect the historical legacy of the Spanish land grant system imposed over centuries of colonialism. Wealthy rural oligarchs often control vast estates and plantations designed primarily for export crops, not domestic consumption. The majority of the rural population, consequently, is landless and must sell its labor, often at very low (below subsistence) wages, and live in serf-like conditions. Shortages of land are accentuated by agricultural mechanization and high population growth. The result is frequent political turmoil, including demands for land redistribution; in much of Latin America, peasant social movements to regain control of the land are very active and violent conflicts over land possession are frequent. Further, rural areas with high natural growth rates and widespread poverty become the source of rural-to-urban waves of migration, which often swamp cities with desperately poor people seeking jobs.

Throughout much of the developing world, land reform, which is vitally important to increased agricultural output, is not carried out because governments are too inept or corrupt to redistribute the land owned by a few wealthy families. For some nations, such as the Philippines, land reform is the single most pressing problem deterring economic development. In contrast, strong action taken by the South Korean government after the Korean War allowed for increased productivity and the development of an industrial and commercial middle class that made South Korea one of the success stories of the Pacific Rim.

Poor Terms of Trade

A country's **terms of trade** are the relative values of its exports and imports. When the terms of trade are good, a country sells, on average, relatively high-valued commodities (e.g., manufactured goods) and imports relatively low-valued ones (e.g., agricultural products). This situation generates foreign revenues that allow it to pay off debt or to reinvest revenues in infrastructure, human capital, or new technologies. Unfortunately, many less developed countries, with economies distorted by centuries of colonialism, in which they became suppliers of minerals and foodstuffs, export low-valued primary-sector goods and must import expensive high-valued, manufactured ones. Many less developed countries rely heavily on exports of goods such as petroleum, copper, tin and iron ore, timber, coffee, and fruits and must import items such as automobiles, pharmaceuticals, office equipment, and machine tools. The worldwide glut in raw materials, including petroleum, agricultural crops, and many mineral ores, has depressed the prices for many less developed countries' exports. Without indigenous manufacturing, many are forced to sell low-valued goods such as bananas for high-valued ones such as computers and pharmaceuticals (Figure 24). This situation makes it difficult to generate foreign revenues and

Medioimages/ Photodisc/Getty Images

FIGURE 24 The famous floating markets of Bangkok, Thailand, are one of several faces of the informal economy prevalent in almost all cities in the developing world and, to a lesser extent, in many Western ones too.

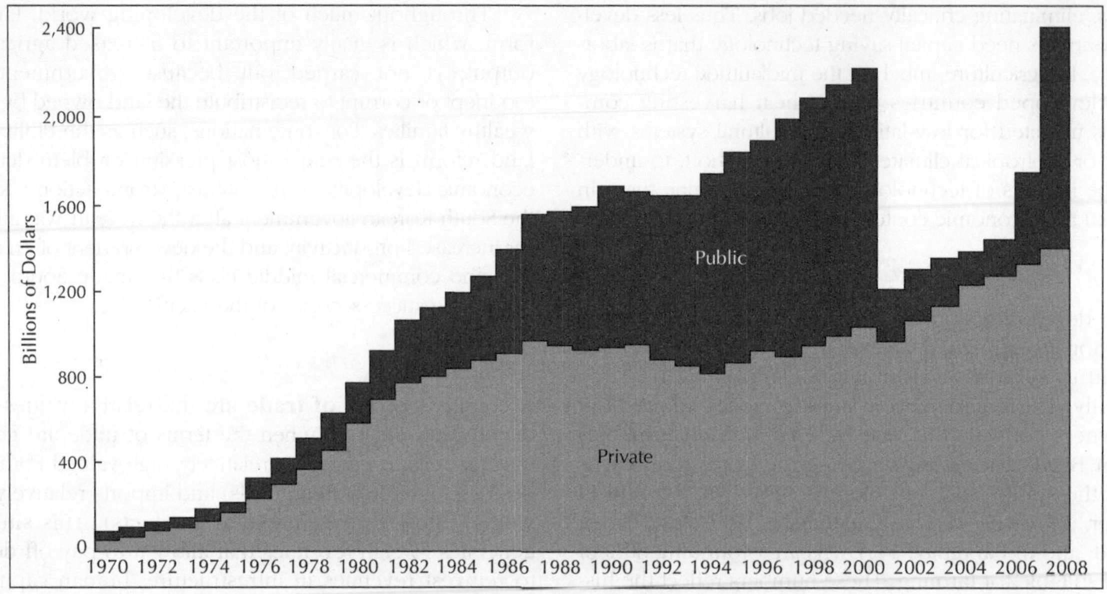

FIGURE 25 The external debt of all developing countries grew substantially between 1970 and 2008. It consists of two parts: the public debt, which is owed to foreign governments, and the private debt, which is owed to private banks.

exacerbates a country's debt problems. Further, poor terms of trade perpetuate a country's cycle of poverty: Low foreign revenues yield little to reinvest, which helps to create a shortage of capital, resulting in low rates of productivity.

Foreign Debt

Much of the developing world is deeply in debt to foreign governments and banks. In 2009, total world debt amounted to more than $56 trillion; in absolute terms, the largest debtors are in the developed world (Figure 25), led by the United States (Figure 26), which owes the world more than $13 trillion (Table 2). However, when we examine this issue in terms of relative debt, or debt as a percent of GDP, quite a different picture emerges: Measured in terms of national income, it is developing countries that owe the most (Figure 27).

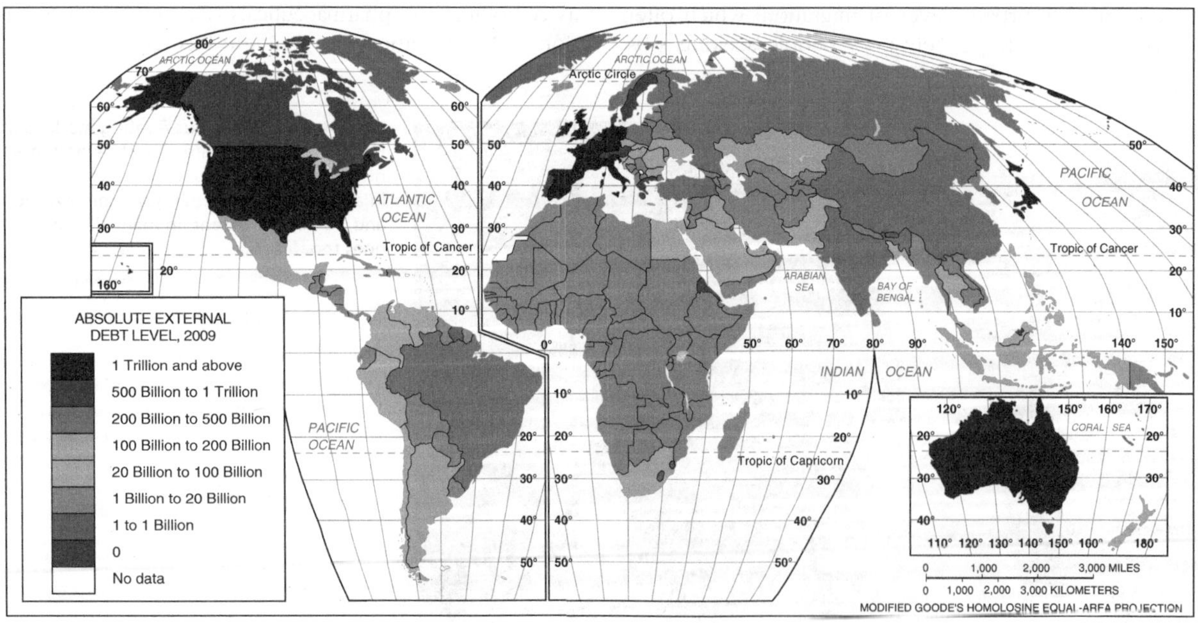

FIGURE 26 Absolute external debt level in 2009. The United States is unquestionably the world's largest debtor, a situation that results from decades of tax cuts and large federal government budget deficits.

TABLE 2 World's Major Debtors, 2009 ($ trillions)

World	56.9
United States	13.45
United Kingdom	9.1
Germany	5.2
France	5.0
Netherlands	3.7
Spain	2.4
Italy	2.3
Ireland	2.3
Japan	2.1
Belgium	1.4
Switzerland	1.3
Australia	0.9
Canada	0.8
Austria	0.8

Some countries such as Sudan or Zimbabwe have foreign debts in excess of 100% of their GDPs. Having borrowed hundreds of billions of dollars, they find themselves unable to repay either the interest or the principal. Debt repayments—annual interest and amortization payments—often consume a large share of their export revenues, which might otherwise be used for development. Indeed, the average developing country spends 50% of its export revenues on debt repayment.

The origins of the debt crisis lie in the 1970s and 1980s, when many less developed countries took out large loans with the expectation of paying them off when their economies improved in the future. A huge influx of petrodollars made Western banks eager for borrowers; developing countries were happy to take advantage of this unaccustomed access to cheap loans with few strings attached. The borrowing enabled them to maintain domestic growth. Even oil-exporting nations such as Mexico, Libya, and Nigeria overborrowed based on their expectations of rapidly inflating oil prices. When oil prices fell significantly in the 1980s, these nations found themselves saddled with debts that they could not afford. In addition, many developing countries faced decades of declining terms of trade and governments that spent heavily on their militaries as well as overly ambitious, unrealistic development schemes.

The mounting debt raised concerns about the stability of the international monetary system. In 1982, Mexico ran into difficulties meeting interest and capital payments on its debts; Brazil and Argentina also appeared ready to default. A collapse of the financial system was forestalled by a series of emergency measures designed to prevent large debtor countries from defaulting on their loans. These measures involved banks, the International Monetary Fund (IMF), the World Bank, and the governments of lending countries in massive bailout exercises that accompanied debt reschedulings.

Debt restructuring policies imposed by the IMF typically tie debt relief to "**structural adjustment**," which includes reductions in government subsidies to the poor (such as for kerosene, cooking oil, or mass transit), a rise in interest rates to attract foreign investors, and devaluations of currencies (which drive up the costs of imports). Such policies lower the quality of life for hundreds of millions, if not

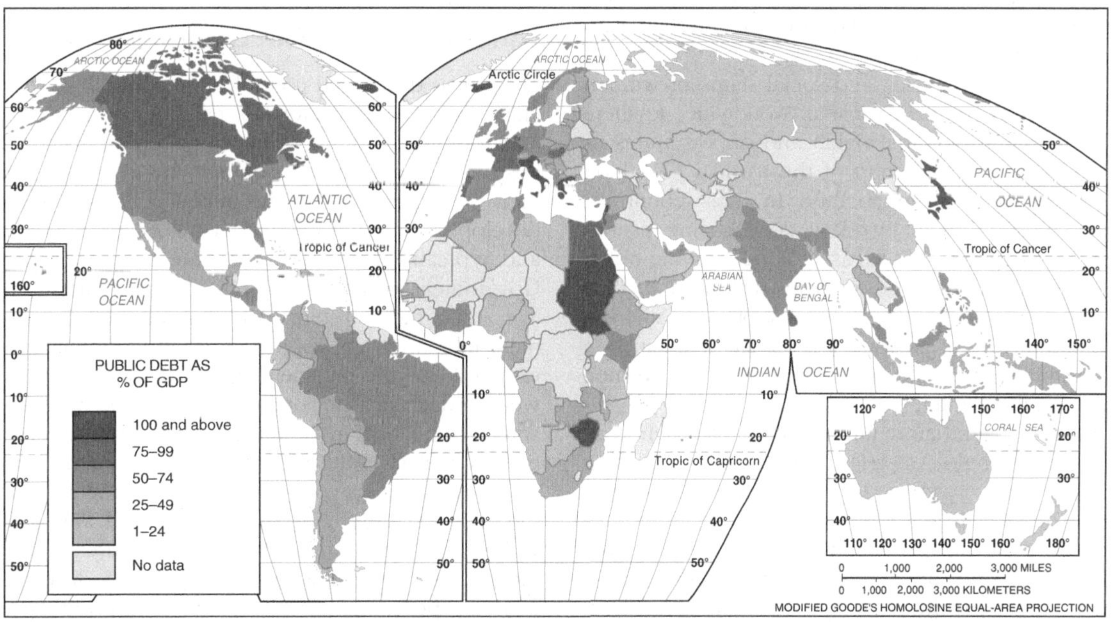

FIGURE 27 External debt as a percent of GDP, 2007. Using GDP as a measure of a country's ability to pay back its debt, rather than absolute debt, points to the enormous debt burden carried by many impoverished developing countries. A large share of government revenues and earnings from exports thus must pay the principal and interest of their debt. Although the poor often did not benefit from these loans, it is their taxes that are often used to pay them back.

billions, of people, and have made the IMF a much hated institution in many countries. For example, a poor mother in a less developed country with a sick infant relies on imported pharmaceuticals when her country lacks a domestic industry; IMF-imposed currency devaluations will drive up the costs of these necessities, often putting them out of reach for the poor, who need them the most. For these reasons, scholars and politicians involved in international development often call for a debt moratorium to allow less developed countries a respite from what are often crushing debt burdens.

As the debt crisis grew in the twentieth century, some less developed countries required foreign banks to rewrite their loans and cancel or *write down* a portion of the principal and interest. The result was a loss of confidence in the future ability of many less developed countries to repay. At the same time, the United States began to generate enormous federal budget deficits; to finance these, it borrowed a large portion of the available investment revenue, driving up interest rates worldwide. Higher oil prices, declining prices for raw material exports, higher world interest rates, an increase in the value of the U.S. dollar, and a decline in public and private lending to less developed countries because of loss of confidence all contributed to the debt crisis that continues today.

FIGURE 28 In Mali, women often must travel several miles a day from their villages to gather firewood. This task is usually a woman's chore in most countries and exemplifies the fact that women almost everywhere work harder and longer than do men.

Restrictive Gender Roles

Deeply entrenched systems of social stratification work against people, especially women, in many developing countries. Gender roles are the socially reproduced differences—including both privileges and obligations—between women and men and permeate every society's allocation of resources and people's life chances. Typically, gender roles work to the advantage of men and the disadvantage of women (Figure 28).

The economic, political, and social status of women around the world is spatially variable. Worldwide, nowhere can women claim the same rights and opportunities as men. Women account for most of the world's 1.5 billion people living in extreme poverty. In most countries, women do most of the field work while still being responsible for household chores such as cooking, carrying water, and raising children. In some countries with more mechanized forms of agriculture, such as most of Latin America, men assume the job of plowing, with female participation strongly diminished in the field. In these cases, women work more in the market.

Women still spend more time working than men in all regions except Anglo America and Australia, and their wages are lower everywhere. In Muslim areas, women are not very economically active outside of the home because of religious prohibitions. In Latin America, the participation of women in the labor force is increasing but mostly outside of the agricultural sector. Sub-Saharan Africa, India, and Southeast Asia are highly dependent on female farm and market income as well as commodified labor in the waged labor market. At a world scale, women generally garner only 60% to 70% of what men earn, often for the same work. This ratio is remarkably widespread and it includes the United States, in which women comprise the bulk of the poor and where women account for less than 3% of the highest levels of management and ownership.

Corrupt and Inefficient Governments

Often, less developed countries are controlled by bureaucrats whose primary interest is catering to the wealthy elite and foreign investors, creating governments that are at best indifferent and often outright hostile to the needs of their own populations. The public policies of many less developed countries are ineffectual or counterproductive and typically ignore the rural areas in favor of cities. Government jobs are frequently allocated through patronage, not a merit system. For example, in Africa, government jobs often go disproportionately to members of the same tribe as the president. The military is often the most well funded and well organized institution and may be a source of political instability from military takeovers of the government in coups d'état. Corruption often is endemic and may become a way of life, generating inefficiency and inequality (Figure 29). During the Cold War, both the United States and the USSR backed oppressive regimes throughout the developing world with subsidies and arms. In many poor countries, dictatorial governments curtail their citizens' civil liberties, censoring the media and imprisoning or executing dissidents. Such types of governments function as kleptocracies, designed to enhance the

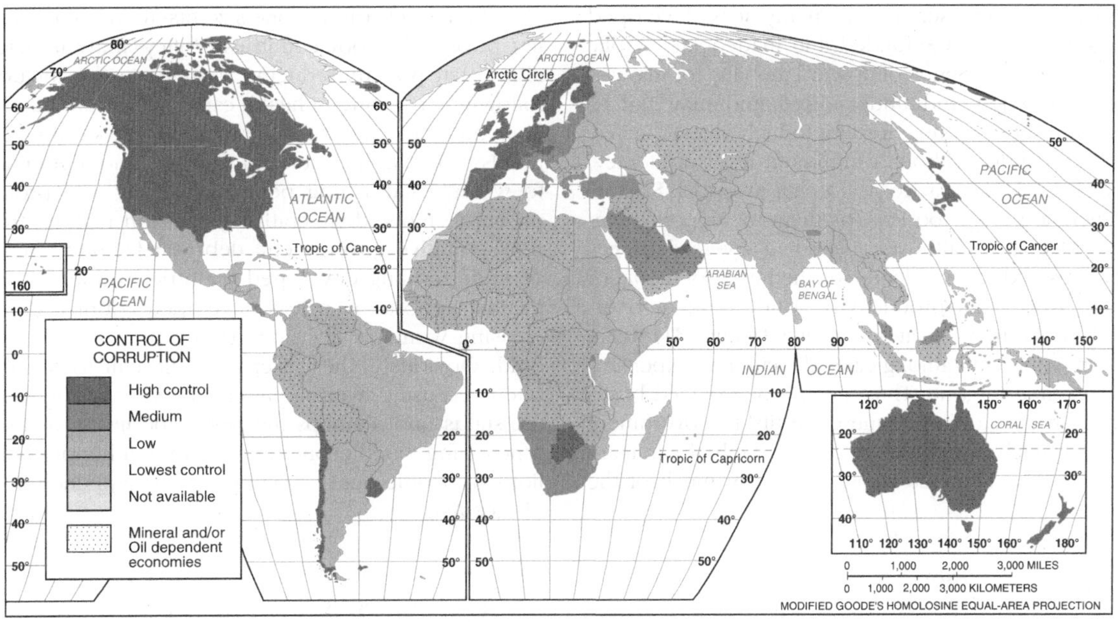

FIGURE 29 Control of corruption around the world. Corruption is a difficult issue to define and may be prone to both underestimation and exaggeration. However, by most measures, countries in the developing world tend to have the highest levels, often taking the forms of bribes and paybacks, which interfere with the efficient functioning of governments.

fortunes of a small, wealthy elite. Attacks on journalists, labor union leaders, student movements, religious groups, and ethnic minorities are common in totalitarian states. Widespread and well-protected civil liberties are more likely under economic development. Understandably, many people feel alienated from oppressive governments and may sympathize with various resistance movements, contributing to frequent political instability.

The governments of less developed countries, often lacking sufficient financial resources, have great difficulty building or maintaining their infrastructures. Roads, bridges, tunnels, and highways may fall into disrepair, driving up transportation costs (Figure 30). Dams, ports, and airports likewise may be neglected. Electrical power stations may not be maintained correctly, leading to frequent brownouts and blackouts. Telephone networks may be missing or unreliable, especially in rural areas. Unsanitary water supplies become major carriers of infectious diseases such as cholera and malaria; indeed, unclean drinking water is one of the world's major public health hazards. Because the infrastructure is essential to economic development, governments that do not reinvest in their nation's transportation, water, and communication lines do not facilitate the process of economic growth.

James Strachan/Robert Harding World Imagery

FIGURE 30 Unpaved road near Kumasi, Ghana, exemplifies a transportation infrastructure suffering from insufficient investment, an obstacle to investment.

Similarly, public services in many less developed countries are often underfunded and inadequate. Public schools—the major avenue of upward mobility for many—are often crowded and understaffed and may fail to educate the nation's young, leading to high illiteracy rates, especially for girls. Without a minimum level of literacy, the labor force is unattractive to foreign investors. Salaries for teachers are often too low for them to support themselves. Inadequate health care, including severe shortages of physicians and nurses, depresses the health of the labor force and lowers its productivity. Underfunded, overcrowded public transportation systems (Figure 31) make circulation within and among cities difficult and expensive in terms of time forgone. Thus, images of crowded buses and trains in developing countries testify to government priorities as much as population growth. Only the military tends to be a well funded public service in much of the world.

Trends and Solutions

Worldwide, the gap between the rich and the poor is widening. The World Bank estimates that in 2008 the global poverty rate was 24%, or about 1.6 billion people. In relative terms, the two regions of the world with the greatest proportions of people living in poverty are Latin America and sub-Saharan Africa. However, in absolute numbers, Asia is home to the greatest number of the world's poor, including vast numbers on the Indian subcontinent. There are indicators that the numbers of the poor in Asia may be declining as economic growth diffuses through the region, while the poverty-stricken population of sub-Saharan Africa is increasing quickly due to war and AIDS.

There is clearly no one-size-fits-all approach to ending poverty in poor countries. Policy analysts generally argue that poverty-reducing strategies must include the following: investing in health care and education; protecting land tenure rights for the rural poor; focusing on rural needs, such as agricultural storage and better access to water; establishing political stability and protecting property rights; democratizing decision making through representative government; debt relief; improving terms of trade; ending developed countries' trade restrictions against imports from less developed countries; and increasing women's decision-making power in the household. Obviously, these steps are difficult to implement, often because powerful, entrenched interests benefit from the status quo. Perhaps the best hope lies in a smoothly functioning world economy that can pull hundreds of millions of people out of poverty, as has happened in much of East Asia.

MAJOR THEORETICAL PERSPECTIVES ON GLOBAL PATTERNS OF DEVELOPMENT

Theories of development have been advanced for a long time. The earliest can be traced to the classical economists of the eighteenth century such as Adam Smith. But in the social sciences discussion of development is fairly recent, flourishing only after World War II. In economic geography, three perspectives hold widely varying assumptions and draw different conclusions on development. These are modernization theory, dependency theory, and world-systems theory. All grapple with the complex question of how the global economy both shapes patterns of opportunities and constrains development in individual countries.

The World Bank\ial

FIGURE 31 The trains are crowded in Bangladesh. Typically, resources for feeding and moving the poor are inadequate. Such systems generate low rates of profit and do not attract large investors. Thus, transport systems in impoverished countries generally are provided by financially strained governments, many of which are reeling under low export prices and neoliberal programs imposed by the IMF. Thus crowding is not simply a result of population growth, but reflects the political economy of public services in developing countries.

Modernization Theory

The first and most widely accepted theory of development is **modernization theory**. This theory starts with the central question: Does the development of the less developed world follow the same historical trajectory as that of the West? The intellectual origins of this line of thought lie with the famous sociologist Max Weber, who attributed the success of northern Europe during the Industrial Revolution to the Protestant ethic. Weber's work underlies the view that economic development and social change are a function of people's ideas, culture, and beliefs rather than their social relations and historical context.

As translated and elaborated by Talcott Parsons, Weber's ideas became enormously popular in the United States following World War II. The American version tended to equate "modern" with Western, denying the possibility of modern, non-Western cultures. During a period of intense competition between the United States and the Soviet Union for influence in the developing world during the Cold War, Parsonian modernization theory explicitly advocated capitalism as the best possible path any country could choose to follow (i.e., the only one that leads to modernity, wealth, and democracy). The theory argued that if less developed countries adopted Western values, market relations, and government institutions, they would become affluent, democratic societies. By changing their way of looking at the world, in particular giving up the traditions that kept them trapped in an irrational past, less developed countries would eventually achieve the same prosperity that Europe and North America enjoyed. Hence, the path to progress from traditional to modern is unidirectional. Rich industrial countries, without rival in social, economic, and political development, are modern, whereas poor countries must undergo the modernization process to acquire these traits. It is worth noting that whereas Weber held that capitalism was an institution unique to Europe, modernization theorists maintained that the triumph of capitalism was inevitable worldwide.

Modernization theory viewed economic development in all economies as occurring through a series of stages, which were to be achieved in a stable and gradual manner, not in revolutionary leaps. These stages, broadly speaking, were, first, agriculture, then manufacturing (i.e., the Industrial Revolution), then services (a postindustrial world). However, although this line of thought is somewhat accurate in terms of the Western world, many developing countries undergo a leapfrogging effect in which many agricultural workers plunge directly into services. This discrepancy points to the dangers of simplistically generalizing from the experience of the West to the developing world, which has a very different historical context and trajectory.

The diffusion of progress was also a major line of thought in modernization theory: New ideas, technologies, and institutions were held to flow from the core to the periphery on both the international and national scales. Internationally, progress spread from the world's core, that is, the developed countries, to the global periphery as multinational corporations invested in less developed countries. Investment would realize a country's comparative advantage and allow it to carve a niche for itself in the world economy; thus modernization theory advocated free trade. Within countries, diffusion occurred from the urban core, that is, cities, which were held to be the foci of modernity, to the rural periphery—the countryside. Thus modernization theory held that urbanization was inherently a good thing, that rural areas were trapped in cycles of tradition and stagnation.

Poverty, in this conception, is the result of the incomplete formation and diffusion of markets, which promote an equalization of standards of living in poor and rich regions through the free flow of capital and labor. Thus modernization theory directly contradicts Marxist claims that capitalism inevitably generates **uneven spatial development**. It likens economic development to a race, in which the rural areas must catch up with the cities, much as less developed countries must catch up with the developed world.

Modernization theory also advocated major social, cultural, and political changes on the road to capitalism. Population growth, for example, must be brought under control or else the demand for resources would exceed the productivity gains of markets. Thus the theory promoted family planning programs, a line of thought similar to the demographic transition. During the Cold War, population control policies were encouraged as a means of raising standards of living and reducing the appeal of the Communist bloc's alternative to capitalism. Culturally, societies (or parts of them) were divided between the traditional and the modern: Tradition was held to be synonymous with irrationality and fear of change, whereas modernity implied the belief that change is good and necessary for the sake of progress. Thus, cities were alleged to be repositories of the modern, and rural areas identified with the traditional, to be transformed and brought into the modern age. The culture of modernity was to be diffused through mass education, the media, and growing literacy rates.

Politically, modernization theory equated capitalism with democracy, arguing that only market-based societies could protect civil liberties. In contrast to the centralized power of many precapitalist societies, markets tend to create multiple centers of power. The rise of a middle class was often held to be a central event in protecting civil liberties: Despotic societies tend to be poor, with a small dictatorial elite in control. By generating an educated, informed citizenry, markets were held to be antithetical to dictatorship. Modernization theorists point to countries such as South Korea as evidence of this thesis.

A major advocate of this approach was Walter Rostow, an influential economist and presidential policy advisor in

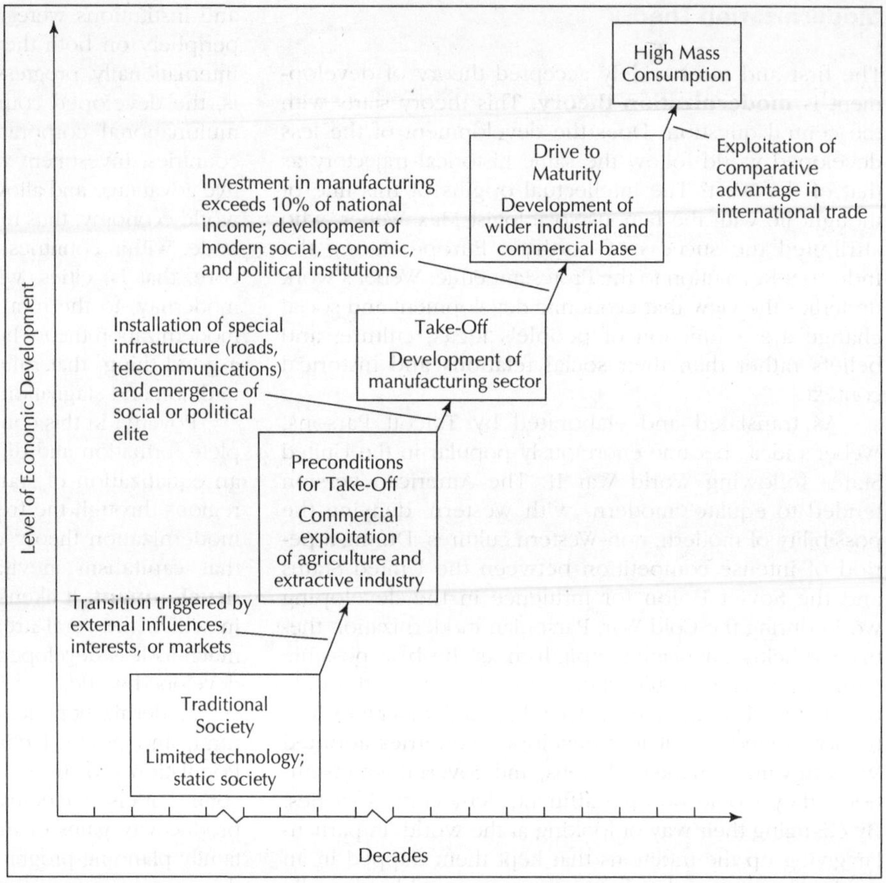

FIGURE 32 Rostow's five stages of the modernization process. The conventional approach to economic development viewed the process as akin to an airplane taking off, in which economies progress through a series of stages culminating in mass consumption. Others have criticized this approach for being ethnocentric (holding up the West as the only model of development) and ignoring the global and historical context of development.

the 1960s who proposed a famous five-stage model of development (Figure 32). Rostow's model, likening economic growth to an airplane taking off, proposed five stages in the modernization process:

1. ***Traditional Society*** This term refers to a country that has not yet started the process of development and is mired in poverty. It includes traditional societies with a very high percentage of people engaged in agriculture and a high percentage of national wealth allocated to what Rostow called "nonproductive" activities, such as the military and religion. Contemporary examples might include Mali or Bhutan.
2. ***Preconditions for Takeoff*** Under the international trade model, the process of development begins when an elite group initiates innovative economic activities. Under the influence of these well-educated leaders, the country starts to invest in new technology and infrastructure, such as water supplies and transportation systems. These projects will ultimately stimulate an increase in productivity. A culture of growth begins to take root. Examples include Thailand, Mexico, and Indonesia.
3. ***The Takeoff*** Rapid growth is generated in a limited number of economic activities, such as textiles or food products. These few takeoff industries achieve technical breakthroughs and become efficiently productive, while other sectors of the economy remain dominated by traditional practices. Examples today include China and Chile.
4. ***The Drive to Maturity*** Modern technology, previously confined to a few takeoff industries, diffuses to a wide variety of industries, which then experience rapid growth comparable to that of the takeoff industries. Workers become more skilled and specialized. Examples include Singapore, Taiwan, and South Korea.
5. ***The Age of Mass Consumption*** The economy shifts from production of heavy industry, such as steel and energy, to consumer goods, like motor vehicles and refrigerators, and advanced services. Examples include the United States and Germany.

Modernization theory assigns every country in the world to one of these stages. Less developed countries reside in the primary stages while developed countries have passed through the preliminary stages and are now in stage 5 or beyond. It is assumed that less developed countries will use this model to follow the steps of the developed countries. International trade keeps countries competitive by forcing the industries to adapt and develop along world standards.

The policy implications of modernization theory are straightforward. Overall, it advocates the formation of unfettered markets. Thus, barriers to trade and investment should be removed; foreign capital in the form of multinational corporations should be welcomed. Urban development should be stressed at the expense of rural areas. Internationally, this view was central in the switch from import substitution to export promotion that occurred in much of East Asia in the 1970s and 1980s. More recently, this theory has been used to justify neoliberal policies of structural adjustment by the International Monetary Fund, including currency devaluations and reductions in government subsidies.

Modernization theory has been soundly criticized on a number of grounds. Critics note that it is ethnocentric: It posits the history of the West as an ideal to be imitated and everyone else's culture as inferior. It offers a simplistic, unidimensional view of history that culminates only in the experience of Europe and North America. Less developed countries are seen as little more than backward versions of the West, not as unique entities with their own distinct cultures and histories. Critics note that hundreds of years of colonialism produced a world economy that greatly advantaged the West and disadvantaged the developing world; development is hardly a fair race when not all countries enjoyed the same starting point. Finally, by uncritically celebrating markets as mechanisms that produce only wealth and not poverty, modernization theory celebrates the benefits but ignores the costs of capitalist development: Markets look fine to the winners, who preach free trade, but appear less rosy to those who have not benefited from them. Finally, modernization theory focuses only on the internal dynamics within countries (i.e., their cultures) and ignores the external context, the colonial legacies, and their position in the global division of labor. In this way it ends up blaming the victim (i.e., attributing poverty to poor people), which is politically easy but not accurate.

Dependency Theory

In the 1960s, numerous scholars from the developing world, particularly Latin America, began to question the promises and assumptions of modernization theory. What appeared as comparative advantage and interdependence to Western scholars, for example, appeared to many in the less developed countries as exploitation. Drawing from the heritage of Marxism, **dependency theorists** claimed that the development of the core countries was intrinsically dependent on the underdevelopment of the periphery countries. The theory suggests that unequal development of the world economy stems directly from the historical experience of colonialism. The development of Europe and North America, especially during the nineteenth and early twentieth centuries, relied on the systematic exploitation of underdeveloped areas by means of unequal terms of trade, abuse of low-skilled and low-paid labor, and profit extraction.

In this view, poverty didn't just happen, but is a historic product (i.e., like a shirt or pair of tennis shoes, poverty is actively made by the dynamics of capitalism, not just something that happens to appear). This process is described as the "development of underdevelopment," which emphasizes that underdevelopment is not a state but an active process. Dependency theorists thus argue that the less developed countries were *made to be poor* by the West. Exploitation of the periphery occurs through the process of uneven exchange, in which less developed countries produce low-valued goods in the primary sector and purchase high-valued goods from the core, a market mechanism that appears like an exchange of opposites but conceals the appropriation of surplus value that low-income workers in developing countries produced.

Unlike modernization theory, therefore, dependency theory focuses on external, not internal, causes of poverty. Because they occupy very different roles in the capitalist world economy, developed and underdeveloped countries are very different animals. Thus, unlike its colonies, the West was undeveloped, but not underdeveloped; less developed countries do not temporarily "lag" behind the West but are mired in poverty produced by the West, a situation naturalized as inevitable by the economics of comparative advantage. Dependency theorists thus argue that development and underdevelopment are twin sides of global capital accumulation: The wealth of developed countries is derived from the labor and resources of the less developed countries. Whereas modernization theory holds that markets eradicate uneven development, dependency theorists maintain markets perpetuate it. Development and underdevelopment are therefore two sides of the same coin, a zero-sum game: Development somewhere (i.e., the West) requires underdevelopment somewhere else (i.e., its colonies). The political and social structures created under this vision of the world economy imply that independent development is impossible. The policy implications of dependency theory, directly opposite of those espoused by modernization theorists, emphasize self-reliance; countries should exclude transnational corporations and practice import substitution to promote domestic production. Some even advocate defaulting on foreign debt.

Like modernization theory, dependency theory is also subject to criticisms. The theory tends to view the global periphery as passive and incapable of taking action; it lumps all less developed countries together as if they were victimized by capitalism to the same degree. It ignores the internal causes of poverty, such as rural aristocracies that inhibit development, and its explanation for the core's wealth is simplistic. For example, dependency theory has not offered an adequate account of technological change and productivity growth in less developed countries. The claim that global capitalism always generates poverty in less developed countries has been unsustainable: Empirically, in the 1970s and 1980s, the rapid growth of the East Asian NICs in particular showed that development on the global periphery is indeed possible and that capitalism does not automatically reduce all less developed countries to an impoverished state.

World-Systems Theory

It is important to keep in mind that the forces driving the world economy are in a continual state of flux. The rise of the United States from a periphery to a core country, the fall of the Soviet Union, the appearance of the NICs in Asia, and the increasing importance of transnational firms exemplify the notion of permanent spatial disequilibrium. A third theory, **world-systems theory**, originating with the sociologist Immanuel Wallerstein, takes into account this disequilibrium in explaining the changing structure of the world economy.

Unlike dependency theory, world-systems theory allows for some mobility within the capitalist world economy. Its focus is on the entire world rather than individual nation-states. Fundamentally, this view maintains that one can't study the internal dynamics of countries without also examining their external ones. Thus the boundary between foreign and domestic effectively disappears. World-systems theory distinguishes between large-scale, precapitalist world empires, such as the Roman, Mongol, or Ottoman empires, which appropriated surplus for the use of the state from their peripheries, and the capitalist world system, which arose in the "long sixteenth century" (1450–1650). Occasional, failed attempts to reassert a world empire included the Holy Roman Empire under the Hapsburgs, the Napoleonic Empire, and the Third Reich in Germany during World Wars I and II. Under global capitalism, there is no single political entity to rule the world (i.e., there is a single market but multiple political centers, meaning there is no effective control over the market). The political geography of capitalism is thus not the nation-state but the interstate system.

World-systems theorists maintain that capitalism takes many forms and uses labor in different ways in different regions. In the core, labor tends to be waged (i.e., organized through labor markets); in developing countries there is considerable use of unfree labor, ranging from slavery to indentured workers to landless peasants working on plantations. The world economy structures places in such a way that high-valued goods are produced in the core and low-valued ones in the periphery. It is the search for profits through low-cost labor that drives the world system forward to expand into uncharted territories.

Both modernization and dependency theories divide countries into two categories: developed and less developed; world-systems theory sees a tripartite division among core, periphery, and semiperiphery countries (Figure 33). The **core and periphery** are the developed

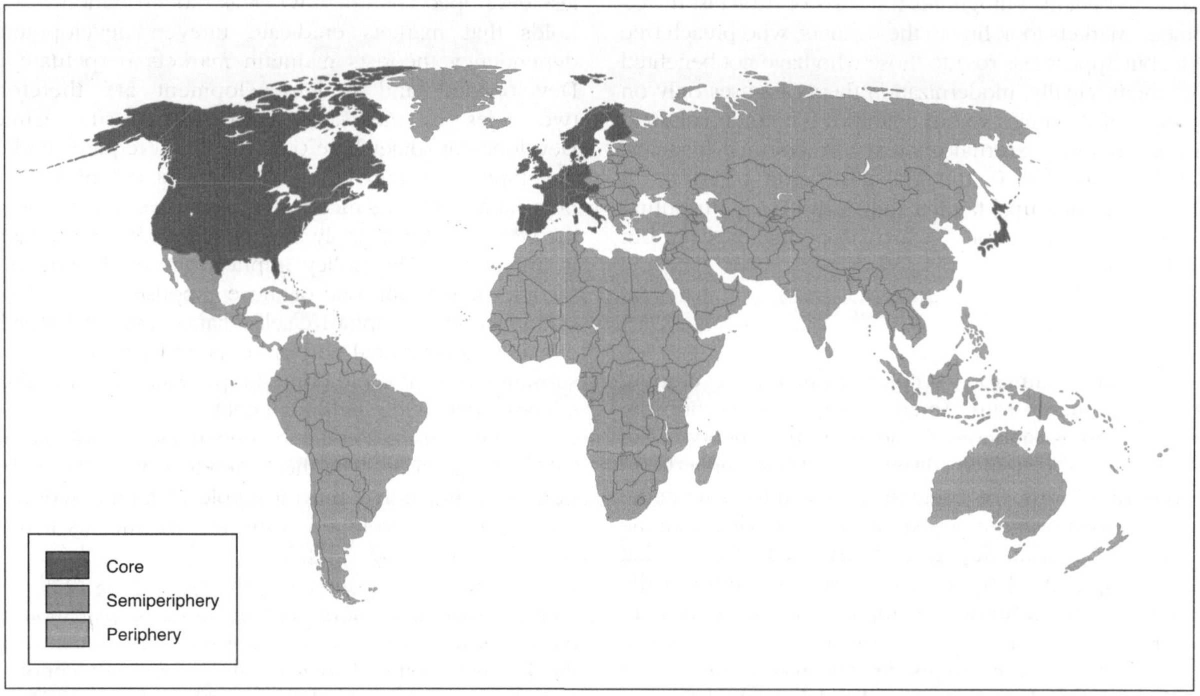

FIGURE 33 The geography of the world system. In world-systems theory, the global network of states and markets is dominated by a hegemonic power and a core of powerful, wealthy countries in Europe, North America, and Japan. The rest of the world is divided into an impoverished periphery, which produces raw materials, and a semiperiphery, including the newly industrialized countries, which has aspects of both the core and periphery. This approach recognizes the movement of states higher and lower in the world system as global capitalism creates new layers of uneven development worldwide.

and undeveloped countries, respectively: One is wealthy, urbanized, industrialized, and democratic; the other rural, impoverished, agriculturally based, and dominated by authoritarian governments. The semiperiphery has characteristics of both the core and the periphery and includes states at the upper tier of the less developed countries, such as the NICs, Brazil, Mexico, Thailand, and Saudi Arabia. Core processes are high wages, high levels of urbanization, industrialism and postindustrialism, the quaternary sector of the economy, advanced technology, and a diversified product mix. The world periphery processes are low wages, low levels of urbanization, preindustrial and industrial technology, and a simple product mix. Consumption is low. In between are states that are part of the semiperiphery where both sets of processes coexist to a greater or lesser degree. The theory suggests that the semiperiphery countries are exploited by the core countries with regard to raw material and product flow while at the same time exploiting periphery countries.

World-systems theory pays particular attention to the role of a single hegemon that dominates the global political and economic system. The hegemon "sets the rules of the game," so to speak. During the period of Spanish dominance in the sixteenth and seventeenth centuries, for example, mercantilism was the dominant ideology. Under the Pax Britannica of the nineteenth century, free trade was the norm. And since the U.S. rise to dominance, especially since World War II, **neocolonialism** has been the typical pattern (although during the Cold War there were two superpowers and a bifurcated world system). Hegemonic powers may overextend themselves militarily, leading to an erosion of their economic base. When powers in the core have conflicts among themselves, they open opportunities both for new hegemons and for countries on the periphery: The Napoleonic wars of the early nineteenth century and World War II in the twentieth were thus openings for nationalist, anticolonial movements worldwide, first in Latin America and later in Asia and Africa.

Hegemony exists when one core power enjoys supremacy in production, commerce, and finance and occupies a position of political leadership. The hegemonic power owns and controls the largest share of the world's production apparatus. It is the leading trading and investment country, its currency is the universal medium of exchange, and its city of primacy is the financial center of the world. Because of political and military superiority, the dominant core country maintains order in the world system and imposes self-interested solutions to international conflicts. Britain played this role in the nineteenth century and the United States has done so since World War II. Consequently, hegemonic situations are characterized by periods of relative peace (e.g., the nineteenth century). During a power's decline from hegemony, rival core states, which can focus on capital accumulation without the burden of maintaining the political and military apparatus of supremacy, catch up and challenge the hegemonic power. Thus, in the early twentieth century, Germany twice challenged Britain for global leadership, with catastrophically violent results.

REGIONAL DISPARITIES WITHIN DEVELOPING COUNTRIES

In addition to the division between the developed and the underdeveloped worlds, which forms the primary axis of the global economy, there is also the predicament of profound spatial differences within developing countries. The geographies of colonialism, including rural-urban differences, are one major dimension of this. Major cities of developing countries operate largely as export platforms linking them and their raw materials to the rich industrial countries. Modern large-scale enterprises tend to concentrate in capital and port cities. Injections of capital into these urban economics attract new migrants from rural areas and provincial towns to principal cities. Migrants, absorbed by the system, are maintained at minimal levels. There is little incentive to decentralize urban economic activities. The markedly hierarchical, authoritarian nature of the political and social institutions retards the diffusion of ideas throughout the urban hierarchy and into rural areas.

Because economic landscapes are produced by social relations, regional inequalities within developing countries are reflections of social inequalities. The class relations in much of the less developed world are often typified by a small, powerful elite and large numbers of poor peasants, with a small middle class. Countries with highly unequal distributions of income, such as Brazil, tend to have highly unequal standards of living among their various regions. Those countries that have achieved more equality economically, such as the NICs of East Asia, tend to have fewer disparities spatially. In this way, economic landscapes mirror and contribute to social bifurcations, revealing how geographies are socially produced and socially producing.

The center-periphery concept echoes the Marxist argument that the center appropriates to itself the surplus of the periphery for its own development. The center–periphery phenomenon may be regarded as a multiple system of nested centers and peripheries. At the world level, the global center (rich industrial countries) drains the global periphery (most of the underdeveloped countries). But within any part of the international system, within any national unit, other centers and peripheries exist. Centers at this level, although considerably less powerful, still have sufficient strength to appropriate to themselves a smaller, yet sizable, fraction of remaining surplus value. A center may be a single urban area or a region encompassing several towns that stand in an advantageous relation to the hinterland. Even in remote peripheral areas local, regional imbalances are likely to exist, with some areas growing and others stagnating or declining.

There are reverse flows from the various centers to the peripheries—to peripheral nations, to peripheral rural areas. Yet these flows, themselves, may further accentuate center–periphery differences. For example, loans from the World Bank, the U.S. Aid for International Development (USAID), and the International Development Association (IDA) support major infrastructure projects such as roads and power stations, which are proven money earners and reinforce the centrality and drawing power of cities and the export sectors. AID supports projects dealing with agriculture, health and family planning, school construction, and road building; industrialization projects are seldom financed.

DEVELOPMENT STRATEGIES

Today, the economically developed countries must come to the aid of less developed countries. How can this be done peacefully? Generally, three methods are cited: (1) expand trade with less developed countries, (2) invest private capital in less developed countries, and (3) provide foreign aid to less developed countries.

Expansion of Trade with Less Developed Countries

Economists suggest that expanding trade with less developed countries is one way to help them. It is true that reducing tariffs and trade barriers will improve the situation of less developed countries somewhat. Eliminating protectionist measures levied against producers in developing countries gives them access to the large, wealthy markets in the United States, Canada, Europe, and Japan, increasing their export volumes and revenues. But free trade can have costs. For example, with the North American Free Trade Agreement (NAFTA), the United States removed all trade barriers with Mexico, and Mexican manufacturing flourished as a result. However, trade liberalization also opened Mexican doors to U.S. imports, particularly heavily subsidized agricultural products, which drove millions of Mexican farmers into bankruptcy.

Private Capital Flows to Less Developed Countries

Less developed countries can be and are recipients of investments from multinational corporations, private banks, and large corporations. Some of this takes the form of foreign direct investment. For example, major U.S. automakers have now built numerous *maquilladora* assembly plants in Brazil and Mexico. Other types of capital flows are purely financial: Citicorp and Chase Manhattan Bank have made loans to the governments of numerous less developed countries. Since the debt crisis of the 1980s and 1990s, however, investments and private capital flows have decreased substantially because of concerns about returns on investment and fears of debt default.

An international trade climate must also be supported by financial and marketing systems, a favorable tax rate, an adequately maintained infrastructure, and a reliable flow of sufficiently skilled labor. Often, however, less developed countries cannot guarantee that a politically stable environment will prevail; many are torn by tribal conflicts, ethnic strife, religious struggles, and civil wars. These obstacles often thwart the major capital flows that might otherwise exist. African states in particular have not been able to tap private capital flows from large corporations and commercial banks because of problems arising from these conditions.

Foreign Aid from Economically Developed Countries

In order to reverse the vicious cycle of poverty in less developed countries, **foreign aid** is needed in the form of direct grants, gifts, and public loans. Capital is necessary to finance companies, build the infrastructure, generate jobs, increase productivity, and retrain the labor pool. In absolute terms, the United States has been a major participant in foreign aid programs. Its foreign aid averages $15 billion per year, for example, but this is less than 1% of GDP, the smallest share among all developed countries (Figure 34). Unfortunately, the contributions of developed countries are much too small to make a meaningful difference to most of the developing world. However, for some countries in Africa, aid can make up as much as 15% of their national output.

Most foreign aid carries stipulations, such as purchase requirements that make the less developed countries patronize the donor country's products and services. Almost 75% of U.S. foreign aid is military in nature, and the vast bulk of it flows to close allies like Israel, Egypt, and Turkey on the basis of political ties rather than economic need. Israel, Egypt, and Turkey each receive nearly $3 billion in aid per year. These nations are neither the most populous nor the most needy, but they occupy strategic areas of the Middle East where vast oil deposits exist and where the United States is struggling against Islamic fundamentalism. In addition, the United States has guaranteed Israel its support in a hostile environment.

INDUSTRIALIZATION IN THE DEVELOPING WORLD

Deindustrialization in the economically developed world did not induce widespread industrialization in the developing world. In 2008, 40 countries accounted for 70% of manufacturing exports from developing countries; the top 15 alone accounted for about 60%. Even more striking is that about one-third of all exports from the less developed countries came from four Southeast Asian countries—Hong Kong, South Korea, Singapore, and Taiwan, the original "tigers" of the East Asian miracle (Figure 35). As the most rapidly growing economies in the world since World War II, East Asian countries have formed a growing

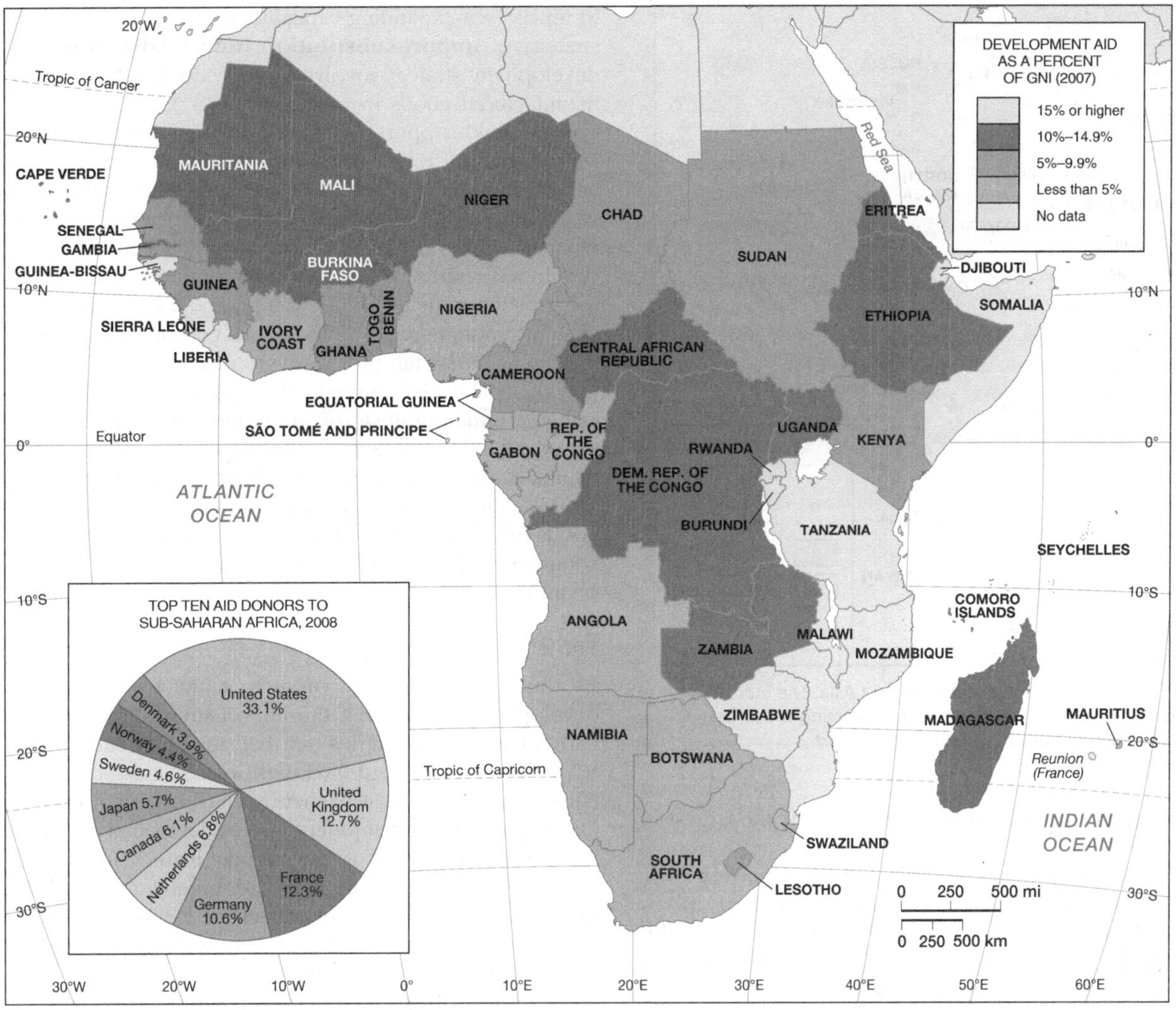

FIGURE 34 Foreign aid donated as a percent of gross national income (GNI). Although the U.S. supply of aid is large in absolute terms, because its economy is huge, the United States donates the least among economically developed countries as a percent of Gross Domestic Product. Most U.S. foreign aid is military in nature, disproportionately contributed to countries such as Israel, Egypt, and Turkey.

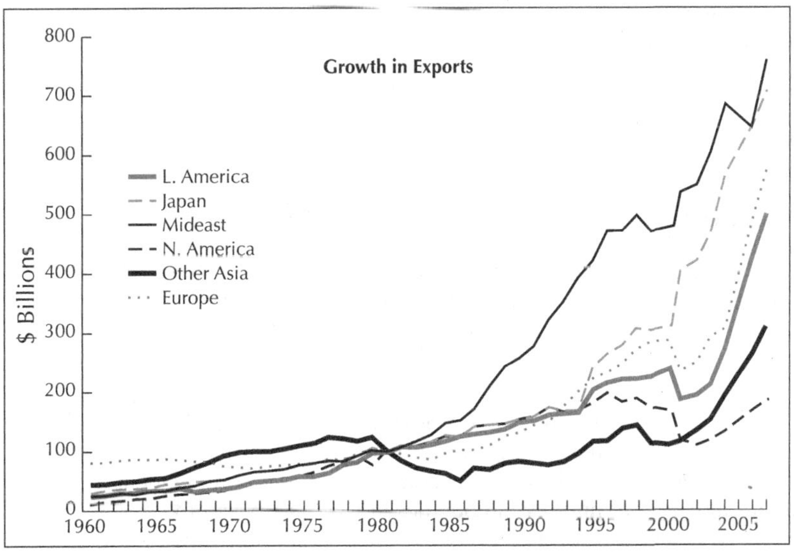

FIGURE 35 Export growth, 1960–2007. The most spectacular rates of export growth have been in East Asia, the world's most rapidly growing area economically, where the NICs have pulled millions out of poverty. China has major development plans for the next century. China is doing big, multibillion-dollar, 25-year-horizon, game-changing investments. China has four going now: (1) building a network of ultramodern airports; (2) building a web of high-speed trains connecting major cities; (3) bioscience, which will launch China's own stem-cell engineering industry, with technology and DNA sequencers being shipped in from San Diego; and (4) Beijing is providing $20 billion in seed money to move China off oil and into the next industrial growth engine: electric cars.

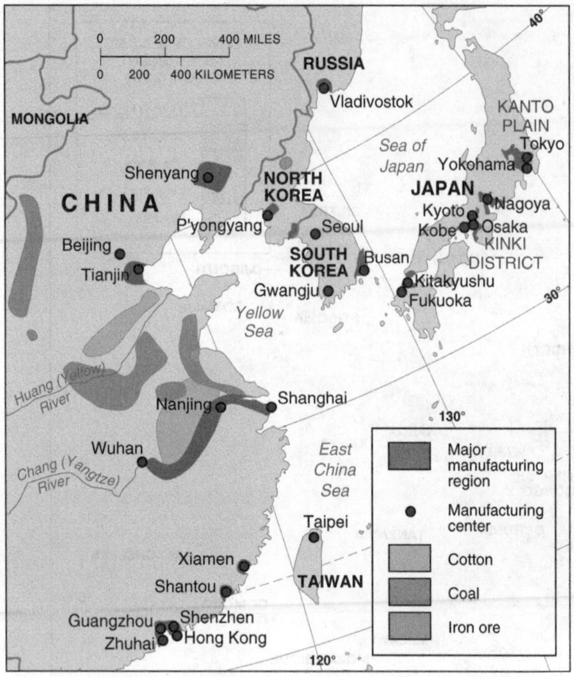

FIGURE 36 Manufacturing centers in East Asia. The industrial districts of China, Korea, and Japan are becoming the largest aggregate manufacturing complex in the world.

belt of manufacturing that may soon become the largest in the world (Figure 36).

Most undeveloped countries saw zero or very little manufacturing growth. Industrialization occurred, therefore, only in selected parts of the developing world. Manufacturing was slowest to take hold in the poorest countries of the periphery, most of which are in Africa. It grew fastest in the newly industrialized countries that made a transition from an industrial strategy based on import substitution to one based on exports. The exporters can be divided into two groups. First, countries such as Mexico, Brazil, Argentina, and India have a relatively large domestic industrial base and established infrastructure. All four of these countries are primarily exporters of traditional manufactured goods, such as furniture, textiles, leather, and footwear. Second, countries such as Hong Kong, Taiwan, South Korea, Malaysia, and Singapore have few natural resources and relatively small domestic markets. But by tailoring their industrial bases to world economic needs, they have become successful exporters to developed countries. These countries emphasize exports in clothing, engineering, metal products, and light manufactures. Their success encouraged other less developed countries to adopt a similar program of export-led industrialization.

Import-Substitution Industrialization

In the post–World War II period, newly independent developing countries sought to break out of their domination by, and dependence on, developed countries. Their goal was to initiate self-expanding capitalist development through a strategy of **import-substitution industrialization**. This development strategy involved the production of domestic manufactured goods to replace imports. Only the middle classes could support a domestic market; thus, industrialization focused on luxuries and consumer durables. The small plants concentrated in existing cities, which increased regional inequalities. These "infant industries" developed behind tariff walls in order to reduce imports from developed countries, but local entrepreneurs had neither the capital nor the technology to expand their domestic industrialization. Foreign multinational corporations came to the rescue. Although projects were often joint ventures involving local capital, "independent" development soon became *dependent industrialization* under the control of foreign capital. Many countries experienced an initial burst in manufacturing growth and a reduction in imports. But after a while, the need to purchase raw materials and capital goods and the heavy repatriation of profits to the home countries of the multinationals dissipated foreign exchange savings.

Export-Led Industrialization

By the 1960s, it was apparent to many leaders of less developed countries that the import-substitution strategy had failed. Only countries that had made an early transition to **export-led industrialization**, such as the Asian NICs, were able to sustain their rates of industrial growth. Once again, the development of less developed countries became strongly linked to the external market. In the past, export-oriented development had involved the export of primary commodities to developed countries. Now, export-oriented development was to be based on the production and export of manufactured goods.

The growth of export-led industrialization coincided with the international economic crisis of the 1970s and 1980s. It took place at a time when the demand for imports in the advanced industrial countries was growing despite the onset of a decline in their industrial bases. It was, in short, a response to the new international division of labor. Export-oriented industrialization tends to concentrate in **export-processing zones**, that usually meet the following four conditions:

1. Import provisions are made for goods used in the production of items for duty-free export, and export duties are waived. There is no foreign exchange control, and there is freedom to repatriate profits.
2. Infrastructure, utilities, factory space, and warehousing are usually provided at subsidized rates.
3. Tax holidays, usually of 5 years, are offered.
4. Abundant, disciplined labor is provided at low wage rates.

The first export-processing zone was not established in the developing world, but in 1958 in Shannon, Ireland, with the local international airport at its core. In the late 1960s, a number of countries in East Asia began to develop export-processing zones, the first being Taiwan's Kaohsiung

FIGURE 37 Export-processing zones in East and Southeast Asia. These platforms, constructed by governments to attract foreign capital and facilitate exports, have become widely adopted throughout the developing world as part of the broader shift from import substitution to export-led industrialization.

Export-Processing Zone, set up in 1965, a strategy soon imitated elsewhere (Figure 37). By 1975, 31 zones existed in 18 countries. By 2008, at least 120 zones had been established in developing countries, mostly in the Caribbean, Central and Latin America, and East Asia.

Export-led industrialization moves work to the workers instead of workers to the work, which was the case during the long postwar boom. In some countries, this form of industrialization has generated substantial employment. For example, since their establishment, export-processing zones have accounted for at least a 60% expansion of manufacturing employment in Malaysia and Singapore. However, in general, the number of workers in the export-processing zones' labor forces is modest. It is unlikely that these zones employ more than a million workers worldwide.

Much of the employment in export-processing zones is in electronics and electrical assembly or in textiles. Young, unmarried women make up the largest part of the workforce in these industries—nearly 90% of zone employment in Sri Lanka, 85% in Malaysia and Taiwan, and 75% in the Philippines. Explanations for this dominance of women in the workforce vary; it is often attributed to sexual stereotyping, in which the docility, patience, manual dexterity, and visual acuity of female labor are presupposed. Of more significance is the fact that women are often paid much less than men are for the same job. Thus the cheap labor so essential to the labor-intensive industries of the global assembly line also sustains the patriarchal relations that keep many women in particularly low-paying jobs.

Export-led industrialization may lead to an imposed economic system at odds with the cultural and political institutions of the people that it exploits. Often people produce things that are of no use to them. How they produce has no relation to how they formerly produced. Workers are often flung into an alien labor process that violates their traditional customs and codes. For example, female factory workers often pay a high price for their escape from family and home production, especially in Asia, which has traditionally emphasized the family roles of women. Because of their relative independence, Westernized dress, and changed lifestyles, women may be rejected by their clan and find it hard to reassimilate when they can no longer find employment on the assembly line.

Economic stagnation in developed countries is a major concern of developing countries that have enjoyed success with the export-led industrialization strategy. In developed countries, which are seeing an outflow of production and investment, purchasing power will be lost. The consequent spiraling down of general economic activity will choke off dependent industrialization and increase poverty and suffering for workers and peasants in less developed countries.

Sweatshops

Some Guess clothing is made by suppliers who use underpaid Latino immigrants in Los Angeles, sometimes in their own homes. Mattel makes tens of millions of toys each year in China, where young female Chinese workers who have migrated hundreds of miles from home are alleged to earn less than the minimum wage of $1.99 a day. Nike is criticized for manufacturing many of its shoes under tough labor conditions in Indonesia, and some of Disney's hottest seasonal products are being made by suppliers in Sri Lanka and Haiti—countries with unsavory reputations for labor and human rights. Soccer balls are sewn together by child laborers in Pakistan, who work up to 12 hours per day. In an era of the global economy, it is impossible for consumers to avoid products made under less than ideal labor conditions. Moreover, what may appear to be horrific working environments to most citizens in the world's richest nation are not just acceptable but actually attractive to others who live overseas or even in Third World pockets of the United States. Anyone even casually familiar with how some Americans employ their (usually immigrant) housekeepers understands their desire to work and support their families.

One icon of American culture whose manufacturing practices seem out of sync with its brand name is Disney. Disney maintains almost 4000 contracts with other companies that grant them the right to manufacture Disney paraphernalia, some of which is then sold in Disney stores. These licensees go to some of the world's lowest-cost-labor countries, including Sri Lanka and Indonesia, to produce stuffed animals and clothes. Disney itself rarely takes a direct hand in manufacturing. Sears, which carries 200,000 products from manufacturers operating in virtually every country, is tightening up on buying goods from suppliers with dubious records. The Gap, after enduring criticism, also has become a model for manufacturing and sourcing products abroad ethically. In contrast to Disney, Mattel does most of its own manufacturing. It makes a staggering 100 million Barbie dolls a year in four factories, two in China and one each in Malaysia and Indonesia.

Does a global economy mean consumers face no choice but to buy products made under conditions Americans don't want to think about? A number of U.S. companies say that intense global competition is no excuse for keeping working standards at the lowest possible level. Levi Strauss, for example, imposes its own "terms of engagement" on manufacturers who make its jeans products in 50 countries.

The East Asian Economic Miracle

What does it take to turn a less developed country into a developed nation? Who is marching successfully forward? The most successful examples have been the trading states of East Asia: South Korea, Taiwan, Singapore, and Hong Kong, following the path pioneered by nearby Japan, which began to modernize in the nineteenth century. In the 1980s and 1990s, Malaysia, Thailand, Indonesia, and the Pacific coast of China headed down the same road. Japan's economy was in ruins after the devastation of World War II. Today, Japan's economy is the second largest in the world, two-thirds the size of the U.S. economy (Japan's population is only half as large). Similarly, in the mid-1960s, South Korea was a land of traditional rice farmers who made up over 70% of the country's workforce. It's GDP per capita—$230—was the same as Ghana's; today, South Korea's GDP per capita is 20 times larger, and over 70% of its people live and work in urban areas rather than on farms. South Korea's economy is now the eighth largest in the world, ahead of Sweden, the Netherlands, and Australia. It has become the world's largest shipbuilding nation and the world's fifth-largest auto manufacturer. Its iron, steel, and chemical industries are thriving, and with the largest number of PhDs per capita in the world, South Korea has become a formidable competitor in research and development of semiconductors, biotechnology, information processing, telecommunications, and civilian nuclear energy. Few other countries have achieved as much economically in so short a time.

How has East Asia emerged as the hub of the increasingly prosperous Pacific Rim? There are several characteristics that these societies share, which, taken together along with a unique combination of circumstances in the world political economy, help to explain their sustained economic growth. First, East Asian countries have made an enormous commitment to education. Some of this may derive from the Confucian respect for learning and scholarship. East Asian educational mores encourage docile, well-trained workforces, often consisting of easily exploited, low-wage female labor, which means that these countries lose relatively few days to strikes and labor unrest.

Second is a high level of national savings. East Asian governments have encouraged personal savings by restricting the movement of capital abroad, maintaining low tax rates while keeping interest rates above the rate of inflation, and limiting the importation of foreign luxury goods. The result has been the accumulation of large amounts of low-interest capital that have allowed Asian countries to finance education, infrastructure, manufacturing, and commerce. Many countries in East Asia save one-third or more of their gross domestic product (in America, in contrast, the domestic savings rate is between zero and 3% of GDP). Only recently, after their economic takeoff was well underway, have East Asian governments loosened financial policies to allow increased consumption and capital investment in consumer durables like new homes. Such purchases, rather than savings, may finance future Asian prosperity. As Asian economies mature and their populations age, it is possible that their saving rates will decline and imports may increase. Will East Asians tend more and more to consume rather than create wealth, as Americans have tended to do?

Third, East Asia has enjoyed a strong political framework within which economic growth has been fostered. Industries targeted for growth were given a variety of supports—export subsidies, training grants, and tariff protection from foreign competitors. Low taxes and energy subsidies assisted the business sector. Trade unions were restricted and democracy was constrained. In Japan, powerful government bureaucrats largely beyond public control promoted industrial expansion with little regard for the opinions or needs of Japanese consumers. Military governments in South Korea and Taiwan dealt harshly with industrial unrest and political dissidents. Authoritarian regimes long ruled Singapore. In no way did Asian governments follow a laissez-faire model; it is a common myth that the Asian NICs demonstrate a "free market" in operation. It is only lately that multiparty politics have been permitted outside Japan.

Fourth, the NICs engaged in widespread land reform in the 1950s. In part, this was made possible by the turmoil of the Japanese occupation (1895–1945) and the Korean War (1950–1953), which dislodged the rural aristocracies that owned much of the arable land. As a result, rural land ownership in these nations is relatively egalitarian, with high rates of productivity, unlike Latin America, which still suffers from the legacy of the Spanish land grant system. Besides a thriving agricultural economy, which has dwindled in the face of rapid industrialization, the NICs have benefited from comparatively low rates of rural-to-urban migration, which helps to prevent the urban areas from being flooded by desperate peasants seeking to escape poverty.

Fifth, East Asian NIC countries exhibited a sustained commitment to exports (export-led industrialization) in contrast to the import-substitution policies that characterized India, Africa, and Latin America until very recently. Instead of encouraging industrialists with low labor costs to target foreign markets and compete there, governments in India, Latin America, and Africa decided to protect their economies rather than open them to international competition. They shielded firms from foreign competition with protective tariffs, government subsidies, and tax breaks. As a result, their products became less competitive abroad. While it was relatively easy to create a basic iron and steel industry, it proved harder to establish high-tech industries like computers, aerospace, machine tools, and pharmaceuticals.

Most import-substituting states depend on imported manufactured goods, whereas exports still chiefly consist of raw materials such as oil, coffee, and soybeans, a condition that creates poor terms of trade. A country that relies on the export of raw materials—mineral and agricultural commodities, "rocks and crops"—with little or no value added to the products through finishing or manufacturing—and then has to import expensive (because of the immense value added) high-tech products is not headed toward development unless the country commits its earnings to investment in quality exports and competitive high-tech manufacturing. Such countries need to get the fundamentals right: Keep inflation low and fiscal policies prudent, maintain high savings and investment rates, improve the educational level of the population, trade with the outside world, and encourage foreign investment.

Throughout Latin America and Africa that sort of economic strategy was often missing. Governments poured money into state-owned enterprises, large bureaucracies, and oversized armed forces, paying for them by printing money and raising loans from Western banks and international agencies. Public spending soared, price inflation accelerated, domestic capital took flight to safe deposits in American and European banks, and indebtedness skyrocketed. Just when Latin America and Africa needed capital for economic growth, countries there found themselves overwhelmed by debt, starved of foreign funds for investment, with currencies made worthless by hyperinflation. Land as well as people paid the price. Forests have been recklessly logged, mineral deposits carelessly mined, fragile lands put to the plow, and fisheries overexploited in a desperate effort to escape the poverty trap.

In addition to their domestic organization, the position of the Asian NICs in the world economy also played a role in their growth. Thus, not all of the success of the NICs is due to internal, domestic factors. For example, putting aside the brutality of war and the untold suffering it caused, many NICs benefited from the legacy of their occupation by Japan between 1895 and 1945. Japan initiated the steel, chemicals, and textiles industries in Korea and Taiwan, as well as in Manchuria. The Japanese also built much of the NICs' industrial infrastructure—roads, bridges, tunnels, ports, and airports—which, while designed to maximize the extraction of raw materials such as coal, nonetheless became important after the war. Japan also centralized the state bureaucracies of these countries and displaced the reactionary rural aristocracies that had controlled much of these economies and hindered growth. Subsequent to the war, many NICs, particularly South Korea, imitated the Japanese interlocking corporate giants known as *zaibatsu* (e.g., the Korean *chaebol*), many of which were closely linked to banks and obtained easy credit. Finally, current Japanese foreign investment in East Asia, which exceeds that of the United States, also has accelerated the industrialization of these countries.

The United States also is partly responsible for the success of the NICs. Throughout the Cold War (1945–1991), it provided generous economic and military aid, freeing resources that could be applied to economic development. The Containment Doctrine, also known as the Truman Doctrine, positioned Japan, South Korea, Taiwan, and, to a lesser extent, other countries, as frontline states in the war against communism. The United States gave copious grants and subsidies to the NICs and awarded them preferential trade status, such as exemptions from tariffs, which allowed them easy access to the American market and significant export earnings. The roles of the United States and Japan have been important in a theoretical sense as well: To some extent, the geopolitical conditions that enabled the NICs to take off were a unique constellation of circumstances that could not be duplicated elsewhere.

In the 1990s, the original Four Tigers of South Korea, Taiwan, Hong Kong, and Singapore were joined by a new group that included Thailand, Malaysia, and Indonesia. Each of them replicated the experience of the original NICs to one extent or another. Thailand, which long enjoyed close economic and military ties to the United States, underwent a wave of investment in textiles, automobiles, toys, and tourism, propelling Bangkok to become a prosperous but crowded urban center. Malaysia, under an authoritarian government, launched its "Plan 2020" to become a fully industrialized country by the year 2020. Already this strategy has succeeded, at least in the western half of the country. The country is the world's largest exporter of refrigerators and semiconductors and has close ties to banks and firms in Singapore as well. Kuala Lumpur has become a thoroughly modernized city. Indonesia, struggling to escape decades of poverty, has enjoyed some of the growth experienced in the Singapore–Kuala Lumpur

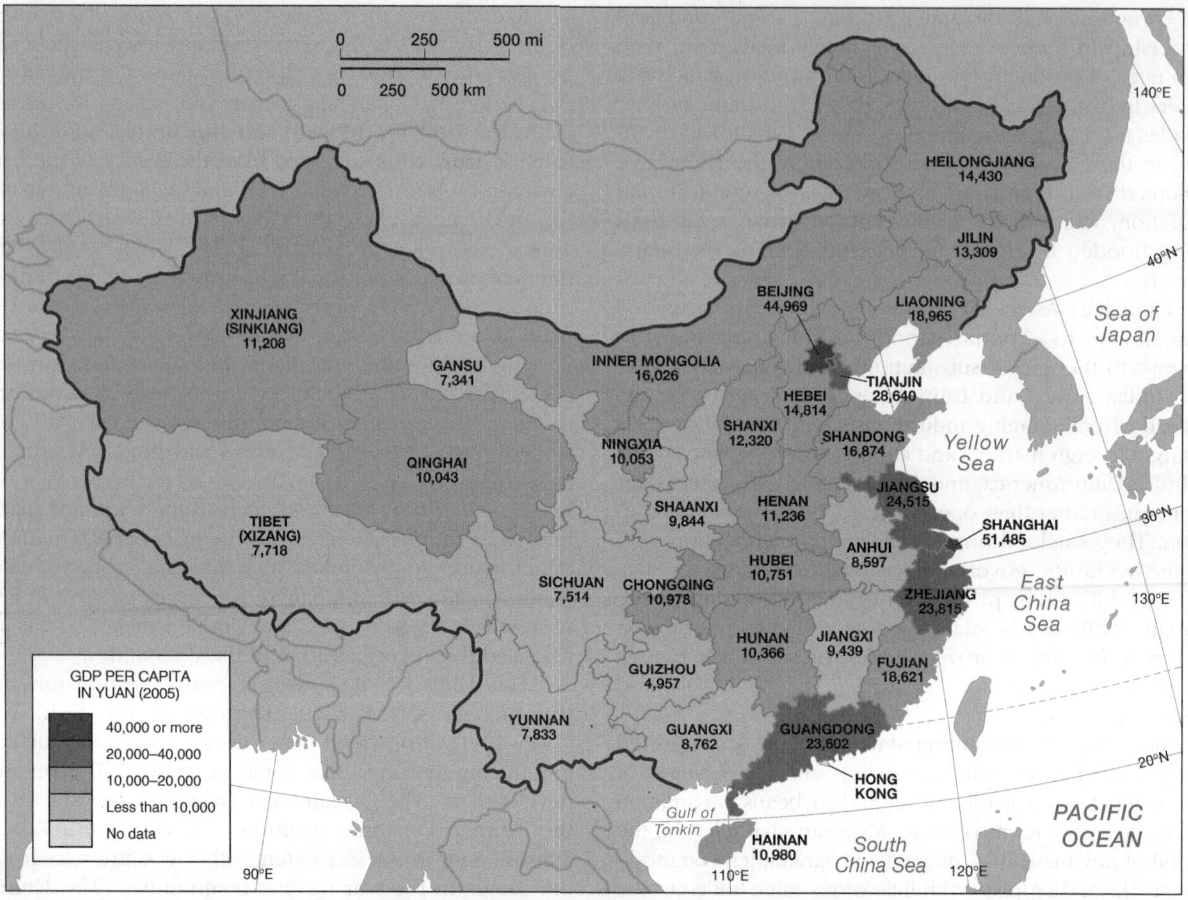

FIGURE 38 Unequal development in China as measured by GDP per capita, 2005. As China reentered the global economy in the late twentieth century and rapidly adopted market-based economic systems, it experienced a surge in social and spatial inequality. The most rapid growth has been along the coasts, most famously in the southern province of Guangdong, while much of the country's interior remains mired in poverty. Not surprisingly, the labor force has responded to these geographic differentials by moving in large numbers to the cities, where recent migrants are often vulnerable to exploitation.

corridor, including textiles, shoes, electronics, and even aerospace.

Finally, no discussion of the NICs could be complete without mentioning China, the "800-pound gorilla" looming behind all of the NICs. Following decades of isolation under the Communists, China began to open itself to the world economy in the late 1970s under the leadership of Deng Xiaoping. The policies implemented in the 1980s revolutionized the economy and society of the most populous nation in the world, encouraging the growth of private property and markets. Whereas the Communists shunned contacts with the West, the Chinese in the last 20 years have welcomed it, making the country the single largest recipient of foreign direct investment in the developing world. The government targeted coastal areas in particular, designating them special economic zones in which investors were showered with subsidies and tax breaks. These regions include Guangdong province, located near and benefiting greatly from its ties to Hong Kong (e.g., cities such as Shenzen); Fujian province across the straits from Taiwan; and Pudong, the financial center near Shanghai (Figure 38). In regions like these, the electronics, toys, and garment industries and other types of light industry have exploded. Indeed, China's economy has grown an average of 8% annually for the past three decades, one of the highest rates in the world, and China enjoys large trade surpluses with most of the developed world. This growth has been geographically uneven (i.e., as a form of uneven spatial development), and in response it has attracted waves of peasants from less densely inhabited, poorer regions in the interior to the prosperous coasts.

Although India, the world's second most populous country (with 1.0 billion people), lags far behind China in its level of economic development, it, too, has seen its pace of economic growth increase recently. Much of this growth is located on the western half of the Indian peninsula (Figure 39), including financial centers such as Mumbai (formerly Bombay) (Figure 40), one of the world's largest cities, and the famous software complex in Bangalore (Figure 41), India's answer to Silicon Valley Indeed, India today is the world's largest producer of software as well as films.

FIGURE 39 Uneven development in India. As India deregulated its economy and lured foreign investment, it, too, experienced uneven growth. The western parts of the country (including the financial and media center of Mumbai, formerly Bombay) and the software district of Bangalore have done much better than the eastern half. India's problems are painfully visible. The roads are dangerous, public transit is poor, and their firms are hobbled by the costs of building their own infrastructure: back-up generators, water treatment plants, and fleets of buses to transport staff to work. India's state may appear weak, but its private companies are strong. Indian capitalism is driven by millions of entrepreneurs, all doing their own thing. In the early 1990s, India dismantled their closed-door policy to trade, and Indian business has boomed ever since. Besides the well-known call centers, the country now boasts legions of thriving small businesses and a number of world-class ones, whose English-speaking managers network with the global business elite. They are less dependent on state patronage than the Chinese firms, and are often more innovative. Indian firms have pioneered the $2000 car and the $500 heart operation, and knowledge-based industries, such as software, love India while shunning the rampant property rights and patent piracy of the Middle Kingdom.

SUSTAINABLE DEVELOPMENT

Economic and population growth, the availability and depletion of resources, and the carrying capacity of the earth and the atmosphere are the core concerns of sustainable development. In a fundamental way, the Western development model is at odds with sustainability of the earth's resource base and carrying capacity. For example, 69% of the world's primary commercial fish species are in decline, mineral resources are becoming scarce and declining in quality, and projections of future freshwater availability indicate that nearly 50% of the world's population will experience water shortages in the next 50 years. The economy is an open subsystem of the earth's network

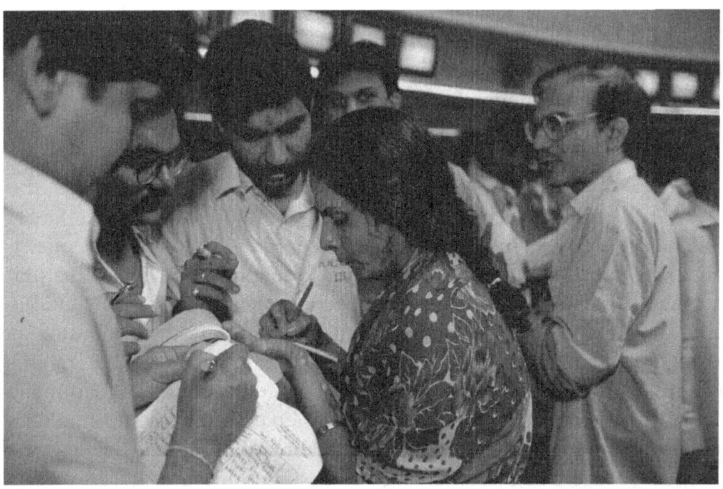

FIGURE 40 The Mumbai (Bombay), India, stock exchange provides evidence for this city's central role in the rapidly globalizing South Asian economy. Despite the world news, India is doing rather well with its development. Its economy is expected to expand by 9% into the future. It does have a long way to go before it is as rich as China. The Chinese economy is four times as large, but India's growth rate could overtake China's very soon. Rapid growth in a country of 1.2 billion people has far-reaching implications.

FIGURE 41 Bangalore, India's Silicon Plateau. With a growing population of well-educated workers, India has become the largest software producer in the world. Many of California's Silicon Valley firms use engineers and programmers in India. There are several reasons why India may soon outpace China. One is its ties to the developed world of Europe and the United States, from its history of English as the lingua franca. Another is demography. China's workforce will shortly start aging, and in a few years' time, it will start shrinking because of the one-child policy. India's dependency ratio—the proportion of children and older people to working adults—is much better than China's and is one of the best in the world. India will benefit from its "demographic dividend" that has propelled many of the Asian economic miracles.

of ecosystems, which is finite and materially closed. As the economic subsystem grows, it incorporates an ever-greater proportion of the total ecosystem. Thus, economic growth is unsustainable, as nature's resources are used up more quickly than they are being replenished. In effect, unsustainable development is intertwined with the loss of carrying capacity and the limits to growth. The long-term result of unsustainable development is the inability to sustain human life; such degradation on a global scale could imply extinction for humanity.

It is for this reason that sustainable development has become an increasingly important issue. **Sustainable development** is defined as development that meets the needs of the present without compromising the ability of future generations to meet their own needs while stressing the crucial need to ease poverty and work within social and technological limits. Those who recognize the challenges of the biophysical limits to growth on the one hand and business-as-usual approaches to development on the other argue that profound transformations of contemporary capitalism are needed sooner rather than later if we are to avoid dramatic disruptions of life on earth. As the global economy is being driven by the profit motive, and as we approach the global limits to growth, a less equitable distribution of wealth and livelihood security has resulted. The implication of this trend is that sustainable development efforts require a restructuring of societal relationships, which have made discussions about sustainability and sustainable development highly politicized.

That capitalism in its present form is unsustainable has become one of the major points of discussion among critics of globalization, including political ecologists. Since many negative effects of global production and trade are not confined within national boundaries but spill over to affect other parts of the world (e.g., greenhouse gas emissions), and since these effects are often more severe in developing countries than in the developed world, some environmentalists and ecologists argue that we should pursue a sustainable development course that would promote fair-trade and fair-share principles. Capitalist economic growth is about quantitative expansion and limitless transformation of resources into manufactured capital. Sustainable development, on the other hand, is about qualitative improvement in livelihoods, permitting increased economic activity only insofar as it does not exceed the capacity of the ecosystem. Thus, sustainable development emphasizes physical parameters such as resources and carrying capacity, along with an acceptance that economic variables like income and profit will have to be adjusted accordingly.

To many, it has become abundantly clear that the world's poor will bear the brunt of the damage to the world's ecological systems and will carry most of the costs associated with it, while they are hardly to blame or in a position to make any changes themselves. New core values like interdependence, empathy, equity, personal responsibility, and intergenerational justice will have to become the guiding principles and the basis for the transformation necessary to meet the sustainability challenges. This process involves a shift in focus *away from* understandings of nature as a free good and a sink for wastes and *toward* the understanding that it is a public good and resource and *toward* anticipatory techniques of environmental policy making, new roles for science in policy, and new legislative and regulatory frameworks emphasizing risk and uncertainty.

Applying laws from the physical sciences to the human economy, as a part contained within a larger ecosystem filled with finite resources, ecological economics has emerged in part as a reaction to the view of natural resources in traditional economics as fully substitutable with other standard input factors to production, such as capital and labor. Ecological economics, however, stresses the relationship between resource flows and human-made production funds as complementary and not substitutable. Development requires little or no additional throughput, the design of existing commodities and institutions can be adjusted to improve the scale, allocation, and distribution of resources. Ecological economics suggests that some problems can be solved by assigning prices to natural resources and taxing polluters. It emphasizes the intrinsic and moral value of ecosystems and their role in sustaining human societies. Ecological economics places importance on the fair distribution of resources, wealth, and income for sustainability and responsible use of resources, and even the fair distribution of resource flows between humans and other species.

Ecological modernization is the view that it is possible to maintain or increase the rate of economic growth *and* protect the environment and that ecological harm may be diminished by policy correctives and technological fixes that design environmental criteria into economic systems. Its appeal lies in a capacity to generate positive-sum solutions to problems conceived as zero-sum; to move beyond the remedial and regulatory environmental strategies of the 1970s; and to avoid structural changes seen as intractably difficult.

Summary

This chapter began by noting how slippery the term *development* is and that there are a variety of ways to measure it. We discussed goals for development by listing objectives that are by and large universally endorsed. We then surveyed the locations of underdevelopment in various regions of the developing world. Next we explored typical characteristics and development problems of less developed countries—overpopulation, lack of resources, capital shortages, insufficient foreign revenues and poor terms of trade, unequal land distribution, inadequate infrastructures and public services, and corrupt governments.

In discussing major perspectives on development, we saw how modernization theories, which stress economic growth and Westernization, have obscured important aspects of underdevelopment, particularly the long history and effects of colonialism and neocolonialism. Two major views derived from political economy—dependency theory and world-systems theory—explain why the Third World does not develop, attributing this lack to the role played by the planet's core in the domination and exploitation of the periphery.

We examined several development strategies that may promote growth in the developing world. A small but growing number of countries is moving from "have-not" to "have" status, or, in the framework of world-systems theory, from the global periphery to the semiperiphery, although many more remain behind. The key factors at play here include the history of colonialism, position in the commodity chains of the global economy, cultural attitudes toward education, political stability, and capacity to carry out long-term plans, all of which shape economic performance from one country to another. The race to develop will as always surely have its winners and losers. Only this time, modern communications will continually remind us of the growing disparity.

We noted how industrialization in the developing world started under policies of import substitution, which rapidly gave way to the far more successful tactic of export promotion. This shift was most successful in the East Asian newly industrialized countries. Nonetheless, industrialization has brought with it brutal exploitation, as in the case of sweatshops. The comfortable lives of many in the West often depend on the misery of workers in many poor countries.

Finally, we turned to the question of sustainable development. As the limitations of the earth's resources and ecosystems have become more obvious, it has become increasingly evident to many observers that global capitalism cannot carry on in its present form into the indefinite future. Rather than a growth-at-all-costs strategy, sustainable development promotes ways of sustaining the lifestyles of the present without jeopardizing the opportunities of future generations. Such a shift in worldview requires thinking about the natural environment as something more than a set of inputs to be used in production. Energy con servation, recycling, smart growth, and a point of view that takes the concerns of the planet seriously are all necessary to effectuate such a transformation.

Key Terms

brain drain
capital flight
core and periphery
dependency theory
development
export-led industrialization
export-processing zones
foreign aid
foreign direct investment (FDI)
human capital
import-substitution industrialization
informal economy
less developed countries
modernization theory
neocolonialism
newly industrializing countries (NICs)
primary sector
secondary sector
squatter settlements
structural adjustment
sustainable development
terms of trade
tertiary sector
underdevelopment
underemployment
unemployment
uneven development
world-systems theory

Study Questions

1. Why aren't poor countries called "primitive"?
2. What are five common characteristics of less developed countries?
3. What were the origins of the world debt crises?
4. What are four ways to measure economic development?
5. Contrast the modernization, dependency, and world-systems theories of development.
6. How does development relate to regional disparities within countries?
7. What were the major cycles of colonialism? What did they have to do with the change in relations between the world's core and periphery countries?
8. How can the First World meaningfully assist the Third?
9. What is sustainable development, and why is it important?

Suggested Readings

Bhagwati, J. 2004. *In Defense of Globalization*. Oxford: Oxford University Press.

Chase-Dunn, C. 1998. *Global Formation: Structures of the World-Economy,* 2nd ed. Lanham, MD: Rowman and Littlefield.

Davis, M. 2007. *Planet of Slums*. London: Verso.

Herod, A. 2009. *Geographies of Globalization*. Chichester, UK: Wiley-Blackwell.

So, A. 1990. *Social Change and Development: Modernization, Dependency, and World-System Theories*. Newbury Park, CA: Sage.

Stiglitz, J. 2002. *Globalization and Its Discontents*. New York: W. W. Norton.

Wallerstein, I. 1979 *The Capitalist World-Economy*. Cambridge: Cambridge University Press.

Web Resources

Terra: Brazil's Landless Movement
http://www.nytimes.com/specials/salgado/home/

The *New York Times* opened this Web special by documenting in both words and pictures the plight of Brazil's landless.

Atlapedia
http://www.atlapedia.com/index.html

Atlapedia Online contains key information on every country of the world. Each country profile provides facts and data on geography, climate, people, religion, language, history, and economy, making it ideal for students of all ages.

United Nations Human Development
http://www.undp.org

Report includes excellent information for developing countries.

CIA World Factbook
https://www.cia.gov/library/publications/the-world-factbook/index.html

This site includes helpful data, including a map, about every country in the world.

REFERENCES

Abu-Lughod, J. 1989. *Before European Hegemony: The World System A.D. 1250–1350.* New York: Oxford University Press.

Bell, D., and G. Valentine. 1997. *Consuming Geographies: We Are Where We Eat.* London: Routledge.

Bhagwati, J. 2004. *In Defense of Globalization.* Oxford: Oxford University Press.

Bluestone, B., and B. Harrison. 1982. *The Deindustrialization of America: Plant Closings, Community Abandonment, and the Dismantling of Basic Industry.* New York: Basic Books.

Boserup, E. 1965. *The Conditions of Agricultural Growth.* London: Earthscan Publications.

Brunn, S., and T. Leinbach, eds. 2001. *Wired Worlds of Electronic Commerce.* London: John Wiley.

Bryson, J., P. Daniels, and B. Warf. 2004. *Service Worlds: People, Organizations, Technologies.* London: Routledge.

Castree, N., and B. Braun, eds. 2001. *Social Nature: Theory, Practice, and Politics.* London: Blackwell.

Chase-Dunn, C. 1998. *Global Formation: Structures of the World-Economy,* 2nd ed. Lanham, MD: Rowman and Littlefield.

Chirot, D. 1985. "The Rise of the West." *American Sociological Review* 50:181–195.

Coe, N., P. Kelly, and H. Yeung. 2007. *Economic Geography: A Contemporary Introduction.* Oxford: Blackwell.

Davis, M. 2007. *Planet of Slums.* London: Verso.

De Graaf, J., D. Wann, and T. Naylor. 2001. *Affluenza: The All-Consuming Epidemic.* San Francisco: Berrett-Koehler Publication.

Diamond, J. 1999. *Guns, Germs, and Steel.* New York: W. W. Norton.

Dicken, P. 2010. *Global Shift: Mapping the Changing Contours of the World Economy,* 6th ed. New York: Guilford Press.

Dodge, M., and R. Kitchin. 2001. *Mapping Cyberspace.* London: Routledge.

Ellis, R. 2003. *The Empty Ocean: Plundering the World's Marine Life.* New York: Shearwear Books.

Falola, T., and A. Genova. 2005. *The Politics of the Global Oil Industry.* New York: Praeger.

Florida, R. 2004. *The Rise of the Creative Class.* New York: Basic Books.

Foster, J., and F. Magdoff. 2009. *The Great Financial Crisis: Causes and Consequences.* New York: Monthly Review Press.

Friedman, T. 2005. *The World Is Flat: A Brief History of the Twenty-First Century.* New York: Farrar, Straus and Giroux.

Goss, J. 1999. Once upon a time in the commodity world: An unofficial guide to the Mall of America. *Annals of the Association of American Geographers* 98: 45–75.

Graham, S., and S. Marvin. 2001. *Splintering Urbanism: Networked Infrastructures, Technological Mobilities and the Urban Condition.* London: Routledge.

Grant, R. 2000. "The Economic Geography of Global Trade." In *A Companion to Economic Geography.* Oxford: Blackwell.

Hardin, G. 1968. The tragedy of the commons. *Science* 162(859):1243–1248.

Hart, J. 2003. *The Changing Scale of American Agriculture.* Charlottesville: University of Virginia Press.

Hartwick, E. 2000. Towards a geographical politics of consumption. *Environment and Planning A* 32:1177–1192.

Harvey, D. 2006. *A Brief History of Neoliberalism.* Oxford: Oxford University Press.

Heckscher, E., and B. Ohlin. 1991. *Heckscher-Ohlin Trade Theory.* Cambridge, MA: MIT Press.

Herod, A. 1998. *Organizing the Landscape: Geographical Perspectives on Labor Unionism.* Minneapolis: University of Minnesota Press.

Herod, A. 2009. *Geographies of Globalization.* Chichester, UK: Wiley-Blackwell.

Hindess, B., and P. Hirst. 1975. *Pre-Capitalist Modes of Production.* London: Routledge and Kegan Paul.

Hiscox, M. 2001. *International Trade and Political Conflict: Commerce, Coalitions, and Mobility.* Princeton: Princeton University Press.

Hugill, P. 1993. *World Trade Since 1431: Geography, Technology, and Capitalism.* Baltimore: Johns Hopkins University Press.

Jones, E. 1981. *The European Miracle.* Cambridge, UK: Cambridge University Press.

Kellerman, A. 2002. *The Internet on Earth: A Geography of Information.* Hoboken, NJ: John Wiley.

Kirk, D. 1996. Demographic transition theory. *Population Studies* 50:361–88.

Klare, M. 2002. *Resource Wars: The New Landscape of Global Conflict.* New York: Owl Books.

Knox, P., J. Agnew, and L. McCarthy. 2008. *The Geography of the World Economy,* 5th ed. London: Edward Arnold.

Knox, P., and L. McCarthy. 2005. *Urbanization: An Introduction to Urban Geography,* 2nd ed. Upper Saddle River, NJ: Prentice Hall.

Knox, P., and S. Pinch. 2005. *Urban Social Geography: An Introduction,* 5th ed. Upper Saddle River, NJ: Prentice Hall.

Kondratiev, N. 1935/2010. *Long Waves in Economic Life.* Whitefish, MT: Kessinger.

Krugman, P. 1991. *Geography and Trade.* Cambridge, MA: MIT Press.

Krugman, P. 2008. *The Return of Depression Economics and the Crisis of 2008*. New York: W. W. Norton.

Kuznets, S. 1965. *Toward a Theory of Economic Growth*. Toronto: George McLeod.

Landes, D. 1969. *The Unbound Prometheus: Technological Change and Industrial Development in Western Europe from 1750 to the Present*. Cambridge, UK: Cambridge University Press.

Linkon, S., and J. Russo. 2002. *Steel-Town U.S.A.: Work & Memory in Youngstown*. Lawrence: University Press of Kansas.

Malthus, T. R. 1798/2006. *An Essay on the Principle of Population Growth*. New York: Cosimo Classics.

Mann, M. 1986. *The Sources of Social Power*. Cambridge, UK: Cambridge University Press.

Mansvelt, J. 2005. *Geographies of Consumption*. Thousand Oaks, CA: Sage.

Meadows, D.H., D.L. Meadows, and J. Randers. 1972. *The Limits to Growth*. New York: Universe Books.

Mielants, E. 2007. *The Origins of Capitalism and the "Rise of the West."* Philadelphia: Temple University Press.

Newbold, K. 2006. *Six Billion Plus: Population Issues in the Twenty-First Century*. Boulder, CO: Rowman and Littlefield.

Pacione, M. 2001. *Urban Geography: A Global Perspective*. London: Routledge.

Parsons, T. 1964. *Essays in Sociological Theory*. New York: Free Press.

Peet, R. 2003. *Unholy Trinity: The IMF, World Bank, and WTO*. New York: Zed Books.

Perelman, M. 2000. *The Invention of Capitalism: Classical Political Economy and the Secret History of Primitive Accumulation*. Durham, NC: Duke University Press.

Perkins, J. 2005. *Confessions of an Economic Hit Man*. New York: Plume.

Peters, G., and R. Larkin. 2005. *Population Geography: Problems, Concepts and Prospects*. New York: Kendall/Hunt.

Porter, M. 1990. *The Competitive Advantage of Nations*. New York: Free Press.

Ricardo, D. 1891/2010. *Principles of Political Economy and Taxation*. New York: CreateSpace.

Rostow, W. 1960/1991. *The Stages of Economic Growth: A Non-Communist Manifesto*. Cambridge, UK: Cambridge University Press.

Rumney, T. 2005. *The Study of Agricultural Geography*. Lanham, MD: Rowman and Littlefield.

Sassen, S. 1991. *The Global City: New York, London, Tokyo*. Princeton, NJ: Princeton University Press.

Schiller, D. 1999. *Digital Capitalism: Networking the Global Market System*. Cambridge, MA: MIT Press.

Schumpeter, J. 1947/2010. *Capitalism, Socialism and Democracy*. Whitefish, MT: Kessinger.

Scott, A.J. 2006. *Geography and Economy*. New York: Oxford University Press.

Shaw, G., and A. Williams. 2004. *Tourism and Tourism Spaces*. London: Sage Publications.

Shepard, E. and T. Barnes, eds. 2002. *A Companion to Economic Geography*. New York: Wiley.

Smith, A. 1776/2008. *An Inquiry into the Nature and Causes of the Wealth of Nations*. New York: Management Laboratory Press.

So, A. 1990. *Social Change and Development: Modernization, Dependency, and World-System Theories*. Newbury Park, CA: Sage.

Solomon, R. 1999. *Money on the Move: The Revolution in International Finance Since 1980*. Princeton, NJ: Princeton University Press.

Standage, T. 1998. *The Victorian Internet*. New York: Walker and Company.

Stearns, P. 2001. *Consumerism in World History: The Global Transformation of Desire*. London: Routledge.

Stiglitz, G. 2002. *Globalization and Its Discontents*. New York: W. W. Norton.

Taylor, P. 2003. *World City Network: A Global Urban Analysis*. London: Routledge.

Thompson, E. P. 1966. *The Making of the English Working Class*. New York: Vintage Press.

Van Marrewijk, C. 2002. *International Trade and the World Economy*. Oxford: Oxford University Press.

von Thünen, J. 1783–1850/1966. *The Isolated State*. Translated by Carla Wartenberg. New York: Oxford Unversity Press.

Wallerstein, I. 1979. *The Capitalist World-Economy*. Cambridge, UK: Cambridge University Press.

Warf, B. 2008. *Time-Space Compression: Historical Geographies*. London: Routledge.

Watts, M. 1996. The global agrofood system and late twentieth-century development. *Progress in Human Geography* 20:230–245.

Weber, M. 1905/2008. *The Protestant Ethic and the Spirit of Capitalism*. New York: BN Publishing.

Whybrow, P. 2006. *American Mania: When More Is Not Enough*. New York: Norton.

World Resources Institute. 2004. *World Resources*. New York: Oxford University Press.

Zimmerer, K. 2006. *Globalization and New Geographies of Conservation*. Chicago: University of Chicago Press.

World Political Map

World – Political

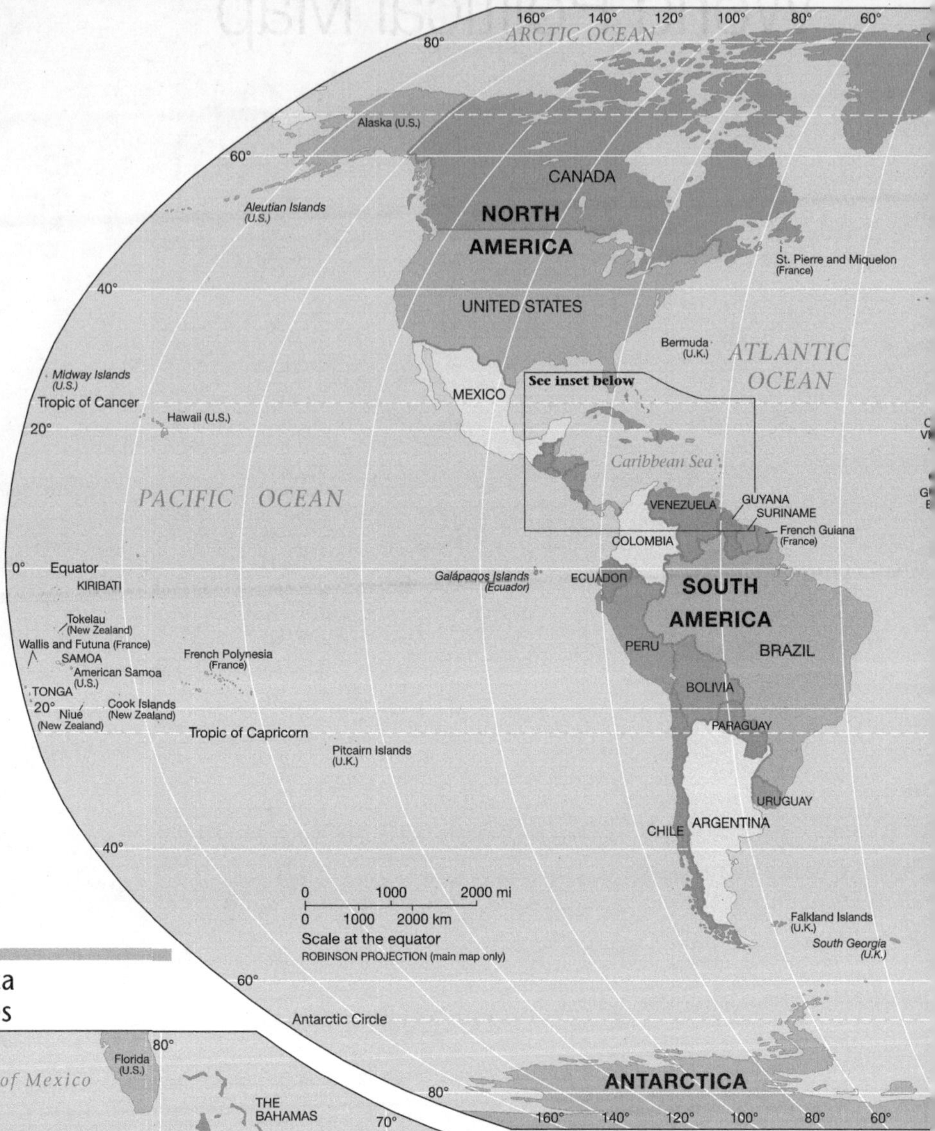

Central America and West Indies

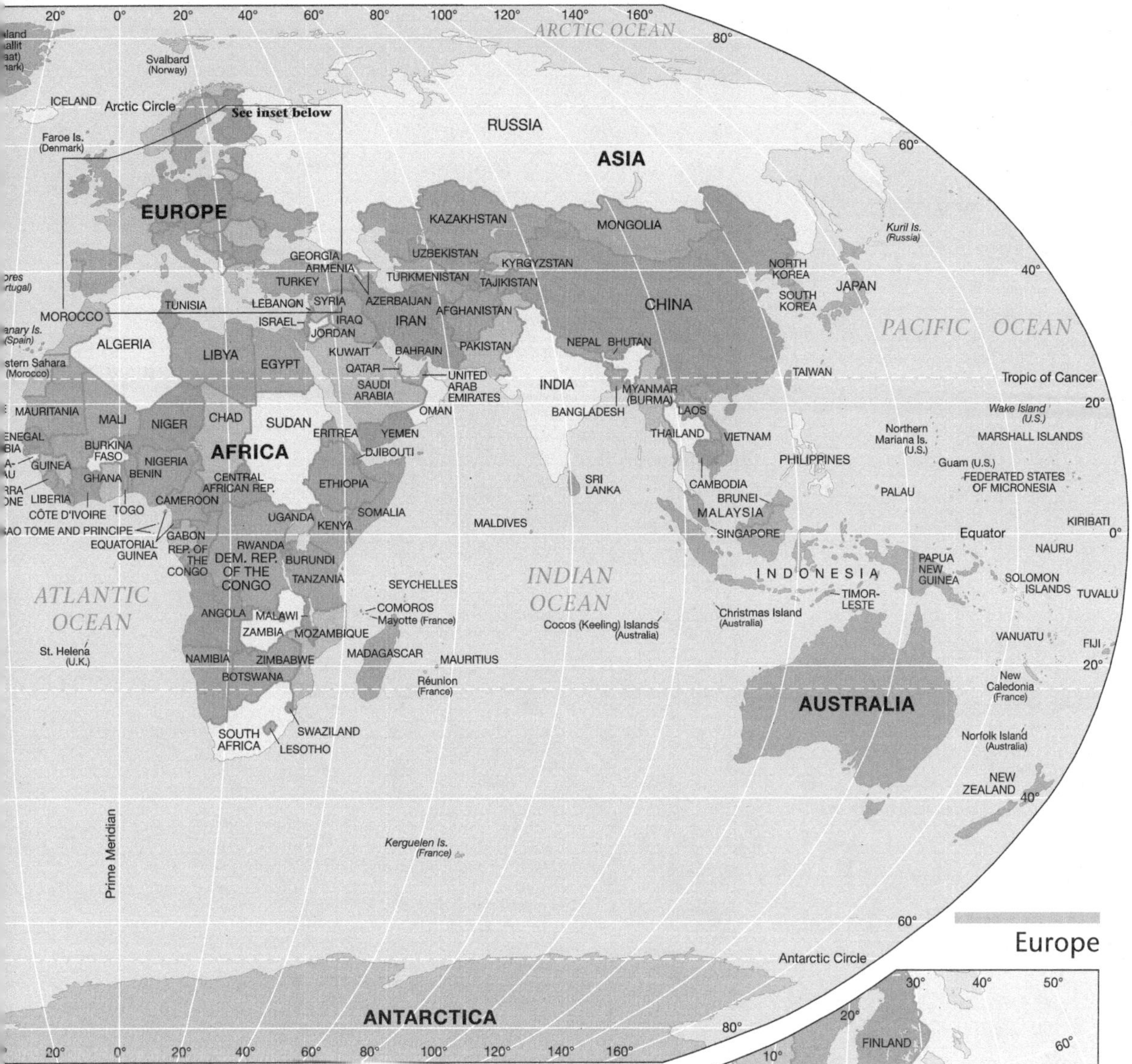

ARCTIC OCEAN
Svalbard (Norway)
ICELAND
Arctic Circle
See inset below
Faroe Is. (Denmark)
RUSSIA
ASIA
EUROPE
KAZAKHSTAN
MONGOLIA
Kuril Is. (Russia)
GEORGIA
ARMENIA
UZBEKISTAN
KYRGYZSTAN
TURKEY
TURKMENISTAN
TAJIKISTAN
NORTH KOREA
SOUTH KOREA
JAPAN
TUNISIA
LEBANON
SYRIA
AZERBAIJAN
MOROCCO
ISRAEL
IRAQ
JORDAN
IRAN
AFGHANISTAN
CHINA
PACIFIC OCEAN
ALGERIA
LIBYA
EGYPT
KUWAIT
BAHRAIN
PAKISTAN
NEPAL
BHUTAN
QATAR
UNITED ARAB EMIRATES
SAUDI ARABIA
INDIA
TAIWAN
Tropic of Cancer
MYANMAR (BURMA)
BANGLADESH
LAOS
MAURITANIA
MALI
NIGER
CHAD
SUDAN
ERITREA
YEMEN
OMAN
Wake Island (U.S.)
Northern Mariana Is. (U.S.)
MARSHALL ISLANDS
THAILAND
VIETNAM
BURKINA FASO
NIGERIA
AFRICA
DJIBOUTI
PHILIPPINES
GUINEA
GHANA
BENIN
CENTRAL AFRICAN REP.
ETHIOPIA
SRI LANKA
CAMBODIA
Guam (U.S.)
FEDERATED STATES OF MICRONESIA
PALAU
LIBERIA
CÔTE D'IVOIRE
TOGO
CAMEROON
UGANDA
SOMALIA
BRUNEI
MALAYSIA
SÃO TOME AND PRINCIPE
KENYA
MALDIVES
SINGAPORE
Equator
KIRIBATI
GABON
EQUATORIAL GUINEA
REP. OF THE CONGO
RWANDA
DEM. REP. OF THE CONGO
BURUNDI
TANZANIA
NAURU
PAPUA NEW GUINEA
SEYCHELLES
INDIAN OCEAN
INDONESIA
SOLOMON ISLANDS
TUVALU
ATLANTIC OCEAN
ANGOLA
MALAWI
ZAMBIA
MOZAMBIQUE
COMOROS
Mayotte (France)
Cocos (Keeling) Islands (Australia)
Christmas Island (Australia)
TIMOR-LESTE
VANUATU
FIJI
St. Helena (U.K.)
NAMIBIA
ZIMBABWE
BOTSWANA
MADAGASCAR
MAURITIUS
Réunion (France)
New Caledonia (France)
AUSTRALIA
SOUTH AFRICA
SWAZILAND
LESOTHO
Norfolk Island (Australia)
NEW ZEALAND
Prime Meridian
Kerguelen Is. (France)
Antarctic Circle
ANTARCTICA

Europe

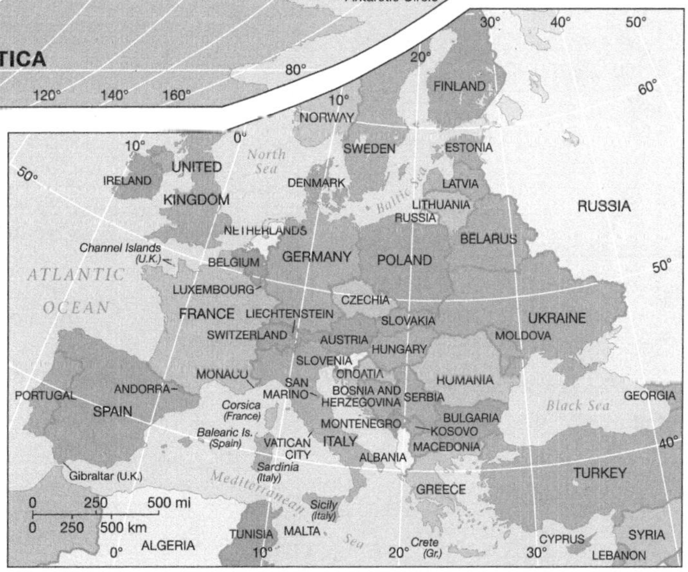

FINLAND
NORWAY
SWEDEN
ESTONIA
North Sea
IRELAND
UNITED KINGDOM
DENMARK
LATVIA
LITHUANIA
RUSSIA
Baltic Sea
NETHERLANDS
BELARUS
Channel Islands (U.K.)
BELGIUM
GERMANY
POLAND
ATLANTIC OCEAN
LUXEMBOURG
CZECHIA
FRANCE
LIECHTENSTEIN
SLOVAKIA
UKRAINE
SWITZERLAND
AUSTRIA
MOLDOVA
HUNGARY
SLOVENIA
CROATIA
MONACO
ROMANIA
SAN MARINO
BOSNIA AND HERZEGOVINA
SERBIA
PORTUGAL
ANDORRA
SPAIN
Corsica (France)
Black Sea
GEORGIA
BULGARIA
MONTENEGRO
KOSOVO
Balearic Is. (Spain)
VATICAN CITY
ITALY
ALBANIA
MACEDONIA
Sardinia (Italy)
TURKEY
Gibraltar (U.K.)
Mediterranean Sea
GREECE
0 250 500 mi
0 250 500 km
Sicily (Italy)
TUNISIA
MALTA
ALGERIA
Crete (Gr.)
CYPRUS
SYRIA
LEBANON

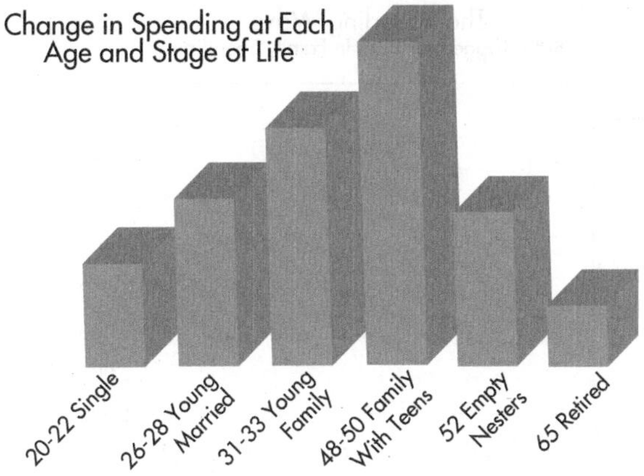

Figure A. Changes in Spending at Each Age and Stage of Life. Spending patterns increase with age and with family size. Total spending increases up to age 48, according to the U.S. Bureau of Labor Statistics, after which it starts to decline. Six stages are shown, starting with individuals entering the workforce at about age 18 to 22; to getting married, which requires more spending, from approximately age 22 to 30; to incurring the expenses of starting a family and the need for a larger place of residence, perhaps purchasing a house, from about age 31 to 42. The average total household spending continues to increase to about age 48 as children reach the teen years and larger houses, durable consumer goods spending, clothes, college tuition, and vacation expenses tend to peak. Beyond about age 50, individuals and households start to have less housing and consumer spending needs, become empty nesters, still have fixed costs, but lower variable costs, and spend less as they pay down debt and start thinking of saving for retirement. People are not making less income, but are spending less. Even though war, terrorism, natural disasters, and economic downturns occur unexpectedly, spending trends do not change much from the adjacent diagram. Even what people buy and how much they buy is surprisingly predictable. This information can be used to forecast how spending will alter the economy in the years to come.

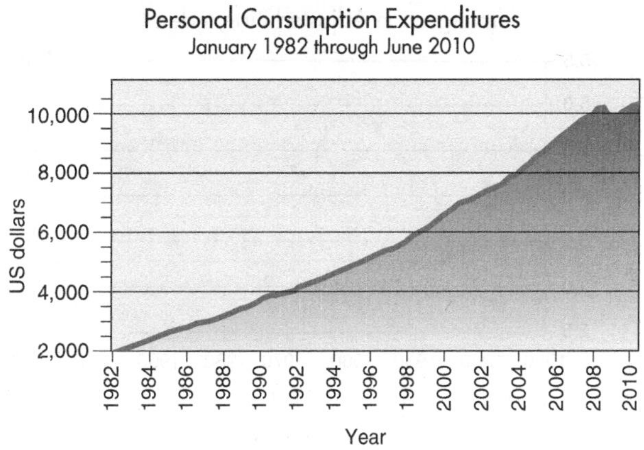

Figure B. Personal Consumption Expenditures. Personal consumption represents about 70% of the U.S. gross domestic product. This fact means that consumer spending is the largest single factor of our economic health. As consumers age and have greater household need, they spend more. In large groups, like the baby boom cohort (those born between the years 1946 and 1964) the economy grows and expands at a rapid rate. When large groups of the population pass their peak spending years, the economy starts to contract, as we are seeing now, in the post-2010 period. The U.S. Department of Commerce conducts the Consumer Expenditure Survey every year, and from this survey researchers have determined how money is spent at different ages and on which products and services. For example, the peak demand for potato chips is age 42. It is at this age that most parents are buying potato chips for their teenage children. The demand for health care, a popular topic of discussion these days in America, slowly increases from age 30; but after age 52, it takes on a dramatic increase. Since there are many baby boomers about to enter retirement age, total health care costs, demand, and employment is set to rise dramatically. Naturally, Medicare, Medicaid, and Social Security programs will be overtaxed.

Please go to http://www.frederickstutz.com for further discussion and videos on the economic crisis and real estate values in the United States.

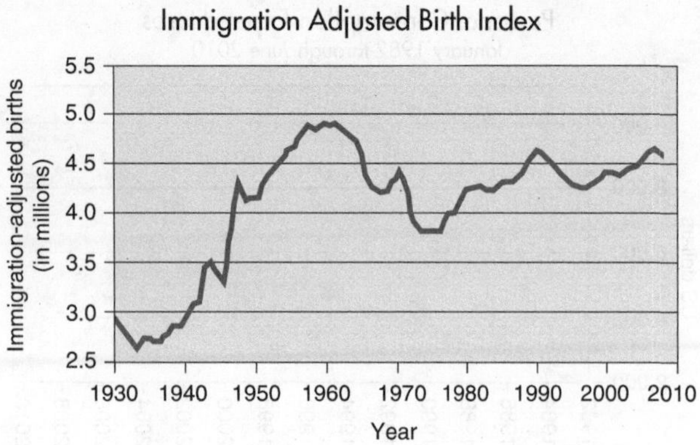

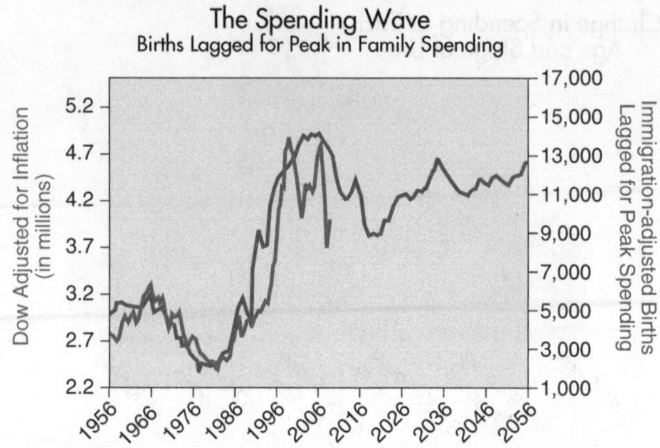

Figure C. U.S. Immigration Adjusted Birth Index in the 20th century. The number of children born reflects the size of their parents' generation, their economic circumstances, and cultural attitudes toward children and larger families. Our interest here is based on the fact that seasons of economic boom and bust are determined by demographic cycles. Births peaked in the late 1950s with what has been called the baby boom in America. New generation bulges occur about every 40 years. And as these bulges move through the predictable age, earning, spending, and growth cycles shown above, the peaks and troughs of such patterns can be forecast by moving the diagram pictured here, which is the births in America (adjusted for the births of all past and future projected immigrants) forward 48 years into the peak spending cycle of the average American. Peak spending and peak economic growth in America should, therefore, correspond with this baby-boom demographic by viewing this bulge in population, the spenders, into their peak spending years, some 48 years after they were born. Thus, the years in and about 2006 should have represented the peak growth of the economy. After that point, the economy should decline because of an aging population that tends to spend less and less. During the Great Depression, fewer babies were born in America. Therefore, we would expect that 48 years later, there would be a relative trough in economic growth and vitality in America. Economic boom cycles are associated with large and growing demographic bulges of the late forties population and bust times are associated with small and declining troughs in these late 40s demographics. The variation in the number of people in their mid to late 40s (at their peak spending age) is called the "spending wave."

Figure D. The Spending Wave. When U.S. births are lagged forward 48 years (red curve) and are compared with the best measurement of the performance of the U.S. economy (green curve) there is a remarkable correlation. The spending wave allows one to see when the stock market will rise and fall many years in advance of the actual event. Equities, especially large company stocks, were the investment tool when increasing numbers of people moved into their peak spending years. During the late 1990s and the first decade of the twenty-first century, from 2000–2010, good things happened to the economy, because private sector spending drives almost three-quarters of the economy. The economy has now moved ahead for roughly 30 years. But looking back to 1970, the stock market began to decline as the number of 48-year-olds declined (children of the Great Depression). In the 1980s, the number of big spenders began to increase; up went stocks, creating the largest bull market in U.S. history. But look what is now coming. The market is set to decline precipitously, as the demographics suggest a spending collapse, set off by the baby boomers reaching their retirement years in mass. The Great Recession set off by bank failure and the sub-prime meltdown of mortgage loans was the beginning. But lower demand for all goods and services, especially houses and major consumer durables, like cars, computers, and refrigerators, and so forth, will now signal slower growth. An increase in demand for retirement and health care services will buoy the economy somewhat, but the overall recovery that everyone seeks will not occur until 2025, when the children of the baby boomers hit their peak spending years.

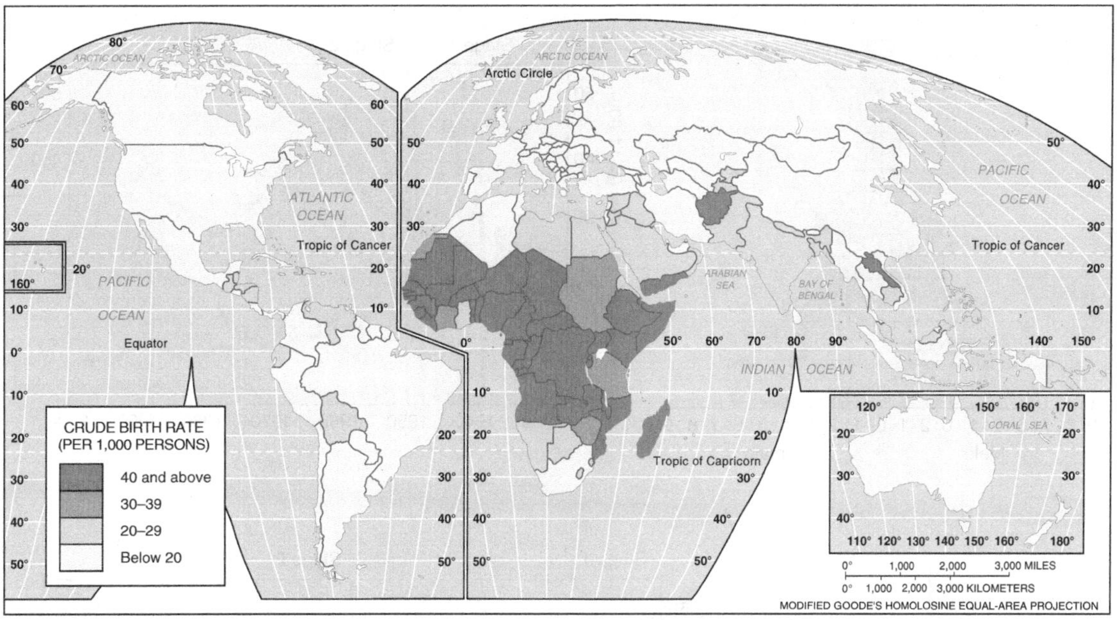

Crude birth rates measure the number of babies born per 1000 people per year. Typically, the poorer a country is, the higher is its birth rate, as rural families need child labor, have many children in response to high infant mortality rates, and require offspring to care for them in old age. Thus, countries in sub-Saharan Africa have the highest birth rates in the world. Economically developed countries, in contrast, in Europe, Japan, North America, and Australia and New Zealand, tend to have lower birth rates, a reflection of the costs and benefits of having children in urbanized contexts in which many women work outside the home.

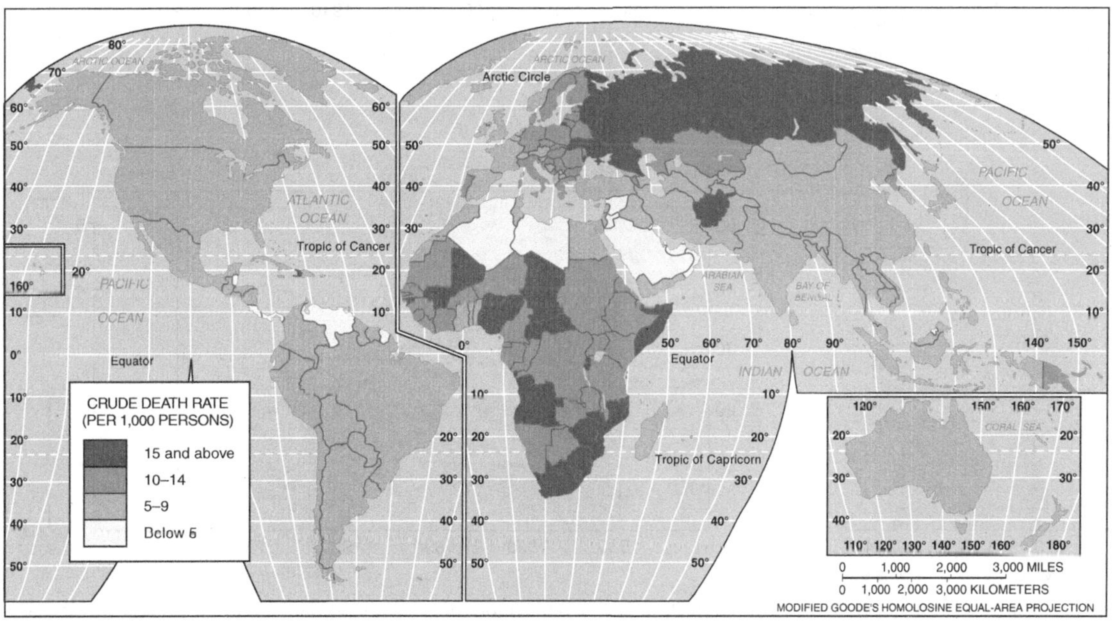

Crude death rates measure the number of deaths per 1000 people per year. In economically underdeveloped countries, malnutrition, unclean drinking water, and infectious diseases are responsible for relatively high death rates and correspondingly low life expectancies. In economically developed countries, in contrast, death rates are lower and life expectancies are higher. Moreover, the causes of death shift as countries urbanize and develop; most deaths are "lifestyle" or behaviorally related ones associated with smoking, heart disease, obesity, and cancers of various sorts, followed by homicide, suicide, and accidents. (Note that the high death rate in Haiti is due to the 2010 earthquake in Port-au-Prince, which killed an estimated 230,000 people.)

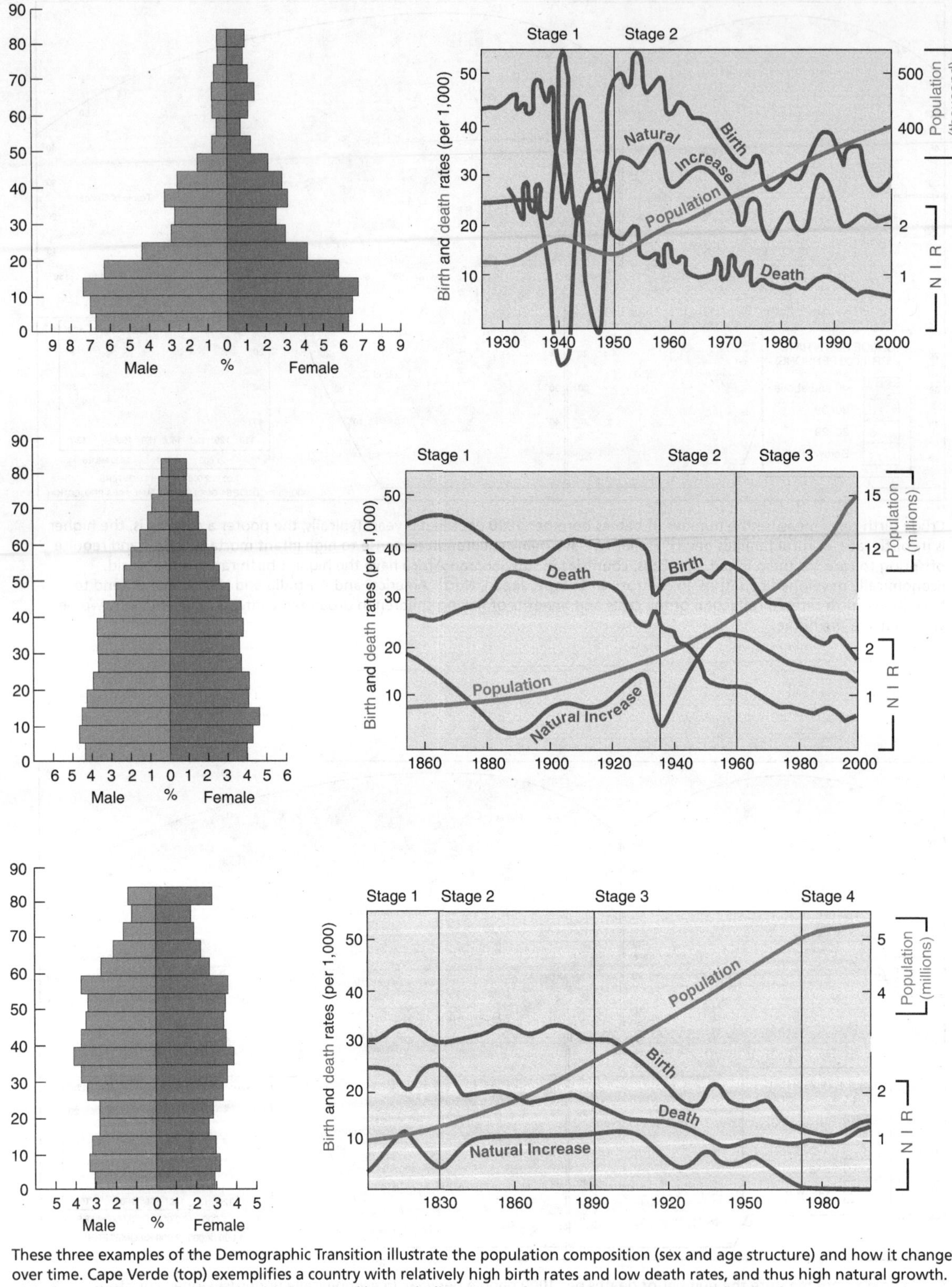

These three examples of the Demographic Transition illustrate the population composition (sex and age structure) and how it changes over time. Cape Verde (top) exemplifies a country with relatively high birth rates and low death rates, and thus high natural growth. Chile (center) illustrates a declining birth rate and hence declining rate of natural increase. Denmark (bottom) is an example where births and deaths are roughly equal, leading to zero natural growth.

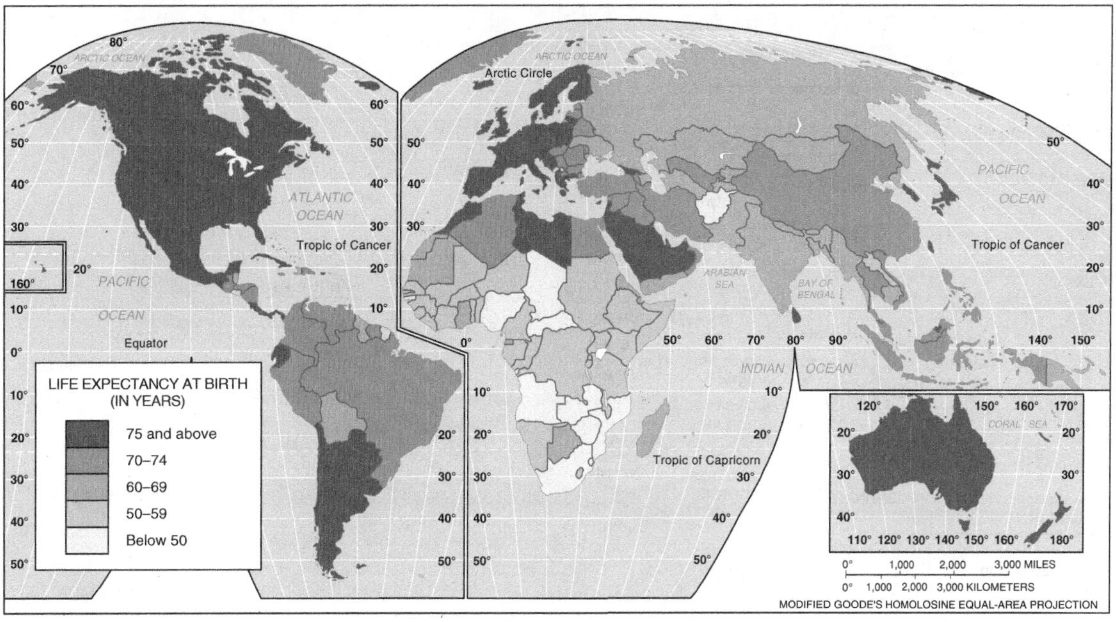

How long the average person can expect to live at birth is largely a function of a country's level of economic development, including access to an adequate diet, clean water, and health care. In the economically developed world, most people can expect to see their 75th birthday (although there are significant variations in terms of gender, class, and ethnicity). In the poorest countries in Africa, in contrast, most people will die before they reach the age of 50, a reflection of their poverty, malnutrition, the widespread wars on the continent, and the growing incidence of diseases, particularly AIDS.

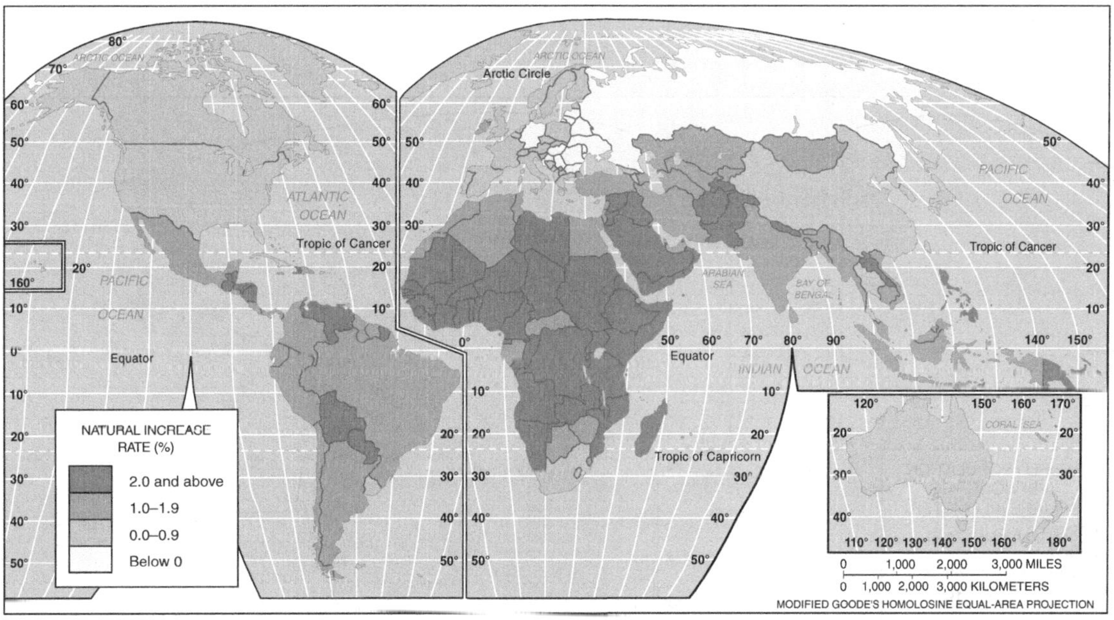

The rate of natural increase is the difference between birth and death rates. Almost always, the poorest countries grow the most rapidly, a reflection of the need to have many children and large families in agricultural contexts. Thus, Africa reveals the highest rates of natural growth, often 3.0% annually or higher, although the Arab world is also growing quickly. In contrast, economically wealthy countries in North America, Europe, Japan, and Australia and New Zealand, in which both birth and death rates are low, tend to hover around zero population growth. Russia and several countries in Eastern Europe have death rates exceeding birth rates, indicating they are losing people and in a state of population decline.

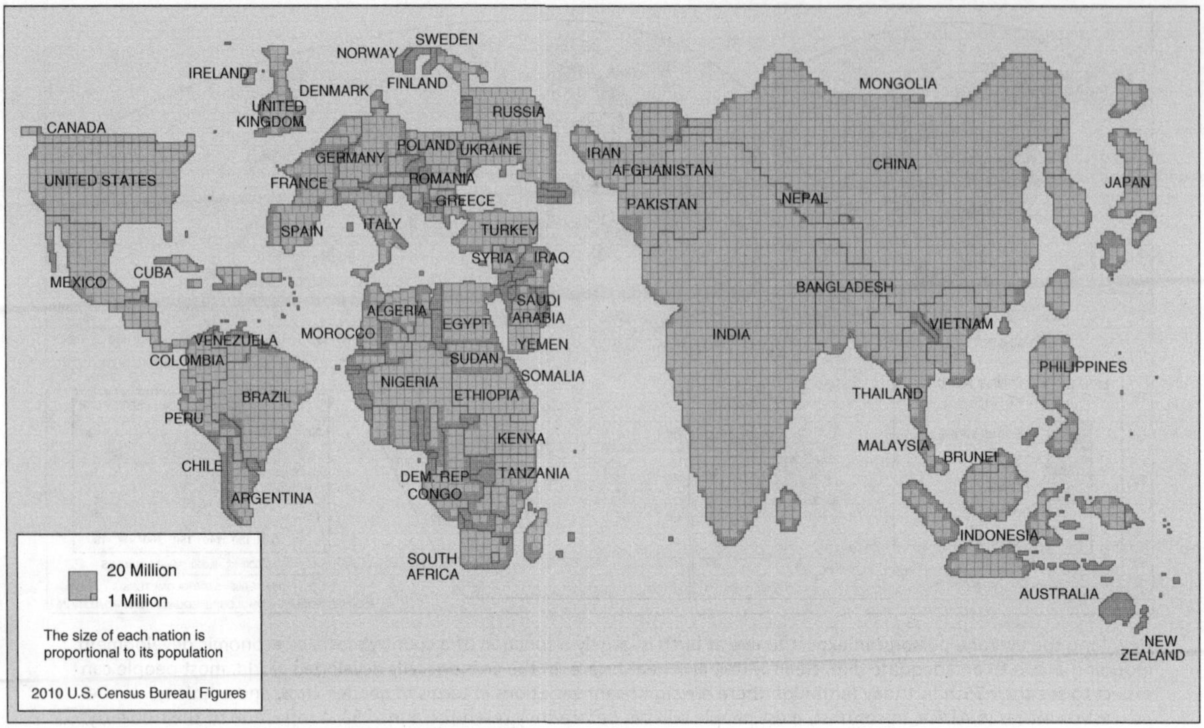

A cartogram is a map that deliberately distorts the size of units (in this case, countries) in proportion to a given variable (in this case, total population). Such a map reveals the vast numbers of people who live in East and South Asia, and the relatively small proportions in Africa and North and South America.

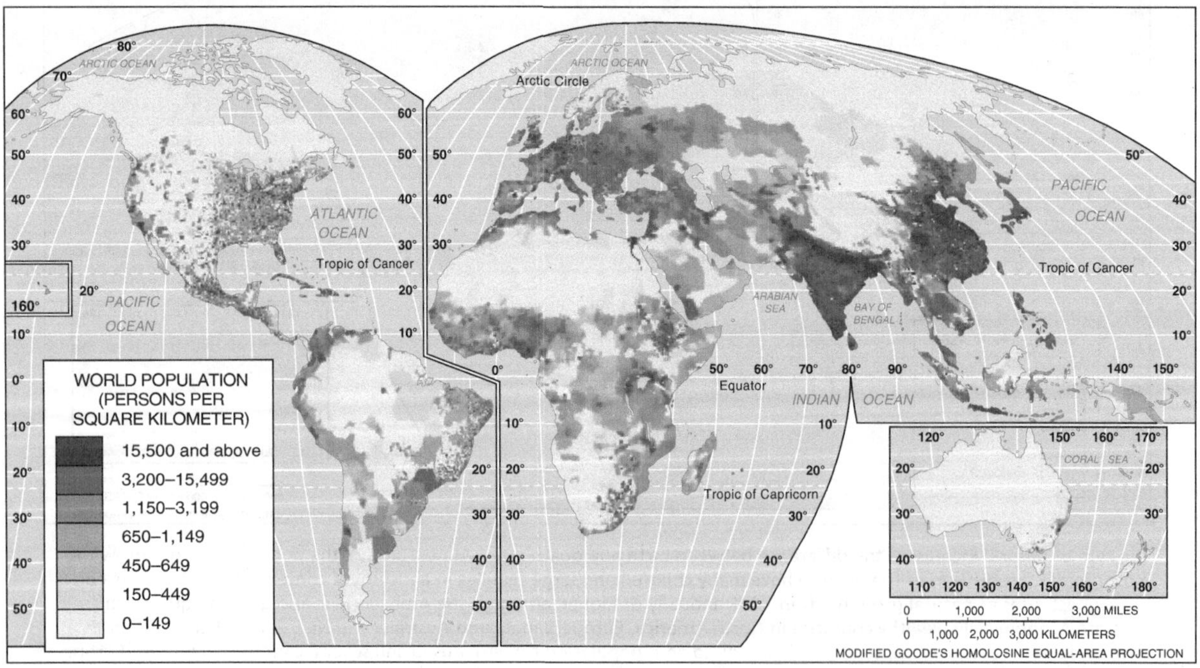

The geography of the world's 6.3 billion people is fundamental to its economic activity. Three major clusters stand out—East Asia, South Asia, and Europe. Vast areas in central Asia, Northern Africa, Australia, and the interior of South America are too hot, too cold, or too dry to sustain dense populations.

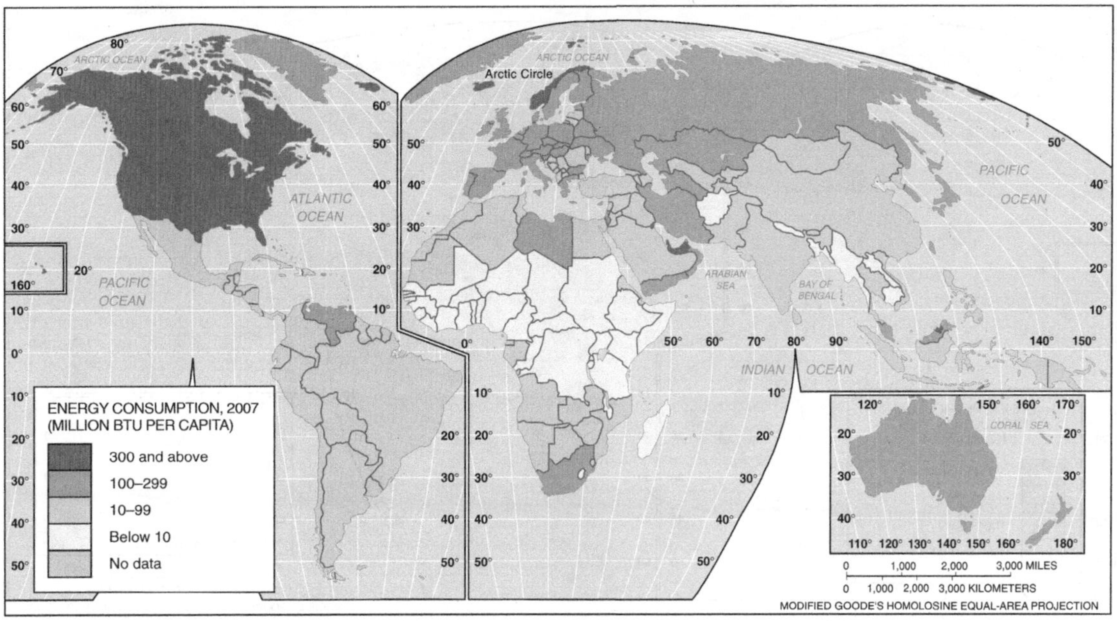

The average rate of energy consumption per person is a reliable measure of economic development. Urbanized, industrialized economies in North America, Europe, Japan, and Australia and New Zealand consume vast quantities for manufacturing, transportation, electricity generation, and heating. In contrast, in much of the developing world, energy consumption levels—and corresponding standard of living—are relatively low.

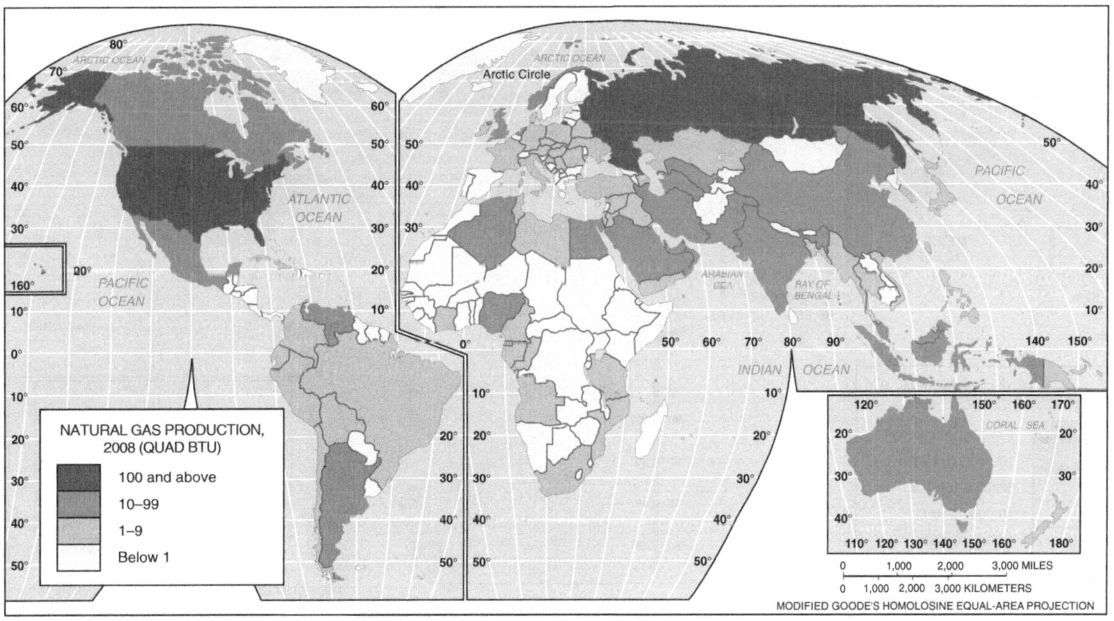

Along with petroleum and coal, natural gas is an important fossil fuel, and one that produces virtually no pollution. The world's major suppliers are in North America, Russia, Britain, and parts of Asia.

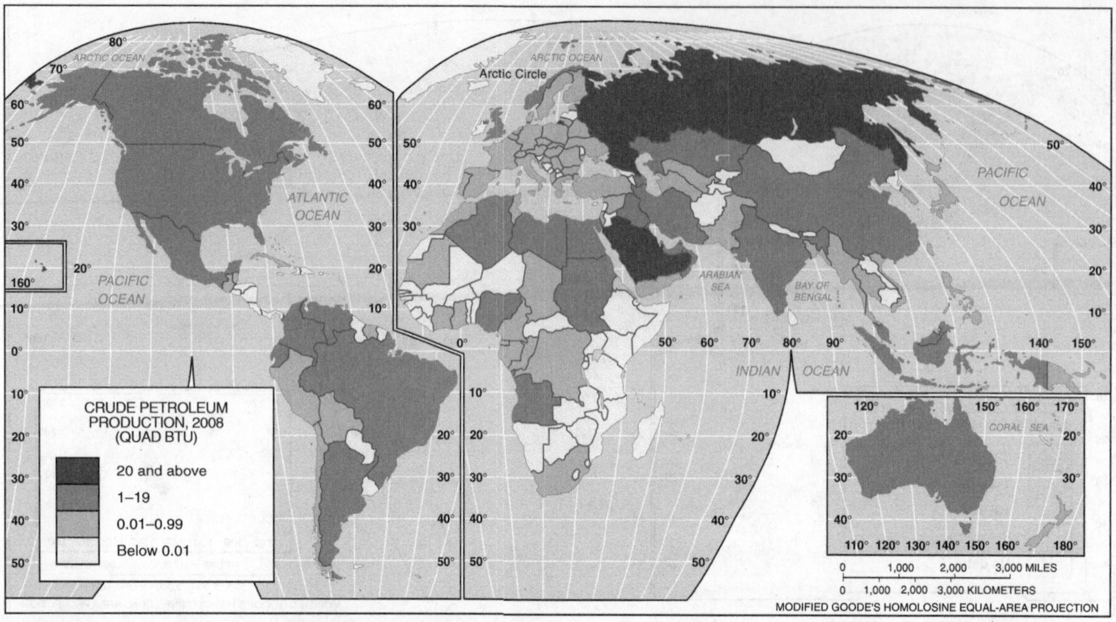

Petroleum—the lifeblood of the industrial world's economies—is produced in a select group of countries. Some of these, including those in the Middle East as well as Venezuela and Indonesia, belong to the Organization of Petroleum Exporting Countries (OPEC). Other producers, which do not belong to OPEC, include Russia, the United States, Mexico, Britain, and Norway.

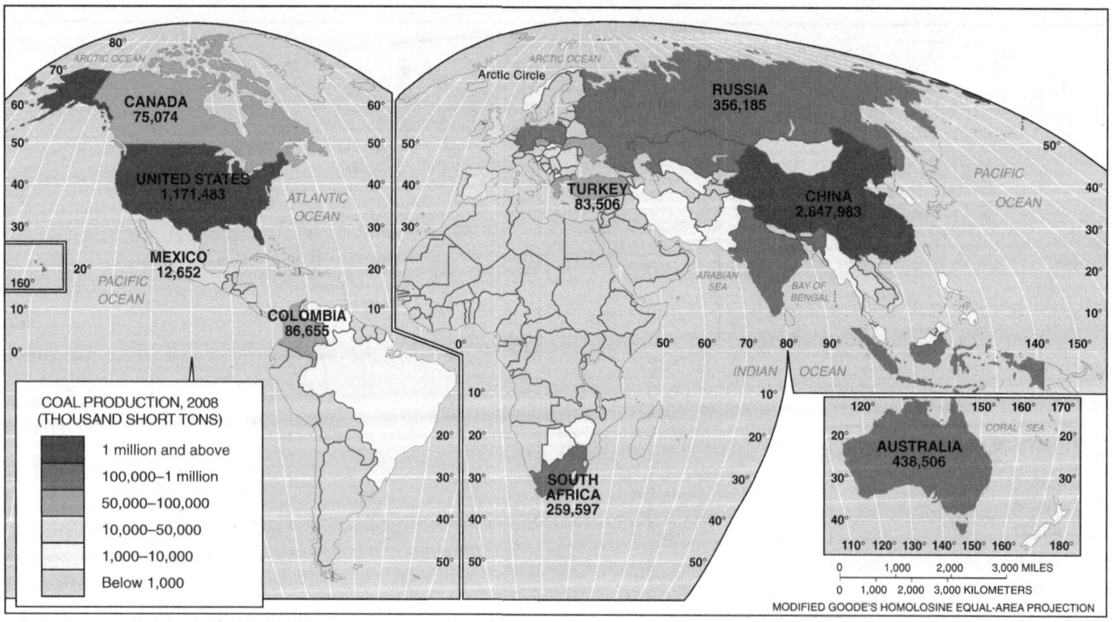

Coal, which made the Industrial Revolution possible, continues to be used in large quantities today. It is a major source of energy for the United States, and the largest source for China, the world's major producer and consumer. However, the burning of coal has serious environmental consequences, including urban air pollution; the release of vast quantities of carbon dioxide, which contributes to global warming; and acid rain, which destroys forests.

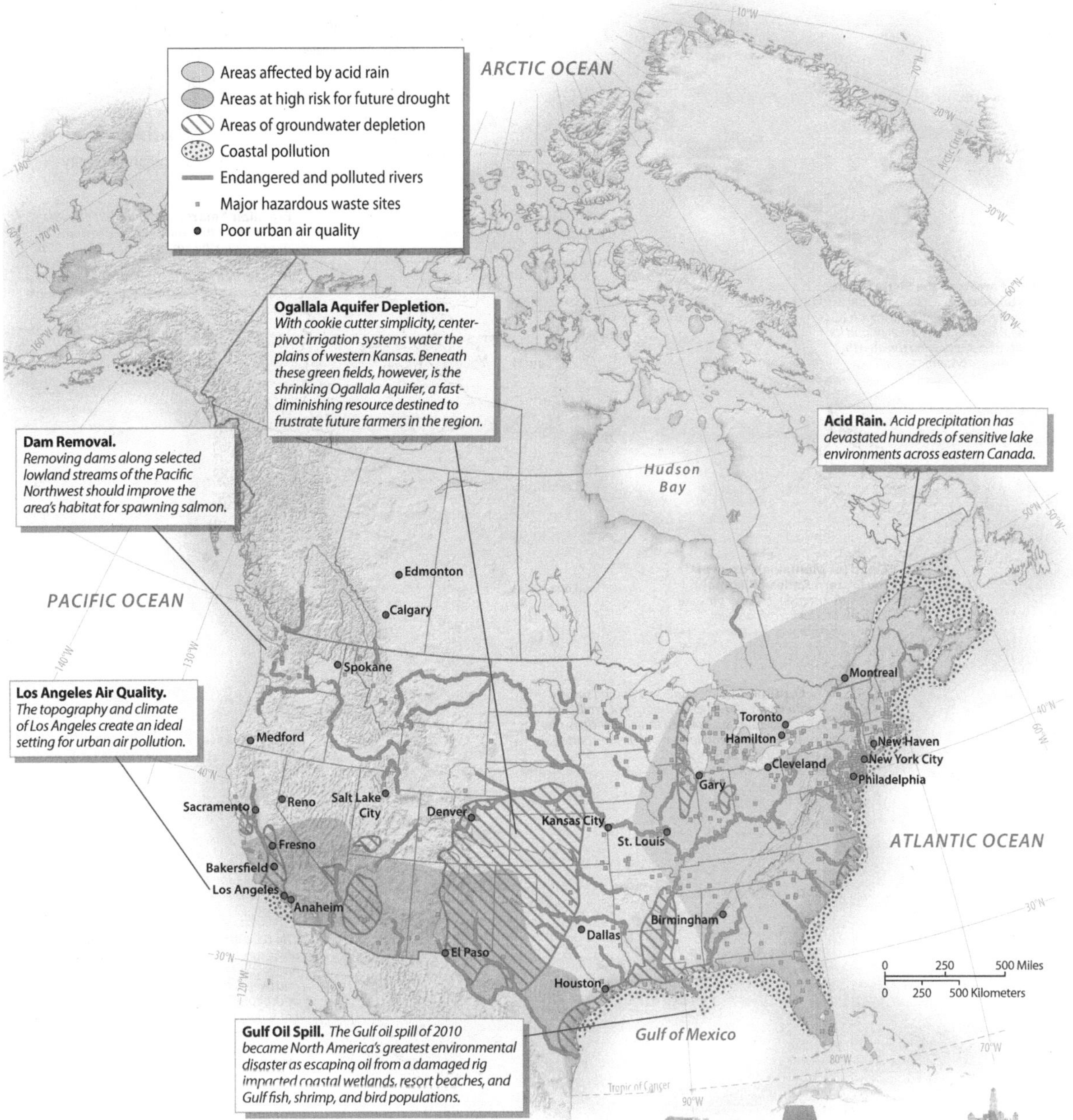

Jeremy Woodhouse/Photodisc/Getty Images

Top: Many environmental issues threaten North America. Acid rain damage is widespread in regions downwind from industrial source areas. Elsewhere, widespread water pollution, cities with high levels of air pollution, and zones of accelerating groundwater depletion pose health dangers and economic costs to residents of the region. Since 1970, however, both Americans and Canadians have become increasingly responsive to the dangers posed by these environmental challenges. **Right:** Rapid job creation has transformed Sun Belt cities such as Dallas, Texas. Healthy growth in office space, specialty retailing, and entertainment districts have fueled downtown Dallas's expansion and reshaped the look of the central-city skyline.

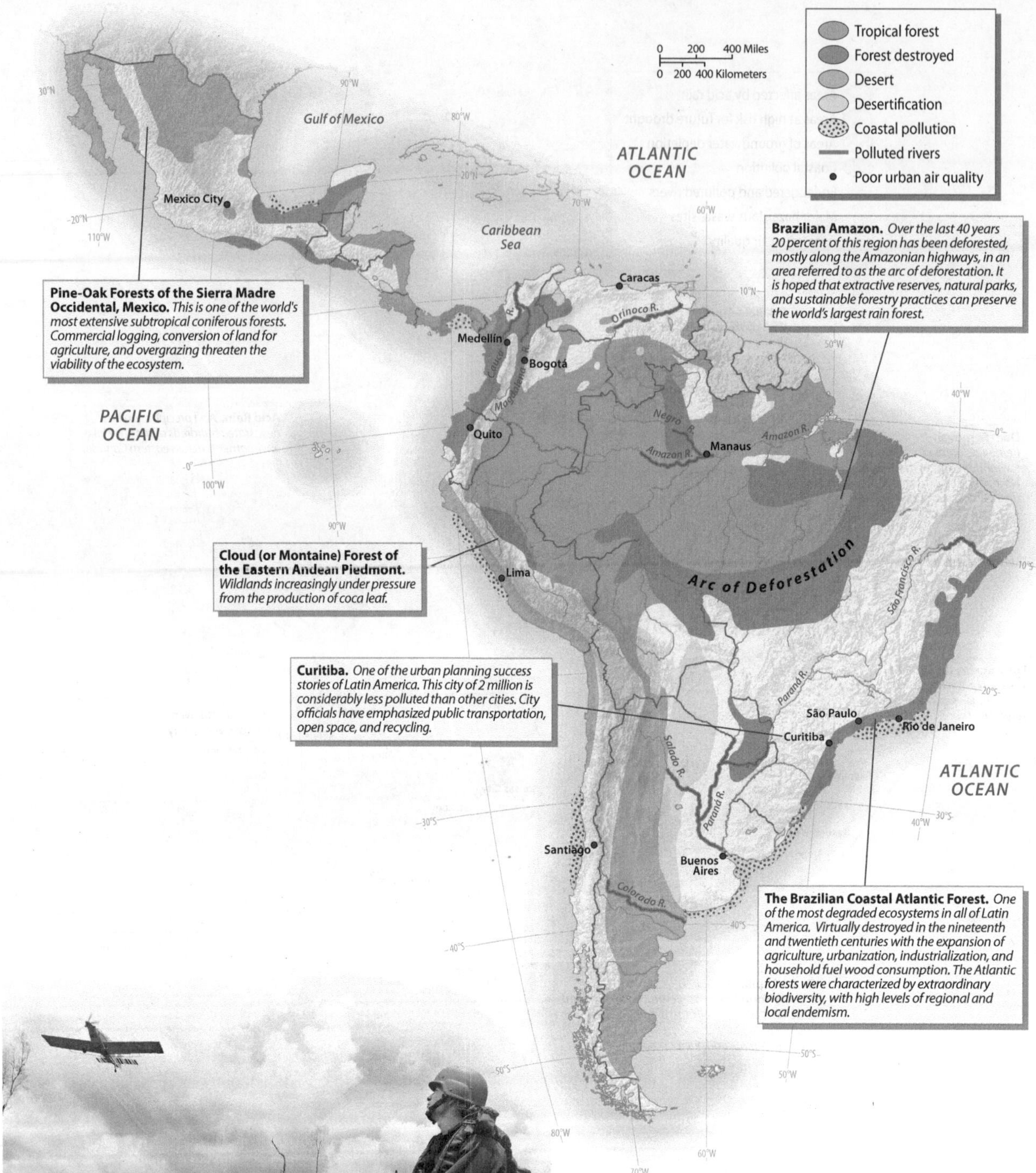

Top: Tropical forest destruction, desertification, water pollution, and poor urban air quality are some of the pressing environmental problems facing Latin America. Still present, however, are vast areas of tropical forest, supporting a wealth of genetic and biological diversity. **Left:** A member of an elite antidrug unit of Colombia's National Police watches as an aircraft dumps herbicide on an illicit coca crop in Colombia. Coca leaf, the raw material for cocaine, has become the leading cash crop in the area. With financial assistance from the United States, Colombians use aerial spraying in an effort to eradicate the crop.

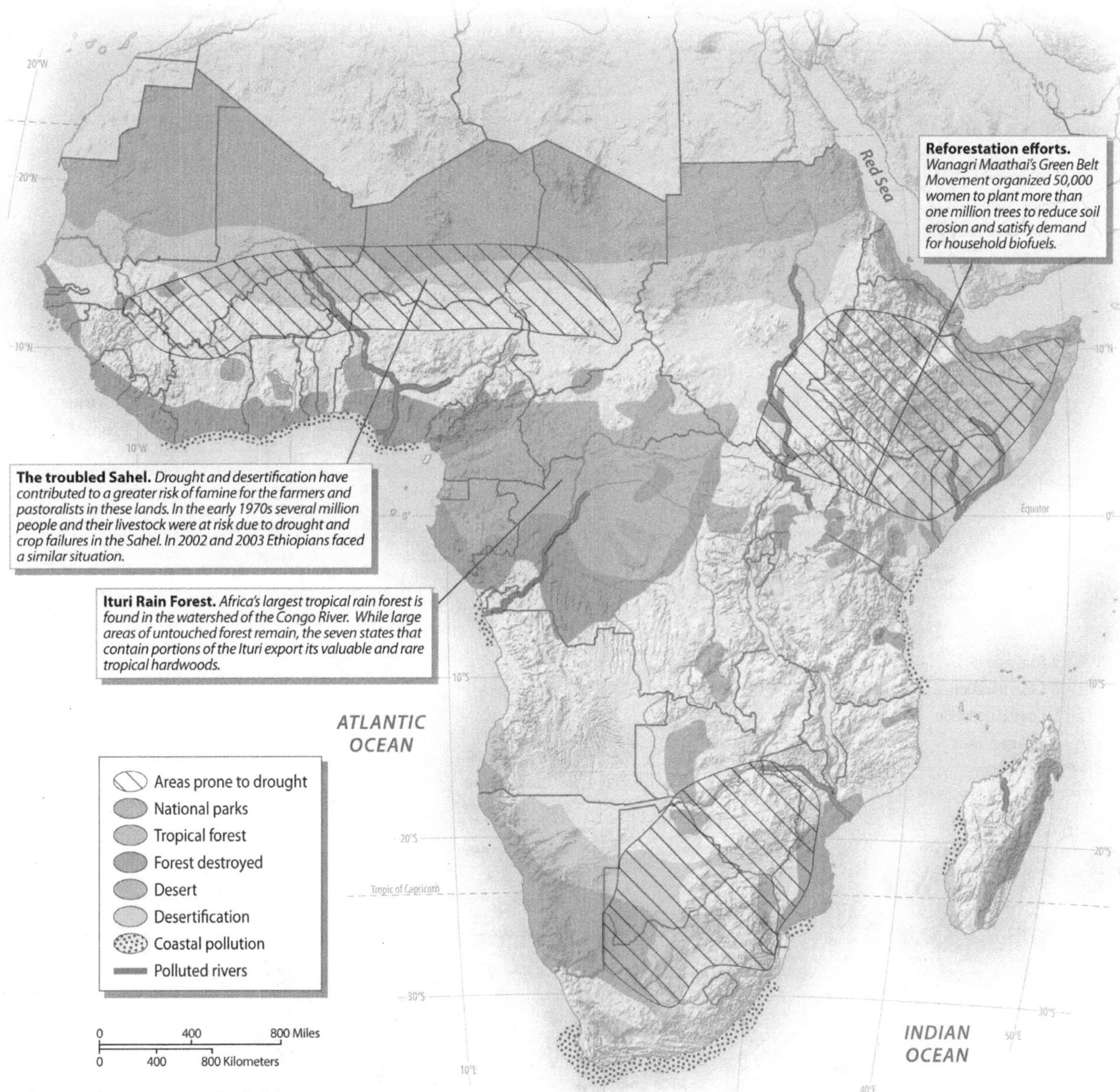

Steve Robinson/Balance/Photoshot./Newscom

Top: Given the immense size of sub-Saharan Africa, it is difficult to generalize about environmental problems. Dependence on trees for fuel places strains on forests and wooded savannas throughout the region. In semiarid regions, such as the Sahel and Horn of Africa, population pressures, climate change, and land-use practices seem to have exacerbated desertification. Yet sub-Saharan Africa also supports the most impressive array of wildlife, especially large mammals, on earth. **Right:** Heavy industry, such as this chemical plant in Kafue, Zambia, has failed to deliver sub-Saharan Africa from poverty. In the worst cases, these industrial enterprises were unable to produce competitive products for world and domestic markets.

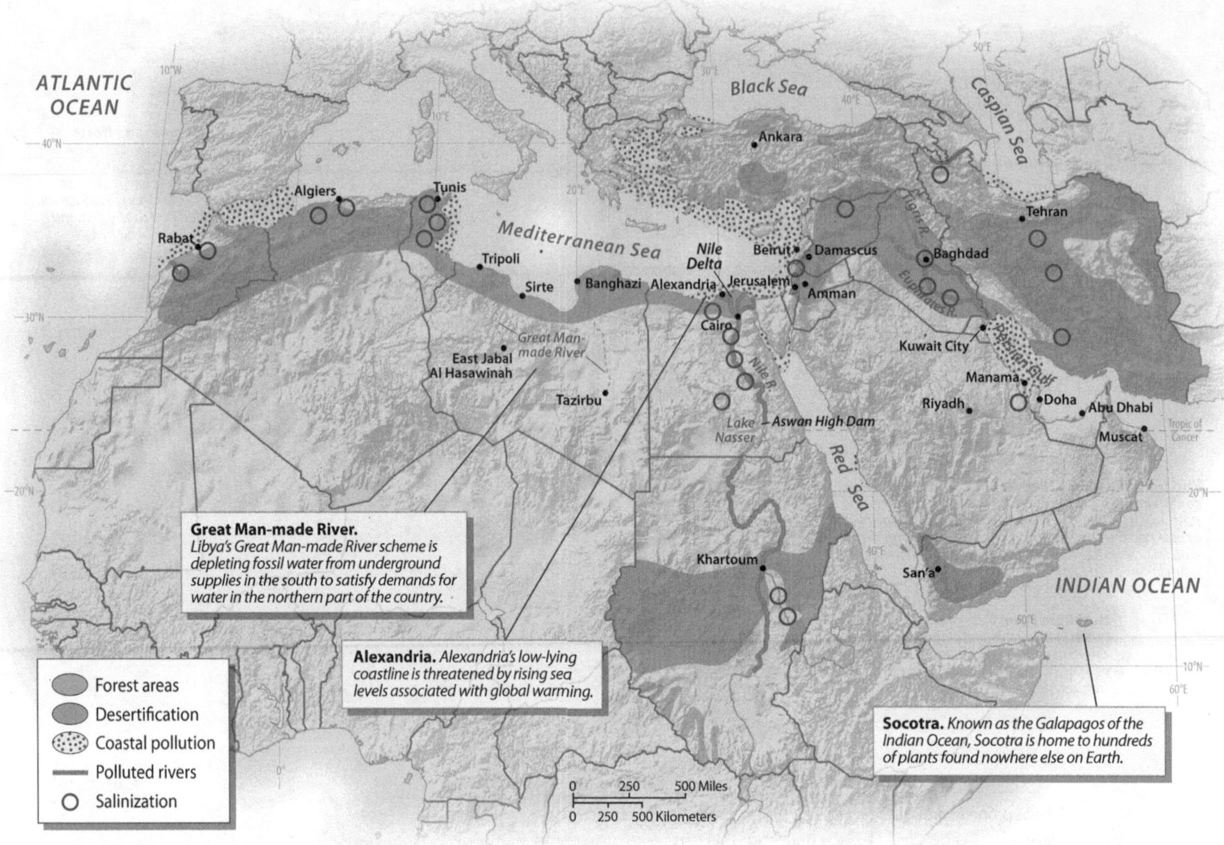

Top: Growing populations, pressures for economic development, and pervasive aridity combine to create environmental hazards across Southwest Asia and North Africa. Long human occupance has contributed to deforestation, irrigation-induced salinization, and expanding desertification. Saudi Arabia's deep-water wells, Egypt's Aswan High Dam, and Libya's Great Manmade River are all recent technological attempts to expand settlement, but they may carry a high long-term environmental price tag.
Left: These Israeli workers are employees of the Telrad factory, a high-tech division of the Koor conglomerate.
Right: Three Iranian teenagers ponder the latest computers at the Paytakht Mall, a popular shopping spot in northern Tehran.

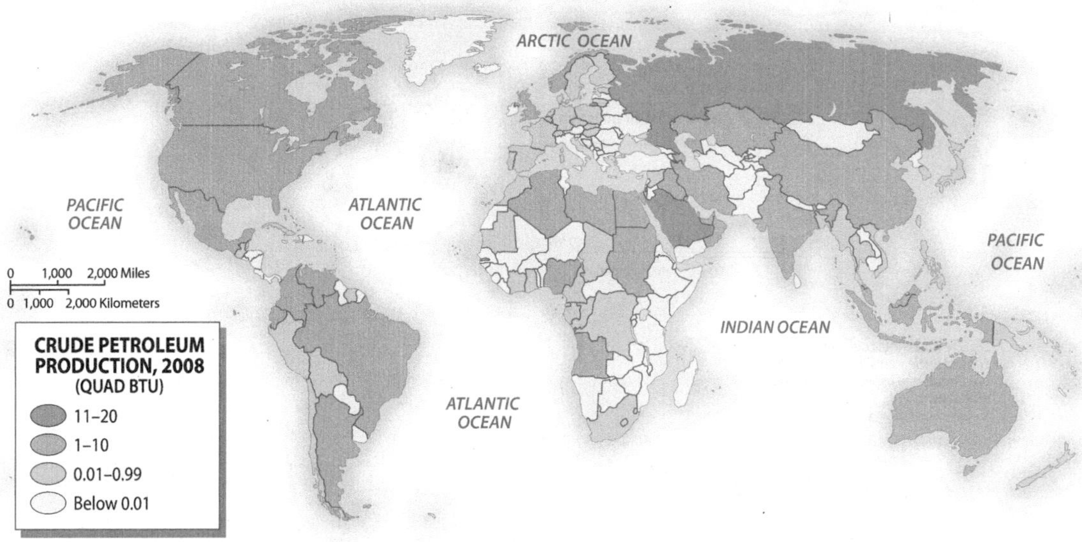

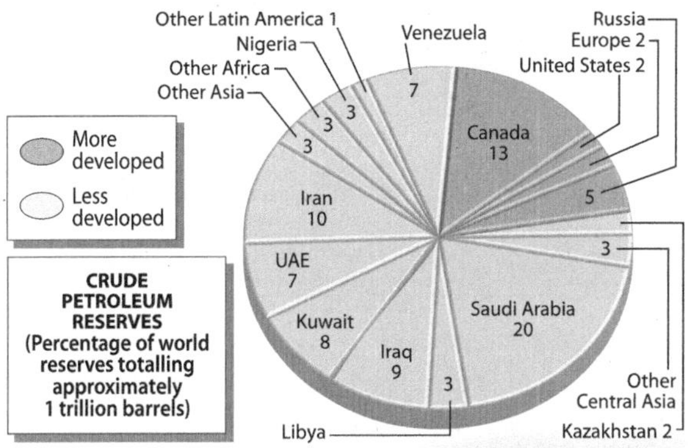

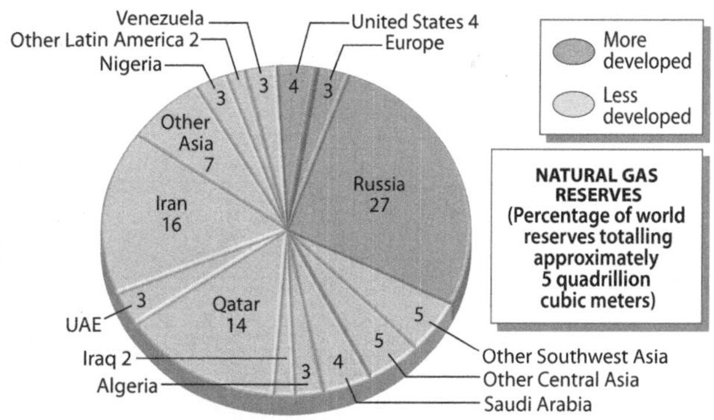

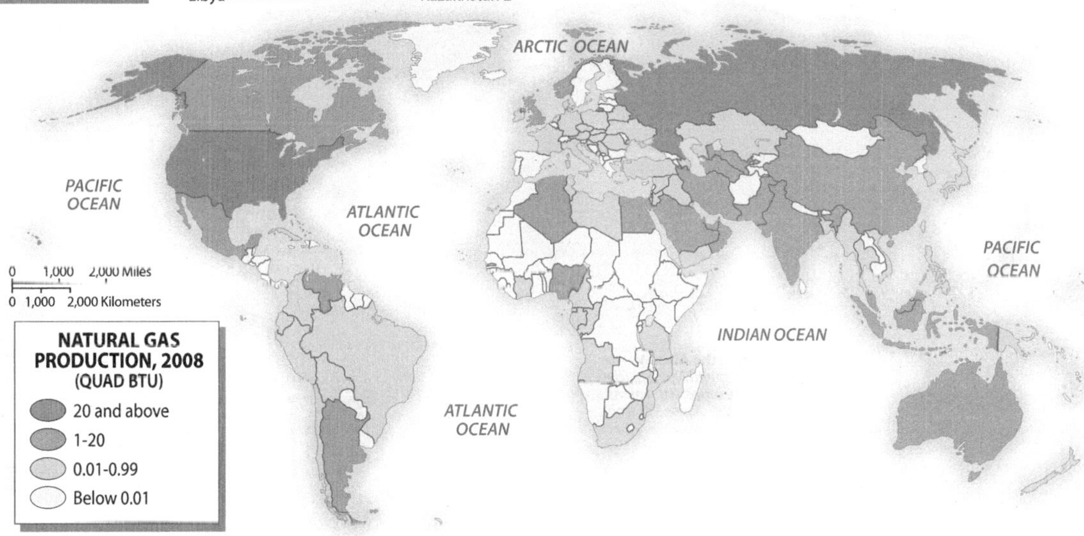

Saudi Arabia and Russia are the largest producers of petroleum. The United States and Russia are the largest producers of natural gas.

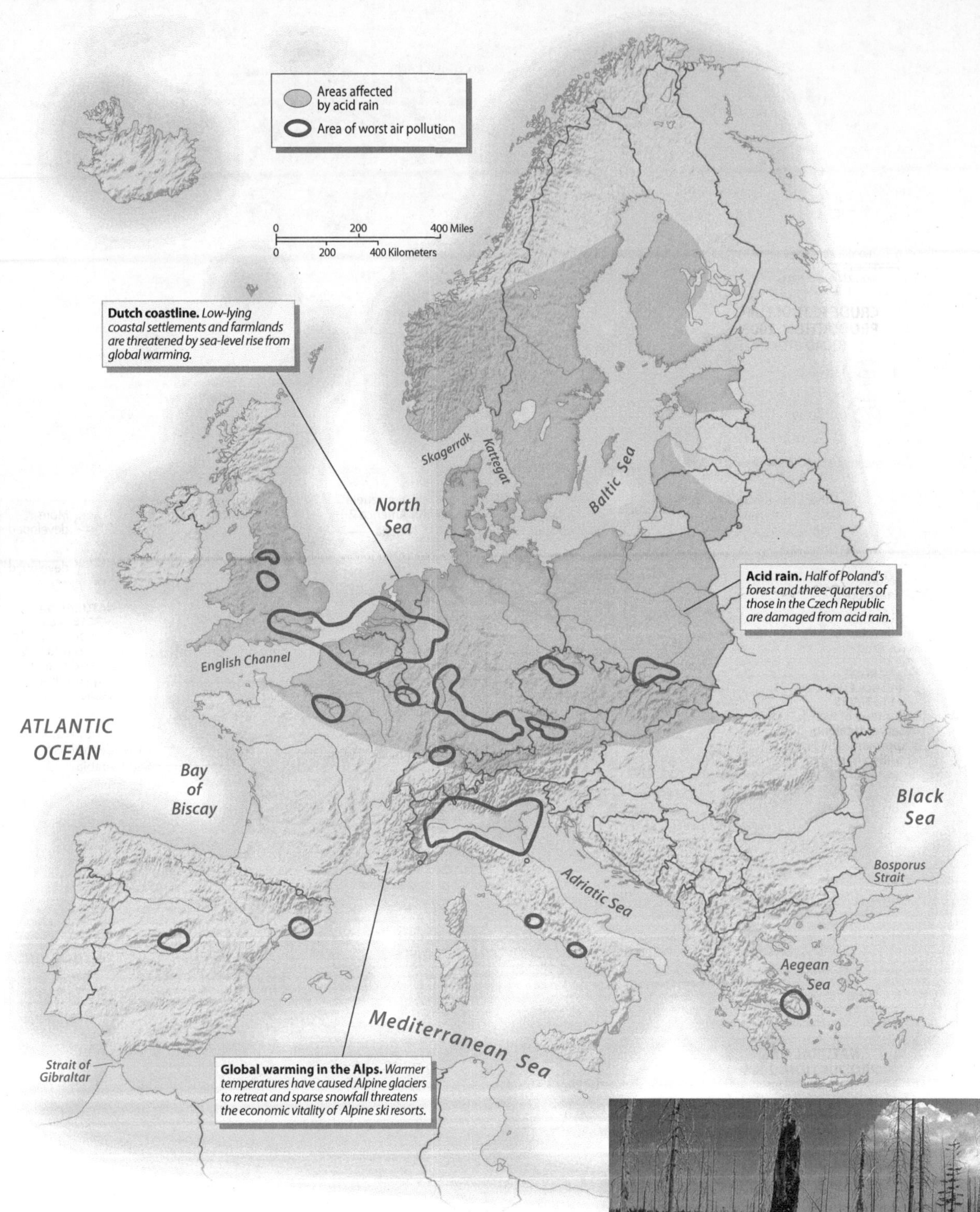

Karol Kallay/Agentur Bilderburg GmbH

Top: While Western Europe has worked energetically over the past 50 years to solve environmental problems such as air and water pollution, Eastern Europe lags behind because environmental protection was not a high priority during the postwar communist period. Current efforts, though, show great promise.
Right: Acid precipitation has taken a devastating toll on eastern European forests, such as those shown here in Bohemia, Czech Republic. In this country, three-quarters of the forests have been damaged by acid precipitation, which is caused by industrial and auto emissions both in and outside the country.

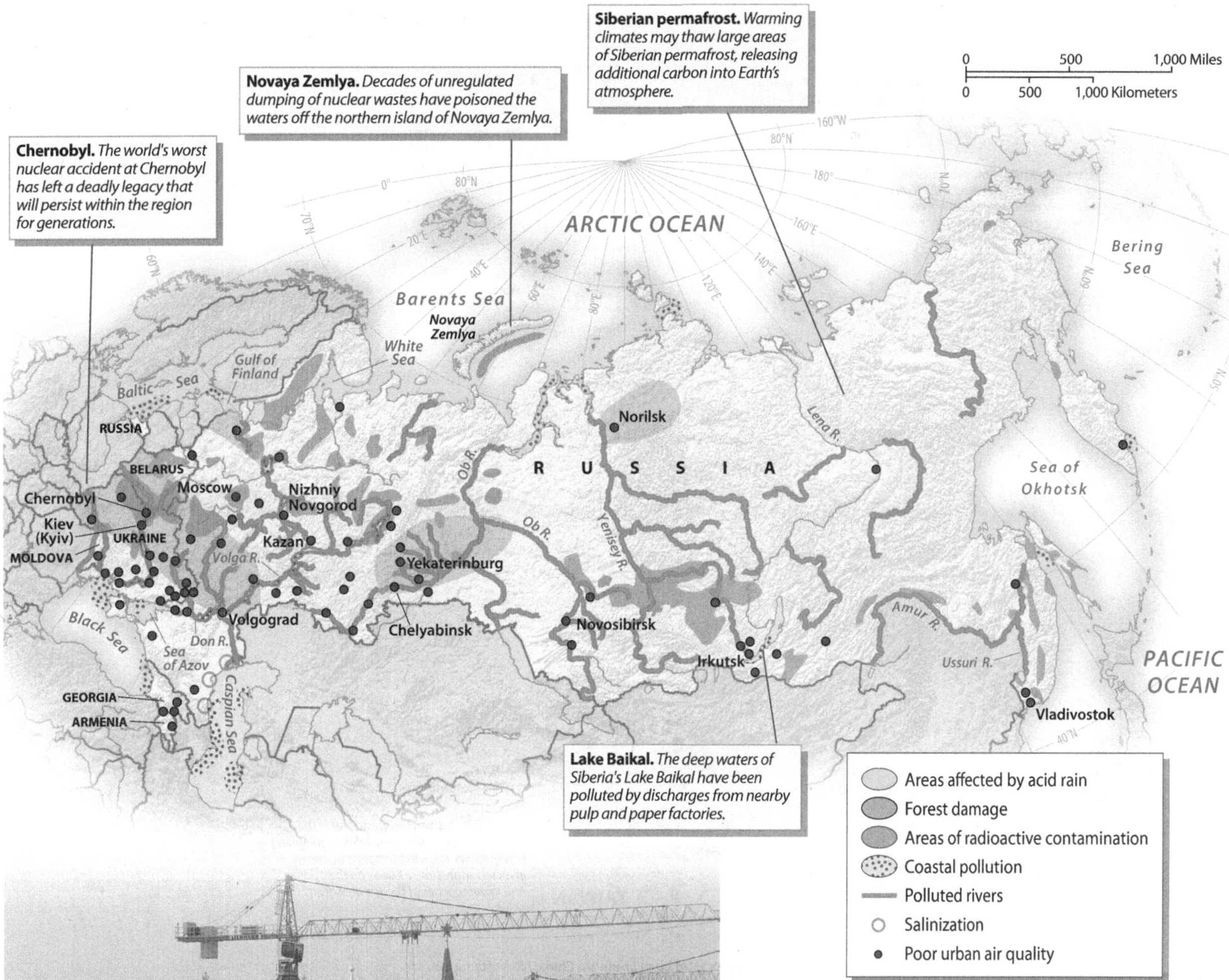

© Grigory Dukor/Reuters/Corbis

ITAR-TASS/Yuri Belinsky/Landov

Top: Varied environmental hazards have left a devastating legacy across the region. The landscape has been littered with nuclear waste, heavy metals, and air pollution. Fouled lakes and rivers pose additional problems in many localities. Present economic difficulties and political uncertainties only add to the costly challenge of improving the region's environmental quality in the twenty-first century.
Left: The city's downtown landscape has been rapidly transformed in the recent economic boom. It remains a complex and fascinating mix of imperial, Soviet, and post-Soviet influences.
Right: A slight rise in Russian birth rates may help slow that country's population decline. Still, high death rates are a major issue, particularly for males throughout the region, and they are the product of many environmental and lifestyle variables.

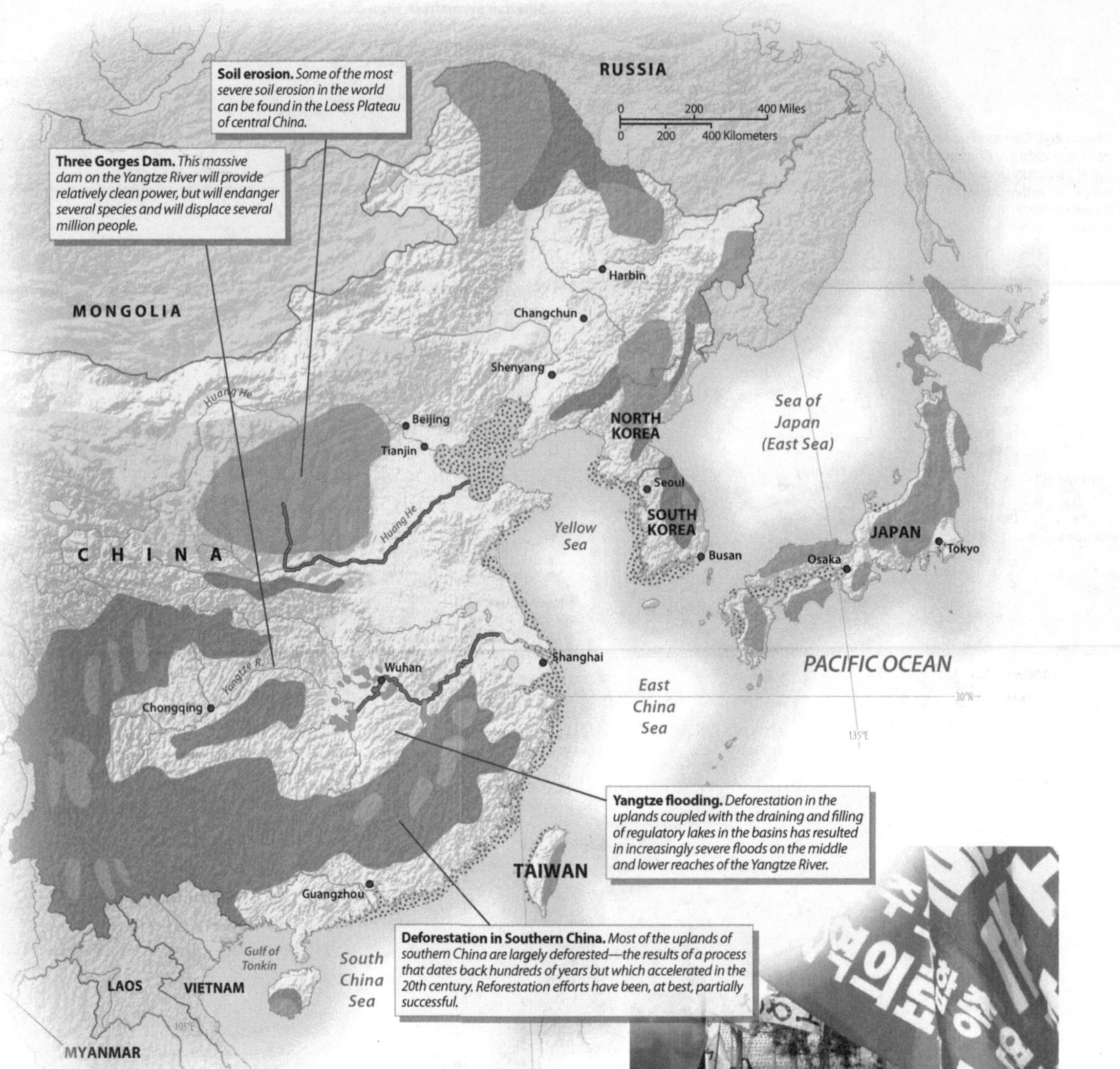

Jodi Cobb/National Geographic Stock

AP Photo/Yun Jai-hyoung

Top: This vast world region has been almost completely transformed from its natural state and continues to have serious environmental problems. In China, some of the more pressing environmental issues involve deforestation, flooding, water control, and soil erosion.
Left: Part of Japan's economic success has resulted from automation of its factory assembly lines. Here Mazda automobiles are assembled in a Hiroshima plant. These cars are destined for the East Coast of the United States.
Right: South Korean students have long been noted for their militant protests against their government, international organizations such as the IMF, and, on occasion, the United States. This rally was in the capital city, Seoul, in 1997, to protest government financial and trade policies.

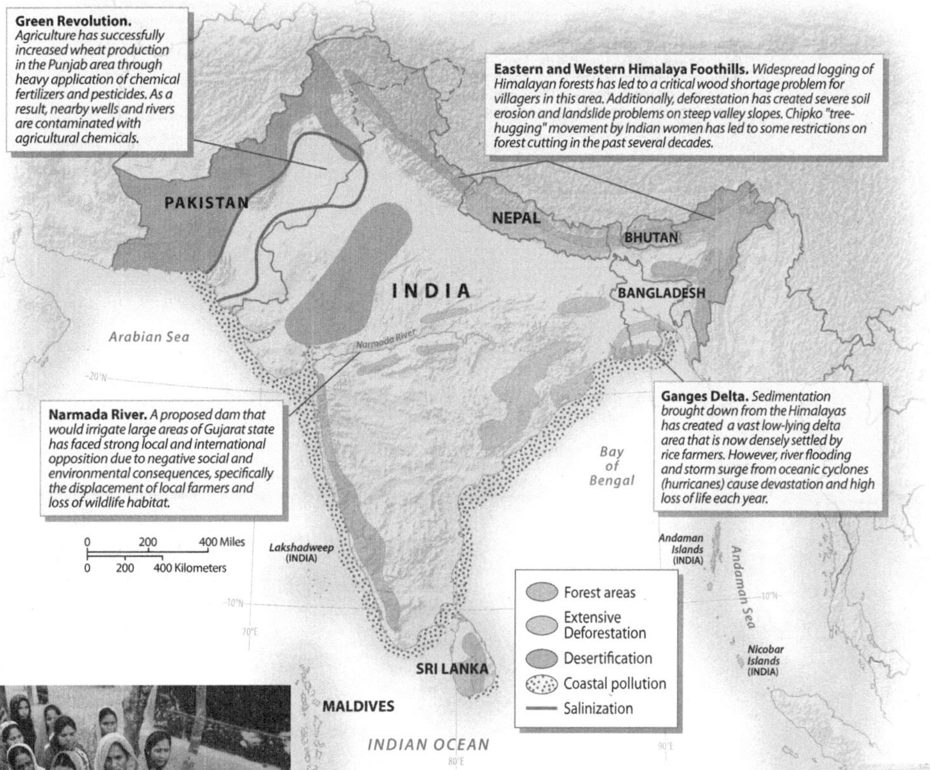

Dinodia/ Sharad J. Devare

© Derek Hall; Frank Lane Picture Agency/CORBIS

© John Van Hasselt/Sygma/Corbis

Balan Madhavan/Hombil Images Pvt. Ltd.

Top: As might be expected in a highly diverse and densely populated region, South Asia suffers from a number of environmental problems. These range from salinization of irrigated lands in the dry lands of Pakistan and Western India to groundwater pollution from Green Revolution fertilizers and pesticides. Additionally, deforestation and erosion are widespread in upland areas.
Middle left: Grameen Bank is an innovative institution that lends money to rural women so they can buy land, purchase homes, or start cottage industries. In this photo, taken in Bangladesh, women proudly repay their loans to a bank official as testimony to their success.
Middle right: During the summer monsoon, some Indian cities such as Mumbai (Bombay) receive more than 70 inches (178 centimeters) of rain in just 3 months. These daily torrents cause floods, power outages, and daily inconvenience. However, the monsoon rains are crucial to India's agriculture. If the rains are late or abnormally weak, crop failure often results.
Bottom left: India's seven government-run institutes of technology provide world-class training for the country's top students in science and engineering, helping India develop globally oriented information technology industries.
Bottom right: A large amount of irrigation water is needed to grow rice, as is apparent from this photo from Sri Lanka. Rice also is the main crop in the lower Ganges Valley and Delta, along the lower Indus River of Pakistan, and on India's coastal plains.

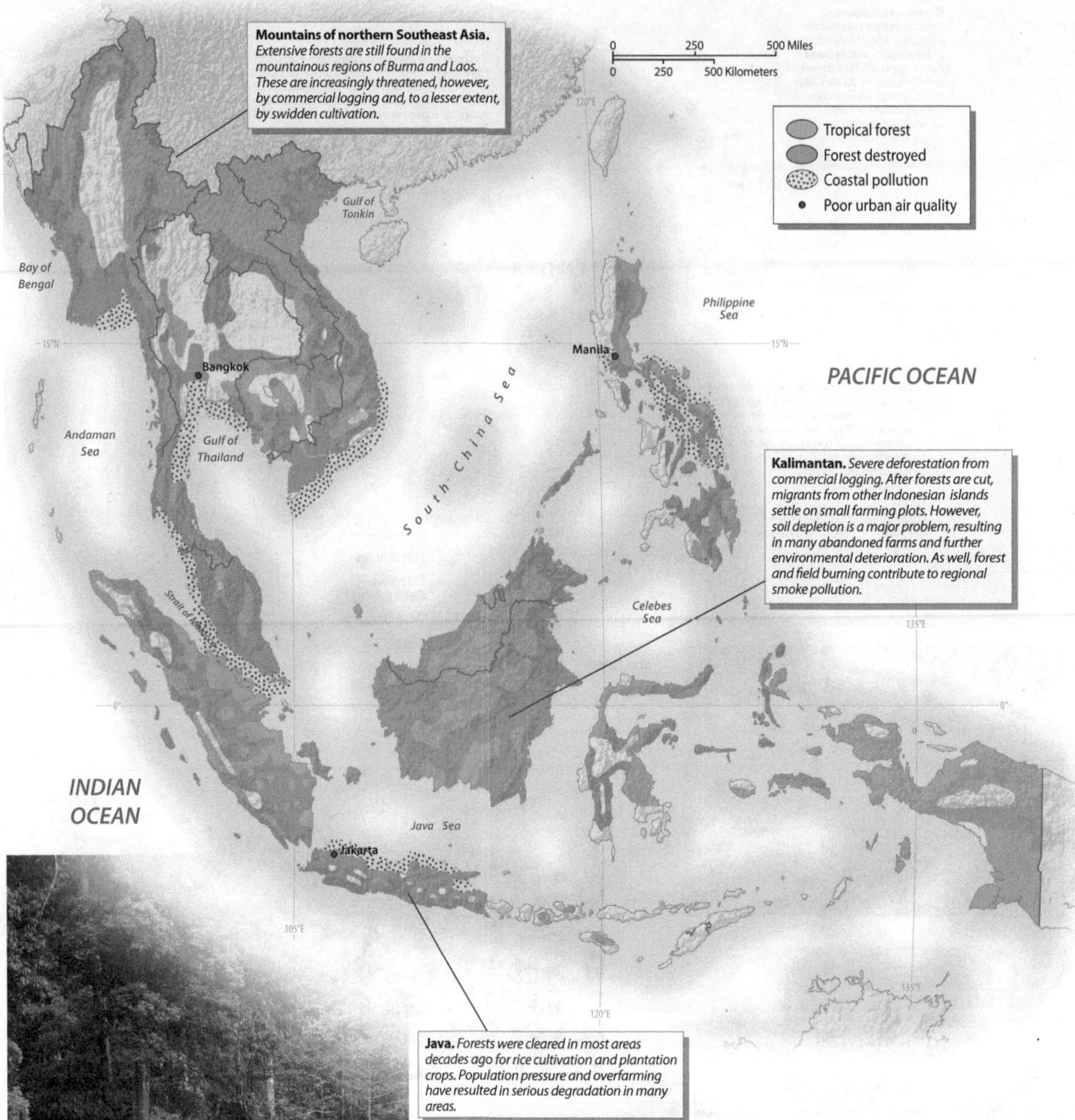

Barry Calhoun/Redux Pictures

© Peter Bowater/Alamy

Top: Southeast Asia was once one of the most heavily forested regions of the world. Most of the tropical forests of Thailand, the Philippines, peninsular Malaysia, Sumatra, and Java, however, have been destroyed by a combination of commercial logging and agricultural settlement. The forests of Borneo (Kalimantan), Burma (Myanmar), Laos, and Vietnam, moreover, are now being rapidly cleared. Water and urban air pollution, as well as soil erosion, are also widespread in Southeast Asia.
Left: Southeast Asia has long been the world's most important supplier of tropical hardwoods. The logging process has destroyed most of the tropical forests of the Philippines and Thailand, as well as on the Indonesian islands of Java and Sumatra.
Right: Plantation crops, such as tea, are major sources of exports for several Southeast Asian countries. Coconut, rubber, oil palms, and coffee are other major cash crops. Many of these crops require large amounts of labor, particularly at harvest time.

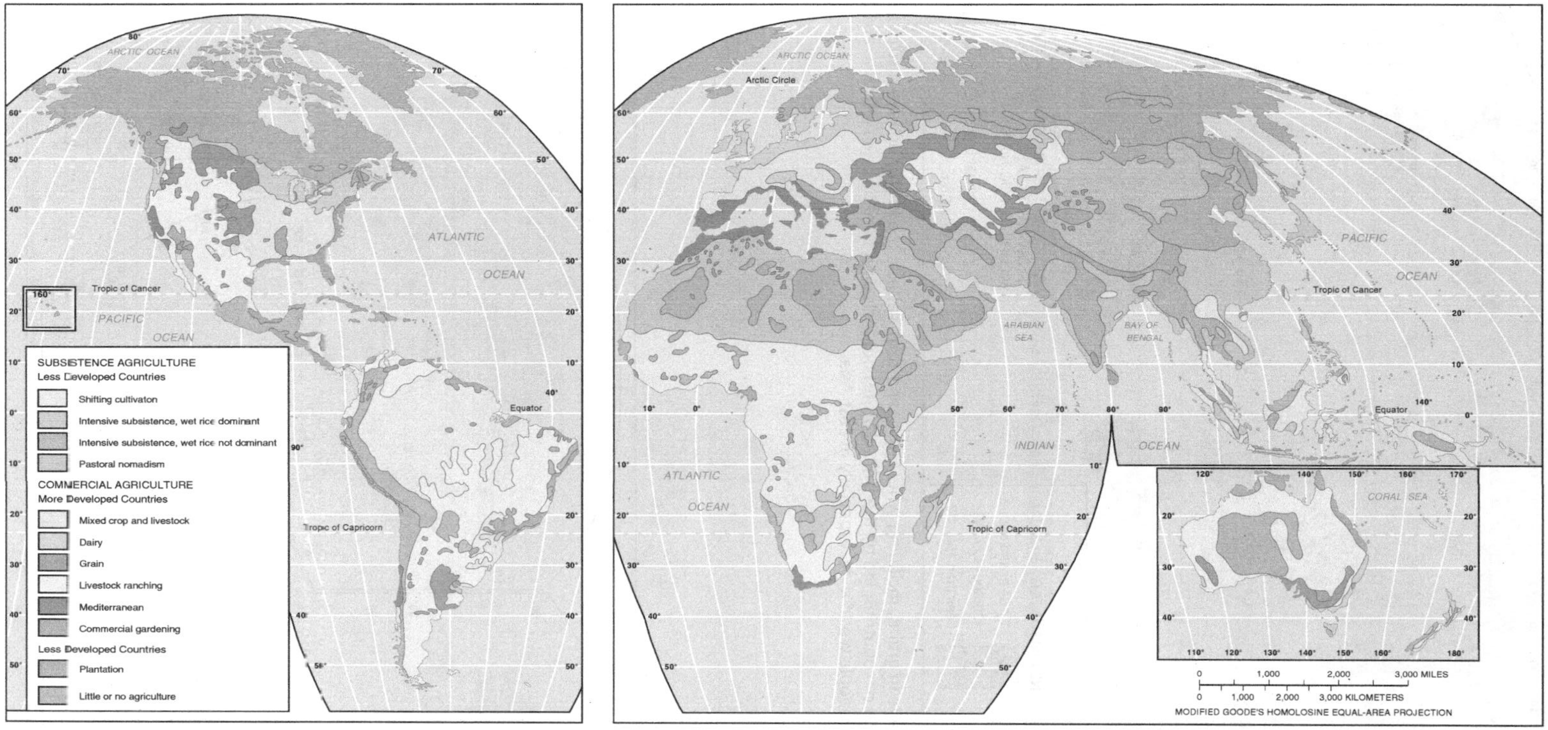

The geography of agriculture worldwide reflects an enormous number of forces, including, among others, climatic patterns and soil fertility, topography, the availability of water, as well as social forces such as the impacts of colonialism, the legal and cultural organization of property ownership, population growth, the market prices of crops, government subsidies, and the role of multinational corporations. Agriculture thus reflects the complex intersections of nature and society in such a way that they cannot be separated. There are a large variety of forms of agriculture, ranging from pre-industrial, subsistence production (such as shifting cultivation and intensive rice cropping) to highly industrialized, capital- and energy-intensive corporate agriculture such as is found in the United States.

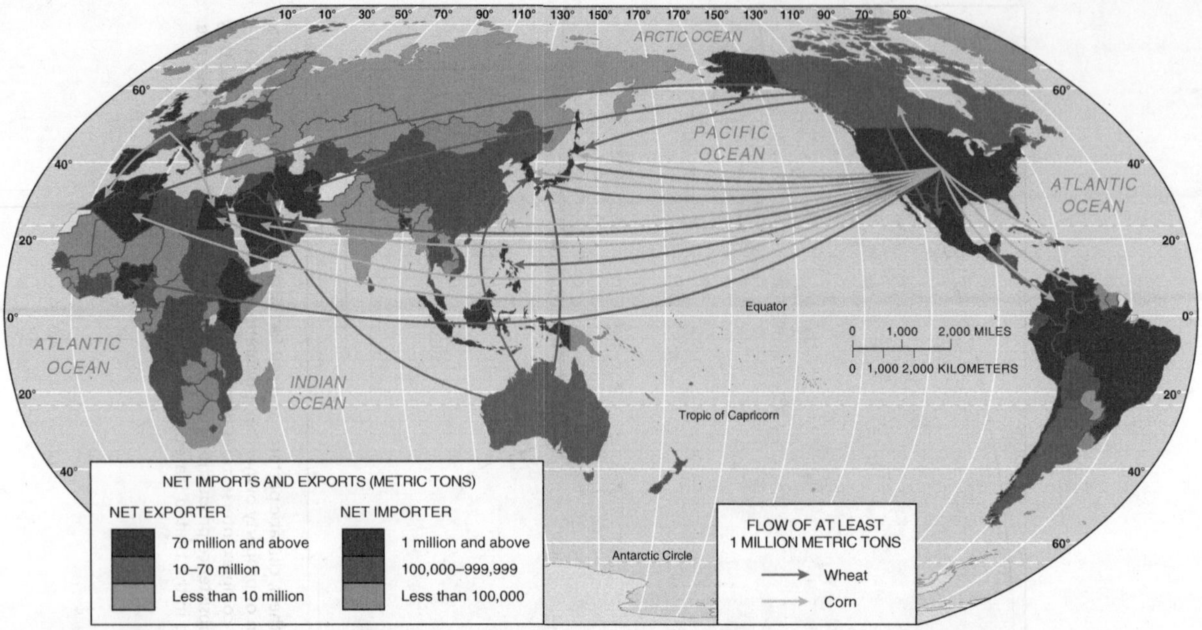

The geography of international trade in grains, including wheat, rice, and corn, reflects the highly uneven distribution of supply and demand over space. A small handful of countries, led by the United States but including Canada and Australia, are the major exporters, supplying food to vast numbers of people worldwide, especially in East Asia.

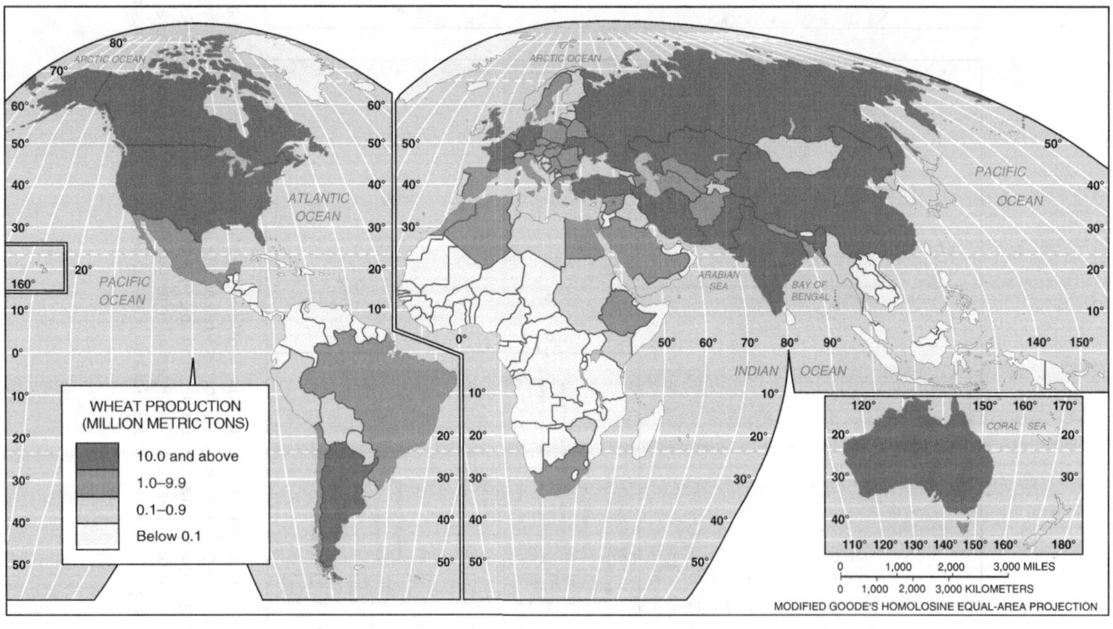

Wheat is one of the world's premier agricultural crops. While wheat is bought and sold extensively on the international market, much of it is consumed domestically. Wheat production levels tend to be highest in midlatitude countries with vast open areas, suitable climate and rainfall, and rich soils, including those in North America, Argentina, Russia and Ukraine, as well as India and northern China.

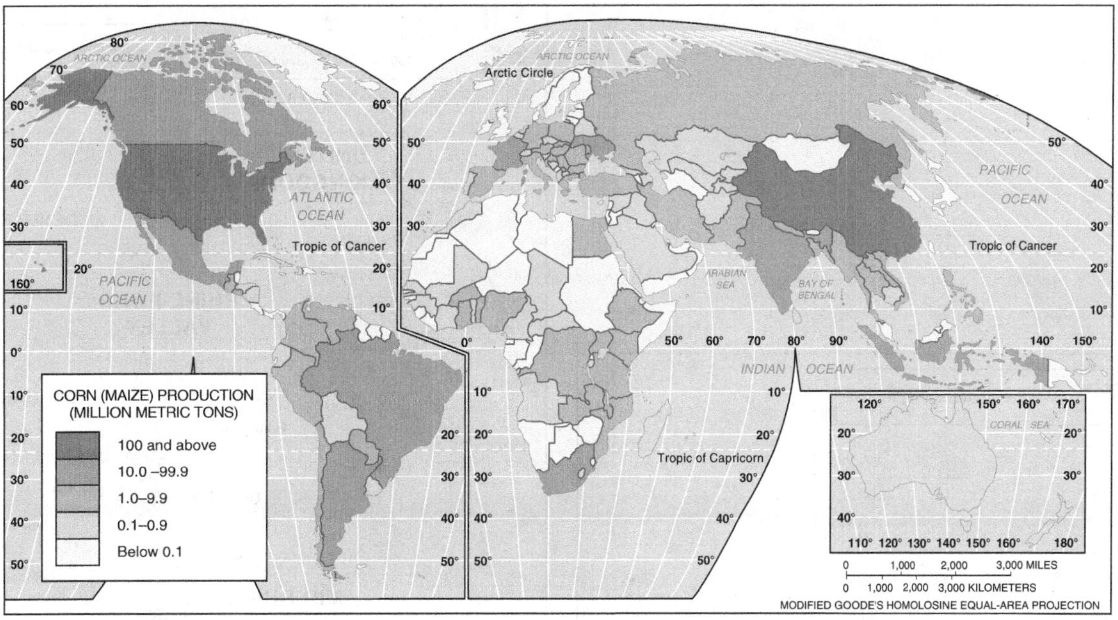

Corn is an important crop that is used to feed not only people but often animals as well. China and the United States dominate world corn production. In the United States, the bulk of corn is used to fatten cattle, as well as to produce sweeteners.

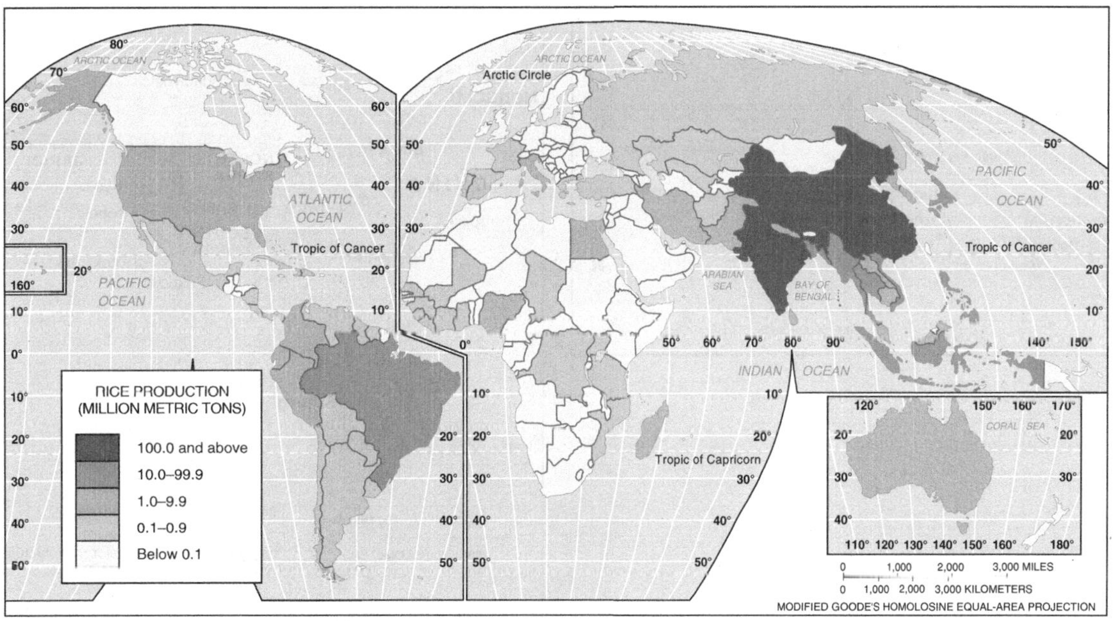

Rice is the most commonly grown crop in the world. In East Asia and Eastern India, where rice is the staple food for one-third of humanity, it has long been grown in the labor-intensive form of irrigated rice paddies, often carved out of hillsides, where hundreds of millions of people engage in arduous tasks such as planting and harvesting. Because rice requires large amounts of water, this type of agriculture involves intricate irrigation systems.

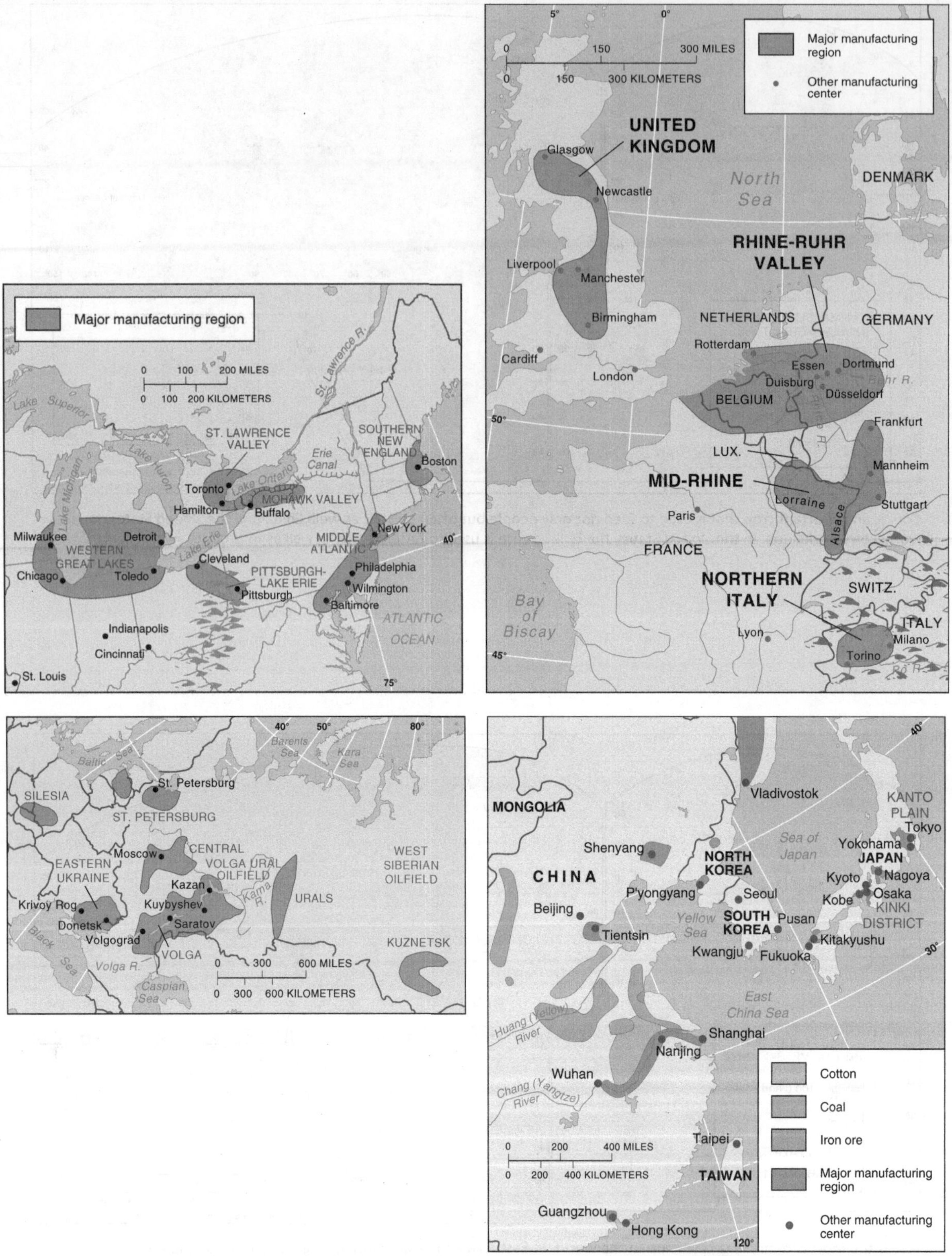

Manufacturing and industrial activity tends to be tightly concentrated in a few, select urban regions, where firms have access to agglomeration economies, specialized pools of labor, information, and infrastructure. In North America, several industrial districts form the Manufacturing Belt of the Northeast and Midwest. In Western Europe, a broad north-south belt extends from central Britain to Western Germany to Northern Italy. In Eastern Europe and Russia, the complexes of Silesia, St. Petersburg, the Donetsk basin in Ukraine, and the Volga River basin stand out. In East Asia, manufacturing is clustered in central Japan, South Korea, and the rapidly growing districts of China, particularly along the southern Yangtze River basin.

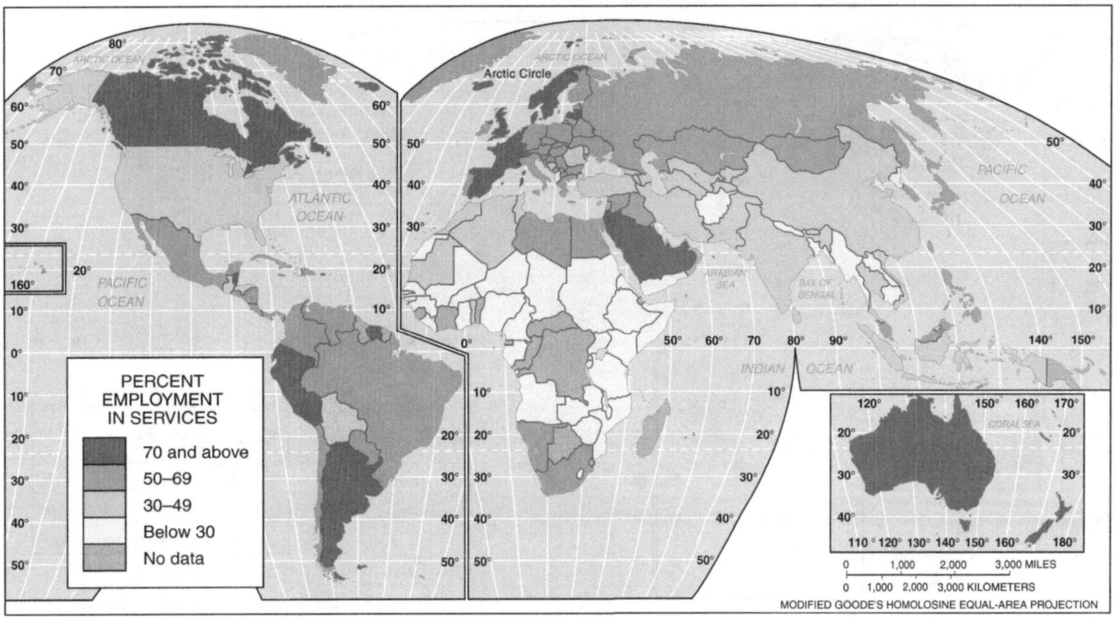

The proportion of workers in service occupations is closely associated with the degree of national development. Generally, in the First World (Europe, North America, Japan, Australia and New Zealand), the vast bulk of workers are involved in the production of intangibles, while in the developing world this proportion tends to be relatively low. The category "services," however, encompasses a vast array of different jobs ranging from prostitutes to professors.

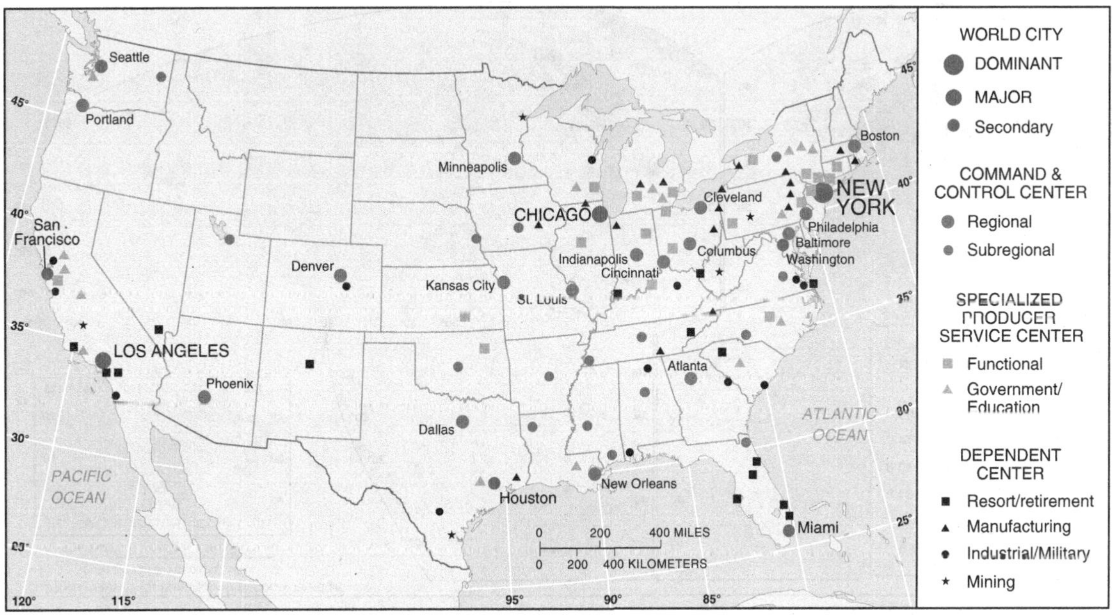

Business services, which cater to other firms rather than households, include firms that produce specialized expertise, such as legal services, advertising, accounting, engineering and architecture, marketing, and corporate research and development. These activities agglomerate in large cities, where they rely upon dense webs of connections to other firms, often in the form of face-to-face contact.

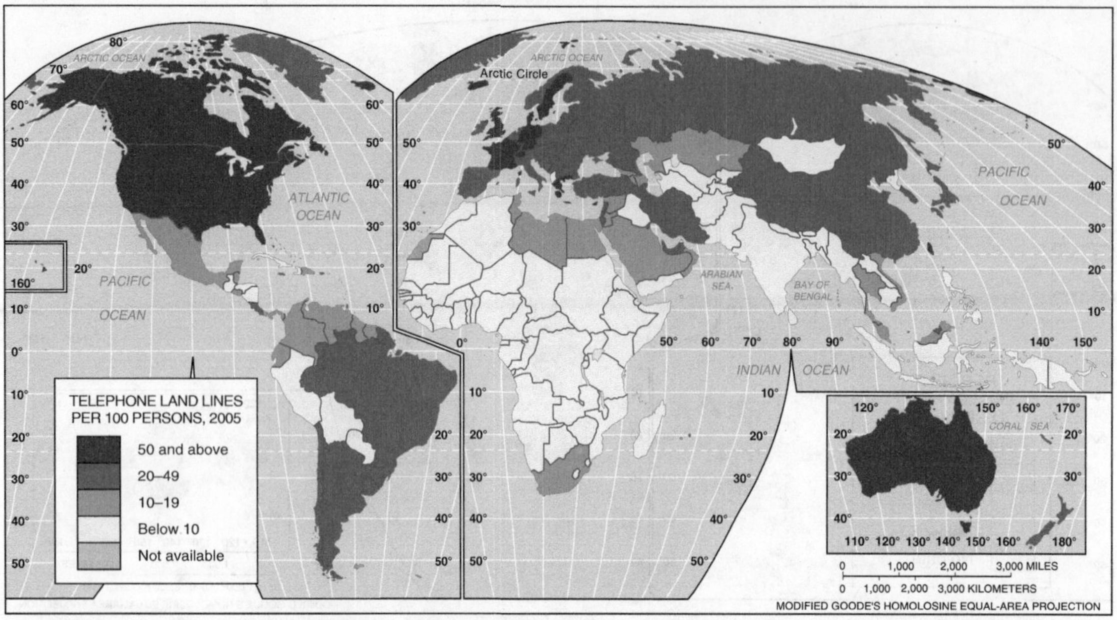

The workhorse of telecommunications is the telephone. While the ratio of people in the First World is relatively low (often as low as 2 per phone), in the developing world there may be 50 or more people per phone. One-half of the world has never made a phone call.

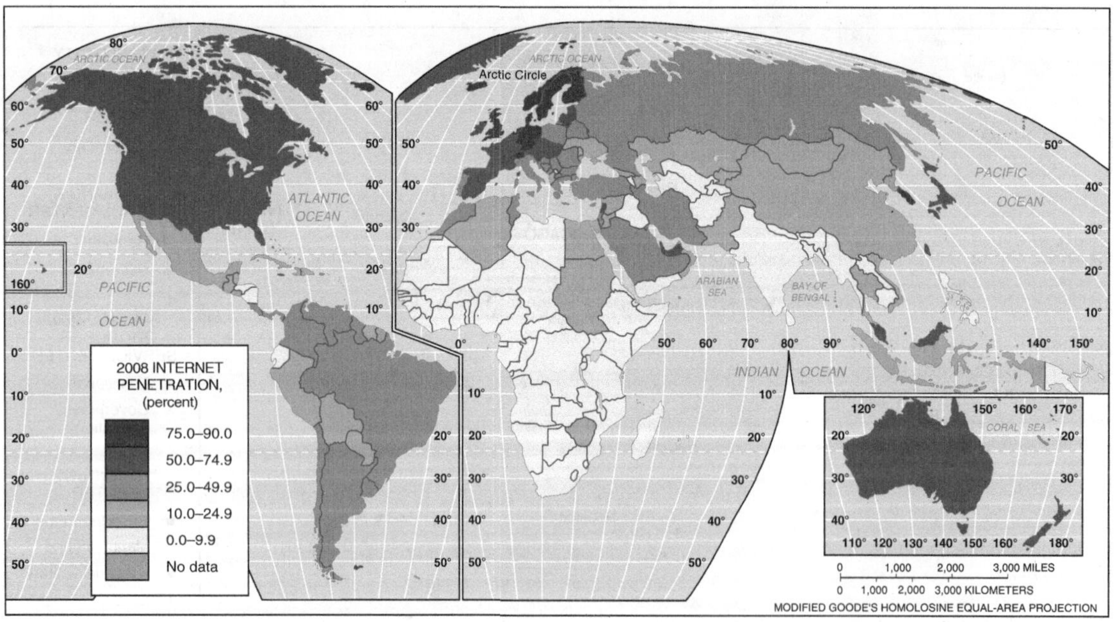

About 600 million people—roughly 10% of the planet—is now wired into the Internet. Yet as this map shows, access to the Internet is in many ways a map of the world's wealth and poverty. Internet penetration rate (% of people with access) range from as little as 0.9% in Africa to as high as 62% in North America. While citizens in the developed world suffer from information overabundance, vast numbers in impoverished countries will never get online.

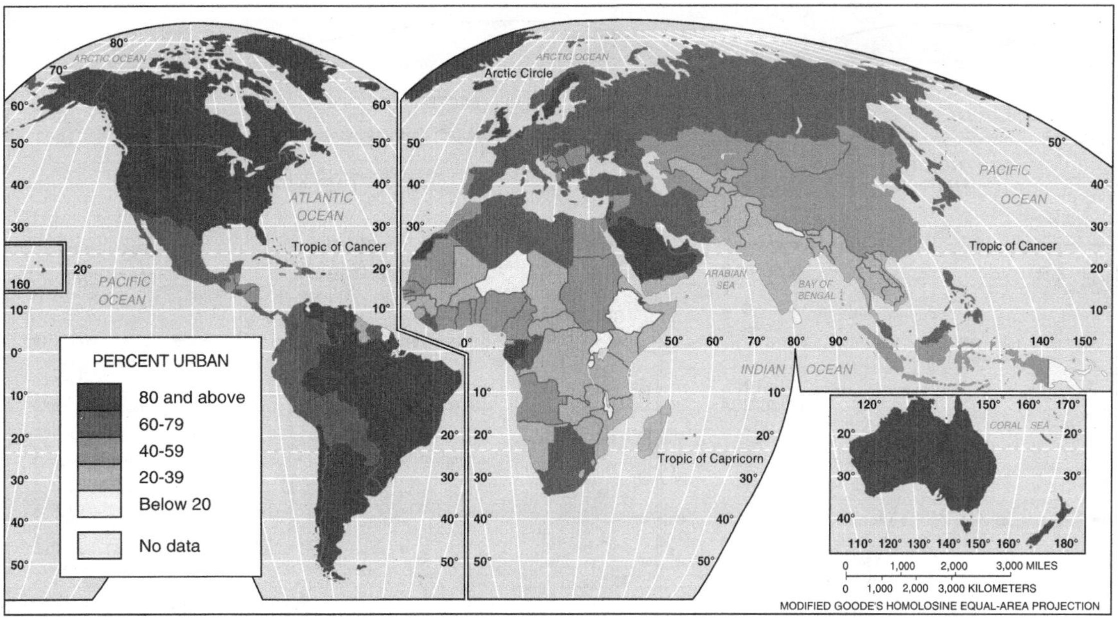

Today, one-half of the world's 6 billion people live in cities. Definitions of "urban" vary widely among countries. The proportion of urbanites is closely associated with countries' wealth, labor force and occupational structure, and degree of rural development, factors that in turn reflect their historical circumstances and position in the world economy. Generally, wealthy, industrialized countries are more heavily urbanized than poorer, pre-industrial ones.

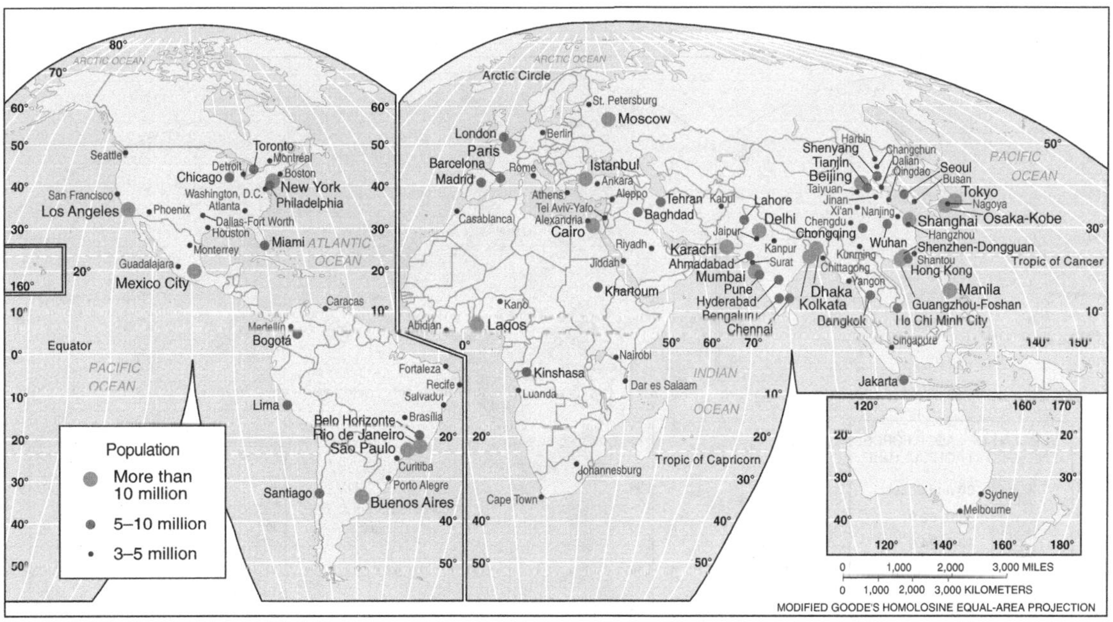

There are dozens of large cities around the world, with some reaching levels as high as 20 million people (e.g., Mexico City, Sao Paolo). Most urban growth, and most of the world's largest cities, are found in the developing world, where cities are swollen from relatively high natural rates of growth and rural-to-urban migration.

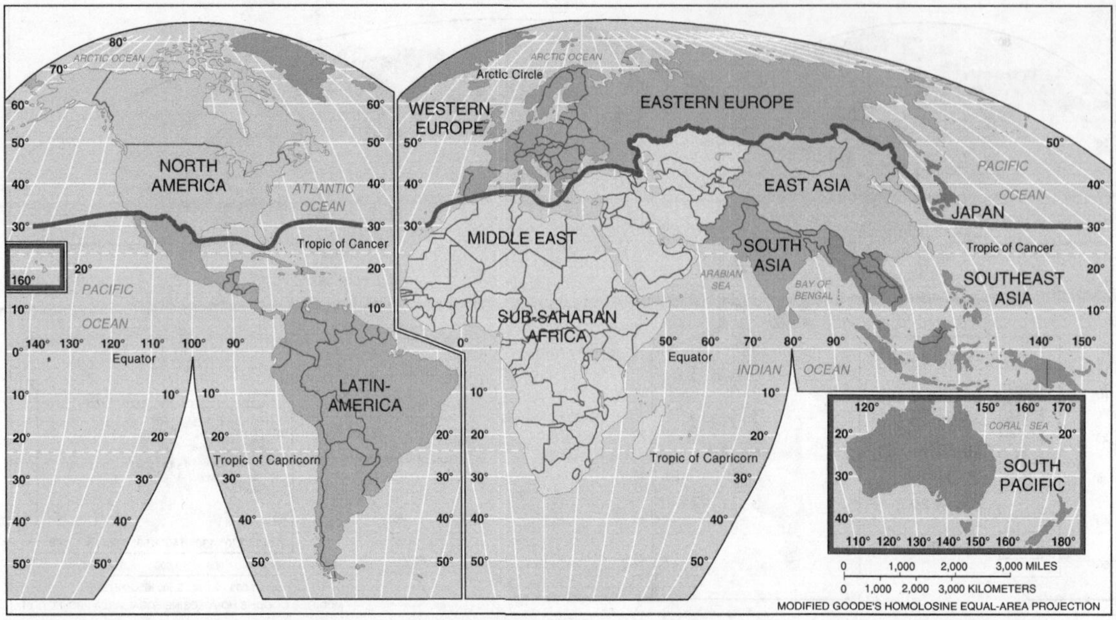

Perhaps the defining aspect of the geography of the global economy is the division of the world into two major parts, the economically developed and underdeveloped regions. During the Cold War, the term First World was used to describe the economically advanced countries of Western Europe, Japan, Australia and New Zealand, and the United States and Canada. The Third World consisted of the former European colonies in Latin America, Africa, and Asia (excluding Japan). The Second World consisted of the Soviet Union and its allies in Eastern Europe; with the disintegration of the USSR in 1991, however, this world has disappeared, divided into the other two.

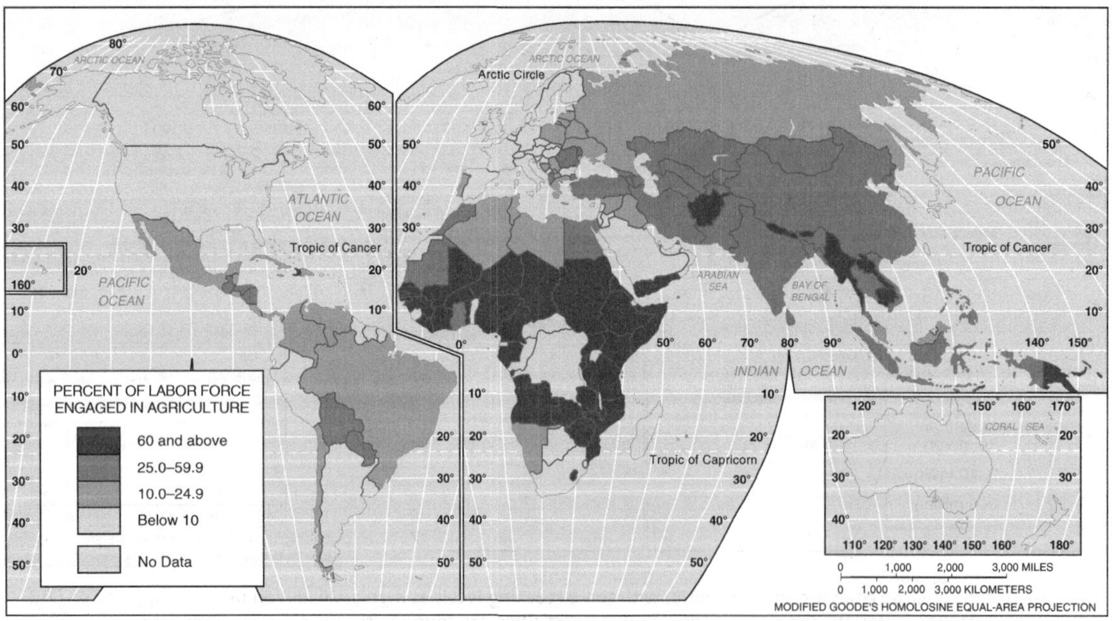

The primary economic sector consists of workers in extractive activities—agriculture, forestry, fishing, and mining. The relative size of this sector is a measure of the level of economic development of a country. In economically advanced countries, relatively few workers are employed in this way (in the United States it is roughly 5%). In developing countries, in contrast, the bulk of workers are often farmers, as in most of Africa and Asia.

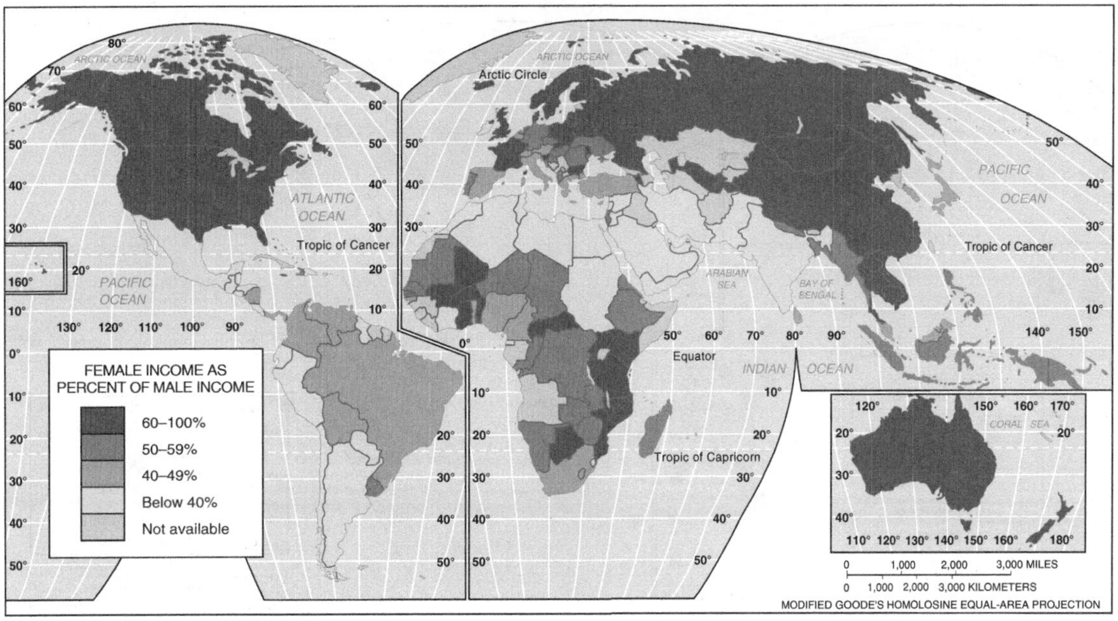

Gender is a major dimension of economic structure in time and space, and intersects with class and ethnicity in complex ways. In almost all societies, men enjoy a privileged social and economic status compared to women, with higher incomes and positions of responsibility. The gap between men's and women's incomes is thus a measure of the geography of patriarchy. Economically advanced countries tend, in general, to exhibit comparatively greater equality between men and women in this regard, although there are notable exceptions (e.g., Japan). In the developing world, men frequently enjoy much greater advantages than do women, although this relation is mediated by culture. In the Muslim world of North Africa and the Middle East, for example, the gap between men's and women's incomes is the world's greatest.

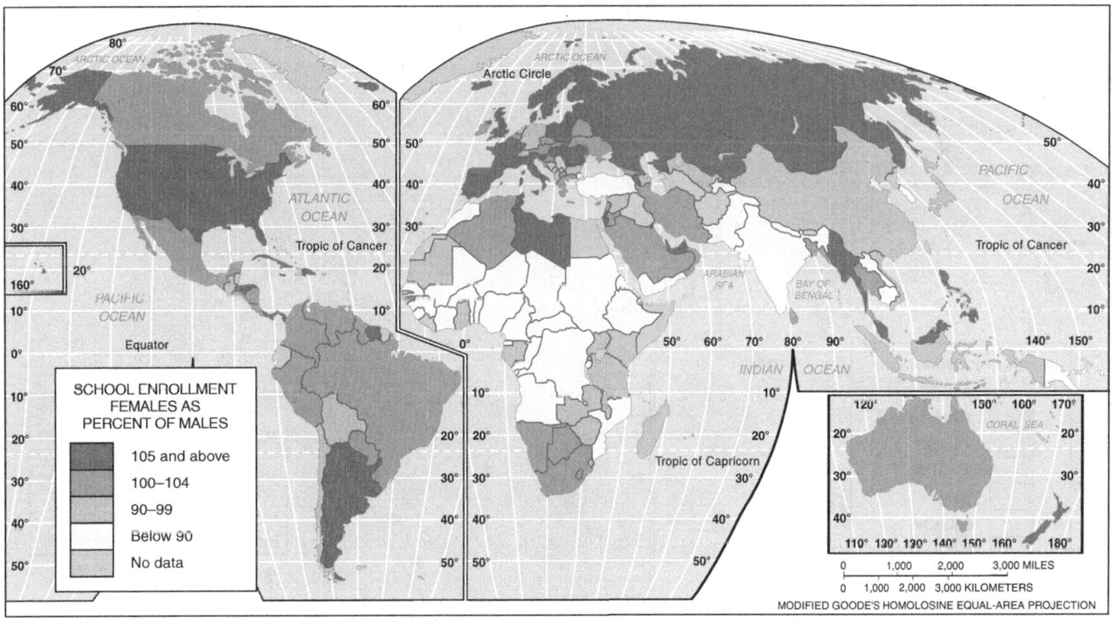

One of the resources that males typically enjoy more abundantly compared to women is education. In many developing countries, girls are much less likely to go to school than are boys, and female literacy rates are typically much lower. Low education levels handicap women's participation in the economy and public life, and raising women's literacy levels is a significant step in the process of economic development.

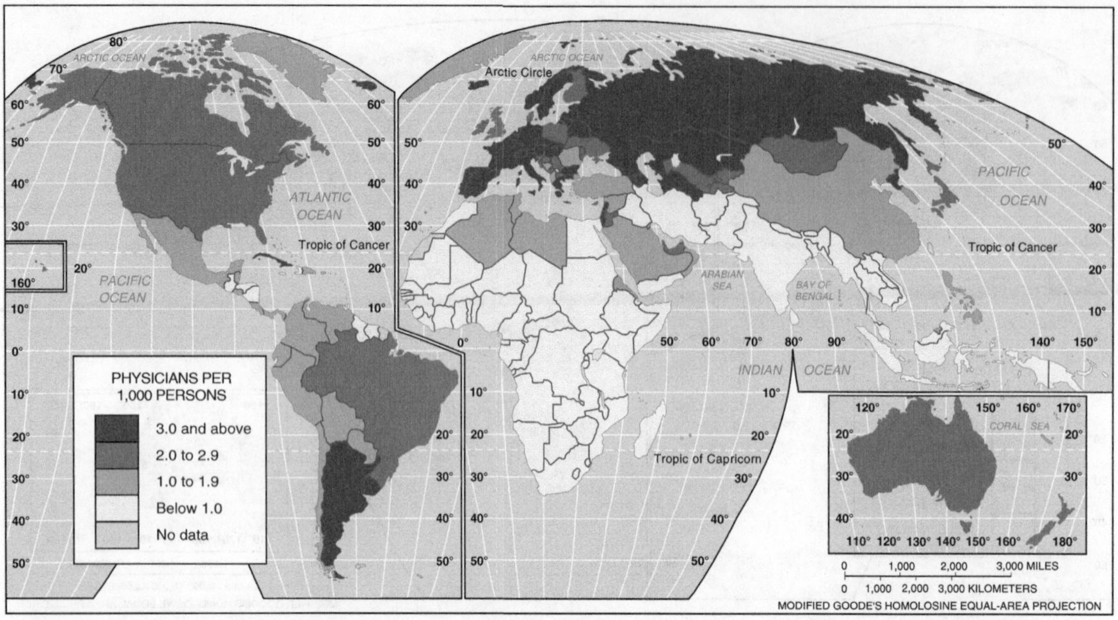

Access to health care, here measured in terms of the number of people per physician, is an important dimension of quality of life and economic development. In the economically developed world, this ratio is generally low, that is, often 500 or less, meaning that most people can see a doctor when they need one (although accessibility varies widely within countries as well, depending on income and the form of national health care available). In most of the developing world, in contrast, population/physician ratios are high—in sub-Saharan Africa they exceed 10,000—meaning that few people have access to health care, a number that translates into high mortality rates (including infants), low life expectancy, and diminished quality of life.

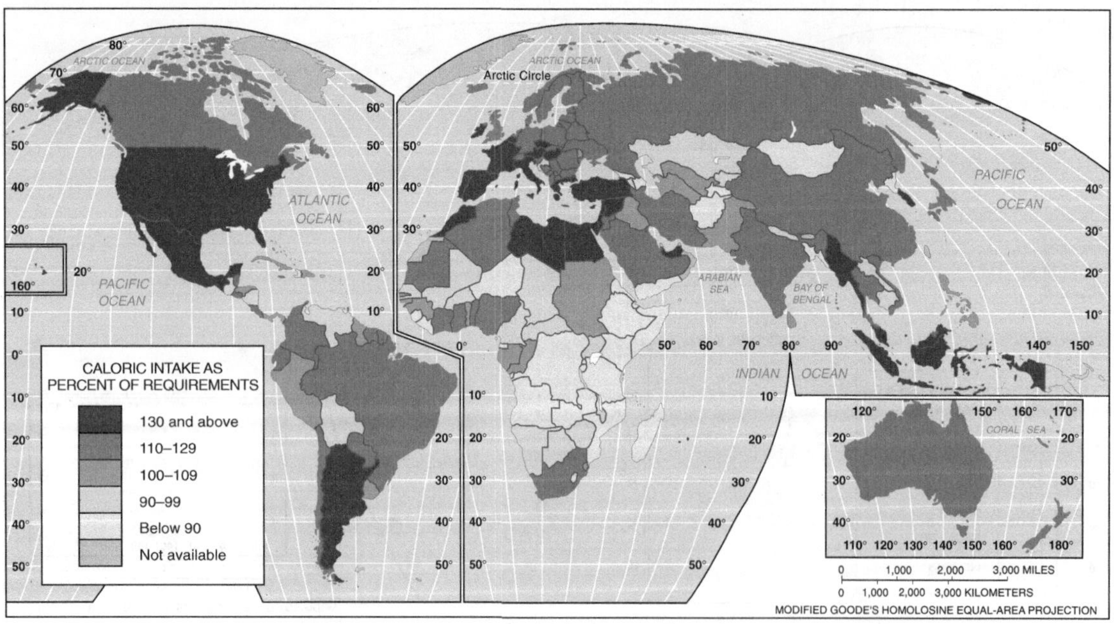

Caloric intake per capita reflects the quantity and quality of a nation's food supply as well as the demands placed upon it by its population. Generally, in wealthy, industrialized countries, relatively few people go hungry or are malnourished; indeed, in the United States, the more pressing problem is too much food and associated obesity rates. In the developing world, access to food varies widely, and many people suffer from less than 100% of their body's caloric requirements, leading to widespread malnutrition. It is worst in the impoverished nations of sub-Saharan Africa, site of most of the world's famines.

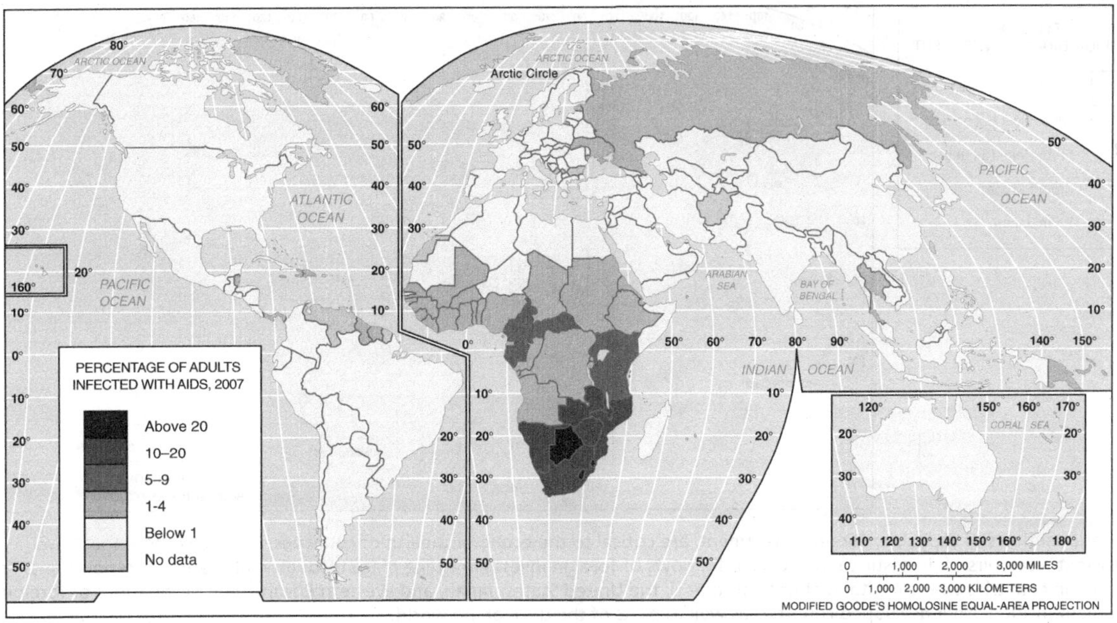

The human immunodeficiency virus (HIV), which causes acquired immune deficiency syndrome (AIDS), has reached the levels of a global epidemic. More than 42 million people worldwide are infected with HIV. With a long lead time before AIDS symptoms appear, HIV is transmitted primarily through sexual contact. Sub-Saharan Africa is the epicenter of the world's AIDS pandemic, where in many countries HIV infection rates have reached 40%. Vast numbers of villages have lost adults in their prime working years and millions of children are orphaned as a result. The virus is just now beginning to penetrate the huge masses of people in Southern and Eastern Asia.

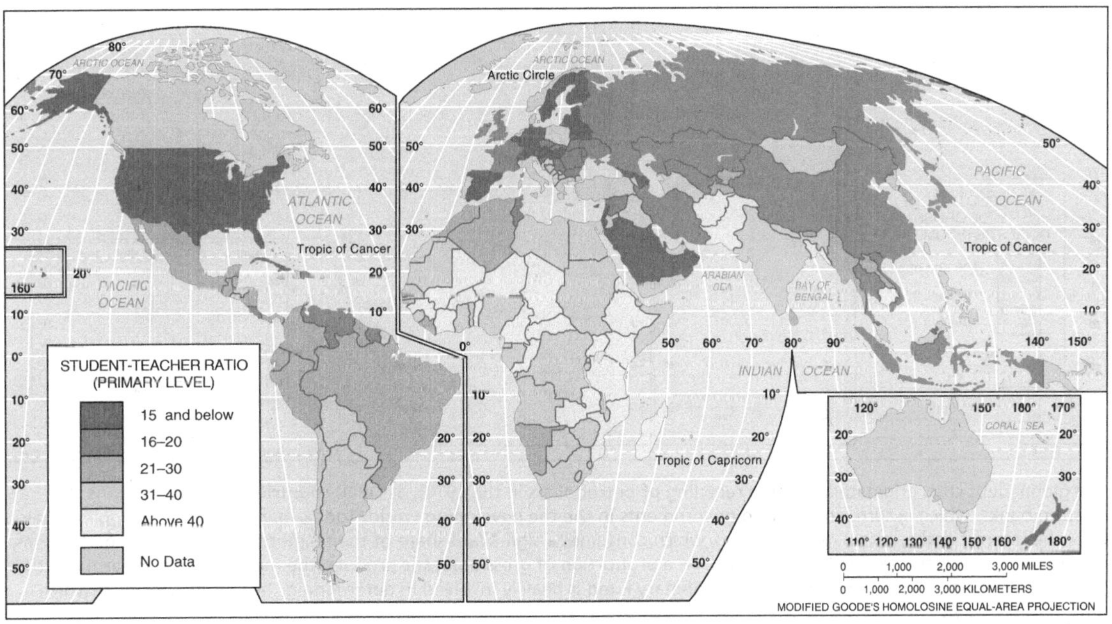

Access to education is often measured by the ratio of students to teachers. In the economically developed world, where education generally receives adequate (if just barely) funding, student/teacher ratios are relative small (i.e., below 15). In most developing countries, with government budgets strapped and priorities focused on the military, schools are badly underfunded, teachers underpaid, and student/teacher ratios high, a well known variable that inhibits effective education.

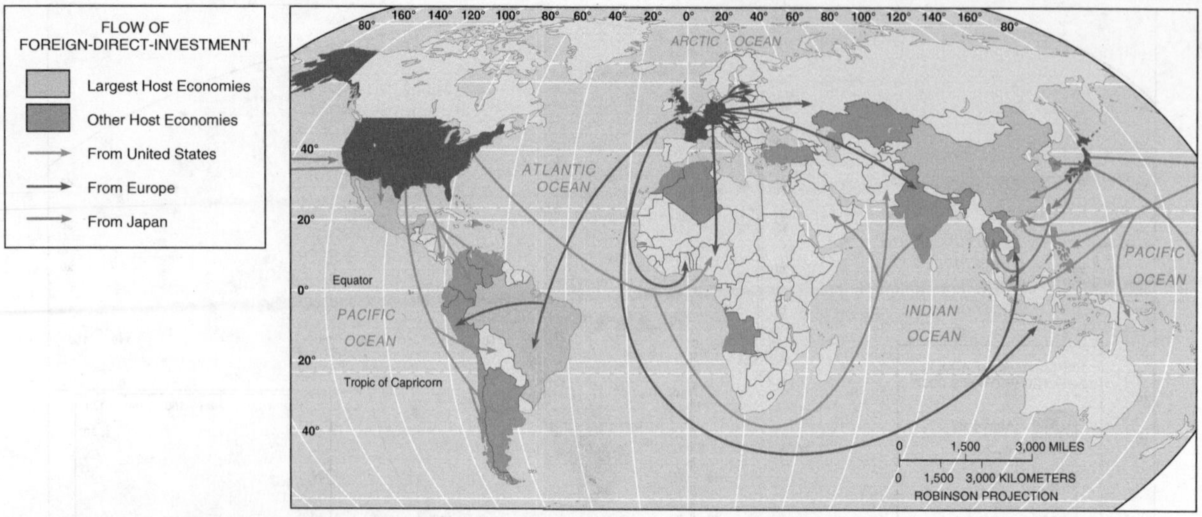

International flows of capital, or foreign investment, are critical to the economic health of countries, especially those that have insufficient reservoirs of domestic capital. Generally, flows of foreign investment take place through multinational corporations. They originate in countries where capital is abundant, that is, the United States, Japan, and Western Europe. The destinations of foreign investment include other developed countries as well as parts of the developing world.

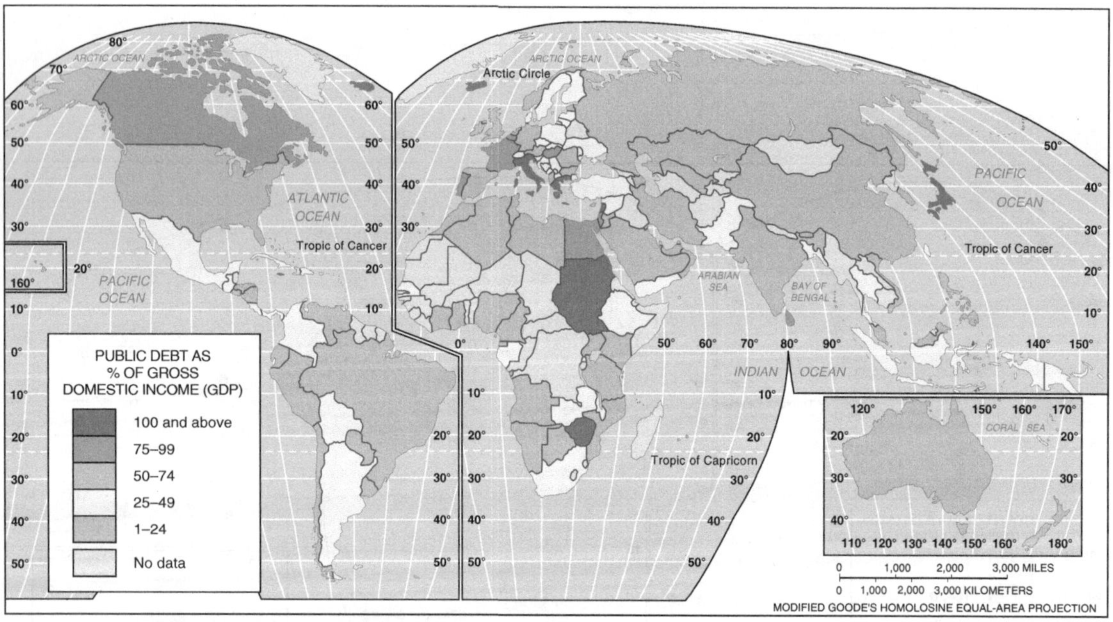

The global debt crisis originated with the recycling of petrodollars in the 1970s, as OPEC countries deposited billions in western banks, which in turn lent them to governments in the the developing world. Today, many countries struggle to pay back the interest and principal on their loans, which consumes a significant share of their export revenues. In absolute terms, the United States is the world's largest debtor. As a proportion of gross domestic product, the most indebted countries are in sub-Saharan Africa. Often the International Monetary Fund is heavily involved in determining the conditions under which debt is to be repaid.

© Catherine Karnow/CORBIS

© Paul Conklin/PhotoEdit

Frederick P. Stutz

© Michael Brennan/CORBIS

© Dave G. Houser/Corbis

Top: With more than one-half of the world's 6.3 billion people living in cities, urban areas have become critical centers of social, political, and economic life. Most of the planet's largest cities are in the developing world, where they are often swollen through rural-to-urban migration and high rates of natural population growth. Large numbers of people live in intolerable conditions marked by substandard housing, poor infrastructure, inadequate employment opportunities, and little access to health and educational services. Many cities in the developing world are ringed by huge shantytown districts, sometimes with millions of inhabitants apiece. Improving the lives and hopes of people caught under these circumstances will require urban as well as rural development.

Left: A woman inspects a fair trade coffee label. Fair Trade is an organized social movement and market-based approach that aims to help producers in developing countries obtain better trading conditions and prices, leading toward sustainability. Fair Trade advocates work for higher prices for producers as well as improved environmental standards. Fair Trade focuses on exports from developing countries which are sent to developed countries. While Fair Trade is still a small proportion of international shipments, certain types of goods, especially primary products, like coffee, cocoa, sugar, tea, fruit, cotton, and fabrics, comprise from 20 to 40% of their respective total world shipments. Many farmers, without other means of subsistence, are obliged to produce more and more, no matter how low the prices are. Research has shown that those who suffer most from declines in commodity prices are the rural poor. Basic agriculture employs over 50% of the people in developing countries, and accounts for 33% of their GDP.

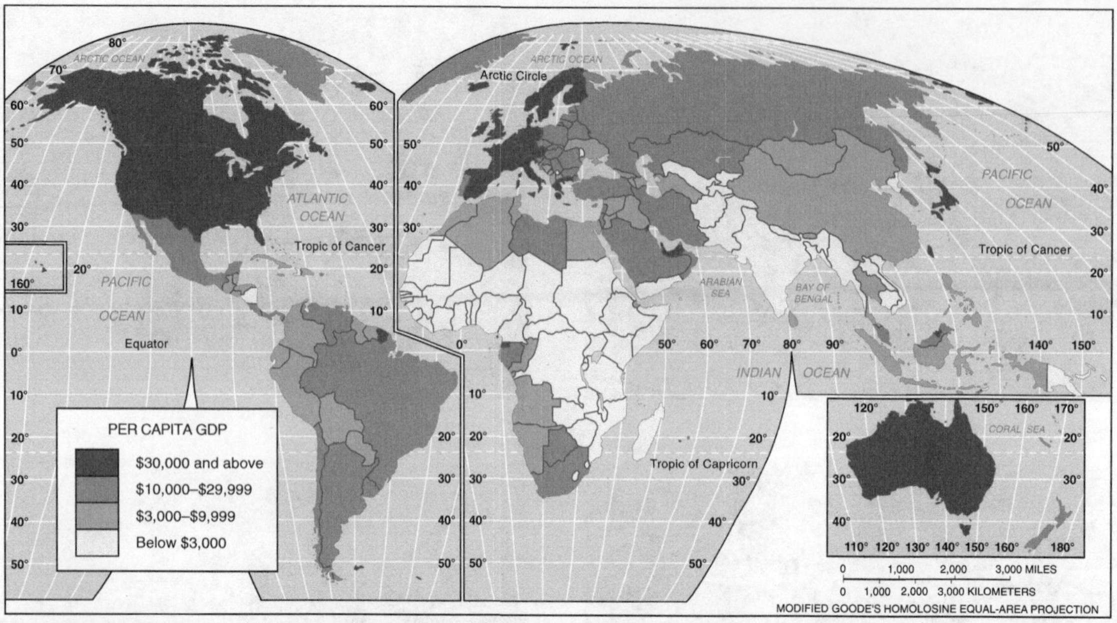

Annual gross domestic product (GDP) per capita is the most widely used measure of economic development, or the lack thereof. It includes the sum total of a country's economic output (goods and services) divided by its population, and ranges from less than $5000 in most of the developing world to more than $30,000 per person in North America, Japan, and parts of Europe. However, GDP per capita is an imperfect measure, failing to capture noncommodified forms of production (e.g., household labor, subsistence production) and being sensitive to exchange rate fluctuations.

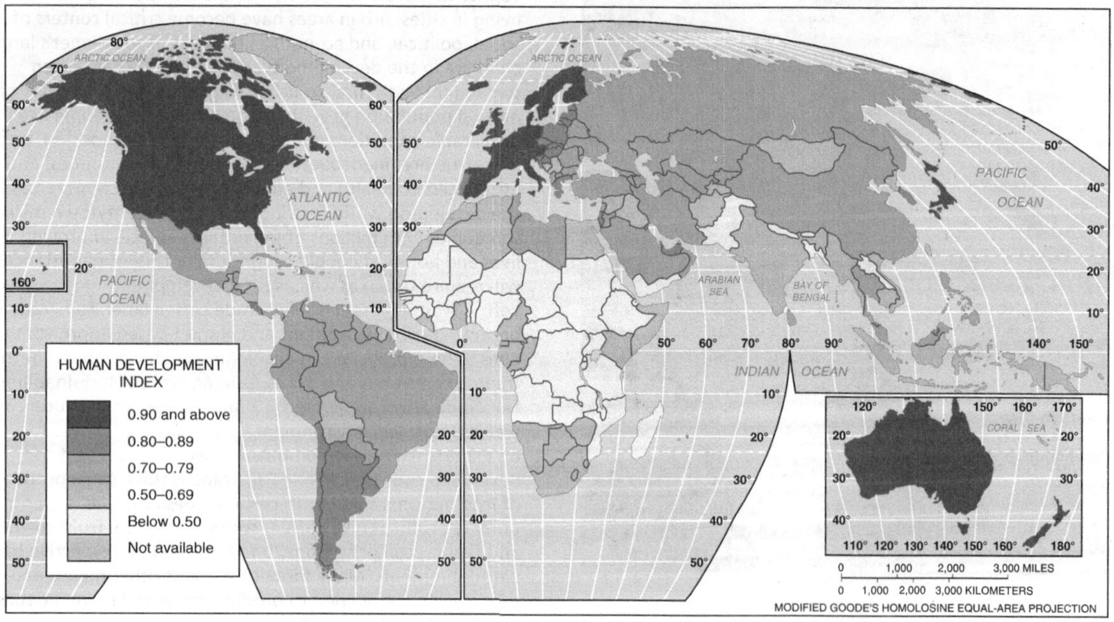

The Human Development Index (HDI), created by the United Nations, combines multiple measures of social and economic development, including gross domestic product per capita, life expectancy at birth, and literacy rates. It thus encapsulates the historical forces that surround people's lives as they play out unevenly over space. The geography of the HDI reflects the world's sharp bifurcation into a small group of relatively well off countries (many of which contain internal pockets of poverty) and large numbers of poor countries in which the HDI is low.

Index

L

M

N

O

P

T